CONVERSION FACTORS

Length

1 m = 39.37 in. = 3.281 ft
1 in. = 2.54 cm
1 km = 0.621 mi
1 mi = 5 280 ft = 1.609 km
1 light year (ly) = 9.461×10^{15} m
1 angstrom (Å) = 10^{-10} m

Mass

1 kg = 10^3 g = 6.85×10^{-2} slug
1 slug = 14.59 kg
1 u = 1.66×10^{-27} kg = 931.5 MeV/c^2

Time

1 min = 60 s
1 h = 3 600 s
1 day = 8.64×10^4 s
1 yr = 365.242 days = 3.156×10^7 s

Volume

1 L = 1 000 cm^3 = 3.531×10^{-2} ft^3
1 ft^3 = 2.832×10^{-2} m^3
1 gal = 3.786 L = 231 in.^{3}

Angle

180° = π rad
1 rad = 57.30°
1° = 60 min = 1.745×10^{-2} rad

Speed

1 km/h = 0.278 m/s = 0.621 mi/h
1 m/s = 2.237 mi/h = 3.281 ft/s
1 mi/h = 1.61 km/h = 0.447 m/s = 1.47 ft/s

Force

1 N = 0.224 8 lb = 10^5 dynes
1 lb = 4.448 N
1 dyne = 10^{-5} N = 2.248×10^{-6} lb

Work and energy

1 J = 10^7 erg = 0.738 ft·lb = 0.239 cal
1 cal = 4.186 J
1 ft·lb = 1.356 J
1 Btu = 1.054×10^3 J = 252 cal
1 J = 6.24×10^{18} eV
1 eV = 1.602×10^{-19} J
1 kWh = 3.60×10^6 J

Pressure

1 atm = 1.013×10^5 N/m^2 (or Pa) = 14.70 lb/in.^{2}
1 Pa = 1 N/m^2 = 1.45×10^{-4} lb/in.^{2}
1 lb/in.^{2} = 6.895×10^3 N/m^2

Power

1 hp = 550 ft·lb/s = 0.746 kW
1 W = 1 J/s = 0.738 ft·lb/s
1 Btu/h = 0.293 W

College Physics

ENHANCED · *seventh edition*

VOLUME 2 / CHAPTERS 15–30

Raymond A. Serway
Emeritus, James Madison University

Jerry S. Faughn
Emeritus, Eastern Kentucky University

Chris Vuille
Embry-Riddle Aeronautical University

Charles A. Bennett
The University of North Carolina, Asheville
Contributing Author, Media

THOMSON

BROOKS/COLE

Australia · Canada · Mexico · Singapore · Spain · United Kingdom · United States

THOMSON

BROOKS/COLE

Physics Acquisitions Editor: Chris Hall
Publisher: David Harris
Vice President, Editor-in-Chief, Sciences: Michelle Julet
Development Editor: Ed Dodd
Assistant Editor: Annie Mac
Editorial Assistant: Kyra Engelberg
Technology Project Manager: Sam Subity
Marketing Manager: Erik Evans, Julie Conover
Marketing Assistant: Leyla Jowza
Advertising Project Manager: Stacey Purviance
Project Manager, Editorial Production: Teri Hyde
Print/Media Buyer: Judy Inouye

Permissions Editor: Audrey Pettengill
Production Service: Progressive Publishing Alternatives
Text Designer: Patrick Devine
Photo Researcher: Jane Sanders
Copy Editor: Progressive Publishing Alternatives
Illustrator: Rolin Graphics, Progressive Information Technologies
Cover Designer: Patrick Devine
Cover Image: ML Sinibaldi/CORBIS
Cover Printer: Phoenix Color Corp., MD
Compositor: Progressive Information Technologies
Printer: R.R. Donnelley/Willard

Printed in the United States of America
1 2 3 4 5 6 7 09 08 07 06

For more information about our products, contact us at:
Thomson Learning Academic Resource Center
1-800-423-0563

For permission to use material from this text or product, submit a request online at **http://www.thomsonrights.com**.
Any additional questions about permissions can be submitted by email to **thomsonrights@thomson.com**.

Thomson Brooks/Cole
10 Davis Drive
Belmont, CA 94002
USA

Library of Congress Control Number: 2004113839

Student Edition: ISBN 0-495-11376-x
Instructor's Edition: ISBN 0-534-99724-4

Asia
Thomson Learning
5 Shenton Way #01-01
UIC Building
Singapore 068808

Australia/New Zealand
Thomson Learning
102 Dodds Street
Southbank, Victoria 3006
Australia

Canada
Nelson
1120 Birchmount Road
Toronto, Ontario M1K 5G4
Canada

Europe/Middle East/Africa
Thomson Learning
High Holborn House
50/51 Bedford Row
London WC1R 4LR
United Kingdom

Latin America
Thomson Learning
Seneca, 53
Colonia Polanco
11560 Mexico D.F.
Mexico

Spain/Portugal
Paraninfo
Calle Magallanes, 25
28015 Madrid, Spain

Your quick start guide to Physics⊙Now

Welcome to **PhysicsNow**, your fully integrated system for physics tutorials and self-assessment on the web. To get started, just follow these simple instructions.

Your first visit to PhysicsNow

1. Go to **http://www.cp7e.com** and click the **Register** button.

2. The first time you visit, you will be asked to select your school. Choose your state from the drop-down menu, then type in your school's name in the box provided and click **Search**. A list of schools with names similar to what you entered will show on the right. Find your school and click on it.

3. On the next screen, enter the access code from the card that came with your textbook in the "Content or Course Access Code" box*. Enter your email address in the next box and click **Submit.**

   * PhysicsNow access codes may be purchased separately. Should you need to purchase an access code, go back to http://www.cp7e.com and click the Buy button.

4. On the next screen, choose a password and click **Submit.**

5. Lastly, fill out the registration form and click **Register and Enter iLrn.** This information will only be used to contact you if there is a problem with your account.

6. You should now see the **PhysicsNow** homepage. Select a chapter and begin!

   Note: Your account information will be sent to the email address that you entered in Step 3, so be sure to enter a valid email address. You will use your email address as your username the next time you login.

Second and later visits

1. Go to **http://www.cp7e.com** and click the **Login** button.

2. Enter your user name (the email address you entered when you registered) and your password and then click **Login.**

SYSTEM REQUIREMENTS:
(Please see the System Requirements link at www.ilrn.com for complete list.)
PC: Windows 98 or higher, Internet Explorer 5.5 or higher
Mac: OS X or higher, Mozilla browser 1.2.1 or higher

TECHNICAL SUPPORT:
For online help, click on **Technical Support** in the upper right corner of the screen, or contact us at:

 1-800-423-0563 Monday–Friday • 8:30 A.M. to 6:00 P.M. EST
tl.support@thomson.com

Turn the page to learn more about **PhysicsNow** and how it can help you achieve success in your course!

THOMSON
BROOKS/COLE

PhysicsNow™ Quick Start Guide

iii

What do you need to learn now?

Take charge of your learning with **PhysicsNow**™, a powerful student-learning tool for physics! This interactive resource helps you gauge your unique study needs, then gives you a *Personalized Learning Plan* that will help you focus in on the concepts and problems that will most enhance your understanding. With **PhysicsNow**, you have the resources you need to take charge of your learning!

The access code card included with this new copy of **College Physics** is your ticket to all of the resources in **PhysicsNow.** (See the previous page for login instructions.)

Interact at every turn with the POWER and SIMPLICITY of PhysicsNow!

PhysicsNow combines Serway and Faughn's best-selling *College Physics* with carefully crafted media resources that will help you learn. This dynamic resource and the Seventh Edition of the text were developed in concert, to enhance each other and provide you with a seamless, integrated learning system.

As you work through the text, you will see notes that direct you to the media-enhanced activities in **PhysicsNow**. This precise page-by-page integration means you'll spend less time flipping through pages or navigating websites looking for useful exercises. These multimedia exercises will make all the difference when you're studying and taking exams ... after all, it's far easier to understand physics if it's seen in action, and **PhysicsNow** enables you to become a part of the action!

Begin at http://www.cp7e.com and build your own Personalized Learning Plan now!

Log into **PhysicsNow** at http://www.cp7e.com by using the free access code packaged with the text. You'll immediately notice the system's simple, browser-based format. You can build a complete *Personalized Learning Plan* for yourself by taking advantage of all three powerful components found on **PhysicsNow**:

▶ **What I Know**
▶ **What I Need to Learn**
▶ **What I've Learned**

The best way to maximize the system and optimize your time is to start by taking the *Pre-Test* ▶▶▶

PhysicsNow™ Quick Start Guide

What I Know

▲ You take a *Pre-Test* to measure your level of comprehension after reading a chapter. Each *Pre-Test* includes approximately 15 questions. The *Pre-Test* is your first step in creating your custom-tailored *Personalized Learning Plan*.

An item-by-item analysis gives you feedback on each of your answers.

▲ Once you've completed the "What I Know" *Pre-Test*, you are presented with a detailed *Personalized Learning Plan*, with text references that outline the elements you need to review in order to master the chapter's most essential concepts. This roadmap to concept mastery guides you to exercises designed to improve skills and to increase your understanding of the basic concepts.

At each stage, the *Personalized Learning Plan* refers to *College Physics* to reinforce the connection between text and technology as a powerful learning tool.

What I Need to Learn

Once you've completed the *Pre-Test*, you're ready to work through tutorials and exercises that will help you master the concepts that are essential to your success in the course.

ACTIVE FIGURES

A remarkable bank of more than 200 animated figures helps you visualize physics in action. Taken straight from illustrations in the text, these *Active Figures* help you master key concepts from the book. By interacting with the animations and accompanying quiz questions, you come to an even greater understanding of the concepts you need to learn from each chapter. ▼

Each figure is titled so you can easily identify the concept you are seeing. The final tab features a *Quiz*. The *Explore* tab guides you through the animation so you understand what you should be seeing and learning.

▲ The brief *Quiz* ensures that you mastered the concept played out in the animation—and gives you feedback on each response.

Continued on the next page ▶

PhysicsNow™ Quick Start Guide

What I Need to Learn *Continued*

COACHED PROBLEMS

Engaging *Coached Problems* reinforce the lessons in the text by taking a step-by-step approach to problem-solving methodology. Each *Coached Problem* gives you the option of breaking down a problem from the text into steps with feedback to 'coach' you toward the solution. There are approximately five *Coached Problems* per chapter.

You can choose to work through the *Coached Problems* by inputting an answer directly or working in steps with the program. If you choose to work in steps, the problem is solved with the same problem-solving methodology used in *College Physics* to reinforce these critical skills. Once you've worked through the problem, you can click **Try Another** to change the variables in the problem for more practice.

▲ Also built into each *Coached Problem* is a link to Brooks/Cole's exclusive **vMentor**™ web-based tutoring service site that lets you interact directly with a live physics tutor. If you're stuck on math, a *MathAssist* link on each *Coached Problem* launches tutorials on math specific to that problem.

INTERACTIVE EXAMPLES

You'll strengthen your problem-solving and visualization skills with *Interactive Examples*. Extending selected examples from the text, *Interactive Examples* utilize the proven and trusted problem-solving methodology presented in **College Physics.** These animated learning modules give you all the tools you need to solve a problem type—you're then asked to apply what you have learned to different scenarios. You will find approximately two *Interactive Examples* for each chapter of the text.▼

You're guided through the steps to solve the problem and then asked to input an answer in a simulation to see if your result is correct. Feedback is instantaneous.

What I've Learned

▶ After working through the problems highlighted in your *Personalized Learning Plan*, you move on to a *Post Test*, about 15 questions per chapter

◀ Once you've completed the *Post Test*, you receive your percentage score and specific feedback on each answer. The *Post Tests* give you a new set of questions with each attempt, so you can take them over and over as you continue to build your knowledge and skills and master concepts.

Also available to help you succeed in your course

Student Solutions Manual and Study Guide
Volume I (Ch. 1–14) ISBN: 0-534-99920-4
Volume II (Ch. 15–30) ISBN: 0-534-99930-1
These manuals contain detailed solutions to approximately 12 problems per chapter. These problems are indicated in the textbook with boxed problem numbers. Each manual also features a skills section, important notes from key sections of the text, and a list of important equations and concepts.

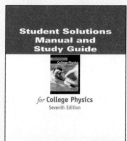

Core Concepts in College Physics CD-ROM, Version 2.0
ISBN: 0-03-033701-1
Explore the core of physics with this powerful CD-ROM/workbook program! Content screens provide in-depth coverage of abstract and often difficult principles, building connections between physical concepts and mathematics. The presentation contains more than 350 movies—both animated and live video—including laboratory demonstrations, real-world examples, graphic models, and step-by-step explanations of essential mathematics. An accompanying workbook contains practical physics problems directly related to the presentation, along with worked solutions. Package includes three discs and a workbook.

Welcome to your MCAT Test Preparation Guide

The **MCAT Test Preparation Guide** makes your copy of *College Physics,* Seventh Edition, the most comprehensive MCAT study tool and classroom resource in introductory physics. The grid, which begins below and continues on the next two pages, outlines twelve concept-based **study courses** for the physics part of your MCAT exam. Use it to prepare for the MCAT, class tests, and your homework assignments.

Vectors

Skill Objectives: To calculate distance, angles between vectors, and magnitudes.

Review Plan:

Distance and Angles:
- Chapter 1, Sections 1.7, 1.8
- Active Figure 1.6
- Chapter Problems 35, 39, 42

Using Vectors:
- Chapter 3, Sections 3.1, 3.2
- Quick Quizzes 3.1–3.3
- Examples 3.1–3.3
- Active Figure 3.3
- Chapter Problems 4, 11, 19

Motion

Skill Objectives: To understand motion in two dimensions, to calculate speed and velocity, centripetal acceleration, and acceleration in free fall problems.

Review Plan:

Motion in 1 Dimension:
- Chapter 2, Sections 2.1–2.6
- Quick Quizzes 2.1–2.8
- Examples 2.1–2.10
- Active Figure 2.15
- Chapter Problems 1, 8, 17, 21, 29, 33, 38, 47, 51

Motion in 2 Dimensions:
- Chapter 3, Sections 3.3, 3.4
- Quick Quizzes 3.4–3.7
- Examples 3.4–3.8
- Active Figures 3.14, 3.15
- Chapter Problems 27, 33

Centripetal Acceleration:
- Chapter 7, Section 7.4
- Quick Quizzes 7.6, 7.7
- Example 7.6

Force

Skill Objectives: To know and understand Newton's Laws, to calculate resultant forces, and weight.

Review Plan:

Newton's Laws:
- Chapter 4, Sections 4.1–4.4
- Quick Quizzes 4.1–4.5
- Examples 4.1–4.4
- Active Figure 4.6
- Chapter Problems 5, 7, 11

Resultant Forces:
- Chapter 4, Section 4.5
- Quick Quizzes 4.6, 4.7
- Examples 4.7, 4.9, 4.10
- Chapter Problems 17, 25, 33

Equilibrium

Skill Objectives: To calculate momentum and impulse, center of gravity, and torque.

Review Plan:

Momentum:
- Chapter 6, Sections 6.1–6.3
- Quick Quizzes 6.2–6.6
- Examples 6.1–6.4, 6.6
- Active Figures 6.7, 6.10, 6.13
- Chapter Problems 7, 16, 21

Torque:
- Chapter 8, Sections 8.1–8.4
- Examples 8.1–8.7
- Chapter Problems 5, 9

Work

Skill Objectives: To calculate friction, work, kinetic energy, potential energy, and power.

Review Plan:

Friction:
- Chapter 4, Section 4.6
- Quick Quizzes 4.8–4.10
- Active Figure 4.19

Work:
- Chapter 5, Section 5.1
- Quick Quiz 5.1
- Example 5.1
- Active Figure 5.5
- Chapter Problems 11, 17

Energy:
- Chapter 5, Sections 5.2, 5.3
- Examples 5.4, 5.5
- Quick Quizzes 5.2, 5.3

Power:
- Chapter 5, Section 5.6
- Examples 5.12, 5.13

Waves

Skill Objectives: To understand interference of waves, to calculate basic properties of waves, properties of springs, and properties of pendulums.

Review Plan:

Wave Properties:
- Chapters 13, Sections 13.1–13.4, 13.7–13.11
- Quick Quizzes 13.1–13.6
- Examples 13.1, 13.6, 13.8–13.10
- Active Figures 13.1, 13.8, 13.12, 13.13, 13.24, 13.26, 13.32, 13.33, 13.34, 13.35
- Chapter Problems 9, 15, 23, 29, 41, 47, 53

Pendulum:
- Chapter 13, Sections 13.5
- Quick Quizzes 13.7–13.9
- Example 13.7
- Active Figures 13.15, 13.16
- Chapter Problem 35

Matter

Skill Objectives: To calculate pressure, density, specific gravity, and flow rates.

Review Plan:

Properties:
- Chapter 9, Sections 9.1–9.3
- Quick Quiz 9.1
- Examples 9.1, 9.3, 9.4
- Active Figure 9.3
- Chapter Problem 7

Pressure:
- Chapter 9, Sections 9.3–9.6
- Quick Quizzes 9.2–9.6
- Examples 9.4–9.9
- Active Figures 9.19, 9.20
- Chapter Problems 13, 23, 39

Flow rates:
- Chapter 9, Sections 9.7, 9.8
- Quick Quiz 9.7
- Examples 9.11–9.14
- Chapter Problem 42

Sound

Skill Objectives: To understand interference of waves, to calculate properties of waves, the speed of sound, Doppler shifts, and intensity.

Review Plan:

Sound Properties:
- Chapter 14, Sections 14.1–14.4, 14.6
- Quick Quizzes 14.1, 14.2
- Examples 14.1, 14.2, 14.4, 14.5
- Active Figures 14.6, 14.11
- Chapter Problems 7, 13,25

Interference/Beats:
- Chapter 14, Sections 14.7, 14.8, 14.11
- Quick Quiz 14.7
- Examples 14.6, 14.11
- Active Figures 14.18, 14.25
- Chapter Problems 33, 37, 51

Light

Skill Objectives: To understand mirrors and lenses, to calculate the angles of reflection, to use the index of refraction, and to find focal lengths.

Review Plan:

Reflection and Refraction:
- Chapter 22, Sections 22.1–22.4
- Quick Quizzes 22.2–22.4
- Examples 22.1–22.4
- Active Figures 22.4, 22.6, 22.7
- Chapter Problems 9, 17, 21, 25

Mirrors and Lenses:
- Chapter 23, Sections 23.1–23.6
- Quick Quizzes 23.1, 23.2, 23.4–23.6
- Examples 23.7, 23.8, 23.9
- Active Figures 23.2, 23.16, 23.25
- Chapter Problems 25, 29, 33, 39

Electrostatics

Skill Objectives: To understand and calculate the electric field, the electrostatic force, and the electric potential.

Review Plan:

Coulomb's Law:
- Chapter 15, Sections 15.1–15.3
- Quick Quiz 15.2
- Examples 15.1–15.3
- Active Figure 15.6
- Chapter Problems 11

Electric Field:
- Chapter 15, Sections 15.4, 15.5
- Quick Quizzes 15.3–15.6
- Examples 15.4, 15.5
- Active Figures 15.11, 15.16
- Chapter Problems 19, 23, 27

Potential:
- Chapter 16, Sections 16.1–16.3
- Quick Quizzes 16.1–16.5
- Examples 16.1, 16.4
- Active Figure 16.7
- Chapter Problems 1, 7, 15

Circuits

Skill Objectives: To understand and calculate current, resistance, voltage, power, and energy, and to use circuit analysis.

Review Plan:

Ohm's Law:
- Chapter 17, Sections 17.1–17.4
- Quick Quizzes 17.1, 17.3
- Examples 17.1, 17.3
- Chapter Problem 15

Power and energy:
- Chapter 17, Section 17.8
- Quick Quizzes 17.6–17.8
- Example 17.6
- Active Figure 17.11
- Chapter Problem 36

Circuits:
- Chapter 18, Sections 18.2, 18.3
- Quick Quizzes 18.1–18.5
- Examples 18.1–18.3
- Active Figures 18.2, 18.6
- Chapter Problems 4, 13

Atoms

Skill Objectives: To calculate half-life, and to understand decay processes and nuclear reactions.

Review Plan:

Atoms:
- Chapter 29, Sections 29.1, 29.2

Radioactive Decay:
- Chapter 29, Sections 29.3–29.5
- Examples 29.3, 29.4, 29.7
- Active Figures 29.6, 29.7
- Chapter Problems 15, 21, 29, 35

Nuclear reactions:
- Chapter 29, Sections 29.6
- Quick Quiz 29.4
- Examples 29.8, 29.9
- Chapter Problems 41, 45

Contents Overview

Contents
Volume 2

About the Authors

Raymond A. Serway received his doctorate at Illinois Institute of Technology and is Professor Emeritus at James Madison University. In 1990, he received the Madison Scholar Award at James Madison University, where he taught for 17 years. Dr. Serway began his teaching career at Clarkson University, where he conducted research and taught from 1967 to 1980. He was the recipient of the Distinguished Teaching Award at Clarkson University in 1977 and of the Alumni Achievement Award from Utica College in 1985. As Guest Scientist at the IBM Research Laboratory in Zurich, Switzerland, he worked with K. Alex Müller, 1987 Nobel Prize recipient. Dr. Serway also was a visiting scientist at Argonne National Laboratory, where he collaborated with his mentor and friend, Sam Marshall. In addition to earlier editions of this textbook, Dr. Serway is the co-author of *Physics for Scientists and Engineers,* Sixth Edition; *Principles of Physics,* Fourth Edition; and *Modern Physics,* Third Edition. He also is the author of the high-school textbook *Physics,* published by Holt, Rinehart, & Winston. In addition, Dr. Serway has published more than 40 research papers in the field of condensed matter physics and has given more than 70 presentations at professional meetings. Dr. Serway and his wife Elizabeth enjoy traveling, golfing, and spending quality time with their four children and six grandchildren.

Jerry S. Faughn earned his doctorate at the University of Mississippi. He is Professor Emeritus and former Chair of the Department of Physics and Astronomy at Eastern Kentucky University. Dr. Faughn has also written a microprocessor interfacing text for upper-division physics students. He is co-author of a non-mathematical physics text and a physical science text for general education students, and (with Dr. Serway) the high-school textbook *Physics,* published by Holt, Reinhart, & Winston. He has taught courses ranging from the lower division to the graduate level, but his primary interest is in students just beginning to learn physics. Dr. Faughn has a wide variety of hobbies, among which are reading, travel, genealogy, and old-time radio. His wife Mary Ann is an avid gardener, and he contributes to her efforts by staying out of the way. His daughter Laura is in family practice and his son David is an attorney.

Chris Vuille is an associate professor of physics at Embry-Riddle Aeronautical University, Daytona Beach, Florida, the world's premier institution for aviation higher education. He received his doctorate in physics at the University of Florida in 1989, moving to Daytona after a year at ERAU's Prescott, Arizona campus. While he has taught courses at all levels, including post-graduate, his primary interest has been the delivery of introductory physics. He has received several awards for teaching excellence, including the Senior Class Appreciation Award (three times). He conducts research in general relativity and quantum theory, and was a participant in the JOVE program, a special three-year NASA grant program during which he studied neutron stars. His work has appeared in a number of scientific journals, and he has been a featured science writer in Analog Science Fiction/Science Fact magazine. Dr. Vuille enjoys tennis, lap swimming, guitar and classical piano, and is a former chess champion of St. Petersburg and Atlanta. In his spare time he writes science fiction and goes to the beach. His wife, Dianne Kowing, is an optometrist for a local VA clinic. Teen daughter Kira Vuille-Kowing is an accomplished swimmer, violinist, and mall shopper. He has two sons, twelve-year-old Christopher, a cellist and electronic game expert, and three-year-old James, master of tinkertoys.

Charles A. Bennett received his doctorate at North Carolina State University, and is Professor of Physics at the University of North Carolina at Asheville. His research interests include quantum and physical optics, and laser applications in environmental and fusion energy research. He has collaborated with Oak Ridge National Laboratory since 1983, where he is currently an adjunct research and development associate of the Advanced Laser and Optical Technology and Development group. In addition to his work in optics, Dr. Bennett has a long record of innovation in educational technology, particularly in the integration of active media into on-line homework. He is a past director of the UNCA Center for Teaching and Learning, and has received UNCA's most prestigious recognition for scholarship: the Ruth and Leon Feldman Professorship for 1996–1997. He is an avid hiker, and enjoys spending time with his wife Karen and four children.

Preface

College Physics is written for a one-year course in introductory physics usually taken by students majoring in biology, the health professions, and other disciplines including environmental, earth, and social sciences, and technical fields such as architecture. The mathematical techniques used in this book include algebra, geometry, and trigonometry, but not calculus.

The main objectives of this introductory textbook are twofold: to provide the student with a clear and logical presentation of the basic concepts and principles of physics, and to strengthen an understanding of the concepts and principles through a broad range of interesting applications to the real world. To meet these objectives, we have emphasized sound physical arguments and problem-solving methodology. At the same time, we have attempted to motivate the student through practical examples that demonstrate the role of physics in other disciplines.

This textbook, which covers the standard topics in classical physics and 20th-century physics, is divided into six parts. Part I (Chapters 1–9) deals with Newtonian mechanics and the physics of fluids; Part II (Chapters 10–12) is concerned with heat and thermodynamics; Part III (Chapters 13–14) covers wave motion and sound; Part IV (Chapters 15–21) develops the concepts of electricity and magnetism; Part V (Chapters 22–25) treats the properties of light and the field of geometric and wave optics; and Part VI (Chapters 26–30) provides an introduction to special relativity, quantum physics, atomic physics, and nuclear physics.

CHANGES TO THE SEVENTH EDITION

A number of new features, changes, and improvements have been added to this edition. Based on comments from users of the sixth edition and reviewers' suggestions, a major effort was made to improve clarity of presentation, precision of language, and accuracy throughout. The new pedagogical features added to this edition are based on current trends in science education. The following represent the major changes in the seventh edition.

Pedagogical Changes

- **Examples** All in-text worked examples have been reconstituted for the seventh edition, and now are presented in a two-column format that better aids student learning and helps to better reinforce physical concepts. Special care has been taken in this new edition to present a range of levels within each chapter's collection of worked examples, so that students are better prepared to solve the end-of-chapter problems. The examples are set off from the text for ease of location and are given titles to describe their content. All worked examples now include the following parts:

  (a) **Goal,** which describes the physical concepts being explored within the Worked Example.
  (b) **Problem,** which presents the problem itself.
  (c) **Strategy,** which helps students analyze the problem and create a framework for working out the solution.
  (d) **Solution,** which uses a two-column format that gives the explanation for each step of the solution in the left-hand column, while giving each accompanying mathematical step in the right-hand column. This layout facilitates matching the idea with its execution, and helps students learn how to organize their work. Another benefit: students can easily use this format as a training tool, covering up the solution on the right and solving the problem using the comments on the left as a guide.

(e) **Remarks,** which follow each Solution and highlight some of the underlying concepts and methodology used in arriving at a correct solution. In addition, the remarks are often used to put the problem into a larger, real world context.

(f) **Exercise/Answer** By popular demand of users and reviewers, every worked example in the new edition is now followed immediately by an exercise with an answer. These exercises allow the students to reinforce their understanding by working a similar or related problem, with the answers giving them instant feedback. At the option of the instructor, the exercises can also be assigned as homework. Students who work through these exercises on a regular basis will find the end-of-chapter problems less intimidating.

■ *PhysicsNow*™ Directly links with *College Physics, Seventh Edition. PhysicsNow*™ and *College Physics, Seventh Edition* were built in concert to enhance each other and provide a seamless, integrated learning system. Throughout the text, readers will see Physics⊗Now™ icons that direct them to media-enhanced activities at the *PhysicsNow*™ Web site (**www.cp7e.com**). The precise page-by-page integration means professors and students will spend less time flipping through pages and navigating Web sites for useful exercises.

■ **Active Figures** Many diagrams from the text have been animated to form Active Figures, part of the new *PhysicsNow*™ integrated Web-based learning system. There are over 70 Active Figures in Volume 2, available at **www.cp7e.com.** By visualizing phenomena and processes that cannot be fully represented on a static page, students greatly increase their conceptual understanding. Active Figures are easily identified by their red figure legend and Physics⊗Now™ icon, and the caption after the icon describes briefly the nature and contents of the animation. In addition to viewing animations of the figures, students can change variables to see the effects, conduct suggested explorations of the principles involved in the figure, and take and receive feedback on quizzes related to the figure.

■ **Interactive Examples** Sixteen of the worked examples in Volume 2 have been identified as interactive. As part of the *PhysicsNow*™ Web-based learning system, students can engage in an extension of the problem solved in the example. This often includes elements of both visualization and calculation, and may also involve prediction and intuition-building. Interactive Examples are available at **www.cp7e.com.**

■ **Quick Quizzes** All of the Quick Quizzes in this edition have been cast in an objective format, including multiple choice, true-false, and ranking. Quick Quizzes provide students with opportunities to test their understanding of the physical concepts presented. The questions require students to make decisions on the basis of sound reasoning, and some of them have been written to help students overcome common misconceptions. Answers to all Quick Quiz questions are found at the end of the textbook, while answers with detailed explanations are provided in the *Instructor's Manual.* Many instructors choose to use Quick Quiz questions in a "peer instruction" teaching style.

■ **Summary** The Summary is now organized by each individual chapter heading, for ease of reference.

■ **Problems and Conceptual Questions** All problems have been carefully worded and have been checked for clarity and accuracy, and approximately 15% of the conceptual questions and problems are new to this edition. Solutions to approximately 20% of the end-of-chapter problems are included in the *Student Solutions Manual and Study Guide.* These problems are identified with a box around the problem number. A smaller subset of problems will be available with coached solutions as part of the *PhysicsNow*™ Web-based learning system and will be accessible to students and instructors using *College Physics.* The Physics⊗Now™ icon identifies these problems.

■ **Line-by-Line Revision** The text has been carefully edited to improve clarity of presentation and precision of language. We hope that the result is a book both accurate and enjoyable to read.

CONTENT CHANGES

Although the overall content and organization of the textbook is similar to that of the sixth edition, several changes were implemented.

- Vectors are now denoted in bold face with arrows over them (for example, $\vec{\mathbf{v}}$), making them easier to recognize.
- In many chapters, one or two examples have been modified so as to combine new concepts with previously studied concepts. This not only reviews prior material, but also integrates different concepts, showing how they work together to enhance our understanding of the physical world.
- In Chapter 4, on Newton's laws, we have doubled the number of examples, from seven in the sixth edition to fourteen in the seventh, to help aid student understanding.
- Chapter 5, on work and energy, has been extensively reorganized, with gravitational potential energy and spring potential energy handled in separate sections. The connection between work done by conservative forces and the definition of the potential energy of a system has been emphasized and clarified.
- The number of examples in Chapter 9, Solids and Fluids, has been increased from thirteen to nineteen.
- Chapter 12, The Laws of Thermodynamics, now covers the important concepts of degrees of freedom and the equipartition of energy, and relates them to determining molar specific heats. The different kinds of thermodynamic processes have been organized into their own subsections, and the work done during an isothermal process is now presented and used in examples. Generic processes are also briefly addressed. The methods for calculating the quantities appearing in the first law are organized into a table that makes clear the similarities and differences between the processes.
- The concepts of electric potential and electric potential energy have been clarified in Chapter 16.
- In Chapter 23, the tables for sign conventions for mirrors and lenses have been reorganized for greater clarity.
- The methods of handling interference in thin films have been organized into a table in Chapter 24, facilitating student understanding and problem-solving in that area.
- Additional detail on prescribing lenses has been added to the example problems of Chapter 25, Optical Instruments.
- In Chapter 26, the Twin Paradox and concepts of general relativity were clarified and enhanced.

TEXTBOOK FEATURES

Most instructors would agree that the textbook assigned in a course should be the student's primary guide for understanding and learning the subject matter. Furthermore, the textbook should be easily accessible and written in a style that facilitates instruction and learning. With this in mind, we have included many pedagogical features that are intended to enhance the textbook's usefulness to both students and instructors. These features are as follows:

STYLE   To facilitate rapid comprehension, we have attempted to write the book in a style that is clear, logical, relaxed, and engaging. The somewhat informal and relaxed writing style is designed to connect better with students and enhance their reading enjoyment. New terms are carefully defined, and we have tried to avoid the use of jargon.

PREVIEWS   All chapters begin with a brief preview that includes a discussion of the chapter's objectives and content.

UNITS   The international system of units (SI) is used throughout the text. The U.S. customary system of units is used only to a limited extent in the chapters on mechanics and thermodynamics.

MARGINAL NOTES   Comments and notes appearing in the margin can be used to locate important statements, equations, and concepts in the text.

PROBLEM-SOLVING STRATEGIES   A general strategy to be followed by the student is outlined at the end of Chapter 1 and provides students with a structured process for solving problems. In most chapters, more specific strategies and suggestions are included for solving the types of problems featured in both the worked examples and end-of-chapter problems. This feature, highlighted by a surrounding box, is intended to help students identify the essential steps in solving problems and increases their skills as problem solvers.

BIOMEDICAL APPLICATIONS   For biology and pre-med students, icons point the way to various practical and interesting applications of physical principles to biology and medicine. Where possible, an effort was made to include more problems that would be relevant to these disciplines.

APPLYING PHYSICS   provide students with an additional means of reviewing concepts presented in that section. Some Applying Physics examples demonstrate the connection between the concepts presented in that chapter and other scientific disciplines. These examples also serve as models for students when assigned the task of responding to the Conceptual Questions presented at the end of each chapter.

TIPS   are placed in the margins of the text and address common student misconceptions and situations in which students often follow unproductive paths. Approximately 40 Tips are provided in Volume 2 to help students avoid common mistakes and misunderstandings.

IMPORTANT STATEMENTS AND EQUATIONS   Most important statements and definitions are set in **boldface** type or are highlighted with a background screen for added emphasis and ease of review. Similarly, important equations are highlighted with a tan background screen to facilitate location.

ILLUSTRATIONS AND TABLES   The readability and effectiveness of the text material, worked examples, and end-of-chapter conceptual questions and problems are enhanced by the large number of figures, diagrams, photographs, and tables. Full color adds clarity to the artwork and makes illustrations as realistic as possible. Three-dimensional effects are rendered with the use of shaded and lightened areas where appropriate. Vectors are color coded, and curves in graphs are drawn in color. Color photographs have been carefully selected, and their accompanying captions have been written to serve as an added instructional tool. A complete description of the pedagogical use of color appears on the inside front cover.

SIGNIFICANT FIGURES   Significant figures in both worked examples and end-of-chapter problems have been handled with care. Most numerical examples and problems are worked out to either two or three significant figures, depending on the accuracy of the data provided. Intermediate results presented in the examples are rounded to the proper number of significant figures, and only those digits are carried forward.

CONCEPTUAL QUESTIONS   are provided at the end of each chapter (almost 300 questions are included in Volume 2). The **Applying Physics** examples presented in the text serve as models for students when conceptual questions are assigned, and show how the concepts can be applied to understanding the physical world. The conceptual questions provide the student with a means of self-testing the concepts presented in the chapter. Some conceptual questions are appropriate for initiating classroom discussions. Answers to all odd-numbered conceptual questions are located in the answer section at the end of the book.

End-of-Chapter Problems   An extensive set of problems is included at the end of each chapter (in all, over 1,000 problems are provided in Volume 2). Answers to odd-numbered problems are given at the end of the book. For the convenience of both the student and instructor, about two thirds of the problems are keyed to specific sections of the chapter. The remaining problems, labeled "Additional Problems," are not keyed to specific sections. There are three levels of problems that are graded according to their difficulty. Straightforward problems are numbered in black, intermediate level problems are numbered in blue, and the most challenging problems are numbered in magenta. The ▨ icon identifies problems dealing with applications to the life sciences and medicine.

Solutions to approximately 20% of the problems in each chapter are in the *Student Solutions Manual and Study Guide*. Among these, selected problems are identified with *PhysicsNow*™ icons and have coached solutions available at **www.cp7e.com.**

Activities   have been included at the end of each chapter's problem set to encourage students to engage in scientific activities outside of the classroom.

Appendices and Endpapers   Several appendices are provided at the end of the textbook. Most of the appendix material represents a review of mathematical concepts and techniques used in the text, including scientific notation, algebra, geometry, trigonometry, differential calculus, and integral calculus. Reference to these appendices is made as needed throughout the text. Most of the mathematical review sections include worked examples and exercises with answers. In addition to the mathematical review, some appendices contain useful tables that supplement textual information. For easy reference, the front endpapers contain a chart explaining the use of color throughout the book and a list of frequently used conversion factors.

TEACHING OPTIONS

This book contains more than enough material for a one-year course in introductory physics. This serves two purposes. First, it gives the instructor more flexibility in choosing topics for a specific course. Second, the book becomes more useful as a resource for students. On the average, it should be possible to cover about one chapter each week for a class that meets three hours per week. Those sections, examples, and end-of-chapter problems dealing with applications of physics to life sciences are identified with the DNA icon ▨. We offer the following suggestions for shorter courses for those instructors who choose to move at a slower pace through the year.

Option A: If you choose to place more emphasis on contemporary topics in physics, you should consider omitting all or parts of Chapter 8 (Rotational Equilibrium and Rotational Dynamics), Chapter 21 (Alternating Current Circuits and Electromagnetic Waves), and Chapter 25 (Optical Instruments).

Option B: If you choose to place more emphasis on classical physics, you could omit all or parts of Part VI of the textbook, which deals with special relativity and other topics in 20th-century physics.

The *Instructor's Manual* offers additional suggestions for specific sections and topics that may be omitted without loss of continuity if time presses.

ANCILLARIES

The ancillary package has been updated substantially and streamlined in response to suggestions from users of the sixth edition. The most essential parts of the student package are the two-volume *Student Solutions Manual and Study Guide* with a tight focus on problem-solving and the Web-based *PhysicsNow*™ learning system. Instructors will find increased support for their teaching efforts with new electronic materials.

STUDENT ANCILLARIES

Thomson • Brooks/Cole offers several items to supplement and enhance the classroom experience. These ancillaries will allow instructors to customize the textbook to their students' needs and to their own style of instruction. One or more of these ancillaries may be shrink-wrapped with the text at a reduced price:

Student Solutions Manual and Study Guide   by John R. Gordon, Charles Teague, and Raymond A. Serway. Now offered in two volumes, this manual features detailed solutions to approximately 12 problems per chapter. Boxed numbers identify those problems in the textbook for which complete solutions are found in the manual. The manual also features a skills section, important notes from key sections of the text, and a list of important equations and concepts. Volume 1 contains Chapters 1–14 and Volume 2 contains Chapters 15–30.

Physics⊗Now™ Students log into *PhysicsNow*™ at **www.cp7e.com** by using the free access code packaged with this text.* The *PhysicsNow*™ system is made up of three interrelated parts:

- How much do you know?
- What do you need to learn?
- What have you learned?

Students maximize their success by starting with the Pre-Test for the relevant chapter. Each Pre-Test is a mix of conceptual and numerical questions. After completing the Pre-Test, each student is presented with a detailed Learning Plan. The Learning Plan outlines elements to review in the text and Web-based media (Active Figures, Interactive Examples, and Coached Problems) in order to master the chapter's most essential concepts. After working through these materials, students move on to a multiple-choice Post-Test presenting them with questions similar to those that might appear on an exam. Results can be e-mailed to instructors.

Physics Laboratory Manual,   2nd edition by David Loyd. This manual supplements the learning of basic physical principles while introducing laboratory procedures and equipment. Each chapter of the manual includes a pre-laboratory assignment, objectives, an equipment list, the theory behind the experiment, experimental procedures, graphs, and questions. A laboratory report is provided for each experiment so the student can record data, calculations, and experimental results. Students are encouraged to apply statistical analysis to their data in order to develop their ability to judge the validity of their results.

WebTutor™ on WebCT and Blackboard WebTutor™ offers students real-time access to a full array of study tools, including PhysicsNow's Active Figures, chapter outlines, summaries, and learning objectives. New to this edition is a bank of end-of-chapter problems from each chapter of the book coded in multiple choice format and ready to assign as quizzes or homework.

The Brooks/Cole Physics Resource Center You'll find additional online quizzes, Web links, and animations at **http://physics.brookscole.com.**

INSTRUCTOR ANCILLARIES

The following ancillaries are available to qualified adopters. Please contact your local Thomson • Brooks/Cole sales representative for details.
Ancillaries offered in two volumes are split as follows: Volume 1 contains Chapters 1–14 and Volume 2 contains Chapters 15–30.

Instructor's Solutions Manual   by Jerry Faughn and Charles Teague. Available in two volumes, this manual consists of complete solutions to all the problems in the text,

*Free pincodes are only available with new copies of *College Physics*, 7th edition.

answers to the even-numbered problems and conceptual questions, and full answers with explanations to the Quick Quizzes.

Test Bank   by Ed Oberhofer. Contains approximately 1 750 multiple-choice problems and questions. Answers are provided in a separate key. It is provided in print form (in two volumes) for the instructor who does not have access to a computer, and instructors may duplicate pages for distribution to students. The questions in the *Test Bank* are also available in electronic format with complete answers and solutions in iLrn's Computerized Testing.

Overhead Transparency Acetates   The collection of transparencies in two volumes consists of approximately 200 full-color figures and photographs from the text to enhance lectures. These transparencies feature large print for easy viewing in the classroom.

Instructor's Manual for Physics Laboratory Manual   2nd edition by David Loyd. Each chapter contains a discussion of the experiment, teaching hints, answers to selected questions from the student laboratory manual, and a post-laboratory quiz with short answers and essay questions. The author has also included a list of the suppliers of scientific equipment and a summary of the equipment needed for all the experiments in the manual.

Personal Response System Content   To support your use of personal response systems in the classroom, we've coded the questions from the Quick Quiz questions for use with the system of your choice. Contact your Thomson representative for more details.

Multimedia Manager Instructor's Resource CD   This easy-to-use lecture tool allows you to quickly assemble art, photos, and multimedia with notes to create fluid lectures. The CD-ROM set (Volume 1, Chapters 1–14; Volume 2, Chapters 15–30) includes a database of animations, video clips and digital art from the text as well as PowerPoint lectures and electronic files of the Instructor's Solutions Manual and **Test Bank.** You'll also find additional content from the textbook like the Quick Quizzes and Conceptual Questions.

Physics⊗Now™   *PhysicsNow*™ **Course Management Tools powered by iLrn**   This extension to the student tutorial environment of PhysicsNow™ allows instructors to deliver online assignments in an environment that is familiar to students. This powerful system is your gateway to managing online homework, testing, and course administration all in one shell with the proven content to make your course a success. To see a demonstration of this powerful system, contact your Thomson representative or go to **www.cp7e.com.**

Physics⊗Now™   *PhysicsNow*™ **Homework Management powered by iLrn**   To further enhance PhysicsNow, contributing author Chuck Bennett has created ready-to-use homework assignments which are based on the in-chapter examples. These assignments feature step-by-step guides through algorithmic versions of the in-text examples paired with more challenging examples to extend your students' problem-solving abilities. The problems include hints and feedback to help students pinpoint their errors. The goal of this set of exercises is to bring students' problem solving abilities up to the level they will need to work the end-of-chapter problems.

   PhysicsNow™ gives you a rich array of problem types and grading options. Its library of assignable questions includes all of the end-of-chapter problems from the text so that you can select the problems you want to include in your online homework assignments. These well-crafted problems are algorithmically generated so that you can assign the same problem with different variables for each student. A flexible grading tolerance feature allows you to specify a percentage range of correct answers so that your students are not penalized for rounding errors. You can give students the option to work an assignment multiple times and record the highest score or limit the times they are able to attempt it. In addition, you can create your own problems to complement the problems from the text. Results flow automatically to an exportable grade book so that instructors are better able to assess student understanding of the material, even prior to class or to an actual test.

Physics⊗Now™ **iLrn Computerized Testing**   Extend the student experience with iLrn into a testing or quizzing environment. The test item file from the text is included to give you a bank of well-crafted questions that you can deliver online or print out. As with the homework problems, you can use the program's friendly interface to craft your own questions to complement the Serway/Faughn questions. You have complete control over grading, deadlines, and availability and can create multiple tests based on the same material.

WebTutor™ on WebCT and Blackboard   With **WebTutor™**'s text-specific, pre-formatted content and total flexibility, instructors can easily create and manage their own personal Web site. WebTutor™'s course management tool gives instructors the ability to provide virtual office hours, post syllabi, set up threaded discussions, track student progress with the quizzing material, and much more. WebTutor™ also provides robust communication tools, such as a course calendar, asynchronous discussion, real-time chat, a whiteboard, and an integrated e-mail system.

Additional Options for Online Homework

WebAssign: A Web-Based Homework System   WebAssign is the most utilized homework system in physics. Designed by physicists for physicists, this system is a trusted companion to your teaching. An enhanced version of WebAssign is available for College Physics. In addition to selected end-of-chapter problems, this enhanced version includes animations with conceptual questions. We've also added tutorial problems with feedback and hints to guide students' content mastery. Take a look at this new innovation from the most trusted name in physics homework. **www.webassign.net**

LON-CAPA   This Web-based course management system developed at Michigan State University includes selected end of chapter problems from *College Physics*. For more information, visit the LON-CAPA Web site at **www.lon-capa.org.**

UT Austin Homework Service   Details about and a demonstration of this service are available at **http://hw.ph.utexas.edu/hw.html.**

ACKNOWLEDGEMENTS

In preparing the seventh edition of this textbook, we have been guided by the expertise of many people who have reviewed manuscript and/or provided pre-revision suggestions. We wish to acknowledge the following reviewers and express our sincere appreciation for their helpful suggestions, criticism, and encouragement.

Seventh edition reviewers:
Gary B. Adams, *Arizona State University*
Ricardo Alarcon, *Arizona State University*
Natalie Batalha, *San Jose State University*
Ken Bolland, *Ohio State University*
Kapila Clara Castoldi, *Oakland University*
Andrew Cornelius, *University of Nevada, Las Vegas*
Yesim Darici, *Florida International University*
N. John DiNardo, *Drexel University*
Steve Ellis, *University of Kentucky*

Hasan Fakhruddin, *Ball State University/ The Indiana Academy*
Emily Flynn
Lewis Ford, *Texas A & M University*
James R. Goff, *Pima Community College*
Yadin Y. Goldschmidt, *University of Pittsburgh*
Steve Hagen, *University of Florida*
Raymond Hall, *California State University, Fresno*
Patrick Hamill, *San Jose State University*
Joel Handley
Grant W. Hart, *Brigham Young University*

James E. Heath, *Austin Community College*
Grady Hendricks, *Blinn College*
Rhett Herman, *Radford University*
Aleksey Holloway, *University of Nebraska at Omaha*
Joey Huston, *Michigan State University*
Mark James, *Northern Arizona University*
Randall Jones, *Loyola College in Maryland*
Joseph Keane, *St. Thomas Aquinas College*
Dorina Kosztin, *University of Missouri-Columbia*
Martha Lietz, *Niles West High School*
Edwin Lo
Rafael Lopez-Mobilia, *University of Texas at San Antonio*
John A. Milsom, *University of Arizona*

Monty Mola, *Humboldt State University*
Charles W. Myles, *Texas Tech University*
Ed Oberhofer, *Lake Sumter Community College*
Chris Pearson, *University of Michigan-Flint*
Alexey A. Petrov, *Wayne State University*
M. Anthony Reynolds, *Embry-Riddle Aeronautical University*
Dubravka Rupnik, *Louisiana State University*
Surajit Sen, *State University of New York at Buffalo*
Marllin L. Simon, *Auburn University*
Matthew Sirocky
George Strobel, *University of Georgia*
James Wanliss, *Embry-Riddle Aeronautical University*

Volume 2 of *College Physics,* seventh edition was carefully checked for accuracy by Ken Bolland (Ohio State University), Randall Jones (Loyola College in Maryland), and Edwin Lo. Though responsibility for any remaining errors rests with us, we thank them for their dedication and vigilance.

We thank the following people for their suggestions and assistance during the preparation of earlier editions of this textbook:

Marilyn Akins, *Broome Community College*
Albert Altman, *University of Lowell*
John Anderson, *University of Pittsburgh*
Lawrence Anderson-Huang, *University of Toledo*
Subhash Antani, *Edgewood College*
Neil W. Ashcroft, *Cornell University*
Charles R. Bacon, *Ferris State University*
Dilip Balamore, *Nassau Community College*
Ralph Barnett, *Florissant Valley Community College*
Lois Barrett, *Western Washington University*
Paul D. Beale, *University of Colorado at Boulder*
Paul Bender, *Washington State University*
David H. Bennum, *University of Nevada at Reno*
Jeffery Braun, *University of Evansville*
John Brennan, *University of Central Florida*
Michael Bretz, *University of Michigan, Ann Arbor*
Michael E. Browne, *University of Idaho*
Joseph Cantazarite, *Cypress College*
Ronald W. Canterna, *University of Wyoming*
Clinton M. Case, *Western Nevada Community College*
Neal M. Cason, *University of Notre Dame*
Roger W. Clapp, *University of South Florida*
Giuseppe Colaccico, *University of South Florida*
Lattie F. Collins, *East Tennessee State University*
Lawrence B. Colman, *University of California, Davis*
Jorge Cossio, *Miami Dade Community College*
Terry T. Crow, *Mississippi State College*
Stephen D. Davis, *University of Arkansas at Little Rock*
John DeFord, *University of Utah*
Chris J. DeMarco, *Jackson Community College*
Michael Dennin, *University of California, Irvine*
N. John DiNardo, *Drexel University*
Robert J. Endorf, *University of Cincinnati*
Paul Feldker, *Florissant Valley Community College*
Leonard X. Finegold, *Drexel University*
Tom French, *Montgomery County Community College*
Albert Thomas Frommhold, Jr., *Auburn University*
Lothar Frommhold, *University of Texas at Austin*
Eric Ganz, *University of Minnesota*
Teymoor Gedayloo, *California Polytechnic State University*

Simon George, *California State University, Long Beach*
John R. Gordon, *James Madison University*
George W. Greenlees, *University of Minnesota*
Wlodzimierz Guryn, *Brookhaven National Laboratory*
James Harmon, *Oklahoma State University*
Grant W. Hart, *Brigham Young University*
Christopher Herbert, *New Jersey City University*
John Ho, *State University of New York at Buffalo*
Murshed Hossain, *Rowan University*
Robert C. Hudson, *Roanoke College*
Fred Inman, *Mankato State University*
Ronald E. Jodoin, *Rochester Institute of Technology*
Drasko Jovanovic, *Fermilab*
George W. Kattawar, *Texas A & M University*
Frank Kolp, *Trenton State University*
Joan P.S. Kowalski, *George Mason University*
Ivan Kramer, *University of Maryland, Baltimore County*
Sol Krasner, *University of Chicago*
Karl F. Kuhn, *Eastern Kentucky University*
David Lamp, *Texas Tech University*
Harvey S. Leff, *California State Polytechnic University*
Joel Levine, *Orange Coast College*
Michael Lieber, *University of Arkansas*
James Linbald, *Saddleback Community College*
Bill Lochslet, *Pennsylvania State University*
Michael LoPresto, *Henry Ford Community College*
Bo Lou, *Ferris State University*
Jeffrey V. Mallow, *Loyola University of Chicago*
David Markowitz, *University of Connecticut*
Steven McCauley, *California State Polytechnic University, Pomona*
Joe McCauley, Jr., *University of Houston*
Ralph V. McGrew, *Broome Community College*
Bill F. Melton, *University of North Carolina at Charlotte*
H. Kent Moore, *James Madison University*
John Morack, *University of Alaska, Fairbanks*
Steven Morris, *Los Angeles Harbor College*
Carl R. Nave, *Georgia State University*
Martin Nikolo, *Saint Louis University*
Blaine Norum, *University of Virginia*

M. E. Oakes, *University of Texas at Austin*
Lewis J. Oakland, *University of Minnesota*
Ed Oberhofer, *Lake Sumter Community College*
Lewis O'Kelly, *Memphis State University*
David G. Onn, *University of Delaware*
J. Scott Payson, *Wayne State University*
T. A. K. Pillai, *University of Wisconsin, La Crosse*
Lawrence S. Pinsky, *University of Houston*
William D. Ploughe, *Ohio State University*
Patrick Polley, *Beloit College*
Brooke M. Pridmore, *Clayton State University*
Joseph Priest, *Miami University*
James Purcell, *Georgia State University*
W. Steve Quon, *Ventura College*
Michael Ram, *State University of New York at Buffalo*
Kurt Reibel, *Ohio State University*
Barry Robertson, *Queen's University*
Virginia Roundy, *California State University, Fullerton*

Larry Rowan, *University of North Carolina, Chapel Hill*
William R. Savage, *The University of Iowa*
Reinhard A. Schumacher, *Carnegie Mellon University*
John Simon, *University of Toledo*
Donald D. Snyder, *Indiana University at Southbend*
Carey E. Stronach, *Virginia State University*
Thomas W. Taylor, *Cleveland State University*
Perry A. Tompkins, *Samford University*
L. L. Van Zandt, *Purdue University*
Howard G. Voss, *Arizona State University*
Larry Weaver, *Kansas State University*
Donald H. White, *Western Oregon State College*
Bernard Whiting, *University of Florida*
George A. Williams, *The University of Utah*
Jerry H. Wilson, *Metropolitan State College*
Robert M. Wood, *University of Georgia*
Clyde A. Zaidins, *University of Colorado at Denver*

Randall Jones generously contributed many of the new end-of-chapter problems, especially those of interest to the life sciences. Edward F. Redish of the University of Maryland graciously allowed us to list some of his problems from the Activity Based Physics Project as both problems and activities. Some of the remaining activities were written by Robert J. Beichner of North Carolina State University.

We are extremely grateful to the publishing team at the Brooks/Cole Publishing Company for their expertise and outstanding work in all aspects of this project. In particular, we'd like to thank Ed Dodd, who tirelessly coordinated and directed our efforts in preparing the manuscript in its various stages, and Annie Mac, who worked behind the scenes securing text and accuracy reviews, and coordinated and transmitted all the print ancillaries. Jane Sanders, the photo researcher, did a great job finding photos of physical phenomena, Sam Subity coordinated the building of the *PhysicsNow*™ Web site, and Rob Hugel helped translate our rough sketches into accurate, compelling art. Donna King of Progressive Publishing Alternatives managed the difficult task of keeping production moving and on schedule. Chris Hall and Teri Hyde also made numerous valuable contributions. Chris provided just the right amount of guidance and vision throughout the project. We would like to thank David Harris, a great team builder and motivator with loads of enthusiasm and an infectious sense of humor.

Finally, we are deeply indebted to our wives and children for their love, support, and long-term sacrifices.

Raymond A. Serway
St. Petersburg, Florida

Jerry S. Faughn
Richmond, Kentucky

Chris Vuille
Daytona Beach, Florida

Charles A. Bennett
Asheville, North Carolina

Applications

Although physics is relevant to so much in our modern lives, this may not be obvious to students in an introductory course. In this seventh edition of *College Physics*, we continue a design feature begun in the previous edition. This feature makes the relevance of physics to everyday life more obvious by pointing out specific applications in the form of a marginal note. Some of these applications pertain to the life sciences and are marked with the DNA icon ⬛. The list below is not intended to be a complete listing of all the applications of the principles of physics found in this textbook. Many other applications are to be found within the text and especially in the worked examples, conceptual questions, and end-of-chapter problems.

To the Student

As a student, it's important that you understand how to use the book most effectively, and how best to go about learning physics. Scanning through the Preface will acquaint you with the various features available, both in the book and online. Awareness of your educational resources and how to use them is essential. Though physics is challenging, it can be mastered with the correct approach.

HOW TO STUDY

Students often ask how best to study physics and prepare for examinations. There's no simple answer to this question, but we'd like to offer some suggestions based on our own experiences in learning and teaching over the years.

First and foremost, maintain a positive attitude toward the subject matter. Like learning a language, physics takes time. Those who keep applying themselves on a daily basis can expect to reach understanding. Keep in mind that physics is the most fundamental of all natural sciences. Other science courses that follow will use the same physical principles, so it is important that you understand and are able to apply the various concepts and theories discussed in the text.

CONCEPTS AND PRINCIPLES

Students often try to do their homework without first studying the basic concepts. It is essential that you understand the basic concepts and principles *before* attempting to solve assigned problems. You can best accomplish this goal by carefully reading the textbook *before* you attend your lecture on the covered material. When reading the text, you should jot down those points that are not clear to you. Also be sure to make a diligent attempt at answering the questions in the Quick Quizzes as you come to them in your reading. We have worked hard to prepare questions that help you judge for yourself how well you understand the material. Pay careful attention to the many Tips throughout the text. These will help you avoid misconceptions, mistakes, and misunderstandings as well as maximize the efficiency of your time by minimizing adventures along fruitless paths. During class, take careful notes and ask questions about those ideas that are unclear to you. Keep in mind that few people are able to absorb the full meaning of scientific material after only one reading.

Be sure to take advantage of the features available in the *PhysicsNow*™ learning system, such as the Active Figures, Interactive Examples, and Coached Problems. Your lectures and laboratory work supplement your textbook and should clarify some of the more difficult material. You should minimize rote memorization of material. Successful memorization of passages from the text, equations, and derivations does not necessarily indicate that you understand the fundamental principles.

Your understanding will be enhanced through a combination of efficient study habits, discussions with other students and with instructors, and your ability to solve the problems presented in the textbook. Ask questions whenever you feel clarification of a concept is necessary.

STUDY SCHEDULE

It is important for you to set up a regular study schedule, preferably a daily one. Make sure you read the syllabus for the course and adhere to the schedule set by your instructor. As a general rule, you should devote about two hours of study time for every hour you are in class. If you are having trouble with the course, seek the advice of the instructor or other students who have taken the course. You may find it necessary to seek further instruction from experienced students. Very often, instructors offer review sessions in addition to regular class periods. It is important that you avoid the practice of delaying study until a day or two before an exam. One hour of study a day for fourteen days is far more effective than fourteen hours the day before the exam. "Cramming" usually produces disastrous results, especially in science. Rather than undertake an all-night study session just before an exam, briefly review the basic concepts and equations and get a good night's rest. If you feel you need additional help in understanding the concepts, in preparing for exams, or in problem-solving, we suggest that you acquire a copy of the *Student Solutions Manual and Study Guide* that accompanies this textbook; this manual should be available at your college bookstore.

USE THE FEATURES

You should make full use of the various features of the text discussed in the preface. For example, marginal notes are useful for locating and describing important equations and concepts, and **boldfaced** type indicates important statements and definitions. Many useful tables are contained in the Appendices, but most tables are incorporated in the text where they are most often referenced. Appendix A is a convenient review of mathematical techniques.

Answers to all Quick Quizzes, as well as odd-numbered conceptual questions and problems, are given at the end of the textbook. Answers to selected end-of-chapter problems are provided in the *Student Solutions Manual and Study Guide.* Problem-Solving Strategies included in selected chapters throughout the text give you additional information about how you should solve problems. The Table of Contents provides an overview of the entire text, while the Index enables you to locate specific material quickly. Footnotes sometimes are used to supplement the text or to cite other references on the subject discussed.

After reading a chapter, you should be able to define any new quantities introduced in that chapter and to discuss the principles and assumptions used to arrive at certain key relations. The chapter summaries and the review sections of the *Student Solutions Manual and Study Guide* should help you in this regard. In some cases, it may be necessary for you to refer to the index of the text to locate certain topics. You should be able to correctly associate with each physical quantity the symbol used to represent that quantity and the unit in which the quantity is specified. Furthermore, you should be able to express each important relation in a concise and accurate prose statement.

PROBLEM-SOLVING

R. P. Feynman, Nobel laureate in physics, once said, "You do not know anything until you have practiced." In keeping with this statement, we strongly advise that you develop the skills necessary to solve a wide range of problems. Your ability to solve problems will be one of the main tests of your knowledge of physics; therefore, you should try to solve as many problems as possible. It is essential that you understand basic concepts and principles before attempting to solve problems. It is good practice to try to find alternate solutions to the same problem. For example, you can solve problems in mechanics using Newton's laws, but very often an alternate method that draws on energy considerations is more direct. You should not deceive yourself into thinking you understand a problem merely because you have seen it solved in class. You must be able to solve the problem and similar problems on your own. We have recast the examples in this book to help you in this regard. After studying an example, see if you can cover up the right-hand side and do it yourself, using only the written descriptions on the left as hints. Once you succeed at that, try solving the example completely on your own. Finally, solve the exercise. Once you have accomplished all this, you will have a good mastery of the problem, its concepts, and mathematical technique.

The approach to solving problems should be carefully planned. A systematic plan is especially important when a problem involves several concepts. First, read the problem several times until you are confident you understand what is being asked. Look for any key words that will help you interpret the problem and perhaps allow you to make certain assumptions. Your ability to interpret a question properly is an integral part of problem-solving. Second, you should acquire the habit of writing down the information given in a problem and those quantities that need to be found; for example, you might construct a table listing both the quantities given and the quantities to be found. This procedure is sometimes used in the worked examples of the textbook. After you have decided on the method you feel is appropriate for a given problem, proceed with your solution. Finally, check your results to see if they are reasonable and consistent with your initial understanding of the problem. General problem-solving strategies of this type are included in the text and are highlighted with a surrounding box. If you follow the steps of this procedure, you will find it easier to come up with a solution and also gain more from your efforts.

Often, students fail to recognize the limitations of certain equations or physical laws in a particular situation. It is very important that you understand and remember the assumptions underlying a particular theory or formalism. For example, certain equations in kinematics apply only to a particle moving with constant acceleration. These equations are not valid for describing motion whose acceleration is not constant, such as the motion of an object connected to a spring or the motion of an object through a fluid.

EXPERIMENTS

Physics is a science based on experimental observations. In view of this fact, we recommend that you try to supplement the text by performing various types of "hands-on" experiments, either at home or in the laboratory. For example, the common Slinky™ toy is excellent for studying traveling waves; a ball swinging on the end of a long string can be used to investigate pendulum motion; various masses attached to the end of a vertical spring or rubber band can be used to determine their elastic nature; an old pair of Polaroid sunglasses and some discarded lenses and a magnifying glass are the components of various experiments in optics; and the approximate measure of the free-fall acceleration can be determined simply by measuring with a stopwatch the time it takes for a ball to drop from a known height. The list of such experiments is endless. When physical models are not available, be imaginative and try to develop models of your own.

Some of the Activities at the end of each chapter can be performed on your own or with the help of a friend and will give you specific guidelines for acquiring "hands-on" experience.

PhysicsNow™ with Active Figures and Interactive Examples

We strongly encourage you to use the *PhysicsNow*™ Web-based learning system that accompanies this textbook. It is far easier to understand physics if you see it in action, and these new materials will enable you to become a part of that action. *PhysicsNow*™ media described in the Preface are accessed at the URL **www.cp7e.com**, and feature a three-step learning process consisting of a Pre-Test, a personalized learning plan, and a Post-Test.

In addition to the Coached Problems identified with icons, *PhysicsNow*™ includes the following Active Figures and Interactive Examples from Volume 2:

Chapter 15

Active Figures 15.6, 15.11, 15.21, and 15.28

Interactive Example 15.2

Chapter 16

Active Figures 16.7, 16.17, and 16.19

Interactive Examples 16.2 and 16.8

Chapter 17

Active Figures 17.4 and 17.11

Interactive Example 17.4

Chapter 18

Active Figures 18.1, 18.6, 18.16, and 18.17

Interactive Examples 18.2 and 18.5

An Invitation to Physics

It is our hope that you too will find physics an exciting and enjoyable experience and that you will profit from this experience, regardless of your chosen profession. Welcome to the exciting world of physics!

To see the World in a Grain of Sand
And a Heaven in a Wild Flower,
Hold infinity in the palm of your hand
And Eternity in an hour.

William Blake, "Auguries of Innocence"

This nighttime view of multiple bolts of lightning was photographed in Tucson, Arizona. During a thunderstorm, a high concentration of electrical charge in a thundercloud creates a higher-than-normal electric field between the thundercloud and the negatively charged Earth's surface. This strong electric field creates an electric discharge between the charged cloud and the ground—an enormous spark. Other discharges that are observed in the sky include cloud-to-cloud discharges and the more frequent intracloud discharges.

© Keith Kent/Science Photo Library/Photo Researchers, Inc.

CHAPTER

15

Electric Forces and Electric Fields

Electricity is the lifeblood of technological civilization and modern society. Without it, we revert to the mid-nineteenth century: no telephones, no television, none of the household appliances that we take for granted. Modern medicine would be a fantasy, and due to the lack of sophisticated experimental equipment and fast computers—and especially the slow dissemination of information—science and technology would grow at a glacial pace.

Instead, with the discovery and harnessing of electric forces and fields, we can view arrangements of atoms, probe the inner workings of the cell, and send spacecraft beyond the limits of the solar system. All this has become possible in just the last few generations of human life, a blink of the eye compared to the million years our kind spent foraging the savannahs of Africa.

Around 700 B.C. the ancient Greeks conducted the earliest known study of electricity. It all began when someone noticed that a fossil material called amber would attract small objects after being rubbed with wool. Since then we have learned that this phenomenon is not restricted to amber and wool, but occurs (to some degree) when almost any two nonconducting substances are rubbed together.

In the current chapter we use the effect of charging by friction to begin an investigation of electric forces. We then discuss Coulomb's law, which is the fundamental law of force between any two stationary charged particles. The concept of an electric field associated with charges is introduced and its effects on other charged particles described. We end with discussions of the Van de Graaff generator and Gauss's law.

15.1 PROPERTIES OF ELECTRIC CHARGES

After running a plastic comb through your hair, you will find that the comb attracts bits of paper. The attractive force is often strong enough to suspend the

Figure 15.1 (a) A negatively charged rubber rod, suspended by a thread, is attracted to a positively charged glass rod. (b) A negatively charged rubber rod is repelled by another negatively charged rubber rod.

(a)   (b)

BENJAMIN FRANKLIN
(1706–1790)

Franklin was a printer, author, physical scientist, inventor, diplomat, and a founding father of the United States. His work on electricity in the late 1740s changed a jumbled, unrelated set of observations into a coherent science.

Like charges repel
Unlike charges attract ▶

paper from the comb, defying the gravitational pull of the entire Earth. The same effect occurs with other rubbed materials, such as glass and hard rubber.

Another simple experiment is to rub an inflated balloon against wool (or across your hair). On a dry day, the rubbed balloon will then stick to the wall of a room, often for hours. These materials have become **electrically charged**. You can give your body an electric charge by vigorously rubbing your shoes on a wool rug or by sliding across a car seat. You can then surprise and annoy a friend or co-worker with a light touch on the arm, delivering a slight shock to both yourself and your victim. (If the co-worker is your boss, don't expect a promotion!) These experiments work best on a dry day because excessive moisture can facilitate a leaking away of the charge.

Experiments also demonstrate that there are two kinds of electric charge, which Benjamin Franklin (1706–1790) named **positive** and **negative**. Figure 15.1 illustrates the interaction of the two charges. A hard rubber (or plastic) rod that has been rubbed with fur is suspended by a piece of string. When a glass rod that has been rubbed with silk is brought near the rubber rod, the rubber rod is attracted toward the glass rod (Fig. 15.1a). If two charged rubber rods (or two charged glass rods) are brought near each other, as in Figure 15.1b, the force between them is repulsive. These observations may be explained by assuming that the rubber and glass rods have acquired different kinds of excess charge. We use the convention suggested by Franklin, where the excess electric charge on the glass rod is called positive and that on the rubber rod is called negative. On the basis of observations such as these, we conclude that **like charges repel one another and unlike charges attract one another**. Objects usually contain equal amounts of positive and negative charge—electrical forces between objects arise when those objects have net negative or positive charges.

Nature's basic carriers of positive charge are protons, which, along with neutrons, are located in the nuclei of atoms. The nucleus, about 10^{-15} m in radius, is surrounded by a cloud of negatively charged electrons about ten thousand times larger in extent. An electron has the same magnitude charge as a proton, but the opposite sign. In a gram of matter there are approximately 10^{23} positively charged protons and just as many negatively charged electrons, so the net charge is zero. Because the nucleus of an atom is held firmly in place inside a solid, protons never move from one material to another. Electrons are far lighter than protons and hence more easily accelerated by forces. Furthermore, they occupy the outer regions of the atom. Consequently, objects become charged by gaining or losing electrons.

Charge transfers readily from one type of material to another. Rubbing the two materials together serves to increase the area of contact, facilitating the transfer process.

An important characteristic of charge is that **electric charge is always conserved**. Charge isn't *created* when two neutral objects are rubbed together; rather, the objects become charged because **negative charge is transferred from one object to the other**. One object gains a negative charge while the other loses an equal amount of negative charge and hence is left with a net positive charge. When a glass rod is rubbed with silk, as in Figure 15.2, electrons are transferred from the rod to the silk. As a result, the glass rod carries a net positive charge, the silk a net negative charge. Likewise, when rubber is rubbed with fur, electrons are transferred from the fur to the rubber.

In 1909 Robert Millikan (1886–1953) discovered that if an object is charged, its charge is always a multiple of a fundamental unit of charge, designated by the symbol e. In modern terms, the charge is said to be **quantized**, meaning that charge occurs in discrete chunks that can't be further subdivided. An object may have a charge of $\pm e$, $\pm 2e$, $\pm 3e$, and so on, but never[1] a fractional charge of $\pm 0.5e$ or $\pm 0.22e$. Other experiments in Millikan's time showed that the electron has a charge of $-e$ and the proton has an equal and opposite charge of $+e$. Some particles, such as a neutron, have no net charge. A neutral atom (an atom with no net charge) contains as many protons as electrons. The value of e is now known to be $1.602\ 19 \times 10^{-19}$ C. (The SI unit of electric charge is the **coulomb** [C].)

◀ Charge is conserved; charge is quantized

Figure 15.2   When a glass rod is rubbed with silk, electrons are transferred from the glass to the silk. Because of conservation of charge, each electron adds negative charge to the silk, and an equal positive charge is left behind on the rod. Also, because the charges are transferred in discrete bundles, the charges on the two objects are $\pm e$, $\pm 2e$, $\pm 3e$, and so on.

15.2   INSULATORS AND CONDUCTORS

Substances can be classified in terms of their ability to conduct electric charge.

> In **conductors**, electric charges move freely in response to an electric force. All other materials are called **insulators**.

Glass and rubber are insulators. When such materials are charged by rubbing, only the rubbed area becomes charged, and there is no tendency for the charge to move into other regions of the material. In contrast, materials such as copper, aluminum, and silver are good conductors. When such materials are charged in some small region, the charge readily distributes itself over the entire surface of the material. If you hold a copper rod in your hand and rub the rod with wool or fur, it will not attract a piece of paper. This might suggest that a metal can't be charged. However, if you hold the copper rod with an insulator and then rub it with wool or fur, the rod remains charged and attracts the paper. In the first case, the electric charges produced by rubbing readily move from the copper through your body and finally to ground. In the second case, the insulating handle prevents the flow of charge to ground.

Semiconductors are a third class of materials, and their electrical properties are somewhere between those of insulators and those of conductors. Silicon and germanium are well-known semiconductors that are widely used in the fabrication of a variety of electronic devices.

Charging by Conduction

Consider a negatively charged rubber rod brought into contact with an insulated neutral conducting sphere. The excess electrons on the rod repel electrons on the sphere, creating local positive charges on the neutral sphere. On contact, some electrons on the rod are now able to move onto the sphere, as in Figure 15.3, neutralizing the positive charges. When the rod is removed, the sphere is left with a net negative charge. This process is referred to as charging by **conduction.** The object being charged in such a process (the sphere) is always left with a charge having the same sign as the object doing the charging (the rubber rod).

Rubber

(a) Before

(b) Contact

(c) After breaking contact

Figure 15.3   Charging a metallic object by conduction. (a) Just before contact, the negative rod repels the sphere's electrons, inducing a localized positive charge. (b) After contact, electrons from the rod flow onto the sphere, neutralizing the local positive charges. (c) When the rod is removed, the sphere is left with a negative charge.

---

[1]There is strong evidence for the existence of fundamental particles called **quarks** that have charges of $\pm e/3$ or $\pm 2e/3$. The charge is *still* quantized, but in units of $\pm e/3$ rather than $\pm e$. A more complete discussion of quarks and their properties is presented in Chapter 30.

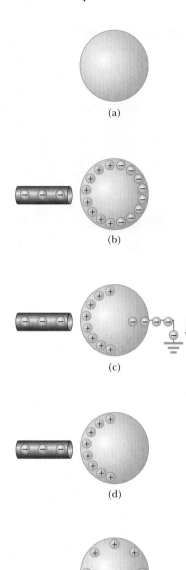

Figure 15.4 Charging a metallic object by *induction*. (a) A neutral metallic sphere with equal numbers of positive and negative charges. (b) The charge on a neutral metal sphere is redistributed when a charged rubber rod is placed near the sphere. (c) When the sphere is grounded, some of the electrons leave it through the ground wire. (d) When the ground connection is removed, the nonuniformly charged sphere is left with excess positive charge. (e) When the rubber rod is moved away, the charges on the sphere redistribute themselves until the sphere's surface becomes uniformly charged.

Charging by Induction

An object connected to a conducting wire or copper pipe buried in the Earth is said to be **grounded**. The Earth can be considered an infinite reservoir for electrons; in effect, it can accept or supply an unlimited number of electrons. With this idea in mind, we can understand the charging of a conductor by a process known as **induction**.

Consider a negatively charged rubber rod brought near a neutral (uncharged) conducting sphere that is insulated, so there is no conducting path to ground (Fig. 15.4). Initially the sphere is electrically neutral (Fig. 15.4a). When the negatively charged rod is brought close to the sphere, the repulsive force between the electrons in the rod and those in the sphere causes some electrons to move to the side of the sphere farthest away from the rod (Fig. 15.4b). The region of the sphere nearest the negatively charged rod has an excess of positive charge because of the migration of electrons away from that location. If a grounded conducting wire is then connected to the sphere, as in Figure 15.4c, some of the electrons leave the sphere and travel to ground. If the wire to ground is then removed (Fig. 15.4d), the conducting sphere is left with an excess of induced positive charge. Finally, when the rubber rod is removed from the vicinity of the sphere (Fig. 15.4e), the induced positive charge remains on the ungrounded sphere. Even though the positively charged atomic nuclei remain fixed, this excess positive charge becomes uniformly distributed over the surface of the ungrounded sphere because of the repulsive forces among the like charges and the high mobility of electrons in a metal.

In the process of inducing a charge on the sphere, the charged rubber rod doesn't lose any of its negative charge because it never comes in contact with the sphere. Furthermore, the sphere is left with a charge opposite that of the rubber rod. **Charging an object by induction requires no contact with the object inducing the charge**.

A process similar to charging by induction in conductors also takes place in insulators. In most neutral atoms or molecules, the center of positive charge coincides with the center of negative charge. However, in the presence of a charged object, these centers may separate slightly, resulting in more positive charge on one side of the molecule than on the other. This effect is known as **polarization.** The realignment of charge within individual molecules produces an induced charge on the surface of the insulator, as shown in Figure 15.5a. This explains why a balloon charged through rubbing will stick to an electrically neutral wall, or the comb you just used on your hair attracts tiny bits of neutral paper.

Quick Quiz 15.1

A suspended object A is attracted to a neutral wall. It's also attracted to a positively charged object B. Which of the following is true about object A? (a) It is uncharged. (b) It has a negative charge. (c) It has a positive charge. (d) It may be either charged or uncharged.

15.3 COULOMB'S LAW

In 1785 Charles Coulomb (1736–1806) experimentally established the fundamental law of electric force between two stationary charged particles.

An **electric force** has the following properties:

1. It is directed along a line joining the two particles and is inversely proportional to the square of the separation distance r, between them.
2. It is proportional to the product of the magnitudes of the charges, $|q_1|$ and $|q_2|$, of the two particles.
3. It is attractive if the charges are of opposite sign and repulsive if the charges have the same sign.

Figure 15.5 (a) The charged object on the left induces charges on the surface of an insulator. (b) A charged comb attracts bits of paper because charges are displaced in the paper.

(a)   (b)

From these observations, Coulomb proposed the following mathematical form for the electric force between two charges:

◀ Coulomb's law

> The magnitude of the electric force F between charges q_1 and q_2 separated by a distance r is given by
>
> $$F = k_e \frac{|q_1||q_2|}{r^2} \qquad \text{[15.1]}$$
>
> where k_e is a constant called the *Coulomb constant*.

Equation 15.1, known as **Coulomb's law**, applies exactly only to point charges and to spherical distributions of charges, in which case r is the distance between the two centers of charge. Electric forces between unmoving charges are called *electrostatic* forces. Moving charges, in addition, create magnetic forces, studied in Chapter 19.

The value of the Coulomb constant in Equation 15.1 depends on the choice of units. The SI unit of charge is the **coulomb** (C). From experiment, we know that the **Coulomb constant** in SI units has the value

$$k_e = 8.9875 \times 10^9 \, \text{N} \cdot \text{m}^2/\text{C}^2 \qquad \text{[15.2]}$$

This number can be rounded, depending on the accuracy of other quantities in a given problem. We'll use either two or three significant digits, as usual.

The charge on the proton has a magnitude of $e = 1.6 \times 10^{-19}$ C. Therefore, it would take $1/e = 6.3 \times 10^{18}$ protons to create a total charge of $+ 1.0$ C. Likewise, 6.3×10^{18} electrons would have a total charge of $- 1.0$ C. Compare this with the number of free electrons in 1 cm^3 of copper, which is on the order of 10^{23}. Even so, 1.0 C is a very large amount of charge. In typical electrostatic experiments in which a rubber or glass rod is charged by friction, there is a net charge on the order of 10^{-6} C ($= 1 \, \mu$C). Only a very small fraction of the total available charge is transferred between the rod and the rubbing material. Table 15.1 lists the charges and masses of the electron, proton, and neutron.

TABLE 15.1

Charge and Mass of the Electron, Proton, and Neutron

Particle	Charge (C)	Mass (kg)
Electron	$- 1.60 \times 10^{-19}$	9.11×10^{-31}
Proton	$+ 1.60 \times 10^{-19}$	1.67×10^{-27}
Neutron	0	1.67×10^{-27}

CHARLES COULOMB
(1736–1806)

Coulomb's major contribution to science was in the field of electrostatics and magnetism. During his lifetime, he also investigated the strengths of materials and identified the forces that affect objects on beams, thereby contributing to the field of structural mechanics.

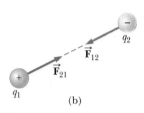

ACTIVE FIGURE 15.6
Two point charges separated by a distance r exert a force on each other given by Coulomb's law. The force on q_1 is equal in magnitude and opposite in direction to the force on q_2. (a) When the charges are of the same sign, the force is repulsive. (b) When the charges are of opposite sign, the force is attractive.

Physics⊗Now™

Log into PhysicsNow at **www.cp7e.com** and go to Active Figure 15.6, where you can move the charges to any position in two-dimensional space and observe the electric forces acting on them.

When using Coulomb's force law, remember that force is a vector quantity and must be treated accordingly. Active Figure 15.6a shows the electric force of repulsion between two positively charged particles. Like other forces, electric forces obey Newton's third law; hence, the forces $\vec{F}_{12}$ and $\vec{F}_{21}$ are equal in magnitude but opposite in direction. (The notation $\vec{F}_{12}$ denotes the force exerted by particle 1 on particle 2; likewise, $\vec{F}_{21}$ is the force exerted by particle 2 on particle 1.) From Newton's third law, F_{12} and F_{21} are always equal regardless of whether q_1 and q_2 have the same magnitude.

Quick Quiz 15.2

Object A has a charge of $+2\ \mu$C, and object B has a charge of $+6\ \mu$C. Which statement is true?

(a) $\vec{F}_{AB} = -3\vec{F}_{BA}$ (b) $\vec{F}_{AB} = -\vec{F}_{BA}$ (c) $3\vec{F}_{AB} = -\vec{F}_{BA}$

The Coulomb force is similar to the gravitational force. Both act at a distance without direct contact. Both are inversely proportional to the distance squared, with the force directed along a line connecting the two bodies. The mathematical form is the same, with the masses m_1 and m_2 in Newton's law replaced by q_1 and q_2 in Coulomb's law and with Newton's constant G replaced by Coulomb's constant k_e. There are two important differences: (1) electric forces can be either attractive or repulsive, but gravitational forces are always attractive, and (2) the electric force between charged elementary particles is far stronger than the gravitational force between the same particles, as the next example shows.

EXAMPLE 15.1   Forces in a Hydrogen Atom

Goal   Contrast the magnitudes of an electric force and a gravitational force.

Problem   The electron and proton of a hydrogen atom are separated (on the average) by a distance of about 5.3×10^{-11} m. Find the magnitudes of the electric force and the gravitational force that each particle exerts on the other, and the ratio of the electric force F_e to the gravitational force F_g.

Strategy   Solving this problem is just a matter of substituting known quantities into the two force laws and then finding the ratio.

Solution

Substitute $|q_1| = |q_2| = e$ and the distance into Coulomb's law to find the electric force:

$$F_e = k_e \frac{|e|^2}{r^2} = \left(8.99 \times 10^9\ \frac{\text{N} \cdot \text{m}^2}{\text{C}^2}\right) \frac{(1.6 \times 10^{-19}\ \text{C})^2}{(5.3 \times 10^{-11}\ \text{m})^2}$$

$$= \boxed{8.2 \times 10^{-8}\ \text{N}}$$

Substitute the masses and distance into Newton's law of gravity to find the gravitational force:

$$F_g = G \frac{m_e m_p}{r^2}$$

$$= \left(6.67 \times 10^{-11}\ \frac{\text{N} \cdot \text{m}^2}{\text{kg}^2}\right) \frac{(9.11 \times 10^{-31}\ \text{kg})(1.67 \times 10^{-27}\ \text{kg})}{(5.3 \times 10^{-11}\ \text{m})^2}$$

$$= \boxed{3.6 \times 10^{-47}\ \text{N}}$$

Find the ratio of the two forces:

$$\frac{F_e}{F_g} = \boxed{2.27 \times 10^{39}}$$

Remarks   The gravitational force between the charged constituents of the atom is negligible compared with the electric force between them. The electric force is so strong, however, that any net charge on an object quickly attracts nearby opposite charges, neutralizing the object. As a result, gravity plays a greater role in the mechanics of moving objects in everyday life.

Exercise 15.1
Find the magnitude of the electric force between two protons separated by one femtometer (10^{-15} m), approximately the distance between two protons in the nucleus of a helium atom. The answer may not appear large, but if not for the strong nuclear force, the two protons would fly apart at an initial acceleration of nearly 7×10^{28} m/s^2!

Answers 2.30×10^2 N

The Superposition Principle

When a number of separate charges act on the charge of interest, each exerts an electric force. These electric forces can all be computed separately, one at a time, then added as vectors. This is another example of the **superposition principle**. The following example illustrates this procedure in one dimension.

INTERACTIVE EXAMPLE 15.2   May the Force Be Zero

Goal   Apply Coulomb's law in one dimension.

Problem   Three charges lie along the x-axis as in Figure 15.7. The positive charge $q_1 = 15\ \mu C$ is at $x = 2.0$ m, and the positive charge $q_2 = 6.0\ \mu C$ is at the origin. Where must a *negative* charge q_3 be placed on the x-axis so that the resultant electric force on it is zero?

Strategy   If q_3 is to the right or left of the other two charges, then the net force on q_3 can't be zero, because then $\vec{\mathbf{F}}_{13}$ and $\vec{\mathbf{F}}_{23}$ act in the same direction. Consequently, q_3 must lie between the two other charges. Write $\vec{\mathbf{F}}_{13}$ and $\vec{\mathbf{F}}_{23}$ in terms of the unknown coordinate position x, sum them and set them equal to zero, solving for the unknown. The solution can be obtained with the quadratic formula.

Figure 15.7   (Example 15.2) Three point charges are placed along the x-axis. The charge q_3 is negative, whereas q_1 and q_2 are positive. If the resultant force on q_3 is zero, then the force $\vec{\mathbf{F}}_{13}$ exerted by q_1 on q_3 must be equal in magnitude and opposite the force $\vec{\mathbf{F}}_{23}$ exerted by q_2 on q_3.

Solution

Write the x-component of $\vec{\mathbf{F}}_{13}$:

$$F_{13x} = + k_e \frac{(15 \times 10^{-6}\ \text{C})|q_3|}{(2.0\ \text{m} - x)^2}$$

Write the x-component of $\vec{\mathbf{F}}_{23}$:

$$F_{23x} = - k_e \frac{(6.0 \times 10^{-6}\ \text{C})|q_3|}{x^2}$$

Set the sum equal to zero:

$$k_e \frac{(15 \times 10^{-6}\ \text{C})|q_3|}{(2.0\ \text{m} - x)^2} - k_e \frac{(6.0 \times 10^{-6}\ \text{C})|q_3|}{x^2} = 0$$

Cancel k_e, 10^{-6} and q_3 from the equation, and rearrange terms (explicit significant figures and units are temporarily suspended for clarity):

$$6(2 - x)^2 = 15x^2$$

Put this equation into standard quadratic form, $ax^2 + bx + c = 0$:

$$6(4 - 4x + x^2) = 15x^2 \quad \rightarrow \quad 2(4 - 4x + x^2) = 5x^2$$
$$3x^2 + 8x - 8 = 0$$

Apply the quadratic formula:

$$x = \frac{-8 \pm \sqrt{64 - (4)(3)(-8)}}{2 \cdot 3} = \frac{-4 \pm 2\sqrt{10}}{3}$$

Only the positive root makes sense:

$$x = \boxed{0.77\ \text{m}}$$

Remarks   Notice that it was necessary to use physical reasoning to choose between the two possible answers for x. This is nearly always the case when quadratic equations are involved.

Exercise 15.2

Three charges lie along the x-axis. The positive charge $q_1 = 10.0\ \mu C$ is at $x = 1.00$ m, and the *negative charge* $q_2 = -2.00\ \mu C$ is at the origin. Where must a *positive* charge q_3 be placed on the x-axis so that the resultant force on it is zero?

Answer $x = -0.809$ m

Physics⊗Now™ You can predict where on the x-axis the electric force is zero for random values of of q_1 and q_2 by logging into PhysicsNow at **www.cp7e.com** and going to Interactive Example 15.2.

EXAMPLE 15.3 A Charge Triangle

Goal Apply Coulomb's law in two dimensions.

Problem Consider three point charges at the corners of a triangle, as shown in Figure 15.8, where $q_1 = 6.00 \times 10^{-9}$ C, $q_2 = -2.00 \times 10^{-9}$ C, and $q_3 = 5.00 \times 10^{-9}$ C. **(a)** Find the components of the force $\vec{F}_{23}$ exerted by q_2 on q_3. **(b)** Find the components of the force $\vec{F}_{13}$ exerted by q_1 on q_3. **(c)** Find the resultant force on q_3, in terms of components and also in terms of magnitude and direction.

Strategy Coulomb's law gives the magnitude of each force, which can be split with right-triangle trigonometry into x- and y-components. Sum the vectors component-wise, and then find the magnitude and direction of the resultant vector.

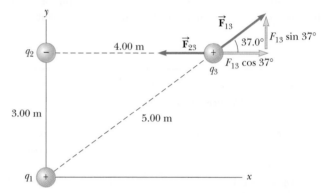

Figure 15.8 (Example 15.3) The force exerted by q_1 on q_3 is $\vec{F}_{13}$. The force exerted by q_2 on q_3 is $\vec{F}_{23}$. The *resultant force* $\vec{F}_3$ exerted on q_3 is the *vector* sum $\vec{F}_{13} + \vec{F}_{23}$.

Solution

(a) Find the components of the force exerted by q_2 on q_3.

Find the magnitude of $\vec{F}_{23}$ with Coulomb's law:

$$F_{23} = k_e \frac{|q_2||q_3|}{r^2}$$

$$= (8.99 \times 10^9\ \text{N·m}^2/\text{C}^2) \frac{(2.00 \times 10^{-9}\ \text{C})(5.00 \times 10^{-9}\ \text{C})}{(4.00\ \text{m})^2}$$

$$F_{23} = 5.62 \times 10^{-9}\ \text{N}$$

Because $\vec{F}_{23}$ is horizontal and points in the negative x-direction, the negative of the magnitude gives the x-component, and the y-component is zero:

$$F_{23x} = \boxed{-5.62 \times 10^{-9}\ \text{N}}$$

$$F_{23y} = \boxed{0}$$

(b) Find the components of the force exerted by q_1 on q_3.

Find the magnitude of $\vec{F}_{13}$:

$$F_{13} = k_e \frac{|q_1||q_3|}{r^2}$$

$$= (8.99 \times 10^9\ \text{N·m}^2/\text{C}^2) \frac{(6.00 \times 10^{-9}\ \text{C})(5.00 \times 10^{-9}\ \text{C})}{(5.00\ \text{m})^2}$$

$$F_{13} = 1.08 \times 10^{-8}\ \text{N}$$

Use the given triangle to find the components:

$$F_{13x} = F_{13} \cos\theta = (1.08 \times 10^{-8}\ \text{N})\cos(37°)$$

$$= \boxed{8.63 \times 10^{-9}\ \text{N}}$$

$$F_{13y} = F_{13} \sin\theta = (1.08 \times 10^{-8}\ \text{N})\sin(37°)$$

$$= \boxed{6.50 \times 10^{-9}\ \text{N}}$$

(c) Find the components of the resultant vector.

Sum the x-components to find the resultant F_x:

$$F_x = -5.62 \times 10^{-9}\,\text{N} + 8.63 \times 10^{-9}\,\text{N}$$
$$= \boxed{3.01 \times 10^{-9}\,\text{N}}$$

Sum the y-components to find the resultant F_y:

$$F_y = 0 + 6.50 \times 10^{-9}\,\text{N} = \boxed{6.50 \times 10^{-9}\,\text{N}}$$

Find the magnitude of the resultant force on the charge q_3, using the Pythagorean theorem:

$$|\vec{\mathbf{F}}| = \sqrt{F_x{}^2 + F_y{}^2}$$
$$= \sqrt{(3.01 \times 10^{-9}\,\text{N})^2 + (6.50 \times 10^{-9}\,\text{N})^2}$$
$$= \boxed{7.16 \times 10^{-9}\,\text{N}}$$

Find the angle the force vector makes with respect to the positive x-axis:

$$\theta = \tan^{-1}\left(\frac{F_y}{F_x}\right) = \tan^{-1}\left(\frac{6.50 \times 10^{-9}\,\text{N}}{3.01 \times 10^{-9}\,\text{N}}\right) = \boxed{65.2°}$$

Remarks   The methods used here are just like those used with Newton's law of gravity in two dimensions.

Exercise 15.3
Using the same triangle, find the vector components of the electric force on q_1 and the vector's magnitude and direction.

Answers $F_x = -8.63 \times 10^{-9}\,\text{N}$, $F_y = 5.50 \times 10^{-9}\,\text{N}$, $F = 1.02 \times 10^{-8}\,\text{N}$, $\theta = 147°$

15.4   THE ELECTRIC FIELD

The gravitational force and the electrostatic force are both capable of acting through space, producing an effect even when there isn't any physical contact between the objects involved. Field forces can be discussed in a variety of ways, but an approach developed by Michael Faraday (1791–1867) is the most practical. In this approach, an **electric field** is said to exist in the region of space around a charged object. The electric field exerts an electric force on any other charged object within the field. This differs from the Coulomb's law concept of a force exerted at a distance, in that the force is now exerted by something—the field—that is in the same location as the charged object.

Figure 15.9 shows an object with a small positive charge q_0 placed near a second object with a much larger positive charge Q.

The electric field $\vec{\mathbf{E}}$ produced by a charge Q at the location of a small "test" charge q_0 is defined as the electric force $\vec{\mathbf{F}}$ exerted by Q on q_0, divided by the test charge q_0:

$$\vec{\mathbf{E}} \equiv \frac{\vec{\mathbf{F}}}{q_0} \qquad\qquad \textbf{[15.4]}$$

SI Unit: newton per coulomb (N/C)

Conceptually and experimentally, the test charge q_0 is required to be very small (arbitrarily small, in fact), so it doesn't cause any significant rearrangement of the charge creating the electric field $\vec{\mathbf{E}}$. Mathematically, however, the size of the test charge makes no difference: the calculation comes out the same, regardless. In view of this, using $q_0 = 1$ C in Equation 15.4 can be convenient if not rigorous.

When a positive test charge is used, the electric field always has the same direction as the electric force on the test charge. This follows from Equation 15.4. Hence in Figure 15.9, the direction of the electric field is horizontal and to the

Figure 15.9   A small object with a positive charge q_0 placed near an object with a larger positive charge Q is subject to an electric field $\vec{\mathbf{E}}$ directed as shown. The magnitude of the electric field at the location of q_0 is defined as the electric force on q_0 divided by the charge q_0.

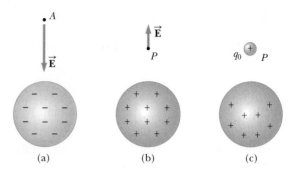

Figure 15.10 (a) The electric field at A due to the negatively charged sphere is downward, toward the negative charge. (b) The electric field at P due to the positively charged conducting sphere is upward, away from the positive charge. (c) A test charge q_0 placed at P will cause a rearrangement of charge on the sphere, unless q_0 is very small compared with the charge on the sphere.

(a)

(b)

ACTIVE FIGURE 15.11
A test charge q_0 at P is a distance r from a point charge q. (a) If q is positive, the electric field at P points radially *outwards* from q. (b) If q is negative, the electric field at P points radially *inwards* toward q.

Physics⊗Now™

Log into PhysicsNow at **www.cp7e.com** and go to Active Figure 15.11, where you can move point P to any position in two-dimensional space and observe the electric field due to q.

right. The electric field at point A in Figure 15.10a is vertical and downward because at that point a positive test charge would be attracted toward the negatively charged sphere.

Once the electric field due to a given arrangement of charges is known at some point, the force on *any* particle with charge q placed at that point can be calculated from a rearrangement of Equation 15.4:

$$\vec{F} = q\vec{E} \qquad [15.5]$$

Here q_0 has been replaced by q, which need not be a mere test charge.

As shown in Active Figure 15.11, the direction of $\vec{E}$ is the direction of the force that acts on a positive test charge q_0 placed in the field. We say that **an electric field exists at a point if a test charge at that point is subject to an electric force there**.

Consider a point charge q located a distance r from a test charge q_0. According to Coulomb's law, the *magnitude* of the electric force of the charge q on the test charge is

$$F = k_e \frac{|q||q_0|}{r^2}$$

Because the magnitude of the electric field at the position of the test charge is defined as $E = F/q_0$, we see that the *magnitude* of the electric field due to the charge q at the position of q_0 is

$$E = k_e \frac{|q|}{r^2} \qquad [15.6]$$

Equation 15.6 points out an important property of electric fields that makes them useful quantities for describing electrical phenomena. As the equation indicates, an electric field at a given point depends only on the charge q on the object setting up the field and the distance r from that object to a specific point in space. As a result, we can say that an electric field exists at point P in Active Figure 15.11 whether or not there is a test charge at P.

The principle of superposition holds when the electric field due to a group of point charges is calculated. We first use Equation 15.6 to calculate the electric field produced by each charge individually at a point and then add the electric fields together as vectors.

It's also important to exploit any symmetry of the charge distribution. For example, if equal charges are placed at $x = a$ and at $x = -a$, the electric field is zero at the origin, by symmetry. Similarly, if the x-axis has a uniform distribution of positive charge, it can be guessed by symmetry that the electric field points away from the x-axis and is zero parallel to that axis.

Quick Quiz 15.3

A test charge of $+3 \ \mu C$ is at a point P where the electric field due to other charges is directed to the right and has a magnitude of 4×10^6 N/C. If the test charge is replaced with a charge of $-3 \ \mu C$, the electric field at P (a) has the same magnitude as before, but changes direction, (b) increases in magnitude and changes direction, (c) remains the same, or (d) decreases in magnitude and changes direction.

Quick Quiz 15.4

A circular ring of charge of radius b has a total charge q uniformly distributed around it. The magnitude of the electric field at the center of the ring is

(a) 0   (b) $k_e q/b^2$ (c) $k_e q^2/b^2$ (d) $k_e q^2/b$ (e) none of these.

Quick Quiz 15.5

A "free" electron and a "free" proton are placed in an identical electric field. Which of the following statements are true? (a) Each particle is acted upon by the same electric force and has the same acceleration. (b) The electric force on the proton is greater in magnitude than the force on the electron, but in the opposite direction. (c) The electric force on the proton is equal in magnitude to the force on the electron, but in the opposite direction. (d) The magnitude of the acceleration of the electron is greater than that of the proton. (e) Both particles have the same acceleration.

EXAMPLE 15.4   Electrified Oil

Goal   Use electric forces and fields together with Newton's second law in a one-dimensional problem.

Problem   Tiny droplets of oil acquire a small negative charge while dropping through a vacuum (pressure = 0) in an experiment. An electric field of magnitude 5.92×10^4 N/C points straight down. **(a)** One particular droplet is observed to remain suspended against gravity. If the mass of the droplet is 2.93×10^{-15} kg, find the charge carried by the droplet. **(b)** Another droplet of the same mass falls 10.3 cm from rest in 0.250 s, again moving through a vacuum. Find the charge carried by the droplet.

Strategy   We use Newton's second law with both gravitational and electric forces. In both parts, the electric field $\vec{E}$ is pointing down, taken as the negative direction, as usual. In part (a), the acceleration is equal to zero. In part (b), the acceleration is uniform, so the kinematic equations yield the acceleration. Newton's law can then be solved for q.

Solution

(a) Find the charge on the suspended droplet.

Apply Newton's second law to the droplet in the vertical direction:

$$(1)\quad ma = \Sigma F = -mg + Eq$$

E points downward, hence is negative.
Set $a = 0$ and solve for q:

$$q = \frac{mg}{E} = \frac{(2.93 \times 10^{-15}\,\text{kg})\,(9.80\,\text{m/s}^2)}{-5.92 \times 10^4\,\text{N/C}}$$

$$= \boxed{-4.85 \times 10^{-19}\,\text{C}}$$

(b) Find the charge on the falling droplet.

Use the kinematic displacement equation to find the acceleration:

$$\Delta y = \tfrac{1}{2} at^2 + v_0 t$$

Substitute $\Delta y = -0.103$ m, $t = 0.250$ s, and $v_0 = 0$:

$$-0.103\,\text{m} = \tfrac{1}{2} a(0.250\,\text{s})^2 \quad \rightarrow \quad a = -3.30\,\text{m/s}^2$$

Solve Equation 1 for q and substitute:

$$q = \frac{m(a + g)}{E}$$

$$= \frac{(2.93 \times 10^{-15}\,\text{kg})(-3.30\,\text{m/s}^2 + 9.80\,\text{m/s}^2)}{-5.92 \times 10^4\,\text{N/C}}$$

$$= \boxed{-3.22 \times 10^{-19}\,\text{C}}$$

Remark This example exhibits features similar to the Millikan Oil-Drop experiment discussed in Section 15.7, which determined the value of the fundamental electric charge e. Notice that in both parts of the example, the charge is very nearly a multiple of e.

Exercise 15.4

Suppose a droplet of unknown mass remains suspended against gravity when $E = -2.70 \times 10^5$ N/C. What is the minimum mass of the droplet?

Answer 4.41×10^{-15} kg

Problem-Solving Strategy
Calculating Electric Forces and Fields

The following procedure is used to calculate electric forces (the same procedure can be used to calculate an electric field, a simple matter of replacing the charge of interest, q, with a convenient test charge and dividing by the test charge at the end):

1. **Draw** a diagram of the charges in the problem.
2. **Identify** the charge of interest, q, and circle it.
3. **Convert all units** to SI, with charges in coulombs and distances in meters, so as to be consistent with the SI value of the Coulomb constant k_e.
4. **Apply Coulomb's law.** For each charge Q, find the electric force on the charge of interest, q. The magnitude of the force can be found using Coulomb's law. The vector direction of the electric force is along the line of the two charges, directed away from Q if the charges have the same sign, toward Q if the charges have the opposite sign. Find the angle θ this vector makes with the positive x-axis. The x-component of the electric force exerted by Q on q will be $F \cos \theta$, and the y-component will be $F \sin \theta$.
5. **Sum all the x-components,** getting the x-component of the resultant electric force.
6. **Sum all the y-components,** getting the y-component of the resultant electric force.
7. **Use the Pythagorean theorem and trigonometry** to find the magnitude and direction of the resultant force if desired.

EXAMPLE 15.5   Electric Field Due to Two Point Charges

Goal   Use the superposition principle to calculate the electric field due to two point charges.

Problem   Charge $q_1 = 7.00$ μC is at the origin, and charge $q_2 = -5.00$ μC is on the x-axis, 0.300 m from the origin (Fig. 15.12). **(a)** Find the magnitude and direction of the electric field at point P, which has coordinates $(0, 0.400)$ m. **(b)** Find the force on a charge of 2.00×10^{-8} C placed at P.

Strategy   Follow the problem-solving strategy, finding the electric field at point P due to each individual charge in terms of x- and y-components, then adding the components of each type to get the x- and y-components of the resultant electric field at P. The magnitude of the force in part (b) can be found by simply multiplying the magnitude of the electric field by the charge.

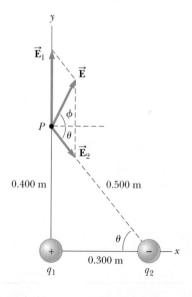

Figure 15.12   (Example 15.5) The resultant electric field $\vec{\mathbf{E}}$ at P equals the vector sum $\vec{\mathbf{E}}_1 + \vec{\mathbf{E}}_2$, where $\vec{\mathbf{E}}_1$ is the field due to the positive charge q_1 and $\vec{\mathbf{E}}_2$ is the field due to the negative charge q_2.

Solution

(a) Calculate the electric field at P.

Find the magnitude of $\vec{E}_1$ with Equation 15.6:

$$E_1 = k_e \frac{|q_1|}{r_1^2} = (8.99 \times 10^9 \text{ N·m}^2/\text{C}^2)\frac{(7.00 \times 10^{-6} \text{ C})}{(0.400 \text{ m})^2}$$

$$= 3.93 \times 10^5 \text{ N/C}$$

The vector $\vec{E}_1$ is vertical, making an angle of $90°$ with respect to the positive x-axis. Use this fact to find its components:

$$E_{1x} = E_1 \cos (90°) = 0$$

$$E_{1y} = E_1 \sin (90°) = 3.93 \times 10^5 \text{ N/C}$$

Next, find the magnitude of $\vec{E}_2$, again with Equation 15.6:

$$E_2 = k_e \frac{|q_2|}{r_2^2} = (8.99 \times 10^9 \text{ N·m}^2/\text{C}^2)\frac{(5.00 \times 10^{-6} \text{ C})}{(0.500 \text{ m})^2}$$

$$= 1.80 \times 10^5 \text{ N/C}$$

Obtain the x-component of $\vec{E}_2$, using the triangle in Figure 15.12 to find $\cos \theta$:

$$\cos \theta = \frac{\text{adj}}{\text{hyp}} = \frac{0.300}{0.500} = 0.600$$

$$E_{2x} = E_2 \cos \theta = (1.80 \times 10^5 \text{ N/C})(0.600)$$

$$= 1.08 \times 10^5 \text{ N/C}$$

Obtain the y-component in the same way, but a minus sign has to be provided for $\sin \theta$ because this component is directed downwards:

$$\sin \theta = \frac{\text{opp}}{\text{hyp}} = \frac{0.400}{0.500} = 0.800$$

$$E_{2y} = E_2 \sin \theta = (1.80 \times 10^5 \text{ N/C})(-0.800)$$

$$= -1.44 \times 10^5 \text{ N/C}$$

Sum the x-components to get the x-component of the resultant vector:

$$E_x = E_{1x} + E_{2x} = 0 + 1.08 \times 10^5 \text{ N/C} = 1.08 \times 10^5 \text{ N/C}$$

Sum the y-components to get the y-component of the resultant vector:

$$E_y = E_{1y} + E_{2y} = 0 + 3.93 \times 10^5 \text{ N/C} - 1.44 \times 10^5 \text{ N/C}$$

$$E_y = 2.49 \times 10^5 \text{ N/C}$$

Use the Pythagorean theorem to find the magnitude of the resultant vector:

$$E = \sqrt{E_x^2 + E_y^2} = 2.71 \times 10^5 \text{ N/C}$$

The inverse tangent function yields the direction of the resultant vector:

$$\phi = \tan^{-1}\left(\frac{E_y}{E_x}\right) = \tan^{-1}\left(\frac{2.49 \times 10^5 \text{ N/C}}{1.08 \times 10^5 \text{ N/C}}\right) = 66.6°$$

(b) Find the force on a charge of 2.00×10^{-8} C placed at P.

Calculate the magnitude of the force (the direction is the same as that of $\vec{E}$ because the charge is positive):

$$F = Eq = (2.71 \times 10^5 \text{ N/C})(2.00 \times 10^{-8} \text{ C})$$

$$= 5.42 \times 10^{-3} \text{ N}$$

Remarks   There were numerous steps to this problem, but each was very short. When attacking such problems, it's important to focus on one small step at a time. The solution comes not from a leap of genius, but from the assembly of a number of relatively easy parts.

Exercise 15.5

(a) Place a charge of $-7.00 \ \mu\text{C}$ at point P and find the magnitude and direction of the electric field at the location of q_2 due to q_1 and the charge at P. **(b)** Find the magnitude and direction of the force on q_2.

Answer   **(a)** 5.84×10^5 N/C, $\phi = 20.2°$ **(b)** $F = 2.92$ N, $\phi = 200.°$

15.5 ELECTRIC FIELD LINES

A convenient aid for visualizing electric field patterns is to draw lines pointing in the direction of the electric field vector at any point. These lines, introduced by Michael Faraday and called **electric field lines**, are related to the electric field in any region of space in the following way:

1. The electric field vector $\vec{E}$ is tangent to the electric field lines at each point.
2. The number of lines per unit area through a surface perpendicular to the lines is proportional to the strength of the electric field in a given region.

Note that $\vec{E}$ is large when the field lines are close together and small when the lines are far apart.

Figure 15.13a shows some representative electric field lines for a single positive point charge. This two-dimensional drawing contains only the field lines that lie in the plane containing the point charge. The lines are actually directed radially outward from the charge in *all* directions, somewhat like the quills of an angry porcupine. Because a positive test charge placed in this field would be repelled by the charge *q*, the lines are directed radially away from the positive charge. The electric field lines for a single negative point charge are directed toward the charge (Fig. 15.13b), because a positive test charge is attracted by a negative charge. In either case, the lines are radial and extend all the way to infinity. Note that the lines are closer together as they get near the charge, indicating that the strength of the field is increasing. Equation 15.6 verifies that this is indeed the case.

The rules for drawing electric field lines for any charge distribution follow directly from the relationship between electric field lines and electric field vectors:

1. The lines for a group of point charges must begin on positive charges and end on negative charges. In the case of an excess of charge, some lines will begin or end infinitely far away.
2. The number of lines drawn leaving a positive charge or ending on a negative charge is proportional to the magnitude of the charge.
3. No two field lines can cross each other.

Figure 15.14 shows the beautifully symmetric electric field lines for two point charges of equal magnitude but opposite sign. This charge configuration is called an **electric dipole**. Note that the number of lines that begin at the positive charge must equal the number that terminate at the negative charge. At points very near either charge, the lines are nearly radial. The high density of lines between the charges indicates a strong electric field in this region.

TIP 15.1 Electric Field Lines Aren't Paths of Particles

Electric field lines are *not* material objects. They are used only as a pictorial representation of the electric field at various locations. Except in special cases, they *do not* represent the path of a charged particle released in an electric field.

(a)

(b)

(c)

Photo courtesy of Harold M. Waage, Princeton University

Figure 15.13 The electric field lines for a point charge. (a) For a positive point charge, the lines radiate outward. (b) For a negative point charge, the lines converge inward. Note that the figures show only those field lines which lie in the plane containing the charge. (c) The dark lines are small pieces of thread suspended in oil, which align with the electric field produced by a small charged conductor at the center.

(a)

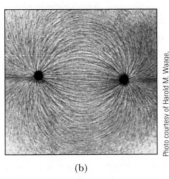
(b)

Photo courtesy of Harold M. Waage. Princeton University

Figure 15.14 (a) The electric field lines for two equal and opposite point charges (an electric dipole). Note that the number of lines leaving the positive charge equals the number terminating at the negative charge. (b) The dark lines are small pieces of thread suspended in oil, which align with the electric field produced by two charged conductors.

Figure 15.15 shows the electric field lines in the vicinity of two equal positive point charges. Again, close to either charge the lines are nearly radial. The same number of lines emerges from each charge because the charges are equal in magnitude. At great distances from the charges, the field is approximately equal to that of a single point charge of magnitude $2q$. The bulging out of the electric field lines between the charges reflects the repulsive nature of the electric force between like charges. Also, the low density of field lines between the charges indicates a weak field in this region, unlike the dipole.

Finally, Active Figure 15.16 is a sketch of the electric field lines associated with the positive charge $+2q$ and the negative charge $-q$. In this case, the number of lines leaving charge $+2q$ is twice the number terminating on charge $-q$. Hence, only half of the lines that leave the positive charge end at the negative charge. The remaining half terminate on negative charges that we assume to be located at infinity. At great distances from the charges (great compared with the charge separation), the electric field lines are equivalent to those of a single charge $+q$.

Quick Quiz 15.6

Rank the magnitudes of the electric field at points A, B, and C in Figure 15.15, with the largest magnitude first.
(a) A, B, C (b) A, C, B (c) C, A, B (d) Can't be determined by visual inspection

(a)

(b)

Photo courtesy of Harold M. Waage, Princeton University

Figure 15.15 (a) The electric field lines for two positive point charges. The points A, B, and C will be discussed in Quick Quiz 15.6. (b) The dark lines are small pieces of thread suspended in oil, which align with the electric field produced by two charged conductors.

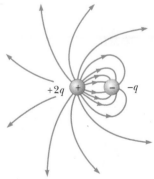

ACTIVE FIGURE 15.16
The electric field lines for a point charge of $+2q$ and a second point charge of $-q$. Note that two lines leave the charge $+2q$ for every line that terminates on $-q$.

Physics Now™
Log into PhysicsNow at
www.cp7e.com and go to Active Figure 15.16, where you can choose the values and signs for the two charges and observe the resulting electric field lines.

Applying Physics 15.1   Measuring Atmospheric Electric Fields

The electric field near the surface of the Earth in fair weather is about 100 N/C downward. Under a thundercloud, the electric field can be very large, on the order of 20 000 N/C. How are these electric fields measured?

Explanation A device for measuring these fields is called the *field mill*. Figure 15.17 shows the fundamental components of a field mill: two metal plates parallel to the ground. Each plate is connected to ground with a wire, with an ammeter (a low-resistance device for measuring the flow of charge, to be discussed in Section 19.6) in one path. Consider first just the lower plate. Because it's connected to ground and the ground carries a negative charge, the plate is negatively charged. The electric field lines, therefore, are directed downward, ending on the plate as in

Figure 15.17a. Now imagine that the upper plate is suddenly moved over the lower plate, as in Figure 15.17b. This plate is also connected to ground and is also negatively charged, so the field lines now end on the upper plate. The negative charges in the lower plate are repelled by those on the upper plate and must pass through the ammeter, registering a flow of charge. The amount of charge that was on the lower plate is related to the strength of the electric field. In this way, the flow of charge through the ammeter can be calibrated to measure the electric field. The plates are normally designed like the blades of a fan, with the upper plate rotating so that the lower plate is alternately covered and uncovered. As a result, charges flow back and forth continually through the ammeter, and the reading can be related to the electric field strength.

(a)   (b)

Figure 15.17   (Applying Physics 15.1) In (a), electric field lines end on negative charges on the lower plate. In (b), the second plate is moved above the lower plate. Electric field lines now end on the upper plate, and the negative charges in the lower plate are repelled through the ammeter.

15.6 CONDUCTORS IN ELECTROSTATIC EQUILIBRIUM

A good electric conductor like copper, though electrically neutral, contains charges (electrons) that aren't bound to any atom and are free to move about within the material. When no net motion of charge occurs within a conductor, the conductor is said to be in **electrostatic equilibrium**. An isolated conductor (one that is insulated from ground) has the following properties:

Properties of an isolated conductor ▶

> **1.** The electric field is zero everywhere inside the conducting material.
> **2.** Any excess charge on an isolated conductor resides entirely on its surface.
> **3.** The electric field just outside a charged conductor is perpendicular to the conductor's surface.
> **4.** On an irregularly shaped conductor, the charge accumulates at sharp points, where the radius of curvature of the surface is smallest.

The first property can be understood by examining what would happen if it were *not* true. If there were an electric field inside a conductor, the free charge

(a)   (b)

Figure 15.18   (a) Negative charges at the surface of a conductor. If the electric field were at an angle to the surface, as shown, an electric force would be exerted on the charges along the surface and they would move to the left. Because the conductor is assumed to be in electrostatic equilibrium, $\vec{E}$ cannot have a component along the surface and hence must be perpendicular to it. (b) The electric field pattern of a charged conducting plate near an oppositely charged conducting cylinder. Small pieces of thread suspended in oil align with the electric field lines. Note that (1) the electric field lines are perpendicular to the conductors and (2) there are no lines inside the cylinder ($\vec{E} = 0$).

there would move and a flow of charge, or current, would be created. However, if there were a net movement of charge, the conductor would no longer be in electrostatic equilibrium.

Property 2 is a direct result of the $1/r^2$ repulsion between like charges described by Coulomb's law. If by some means an excess of charge is placed inside a conductor, the repulsive forces between the like charges push them as far apart as possible, causing them to quickly migrate to the surface. (We won't prove it here, but the excess charge resides on the surface due to the fact that Coulomb's law is an inverse-square law. With any other power law, an excess of charge would exist on the surface, but there would be a distribution of charge, of either the same or opposite sign, inside the conductor.)

Property 3 can be understood by again considering what would happen if it were not true. If the electric field in Figure 15.18a were not perpendicular to the surface, it would have a component along the surface, which would cause the free charges of the conductor to move (to the left in the figure). If the charges moved, however, a current would be created and the conductor would no longer be in electrostatic equilibrium. Therefore, $\vec{E}$ must be perpendicular to the surface.

To see why property 4 must be true, consider Figure 15.19a, which shows a conductor that is fairly flat at one end and relatively pointed at the other. Any excess charge placed on the object moves to its surface. Figure 15.19b shows the forces between two such charges at the flatter end of the object. These forces are predominantly directed parallel to the surface, so the charges move apart until repulsive forces from other nearby charges establish an equilibrium. At the sharp end, however, the forces of repulsion between two charges are directed predominantly away from the surface, as in Figure 15.19c. As a result, there is less tendency for the charges to move apart along the surface here, and the amount of charge per unit area is greater than at the flat end. The cumulative effect of many such outward forces from nearby charges at the sharp end produces a large resultant force directed away from the surface that can be great enough to cause charges to leap from the surface into the surrounding air.

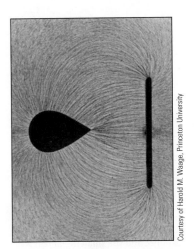

Electric field pattern of a charged conducting plate near an oppositely charged pointed conductor. Small pieces of thread suspended in oil align with the electric field lines. Note that the electric field is most intense near the pointed part of the conductor, where the radius of curvature is the smallest. Also, the lines are perpendicular to the conductors.

A   B   A   B   B

(a)   (b)   (c)

Figure 15.19   (a) A conductor with a flatter end A and a relatively sharp end B. Excess charge placed on this conductor resides entirely at its surface and is distributed so that (b) there is less charge per unit area on the flatter end and (c) there [is a] large charge per unit area on [the] sharper end.

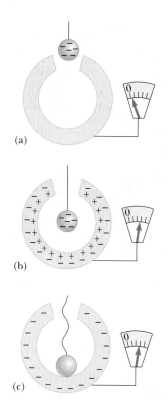

(a)

(b)

(c)

(d)

Figure 15.20 An experiment showing that any charge transferred to a conductor resides on its surface in electrostatic equilibrium. The hollow conductor is insulated from ground, and the small metal ball is supported by an insulating thread.

APPLICATION

Lightning Rods

Many experiments have shown that the net charge on a conductor resides on its surface. One such experiment was first performed by Michael Faraday, and is referred to as *Faraday's ice-pail experiment*. Faraday lowered a metal ball having a negative charge at the end of a silk thread (an insulator) into an uncharged hollow conductor insulated from ground, a metal ice-pail as in Figure 15.20a. As the ball entered the pail, the needle on an electrometer attached to the outer surface of the pail was observed to deflect. (An electrometer is a device used to measure charge.) The needle deflected because the charged ball induced a positive charge on the inner wall of the pail, which left an equal negative charge on the outer wall (Fig. 15.20b).

Faraday next touched the inner surface of the pail with the ball and noted that the deflection of the needle did not change, either when the ball touched the inner surface of the pail (Fig. 15.20c) or when it was removed (Fig. 15.20d). Further, he found that the ball was now uncharged, because when it touched the inside of the pail, the excess negative charge on the ball had been drawn off, neutralizing the induced positive charge on the inner surface of the pail. In this way, Faraday discovered the useful result that *all* the excess charge on an object can be transferred to an already charged metal shell if the object is touched to the *inside* of the shell. As we will see, this is the principle of operation of the Van de Graaff generator.

Faraday concluded that because the deflection of the needle in the electrometer didn't change when the charged ball touched the inside of the pail, the positive charge induced on the inside surface of the pail was just enough to neutralize the negative charge on the ball. As a result of his investigations, he concluded that a charged object suspended inside a metal container rearranged the charge on the container so that the sign of the charge on its inside surface was *opposite* the sign of the charge on the suspended object. This produced a charge on the outside surface of the container of the same sign as that on the suspended object.

Faraday also found that if the electrometer was connected to the inside surface of the pail after the experiment had been run, the needle showed no deflection. Thus, the *excess* charge acquired by the pail when contact was made between ball and pail appeared on the outer surface of the pail.

If a metal rod having sharp points is attached to a house, most of any charge on the house passes through these points, eliminating the induced charge on the house produced by storm clouds. In addition, a lightning discharge striking the house passes through the metal rod and is safely carried to the ground through wires leading from the rod to the Earth. Lightning rods using this principle were first developed by Benjamin Franklin. Some European countries couldn't accept the fact that such a worthwhile idea could have originated in the New World, so they "improved" the design by eliminating the sharp points!

Applying Physics 15.2 Conductors and Field Lines

Suppose a point charge $+Q$ is in empty space. Wearing rubber gloves, you proceed to surround the charge with a concentric spherical conducting shell. What effect does this have on the field lines from the charge?

Explanation When the spherical shell is placed around the charge, the charges in the shell rearrange to satisfy the rules for a conductor in equilibrium. A net charge of $-Q$ moves to the interior surface of the conductor, so that the electric field inside the conductor becomes zero. This means the field lines originating on the $+Q$ charge now terminate on the negative charges. The movement of the negative charges to the inner surface of the sphere leaves a net charge of $+Q$ on the outer surface of the sphere. Then the field lines outside the sphere look just as before: the only change, overall, is the absence of field lines within the conductor.

Applying Physics 15.3   Driver Safety During Electrical Storms

Why is it safe to stay inside an automobile during a lightning storm?

Explanation Many people believe that staying inside the car is safe because of the insulating characteristics of the rubber tires, but in fact this isn't true. Lightning can travel through several kilometers of air, so it can certainly penetrate a centimeter of rubber. The safety of remaining in the car is due to the fact that charges on the metal shell of the car will reside on the outer surface of the car, as noted in property 2 discussed earlier. As a result, an occupant in the automobile touching the inner surfaces is not in danger.

15.7   THE MILLIKAN OIL-DROP EXPERIMENT

From 1909 to 1913, Robert Andrews Millikan (1868–1953) performed a brilliant set of experiments at the University of Chicago in which he measured the elementary charge e of the electron and demonstrated the quantized nature of the electronic charge. The apparatus he used, diagrammed in Active Figure 15.21, contains two parallel metal plates. Oil droplets that have been charged by friction in an atomizer are allowed to pass through a small hole in the upper plate. A horizontal light beam is used to illuminate the droplets, which are viewed by a telescope with axis at right angles to the beam. The droplets then appear as shining stars against a dark background, and the rate of fall of individual drops can be determined.

We assume that a single drop having a mass of m and carrying a charge of q is being viewed and that its charge is negative. If no electric field is present between the plates, the two forces acting on the charge are the force of gravity, $m\vec{g}$, acting downward, and an upward viscous drag force $\vec{D}$ (Fig. 15.22a). The drag force is proportional to the speed of the drop. When the drop reaches its terminal speed, v, the two forces balance each other ($mg = D$).

Now suppose that an electric field is set up between the plates by a battery connected so that the upper plate is positively charged. In this case, a third force, $q\vec{E}$, acts on the charged drop. Because q is negative and $\vec{E}$ is downward, the electric

Oil droplets

Pinhole

Telescope with scale in eyepiece

d

q $\vec{v}$

Illumination

ACTIVE FIGURE 15.21
A schematic view of Millikan's oil-drop apparatus.

Physics⊗Now™
Log into PhysicsNow at **www.cp7e.com** and go to Active Figure 15.21, where you can do a simulation of the experiment, taking data on a number of oil drops and determining the elementary charge from your data.

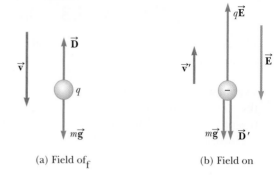

Figure 15.22 The forces on a charged oil droplet in Millikan's experiment.

(a) Field of~~f~~

(b) Field on

force is *upward* as in Figure 15.22b. If this force is great enough, the drop moves upward and the drag force $\vec{\mathbf{D}}'$ acts downward. When the upward electric force, $q\vec{\mathbf{E}}$, balances the sum of the force of gravity and the drag force, both acting downward, the drop reaches a new terminal speed v'.

With the field turned on, a drop moves slowly upward, typically at a rate of *hundredths* of a centimeter per second. The rate of fall in the absence of a field is comparable. Hence, a single droplet with constant mass and radius can be followed for hours as it alternately rises and falls, simply by turning the electric field on and off.

After making measurements on thousands of droplets, Millikan and his co-workers found that, to within about 1% precision, every drop had a charge equal to some positive or negative integer multiple of the elementary charge e,

$$q = ne \qquad n = 0, \pm 1, \pm 2, \pm 3, \ldots \qquad \text{[15.7]}$$

where $e = 1.60 \times 10^{-19}$ C. It was later established that positive integer multiples of e would arise when an oil droplet had lost one or more electrons. Likewise, negative integer multiples of e would arise when a drop had gained one or more electrons. Gains or losses in integral numbers provide conclusive evidence that charge is quantized. In 1923, Millikan was awarded the Nobel prize in physics for this work.

15.8 THE VAN DE GRAAFF GENERATOR

In 1929 Robert J. Van de Graaff (1901–1967) designed and built an electrostatic generator that has been used extensively in nuclear physics research. The principles of its operation can be understood with knowledge of the properties of electric fields and charges already presented in this chapter. Figure 15.23 shows the basic construction of this device. A motor-driven pulley P moves a belt past positively charged comb-like metallic needles positioned at A. Negative charges are attracted to these needles from the belt, leaving the left side of the belt with a net positive charge. The positive charges attract electrons onto the belt as it moves past a second comb of needles at B, increasing the excess positive charge on the dome. Because the electric field inside the metal dome is negligible, the positive charge on it can easily be increased regardless of how much charge is already present. The result is that the dome is left with a large amount of positive charge.

This accumulation of charge on the dome can't continue indefinitely. As more and more charge appears on the surface of the dome, the magnitude of the electric field at that surface is also increasing. Finally, the strength of the field becomes great enough to partially ionize the air near the surface, increasing the conductivity of the air. Charges on the dome now have a pathway to leak off into the air, producing some spectacular "lightning bolts" as the discharge occurs. As noted earlier, charges find it easier to leap off a surface at points where the curvature is great. As a result, one way to inhibit the electric discharge, and to increase the amount of charge that can be stored on the dome, is to increase its radius.

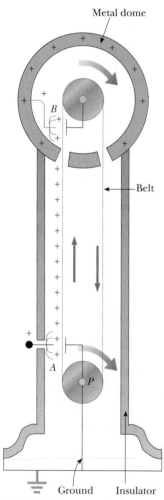

Figure 15.23 A diagram of a Van de Graaff generator. Charge is transferred to the dome by means of a rotating belt. The charge is deposited on the belt at point A and transferred to the dome at point B.

Metal dome

Belt

Ground   Insulator

Another method for inhibiting discharge is to place the entire system in a container filled with a high-pressure gas, which is significantly more difficult to ionize than air at atmospheric pressure.

If protons (or other charged particles) are introduced into a tube attached to the dome, the large electric field of the dome exerts a repulsive force on the protons, causing them to accelerate to energies high enough to initiate nuclear reactions between the protons and various target nuclei.

15.9 ELECTRIC FLUX AND GAUSS'S LAW

Gauss's law is essentially a technique for calculating the average electric field on a closed surface, developed by Karl Friedrich Gauss (1777–1855). When the electric field, because of its symmetry, is constant everywhere on that surface and perpendicular to it, the exact electric field can be found. In such special cases, Gauss's law is far easier to apply than Coulomb's law.

Gauss's law relates the electric flux through a closed surface and the total charge inside that surface. A *closed surface* has an inside and an outside: an example is a sphere. *Electric flux* is a measure of how much the electric field vectors penetrate through a given surface. If the electric field vectors are tangent to the surface at all points, for example, then they don't penetrate the surface and the electric flux through the surface is zero. These concepts will be discussed more fully in the next two subsections. As we'll see, Gauss's law states that the electric flux through a closed surface is proportional to the charge contained *inside* the surface.

Electric Flux

Consider an electric field that is uniform in both magnitude and direction, as in Figure 15.24. The electric field lines penetrate a surface of area A, which is perpendicular to the field. The technique used for drawing a figure such as Figure 15.24 is that the number of lines per unit area, N/A, is proportional to the magnitude of the electric field, or $E \propto N/A$. We can rewrite this as $N \propto EA$, which means that the number of field lines is proportional to the *product* of E and A, called the **electric flux** and represented by the symbol Φ_E:

$$\Phi_E = EA \qquad \qquad \text{[15.8]}$$

Note that Φ_E has SI units of $\text{N} \cdot \text{m}^2/\text{C}$ and is proportional to the number of field lines that pass through some area A oriented perpendicular to the field. (It's called flux by analogy with the term *flux* in fluid flow, which is just the volume of liquid flowing through a perpendicular area per second.). If the surface under consideration is not perpendicular to the field, as in Figure 15.25, the expression for the electric flux is

$$\Phi_E = EA \cos \theta \qquad \qquad \text{[15.9]}$$

where a vector perpendicular to the area A is at an angle θ with respect to the field. This vector is often said to be *normal* to the surface, and we will refer to it as

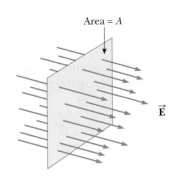

Area = A

$\vec{\mathbf{E}}$

Figure 15.24 Field lines of a uniform electric field penetrating a plane of area A perpendicular to the field. The electric flux Φ_E through this area is equal to EA.

◄ Electric flux

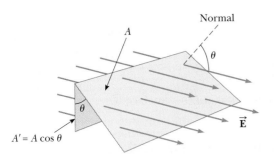

Normal

A

θ

$\vec{\mathbf{E}}$

$A' = A \cos \theta$

Figure 15.25 Field lines for a uniform electric field through an area A that is at an angle of $(90° - \theta)$ to the field. Because the number of lines that go through the shaded area A' is the same as the number that go through A, we conclude that the flux through A' is equal to the flux through A and is given by $\Phi_E = EA \cos \theta$.

"the normal vector to the surface." The number of lines that cross this area is equal to the number that cross the projected area A', which is perpendicular to the field. We see that the two areas are related by $A' = A \cos \theta$. From Equation 15.9, we see that the flux through a surface of fixed area has the maximum value EA when the surface is perpendicular to the field (when $\theta = 0°$) and that the flux is zero when the surface is parallel to the field (when $\theta = 90°$). **By convention, for a closed surface, the flux lines passing into the interior of the volume are negative and those passing out of the interior of the volume are positive.** This convention is equivalent to requiring the normal vector of the surface to point outward when computing the flux through a closed surface.

Quick Quiz 15.7

Calculate the magnitude of the flux of a constant electric field of 5.00 N/C in the z-direction through a rectangle with area 4.00 m^2 in the xy-plane. (a) 0 (b) 10.0 N·m^2/C (c) 20.0 N·m^2/C (d) more information is needed

Quick Quiz 15.8

Suppose the electric field of Quick Quiz 15.7 is tilted 60° away from the positive z-direction. Calculate the magnitude of the flux through the same area. (a) 0 (b) 10.0 N·m^2/C (c) 20.0 N·m^2/C (d) more information is needed

---

EXAMPLE 15.6   Flux Through a Cube

Goal   Calculate the electric flux through a closed surface.

Problem   Consider a uniform electric field oriented in the x-direction. Find the electric flux through each surface of a cube with edges L oriented as shown in Figure 15.26, and the net flux.

Strategy   This problem involves substituting into the definition of electric flux given by Equation 15.9. E and $A = L^2$ are the same in each case; the only difference is the angle θ that the electric field makes with respect to a vector perpendicular to a given surface and pointing outward (the normal vector to the surface). The angles can be determined by inspection. The flux through a surface parallel to the xy-plane will be labeled Φ_{xy} and further designated by position (front, back); others will be labeled similarly: Φ_{xz} top or bottom, and Φ_{yz} left or right.

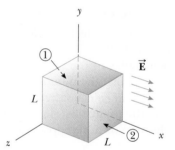

Figure 15.26   (Example 15.6) A hypothetical surface in the shape of a cube in a uniform electric field parallel to the x-axis. The net flux through the surface is zero when the net charge inside the cube is zero.

Solution

The normal vector to the xy-plane points in the negative z-direction. This, in turn, is perpendicular to $\vec{E}$, so $\theta = 90°$. (The opposite side works similarly.)

$$\Phi_{xy} = EA \cos(90°) = \boxed{0} \quad \text{(back and front)}$$

The normal vector to the xz-plane points in the negative y-direction. This, in turn, is perpendicular to $\vec{E}$, so again $\theta = 90°$. (The opposite side works similarly.)

$$\Phi_{xz} = EA \cos(90°) = \boxed{0} \quad \text{(top and bottom)}$$

The normal vector to surface ① (the yz-plane) points in the negative x-direction. This is antiparallel to $\vec{E}$, so $\theta = 180°$.

$$\Phi_{yz} = EA \cos(180°) = \boxed{-EL^2} \quad \text{(surface ①)}$$

Surface ② has normal vector pointing in the positive x-direction, so $\theta = 0°$.

$$\Phi_{yz} = EA \cos(0°) = \boxed{EL^2} \quad \text{(surface ②)}$$

We calculate the net flux by summing:

$$\Phi_{net} = 0 + 0 + 0 + 0 - EL^2 + EL^2 = \boxed{0}$$

Remarks   In doing this calculation, it is necessary to remember that the angle in the definition of flux is measured from the normal vector to the surface and that this vector must point outwards for a closed surface. As a result, the normal vector for the yz-plane on the left points in the negative x-direction, and the normal vector to the plane parallel to the yz-plane on the right points in the positive x-direction. Notice that there aren't any charges in the box. The net electric flux is always zero for closed surfaces that don't contain net charge.

Exercise 15.6
Suppose the constant electric field in Example 15.6 points in the positive y-direction instead. Calculate the flux through the xz-plane and the surface parallel to it. What's the net electric flux through the surface of the cube?

Answers $\Phi_{xz} = -EL^2$ (bottom), $\Phi_{xz} = +EL^2$ (top). The net flux is still zero.

Gauss's Law

Consider a point charge q surrounded by a spherical surface of radius r centered on the charge, as in Figure 15.27a. The magnitude of the electric field everywhere on the surface of the sphere is

$$E = k_e \frac{q}{r^2}$$

Note that the electric field is perpendicular to the spherical surface at all points on the surface. The electric flux through the surface is therefore EA, where $A = 4\pi r^2$ is the surface area of the sphere:

$$\Phi_E = EA = k_e \frac{q}{r^2} (4\pi r^2) = 4\pi k_e q$$

It's sometimes convenient to express k_e in terms of another constant, ϵ_0, as $k_e = 1/(4\pi\epsilon_0)$. The constant ϵ_0 is called the **permittivity of free space** and has the value

$$\epsilon_0 = \frac{1}{4\pi k_e} = 8.85 \times 10^{-12} \, \text{C}^2/\text{N} \cdot \text{m}^2 \qquad \textbf{[15.10]}$$

The use of k_e or ϵ_0 is strictly a matter of taste. The electric flux through the closed spherical surface that surrounds the charge q can now be expressed as

$$\Phi_E = 4\pi k_e q = \frac{q}{\epsilon_0}$$

This result says that the electric flux through a sphere that surrounds a charge q is equal to the charge divided by the constant ϵ_0. Using calculus, this result can be proven for *any* closed surface that surrounds the charge q. For example, if the surface surrounding q is irregular, as in Figure 15.27b, the flux through that surface is also q/ϵ_0. This leads to the following general result, known as Gauss's Law:

> The electric flux Φ_E through any closed surface is equal to the net charge inside the surface, Q_{inside}, divided by ϵ_0:
>
> $$\Phi_E = \frac{Q_{\text{inside}}}{\epsilon_0} \qquad \textbf{[15.11]}$$

Though it's not obvious, Gauss's law describes how charges create electric fields. In principle, it can always be used to calculate the electric field of a system of charges or a continuous distribution of charge. In practice, the technique is useful only in a limited number of cases in which there is a high degree of symmetry, such as spheres, cylinders, or planes. With the symmetry of these special shapes, the charges can be surrounded by an imaginary surface, called a Gaussian surface. This imaginary surface is used strictly for mathematical calculation, and need not be an actual, physical surface. If the imaginary surface is chosen so that the

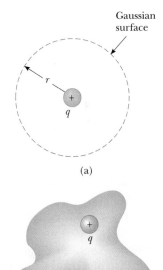

Figure 15.27   (a) The flux through a spherical surface of radius r surrounding a point charge q is $\Phi_E = q/\epsilon_0$. (b) The flux through any arbitrary surface surrounding the charge is also equal to q/ϵ_0.

◀ Gauss's Law

TIP 15.2   Gaussian Surfaces Aren't Real

A Gaussian surface is an imaginary surface, created solely to facilitate a mathematical calculation. It doesn't necessarily coincide with the surface of a physical object.

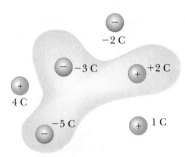

ACTIVE FIGURE 15.28
(Quick Quiz 15.9)

Physics⊗Now™
Log into PhysicsNow at
www.cp7e.com and go to Active
Figure 15.28, where you can change
the size and shape of a closed surface
and see the effect of surrounding
combinations of charge on the electric flux going through that surface.

electric field is constant everywhere on it, then the electric field can be computed
with

$$EA = \Phi_E = \frac{Q_{inside}}{\epsilon_0} \qquad \text{[15.12]}$$

as will be seen in the examples. Though Gauss's law in this form can be used to obtain the electric field only for problems with a lot of symmetry, it can *always* be
used to obtain the *average* electric field on *any* surface.

Quick Quiz 15.9

Find the electric flux through the surface in Active Figure 15.28. (a) $-(3\,C)/\epsilon_0$
(b) $(3\,C)/\epsilon_0$ (c) 0   (d) $-(6\,C)/\epsilon_0$

Quick Quiz 15.10

For a closed surface through which the net flux is zero, each of the following four
statements *could* be true. Which of the statements *must* be true? (There may be
more than one.) (a) There are no charges inside the surface. (b) The net charge
inside the surface is zero. (c) The electric field is zero everywhere on the surface.
(d) The number of electric field lines entering the surface equals the number
leaving the surface.

EXAMPLE 15.7   The Electric Field of a Charged Spherical Shell

Goal   Use Gauss's law to determine electric fields when the symmetry is spherical.

Problem   A spherical conducting shell of inner radius a and outer radius b carries a total charge $+Q$ distributed on
the surface of a conducting shell (Fig. 15.29a). The quantity Q is taken to be positive. **(a)** Find the electric field in the
interior of the conducting shell, for $r < a$, and **(b)** the electric field outside the shell, for $r > b$. **(c)** If an additional
charge of $-Q$ is placed at the center, find the electric field for $r > b$.

Strategy   For each part, draw a spherical Gaussian surface in the region of interest. Add up the charge inside the
Gaussian surface, substitute it and the area into Gauss's law, and solve for the electric field.

Figure 15.29   (Example 15.7)
(a) The electric field inside a uniformly charged spherical shell is *zero*.
It is also zero for the conducting material in the region $a < r < b$. The
field outside is the same as that of a
point charge having a total charge Q
located at the center of the shell.
(b) The construction of a Gaussian
surface for calculating the electric
field *inside* a spherical shell. (c) The
construction of a Gaussian surface
for calculating the electric field
outside a spherical shell.

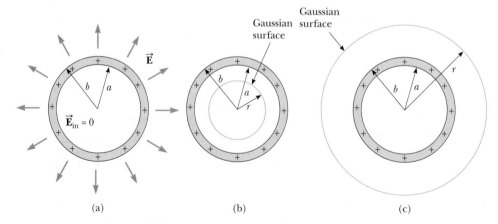

Solution
(a) Find the electric field for $r < a$.

Apply Gauss's law, Equation 15.12, to the Gaussian surface illustrated in Figure 15.29b (note that there isn't
any charge inside this surface):

$$EA = E(4\pi r^2) = \frac{Q_{inside}}{\epsilon_0} = 0 \quad \rightarrow \quad \boxed{E = 0}$$

(b) Find the electric field for $r > b$.

Apply Gauss's law, Equation 15.12, to the Gaussian surface illustrated in Figure 15.29c:

$$EA = E(4\pi r^2) = \frac{Q_{inside}}{\epsilon_0} = \frac{Q}{\epsilon_0}$$

Divide by the area:

$$E = \frac{Q}{4\pi\epsilon_0 r^2}$$

(c) Now an additional charge of $-Q$ is placed at the center of the sphere. Compute the new electric field outside the sphere, for $r > b$.

Apply Gauss's law as in part (b), including the new charge in Q_{inside}:

$$EA = E(4\pi r^2) = \frac{Q_{inside}}{\epsilon_0} = \frac{+Q - Q}{\epsilon_0} = 0$$

$$E = 0$$

Remarks   The important thing to notice is that in each case, the charge is spread out over a region with spherical symmetry or is located at the exact center. This is what allows the computation of a value for the electric field.

Exercise 15.7
Suppose the charge at the center is now increased to $-2Q$, while the surface of the conductor still retains a charge of $+Q$. (a) Find the electric field exterior to the sphere, for $r > b$. (b) What's the electric field inside the conductor, for $a < r < b$?

Answers   (a) $E = -Q/4\pi\epsilon_0 r^2$ (b) $E = 0$, which is always the case when charges are not moving in a conductor.

In Example 15.7, not much was said about the distribution of charge on the conductor. Whenever there is a net nonzero charge, the individual charges will try to get as far away from each other as possible. Hence, charge will reside either on the inside surface or on the outside surface. Because the electric field in the conductor is zero, there will always be enough charge on the inner surface to cancel whatever charge is at the center. In part (b), there is no charge on the inner surface and a charge of $+Q$ on the outer surface. In part (c), with a $-Q$ charge at the center, $+Q$ is on the inner surface, 0 C on the outer surface. Finally, in the exercise, with $-2Q$ in the center, there must be $+2Q$ on the inner surface and $-Q$ on the outer surface. In each case, the total charge on the conductor remains the same, $+Q$; it's just arranged differently.

Problems like Example 15.7 are often said to have "thin nonconducting shells" carrying a uniformly distributed charge. In these cases, no distinction need be made between the outer surface and inner surface of the shell. The next example makes that implicit assumption.

EXAMPLE 15.8   A Nonconducting Plane Sheet of Charge

Goal   Apply Gauss's law to a problem with plane symmetry.

Problem   Find the electric field above and below a nonconducting infinite plane sheet of charge with uniform positive charge per unit area σ (Figure 15.30a).

Strategy   By symmetry, the electric field must be perpendicular to the plane and directed away from it on either side, as shown in Figure 15.30b. For the Gaussian surface, choose a small cylinder with axis perpendicular to the plane, each end having area A_0. No electric field lines pass through the curved surface of the cylinder, only through the two ends, which have total area $2A_0$. Apply Gauss's law, using Figure 15.30b.

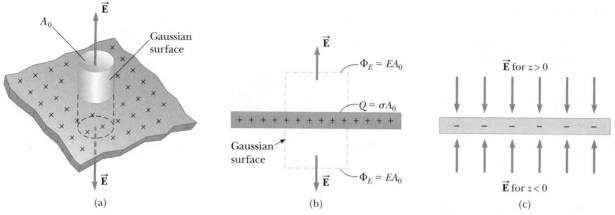

Figure 15.30 (Example 15.8) (a) A cylindrical Gaussian surface penetrating an infinite sheet of charge. (b) A cross section of the same Gaussian cylinder. The flux through each end of the Gaussian surface is EA_0. There is no flux through the cylindrical surface. (c) (Exercise 15.8).

Solution

(a) Find the electric field above and below a plane of uniform charge.

Apply Gauss's law, Equation 15.12:

$$EA = \frac{Q_{inside}}{\epsilon_0}$$

The total charge inside the Gaussian cylinder is the charge density times the cross-sectional area:

$$Q_{inside} = \sigma A_0$$

The electric flux comes entirely from the two ends, each having area A_0. Substitute $A = 2A_0$ and Q_{inside} and solve for E.

$$E = \frac{\sigma A_0}{(2A_0)\epsilon_0} = \boxed{\frac{\sigma}{2\epsilon_0}}$$

This is the *magnitude* of the electric field. Find the z-component of the field above and below the plane. The electric field points away from the plane, so it's positive above the plane and negative below the plane.

$$E_z = \frac{\sigma}{2\epsilon_0} \qquad z > 0$$

$$E_z = -\frac{\sigma}{2\epsilon_0} \qquad z < 0$$

Remarks   Notice here that the plate was taken to be a thin nonconducting shell. If it's made of metal, of course, the electric field inside it is zero, with half the charge on the upper surface and half on the lower surface.

Exercise 15.8

Suppose an infinite non-conducting plane of charge as in Example 15.8 has a uniform negative charge density of $-\sigma$. Find the electric field above and below the plate. Sketch the field.

Answers

$$E_z = \frac{-\sigma}{2\epsilon_0} \quad z > 0; \quad E_z = \frac{\sigma}{2\epsilon_0} \quad z < 0$$

See Figure 15.30c for the sketch.

An important circuit element that will be studied extensively in the next chapter is the parallel plate capacitor. The device consists of a plate of positive charge, as in Example 15.8, with the negative plate of Exercise 15.8 placed above it. The sum of these two fields is illustrated in Figure 15.31. The result is an electric field with double the magnitude in between the two plates:

$$E = \frac{\sigma}{\epsilon_0} \qquad [15.13]$$

Outside the plates, the electric fields cancel.

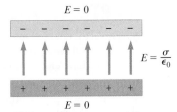

Figure 15.31 Cross section of an idealized parallel-plate capacitor. Electric field vector contributions sum together in between the plates, but cancel outside.

SUMMARY

Physics⊗Now™ Take a practice test by logging into Physics-Now at **www.cp7e.com** and clicking on the Pre-Test link for this chapter.

15.1 Properties of Electric Charges

Electric charges have the following properties:

1. Unlike charges attract one another and like charges repel one another.
2. Electric charge is always conserved.
3. Charge comes in discrete packets that are integral multiples of the basic electric charge $e = 1.6 \times 10^{-19}$ C.
4. The force between two charged particles is proportional to the inverse square of the distance between them.

15.2 Insulators and Conductors

Conductors are materials in which charges move freely in response to an electric field. All other materials are called **insulators**.

15.3 Coulomb's Law

Coulomb's law states that the electric force between two stationary charged particles separated by a distance r has the magnitude

$$F = k_e \frac{|q_1||q_2|}{r^2} \qquad [15.1]$$

where $|q_1|$ and $|q_2|$ are the magnitudes of the charges on the particles in coulombs and

$$k_e \approx 8.99 \times 10^9 \ \text{N} \cdot \text{m}^2/\text{C}^2 \qquad [15.2]$$

is the **Coulomb constant**.

15.4 The Electric Field

An electric field $\vec{E}$ exists at some point in space if a small test charge q_0 placed at that point is acted upon by an electric force $\vec{F}$. The electric field is defined as

$$\vec{E} \equiv \frac{\vec{F}}{q_0} \qquad [15.4]$$

The **direction** of the electric field at a point in space is defined to be the direction of the electric force that would be exerted on a small positive charge placed at that point.

The magnitude of the electric field due to a *point charge* q at a distance r from the point charge is

$$E = k_e \frac{|q|}{r^2} \qquad [15.6]$$

15.5 Electric Field Lines

Electric field lines are useful for visualizing the electric field in any region of space. The electric field vector $\vec{E}$ is tangent to the electric field lines at every point. Furthermore, the number of electric field lines per unit area through a surface perpendicular to the lines is proportional to the strength of the electric field at that surface.

15.6 Conductors in Electrostatic Equilibrium

A **conductor in electrostatic equilibrium** has the following properties:

1. The electric field is zero everywhere inside the conducting material.
2. Any excess charge on an isolated conductor must reside entirely on its surface.
3. The electric field just outside a charged conductor is perpendicular to the conductor's surface.
4. On an irregularly shaped conductor, charge accumulates where the radius of curvature of the surface is smallest, at sharp points.

15.9 Electric Flux and Gauss's Law

Gauss's law states that the electric flux through any closed surface is equal to the net charge Q inside the surface divided by the permittivity of free space, ϵ_0:

$$EA = \Phi_E = \frac{Q_{\text{inside}}}{\epsilon_0} \qquad [15.12]$$

For highly symmetric distributions of charge, Gauss's law can be used to calculate electric fields.

CONCEPTUAL QUESTIONS

1. A glass object is charged to $+3$ nC by rubbing it with a silk cloth. In the rubbing process, have protons been added to the object or have electrons been removed from it?

2. Why must hospital personnel wear special conducting shoes while working around oxygen in an operating room. What might happen if the personnel wore shoes with rubber soles?

3. Two insulated rods are oppositely charged on their ends. They are mounted at the centers so that they are free to rotate, and then held in the position shown in Figure Q15.3 in a view from above. The rods rotate in the plane of the paper. Will the rods stay in those positions when released? If not, into what position(s) will they move? Will the final configuration(s) be stable?

Figure Q15.3

4. Explain from an atomic viewpoint why charge is usually transferred by electrons.

5. Explain how a positively charged object can be used to leave another metallic object with a net negative charge. Discuss the motion of charges during the process.

6. If a suspended object A is attracted to a charged object B, can we conclude that A is charged? Explain.

7. If a metal object receives a positive charge, does its mass increase, decrease, or stay the same? What happens to its mass if the object receives a negative charge?

8. When defining the electric field, why is it necessary to specify that the magnitude of the test charge be very small?

9. In fair weather, there is an electric field at the surface of the Earth, pointing down into the ground. What is the electric charge on the ground in this situation?

10. A student stands on a thick piece of insulating material, places her hand on top of a Van de Graaff generator, and then turns on the generator. Does she receive a shock?

11. An uncharged, metallic-coated Styrofoam ball is suspended in the region between two vertical metal plates. If the two plates are charged, one positively and one negatively, describe the motion of the ball after it is brought into contact with one of the plates.

12. Is it possible for an electric field to exist in empty space? Explain.

13. There are great similarities between electric and gravitational fields. A room can be electrically shielded so that there are no electric fields in the room by surrounding it with a conductor. Can a room be gravitationally shielded? Why or why not? [*Hint:* There are two kinds of charge in nature, but only one kind of mass.]

14. Would life be different if the electron were positively charged and the proton were negatively charged? Does the choice of signs have any bearing on physical and chemical interactions? Explain.

15. Explain why Gauss's law cannot be used to calculate the electric field of (a) a polar molecule consisting of a positive and a negative charge separated by a very small distance, (b) a charged disk, and (c) three point charges at the corner of a triangle.

16. Why should a ground wire be connected to the metal support rod for a television antenna?

17. A balloon negatively charged by rubbing clings to a wall. Does this mean that the wall is positively charged? Why does the balloon eventually fall?

18. A spherical surface surrounds a point charge q. Describe what happens to the total flux through the surface if (a) the charge is tripled, (b) the volume of the sphere is doubled, (c) the surface is changed to a cube, (d) the charge is moved to another location inside the surface, and (e) the charge is moved outside the surface.

19. A charged comb often attracts small bits of dry paper that then fly away when they touch the comb. Explain.

20. Answer the given questions with one of the statements that follow and defend your answer. The answer to each part is either $+Q$, 0, or $-Q$. A spherical conducting object A with a charge of $+Q$ is lowered through a hole into a metal container B that is initially uncharged.
    (a) When A is at the center of B, but not touching it, the charge on the inner surface of B is _____.
    (b) The charge on the outer surface of B is _____.
    (c) Object A is now allowed to touch the inner surface of B. The charge on A is now _____.
    (d) The charge on the inner surface of B is now _____.
    (e) The charge on the outer surface of B is now _____.

21. A positively charged ball hanging from a nonconducting string is brought near a nonconducting object. Based on the behavior of the ball–string combination, the ball is seen to be attracted to the object. From this experiment, it is not possible to determine whether the object is negatively charged or neutral. Why not? What additional experiment would help you decide between these two possibilities?

22. An electron moving horizontally passes between two horizontal plates, the upper charged negatively, the lower positively. A uniform, upward-directed electric field exists in the region between the plates, and this field exerts an electric force downward on the electron. Describe the movement of the electron in this region.

PROBLEMS

1, 2, 3 = straightforward, intermediate, challenging   ☐ = full solution available in *Student Solutions Manual/Study Guide*
Physics ⊗ Now™ = coached solution with hints available at **www.cp7e.com**   🔬 = biomedical application

Section 15.3 Coulomb's Law

1. A charge of 4.5×10^{-9} C is located 3.2 m from a charge of -2.8×10^{-9} C. Find the electrostatic force exerted by one charge on the other.

2. The Moon and Earth are bound together by gravity. If, instead, the force of attraction were the result of each having a charge of the same magnitude but opposite in

sign, find the quantity of charge that would have to be placed on each to produce the required force.

3. An alpha particle (charge $= +2.0e$) is sent at high speed toward a gold nucleus (charge $= +79e$). What is the electrical force acting on the alpha particle when it is 2.0×10^{-14} m from the gold nucleus?

4. Four point charges are situated at the corners of a square with sides of length a, as in Figure P15.4. Find the expression for the resultant force on the positive charge q.

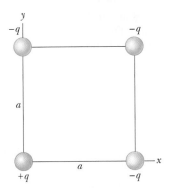

Figure P15.4

5. The nucleus of ^{8}Be, which consists of 4 protons and 4 neutrons, is very unstable and spontaneously breaks into two alpha particles (helium nuclei, each consisting of 2 protons and 2 neutrons). (a) What is the force between the two alpha particles when they are 5.00×10^{-15} m apart, and (b) what will be the magnitude of the acceleration of the alpha particles due to this force? Note that the mass of an alpha particle is 4.0026 u.

6. A molecule of DNA (deoxyribonucleic acid) is 2.17 μm long. The ends of the molecule become singly ionized—negative on one end, positive on the other. The helical molecule acts like a spring and compresses 1.00% upon becoming charged. Determine the effective spring constant of the molecule.

7. Suppose that 1.00 g of hydrogen is separated into electrons and protons. Suppose also that the protons are placed at the Earth's North Pole and the electrons are placed at the South Pole. What is the resulting compressional force on the Earth?

8. An electron is released a short distance above the surface of the Earth. A second electron directly below it exerts an electrostatic force on the first electron just great enough to cancel the gravitational force on it. How far below the first electron is the second?

9. Two identical conducting spheres are placed with their centers 0.30 m apart. One is given a charge of 12×10^{-9} C, the other a charge of -18×10^{-9} C. (a) Find the electrostatic force exerted on one sphere by the other. (b) The spheres are connected by a conducting wire. Find the electrostatic force between the two after equilibrium is reached.

10. Calculate the magnitude and direction of the Coulomb force on each of the three charges shown in Figure P15.10.

Figure P15.10   (Problems 10 and 18)

11. **Physics⊗Now**™ Three charges are arranged as shown in Figure P15.11. Find the magnitude and direction of the electrostatic force on the charge at the origin.

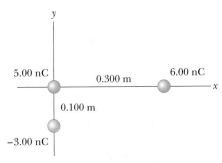

Figure P15.11

12. Three charges are arranged as shown in Figure P15.12. Find the magnitude and direction of the electrostatic force on the 6.00-nC charge.

Figure P15.12

13. Three point charges are located at the corners of an equilateral triangle as in Figure P15.13. Calculate the net electric force on the 7.00-μC charge.

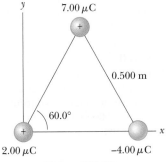

Figure P15.13

14. Two small beads having positive charges $3q$ and q are fixed at the opposite ends of a horizontal insulating rod, extending from the origin to the point $x = d$. As shown in Figure P15.14, a third small charged bead is free to slide on the rod. At what position is the third bead in equilibrium? Can it be in stable equilibrium?

Figure P15.14

Two small metallic spheres, each of mass 0.20 g, are suspended as pendulums by light strings from a common point as shown in Figure P15.15. The spheres are given the same electric charge, and it is found that they come to equilibrium when each string is at an angle of 5.0° with the vertical. If each string is 30.0 cm long, what is the magnitude of the charge on each sphere?

Figure P15.15

16. A charge of 6.00×10^{-9} C and a charge of -3.00×10^{-9} C are separated by a distance of 60.0 cm. Find the position at which a third charge, of 12.0×10^{-9} C, can be placed so that the net electrostatic force on it is zero.

Section 15.4 The Electric Field

17. An object with a net charge of 24 μC is placed in a uniform electric field of 610 N/C, directed vertically. What is the mass of the object if it "floats" in the electric field?

18. (a) Determine the electric field strength at a point 1.00 cm to the left of the middle charge shown in Figure P15.10. (b) If a charge of -2.00 μC is placed at this point, what are the magnitude and direction of the force on it?

19. An airplane is flying through a thundercloud at a height of 2 000 m. (This is a very dangerous thing to do because of updrafts, turbulence, and the possibility of electric discharge.) If there are charge concentrations of $+40.0$ C at a height of 3 000 m within the cloud and -40.0 C at a height of 1 000 m, what is the electric field $\vec{E}$ at the aircraft?

20. An electron is accelerated by a constant electric field of magnitude 300 N/C. (a) Find the acceleration of the electron. (b) Use the equations of motion with constant acceleration to find the electron's speed after 1.00×10^{-8} s, assuming it starts from rest.

21. A Styrofoam® ball covered with a conducting paint has a mass of 5.0×10^{-3} kg and has a charge of 4.0 μC. What electric field directed upward will produce an electric force on the ball that will balance its weight?

22. Each of the protons in a particle beam has a kinetic energy of 3.25×10^{-15} J. What are the magnitude and direction of the electric field that will stop these protons in a distance of 1.25 m?

23. A proton accelerates from rest in a uniform electric field of 640 N/C. At some later time, its speed is 1.20×10^6 m/s. (a) Find the magnitude of the acceleration of the proton. (b) How long does it take the proton to reach this speed? (c) How far has it moved in that interval? (d) What is its kinetic energy at the later time?

24. Three charges are at the corners of an equilateral triangle, as shown in Figure P15.24. Calculate the electric field at a point midway between the two charges on the x-axis.

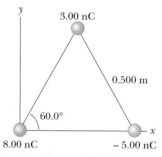

Figure P15.24

25. Three identical charges ($q = -5.0$ μC) lie along a circle of radius 2.0 m at angles of 30°, 150°, and 270°, as shown in Figure P15.25. What is the resultant electric field at the center of the circle?

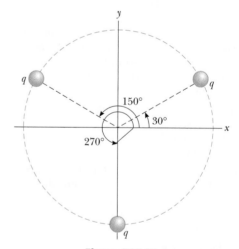

Figure P15.25

26. Two point charges lie along the y-axis. A charge of $q_1 = -9.0$ μC is at $y = 6.0$ m, and a charge of $q_2 = -8.0$ μC is at $y = -4.0$ m. Locate the point (other than infinity) at which the total electric field is zero.

27. Physics⊗Now™ In Figure P15.27, determine the point (other than infinity) at which the total electric field is zero.

Figure P15.27

Section 15.5 Electric Field Lines

Section 15.6 Conductors in Electrostatic Equilibrium

28. Figure P15.28 shows the electric field lines for two point charges separated by a small distance. (a) Determine the ratio q_1/q_2. (b) What are the signs of q_1 and q_2?

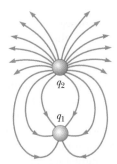

Figure P15.28

29. (a) Sketch the electric field lines around an isolated point charge $q > 0$. (b) Sketch the electric field pattern around an isolated negative point charge of magnitude $-2q$.

30. (a) Sketch the electric field pattern around two positive point charges of magnitude $1\ \mu C$ placed close together. (b) Sketch the electric field pattern around two negative point charges of $-2\ \mu C$, placed close together. (c) Sketch the pattern around two point charges of $+1\ \mu C$ and $-2\ \mu C$, placed close together.

31. Two point charges are a small distance apart. (a) Sketch the electric field lines for the two if one has a charge four times that of the other and both charges are positive. (b) Repeat for the case in which both charges are negative.

32. (a) Sketch the electric field pattern set up by a positively charged hollow sphere. Include regions inside and regions outside the sphere. (b) A conducting cube is given a positive charge. Sketch the electric field pattern both inside and outside the cube.

33. Refer to Figure 15.20. The charge lowered into the center of the hollow conductor has a magnitude of $5\ \mu C$. Find the magnitude and sign of the charge on the inside and outside of the hollow conductor when the charge is as shown in (a) Figure 15.20a; and (b) Figure 15.20b; (c) Figure 15.20c; and (d) Figure 15.20d.

Section 15.8 The Van de Graaff Generator

34. The dome of a Van de Graaff generator receives a charge of 2.0×10^{-4} C. Find the strength of the electric field (a) inside the dome; (b) at the surface of the dome, assuming it has a radius of 1.0 m; and (c) 4.0 m from the center of the dome. [*Hint:* See Section 15.6 to review properties of conductors in electrostatic equilibrium. Also, use the fact that the points on the surface are outside a spherically symmetric charge distribution; the total charge may be considered to be located at the center of the sphere.]

35. If the electric field strength in air exceeds 3.0×10^6 N/C, the air becomes a conductor. Using this fact, determine the maximum amount of charge that can be carried by a metal sphere 2.0 m in radius. (See the hint in Problem 34.)

36. In the Millikan oil drop experiment, an atomizer (a sprayer with a fine nozzle) is used to introduce many tiny droplets of oil between two oppositely charged parallel metal plates. Some of the droplets pick up one or more excess electrons. The charge on the plates is adjusted so that the electric force on the excess electrons exactly balances the weight of the droplet. The idea is to look for a droplet that has the smallest electric force and assume that it has only one excess electron. This strategy lets the observer measure the charge on the electron. Suppose we are using an electric field of 3×10^4 N/C. The charge on one electron is about 1.6×10^{-19} C. Estimate the radius of an oil drop of density $858\ kg/m^3$ for which its weight could be balanced by the electric force of this field on one electron. (Problem 36 is courtesy of E.F. Redish. For more problems of this type, visit http://www.physics.umd.edu/perg/.)

37. A Van de Graaff generator is charged so that the electric field at its surface is 3.0×10^4 N/C. Find (a) the electric force exerted on a proton released at its surface and (b) the acceleration of the proton at that instant of time.

Section 15.9 Electric Flux and Gauss's Law

38. A flat surface having an area of $3.2\ m^2$ is rotated in a uniform electric field of magnitude $E = 6.2 \times 10^5$ N/C. Determine the electric flux through this area (a) when the electric field is perpendicular to the surface and (b) when the electric field is parallel to the surface.

39. An electric field of intensity 3.50 kN/C is applied along the x-axis. Calculate the electric flux through a rectangular plane 0.350 m wide and 0.700 m long if (a) the plane is parallel to the yz-plane; (b) the plane is parallel to the xy-plane; and (c) the plane contains the y-axis, and its normal makes an angle of $40.0°$ with the x-axis.

40. The electric field everywhere on the surface of a thin spherical shell of radius 0.750 m is measured to be equal to 890 N/C and points radially toward the center of the sphere. (a) What is the net charge within the sphere's surface? (b) What can you conclude about the nature and distribution of the charge inside the spherical shell?

41. A 40-cm-diameter loop is rotated in a uniform electric field until the position of maximum electric flux is found. The flux in that position is measured to be 5.2×10^5 N·m²/C. Calculate the electric field strength in this region.

42. A point charge of $+5.00\ \mu C$ is located at the center of a sphere with a radius of 12.0 cm. Determine the electric flux through the surface of the sphere.

43. A point charge q is located at the center of a spherical shell of radius a that has a charge $-q$ uniformly distributed on its surface. Find the electric field (a) for all points outside the spherical shell and (b) for a point inside the shell a distance r from the center.

44. Use Gauss's law and the fact that the electric field inside any closed conductor in electrostatic equilibrium is zero to show that any excess charge placed on the conductor must reside on its surface.

45. **Physics⊗Now™** An infinite plane conductor has charge spread out on its surface as shown in Figure P15.45. Use Gauss's law to show that the electric field at any point outside the conductor is given by $E = \sigma/\epsilon_0$, where σ is the charge per unit area on the conductor. [*Hint:* Choose a Gaussian surface in the shape of a cylinder with one end inside the conductor and one end outside the conductor.]

σ σ

Gaussian surface

Figure P15.45

46. Show that the electric field just outside the surface of a good conductor of any shape is given by $E = \sigma/\epsilon_0$, where σ is the charge per unit area on the conductor. [*Hint:* The electric field just outside the surface of a charged conductor is perpendicular to its surface.]

ADDITIONAL PROBLEMS

47. Two protons in an atomic nucleus are typically separated by a distance of 2×10^{-15} m. The electric repulsion force between the protons is huge, but the attractive nuclear force is even stronger and keeps the nucleus from bursting apart. What is the magnitude of the electrical force between two protons separated by 2.00×10^{-15} m?

48. In the Bohr theory of the hydrogen atom, an electron moves in a circular orbit about a proton. The radius of the orbit is 0.53×10^{-10} m. (a) Find the electrostatic force acting on each particle. (b) If this force causes the centripetal acceleration of the electron, what is the speed of the electron?

49. Three point charges are aligned along the x-axis as shown in Figure P15.49. Find the electric field at the position $x = +2.0$ m, $y = 0$.

Figure P15.49

50. A small 2.00-g plastic ball is suspended by a 20.0-cm-long string in a uniform electric field, as shown in Figure P15.50. If the ball is in equilibrium when the string makes a 15.0° angle with the vertical as indicated, what is the net charge on the ball?

Figure P15.50

51. (a) Two identical point charges $+q$ are located on the y-axis at $y = +a$ and $y = -a$. What is the electric field along the x-axis at $x = b$? (b) A circular ring of charge of radius a has a total positive charge Q distributed uniformly around it. The ring is in the $x = 0$ plane with its center at the origin. What is the electric field along the x-axis at $x = b$ due to the ring of charge? [*Hint:* Consider the charge Q to consist of many pairs of identical point charges positioned at the ends of diameters of the ring.]

52. A positively charged bead having a mass of 1.00 g falls from rest in a vacuum from a height of 5.00 m in a uniform vertical electric field with a magnitude of 1.00×10^4 N/C. The bead hits the ground at a speed of 21.0 m/s. Determine (a) the direction of the electric field (upward or downward), and (b) the charge on the bead.

53. A solid conducting sphere of radius 2.00 cm has a charge of 8.00 μC. A conducting spherical shell of inner radius 4.00 cm and outer radius 5.00 cm is concentric with the solid sphere and has a charge of -4.00 μC. Find the electric field at (a) $r = 1.00$ cm, (b) $r = 3.00$ cm, (c) $r = 4.50$ cm, and (d) $r = 7.00$ cm from the center of this charge configuration.

54. Two small silver spheres, each with a mass of 100 g, are separated by 1.00 m. Calculate the fraction of the electrons in one sphere that must be transferred to the other in order to produce an attractive force of 1.00×10^4 N (about a ton) between the spheres. (The number of electrons per atom of silver is 47, and the number of atoms per gram is Avogadro's number divided by the molar mass of silver, 107.87 g/mol.)

55. A vertical electric field of magnitude 2.00×10^4 N/C exists above the Earth's surface on a day when a thunderstorm is brewing. A car with a rectangular size of 6.00 m by 3.00 m is traveling along a roadway sloping downward at 10.0°. Determine the electric flux through the bottom of the car.

56. A 2.00-μC charged 1.00-g cork ball is suspended vertically on a 0.500-m-long light string in the presence of a uniform downward-directed electric field of magnitude $E = 1.00 \times 10^5$ N/C. If the ball is displaced slightly from the vertical, it oscillates like a simple pendulum. (a) Determine the period of the ball's oscillation. (b) Should gravity be included in the calculation for part (a)? Explain.

57. Physics⊗Now™ Two 2.0-g spheres are suspended by 10.0-cm-long light strings (Fig. P15.57). A uniform electric field is applied in the x-direction. If the spheres have charges of -5.0×10^{-8} C and $+5.0 \times 10^{-8}$ C, determine the electric field intensity that enables the spheres to be in equilibrium at $\theta = 10°$.

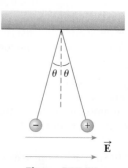

Figure P15.57

58. Two point charges like those in Figure P15.58 are called an electric dipole. Show that the electric field at a distant point along the x-axis is given by $E_x = 4k_e qa/x^3$.

Figure P15.58

59. A charged cork ball of mass 1.00 g is suspended on a light string in the presence of a uniform electric field as in Figure P15.59. When the electric field has an x-component of 3.00×10^5 N/C and a y-component of 5.00×10^5 N/C, the ball is in equilibrium at $\theta = 37.0°$. Find (a) the charge on the ball and (b) the tension in the string.

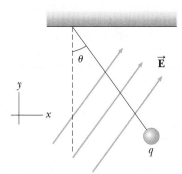

Figure P15.59

60. A point charge of magnitude 5.00 μC is at the origin of a coordinate system, and a charge of -4.00 μC is at the point $x = 1.00$ m. There is a point on the x-axis, at x less than infinity, where the electric field goes to zero. (a) Show by conceptual arguments that this point cannot be located between the charges. (b) Show by conceptual arguments that the point cannot be at any location between $x = 0$ and negative infinity. (c) Show by conceptual arguments that the point must be between $x = 1.00$ m and $x =$ positive infinity. (d) Use the values given to find the point and show that it is consistent with your conceptual argument.

61. Two hard rubber spheres of mass 15 g are rubbed vigorously with fur on a dry day and are then suspended from a rod with two insulating strings of length 5.0 cm. They are observed to hang at equilibrium as shown in Figure P15.61, each at an angle of 10° with the vertical. Estimate the amount of charge that is found on each sphere. (Problem 61 is courtesy of E.F. Redish. For more problems of this type, visit http://www.physics.umd.edu/perg/.)

Figure P15.61

62. Three identical point charges, each of mass $m = 0.100$ kg, hang from three strings, as shown in Figure P15.62. If the lengths of the left and right strings are each $L = 30.0$ cm, and if the angle θ is 45.0°, determine the value of q.

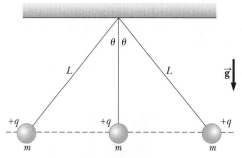

Figure P15.62

63. Each of the electrons in a particle beam has a kinetic energy of 1.60×10^{-17} J. (a) What is the magnitude of the uniform electric field (pointing in the direction of the electrons' movement) that will stop these electrons in a distance of 10.0 cm? (b) How long will it take to stop the electrons? (c) After the electrons stop, what will they do? Explain.

64. Protons are projected with an initial speed $v_0 = 9\,550$ m/s into a region where a uniform electric field $E = 720$ N/C is present (Fig. P15.64). The protons are to hit a target that lies a horizontal distance of 1.27 mm from the point where the protons are launched. Find (a) the two projection angles θ that will result in a hit and (b) the total duration of flight for each of the two trajectories.

Figure P15.64

ACTIVITIES

1. The following are a number of experiments that you can perform to investigate static electricity.

   (a) Attach two inflated balloons to the ends of a light string having a length of about 2 m. Tape the center of the string to the top of an open doorway as in Figure A15.1. Note that the balloons touch each other as they hang freely. Now rub each balloon several times with a wool cloth, and let them hang freely once again. Why are the balloons no longer touching each other, but are now separated?

   (b) Rub an inflated balloon with a piece of wool and press it against a wall. Note that the balloon adheres to the wall. Why?

(c) Rub nylon hose with a plastic dry-cleaner bag and observe how the hose expands like a balloon. Why does it do so?

(d) Tear some paper into very small pieces. Comb your hair and then bring the comb close to the pieces of paper. Notice that they accelerate toward the comb. How does the magnitude of the electric force compare with the magnitude of the gravitational force exerted on the paper? Keep watching, and you might see a few pieces jump away from the comb. They do not fall away; they are clearly repelled. What causes this repulsion?

Figure A15.1

2. For this experiment, you will need two 20-cm strips of transparent tape. (The mass of each is about 65 mg.) Fold about 1 cm of tape over at one end of each strip to create a handle. Press both pieces of tape side by side onto a tabletop, rubbing your finger back and forth across the strips. Quickly pull the strips off the surface so that they become charged. Hold the tape handles together, and the strips will repel each other, forming an inverted "V" shape. Measure the angle between the pieces and estimate the excess charge on each strip. Assume that the charges act as if they were located at the center of mass of each strip.

3. Rub an inflated balloon with a piece of wool cloth and place the balloon near a fine stream of water falling from a faucet. The stream of water will deflect toward the balloon. Why? Vary the distance between the balloon and the stream, and observe the displacement of the water stream for different distances. What is the relationship between the displacement of the stream and the distance of separation?

4. If you have access to a Van de Graaff generator, here are a few interesting things to try: (a) Stack several aluminum pie plates on top of the generator and then turn the generator on. What do you think is going to happen? Why? Try it. (b) Tape one pie plate on top of the generator and pour in some puffed rice or pieces of paper. What do you think will happen when you turn the generator on? Why? Try it and see. (c) Tape a glass beaker to the top of the generator and pour in some puffed rice. What do you think will happen when you turn the generator on? Try it. (d) Bring a fluorescent bulb close to a charged generator and observe what happens. Why does it happen?

Everything in the foreground of this picture is at the same electrical potential of many kilovolts. With no differences in potential, no charge is moving and no one gets a shock.

Electrical Energy and Capacitance

The concept of potential energy was first introduced in Chapter 5 in connection with the conservative forces of gravity and springs. By using the principle of conservation of energy, we were often able to avoid working directly with forces when solving problems. Here we learn that the potential energy concept is also useful in the study of electricity. Because the Coulomb force is conservative, we can define an electric potential energy corresponding to that force. In addition, we define an electric potential—the potential energy per unit charge—corresponding to the electric field.

With the concept of electric potential in hand, we can begin to understand electric circuits, starting with an investigation of a common circuit element called a capacitor. These simple devices store electrical energy and have found uses virtually everywhere, from etched circuits on a microchip to the creation of enormous bursts of power in fusion experiments.

16.1 POTENTIAL DIFFERENCE AND ELECTRIC POTENTIAL

Electric potential energy and electric potential are closely related concepts. The electric potential turns out to be just the electric potential energy per unit charge. This is similar to the relationship between electric force and the electric field, which is the electric force per unit charge.

Work and Electric Potential Energy

Recall from Chapter 5 that the work done by a conservative force $\vec{F}$ on an object depends only on the initial and final positions of the object and not on the path taken between those two points. This, in turn, means that a potential energy

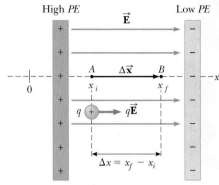

Figure 16.1 When a charge q moves in a uniform electric field $\vec{\mathbf{E}}$ from point A to point B, the work done on the charge by the electric force is $qE_x\Delta x$.

function PE exists. As we have seen, potential energy is a scalar quantity with change equal to the negative of the work done by the conservative force: $\Delta PE = PE_f - PE_i = -W_F$.

Both the Coulomb force law and the universal law of gravity are proportional to $1/r^2$. Because they have the same mathematical form, and because the gravity force is conservative, it follows that **the Coulomb force is also conservative**. As with gravity, an electrical potential energy function can be associated with this force.

To make these ideas more quantitative, imagine a small positive charge placed at point A in a *uniform* electric field $\vec{\mathbf{E}}$, as in Figure 16.1. For simplicity, we first consider only constant electric fields and charges which move parallel to that field in one dimension (taken to be the x-axis). The electric field between equally and oppositely charged parallel plates is an example of a field that is approximately constant. (See Chapter 15.) As the charge moves from point A to point B under the influence of the electric field $\vec{\mathbf{E}}$, the work done on the charge by the electric field is equal to the part of the electric force $q\vec{\mathbf{E}}$ acting parallel to the displacement, times the displacement $\Delta x = x_f - x_i$:

$$W_{AB} = F_x\Delta x = qE_x(x_f - x_i)$$

In this expression, q is the charge and E_x is the vector component of $\vec{\mathbf{E}}$ in the x-direction (*not* the magnitude of $\vec{\mathbf{E}}$). Unlike the magnitude of $\vec{\mathbf{E}}$, the component E_x can be positive or negative depending on the direction of $\vec{\mathbf{E}}$, though in Figure 16.1 E_x is positive. Finally, note that the displacement, like q and E_x, can also be either positive or negative, depending on the direction of the displacement.

The preceding expression for the work done by an electric field on a charge moving in one dimension is valid for both positive and negative charges and for constant electric fields pointing in *any* direction. When numbers are substituted with correct signs, the overall correct sign automatically results. In some books the expression $W = qEd$ is used, instead, where E is the magnitude of the electric field and d the distance the particle travels. The weakness of this formulation is that it doesn't allow, mathematically, for negative electric work on positive charges, nor for positive electric work on negative charges! Nonetheless, the expression is easy to remember and useful for finding magnitudes: the magnitude of the work done by a constant electric field on a charge moving parallel to the field is always given by $|W| = |q|Ed$.

We can substitute our definition of electric work into the work-energy theorem (assume other forces are absent):

$$W = qE_x\Delta x = \Delta KE$$

The electric force is conservative, so the electric work depends only on the end-points of the path, A and B, not on the path taken. Therefore, as the charge accelerates to the right in Figure 16.1, it gains kinetic energy, and loses an equal amount of potential energy. Recall from Chapter 5 that **the work done by a conservative force can be reinterpreted as the negative of the change in a potential energy associated with that force**. This motivates the definition of the change in electric potential energy:

Change in electric potential energy ▶

> The change in the electric potential energy, ΔPE, of a system consisting of an object of charge q moving through a displacement Δx in a constant electric field $\vec{\mathbf{E}}$ is given by
>
> $$\Delta PE = -W_{AB} = -qE_x\Delta x \qquad \text{[16.1]}$$
>
> where E_x is the x-component of the electric field and $\Delta x = x_f - x_i$ is the displacement of the charge along the x-axis.
>
> **SI Unit: joule (J)**

Although potential energy can be defined for any electric field, **Equation 16.1 is valid only for the case of a uniform (i.e., constant) electric field, for a particle that**

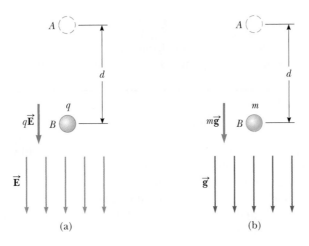

Figure 16.2 (a) When the electric field $\vec{E}$ is directed downward, point B is at a lower electric potential than point A. As a positive test charge moves from A to B, the electric potential energy decreases. (b) An object of mass m moves in the direction of the gravitational field $\vec{g}$, the gravitational potential energy decreases.

undergoes a displacement along a given axis (here called the *x*-axis). Because the electric field is conservative, the change in potential energy doesn't depend on the path. Consequently, it's unimportant whether or not the charge remains on the axis at all times during the displacement: the change in potential energy will be the same. In subsequent sections we will examine situations in which the electric field is not uniform.

Electric and gravitational potential energy can be compared in Figure 16.2. In this figure, the electric and gravitational fields are both directed downwards. We see that positive charge in an electric field acts very much like mass in a gravity field: a positive charge at point A falls in the direction of the electric field, just as a positive mass falls in the direction of the gravity field. Let point B be the zero point for potential energy in both Figure 16.2a and Figure 16.2b. From conservation of energy, in falling from point A to point B the positive charge gains kinetic energy equal in magnitude to the loss of electric potential energy:

$$\Delta KE + \Delta PE_{el} = \Delta KE + (0 - |q|Ed) = 0 \quad \rightarrow \quad \Delta KE = |q|Ed$$

The absolute value signs on q are there only to make explicit that the charge is positive in this case. Similarly, the object in Figure 16.2b gains kinetic energy equal in magnitude to the loss of gravitational potential energy:

$$\Delta KE + \Delta PE_{g} = \Delta KE + (0 - mgd) = 0 \quad \rightarrow \quad \Delta KE = mgd$$

So for positive charges, electric potential energy works very much like gravitational potential energy. In both cases, moving an object opposite the direction of the field results in a gain of potential energy, and upon release, the potential energy is converted to the object's kinetic energy.

Electric potential energy differs significantly from gravitational potential energy, however, in that there are two kinds of electrical charge—positive and negative—whereas gravity has only positive "gravitational charge" (i.e. mass). A negatively charged particle at rest at point A in Figure 16.2a would have to be *pushed* down to point B. To see this, apply the work–energy theorem to a negative charge at rest at point A and assumed to have some speed v on arriving at point B:

$$W = \Delta KE + \Delta PE_{el} = (\tfrac{1}{2}mv^2 - 0) + (0 - (-|q|Ed))$$

$$W = \tfrac{1}{2}mv^2 + |q|Ed$$

Notice that the negative charge, $-|q|$, unlike the positive charge, had a positive change in electric potential energy in moving from point A to point B. If the negative charge has any speed at point B, the kinetic energy corresponding to that speed is also positive. Because both terms on the right-hand side of the work–energy equation are positive, there is no way of getting the negative charge from point A to point B without doing positive work W on it. In fact, if the negative charge is simply released at point A, it will "fall" upwards against the direction of the field!

If an electron is released from rest in a uniform electric field, the electric potential energy of the charge–field system (a) increases, (b) decreases, or (c) remains the same.

EXAMPLE 16.1   Potential Energy Differences in an Electric Field

Goal   Illustrate the concept of electric potential energy.

Problem   A proton is released from rest at $x = -2.00$ cm in a constant electric field with magnitude 1.50×10^3 N/C, pointing in the positive x-direction. **(a)** Calculate the change in the electric potential energy associated with the proton when it reaches $x = 5.00$ cm. **(b)** An electron is now fired in the same direction from the same position. What is its change in electric potential energy associated with the electron if it reaches $x = 12.0$ cm? **(c)** If the direction of the electric field is reversed and an electron is released from rest at $x = 3.00$ cm, by how much has the electric potential energy changed when the electron reaches $x = 7.00$ cm?

Strategy   This problem requires a straightforward substitution of given values into the definition of electric potential energy, Equation 16.1.

Solution
(a) Calculate the change in the electric potential energy associated with the proton.

Apply Equation 16.1:

$$\Delta PE = -qE_x\Delta x = -qE_x(x_f - x_i)$$
$$= -(1.60 \times 10^{-19}\,\text{C})(1.50 \times 10^3\,\text{N/C})$$
$$\times [0.050\,0\,\text{m} - (-0.020\,0\,\text{m})]$$
$$= -1.68 \times 10^{-17}\,\text{J}$$

(b) Find the change in electric potential energy associated with an electron fired from $x = -0.0200$ m and reaching $x = 0.120$ m.

Apply Equation 16.1, but in this case note that the electric charge q is negative:

$$\Delta PE = -qE_x\Delta x = -qE_x(x_f - x_i)$$
$$= -(-1.60 \times 10^{-19}\,\text{C})(1.50 \times 10^3\,\text{N/C})$$
$$\times [(0.120\,\text{m} - (-0.020\,0\,\text{m})]$$
$$= +3.36 \times 10^{-17}\,\text{J}$$

(c) Find the change in potential energy associated with an electron traveling from $x = 3.00$ cm to $x = 7.00$ cm if the direction of the electric field is reversed.

Substitute, but now the electric field points in the negative x-direction, hence carries a minus sign:

$$\Delta PE = -qE_x\Delta x = -qE_x(x_f - x_i)$$
$$= -(-1.60 \times 10^{-19}\,\text{C})(-1.50 \times 10^3\,\text{N/C})$$
$$\times (0.070\,\text{m} - 0.030\,\text{m})$$
$$= -9.60 \times 10^{-18}\,\text{J}$$

Remarks   Notice that the proton (actually the proton–field system) lost potential energy when it moved in the positive x-direction, while the electron gained potential energy when it moved in the same direction. Finding changes in potential energy with the field reversed was just a matter of supplying a minus sign, bringing the total number in this case to three! It's important not to drop any of the signs.

Exercise 16.1
Find the change in electric potential energy associated with the electron in part (b) as it goes on from $x = 0.120$ m to $x = -0.180$ m. (Note that the electron must turn around and go back at some point. The location of the turning point is unimportant, because changes in potential energy depend only on the end points of the path.)

Answer $-7.20 \times 10^{-17}\,\text{J}$

INTERACTIVE EXAMPLE 16.2   Dynamics of Charged Particles

Goal   Use electric potential energy in conservation of energy problems.

Problem   (a) Find the speed of the proton at $x = 0.050\ 0$ m in part (a) of Example 16.1. (b) Find the initial speed of the electron (at $x = -2.00$ cm) in part (b) of Example 16.1, given that its speed has fallen by half when it reaches $x = 0.120$ m.

Strategy   Apply conservation of energy, solving for the unknown speeds. Part (b) involves two equations: the conservation equation, and the condition $v_f = \frac{1}{2}v_i$ for the unknown initial and final speeds. The changes in electric potential energy have already been calculated in Example 16.1.

Solution
(a) Calculate the proton's speed at $x = 0.050$ m.

Use conservation of energy, with an initial speed of zero:

$$\Delta KE + \Delta PE = 0 \quad \rightarrow \quad (\tfrac{1}{2}m_p v^2 - 0) + \Delta PE = 0$$

Solve for v, and substitute the change in potential energy found in Example 16.1a:

$$v^2 = -\frac{2}{m_p}\Delta PE$$

$$v = \sqrt{-\frac{2}{m_p}\Delta PE}$$

$$= \sqrt{-\frac{2}{(1.67 \times 10^{-27}\ \text{kg})}(-1.68 \times 10^{-17}\ \text{J})}$$

$$v = \boxed{1.42 \times 10^5\ \text{m/s}}$$

(b) Find the electron's initial speed, given that its speed has fallen by half at $x = 0.120$ m.

Apply conservation of energy once again, substituting expressions for the initial and final kinetic energies:

$$\Delta KE + \Delta PE = 0$$

$$(\tfrac{1}{2}m_e v_f^2 - \tfrac{1}{2}m_e v_i^2) + \Delta PE = 0$$

Substitute the condition $v_f = \frac{1}{2}v_i$, and subtract the change in potential energy from both sides:

$$\tfrac{1}{2}m_e(\tfrac{1}{2}v_i)^2 - \tfrac{1}{2}m_e v_i^2 = -\Delta PE$$

Combine terms and solve for v_i, the initial speed, and substitute the change in potential energy found in Example 16.1b:

$$-\tfrac{3}{8}m_e v_i^2 = -\Delta PE$$

$$v_i = \sqrt{\frac{8\Delta PE}{3m_e}} = \sqrt{\frac{8(3.36 \times 10^{-17}\ \text{J})}{3(9.11 \times 10^{-31}\ \text{kg})}}$$

$$= \boxed{9.92 \times 10^6\ \text{m/s}}$$

Remarks   While the changes in potential energy associated with the proton and electron were similar in magnitude, the effect on their speeds differed dramatically. The change in potential energy had a proportionately much greater effect on the much lighter electron than on the proton.

Exercise 16.2
Refer to Exercise 16.1. Find the electron's speed at $x = -0.180$ m. The answer is 4.5% of the speed of light.

Answer 1.35×10^7 m/s

Physics⊗Now™   You can predict and observe the speed of the proton as it arrives at the negative plate, for random values of the electric field, by logging into PhysicsNow at **www.cp7e.com** and going to Interactive Example 16.2.

Electric Potential

In Chapter 15, it was convenient to define an electric field $\vec{\mathbf{E}}$ related to the electric force $\vec{\mathbf{F}} = q\vec{\mathbf{E}}$. In this way, the properties of fixed collections of charges could be easily studied, and the force on any particle in the electric field could be obtained simply by multiplying by the particle's charge q. For the same reasons, it's useful to define an *electric potential difference* ΔV related to the potential energy by $\Delta PE = q\Delta V$:

◄ Potential difference between two points

The electric potential difference ΔV between points A and B is the change in electric potential energy as a charge q moves from A to B, divided by the charge q:

$$\Delta V = V_B - V_A = \frac{\Delta PE}{q}$$ [16.2]

SI Unit: joule per coulomb, or volt (J/C, or V)

This definition is completely general, although in many cases calculus would be required to compute the change in potential energy of the system. Because electric potential energy is a scalar quantity, **electric potential is also a scalar quantity**. From Equation 16.2, we see that electric potential difference is a measure of the change in electric potential energy per unit charge. Alternately, the electric potential difference is the work per unit charge that would have to be done by some force to move a charge from point A to point B in the electric field. The SI unit of electric potential is the joule per coulomb, called the volt (V). From the definition of that unit, 1 J of work must be done to move a 1-C charge between two points that are at a potential difference of 1 V. In the process of moving through a potential difference of 1 V, the 1-C charge gains 1 J of energy.

For the special case of a uniform electric field such as that between charged parallel plates, dividing Equation 16.1 by q gives

$$\frac{\Delta PE}{q} = -E_x\Delta x$$

TIP 16.1 Potential and Potential Energy

Electric potential is characteristic of the field only, independent of a test charge that may be placed in that field. On the other hand, potential energy is a characteristic of the charge–field system due to an interaction between the field and a charge placed in the field.

Comparing this equation with Equation 16.2, we find that

$$\Delta V = -E_x\Delta x$$ [16.3]

Equation 16.3 shows that potential difference also has units of electric field times distance. From this it follows that the SI unit of the electric field, the newton per coulomb, can also be expressed as volts per meter:

$$1 \text{ N/C} = 1 \text{ V/m}$$

Because Equation 16.3 is directly related to Equation 16.1, remember that it's valid only for the system consisting of a uniform electric field and a charge moving in one dimension.

Released from rest, positive charges accelerate spontaneously from regions of high potential to low potential. If a positive charge is given some initial velocity in the direction of high potential, it can move in that direction, but will slow and finally turn around, just like a ball tossed upwards in a gravity field. Negative charges do exactly the opposite: Released from rest, they accelerate from regions of low potential toward regions of high potential. Work must be done on negative charges to make them go in the direction of lower electric potential.

Quick Quiz 16.2

Figure 16.3 is a graph of an electric potential as a function of position. If a positively charged particle is placed at point A, what will its subsequent motion be? It will (a) go to the right (b) go to the left (c) remain at point A (d) oscillate around point B.

Now, if a negatively charged particle is placed at point B and given a very small kick to the right, what will its subsequent motion be? It will (a) go to the right, and not return (b) go to the left (c) remain at point B (d) oscillate around point B.

Figure 16.3   (Quick Quiz 16.2)

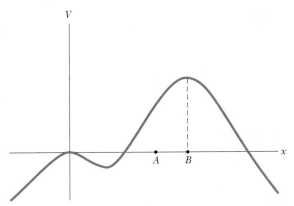

An application of potential difference is the 12-V battery found in an automobile. Such a battery maintains a potential difference across its terminals, with the positive terminal 12 V higher in potential than the negative terminal. In practice, the negative terminal is usually connected to the metal body of the car, which can be considered to be at a potential of zero volts. The battery provides the electrical current necessary to operate headlights, a radio, power windows, motors, and so forth. Now consider a charge of $+1$ C, to be moved around a circuit that contains the battery connected to some of these external devices. As the charge is moved inside the battery from the negative terminal (at 0 V) to the positive terminal (at 12 V), the work done on the charge by the battery is 12 J. Every coulomb of positive charge that leaves the positive terminal of the battery carries an energy of 12 J. As the charge moves through the external circuit toward the negative terminal, it gives up its 12 J of electrical energy to the external devices. When the charge reaches the negative terminal, its electrical energy is zero again. At this point, the battery takes over and restores 12 J of energy to the charge as it is moved from the negative to the positive terminal, enabling it to make another transit of the circuit. The actual amount of charge that leaves the battery each second and traverses the circuit depends on the properties of the external devices, as seen in the next chapter.

APPLICATION
Automobile Batteries

EXAMPLE 16.3   TV Tubes and Atom Smashers

Goal   Relate electric potential to an electric field and conservation of energy.

Problem   In atom smashers (also known as cyclotrons and linear accelerators), charged particles are accelerated in much the same way they are accelerated in TV tubes: through potential differences. Suppose a proton is injected at a speed of 1.00×10^6 m/s between two plates 5.00 cm apart, as shown in Figure 16.4. The proton subsequently accelerates across the gap and exits through the opening. **(a)** What must the electric potential difference be if the exit speed is to be 3.00×10^6 m/s? **(b)** What is the magnitude of the electric field between the plates?

Strategy   Use conservation of energy, writing the change in potential energy in terms of the change in electric potential, ΔV, and solve for ΔV. For part (b), solve Equation 16.3 for the electric field.

Figure 16.4   (Example 16.3) A proton enters a cavity and accelerates from one charged plate toward the other in an electric field $\vec{\mathbf{E}}$.

Solution

(a) Find the electric potential yielding the desired exit speed of the proton.

Apply conservation of energy, writing the potential energy in terms of the electric potential:

$$\Delta KE + \Delta PE = \Delta KE + q\Delta V = 0$$

Solve the energy equation for the change in potential:

$$\Delta V = -\frac{\Delta KE}{q} = -\frac{\frac{1}{2}m_p v_f^2 - \frac{1}{2}m_p v_i^2}{q} = -\frac{m_p}{2q}(v_f^2 - v_i^2)$$

Substitute the given values, obtaining the necessary potential difference:

$$\Delta V = -\frac{(1.67 \times 10^{-27}\ \text{kg})}{2(1.60 \times 10^{-19}\ \text{C})}\left[(3.00 \times 10^6\ \text{m/s})^2\right.$$

$$\left.- (1.00 \times 10^6\ \text{m/s})^2\right]$$

$$\Delta V = \boxed{-4.18 \times 10^4\ \text{V}}$$

(b) What electric field must exist between the plates?

Solve Equation 16.3 for the electric field, and substitute:

$$E = -\frac{\Delta V}{\Delta x} = \frac{4.18 \times 10^4\ \text{V}}{0.0500\ \text{m}} = \boxed{8.36 \times 10^5\ \text{N/C}}$$

Remarks   Systems of such cavities, consisting of alternating positive and negative plates, are used to accelerate charged particles to high speed before smashing them into targets. To prevent a slowing of, say, a positively-charged particle after it passes through the negative plate of one cavity and enters the next, the charges on the plates are reversed. Otherwise, the particle would be traveling from the negative plate to a positive plate in the second cavity, and the kinetic energy gained in the previous cavity would be lost in the second.

Exercise 16.3

Suppose electrons in a TV tube are accelerated through a potential difference of 2.00×10^4 V from the heated cathode (negative electrode), where they are produced, towards the screen, which also serves as the anode (positive electrode), 25.0 cm away. (a) At what speed would the electrons impact the phosphors on the screen? Assume they accelerate from rest, and ignore relativistic effects (Chapter 26). (b) What's the magnitude of the electric field, if it is assumed constant?

Answers   (a) 8.38×10^7 m/s   (b) 8.00×10^4 V/m

16.2   ELECTRIC POTENTIAL AND POTENTIAL ENERGY DUE TO POINT CHARGES

In electric circuits, a point of zero electric potential is often defined by grounding (connecting to Earth) some point in the circuit. For example, if the negative terminal of a 12-V battery were connected to ground, it would be considered to have a potential of zero, while the positive terminal would have a potential of + 12 V. The potential difference created by the battery, however, is only locally defined. In this section we describe the electric potential of a point charge, which is defined throughout space.

The electric field of a point charge extends throughout space, so its electric potential does, also. The zero point of electric potential could be taken anywhere, but is usually taken to be an infinite distance from the charge, far from its influence and the influence of any other charges. With this choice, the methods of calculus can be used to show that the electric potential created by a point charge q at any distance r from the charge is given by

Electric potential created by a point charge ▶

$$V = k_e \frac{q}{r} \qquad\qquad \textbf{[16.4]}$$

Equation 16.4 shows that the electric potential, or work per unit charge, required to move a test charge in from infinity to a distance r from a positive point charge q increases as the positive test charge moves closer to q. A plot of Equation 16.4 in Figure 16.5 shows that the potential associated with a point charge decreases as $1/r$ with increasing r, in contrast to the magnitude of the charge's electric field, which decreases as $1/r^2$.

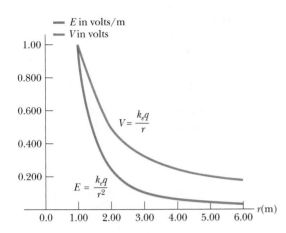

— E in volts/m
— V in volts

Figure 16.5 Electric field and electric potential versus distance from a point charge of 1.11×10^{-10} C. Note that V is proportional to $1/r$, while E proportional to $1/r^2$.

The electric potential of two or more charges is obtained by applying the **superposition principle: the total electric potential at some point P due to several point charges is the algebraic sum of the electric potentials due to the individual charges**. This is similar to the method used in Chapter 15 to find the resultant electric field at a point in space. Unlike electric field superposition, which involves a sum of vectors, the superposition of electric potentials requires evaluating a sum of scalars. As a result, it's much easier to evaluate the electric potential at some point due to several charges than to evaluate the electric field, which is a vector quantity.

◀ Superposition principle

Figure 16.6 is a computer-generated plot of the electric potential associated with an electric dipole, which consists of two charges of equal magnitude but opposite in sign. The charges lie in a horizontal plane at the center of the potential spikes. The value of the potential is plotted in the vertical dimension. The computer program has added the potential of each charge to arrive at total values of the potential.

Just as in the case of constant electric fields, there is a relationship between electric potential and electric potential energy. If V_1 is the electric potential due to charge q_1 at a point P (Active Figure 16.7a, page 540), then the work required to bring charge q_2 from infinity to P without acceleration is q_2V_1. By definition, this work equals the potential energy PE of the two-particle system when the particles are separated by a distance r (Active Fig. 16.7b).

We can therefore express the electrical potential energy of the *pair* of charges as

$$PE = q_2V_1 = k_e \frac{q_1 q_2}{r} \qquad \text{[16.5]}$$

◀ Potential energy of a pair of charges

If the charges are of the *same* sign, PE is positive. This is consistent with the fact that like charges repel, so positive work must be done on the system by an external agent to force the two charges near one another. Conversely, if the charges are of

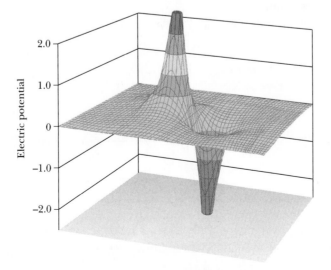

Figure 16.6 The electric potential (in arbitrary units) in the plane containing an electric dipole. Potential is plotted in the vertical dimension.

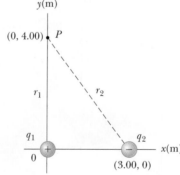

$$P \quad V_1 = \frac{k_e q_1}{r}$$

q_1

(a)

q_2

r

$$PE = \frac{k_e q_1 q_2}{r}$$

q_1

(b)

ACTIVE FIGURE 16.7
(a) The electric potential V_1 at P due to the point charge q_1 is $V_1 = k_e q_1 / r$.
(b) If a second charge, q_2, is brought from infinity to P, the potential energy of the pair is $PE = k_e q_1 q_2 / r$.

Physics⊗Now™

Log into PhysicsNow at **www.cp7e.com** and go to Active Figure 16.7, where you can move one of the charges and see the effect on the electric potential energy, or move a point P to see the effect on the electric potential at P.

opposite sign, the force is attractive and *PE* is negative. This means that negative work must be done to prevent unlike charges from accelerating toward each other as they are brought close together.

Quick Quiz 16.3

Consider a collection of charges in a given region, and suppose all other charges are distant and have a negligible effect. Further, the electric potential is taken to be zero at infinity. If the electric potential at a given point in the region is zero, which of the following statements must be true? (a) The electric field is zero at that point. (b) The electric potential energy is a minimum at that point. (c) There is no net charge in the region. (d) Some charges in the region are positive and some are negative. (e) The charges have the same sign and are symmetrically arranged around the given point.

Quick Quiz 16.4

A spherical balloon contains a positively charged particle at its center. As the balloon is inflated to a larger volume while the charged particle remains at the center, which of the following are true? (a) The electric potential at the surface of the balloon increases. (b) The magnitude of the electric field at the surface of the balloon increases. (c) The electric flux through the balloon remains the same. (d) none of these.

Problem-Solving Strategy   Electric Potential

1. Draw a diagram of all charges, and circle the point of interest.
2. Calculate the distance from each charge to the point of interest, labeling it on the diagram.
3. For each charge q, calculate the scalar quantity $V = \dfrac{k_e q}{r}$. *The sign of each charge must be included in your calculations!*
4. Sum all the numbers found in the previous step, obtaining the electric potential at the point of interest.

EXAMPLE 16.4   Finding the Electric Potential

Goal   Calculate the electric potential due to a collection of point charges.

Problem   A 5.00-μC point charge is at the origin, and a point charge $q_2 = -2.00\ \mu$C is on the x-axis at $(3.00, 0)$ m, as in Figure 16.8. **(a)** If the electric potential is taken to be zero at infinity, find the total electric potential due to these charges at point P with coordinates $(0, 4.00)$ m. **(b)** How much work is required to bring a third point charge of 4.00 μC from infinity to P?

Strategy   **(a)** The electric potential at P due to each charge can be calculated from $V = k_e q / r$. The total electric potential at P is the sum of these two numbers. **(b)** Use the work–energy theorem, together with Equation 16.5, recalling that the potential at infinity is taken to be zero.

Figure 16.8   (Example 16.4) The electric potential at point P due to the point charges q_1 and q_2 is the algebraic sum of the potentials due to the individual charges.

Solution

(a) Find the electric potential at point P.

Calculate the electric potential at P due to the 5.00-μC charge:

$$V_1 = k_e \frac{q_1}{r_1} = \left(8.99 \times 10^9\ \frac{\text{N} \cdot \text{m}^2}{\text{C}^2}\right)\left(\frac{5.00 \times 10^{-6}\ \text{C}}{4.00\ \text{m}}\right)$$

$$= 1.12 \times 10^4\ \text{V}$$

Find the electric potential at P due to the -2.00-μC charge:

$$V_2 = k_e \frac{q_2}{r_2} = \left(8.99 \times 10^9 \frac{\text{N} \cdot \text{m}^2}{\text{C}^2}\right)\left(\frac{-2.00 \times 10^{-6} \text{ C}}{5.00 \text{ m}}\right)$$

$$= -0.360 \times 10^4 \text{ V}$$

Sum the two numbers to find the total electric potential at P:

$$V_P = V_1 + V_2 = 1.12 \times 10^4 \text{ V} + (-0.360 \times 10^4 \text{ V})$$

$$= 7.60 \times 10^3 \text{ V}$$

(b) Find the work needed to bring the 4.00-μC charge from infinity to P.

Apply the work–energy theorem, with Equation 16.5:

$$W = \Delta PE = q_3 \Delta V = q_3(V_P - V_\infty)$$

$$= (4.00 \times 10^{-6} \text{ C})(7.60 \times 10^3 \text{ V} - 0)$$

$$W = 3.04 \times 10^{-2} \text{ J}$$

Exercise 16.4

Suppose a charge of -2.00 μC is at the origin and a charge of 3.00 μC is at the point $(0, 3.00)$ m. (a) Find the electric potential at $(4.00, 0)$ m, assuming the electric potential is zero at infinity, and (b) find the work necessary to bring a 4.00 μC charge from infinity to the point $(4.00, 0)$ m.

Answers   (a) 8.99×10^2 V   (b) 3.60×10^{-3} J

16.3 POTENTIALS AND CHARGED CONDUCTORS

The electric potential at all points on a charged conductor can be determined by combining Equations 16.1 and 16.2. From Equation 16.1, we see that the work done on a charge by electric forces is related to the change in electrical potential energy of the charge by

$$W = -\Delta PE$$

From Equation 16.2, we see that the change in electric potential energy between two points A and B is related to the potential difference between those points by

$$\Delta PE = q(V_B - V_A)$$

Combining these two equations, we find that

$$W = -q(V_B - V_A) \qquad \text{[16.6]}$$

Using this equation, we obtain the following general result: **No net work is required to move a charge between two points that are at the same electric potential.** In mathematical terms, this says that $W = 0$ whenever $V_B = V_A$.

In Chapter 15 we found that when a conductor is in electrostatic equilibrium, a net charge placed on it resides entirely on its surface. Further, we showed that the electric field just outside the surface of a charged conductor in electrostatic equilibrium is perpendicular to the surface and that the field inside the conductor is zero. We now show that **all points on the surface of a charged conductor in electrostatic equilibrium are at the same potential**.

Consider a surface path connecting any points A and B on a charged conductor, as in Figure 16.9. The charges on the conductor are assumed to be in equilibrium with each other, so none are moving. In this case, the electric field $\vec{E}$ is always perpendicular to the displacement along this path. This must be so, for otherwise the part of the electric field tangent to the surface would move the charges. Because $\vec{E}$ is perpendicular to the path, no work is done if a charge is moved between the given two points. From Equation 16.6 we see that if the work done is zero, the difference in electric potential, $V_B - V_A$, is also zero. It

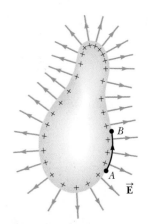

Figure 16.9 An arbitrarily shaped conductor with an excess positive charge. When the conductor is in electrostatic equilibrium, all of the charge resides at the surface, $\vec{E} = 0$ inside the conductor, and the electric field just outside the conductor is perpendicular to the surface. The potential is constant inside the conductor and is equal to the potential at the surface.

follows that **the electric potential is a constant everywhere on the surface of a charged conductor in equilibrium**. Further, because the electric field inside a conductor is zero, no work is required to move a charge between two points inside the conductor. Again, Equation 16.6 shows that if the work done is zero, the difference in electric potential between any two points inside a conductor must also be zero. We conclude that the electric potential is constant everywhere inside a conductor.

Finally, because one of the points inside the conductor could be arbitrarily close to the surface of the conductor, we conclude that **the electric potential is constant everywhere inside a conductor and equal to that same value at the surface**. As a consequence, no work is required to move a charge from the interior of a charged conductor to its surface. (It's important to realize that the potential inside a conductor is not necessarily zero, even though the interior electric field is zero.)

The Electron Volt

An appropriately-sized unit of energy commonly used in atomic and nuclear physics is the electron volt (eV). For example, electrons in normal atoms typically have energies of tens of eV's, excited electrons in atoms emitting x-rays have energies of thousands of eV's, and high energy gamma rays (electromagnetic waves) emitted by the nucleus have energies of millions of eV's.

Definition of the electron volt ▶

> The **electron volt** is defined as the kinetic energy that an electron gains when accelerated through a potential difference of 1 V.

Because $1\ V = 1\ J/C$ and because the magnitude of the charge on the electron is 1.60×10^{-19} C, we see that the electron volt is related to the joule by

$$1\ eV = 1.60 \times 10^{-19}\ C \cdot V = 1.60 \times 10^{-19}\ J \qquad [16.7]$$

Quick Quiz 16.5

An electron initially at rest accelerates through a potential difference of 1 V, gaining kinetic energy KE_e, while a proton, also initially at rest, accelerates through a potential difference of -1 V, gaining kinetic energy KE_p. Which of the following relationships holds? (a) $KE_e = KE_p$ (b) $KE_e < KE_p$. (c) $KE_e > KE_p$ (d) cannot be determined from the given information.

16.4 EQUIPOTENTIAL SURFACES

A surface on which all points are at the same potential is called an **equipotential surface**. The potential difference between any two points on an equipotential surface is zero. Hence, **no work is required to move a charge at constant speed on an equipotential surface**.

Equipotential surfaces have a simple relationship to the electric field: **The electric field at every point of an equipotential surface is perpendicular to the surface**. If the electric field $\vec{E}$ had a component parallel to the surface, that component would produce an electric force on a charge placed on the surface. This force would do work on the charge as it moved from one point to another, in contradiction to the definition of an equipotential surface.

Equipotential surfaces can be represented on a diagram by drawing equipotential contours, which are two-dimensional views of the intersections of the equipotential surfaces with the plane of the drawing. These equipotential contours are generally referred to simply as **equipotentials**. Figure 16.10a shows the equipotentials (in blue) associated with a positive point charge. Note that the equipotentials are perpendicular to the electric field lines (in red) at all points. Recall that the electric potential created by a point charge q is given by $V = k_e q/r$. This relation shows that, for a single point charge, the potential is constant on any surface on which r is constant. It follows that the equipotentials of a point charge are a family

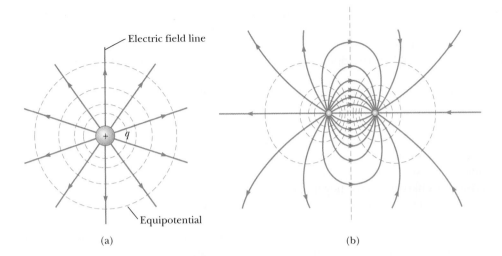

Figure 16.10 Equipotentials (dashed blue lines) and electric field lines (red lines) for (a) a positive point charge and (b) two point charges of equal magnitude and opposite sign. In all cases, the equipotentials are *perpendicular* to the electric field lines at every point.

of spheres centered on the point charge. Figure 16.10b shows the equipotentials associated with two charges of equal magnitude but opposite sign.

16.5 APPLICATIONS

The Electrostatic Precipitator

One important application of electric discharge in gases is a device called an *electrostatic precipitator*. This device removes particulate matter from combustion gases, thereby reducing air pollution. It's especially useful in coal-burning power plants and in industrial operations that generate large quantities of smoke. Systems currently in use can eliminate approximately 90% by mass of the ash and dust from the smoke. Unfortunately, a very high percentage of the lighter particles still escape, and these contribute significantly to smog and haze.

Figure 16.11 illustrates the basic idea of the electrostatic precipitator. A high voltage (typically 40 kV to 100 kV) is maintained between a wire running down the center of a duct and the outer wall, which is grounded. The wire is maintained at a negative electric potential with respect to the wall, so the electric field is directed toward the wire. The electric field near the wire reaches a high enough value to cause a discharge around the wire and the formation of positive ions, electrons, and negative ions, such as O_2^-. As the electrons and negative ions are accelerated

APPLICATION

The Electrostatic Precipitator

(a)   (b)   (c)

Figure 16.11 (a) A schematic diagram of an electrostatic precipitator. The high voltage maintained on the central wires creates an electric discharge in the vicinity of the wire. Compare the air pollution when the precipitator is (b) operating, and (c) turned off.

toward the outer wall by the nonuniform electric field, the dirt particles in the streaming gas become charged by collisions and ion capture. Because most of the charged dirt particles are negative, they are also drawn to the outer wall by the electric field. When the duct is shaken, the particles fall loose and are collected at the bottom.

In addition to reducing the amounts of harmful gases and particulate matter in the atmosphere, the electrostatic precipitator recovers valuable metal oxides from the stack.

A similar device called an *electrostatic air cleaner* is used in homes to relieve the discomfort of allergy sufferers. Air laden with dust and pollen is drawn into the device across a positively charged mesh screen. The airborne particles become positively charged when they make contact with the screen, and then pass through a second, negatively charged mesh screen. The electrostatic force of attraction between the positively charged particles in the air and the negatively charged screen causes the particles to precipitate out on the surface of the screen, removing a very high percentage of contaminants from the air stream.

APPLICATION

The Electrostatic Air Cleaner

APPLICATION

Xerographic Copiers

Xerography and Laser Printers

Xerography is widely used to make photocopies of printed materials. The basic idea behind the process was developed by Chester Carlson, who was granted a patent for his invention in 1940. In 1947 the Xerox Corporation launched a full-scale program to develop automated duplicating machines using Carlson's process. The huge success of that development is evident: today, practically all offices and libraries have one or more duplicating machines, and the capabilities of these machines continue to evolve.

Some features of the xerographic process involve simple concepts from electrostatics and optics. However, the one idea that makes the process unique is the use of photoconductive material to form an image. A photoconductor is a material that is a poor conductor of electricity in the dark, but a reasonably good conductor when exposed to light.

Figure 16.12 illustrates the steps in the xerographic process. First, the surface of a plate or drum is coated with a thin film of the photoconductive material (usually selenium or some compound of selenium), and the photoconductive surface is given a positive electrostatic charge in the dark (Fig. 16.12a). The page to be copied is then projected onto the charged surface (Fig. 16.12b). The photoconducting surface becomes conducting only in areas where light strikes; there the light produces charge carriers in the photoconductor which neutralize the positively charged surface. The charges remain on those areas of the photoconductor not exposed to light, however, leaving a hidden image of the object in the form of a positive distribution of surface charge.

Next, a negatively charged powder called a *toner* is dusted onto the photoconducting surface (Fig. 16.12c). The charged powder adheres only to the areas that contain the positively charged image. At this point, the image becomes visible. It is then transferred to the surface of a sheet of positively charged paper. Finally, the toner is "fixed" to the surface of the paper by heat (Fig. 16.12d), resulting in a permanent copy of the original.

APPLICATION

Laser Printers

The steps for producing a document on a laser printer are similar to those used in a photocopy machine, in that parts (a), (c), and (d) of Figure 16.12 remain essentially the same. The difference between the two techniques lies in the way the image is formed on the selenium-coated drum. In a laser printer, the command to print the letter O, for instance, is sent to a laser from the memory of a computer. A rotating mirror inside the printer causes the beam of the laser to sweep across the selenium-coated drum in an interlaced pattern (Fig. 16.12e). Electrical signals generated by the printer turn the laser beam on and off in a pattern that traces out the letter "O" in the form of positive charges on the selenium. Toner is then applied to the drum, and the transfer to paper is accomplished as in a photocopy machine.

(a) Charging the drum   (b) Imaging the document   (c) Applying the toner

(d) Transferring the
toner to the paper

(e) Laser printer drum

Figure 16.12 The xerographic process. (a) The photoconductive surface is positively charged. (b) Through the use of a light source and a lens, a hidden image is formed on the charged surface in the form of positive charges. (c) The surface containing the image is covered with a negatively charged powder, which adheres only to the image area. (d) A piece of paper is placed over the surface and given a charge. This transfers the image to the paper, which is then heated to "fix" the powder to the paper. (e) The image on the drum of a laser printer is produced by turning a laser beam on and off as it sweeps across the selenium-coated drum.

16.6 CAPACITANCE

A **capacitor** is a device used in a variety of electric circuits—for example, to tune the frequency of radio receivers, eliminate sparking in automobile ignition systems, or store short-term energy for rapid release in electronic flash units. Figure 16.13 shows a typical design for a capacitor. It consists of two parallel metal plates separated by a distance d. Used in an electric circuit, the plates are connected to the positive and negative terminals of a battery or some other voltage source. When this connection is made, electrons are pulled off one of the plates, leaving it with a charge of $+Q$, and are transferred through the battery to the other plate, leaving it with a charge of $-Q$, as shown in the figure. The transfer of charge stops when the potential difference across the plates equals the potential difference of the battery. A charged capacitor is a device that stores energy that can be reclaimed when needed for a specific application.

Figure 16.13 A parallel-plate capacitor consists of two parallel plates, each of area A, separated by a distance d. The plates carry equal and opposite charges.

◀ Capacitance of a pair of conductors

The capacitance C of a capacitor is the ratio of the magnitude of the charge on either conductor (plate) to the magnitude of the potential difference between the conductors (plates):

$$C \equiv \frac{Q}{\Delta V} \qquad [16.8]$$

SI Unit: farad (F) = coulomb per volt (C/V)

The quantities Q and ΔV are always taken to be positive when used in Equation 16.8. For example, if a 3.0-μF capacitor is connected to a 12-V battery, the magnitude of the charge on each plate of the capacitor is

$$Q = C\Delta V = (3.0 \times 10^{-6} \text{ F})(12 \text{ V}) = 36 \text{ } \mu\text{C}$$

From Equation 16.8, we see that a large capacitance is needed to store a large amount of charge for a given applied voltage. The farad is a very large unit of capacitance. In practice, most typical capacitors have capacitances ranging from microfarads (1 μF = 1 $\times$ 10^{-6} F) to picofarads (1 pF = 1 $\times$ 10^{-12} F).

TIP 16.2 Potential Difference Is ΔV, Not V

Use the symbol ΔV for the potential difference across a circuit element or a device (many other books use simply V for potential difference.) The dual use of V to represent potential in one place and a potential difference in another can lead to unnecessary confusion.

16.7 THE PARALLEL-PLATE CAPACITOR

The capacitance of a device depends on the geometric arrangement of the conductors. The capacitance of a parallel-plate capacitor with plates separated by air (see Fig. 16.13) can be easily calculated from three facts. First, recall from Chapter 15 that the magnitude of the electric field between two plates is given by $E = \sigma/\epsilon_0$, where σ is the magnitude of the charge per unit area on each plate. Second, we found earlier in this chapter that the potential difference between two plates is $\Delta V = Ed$, where d is the distance between the plates. Third, the charge on one plate is given by $q = \sigma A$, where A is the area of the plate. Substituting these three facts into the definition of capacitance gives the desired result:

$$C = \frac{q}{\Delta V} = \frac{\sigma A}{Ed} = \frac{\sigma A}{(\sigma/\epsilon_0)\, d}$$

Canceling the charge per unit area, σ, yields

◀ Capacitance of a parallel-plate capacitor

$$C = \epsilon_0 \frac{A}{d} \qquad\qquad \textbf{[16.9]}$$

where A is the area of one of the plates, d is the distance between the plates, and ϵ_0 is the permittivity of free space.

From Equation 16.9, we see that plates with larger area can store more charge. The same is true for a small plate separation d, because then the positive charges on one plate exert a stronger force on the negative charges on the other plate, allowing more charge to be held on the plates.

Figure 16.14 shows the electric field lines of a more realistic parallel-plate capacitor. The electric field is very nearly constant in the center between the plates, but becomes less so when approaching the edges. For most purposes, however, the field may be taken as constant throughout the region between the plates.

APPLICATION

Camera Flash Attachments

One practical device that uses a capacitor is the flash attachment on a camera. A battery is used to charge the capacitor, and the stored charge is then released when the shutter-release button is pressed to take a picture. The stored charge is delivered to a flash tube very quickly, illuminating the subject at the instant more light is needed.

APPLICATION

Computer Keyboards

Computers make use of capacitors in many ways. For example, one type of computer keyboard has capacitors at the bases of its keys, as in Figure 16.15. Each key is connected to a movable plate, which represents one side of the capacitor; the fixed plate on the bottom of the keyboard represents the other side of the capacitor. When a key is pressed, the capacitor spacing decreases, causing an increase in capacitance. External electronic circuits recognize each key by the *change* in its capacitance when it is pressed.

(a)

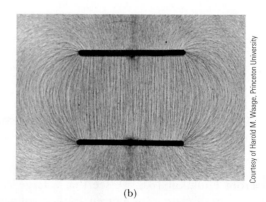

(b)

Courtesy of Harold M. Waage, Princeton University

Figure 16.14 (a) The electric field between the plates of a parallel-plate capacitor is uniform near the center, but nonuniform near the edges. (b) Electric field pattern of two oppositely charged conducting parallel plates. Small pieces of thread on an oil surface align with the electric field.

Key

Movable plate

Dielectric
Fixed plate

Figure 16.15   When the key of one type of keyboard is pressed, the capacitance of a parallel-plate capacitor increases as the plate spacing decreases. The substance labeled "dielectric" is an insulating material, as described in Section 16.10.

Capacitors are useful for storing a large amount of charge that needs to be delivered quickly. A good example on the forefront of fusion research is electrostatic confinement. In this role, capacitors discharge their electrons through a grid. The negatively charged electrons in the grid draw positively charged particles to them and therefore to each other, causing some particles to fuse and release energy in the process.

APPLICATION

Electrostatic Confinement

EXAMPLE 16.5   A Parallel-Plate Capacitor

Goal   Calculate fundamental physical properties of a parallel-plate capacitor.

Problem   A parallel-plate capacitor has an area $A = 2.00 \times 10^{-4}\ \text{m}^2$ and a plate separation $d = 1.00 \times 10^{-3}\ \text{m}$. (a) Find its capacitance. (b) How much charge is on the positive plate if the capacitor is connected to a 3.00-V battery? (c) Calculate the charge density on the positive plate, assuming the density is uniform, and (d) the magnitude of the electric field between the plates.

Strategy   Parts (a) and (b) can be solved by substituting into the basic equations for capacitance. In part (c), use the fact that the voltage difference equals the electric field times the distance.

Solution

(a) Find the capacitance.

Substitute into Equation 16.9:

$$C = \epsilon_0 \frac{A}{d} = (8.85 \times 10^{-12}\ \text{C}^2/\text{N} \cdot \text{m}^2)\left(\frac{2.00 \times 10^{-4}\ \text{m}^2}{1.00 \times 10^{-3}\ \text{m}}\right)$$

$$C = \boxed{1.77 \times 10^{-12}\ \text{F} = 1.77\ \text{pF}}$$

(b) Find the charge on the positive plate after the capacitor is connected to a 3.00-V battery.

Substitute into Equation 16.8:

$$C = \frac{Q}{\Delta V} \quad \rightarrow \quad Q = C\Delta V = (1.77 \times 10^{-12}\ \text{F})(3.00\ \text{V})$$

$$= 5.31 \times 10^{-12}\ \text{C}$$

(c) Calculate the charge density on the positive plate.

Charge density is charge divided by area:

$$\sigma = \frac{Q}{A} = \frac{5.31 \times 10^{-12}\ \text{C}}{2.00 \times 10^{-4}\ \text{m}^2} = \boxed{2.66 \times 10^{-8}\ \text{C}/\text{m}^2}$$

(d) Calculate the magnitude of the electric field between the plates.

Apply Equation 15.13:

$$E = \frac{\sigma}{\epsilon_0} = \frac{2.66 \times 10^{-8}\ \text{C}/\text{m}^2}{8.85 \times 10^{-12}\ \text{C}^2/\text{N} \cdot \text{m}^2} = \boxed{3.01 \times 10^3\ \text{N}/\text{C}}$$

Remarks   The answer to part (d) could also have been obtained from the electric potential, which is $\Delta V = Ed$ for a parallel-plate capacitor.

Exercise 16.5

Two plates, each of area 3.00×10^{-4} m^2, are used to construct a parallel-plate capacitor with capacitance 1.00 pF. (a) Find the necessary separation distance. (b) If the positive plate is to hold a charge of 5.00×10^{-12} C, find the charge density. (c) Find the electric field between the plates. (d) What voltage battery should be attached to the plate to obtain the preceding results?

Answers   (a) 2.66×10^{-3} m   (b) 1.67×10^{-8} C/m^2 (c) 1.89×10^3 N/C   (d) 5.00 V

Figure 16.16 (a) A real circuit and (b) its equivalent circuit diagram.

Symbols for Circuit Elements and Circuits

The symbol that is commonly used to represent a capacitor in a circuit is, ———||——— or sometimes ———|(———. Don't confuse either of these symbols with the circuit symbol, ——+|⊢—— which is used to designate a battery (or any other source of direct current). The positive terminal of the battery is at the higher potential and is represented by the longer vertical line in the battery symbol. In the next chapter we discuss another circuit element, called a resistor, represented by the symbol ———WW———. When wires in a circuit don't have appreciable resistance compared with the resistance of other elements in the circuit, the wires are represented by straight lines.

It's important to realize that a circuit is a collection of real objects, usually containing a source of electrical energy (such as a battery) connected to elements that convert electrical energy to other forms (light, heat, sound) or store the energy in electric or magnetic fields for later retrieval. A real circuit and its schematic diagram are sketched side by side in Figure 16.16. The circuit symbol for a lightbulb shown in Figure 16.16b is ——Ⓧ—— .

If you are not familiar with circuit diagrams, trace the path of the real circuit with your finger to see that it is equivalent to the geometrically regular schematic diagram.

16.8 COMBINATIONS OF CAPACITORS

Two or more capacitors can be combined in circuits in several ways, but most reduce to two simple configurations, called *parallel* and *series*. The idea, then, is to find the single equivalent capacitance due to a combination of several different capacitors that are in parallel or in series with each other. Capacitors are manufactured with a number of different standard capacitances, and by combining them in different ways, any desired value of the capacitance can be obtained.

Capacitors in Parallel

Two capacitors connected as shown in Active Figure 16.17a are said to be in *parallel*. The left plate of each capacitor is connected to the positive terminal of the battery by a conducting wire, so the left plates are at the same potential. In the same way, the right plates, both connected to the negative terminal of the battery, are also at the same potential. This means that **capacitors in parallel both have the same potential difference ΔV across them**. Capacitors in parallel are illustrated in Active Figure 16.17b.

When the capacitors are first connected in the circuit, electrons are transferred from the left plates through the battery to the right plates, leaving the left plates positively charged and the right plates negatively charged. The energy source for this transfer of charge is the internal chemical energy stored in the

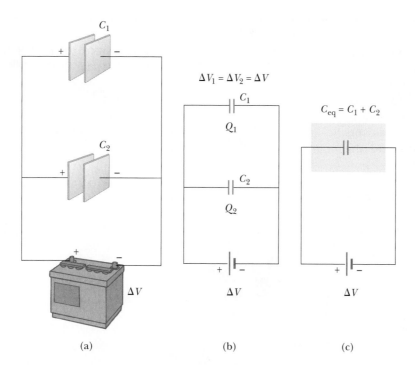

ACTIVE FIGURE 16.17
(a) A parallel connection of two capacitors. (b) The circuit diagram for the parallel combination. (c) The potential differences across the capacitors are the same, and the equivalent capacitance is $C_{eq} = C_1 + C_2$.

Physics⊗Now™
Log into PhysicsNow at **www.cp7e.com** and go to Active Figure 16.17, where you can adjust the battery voltage and the individual capacitances to see the resulting charges and voltages on the capacitors. You can combine up to four capacitors in parallel.

battery, which is converted to electrical energy. The flow of charge stops when the voltage across the capacitors equals the voltage of the battery, at which time the capacitors have their maximum charges. If the maximum charges on the two capacitors are Q_1 and Q_2, respectively, then the *total charge*, Q, stored by the two capacitors is

$$Q = Q_1 + Q_2 \qquad \textbf{[16.10]}$$

We can replace these two capacitors with one equivalent capacitor having a capacitance of C_{eq}. This equivalent capacitor must have exactly the same external effect on the circuit as the original two, so it must store Q units of charge and have the same potential difference across it. The respective charges on each capacitor are

$$Q_1 = C_1\Delta V \qquad \text{and} \qquad Q_2 = C_2\Delta V$$

The charge on the equivalent capacitor is

$$Q = C_{eq}\Delta V$$

Substituting these relationships into Equation 16.10 gives

$$C_{eq}\Delta V = C_1\Delta V + C_2\Delta V$$

or

$$C_{eq} = C_1 + C_2 \qquad \begin{pmatrix} \text{parallel} \\ \text{combination} \end{pmatrix} \qquad \textbf{[16.11]}$$

If we extend this treatment to three or more capacitors connected in parallel, the equivalent capacitance is found to be

$$C_{eq} = C_1 + C_2 + C_3 + \cdots \qquad \begin{pmatrix} \text{parallel} \\ \text{combination} \end{pmatrix} \qquad \textbf{[16.12]}$$

We see that **the equivalent capacitance of a parallel combination of capacitors is greater than any of the individual capacitances**.

TIP 16.3   Voltage is the Same as Potential Difference

A voltage *across* a device, such as a capacitor, has the same meaning as the potential difference across the device. For example, if we say that the voltage across a capacitor is 12 V, we mean that the potential difference between its plates is 12 V.

EXAMPLE 16.6   Four Capacitors Connected in Parallel

Goal   Analyze a circuit with several capacitors in parallel.

Problem   (a) Determine the capacitance of the single capacitor that is equivalent to the parallel combination of capacitors shown in Figure 16.18, and (b) find the charge on the 12.0-μF capacitor.

Strategy   (a) Add the individual capacitances. (b) Apply the formula $C = Q/\Delta V$ to the 12.0-μF capacitor. The voltage difference is the same as the difference across the battery.

Figure 16.18   (Example 16.6) Four capacitors connected in parallel.

3.00 μF

6.00 μF

12.0 μF

24.0 μF

18.0 V

Solution

(a) Find the equivalent capacitance.

Apply Equation 16.12:

$$C_{eq} = C_1 + C_2 + C_3 + C_4$$
$$= 3.00 \ \mu F + 6.00 \ \mu F + 12.0 \ \mu F + 24.0 \ \mu F$$
$$= \boxed{45.0 \ \mu F}$$

(b) Find the charge on the 12-μF capacitor.

Solve the capacitance equation for Q and substitute:

$$Q = C\Delta V = (12.0 \times 10^{-6} \ F)(18.0 \ V) = 216 \times 10^{-6} \ C$$
$$= \boxed{216 \ \mu C}$$

Remarks   The charge on any one of the parallel capacitors can be found as in part (b), since the potential difference is the same.

Exercise 16.6

Find the charge on the 24.0-μF capacitor.

Answer   432 μC

Capacitors in Series

Q is the same for all capacitors connected in series ▶

Now consider two capacitors connected in *series*, as illustrated in Active Figure 16.19a. **For a series combination of capacitors, the magnitude of the charge must be the same on all the plates.** To understand this principle, consider the charge transfer process in some detail. When a battery is connected to the circuit, electrons with total charge $-Q$ are transferred from the left plate of C_1 to the right plate of C_2 through the battery, leaving the left plate of C_1 with a charge of $+Q$. As a consequence, the magnitudes of the charges on the left plate of C_1 and the right plate of C_2 must be the same. Now consider the right plate of C_1 and the left plate of C_2, in the middle. These plates are not connected to the battery (because of the gap across the plates) and, taken together, are electrically neutral. The charge of $+Q$ on the left plate of C_1, however, attracts negative charges to the right plate of C_1. These charges will continue to accumulate until the left and right plates of C_1, taken together, become electrically neutral, which means the charge on the right plate of C_1 is $-Q$. This negative charge could only have come from the left plate of C_2, so C_2 has a charge of $+Q$.

ACTIVE FIGURE 16.19
A series combination of two
capacitors. The charges on the
capacitors are the same, and the
equivalent capacitance can be calculated from the reciprocal relationship
$1/C_{eq} = (1/C_1) + (1/C_2)$.

Physics⊗Now™
Log into PhysicsNow at
www.cp7e.com and go to Active
Figure 16.19, where you can adjust
the battery voltage and the individual
capacitances to see the resulting
charges and voltages on the
capacitors. You can combine up to
four capacitors in series.

Therefore, regardless of how many capacitors are in series or what their capacitances are, **all of the right plates gain charges of $-Q$ and all the left plates have charges of $+Q$.** (This is a consequence of the conservation of charge.)

After an equivalent capacitor for a series of capacitors is fully charged, **the equivalent capacitor must end up with a charge of $-Q$ on its right plate and a charge of $+Q$ on its left plate.** Applying the definition of capacitance to the circuit in Active Figure 16.19b, we have

$$\Delta V = \frac{Q}{C_{eq}}$$

where ΔV is the potential difference between the terminals of the battery and C_{eq} is the equivalent capacitance. Because $Q = C\Delta V$ can be applied to each capacitor, the potential differences across them are given by

$$\Delta V_1 = \frac{Q}{C_1} \qquad \Delta V_2 = \frac{Q}{C_2}$$

From Active Figure 16.19a, we see that

$$\Delta V = \Delta V_1 + \Delta V_2 \qquad \qquad \textbf{[16.13]}$$

where ΔV_1 and ΔV_2 are the potential differences across capacitors C_1 and C_2. (This is a consequence of the conservation of energy.)

The potential difference across any number of capacitors (or other circuit elements) in series equals the sum of the potential differences across the individual capacitors. Substituting these expressions into Equation 16.13, and noting that $\Delta V = Q/C_{eq}$, we have

$$\frac{Q}{C_{eq}} = \frac{Q}{C_1} + \frac{Q}{C_2}$$

Canceling Q, we arrive at the following relationship:

$$\frac{1}{C_{eq}} = \frac{1}{C_1} + \frac{1}{C_2} \qquad \begin{pmatrix} \text{series} \\ \text{combination} \end{pmatrix} \qquad \textbf{[16.14]}$$

If this analysis is applied to three or more capacitors connected in series, the equivalent capacitance is found to be

$$\frac{1}{C_{eq}} = \frac{1}{C_1} + \frac{1}{C_2} + \frac{1}{C_3} + \cdots \qquad \begin{pmatrix} \text{series} \\ \text{combination} \end{pmatrix} \qquad \textbf{[16.15]}$$

As we will show in Example 16.7, Equation 16.15 implies that **the equivalent capacitance of a series combination is always less than any individual capacitance in the combination.**

A capacitor is designed so that one plate is large and the other is small. If the plates are connected to a battery, (a) the large plate has a greater charge than the small plate, (b) the large plate has less charge than the small plate, or (c) the plates have equal, but opposite, charge.

EXAMPLE 16.7   Four Capacitors Connected in Series

Goal Find an equivalent capacitance of capacitors in series, and the charge and voltage on each capacitor.

Problem Four capacitors are connected in series with a battery, as in Figure 16.20. **(a)** Calculate the capacitance of the equivalent capacitor. **(b)** Compute the charge on the 12-μF capacitor. **(c)** Find the voltage drop across the 12-μF capacitor.

Strategy Combine all the capacitors into a single, equivalent capacitor using Equation 16.15. Find the charge on this equivalent capacitor using $C = Q/\Delta V$. This charge is the same as on the individual capacitors. Use this same equation again to find the voltage drop across the 12-μF capacitor.

Figure 16.20 (Example 16.7) Four capacitors connected in series.

Solution

(a) Calculate the equivalent capacitance of the series.

Apply Equation 16.15:

$$\frac{1}{C_{eq}} = \frac{1}{3.0\ \mu F} + \frac{1}{6.0\ \mu F} + \frac{1}{12\ \mu F} + \frac{1}{24\ \mu F}$$

$$C_{eq} = \boxed{1.6\ \mu F}$$

(b) Compute the charge on the 12-μF capacitor.

The desired charge equals the charge on the equivalent capacitor:

$$Q = C_{eq}\Delta V = (1.6 \times 10^{-6}\ F)(18\ V) = \boxed{29\ \mu C}$$

(c) Find the voltage drop across the 12-μF capacitor.

Apply the basic capacitance equation:

$$C = \frac{Q}{\Delta V} \quad \rightarrow \quad \Delta V = \frac{Q}{C} = \frac{29\ \mu C}{12\ \mu F} = \boxed{2.4\ V}$$

Remarks Notice that the equivalent capacitance is less than that of any of the individual capacitors. The relationship $C = Q/\Delta V$ can be used to find the voltage drops on the other capacitors, just as in part (c).

Exercise 16.7

The 24-μF capacitor is removed from the circuit, leaving only three capacitors in series. Find (a) the equivalent capacitance, (b) the charge on the 6-μF capacitor, and (c) the voltage drop across the 6-μF capacitor.

Answers   (a) 1.7 μF   (b) 31 μC   (c) 5.2 V

Problem-Solving Strategy
Complex Capacitor Combinations

1. **Combine** capacitors that are in series or in parallel, following the derived formulas.
2. **Redraw** the circuit after every combination.
3. **Repeat** the first two steps until there is only a single equivalent capacitor.
4. **Find the charge** on the single equivalent capacitor, using $C = Q/\Delta V$.

5. **Work backwards** through the diagrams to the original one, finding the charge and voltage drop across each capacitor along the way. To do this, use the following collection of facts:

   A. The capacitor equation: $C = Q/\Delta V$
   B. Capacitors in parallel: $C_{eq} = C_1 + C_2$
   C. Capacitors in parallel all have the same voltage difference, ΔV, as does their equivalent capacitor.
   D. Capacitors in series: $\dfrac{1}{C_{eq}} = \dfrac{1}{C_1} + \dfrac{1}{C_2}$
   E. Capacitors in series all have the same charge, Q, as does their equivalent capacitor.

INTERACTIVE EXAMPLE 16.8   Equivalent Capacitance

Goal   Solve a complex combination of series and parallel capacitors.

Problem   **(a)** Calculate the equivalent capacitance between a and b for the combination of capacitors shown in Figure 16.21a. All capacitances are in microfarads. **(b)** If a 12-V battery is connected across the system between points a and b, find the charge on the 4.0-μF capacitor in the first diagram and the voltage drop across it.

Figure 16.21   (Example 16.8) To find the equivalent capacitance of the circuit in (a), use the series and parallel rules described in the text to successively reduce the circuit as indicated in (b), (c), and (d).

Strategy   **(a)** Using Equations 16.12 and 16.15, we reduce the combination step by step, as indicated in the figure. **(b)** To find the charge on the 4.0-μF capacitor, start with Figure 16.21c, finding the charge on the 2.0-μF capacitor. This same charge is on each of the 4.0-μF capacitors in the second diagram, by fact 5E of the Problem-Solving Strategy. One of these 4.0-μF capacitors in the second diagram is just the original 4.0-μF capacitor in the first diagram.

Solution

(a) Calculate the equivalent capacitance.

Find the equivalent capacitance of the parallel 1.0-μF and 3.0-μF capacitors in Figure 16.21a:

$$C_{eq} = C_1 + C_2 = 1.0\ \mu F + 3.0\ \mu F = 4.0\ \mu F$$

Find the equivalent capacitance of the parallel 2.0-μF and 6.0-μF capacitors in Figure 16.21a:

$$C_{eq} = C_1 + C_2 = 2.0\ \mu F + 6.0\ \mu F = 8.0\ \mu F$$

Combine the two series 4.0-μF capacitors in Figure 16.21b:

$$\frac{1}{C_{eq}} = \frac{1}{C_1} + \frac{1}{C_2} = \frac{1}{4.0\ \mu F} + \frac{1}{4.0\ \mu F}$$

$$= \frac{1}{2.0\ \mu F} \quad \rightarrow \quad C_{eq} = 2.0\ \mu F$$

Combine the two series 8.0-μF capacitors in Figure 16.21b:

$$\frac{1}{C_{eq}} = \frac{1}{C_1} + \frac{1}{C_2} = \frac{1}{8.0\ \mu F} + \frac{1}{8.0\ \mu F}$$

$$= \frac{1}{4.0\ \mu F} \quad \rightarrow \quad C_{eq} = 4.0\ \mu F$$

Finally, combine the two parallel capacitors in Figure 16.21c to find the equivalent capacitance between a and b:

$$C_{eq} = C_1 + C_2 = 2.0\ \mu F + 4.0\ \mu F = \boxed{6.0\ \mu F}$$

(b) Find the charge on the 4.0-μF capacitor and the voltage drop across it.

Compute the charge on the 2.0-μF capacitor in Figure 16.21c, which is the same as the charge on the 4.0-μF capacitor in Figure 16.21a:

$$C = \frac{Q}{\Delta V} \quad \rightarrow \quad Q = C\Delta V = (2.0\ \mu\text{F})(12\ \text{V}) = \boxed{24\ \mu\text{C}}$$

Use the basic capacitance equation to find the voltage drop across the 4.0-μF capacitor in Figure 16.21a:

$$C = \frac{Q}{\Delta V} \quad \rightarrow \quad \Delta V = \frac{Q}{C} = \frac{24\ \mu\text{C}}{4.0\ \mu\text{F}} = \boxed{6.0\ \text{V}}$$

Remarks To find the rest of the charges and voltage drops, it's just a matter of using $C = Q/\Delta V$ repeatedly, together with facts 5C and 5E in the Problem-Solving Strategy. The voltage drop across the 4.0-μF capacitor could also have been found by noticing, in Figure 16.21b, that both capacitors had the same value and so by symmetry would split the total drop of 12 volts between them.

Exercise 16.8
(a) In Example 16.8, find the charge on the 8.0-μF capacitor in Figure 16.21a and the voltage drop across it. (b) Do the same for the 6.0-μF capacitor in Figure 16.21a.

Answers (a) 48 μC, 6.0 V   (b) 36 μC, 6.0 V

Physics⊗Now™ You can practice reducing a combination of capacitors to a single equivalent capacitor by logging into PhysicsNow at **www.cp7e.com** and going to Interactive Example 16.8.

16.9 ENERGY STORED IN A CHARGED CAPACITOR

Almost everyone who works with electronic equipment has at some time verified that a capacitor can store energy. If the plates of a charged capacitor are connected by a conductor such as a wire, charge transfers from one plate to the other until the two are uncharged. The discharge can often be observed as a visible spark. If you accidentally touched the opposite plates of a charged capacitor, your fingers would act as a pathway by which the capacitor could discharge, inflicting an electric shock. The degree of shock would depend on the capacitance and voltage applied to the capacitor. *Where high voltages and large quantities of charge are present, as in the power supply of a television set, such a shock can be fatal.*

Capacitors store electrical energy, and that energy is the same as the work required to move charge onto the plates. If a capacitor is initially uncharged (both plates are neutral), so that the plates are at the same potential, very little work is required to transfer a small amount of charge ΔQ from one plate to the other. However, once this charge has been transferred, a small potential difference $\Delta V = \Delta Q/C$ appears between the plates, so work must be done to transfer additional charge against this potential difference. From Equation 16.6, if the potential difference at any instant during the charging process is ΔV, then the work ΔW required to move more charge ΔQ through this potential difference is given by

$$\Delta W = \Delta V \Delta Q$$

We know that $\Delta V = Q/C$ for a capacitor that has a total charge of Q. Therefore, a plot of voltage versus total charge gives a straight line with a slope of $1/C$, as shown in Figure 16.22. The work ΔW, for a particular ΔV, is the area of the shaded rectangle. Adding up all the rectangles gives an approximation of the total work needed to fill the capacitor. In the limit as ΔQ is taken to be infinitesimally small, the total work needed to charge the capacitor to a final charge Q and voltage ΔV is the area under the line. This is just the area of a triangle, one-half the base times the height, so it follows that

$$W = \tfrac{1}{2} Q\Delta V \qquad \qquad [16.16]$$

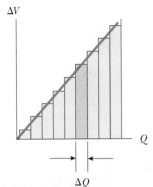

Figure 16.22 A plot of voltage versus charge for a capacitor is a straight line with slope $1/C$. The work required to move a charge of ΔQ through a potential difference of ΔV across the capacitor plates is $\Delta W = \Delta V \Delta Q$, which equals the area of the blue rectangle. The *total work* required to charge the capacitor to a final charge of Q is the area under the straight line, which equals $Q\Delta V/2$.

As previously stated, W is also the energy stored in the capacitor. From the definition of capacitance, we have $Q = C\Delta V$; hence, we can express the energy stored three different ways:

$$\text{Energy stored} = \tfrac{1}{2}Q\Delta V = \tfrac{1}{2}C(\Delta V)^2 = \frac{Q^2}{2C} \qquad \textbf{[16.17]}$$

For example, the amount of energy stored in a 5.0-μF capacitor when it is connected across a 120-V battery is

$$\text{Energy stored} = \tfrac{1}{2}C(\Delta V)^2 = \tfrac{1}{2}(5.0 \times 10^{-6}\,\text{F})(120\,\text{V})^2 = 3.6 \times 10^{-2}\,\text{J}$$

In practice, there is a limit to the maximum energy (or charge) that can be stored in a capacitor. At some point, the Coulomb forces between the charges on the plates become so strong that electrons jump across the gap, discharging the capacitor. For this reason, capacitors are usually labeled with a maximum operating voltage. (This physical fact can actually be exploited to yield a circuit with a regularly blinking light).

Large capacitors can store enough electrical energy to cause severe burns or even death if they are discharged so that the flow of charge can pass through the heart. Under the proper conditions, however, they can be used to *sustain* life by stopping cardiac fibrillation in heart attack victims. When fibrillation occurs, the heart produces a rapid, irregular pattern of beats. A fast discharge of electrical energy through the heart can return the organ to its normal beat pattern. Emergency medical teams use portable defibrillators that contain batteries capable of charging a capacitor to a high voltage. (The circuitry actually permits the capacitor to be charged to a much higher voltage than the battery.) In this case and others (camera flash units and lasers used for fusion experiments), capacitors serve as energy reservoirs that can be slowly charged and then quickly discharged to provide large amounts of energy in a short pulse. The stored electrical energy is released through the heart by conducting electrodes, called paddles, placed on both sides of the victim's chest. The paramedics must wait between applications of electrical energy due to the time it takes for the capacitors to become fully charged. The high voltage on the capacitor can be obtained from a low-voltage battery in a portable machine through the phenomenon of *electromagnetic induction*, to be studied in Chapter 20.

APPLICATION
Defibrillators

EXAMPLE 16.9   Typical Voltage, Energy, and Discharge Time for a Defibrillator

Goal   Apply energy and power concepts to a capacitor.

Problem   A fully charged defibrillator contains 1.20 kJ of energy stored in a 1.10×10^{-4}-F capacitor. In a discharge through a patient, 6.00×10^2 J of electrical energy are delivered in 2.50 ms. **(a)** Find the voltage needed to store 1.20 kJ in the unit. **(b)** What average power is delivered to the patient?

Strategy   (a) Because we know the energy stored and the capacitance, we can use Equation 16.17 to find the required voltage. (b) Dividing the energy delivered by the time gives the average power.

Solution
(a) Find the voltage needed to store 1.20 kJ in the unit.

Solve Equation 16.17 for ΔV:

$$\text{Energy stored} = \tfrac{1}{2}C\Delta V^2$$

$$\Delta V = \sqrt{\frac{2 \times (\text{Energy stored})}{C}}$$

$$= \sqrt{\frac{2(1.20 \times 10^3\,\text{J})}{1.10 \times 10^{-4}\,\text{F}}}$$

$$= \boxed{4.67 \times 10^3\,\text{V}}$$

(b) What average power is delivered to the patient?

Divide the energy delivered by the time:

$$\mathcal{P}_{av} = \frac{\text{Energy delivered}}{\Delta t} = \frac{6.00 \times 10^2 \text{ J}}{2.50 \times 10^{-3} \text{ s}}$$

$$= \boxed{2.40 \times 10^5 \text{ W}}$$

Remarks   The power delivered by a draining capacitor isn't constant, as we'll find in the study of *RC* circuits in Chapter 18. For that reason, we were able to find only an average power. Capacitors are necessary in defibrillators because they can deliver energy far more quickly than batteries. Batteries provide current through relatively slow chemical reactions, whereas capacitors release charge that has already been produced and stored.

Exercise 16.9
(a) Find the energy contained in a 2.50×10^{-5}-F parallel-plate capacitor if it holds 1.75×10^{-3} C of charge. (b) What's the voltage between the plates? (c) What new voltage will result in a doubling of the stored energy?

Answers   (a) 6.13×10^{-2} J   (b) 70.0 V   (c) 99.0 V

Applying Physics 16.1   Maximum Energy Design

How should three capacitors and two batteries be connected so that the capacitors will store the maximum possible energy?

Explanation   The energy stored in the capacitor is proportional to the capacitance and the square of the

potential difference, so we would like to maximize each of these quantities. If the three capacitors are put in parallel, their capacitances add, and if the batteries are in series, their potential differences, similarly, also add together.

Quick Quiz 16.7

A parallel-plate capacitor is disconnected from a battery, and the plates are pulled a small distance further apart. Do the following quantities increase, decrease, or stay the same?
(a) C (b) Q (c) E between the plates   (d) ΔV (e) energy stored in the capacitor

16.10   CAPACITORS WITH DIELECTRICS

A **dielectric** is an insulating material, such as rubber, plastic, or waxed paper. When a dielectric is inserted between the plates of a capacitor, the capacitance increases. If the dielectric completely fills the space between the plates, the capacitance is multiplied by the factor κ, called the **dielectric constant**.

The following experiment illustrates the effect of a dielectric in a capacitor. Consider a parallel-plate capacitor of charge Q_0 and capacitance C_0 in the absence of a dielectric. The potential difference across the capacitor plates can be measured, and is given by $\Delta V_0 = Q_0/C_0$ (Fig. 16.23a). Because the capacitor is not connected to an external circuit, there is no pathway for charge to leave or be added to the plates. If a dielectric is now inserted between the plates as in Figure 16.23b, the voltage across the plates is *reduced* by the factor κ to the value

$$\Delta V = \frac{\Delta V_0}{\kappa}$$

Because $\kappa > 1$, ΔV is less than ΔV_0. Because the charge Q_0 on the capacitor doesn't change, we conclude that the capacitance in the presence of the dielectric must change to the value

$$C = \frac{Q_0}{\Delta V} = \frac{Q_0}{\Delta V_0/\kappa} = \frac{\kappa Q_0}{\Delta V_0}$$

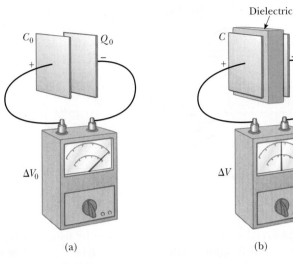

Figure 16.23 (a) With air between the plates, the voltage across the capacitor is ΔV_0, the capacitance is C_0, and the charge is Q_0. (b) With a dielectric between the plates, the charge remains at Q_0, but the voltage and capacitance change.

(a)   (b)

or

$$C = \kappa C_0 \qquad \text{[16.18]}$$

According to this result, the capacitance is *multiplied* by the factor κ when the dielectric fills the region between the plates. For a parallel-plate capacitor, where the capacitance in the absence of a dielectric is $C_0 = \epsilon_0 A/d$, we can express the capacitance in the presence of a dielectric as

$$C = \kappa \epsilon_0 \frac{A}{d} \qquad \text{[16.19]}$$

From this result, it appears that the capacitance could be made very large by decreasing d, the separation between the plates. In practice, the lowest value of d is limited by the electric discharge that can occur through the dielectric material separating the plates. For any given plate separation, there is a maximum electric field that can be produced in the dielectric before it breaks down and begins to conduct. This maximum electric field is called the **dielectric strength**, and for air its value is about 3×10^6 V/m. Most insulating materials have dielectric strengths greater than that of air, as indicated by the values listed in Table 16.1. Figure 16.24 shows an instance of dielectric breakdown in air.

© Loren Winters/Visuals Unlimited

Figure 16.24 Dielectric breakdown in air. Sparks are produced when a large alternating voltage is applied across the wires by a high-voltage induction coil power supply.

TABLE 16.1

Dielectric Constants and Dielectric Strengths of Various Materials at Room Temperature

Material	Dielectric Constant κ	Dielectric Strength (V/m)
Vacuum	1.000 00	—
Air	1.000 59	3×10^6
Bakelite®	4.9	24×10^6
Fused quartz	3.78	8×10^6
Pyrex® glass	5.6	14×10^6
Polystyrene	2.56	24×10^6
Teflon®	2.1	60×10^6
Neoprene rubber	6.7	12×10^6
Nylon	3.4	14×10^6
Paper	3.7	16×10^6
Strontium titanate	233	8×10^6
Water	80	—
Silicone oil	2.5	15×10^6

Figure 16.25 Three commercial capacitor designs. (a) A tubular capacitor whose plates are separated by paper and then rolled into a cylinder. (b) A high-voltage capacitor consisting of many parallel plates separated by oil. (c) An electrolytic capacitor.

(a)   (b)   (c)

(a)

(b)

Figure 16.26 (a) A collection of capacitors used in a variety of applications. (b) A variable capacitor. When one set of metal plates is rotated so as to lie between a fixed set of plates, the capacitance of the device changes.

Commercial capacitors are often made by using metal foil interlaced with thin sheets of paraffin-impregnated paper or Mylar®, which serves as the dielectric material. These alternate layers of metal foil and dielectric are rolled into a small cylinder (Fig. 16.25a). One type of a high-voltage capacitor consists of a number of interwoven metal plates immersed in silicone oil (Fig. 16.25b). Small capacitors are often constructed from ceramic materials. Variable capacitors (typically 10 pF to 500 pF) usually consist of two interwoven sets of metal plates, one fixed and the other movable, with air as the dielectric.

An electrolytic capacitor (Fig. 16.25c) is often used to store large amounts of charge at relatively low voltages. It consists of a metal foil in contact with an electrolyte—a solution that conducts charge by virtue of the motion of the ions contained in it. When a voltage is applied between the foil and the electrolyte, a thin layer of metal oxide (an insulator) is formed on the foil, and this layer serves as the dielectric. Enormous capacitances can be attained because the dielectric layer is very thin.

Figure 16.26 shows a variety of commercially available capacitors. Variable capacitors are used in radios to adjust the frequency.

When electrolytic capacitors are used in circuits, the polarity (the plus and minus signs on the device) must be observed. If the polarity of the applied voltage is opposite that intended, the oxide layer will be removed and the capacitor will conduct rather than store charge. Further, reversing the polarity can result in such a large current that the capacitor may either burn or produce steam and explode.

Applying Physics 16.2   Stud Finders

If you have ever tried to hang a picture on a wall securely, you know that it can be difficult to locate a wooden stud in which to anchor your nail or screw. The principles discussed in this section can be used to detect a stud electronically. The primary element of an electronic stud finder is a capacitor with its plates arranged side by side instead of facing one another, as in Figure 16.27. How does this device work?

Explanation As the detector is moved along a wall, its capacitance changes when it passes across a stud because the dielectric constant of the material "between" the plates changes. The change in capacitance can be used to cause a light to come on, signaling the presence of the stud.

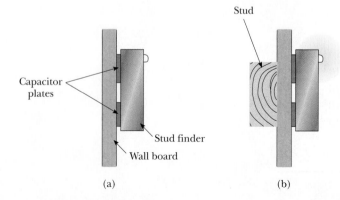

(a)   (b)

Figure 16.27 (Applying Physics 16.2) A stud finder. (a) The materials between the plates of the capacitor are the drywall and the air behind it. (b) The materials become drywall and wood when the detector moves across a stud in the wall. The change in the dielectric constant causes a signal light to illuminate.

A fully charged parallel-plate capacitor remains connected to a battery while you slide a dielectric between the plates. Do the following quantities increase, decrease, or stay the same? (a) C (b) Q (c) E between the plates (d) ΔV (e) energy stored in the capacitor.

EXAMPLE 16.10   A Paper-Filled Capacitor

Goal   Calculate fundamental physical properties of a parallel-plate capacitor with a dielectric.

Problem   A parallel-plate capacitor has plates 2.0 cm by 3.0 cm. The plates are separated by a 1.0-mm thickness of paper. Find **(a)** the capacitance of this device, and **(b)** the maximum charge that can be placed on the capacitor.

Strategy   (a) Obtain the dielectric constant for paper from Table 16.1, and substitute, with other given quantities, into Equation 16.19. (b) Table 16.1 also gives the dielectric strength of paper, which is the maximum electric field that can be applied before electrical breakdown occurs. Use Equation 16.3, $\Delta V = Ed$, to obtain the maximum voltage, and substitute into the basic capacitance equation.

Solution
(a) Find the capacitance of this device.

Substitute into Equation 16.19:

$$C = \kappa\epsilon_0 \frac{A}{d}$$

$$= 3.7 \left(8.85 \times 10^{-12}\, \frac{C^2}{N \cdot m^2} \right)\left(\frac{6.0 \times 10^{-4}\, m^2}{1.0 \times 10^{-3}\, m} \right)$$

$$= 2.0 \times 10^{-11}\, F$$

(b) Find the maximum charge that can be placed on the capacitor.

Calculate the maximum applied voltage, using the dielectric strength of paper, E_{max}.

$$\Delta V_{max} = E_{max}d = (16 \times 10^6\, V/m)(1.0 \times 10^{-3}\, m)$$
$$= 1.6 \times 10^4\, V$$

Solve the basic capacitance equation for Q_{max}, and substitute ΔV_{max} and C:

$$Q_{max} = C\Delta V_{max} = (2.0 \times 10^{-11}\, F)(1.6 \times 10^4\, V)$$
$$= 0.32\, \mu C$$

Remarks   Dielectrics allow κ times as much charge to be stored on a capacitor for a given voltage. They also allow an increase in the applied voltage by increasing the threshold of electrical breakdown.

Exercise 16.10
A parallel-plate capacitor has plate area of $2.50 \times 10^{-3}\, m^2$ and distance between the plates of 2.00 mm. (a) Find the maximum charge that can be placed on the capacitor if air is between the plates. (b) Find the maximum charge if the air is replaced by polystyrene.

Answers   (a) $7 \times 10^{-8}\, C$ (b) $1.4 \times 10^{-6}\, C$

An Atomic Description of Dielectrics

The explanation of why a dielectric increases the capacitance of a capacitor is based on an atomic description of the material, which in turn involves a property of some molecules called **polarization**. A molecule is said to be polarized when there is a separation between the average positions of its negative charge and its positive charge. In some molecules, such as water, this condition is always present. To see why, consider the geometry of a water molecule (Fig. 16.28, page 560).

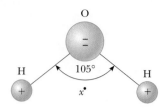

Figure 16.28 The water molecule, H_2O, has a permanent polarization resulting from its bent geometry. The point labeled x is the center of positive charge.

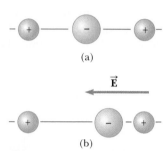

(a)

$\vec{E}$

(b)

Figure 16.29 (a) A symmetric molecule has no permanent polarization. (b) An external electric field induces a polarization in the molecule.

The molecule is arranged so that the negative oxygen atom is bonded to the positively charged hydrogen atoms with a 105° angle between the two bonds. The center of negative charge is at the oxygen atom, and the center of positive charge lies at a point midway along the line joining the hydrogen atoms (point x in the diagram). Materials composed of molecules that are permanently polarized in this way have large dielectric constants, and indeed, Table 16.1 shows that the dielectric constant of water is large ($\kappa = 80$) compared to other common substances.

A symmetric molecule (Fig. 16.29a) can have no permanent polarization, but a polarization can be induced in it by an external electric field. A field directed to the left, as in Figure 16.29b, would cause the center of positive charge to shift to the left from its initial position and the center of negative charge to shift to the right. This *induced polarization* is the effect that predominates in most materials used as dielectrics in capacitors.

To understand why the polarization of a dielectric can affect capacitance, consider the slab of dielectric shown in Figure 16.30. Before placing the slab in between the plates of the capacitor, the polar molecules are randomly oriented (Fig. 16.30a). The polar molecules are dipoles, and each creates a dipole electric field, but because of their random orientation, this field averages to zero.

After insertion of the dielectric slab into the electric field $\vec{E}_0$ between the plates (Fig. 16.30b), the positive plate attracts the negative ends of the dipoles, while the negative plate attracts the positive ends of the dipoles. These forces exert a torque on the molecules making up the dielectric, reorienting them so that on the average the negative pole is more inclined toward the positive plate and the positive pole more aligned toward the negative plate. The positive and negative charges in the middle still cancel each other, but there is a net accumulation of negative charge in the dielectric next to the positive plate and a net accumulation of positive charge next to the negative plate. This configuration can be modeled as an additional pair of charged plates, as in Figure 16.30c, creating an induced electric field $\vec{E}_{ind}$ that partly cancels the original electric field $\vec{E}_0$. If the battery is not connected when the dielectric is inserted, the potential difference ΔV_0 across the plates is reduced to $\Delta V_0/\kappa$.

If the capacitor is still connected to the battery, however, the negative poles push more electrons off the positive plate, making it more positive. Meanwhile, the positive poles attract more electrons onto the negative plate. This situation continues until the potential difference across the battery reaches its original magnitude, equal to the potential gain across the battery. The net effect is an increase in the amount of charge stored on the capacitor. Because the plates can store more charge for a given voltage, it follows from $C = Q\Delta V$ that the capacitance must increase.

Quick Quiz 16.9

Consider a parallel-plate capacitor with a dielectric material between the plates. If the temperature of the dielectric increases, the capacitance (a) decreases (b) increases (c) remains the same.

Figure 16.30 (a) In the absence of an external electric field, polar molecules are randomly oriented. (b) When an external electric field is applied, the molecules partially align with the field. (c) The charged edges of the dielectric can be modeled as an additional pair of parallel plates establishing an electric field $\vec{E}_{ind}$ in the direction opposite to $\vec{E}_0$.

(a)   (b)   (c)

SUMMARY

Physics⊗Now™ Take a practice test by logging into PhysicsNow at **www.cp7e.com** and clicking on the Pre-Test link for this chapter.

16.1 Potential Difference and Electric Potential

The change in the electric potential energy of a system consisting of an object of charge q moving through a displacement Δx in a constant electric field $\vec{\mathbf{E}}$ is given by

$$\Delta PE = -W_{AB} = -qE_x\Delta x \qquad \textbf{[16.1]}$$

where E_x is the component of the electric field in the x-direction and $\Delta x = x_f - x_i$. The **difference in electric potential** between two points A and B is

$$\Delta V = V_B - V_A \equiv \frac{\Delta PE}{q} \qquad \textbf{[16.2]}$$

where ΔPE is the *change* in electrical potential energy as a charge q moves between A and B. The units of potential difference are joules per coulomb, or **volts**; $1\,\text{J/C} = 1\,\text{V}$.

The **electric potential difference** between two points A and B in a *uniform* electric field $\vec{\mathbf{E}}$ is

$$\Delta V = -E_x\Delta x \qquad \textbf{[16.3]}$$

where $\Delta x = x_f - x_i$ is the displacement between A and B and E_x is the x-component of the electric field in that region.

16.2 Electric Potential and Potential Energy Due to Point Charges

The **electric potential** due to a point charge q at distance r from the point charge is

$$V = k_e\frac{q}{r} \qquad \textbf{[16.4]}$$

The **electric potential energy** of a pair of point charges separated by distance r is

$$PE = k_e\frac{q_1 q_2}{r} \qquad \textbf{[16.5]}$$

These equations can be used in the solution of conservation of energy problems and in the work–energy theorem.

16.3 Potentials and Charged Conductors
16.4 Equipotential Surfaces

Every point on the surface of a charged conductor in electrostatic equilibrium is at the same potential. Further, the potential is constant everywhere inside the conductor and equals its value on the surface.

The **electron volt** is defined as the energy that an electron (or proton) gains when accelerated through a potential difference of 1 V. The conversion between electron volts and joules is

$$1\,\text{eV} = 1.60 \times 10^{-19}\,\text{C} \cdot \text{V} = 1.60 \times 10^{-19}\,\text{J} \qquad \textbf{[16.7]}$$

Any surface on which the potential is the same at every point is called an equipotential surface. The electric field is always oriented perpendicular to an equipotential surface.

16.6 Capacitance

A capacitor consists of two metal plates with charges that are equal in magnitude but opposite in sign. The capacitance C of any capacitor is the ratio of the magnitude of the charge Q on either plate to the magnitude of potential difference ΔV between them:

$$C \equiv \frac{Q}{\Delta V} \qquad \textbf{[16.8]}$$

Capacitance has the units coulombs per volt, or farads; $1\,\text{C/V} = 1\,\text{F}$.

16.7 The Parallel-Plate Capacitor

The capacitance of two parallel metal plates of area A separated by distance d is

$$C = \epsilon_0\frac{A}{d} \qquad \textbf{[16.9]}$$

where $\epsilon_0 = 8.85 \times 10^{-12}\,\text{C}^2/\text{N} \cdot \text{m}^2$ is a constant called the **permittivity of free space**.

16.8 Combinations of Capacitors

The **equivalent capacitance of a parallel combination** of capacitors is

$$C_{\text{eq}} = C_1 + C_2 + C_3 + \cdots \qquad \textbf{[16.12]}$$

If two or more capacitors are connected in series, the **equivalent capacitance of the series combination** is

$$\frac{1}{C_{\text{eq}}} = \frac{1}{C_1} + \frac{1}{C_2} + \frac{1}{C_3} + \cdots \qquad \textbf{[16.15]}$$

Problems involving a combination of capacitors can be solved by applying Equations 16.12 and 16.13 repeatedly to a circuit diagram, simplifying it as much as possible. This is followed by working backwards to the original diagram, applying $C = Q/\Delta V$, the fact that parallel capacitors have the same voltage drop, and the fact that series capacitors have the same charge.

16.9 Energy Stored in a Charged Capacitor

Three equivalent expressions for calculating the **energy stored** in a charged capacitor are

$$\text{Energy stored} = \tfrac{1}{2}Q\Delta V = \tfrac{1}{2}C(\Delta V)^2 = \frac{Q^2}{2C} \qquad \textbf{[16.17]}$$

16.10 Capacitors with Dielectrics

When a nonconducting material, called a **dielectric**, is placed between the plates of a capacitor, the capacitance is multiplied by the factor κ, which is called the **dielectric constant**, a property of the dielectric material. The capacitance of a parallel-plate capacitor filled with a dielectric is

$$C = \kappa\epsilon_0\frac{A}{d} \qquad \textbf{[16.19]}$$

CONCEPTUAL QUESTIONS

1. (a) Describe the motion of a proton after it is released from rest in a uniform electric field. (b) Describe the changes (if any) in its kinetic energy and the electric potential energy associated with the proton.

2. Describe how you can increase the maximum operating voltage of a parallel-plate capacitor for a fixed plate separation.

3. A parallel-plate capacitor is charged by a battery, and the battery is then disconnected from the capacitor. Because the charges on the capacitor plates are opposite in sign, they attract each other. Hence, it takes positive work to increase the plate separation. Show that the external work done when the plate separation is increased leads to an increase in the energy stored in the capacitor.

4. Distinguish between electric potential and electrical potential energy.

5. Suppose you are sitting in a car and a 20-kV power line drops across the car. Should you stay in the car or get out? The power line potential is 20 kV compared to the potential of the ground.

6. Why is it important to avoid sharp edges or points on conductors used in high-voltage equipment?

7. Explain why, under static conditions, all points in a conductor must be at the same electric potential.

8. If you are given three different capacitors C_1, C_2, and C_3, how many different combinations of capacitance can you produce, using all capacitors in your circuits.

9. Why is it dangerous to touch the terminals of a high-voltage capacitor even after the voltage source that charged the battery is disconnected from the capacitor? What can be done to make the capacitor safe to handle after the voltage source has been removed?

10. The plates of a capacitor are connected to a battery. What happens to the charge on the plates if the connecting wires are removed from the battery? What happens to the charge if the wires are removed from the battery and connected to each other?

11. Can electric field lines ever cross? Why or why not? Can equipotentials ever cross? Why or why not?

12. Is it always possible to reduce a combination of capacitors to one equivalent capacitor with the rules developed in this chapter? Explain.

13. If you were asked to design a capacitor for which a small size and a large capacitance were required, what factors would be important in your design?

14. Explain why a dielectric increases the maximum operating voltage of a capacitor even though the physical size of the capacitor doesn't change.

15. (a) Capacitors connected in parallel all have the same (i) charge on them, (ii) potential difference across them, or (iii) neither of the above. (b) Capacitors connected in series all have the same (i) charge on them, (ii) potential difference across them, or (iii) neither of the above.

16. (a) The equivalent capacitance for a group of capacitors connected in parallel is (i) greater than the capacitance of any of the capacitors in the group, (ii) less than the capacitance of any of the capacitors in the group, or (iii) neither of the above. (b) The equivalent capacitance for a group of capacitors connected in series is (i) greater than the capacitance of any of the capacitors in the group, (ii) less than the capacitance of any of the capacitors in the group, or (iii) neither of the above.

17. Suppose scientists had chosen to measure small energies in proton volts rather than electron volts. What difference would this make?

18. (a) Under what conditions can the equation $V_B - V_A = -E_x \Delta x$ be used? (b) Can the equation be used to find the difference in potential between two points in an electric field set up by a point charge? (c) Can the equation be used to find the difference in potential between two points in the electric field between the plates of a charged parallel-plate capacitor?

PROBLEMS

1, 2, 3 = straightforward, intermediate, challenging   □ = full solution available in *Student Solutions Manual/Study Guide*

Physics⊗Now™ = coached problem with hints available at **www.cp7e.com**   = biomedical application

Section 16.1 Potential Difference and Electric Potential

1. A proton moves 2.00 cm parallel to a uniform electric field of $E = 200$ N/C. (a) How much work is done by the field on the proton? (b) What change occurs in the potential energy of the proton? (c) What potential difference did the proton move through?

2. A uniform electric field of magnitude 250 V/m is directed in the positive x-direction. A 12-μC charge moves from the origin to the point $(x, y) = (20$ cm, 50 cm$)$. (a) What was the change in the potential energy of this charge? (b) Through what potential difference did the charge move?

3. A potential difference of 90 mV exists between the inner and outer surfaces of a cell membrane. The inner surface is negative relative to the outer surface. How much work is required to eject a positive sodium ion (Na^+) from the interior of the cell?

4. An ion accelerated through a potential difference of 60.0 V has its potential energy decreased by 1.92×10^{-17} J. Calculate the charge on the ion.

5. The potential difference between the accelerating plates of a TV set is about 25 kV. If the distance between the plates is 1.5 cm, find the magnitude of the uniform electric field in the region between the plates.

6. To recharge a 12-V battery, a battery charger must move 3.6×10^5 C of charge from the negative terminal to the positive terminal. How much work is done by the charger? Express your answer in joules.

7. **Physics⊗Now™** Oppositely charged parallel plates are separated by 5.33 mm. A potential difference of 600 V exists between the plates. (a) What is the magnitude of the electric field between the plates? (b) What is the magnitude of the force on an electron between the plates? (c) How much work must be done on the electron to move it to the negative plate if it is initially positioned 2.90 mm from the positive plate?

8. Calculate the speed of a proton that is accelerated from rest through a potential difference of 120 V. (b) Calculate the speed of an electron that is accelerated through the same potential difference.

9. A 4.00-kg block carrying a charge $Q = 50.0 \mu C$ is connected to a spring for which $k = 100$ N/m. The block lies on a frictionless horizontal track, and the system is immersed in a uniform electric field of magnitude $E = 5.00 \times 10^5$ V/m directed as in Figure P16.9. (a) If the block is released at rest when the spring is unstretched (at $x = 0$), by what maximum amount does the spring expand? (b) What is the equilibrium position of the block?

Figure P16.9

10. On planet Tehar, the free-fall acceleration is the same as that on Earth, but there is also a strong downward electric field that is uniform close to the planet's surface. A 2.00-kg ball having a charge of 5.00 μC is thrown upward at a speed of 20.1 m/s. It hits the ground after an interval of 4.10 s. What is the potential difference between the starting point and the top point of the trajectory?

Section 16.2 Electric Potential and Potential Energy Due to Point Charges
Section 16.3 Potentials and Charged Conductors
Section 16.4 Equipotential Surfaces

11. (a) Find the electric potential 1.00 cm from a proton. (b) What is the electric potential difference between two points that are 1.00 cm and 2.00 cm from a proton?

12. Two point charges are on the y-axis, one of magnitude 3.0×10^{-9} C at the origin and a second of magnitude 6.0×10^{-9} C at the point $y = 30$ cm. Calculate the potential at $y = 60$ cm.

13. (a) Find the electric potential, taking zero at infinity, at the upper right corner (the corner without a charge) of the rectangle in Figure P16.13. (b) Repeat if the 2.00-μC charge is replaced with a charge of $-2.00 \mu C$.

Figure P16.13   (Problems 13 and 14)

14. Three charges are situated at corners of a rectangle as in Figure P16.13. How much energy would be expended in moving the 8.00-μC charge to infinity?

15. Two point charges $Q_1 = +5.00$ nC and $Q_2 = -3.00$ nC are separated by 35.0 cm. (a) What is the electric potential at a point midway between the charges? (b) What is the potential energy of the pair of charges? What is the significance of the algebraic sign of your answer?

16. A point charge of 9.00×10^{-9} C is located at the origin. How much work is required to bring a positive charge of 3.00×10^{-9} C from infinity to the location $x = 30.0$ cm?

17. The three charges in Figure P16.17 are at the vertices of an isosceles triangle. Let $q = 7.00$ nC, and calculate the electric potential at the midpoint of the base.

Figure P16.17

18. An electron starts from rest 3.00 cm from the center of a uniformly charged sphere of radius 2.00 cm. If the sphere carries a total charge of 1.00×10^{-9} C, how fast will the electron be moving when it reaches the surface of the sphere?

19. **Physics⊗Now™** In Rutherford's famous scattering experiments that led to the planetary model of the atom, alpha particles (having charges of $+2e$ and masses of 6.64×10^{-27} kg) were fired toward a gold nucleus with charge $+79e$. An alpha particle, initially very far from the gold nucleus, is fired at 2.00×10^7 m/s directly toward the nucleus, as in Figure P16.19. How close does the alpha particle get to the gold nucleus before turning around? Assume the gold nucleus remains stationary.

Figure P16.19

20. Starting with the definition of work, prove that the surface must be perpendicular to the local electric field at every point on an equipotential surface.

21. A small spherical object carries a charge of 8.00 nC. At what distance from the center of the object is the potential equal to 100 V? 50.0 V? 25.0 V? Is the spacing of the equipotentials proportional to the change in voltage?

Section 16.6 Capacitance
Section 16.7 The Parallel-Plate Capacitor

22. (a) How much charge is on each plate of a 4.00-μF capacitor when it is connected to a 12.0-V battery? (b) If this same capacitor is connected to a 1.50-V battery, what charge is stored?

23. Consider the Earth and a cloud layer 800 m above the planet to be the plates of a parallel-plate capacitor. (a) If the cloud layer has an area of 1.0 km^2 = 1.0 × 10^6 m^2, what is the capacitance? (b) If an electric field strength greater than 3.0 × 10^6 N/C causes the air to break down and conduct charge (lightning), what is the maximum charge the cloud can hold?

24. The potential difference between a pair of oppositely charged parallel plates is 400 V. (a) If the spacing between the plates is doubled without altering the charge on the plates, what is the new potential difference between the plates? (b) If the plate spacing is doubled while the potential difference between the plates is kept constant, what is the ratio of the final charge on one of the plates to the original charge?

25. An air-filled capacitor consists of two parallel plates, each with an area of 7.60 cm^2 and separated by a distance of 1.80 mm. If a 20.0-V potential difference is applied to these plates, calculate (a) the electric field between the plates, (b) the capacitance, and (c) the charge on each plate.

26. A 1-megabit computer memory chip contains many 60.0 × 10^{-15}-F capacitors. Each capacitor has a plate area of 21.0 × 10^{-12} m^2. Determine the plate separation of such a capacitor. (Assume a parallel-plate configuration). The diameter of an atom is on the order of 10^{-10} m = 1 Å. Express the plate separation in angstroms.

27. A parallel-plate capacitor has an area of 5.00 cm^2, and the plates are separated by 1.00 mm with air between them. The capacitor stores a charge of 400 pC. (a) What is the potential difference across the plates of the capacitor? (b) What is the magnitude of the uniform electric field in the region between the plates?

28. A small object with a mass of 350 mg carries a charge of 30.0 nC and is suspended by a thread between the vertical plates of a parallel-plate capacitor. The plates are separated by 4.00 cm. If the thread makes an angle of 15.0° with the vertical, what is the potential difference between the plates?

Section 16.8 Combinations of Capacitors

29. A series circuit consists of a 0.050-μF capacitor, a 0.100-μF capacitor, and a 400-V battery. Find the charge (a) on each of the capacitors and (b) on each of the capacitors if they are reconnected in parallel across the battery.

30. Three capacitors, C_1 = 5.00 μF, C_2 = 4.00 μF, and C_3 = 9.00 μF, are connected together. Find the effective capacitance of the group (a) if they are all in parallel, and (b) if they are all in series.

31. (a) Find the equivalent capacitance of the capacitors in Figure P16.31. (b) Find the charge on each capacitor and the potential difference across it.

Figure P16.31

32. Two capacitors give an equivalent capacitance of 9.00 pF when connected in parallel and an equivalent capacitance of 2.00 pF when connected in series. What is the capacitance of each capacitor?

33. Four capacitors are connected as shown in Figure P16.33. (a) Find the equivalent capacitance between points a and b. (b) Calculate the charge on each capacitor if a 15.0-V battery is connected across points a and b.

Figure P16.33

34. Consider the combination of capacitors in Figure P16.34. (a) What is the equivalent capacitance of the group? (b) Determine the charge on each capacitor.

Figure P16.34

35. Find the charge on each of the capacitors in Figure P16.35.

Figure P16.35

36. To repair a power supply for a stereo amplifier, an electronics technician needs a 100-μF capacitor capable of withstanding a potential difference of 90 V between its plates. The only available supply is a box of five 100-μF capacitors, each having a maximum voltage capability of 50 V. Can the technician substitute a combination of these capacitors that has the proper electrical characteristics, and if so, what will be the maximum voltage across any of the capacitors used? [*Hint:* The technician may not have to use all the capacitors in the box.]

37. **Physics⊗Now™** A 25.0-μF capacitor and a 40.0-μF capacitor are charged by being connected across separate 50.0-V batteries. (a) Determine the resulting charge on each capacitor. (b) The capacitors are then disconnected from their batteries and connected to each other, with each negative plate connected to the other positive plate. What is the final charge of each capacitor, and what is the final potential difference across the 40.0-μF capacitor?

38. A 10.0-μF capacitor is fully charged across a 12.0-V battery. The capacitor is then disconnected from the battery and connected across an initially uncharged capacitor with capacitance C. The resulting voltage across each capacitor is 3.00 V. What is the value of C?

39. A 1.00-μF capacitor is charged by being connected across a 10.0-V battery. It is then disconnected from the battery and connected across an uncharged 2.00-μF capacitor. Determine the resulting charge on each capacitor.

40. Find the equivalent capacitance between points a and b for the group of capacitors connected as shown in Figure P16.40 if $C_1 = 5.00$ μF, $C_2 = 10.0$ μF, and $C_3 = 2.00$ μF.

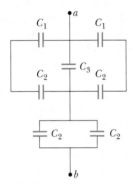

Figure P16.40   (Problems 40 and 41)

41. For the network described in the previous problem and in Figure P16.40, if the potential between points a and b is 60.0 V, what charge is stored on C_3?

42. Find the equivalent capacitance between points a and b in the combination of capacitors shown in Figure P16.42.

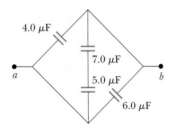

Figure P16.42

Section 16.9 Energy Stored in a Charged Capacitor

43. A parallel-plate capacitor has 2.00-cm^2 plates that are separated by 5.00 mm with air between them. If a 12.0-V battery is connected to this capacitor, how much energy does it store?

44. Two capacitors, $C_1 = 25.0$ μF and $C_2 = 5.00$ μF, are connected in parallel and charged with a 100-V power supply. (a) Calculate the total energy stored in the two capacitors. (b) What potential difference would be required across the same two capacitors connected in *series* in order that the combination store the same energy as in (a)?

45. Consider the parallel-plate capacitor formed by the Earth and a cloud layer as described in Problem 16.23. Assume this capacitor will discharge (i.e., produce lightning) when the electric field strength between the plates reaches 3.0×10^6 N/C. What is the energy released if the capacitor discharges completely during a lightning strike?

46. A certain storm cloud has a potential difference of 1.00×10^8 V relative to a tree. If, during a lightning storm, 50.0 C of charge is transferred through this potential difference and 1.00% of the energy is absorbed by the tree, how much water (sap in the tree) initially at 30.0°C can be boiled away? Water has a specific heat of 4 186 J/kg°C, a boiling point of 100°C, and a heat of vaporization of 2.26×10^6 J/kg.

Section 16.10 Capacitors with Dielectrics

47. A capacitor with air between its plates is charged to 100 V and then disconnected from the battery. When a piece of glass is placed between the plates, the voltage across the capacitor drops to 25 V. What is the dielectric constant of the glass? (Assume the glass completely fills the space between the plates.)

48. Two parallel plates, each of area 2.00 cm^2, are separated by 2.00 mm with purified nonconducting water between them. A voltage of 6.00 V is applied between the plates. Calculate (a) the magnitude of the electric field between the plates, (b) the charge stored on each plate, and (c) the charge stored on each plate if the water is removed and replaced with air.

49. Determine (a) the capacitance and (b) the maximum voltage that can be applied to a Teflon®-filled parallel-plate capacitor having a plate area of 175 cm^2 and an insulation thickness of 0.040 0 mm.

50. A commercial capacitor is constructed as in Figure 16.25a. This particular capacitor is made from a strip of aluminum foil separated by two strips of paraffin-coated paper. Each strip of foil and paper is 7.00 cm wide. The foil is 0.004 00 mm thick, and the paper is 0.025 0 mm thick and has a dielectric constant of 3.70. What length should the strips be if a capacitance of 9.50×10^{-8} F is desired before the capacitor is rolled up? (Use the parallel-plate formula. Adding a second strip of paper and rolling up the capacitor doubles its capacitance by allowing both surfaces of each strip of foil to store charge.)

51. A model of a red blood cell portrays the cell as a spherical capacitor—a positively charged liquid sphere of surface area A separated from the surrounding negatively charged fluid by a membrane of thickness t. Tiny electrodes introduced into the interior of the cell show a potential difference of 100 mV across the membrane. The membrane's thickness is estimated to be 100 nm and whose dielectric constant is 5.00. (a) If an average red blood cell has a mass of 1.00×10^{-12} kg, estimate the volume of the cell and thus find its surface area. The density of blood is 1 100 kg/m^3. (b) Estimate the capacitance of the cell. (c) Calculate the charge on the surface of the membrane. How many electronic charges does the surface charge represent?

ADDITIONAL PROBLEMS

52. Three parallel-plate capacitors are constructed, each having the same plate spacing d and with C_1 having plate area A_1, C_2 having area A_2, and C_3 having area A_3. Show that the total capacitance C of the three capacitors connected in parallel is the same as that of a capacitor having plate spacing d and plate area $A = A_1 + A_2 + A_3$.

53. Physics⊗Now™ Three parallel-plate capacitors are constructed, each having the same plate area A and with C_1 having plate spacing d_1, C_2 having plate spacing d_2, and C_3 having plate spacing d_3. Show that the total capacitance C of the three capacitors connected in series is the same as a capacitor of plate area A and with plate spacing $d = d_1 + d_2 + d_3$.

54. Two capacitors give an equivalent capacitance of C_p when connected in parallel and an equivalent capacitance of C_s when connected in series. What is the capacitance of each capacitor?

55. An isolated capacitor of unknown capacitance has been charged to a potential difference of 100 V. When the charged capacitor is disconnected from the battery and then connected in parallel to an uncharged 10.0-μF capacitor, the voltage across the combination is measured to be 30.0 V. Calculate the unknown capacitance.

56. Two charges of 1.0 μC and -2.0 μC are 0.50 m apart at two vertices of an equilateral triangle as in Figure P16.56. (a) What is the electric potential due to the 1.0-μC charge at the third vertex, point P? (b) What is the electric potential due to the -2.0-μC charge at P? (c) Find the total electric potential at P. (d) What is the work required to move a 3.0-μC charge from infinity to P?

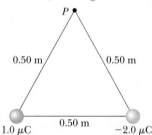

Figure P16.56

57. Find the equivalent capacitance of the group of capacitors shown in Figure P16.57.

Figure P16.57

58. A spherical capacitor consists of a spherical conducting shell of radius b and charge $-Q$ concentric with a smaller conducting sphere of radius a and charge Q. (a) Find the capacitance of this device. (b) Show that as the radius b of the outer sphere approaches infinity, the capacitance approaches the value $a/k_e = 4\pi\epsilon_0 a$.

59. The immediate cause of many deaths is ventricular fibrillation, an uncoordinated quivering of the heart, as opposed to proper beating. An electric shock to the chest can cause momentary paralysis of the heart muscle, after which the heart will sometimes start organized beating again. A *defibrillator* is a device that applies a strong electric shock to the chest over a time of a few milliseconds. The device contains a capacitor of a few microfarads, charged to several thousand volts. Electrodes called paddles, about 8 cm across and coated with conducting paste, are held against the chest on both sides of the heart. Their handles are insulated to prevent injury to the operator, who calls, "Clear!" and pushes a button on one paddle to discharge the capacitor through the patient's chest. Assume that an energy of 300 W·s is to be delivered from a 30.0-μF capacitor. To what potential difference must it be charged?

60. When a certain air-filled parallel-plate capacitor is connected across a battery, it acquires a charge of 150 μC on each plate. While the battery connection is maintained, a dielectric slab is inserted into, and fills, the region between the plates. This results in the accumulation of an additional charge of 200 μC on each plate. What is the dielectric constant of the slab?

61. Capacitors $C_1 = 6.0$ μF and $C_2 = 2.0$ μF are charged as a parallel combination across a 250-V battery. The capacitors are disconnected from the battery and from each other. They are then connected positive plate to negative plate and negative plate to positive plate. Calculate the resulting charge on each capacitor.

62. Capacitors $C_1 = 4.0$ μF and $C_2 = 2.0$ μF are charged as a series combination across a 100-V battery. The two capacitors are disconnected from the battery and from each other. They are then connected positive plate to positive plate and negative plate to negative plate. Calculate the resulting charge on each capacitor.

63. The charge distribution shown in Figure P16.63 is referred to as a *linear quadrupole*. (a) Show that the electric potential at a point on the x-axis where $x > d$ is

$$V = \frac{2k_e Qd^2}{x^3 - xd^2}$$

(b) Show that the expression obtained in (a) when $x \gg d$ reduces to

$$V = \frac{2k_e Qd^2}{x^3}$$

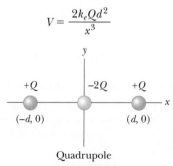

Quadrupole

Figure P16.63

64. The energy stored in a 52.0-μF capacitor is used to melt a 6.00-mg sample of lead. To what voltage must the capacitor be initially charged, assuming that the initial temperature of the lead is 20.0°C? Lead has a specific heat of 128 J/kg·°C, a melting point of 327.3°C, and a latent heat of fusion of 24.5 kJ/kg.

65. Consider a parallel-plate capacitor with charge Q and area A, filled with dielectric material having dielectric constant κ. It can be shown that the magnitude of the attractive force exerted on each plate by the other is $F = Q^2/(2\kappa\epsilon_0 A)$. When a potential difference of 100 V exists between the plates of an air-filled 20-μF parallel-plate capacitor, what force does each plate exert on the other if they are separated by 2.0 mm?

66. An electron is fired at a speed $v_0 = 5.6 \times 10^6$ m/s and at an angle $\theta_0 = -45°$ between two parallel conducting plates that are $D = 2.0$ mm apart, as in Figure P16.66. If the voltage difference between the plates is $\Delta V = 100$ V, determine (a) how close, d, the electron will get to the bottom plate and (b) where the electron will strike the top plate.

Figure P16.66

ACTIVITIES

1. It takes an electric field of about 30 kV/cm to cause a spark in dry air. Shuffle across a rug and reach toward a doorknob. By estimating the length of the spark, determine the electric potential difference that existed between your finger and the doorknob just before you touched the knob. Try this experiment again on a very humid day, and you will find that the spark is much shorter or is imperceptible. Why?

2. Suppose you are given a battery, a capacitor, two switches, a lightbulb, and several pieces of connecting wire. On a sheet of paper, design a circuit that will do the following: (1) When switch 1 is closed and switch 2 is open, the capacitor charges, but no current moves through the lightbulb. (2) Then, when switch 1 is opened and switch 2 closed, the lightbulb is connected to the capacitor, but not to the battery. Describe the motion of charge in the circuit when switch 1 is closed and switch 2 is open. Is energy being stored in the capacitor? What measurements would you have to make to determine how much energy, if any, is stored? What happens to the lightbulb when switch 1 is opened after the capacitor has charged and switch 2 is then closed? Will the bulb light and stay lit? What happens to the charge on the capacitor when switch 2 is closed in this way?

These power lines transfer energy from the power company to homes and businesses. The energy is transferred at a very high voltage, possibly hundreds of thousands of volts in some cases. The high voltage results in less loss of power due to resistance in the wires, so it is used despite the fact that it makes power lines very dangerous.

Current and Resistance

Many practical applications and devices are based on the principles of static electricity, but electricity was destined to become an inseparable part of our daily lives when scientists learned how to produce a continuous flow of charge for relatively long periods of time using batteries. The battery or voltaic cell was invented in 1800 by the Italian physicist Alessandro Volta. Batteries supplied a continuous flow of charge at low potential, in contrast to earlier electrostatic devices that produced a tiny flow of charge at high potential for brief periods. This steady source of electric current allowed scientists to perform experiments to learn how to control the flow of electric charges in circuits. Today, electric currents power our lights, radios, television sets, air conditioners, computers, and refrigerators. They ignite the gasoline in automobile engines, travel through miniature components making up the chips of microcomputers, and provide the power for countless other invaluable tasks.

In this chapter we define current and discuss some of the factors that contribute to the resistance to the flow of charge in conductors. We also discuss energy transformations in electric circuits. These topics will be the foundation for additional work with circuits in later chapters.

17.1 ELECTRIC CURRENT

In Figure 17.1, charges move in a direction perpendicular to a surface of area A. (The area could be the cross-sectional area of a wire, for example.) **The current is the rate at which charge flows through this surface.**

Suppose ΔQ is the amount of charge that flows through an area A in a time interval Δt and that the direction of flow is perpendicular to the area. Then the current I is equal to the amount of charge divided by the time interval:

$$I \equiv \frac{\Delta Q}{\Delta t} \qquad [17.1]$$

SI unit: coulomb/second (C/s), or the ampere (A).

TIP 17.1   *Current Flow* is Redundant

The phrases *flow of current* and *current flow* are commonly used, but here the word *flow* is redundant because current is already defined as a flow (of charge). Avoid this construction!

One ampere of current is equivalent to one coulomb of charge passing through the cross-sectional area in a time interval of 1 s.

When charges flow through a surface as in Figure 17.1, they can be positive, negative, or both. **The direction of conventional current used in this book is the direction positive charges flow.** (This historical convention originated about 200 years ago, when the ideas of positive and negative charges were introduced.) In a common conductor such as copper, the current is due to the motion of negatively charged electrons, so the direction of the current is opposite the direction of motion of the electrons. On the other hand, for a beam of positively-charged protons in an accelerator, the current is in the same direction as the motion of the protons. In some cases—gases and electrolytes, for example—the current is the result of the flows of both positive and negative charges. Moving charges, whether positive or negative, are referred to as *charge carriers*. In a metal, for example, the charge carriers are electrons.

In electrostatics, where charges are stationary, the electric potential is the same everywhere in a conductor. This is no longer true for conductors carrying current: as charges move along a wire, the electric potential is continually decreasing (except in the special case of superconductors).

◀ Direction of current

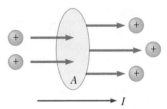

Figure 17.1   Charges in motion through an area A. The time rate of flow of charge through the area is defined as the current I. The direction of the current is the direction of flow of positive charges.

EXAMPLE 17.1   Turn on the Light

Goal   Apply the concept of current.

Problem   The amount of charge that passes through the filament of a certain lightbulb in 2.00 s is 1.67 C. Find **(a)** the current in the bulb and **(b)** the number of electrons that pass through the filament in 5.00 s.

Strategy   Substitute into Equation 17.1 for part (a), then multiply the answer by the time given in part (b) to get the total charge that passes in that time. The total charge equals the number N of electrons going through the circuit times the charge per electron.

Solution
(a) Compute the current in the lightbulb.

Substitute the charge and time into Equation 17.1:

$$I = \frac{\Delta Q}{\Delta t} = \frac{1.67 \text{ C}}{2.00 \text{ s}} = \boxed{0.835 \text{ A}}$$

(b) Find the number of electrons passing through the filament in 5.00 s.

The total number N of electrons times the charge per electron equals the total charge, $I \Delta t$:

$$(1) \quad Nq = I \Delta t$$

Substitute and solve for N:

$$N(1.60 \times 10^{-19} \text{ C/electron}) = (0.835 \text{ A})(5.00 \text{ s})$$

$$N = \boxed{2.61 \times 10^{19} \text{ electrons}}$$

Remarks   In developing the solution, it was important to use units to ensure the correctness of equations such as Equation (1). Notice the enormous number of electrons passing through a given point in a typical circuit.

Exercise 17.1
Suppose 6.40×10^{21} electrons pass through a wire in 2.00 min. Find the current.

Answer   8.53 A

Consider positive and negative charges moving horizontally through the four regions in Figure 17.2. Rank the magnitudes of the currents in these four regions from lowest to highest. (I_a is the current in Figure 17.2a, I_b the current in Figure 17.2b, etc.) (a) I_d, I_a, I_c, I_b (b) I_a, I_c, I_b, I_d (c) I_c, I_a, I_d, I_b (d) I_d, I_b, I_c, I_a (e) I_a, I_b, I_c, I_d (f) none of these

Figure 17.2 (Quick Quiz 17.1)

(a)   (b)   (c)   (d)

Figure 17.3 A section of a uniform conductor of cross-sectional area A. The charge carriers move with a speed v_d, and the distance they travel in time Δt is given by $\Delta x = v_d \Delta t$. The number of mobile charge carriers in the section of length Δx is given by $nAv_d\Delta t$, where n is the number of mobile carriers per unit volume.

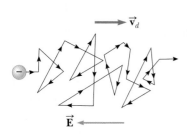

ACTIVE FIGURE 17.4
A schematic representation of the zigzag motion of a charge carrier in a conductor. The sharp changes in direction are due to collisions with atoms in the conductor. Note that the net motion of electrons is opposite the direction of the electric field.

Physics⊗Now™
Log into PhysicsNow at **www.cp7e.com** and go to Active Figure 17.4, where you can observe the random zigzag motion of a charge carrier and see how the motion is affected by an electric field.

TIP 17.2 Electrons are Everywhere in the Circuit

Electrons don't have to travel from the light switch to the light for the light to operate. Electrons already in the filament of the lightbulb move in response to the electric field set up by the battery. Also, the battery does *not* provide electrons to the circuit; it provides *energy* to the existing electrons.

17.2 A MICROSCOPIC VIEW: CURRENT AND DRIFT SPEED

Macroscopic currents can be related to the motion of the microscopic charge carriers making up the current. It turns out that current depends on the average speed of the charge carriers in the direction of the current, the number of charge carriers per unit volume, and the size of the charge carried by each.

Consider identically charged particles moving in a conductor of cross-sectional area A (Fig. 17.3). The volume of an element of length Δx of the conductor is $A \Delta x$. If n represents the number of mobile charge carriers per unit volume, then the number of carriers in the volume element is $nA \Delta x$. The mobile charge ΔQ in this element is therefore

$$\Delta Q = \text{number of carriers} \times \text{charge per carrier} = (nA \Delta x)q$$

where q is the charge on each carrier. If the carriers move with a constant average speed called the **drift speed** v_d, the distance they move in the time interval Δt is $\Delta x = v_d \Delta t$. We can therefore write

$$\Delta Q = (nAv_d \Delta t)q$$

If we divide both sides of this equation by Δt, we see that the current in the conductor is

$$I = \frac{\Delta Q}{\Delta t} = nqv_dA \qquad \textbf{[17.2]}$$

To understand the meaning of drift speed, consider a conductor in which the charge carriers are free electrons. If the conductor is isolated, these electrons undergo random motion similar to the motion of the molecules of a gas. The drift speed is normally much smaller than the free electrons' average speed between collisions with the fixed atoms of the conductor. When a potential difference is applied between the ends of the conductor (say, with a battery), an electric field is set up in the conductor, creating an electric force on the electrons and hence a current. In reality, the electrons don't simply move in straight lines along the conductor. Instead, they undergo repeated collisions with the atoms of the metal, and the result is a complicated zigzag motion with only a small average drift speed along the wire (Active Fig. 17.4). The energy transferred from the electrons to the metal atoms during a collision increases the vibrational energy of the atoms and causes a corresponding increase in the temperature of the conductor. Despite the collisions, however, the electrons move slowly along the conductor in a direction opposite $\vec{\mathbf{E}}$ with the drift velocity $\vec{\mathbf{v}}_d$.

EXAMPLE 17.2   Drift Speed of Electrons

Goal   Calculate a drift speed and compare it with the rms speed of an electron gas.

Problem   A copper wire of cross-sectional area 3.00×10^{-6} m^2 carries a current of 10.0 A. **(a)** Assuming that each copper atom contributes one free electron to the metal, find the drift speed of the electrons in this wire. **(b)** Use the ideal gas model to compare the drift speed with the random rms speed an electron would have at 20.0°C. The density of copper is 8.92 g/cm^3, and its atomic mass is 63.5 u.

Strategy   All the variables in Equation 17.2 are known except for n, the number of free charge carriers per unit volume. We can find n by recalling that one mole of copper contains an Avogadro's number (6.02×10^{23}) of atoms and each atom contributes one charge carrier to the metal. The volume of one mole can be found from copper's known density and atomic mass. The atomic mass is the same, numerically, as the number of grams in a mole of the substance.

Solution
(a) Find the drift speed of the electrons.

Calculate the volume of one mole of copper from its density and its atomic mass:

$$V = \frac{m}{\rho} = \frac{63.5 \text{ g}}{8.92 \text{ g/cm}^3} = 7.12 \text{ cm}^3$$

Convert the volume from cm^3 to m^3:

$$7.12 \text{ cm}^3 \left(\frac{1 \text{ m}}{10^2 \text{ cm}} \right)^3 = 7.12 \times 10^{-6} \text{ m}^3$$

Divide Avogadro's number (the number of electrons in one mole) by the volume per mole to obtain the number density:

$$n = \frac{6.02 \times 10^{23} \text{ electrons/mole}}{7.12 \times 10^{-6} \text{ m}^3/\text{mole}}$$
$$= 8.46 \times 10^{28} \text{ electrons/m}^3$$

Solve Equation 17.2 for the drift speed, and substitute:

$$v_d = \frac{I}{nqA}$$

$$= \frac{10.0 \text{ C/s}}{(8.46 \times 10^{28} \text{ electrons/m}^3)(1.60 \times 10^{-19} \text{ C})(3.00 \times 10^{-6} \text{ m}^2)}$$

$$v_d = \boxed{2.46 \times 10^{-4} \text{ m/s}}$$

(b) Find the rms speed of a gas of electrons at 20.0°C.

Apply Equation 10.18:

$$v_{\text{rms}} = \sqrt{\frac{3 k_B T}{m_e}}$$

Convert the temperature to the Kelvin scale, and substitute values:

$$v_{\text{rms}} = \sqrt{\frac{3(1.38 \times 10^{-23} \text{ J/K})(293 \text{ K})}{9.11 \times 10^{-31} \text{ kg}}}$$

$$= \boxed{1.15 \times 10^5 \text{ m/s}}$$

Remarks   The drift speed of an electron in a wire is very small—only about one-billionth of its random thermal speed.

Exercise 17.2
What current in a copper wire with a cross-sectional area of 7.50×10^{-7} m^2 would result in a drift speed of 5.00×10^{-4} m/s?

Answer   5.08 A

Example 17.2 shows that drift speeds are typically very small. In fact, the drift speed is much smaller than the average speed between collisions. Electrons traveling at 2.46×10^{-4} m/s, as in the example, would take about 68 min to travel 1 m!

In view of this low speed, you might wonder why a light turns on almost instantaneously when a switch is thrown. Think of the flow of water through a pipe. If a drop of water is forced into one end of a pipe that is already filled with water, a drop must be pushed out the other end of the pipe. Although it may take an individual drop a long time to make it through the pipe, a flow initiated at one end produces a similar flow at the other end very quickly. Another familiar analogy is the motion of a bicycle chain. When the sprocket moves one link, the other links all move more or less immediately, even though it takes a given link some time to make a complete rotation. In a conductor, the electric field driving the free electrons travels at a speed close to that of light, so when you flip a light switch, the message for the electrons to start moving through the wire (the electric field) reaches them at a speed on the order of 10^8 m/s!

Quick Quiz 17.2

Suppose a current-carrying wire has a cross-sectional area that gradually becomes smaller along the wire, so that the wire has the shape of a very long cone. How does the drift speed vary along the wire? (a) It slows down as the cross section becomes smaller. (b) It speeds up as the cross section becomes smaller. (c) It doesn't change. (d) More information is needed.

17.3 CURRENT AND VOLTAGE MEASUREMENTS IN CIRCUITS

To study electric current in circuits, we need to understand how to measure currents and voltages.

The circuit shown in Figure 17.5a is a drawing of the actual circuit necessary for measuring the current in Example 17.1. Figure 17.5b shows a stylized figure called a circuit diagram which represents the actual circuit of Figure 17.5a. This circuit consists of only a battery and a lightbulb. The word "circuit" means "a closed loop

Figure 17.5 (a) A sketch of an actual circuit used to measure the current in a flashlight bulb and the potential difference across it. (b) A schematic diagram of the circuit shown in part (a). (c) A digital multimeter can be used to measure both currents and potential differences. Here, the meter is measuring the potential difference across a 9-V battery.

of some sort around which current circulates." The battery pumps charge through the bulb and around the loop. No charge would flow without a complete conducting path from the positive terminal of the battery into one side of the bulb, out the other side, and through the copper conducting wires back to the negative terminal of the battery. The most important quantities that characterize how the bulb works in different situations are the current I in the bulb and the potential difference ΔV across the bulb. To measure the current in the bulb, we place an ammeter, the device for measuring current, in line with the bulb so there is no path for the current to bypass the meter; all of the charge passing through the bulb must also pass through the ammeter. The voltmeter measures the potential difference, or voltage, between the two ends of the bulb's filament. If we use two meters simultaneously as in Figure 17.5a, we can remove the voltmeter and see if its presence affects the current reading. Figure 17.5c shows a digital multimeter—a convenient device, with a digital readout, that can be used to measure voltage, current, or resistance. An advantage of using a digital multimeter as a voltmeter is that it will usually not affect the current, since a digital meter has enormous resistance to the flow of charge in the voltmeter mode.

At this point, you can measure the current as a function of voltage (an I–ΔV curve) of various devices in the lab. All you need is a variable voltage supply (an adjustable battery) capable of supplying potential differences from about -5 V to $+5$ V, a bulb, a resistor, some wires and alligator clips, and a couple of multimeters. Be sure to always start your measurements using the highest multimeter scales (say, 10 A and 1 000 V), and increase the sensitivity one scale at a time to obtain the highest accuracy without overloading the meters. (Increasing the sensitivity means lowering the maximum current or voltage that the scale reads.) Note that the meters must be connected with the proper polarity with respect to the voltage supply, as shown in Figure 17.5b. Finally, follow your instructor's directions carefully to avoid damaging the meters and incurring a soaring lab fee.

Quick Quiz 17.3

Look at the four "circuits" shown in Figure 17.6 and select those that will light the bulb.

(a)   (b)   (c)   (d)

Figure 17.6   (Quick Quiz 17.3)

Figure 17.7   A uniform conductor of length l and cross-sectional area A. The current I in the conductor is proportional to the applied voltage $\Delta V = V_b - V_a$. The electric field $\vec{E}$ set up in the conductor is also proportional to the current.

17.4   RESISTANCE AND OHM'S LAW

When a voltage (potential difference) ΔV is applied across the ends of a metallic conductor as in Figure 17.7, the current in the conductor is found to be proportional to the applied voltage; $I \propto \Delta V$. If the proportionality holds, we can write $\Delta V = IR$, where the proportionality constant R is called the *resistance* of the conductor. In fact, we define the **resistance** as the ratio of the voltage across the conductor to the current it carries:

$$R \equiv \frac{\Delta V}{I}$$ [17.3]   ◀ Resistance

GEORG SIMON OHM
(1787–1854)

A high school teacher in Cologne and later a professor at Munich, Ohm formulated the concept of resistance and discovered the proportionalities expressed in Equation 17.5.

Ohm's law ▶

Resistance has SI units of volts per ampere, called **ohms** (Ω). If a potential difference of 1 V across a conductor produces a current of 1 A, the resistance of the conductor is 1 Ω. For example, if an electrical appliance connected to a 120-V source carries a current of 6 A, its resistance is 20 Ω.

The concepts of electric current, voltage, and resistance can be compared with the flow of water in a river. As water flows downhill in a river of constant width and depth, the flow rate (water current) depends on the steepness of descent of the river and the effects of rocks, the riverbank, and other obstructions. The voltage difference is analogous to the steepness, and the resistance to the obstructions. Based on this analogy, it seems reasonable that increasing the voltage applied to a circuit should increase the current in the circuit, just as increasing the steepness of descent increases the water current. Also, increasing the obstructions in the river's path will reduce the water current, just as increasing the resistance in a circuit will lower the electric current. Resistance in a circuit arises due to collisions between the electrons carrying the current with fixed atoms inside the conductor. These collisions inhibit the movement of charges in much the same way as would a force of friction. For many materials, including most metals, experiments show that **the resistance remains constant over a wide range of applied voltages or currents**. This statement is known as **Ohm's law**, after Georg Simon Ohm (1789–1854), who was the first to conduct a systematic study of electrical resistance.

Ohm's law is given by

$$\Delta V = IR \qquad\qquad [17.4]$$

where R is understood to be independent of ΔV, the potential drop across the resistor, and I, the current in the resistor. We will continue to use this traditional form of Ohm's law when discussing electrical circuits. A **resistor** is a conductor that provides a specified resistance in an electric circuit. The symbol for a resistor in circuit diagrams is a zigzag line: ——⋀⋀⋀——.

Ohm's law is an empirical relationship valid only for certain materials. Materials that obey Ohm's law, and hence have a constant resistance over a wide range of voltages, are said to be **ohmic**. Materials having resistance that changes with voltage or current are **nonohmic**. Ohmic materials have a linear current–voltage relationship over a large range of applied voltages (Fig. 17.8a). Nonohmic materials have a nonlinear current–voltage relationship (Fig. 17.8b). One common semiconducting device that is nonohmic is the *diode*, a circuit element that acts like a one-way valve for current. Its resistance is small for currents in one direction (positive ΔV) and large for currents in the reverse direction (negative ΔV). Most modern electronic devices, such as transistors, have nonlinear current–voltage relationships; their operation depends on the particular ways in which they violate Ohm's law.

Figure 17.8   (a) The current–voltage curve for an ohmic material. The curve is linear, and the slope gives the resistance of the conductor. (b) A nonlinear current–voltage curve for a semiconducting diode. This device doesn't obey Ohm's law.

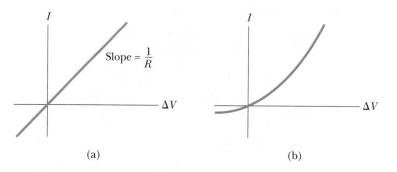

(a)   (b)

In Figure 17.8b, does the resistance of the diode (a) increase or (b) decrease as the positive voltage ΔV increases?

An assortment of resistors used for a variety of applications in electronic circuits.

EXAMPLE 17.3   Resistance of a Steam Iron

Goal   Use Ohm's law to calculate a resistance.

Problem   All electric devices are required to have identifying plates that specify their electrical characteristics. The plate on a certain steam iron states that the iron carries a current of 6.40 A when connected to a source of 1.20×10^2 V. What is the resistance of the steam iron?

Strategy   Substitute into Ohm's law.

Solution
Apply Equation 17.3:

$$R = \frac{\Delta V}{I} = \frac{1.20 \times 10^2 \text{ V}}{6.40 \text{ A}} = \boxed{18.8 \ \Omega}$$

Exercise 17.3
The resistance of a hot plate is 48.0 Ω. How much current does the plate carry when connected to a 1.20×10^2-V source?

Answer   2.50 A

17.5   RESISTIVITY

Electrons don't move in straight-line paths through a conductor. Instead, they undergo repeated collisions with the metal atoms. Consider a conductor with a voltage applied across its ends. An electron gains speed as the electric force associated with the internal electric field accelerates it, giving it a velocity in the direction opposite that of the electric field. A collision with an atom randomizes the electron's velocity, reducing it in the direction opposite the field. The process then repeats itself. Together, these collisions affect the electron somewhat as a force of internal friction would. This is the origin of a material's resistance.

The resistance of an ohmic conductor increases with length, which makes sense because the electrons going through it must undergo more collisions in a longer conductor. A smaller cross-sectional area also increases the resistance of a conductor, just as a smaller pipe slows the fluid moving through it. The resistance, then, is proportional to the conductor's length l and inversely proportional to its cross-sectional area A,

$$R = \rho \frac{l}{A} \qquad \text{[17.5]}$$

where the constant of proportionality, ρ, is called the **resistivity** of the material.[1] Every material has a characteristic resistivity that depends on its electronic structure and on temperature. Good electric conductors have very low resistivities, and good insulators have very high resistivities. Table 17.1 lists the resistivities of various materials at 20°C. Because resistance values are in ohms, resistivity values must be in ohm-meters ($\Omega \cdot$ m).

---

[1]The symbol ρ used for resistivity shouldn't be confused with the same symbol used earlier in the book for density. Often, a single symbol is used to represent different quantities.

TABLE 17.1

Resistivities and Temperature Coefficients of Resistivity for Various Materials (at 20°C)

Material	Resistivity ($\Omega \cdot m$)	Temperature Coefficient of Resistivity $[(°C)^{-1}]$
Silver	1.59×10^{-8}	3.8×10^{-3}
Copper	1.7×10^{-8}	3.9×10^{-3}
Gold	2.44×10^{-8}	3.4×10^{-3}
Aluminum	2.82×10^{-8}	3.9×10^{-3}
Tungsten	5.6×10^{-8}	4.5×10^{-3}
Iron	10.0×10^{-8}	5.0×10^{-3}
Platinum	11×10^{-8}	3.92×10^{-3}
Lead	22×10^{-8}	3.9×10^{-3}
Nichrome[a]	150×10^{-8}	0.4×10^{-3}
Carbon	3.5×10^{5}	-0.5×10^{-3}
Germanium	0.46	-48×10^{-3}
Silicon	640	-75×10^{-3}
Glass	$10^{10}-10^{14}$	
Hard rubber	$\approx 10^{13}$	
Sulfur	10^{15}	
Quartz (fused)	75×10^{16}	

[a]A nickel–chromium alloy commonly used in heating elements.

Applying Physics 17.1   Dimming of Aging Lightbulbs

As a lightbulb ages, why does it gives off less light than when new?

Explanation   There are two reasons for the lightbulb's behavior, one electrical and one optical, but both are related to the same phenomenon occurring within the bulb. The filament of an old lightbulb is made of a tungsten wire that has been kept at a high temperature for many hours. High temperatures evaporate tungsten from the filament, decreasing its radius. From $R = \rho l / A$, we see that a decreased cross-sectional area leads to an increase in the resistance of the filament.

This increasing resistance with age means that the filament will carry less current for the same applied voltage. With less current in the filament, there is less light output, and the filament glows more dimly.

At the high operating temperature of the filament, tungsten atoms leave its surface, much as water molecules evaporate from a puddle of water. The atoms are carried away by convection currents in the gas in the bulb and are deposited on the inner surface of the glass. In time, the glass becomes less transparent because of the tungsten coating, which decreases the amount of light that passes through the glass.

INTERACTIVE EXAMPLE 17.4   The Resistance of Nichrome Wire

Goal   Combine the concept of resistivity with Ohm's law.

Problem   (a) Calculate the resistance per unit length of a 22-gauge nichrome wire of radius 0.321 mm. (b) If a potential difference of 10.0 V is maintained across a 1.00-m length of the nichrome wire, what is the current in the wire? (c) The wire is melted down and recast with twice its original length. Find the new resistance R_N as a multiple of the old resistance R_O.

Strategy   Part (a) requires substitution into Equation 17.5, after calculating the cross-sectional area, while part (b) is a matter of substitution into Ohm's law. Part (c) requires some algebra. The idea is to take the expression for the new resistance and substitute expressions for l_N and A_N, the new length and cross-sectional area, in terms of the old length and cross-section. For the area substitution, use the fact that the volumes of the old and new wires are the same.

Solution

(a) Calculate the resistance per unit length.

Find the cross-sectional area of the wire:

$$A = \pi r^2 = \pi(0.321 \times 10^{-3} \text{ m})^2 = 3.24 \times 10^{-7} \text{ m}^2$$

Obtain the resistivity of nichrome from Table 17.1, solve Equation 17.5 for R/l, and substitute:

$$\frac{R}{l} = \frac{\rho}{A} = \frac{1.5 \times 10^{-6} \, \Omega \cdot \text{m}}{3.24 \times 10^{-7} \text{ m}^2} = \boxed{4.6 \, \Omega/\text{m}}$$

(b) Find the current in a 1.00-m segment of the wire if the potential difference across it is 10.0 V.

Substitute given values into Ohm's law:

$$I = \frac{\Delta V}{R} = \frac{10.0 \text{ V}}{4.6 \, \Omega} = \boxed{2.2 \text{ A}}$$

(c) If the wire is melted down and recast with twice its original length, find the new resistance as a multiple of the old.

Find the new area A_N in terms of the old area A_O, using the fact the volume doesn't change and $l_N = 2l_O$:

$$V_N = V_O \;\rightarrow\; A_N l_N = A_O l_O \;\rightarrow\; A_N = A_O(l_O/l_N)$$
$$A_N = A_O(l_O/2l_O) = A_O/2$$

Substitute into Equation 17.5:

$$R_N = \frac{\rho l_N}{A_N} = \frac{\rho(2l_O)}{(A_O/2)} = 4\frac{\rho l_O}{A_O} = \boxed{4R_O}$$

Remarks   From Table 17.1, the resistivity of nichrome is about 100 times that of copper, a typical good conductor. Therefore, a copper wire of the same radius would have a resistance per unit length of only 0.052 Ω/m, and a 1.00-m length of copper wire of the same radius would carry the same current (2.2 A) with an applied voltage of only 0.115 V.

  Because of its resistance to oxidation, nichrome is often used for heating elements in toasters, irons, and electric heaters.

Exercise 17.4
What is the resistance of a 6.0-m length of nichrome wire that has a radius 0.321 mm? How much current does it carry when connected to a 120-V source?

Answer   28 Ω; 4.3 A

Physics⊗Now™ You can explore the resistance of different materials by logging into PhysicsNow at **www.cp7e.com** and going to Interactive Example 17.4.

Quick Quiz 17.5

Suppose an electrical wire is replaced with one having every linear dimension doubled (i.e. the length and radius have twice their original values). Does the wire now have (a) more resistance than before, (b) less resistance, or (c) the same resistance?

17.6   TEMPERATURE VARIATION OF RESISTANCE

The resistivity ρ, and hence the resistance, of a conductor depends on a number of factors. One of the most important is the temperature of the metal. For most metals, resistivity increases with increasing temperature. This correlation can be understood as follows: as the temperature of the material increases, its constituent atoms vibrate with greater amplitudes. As a result, the electrons find it more diffi-cult to get by those atoms, just as it is more difficult to weave through a crowded

In an old-fashioned carbon filament incandescent lamp, the electrical resistance is typically 10 Ω, but changes with temperature.

room when the people are in motion than when they are standing still. The increased electron scattering with increasing temperature results in increased resistivity. Technically, thermal expansion also affects resistance; however, this is a very small effect.

Over a limited temperature range, the resistivity of most metals increases linearly with increasing temperature according to the expression

$$\rho = \rho_0[1 + \alpha(T - T_0)] \qquad\qquad \text{[17.6]}$$

where ρ is the resistivity at some temperature T (in Celsius degrees), ρ_0 is the resistivity at some reference temperature T_0 (usually taken to be 20°C), and α is a parameter called the **temperature coefficient of resistivity**. Temperature coefficients for various materials are provided in Table 17.1. The interesting negative values of α for semiconductors arise because these materials possess weakly bound charge carriers that become free to move and contribute to the current as the temperature rises.

Because the resistance of a conductor with a uniform cross section is proportional to the resistivity according to Equation 17.5 ($R = \rho l/A$), the temperature variation of resistance can be written

$$R = R_0[1 + \alpha(T - T_0)] \qquad\qquad \text{[17.7]}$$

Precise temperature measurements are often made using this property, as shown by the following example.

EXAMPLE 17.5   A Platinum Resistance Thermometer

Goal   Apply the temperature dependence of resistance.

Problem   A resistance thermometer, which measures temperature by measuring the change in resistance of a conductor, is made of platinum and has a resistance of 50.0 Ω at 20.0°C. **(a)** When the device is immersed in a vessel containing melting indium, its resistance increases to 76.8 Ω. From this information, find the melting point of indium. **(b)** The indium is heated further until it reaches a temperature of 235°C. What is the ratio of the new current in the platinum to the current I_{mp} at the melting point?

Strategy   In part (a), solve Equation 17.7 for $T - T_0$ and get α for platinum from Table 17.1, substituting known quantities. For part (b), use Ohm's law in Equation 17.7.

Solution

(a) Find the melting point of indium.

Solve Equation 17.7 for $T - T_0$:

$$T - T_0 = \frac{R - R_0}{\alpha R_0} = \frac{76.8\ \Omega - 50.0\ \Omega}{[3.92 \times 10^{-3}\ (°C)^{-1}][50.0\ \Omega]}$$
$$= 137°C$$

Substitute $T_0 = 20.0°C$ and obtain the melting point of indium:

$$T = \boxed{157°C}$$

(b) Find the ratio of the new current to the old when the temperature rises from 157°C to 235°C.

Write Equation 17.7, with R_0 and T_0 replaced by R_{mp} and T_{mp}, the resistance and temperature at the melting point.

$$R = R_{mp}[1 + \alpha(T - T_{mp})]$$

According to Ohm's law, $R = \Delta V/I$ and $R_{mp} = \Delta V/I_{mp}$. Substitute these expressions into Equation 17.7:

$$\frac{\Delta V}{I} = \frac{\Delta V}{I_{mp}}[1 + \alpha(T - T_{mp})]$$

Cancel the voltage differences, invert the two expressions, and then divide both sides by I_{mp}:

$$\frac{I}{I_{mp}} = \frac{1}{1 + \alpha(T - T_{mp})}$$

Substitute $T = 235°C$, $T_{mp} = 157°C$, and the value for α, obtaining the desired ratio:

$$\frac{I}{I_{mp}} = \boxed{0.766}$$

Exercise 17.5

Suppose a wire made of an unknown alloy and having a temperature of 20.0°C carries a current of 0.450 A. At 52.0°C the current is 0.370 A for the same potential difference. Find the temperature coefficient of resistivity of the alloy.

Answer 6.76×10^{-3} $(°C)^{-1}$

17.7 SUPERCONDUCTORS

There is a class of metals and compounds with resistances that fall virtually to *zero* below a certain temperature T_c called the *critical temperature*. These materials are known as **superconductors**. The resistance–temperature graph for a superconductor follows that of a normal metal at temperatures above T_c (Fig. 17.9). When the temperature is at or below T_c, however, the resistance suddenly drops to zero. This phenomenon was discovered in 1911 by the Dutch physicist H. Kamerlingh Onnes as he and a graduate student worked with mercury, which is a superconductor below 4.1 K. Recent measurements have shown that the resistivities of superconductors below T_c are less than 4×10^{-25} $\Omega \cdot m$—around 10^{17} times smaller than the resistivity of copper and in practice considered to be zero.

Today thousands of superconductors are known, including such common metals as aluminum, tin, lead, zinc, and indium. Table 17.2 lists the critical temperatures of several superconductors. The value of T_c is sensitive to chemical composition, pressure, and crystalline structure. Interestingly, copper, silver, and gold, which are excellent conductors, don't exhibit superconductivity.

One of the truly remarkable features of superconductors is the fact that once a current is set up in them, it persists *without any applied voltage* (because $R = 0$). In fact, steady currents in superconducting loops have been observed to persist for years with no apparent decay!

An important development in physics that created much excitement in the scientific community was the discovery of high-temperature copper-oxide-based superconductors. The excitement began with a 1986 publication by J. Georg Bednorz and K. Alex Müller, scientists at the IBM Zurich Research Laboratory in Switzerland, in which they reported evidence for superconductivity at a temperature near 30 K in an oxide of barium, lanthanum, and copper. Bednorz and Müller were awarded the Nobel Prize for physics in 1987 for their important discovery. The discovery was remarkable in view of the fact that the critical temperature was significantly higher than those of any previously known superconductors. Shortly thereafter a new family of compounds was investigated, and research activity in the field of superconductivity proceeded vigorously. In early 1987, groups at the University of Alabama at Huntsville and the University of Houston announced the discovery of superconductivity at about 92 K in an oxide of yttrium, barium, and copper ($YBa_2Cu_3O_7$), shown as the gray disk in Figure 17.10. Late in 1987, teams of scientists from Japan and the United States reported superconductivity at 105 K in an oxide of bismuth, strontium, calcium, and copper. More recently, scientists have reported superconductivity at temperatures as high as 150 K in an oxide containing mercury. The search for novel superconducting materials continues, with the hope of someday obtaining a room-temperature superconducting material. This research is important both for scientific reasons and for practical applications.

An important and useful application is the construction of superconducting magnets in which the magnetic field intensities are about ten times greater than those of the best normal electromagnets. Such magnets are being considered as a means of storing energy. The idea of using superconducting power lines to transmit

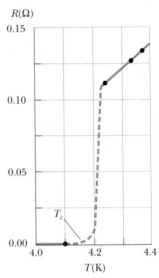

Figure 17.9 Resistance versus temperature for a sample of mercury (Hg). The graph follows that of a normal metal above the critical temperature T_c. The resistance drops to zero at the critical temperature, which is 4.2 K for mercury, and remains at zero for lower temperatures.

TABLE 17.2

Critical Temperatures for Various Superconductors

Material	T_c (K)
Zn	0.88
Al	1.19
Sn	3.72
Hg	4.15
Pb	7.18
Nb	9.46
Nb_3Sn	18.05
Nb_3Ge	23.2
$YBa_2Cu_3O_7$	90
Bi–Sr–Ca–Cu–O	105
Tl–Ba–Ca–Cu–O	125
$HgBa_2Ca_2Cu_3O_8$	134

Figure 17.10 A small permanent magnet floats freely above a ceramic disk made of the superconductor $YBa_2Cu_3O_7$, cooled by liquid nitrogen at 77 K. The superconductor has zero electric resistance at temperatures below 92 K and expels any applied magnetic field.

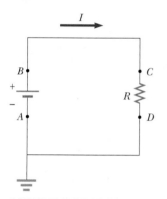

ACTIVE FIGURE 17.11
A circuit consisting of a battery and a resistance R. Positive charge flows clockwise from the positive to the negative terminal of the battery. Point A is grounded.

Physics Now™
Log into PhysicsNow at **www.cp7e.com** and go to Active Figure 17.11, where you can adjust the battery voltage and the resistance, and see the resulting current in the circuit and the power dissipated as heat by the resistor.

power efficiently is also receiving serious consideration. Modern superconducting electronic devices consisting of two thin-film superconductors separated by a thin insulator have been constructed. Among these devices are magnetometers (magnetic-field measuring devices) and various microwave devices.

17.8 ELECTRICAL ENERGY AND POWER

If a battery is used to establish an electric current in a conductor, chemical energy stored in the battery is continuously transformed into kinetic energy of the charge carriers. This kinetic energy is quickly lost as a result of collisions between the charge carriers and fixed atoms in the conductor, causing an increase in the temperature of the conductor. In this way, the chemical energy stored in the battery is continuously transformed into thermal energy.

In order to understand the process of energy transfer in a simple circuit, consider a battery with terminals connected to a resistor (Active Fig. 17.11; remember that the positive terminal of the battery is always at the higher potential). Now imagine following a quantity of positive charge ΔQ around the circuit from point A, through the battery and resistor, and back to A. Point A is a reference point that is grounded (the ground symbol is ⏚), and its potential is taken to be zero. As the charge ΔQ moves from A to B through the battery, the electrical potential energy of the system increases by the amount $\Delta Q \, \Delta V$, and the chemical potential energy in the battery decreases by the same amount. (Recall from Chapter 16 that $\Delta PE = q \, \Delta V$.) However, as the charge moves from C to D through the resistor, it loses this electrical potential energy during collisions with atoms in the resistor. In the process, the energy is transformed to internal energy corresponding to increased vibrational motion of those atoms. Because we can ignore the very small resistance of the interconnecting wires, no energy transformation occurs for paths BC and DA. When the charge returns to point A, the net result is that some of the chemical energy in the battery has been delivered to the resistor and has caused its temperature to rise.

The charge ΔQ loses energy $\Delta Q \, \Delta V$ as it passes through the resistor. If Δt is the time it takes the charge to pass through the resistor, then the rate at which it loses electric potential energy is

$$\frac{\Delta Q}{\Delta t} \Delta V = I \Delta V$$

where I is the current in the resistor and ΔV is the potential difference across it. Of course, the charge regains this energy when it passes through the battery, at the expense of chemical energy in the battery. The rate at which the system loses potential energy as the charge passes through the resistor is equal to the rate at which the system gains internal energy in the resistor. Therefore, the power $\mathcal{P}$, representing the rate at which energy is delivered to the resistor, is

Power ▶
$$\mathcal{P} = I \Delta V \qquad \qquad \textbf{[17.8]}$$

While this result was developed by considering a battery delivering energy to a resistor, Equation 17.8 can be used to determine the power transferred from a voltage source to *any* device carrying a current I and having a potential difference ΔV between its terminals.

Using Equation 17.8 and the fact that $\Delta V = IR$ for a resistor, we can express the power delivered to the resistor in the alternate forms

Power delivered to a resistor ▶
$$\mathcal{P} = I^2 R = \frac{\Delta V^2}{R} \qquad \qquad \textbf{[17.9]}$$

When I is in amperes, ΔV in volts, and R in ohms, the SI unit of power is the watt (introduced in Chapter 5). The power delivered to a conductor of resistance R is often referred to as an I^2R *loss*. Note that Equation 17.9 applies only to resistors and not to nonohmic devices such as lightbulbs and diodes.

Regardless of the ways in which you use electrical energy in your home, you ultimately must pay for it or risk having your power turned off. The unit of energy used by electric companies to calculate consumption, the **kilowatt-hour**, is defined in terms of the unit of power and the amount of time it's supplied. One kilowatt-hour (kWh) is the energy converted or consumed in 1 h at the constant rate of 1 kW. It has the numerical value

$$1 \text{ kWh} = (10^3 \text{ W})(3600 \text{ s}) = 3.60 \times 10^6 \text{ J} \qquad \text{[17.10]}$$

On an electric bill, the amount of electricity used in a given period is usually stated in multiples of kilowatt-hours.

Applying Physics 17.2   Lightbulb Failures

Why do lightbulbs fail so often right after they're turned on?

Explanation Once the switch is closed, the line voltage is applied across the bulb. As the voltage is applied across the cold filament when the bulb is first turned on, the resistance of the filament is low, the current is high, and a relatively large amount of power is delivered to the bulb. This current spike at the beginning of operation is the reason why lightbulbs often fail just after they are turned on. As the filament warms, its resistance rises and the current decreases. As a result, the power delivered to the bulb decreases, and the bulb is less likely to burn out.

Quick Quiz 17.6

A voltage ΔV is applied across the ends of a nichrome heater wire having a cross-sectional area A and length L. The same voltage is applied across the ends of a second heater wire having a cross-sectional area A and length $2L$. Which wire gets hotter? (a) the shorter wire, (b) the longer wire, or (c) more information is needed.

Quick Quiz 17.7

For the two resistors shown in Figure 17.12, rank the currents at points a through f from largest to smallest.

(a) $I_a = I_b > I_e = I_f > I_c = I_d$
(b) $I_a = I_b > I_c = I_d > I_e = I_f$
(c) $I_e = I_f > I_c = I_d > I_a = I_b$

Figure 17.12 (Quick Quiz 17.7)

Quick Quiz 17.8

Two resistors, A and B, are connected in a series circuit with a battery. The resistance of A is twice that of B. Which resistor dissipates more power? (a) resistor A (b) resistor B (c) More information is needed.

Example 17.6   The Cost of Lighting Up Your Life

Goal   Apply the electric power concept, and calculate the cost of power usage using kilowatt-hours.

Problem   A circuit provides a maximum current of 20.0 A at an operating voltage of 1.20×10^2 V. **(a)** How many 75 W bulbs can operate with this voltage source? **(b)** At \$0.120 per kilowatt-hour, how much does it cost to operate these bulbs for 8.00 h?

Strategy   Find the necessary power with $\mathscr{P} = I\Delta V$, then divide by 75.0 W per bulb to get the total number of bulbs. To find the cost, convert power to kilowatts and multiply by the number of hours, then multiply by the cost per kilowatt-hour.

Solution
(a) Find the number of bulbs that can be lighted.

Substitute into Equation 17.8 to get the total power:

$$\mathscr{P}_{total} = I\Delta V = (20.0 \text{ A})(1.20 \times 10^2 \text{ V}) = 2.40 \times 10^3 \text{ W}$$

Divide the total power by the power per bulb to get the number of bulbs.

$$\text{Number of bulbs} = \frac{\mathscr{P}_{total}}{\mathscr{P}_{bulb}} = \frac{2.40 \times 10^3 \text{ W}}{75.0 \text{ W}} = \boxed{32.0}$$

(b) Calculate the cost of this electricity for an 8.00-h day.

Find the energy in kilowatt-hours:

$$\text{Energy} = \mathscr{P}t = (2.40 \times 10^3 \text{ W})\left(\frac{1.00 \text{ kW}}{1.00 \times 10^3 \text{ W}}\right)(8.00 \text{ h})$$

$$= 19.2 \text{ kWh}$$

Multiply by the cost per kilowatt-hour:

$$\text{Cost} = (19.2 \text{ kWh})(\$0.12/\text{kWh}) = \boxed{\$2.30}$$

Remarks   This amount of energy might correspond to what a small office uses in a working day, taking into account all power requirements (not just lighting). In general, resistive devices can have variable power output, depending on how the circuit is wired. Here, power outputs were specified, so such considerations were unnecessary.

Exercise 17.6
(a) How many Christmas tree lights drawing 5.00 W of power each could be run on a circuit operating at 1.20×10^2 V and providing 15.0 A of current? **(b)** Find the cost to operate one such string 24.0 h per day for the Christmas season (two weeks), using the rate \$0.12/kWh.

Answers   (a) 3.60×10^2 bulbs   (b) \$72.60

EXAMPLE 17.7   The Power Converted by an Electric Heater

Goal   Calculate an electrical power output, and link to its effect on the environment through the first law of thermodynamics.

Problem   An electric heater is operated by applying a potential difference of 50.0 V to a nichrome wire of total resistance 8.00 Ω. **(a)** Find the current carried by the wire and the power rating of the heater. **(b)** Using this heater, how long would it take to heat 2.50×10^3 moles of diatomic gas (e.g., a mixture of oxygen and nitrogen—air) from a chilly 10.0°C to 25.0°C? Take the molar specific heat at constant volume of air to be $\frac{5}{2}R$.

Strategy   For part (a), find the current with Ohm's law and substitute into the expression for power. Part (b) is an isovolumetric process, so the thermal energy provided by the heater all goes into the change in internal energy, ΔU. Calculate this quantity using the first law of thermodynamics, and divide by the power to get the time.

Solution
(a) Compute the current and power output.

Apply Ohm's law to get the current:

$$I = \frac{\Delta V}{R} = \frac{50.0 \text{ V}}{8.00 \text{ Ω}} = \boxed{6.25 \text{ A}}$$

Substitute into Equation 17.9 to find the power: $\mathscr{P} = I^2R = (6.25 \text{ A})^2(8.00 \text{ }\Omega) = \boxed{313 \text{ W}}$

(b) How long does it take to heat the gas?

Calculate the thermal energy transfer from the first law. Note that $W = 0$ because the volume doesn't change.

$$Q = \Delta U = nC_v\Delta T$$
$$= (2.50 \times 10^3 \text{ mol})(\tfrac{5}{2} \cdot 8.31 \text{ J/mol} \cdot \text{K})(298 \text{ K} - 283 \text{ K})$$
$$= 7.79 \times 10^5 \text{ J}$$

Divide the thermal energy by the power, to get the time: $t = \dfrac{Q}{\mathscr{P}} = \dfrac{7.79 \times 10^5 \text{ J}}{313 \text{ W}} = \boxed{2.49 \times 10^3 \text{ s}}$

Remarks   The number of moles of gas given here is approximately what would be found in a bedroom. Warming the air with this space heater requires only about forty minutes. However, the calculation doesn't take into account conduction losses. Recall that a 20-cm-thick concrete wall, as calculated in Chapter 11, permitted the loss of over two megajoules an hour by conduction!

Exercise 17.7
A hot-water heater is rated at 4.50×10^3 W and operates at 2.40×10^2 V. (a) Find the resistance in the heating element, and the current. (b) How long does it take to heat 125 L of water from 20.0°C to 50.0°C, neglecting conduction and other losses?

Answers   (a) 12.8 Ω, 18.8 A   (b) 3.49×10^3 s

17.9   ELECTRICAL ACTIVITY IN THE HEART
Electrocardiograms

Every action involving the body's muscles is initiated by electrical activity. The voltages produced by muscular action in the heart are particularly important to physicians. Voltage pulses cause the heart to beat, and the waves of electrical excitation that sweep across the heart associated with the heartbeat are conducted through the body via the body fluids. These voltage pulses are large enough to be detected by suitable monitoring equipment attached to the skin. A sensitive voltmeter making good electrical contact with the skin by means of contacts attached with conducting paste can be used to measure heart pulses, which are typically of the order of 1 mV at the surface of the body. The voltage pulses can be recorded on an instrument called an **electrocardiograph**, and the pattern recorded by this instrument is called an **electrocardiogram** (EKG). In order to understand the information contained in an EKG pattern, it is necessary first to describe the underlying principles concerning electrical activity in the heart.

The right atrium of the heart contains a specialized set of muscle fibers called the SA (sinoatrial) node that initiates the heartbeat (Fig. 17.13). Electric impulses that originate in these fibers gradually spread from cell to cell throughout the right and left atrial muscles, causing them to contract. The pulse that passes through the muscle cells is often called a *depolarization wave* because of its effect on individual cells. If an individual muscle cell were examined in its resting state, a double-layer electric charge distribution would be found on its surface, as shown in Figure 17.14a (page 584). The impulse generated by the SA node momentarily and locally allows positive charge on the outside of the cell to flow in and neutralize the negative charge on the inside layer. This effect changes the cell's charge distribution to that shown in Figure 17.14b. Once the depolarization wave has passed through an individual heart muscle cell, the cell recovers the resting-state charge distribution (positive out, negative in) shown in Figure 17.14a in about 250 ms. When the impulse reaches the atrioventricular (AV) node (Fig. 17.13), the muscles of the atria begin to relax, and the pulse is directed to the ventricular

APPLICATION
 Electrocardiograms

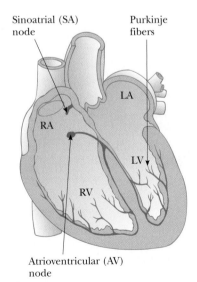

Sinoatrial (SA) node

Purkinje fibers

LA

RA

LV

RV

Atrioventricular (AV) node

Figure 17.13   The electrical conduction system of the human heart. (RA: right atrium; LA: left atrium; RV: right ventricle; LV: left ventricle.)

Figure 17.14 (a) Charge distribution of a muscle cell in the atrium before a depolarization wave has passed through the cell. (b) Charge distribution as the wave passes.

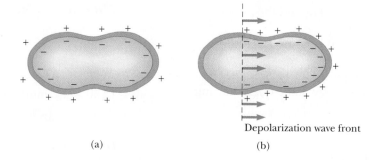

Depolarization wave front

(a)   (b)

Figure 17.15 An EKG response for a normal heart.

APPLICATION

 Cardiac Pacemakers

muscles by the AV node. The muscles of the ventricles contract as the depolarization wave spreads through the ventricles along a group of fibers called the *Purkinje fibers*. The ventricles then relax after the pulse has passed through. At this point, the SA node is again triggered and the cycle is repeated.

A sketch of the electrical activity registered on an EKG for one beat of a normal heart is shown in Figure 17.15. The pulse indicated by *P* occurs just before the atria begin to contract. The *QRS* pulse occurs in the ventricles just before they contract, and the *T* pulse occurs when the cells in the ventricles begin to recover. EKGs for an abnormal heart are shown in Figure 17.16. The *QRS* portion of the pattern shown in Figure 17.16a is wider than normal, indicating that the patient may have an enlarged heart. (Why?) Figure 17.16b indicates that there is no constant relationship between the *P* pulse and the *QRS* pulse. This suggests a blockage in the electrical conduction path between the SA and AV nodes which results in the atria and ventricles beating independently and inefficient heart pumping. Finally, Figure 17.16c shows a situation in which there is no *P* pulse and an irregular spacing between the *QRS* pulses. This is symptomatic of irregular atrial contraction, which is called *fibrillation*. In this condition, the atrial and ventricular contractions are irregular.

As noted previously, the sinoatrial node directs the heart to beat at the appropriate rate, usually about 72 beats per minute. However, disease or the aging process can damage the heart and slow its beating, and a medical assist may be necessary in the form of a *cardiac pacemaker* attached to the heart. This matchbox-sized electrical device implanted under the skin has a lead that is connected to the wall of the right ventricle. Pulses from this lead stimulate the heart to maintain its proper rhythm. In general, a pacemaker is designed to produce pulses at a rate of about 60 per minute, slightly slower than the normal number of beats per minute,

Figure 17.16 Abnormal EKGs.

but sufficient to maintain life. The circuitry basically consists of a capacitor charging up to a certain voltage from a lithium battery and then discharging. The design of the circuit is such that, if the heart is beating normally, the capacitor is not allowed to charge completely and send pulses to the heart.

An Emergency Room in Your Chest

In June 2001, an operation on Vice President Dick Cheney focused attention on the progress in treating heart problems with tiny implanted electrical devices. Aptly termed "an emergency room in your chest" by Cheney's attending physician, devices called *Implanted Cardioverter Defibrillators* (**ICD's**) can monitor, record, and logically process heart signals and then supply different corrective signals to hearts beating too slowly, too rapidly, or irregularly. ICD's can even monitor and send signals to the atria and ventricles independently! Figure 17.17a shows a sketch of an ICD with conducting leads that are implanted in the heart. Figure 17.17b shows an actual titanium-encapsulated dual-chamber ICD.

The latest ICD's are sophisticated devices capable of a number of functions:

1. monitoring both atrial and ventricular chambers to differentiate between atrial and potentially fatal ventricular arrhythmias, which require prompt regulation;
2. storing about a half hour of heart signals that can easily be read out by a physician;
3. being easily reprogrammed with an external magnetic wand;
4. performing complicated signal analysis and comparison;
5. supplying either 0.25- to 10-V repetitive pacing signals to speed up or slow down a malfunctioning heart, or a high-voltage pulse of about 800 V to halt the potentially fatal condition of ventricular fibrillation, in which the heart quivers rapidly rather than beats (people who have experienced such a high-voltage jolt say that it feels like a kick or a bomb going off in the chest);
6. automatically adjusting the number of pacing pulses per minute to match the patient's activity.

ICD's are powered by lithium batteries and have implanted lifetimes of 4–6 years. Some basic properties of these adjustable ICD's are given in Table 17.3 (page 586). In the table, *tachycardia* means "rapid heartbeat" and *bradycardia*

(a)   (b)

FIGURE 17.17   (a) A dual-chamber ICD with leads in the heart. One lead monitors and stimulates the right atrium, and the other monitors and stimulates the right ventricle. (b) Medtronic Dual Chamber ICD.

TABLE 17.3

Properties of Implanted Cardioverter Defibrillators[a]

Physical Specifications	
Mass (g)	85
Size (cm)	7.3 × 6.2 × 1.3 (about five stacked silver dollars)
Antitachycardia Pacing	ICD delivers a burst of critically timed low-energy pulses
Number of Bursts	1–15
Burst Cycle Length (ms)	200–552
Number of Pulses per Burst	2–20
Pulse Amplitude (V)	7.5 or 10
Pulse Width (ms)	1.0 or 1.9
High-Voltage Defibrillation	
Pulse energy (J)	37 stored/33 delivered
Pulse Amplitude (V)	801
Bradycardia Pacing	A dual-chamber ICD can steadily deliver repetitive pulses to both the atrium and the ventricle
Base Frequency (beats/minute)	40–100
Pulse Amplitude (V)	0.25–7.5
Pulse Width (ms)	0.05, 0.1–1.5, 1.9

[a]For more information see **www.photonicd.com/specs.html**.

means "slow heartbeat." A key factor in developing tiny electrical implants that serve as defibrillators is the development of capacitors with relatively large capacitance (125 μf) and small physical size.

SUMMARY

Physics⊗Now™ Take a practice test by logging into Physics-Now at **www.cp7e.com** and clicking on the Pre-Test link for this chapter.

17.1 Electric Current

The **electric current** I in a conductor is defined as

$$I \equiv \frac{\Delta Q}{\Delta t} \qquad [17.1]$$

where ΔQ is the charge that passes through a cross section of the conductor in time Δt. The SI unit of current is the **ampere** (A); 1 A = 1 C/s. By convention, the direction of current is the direction of flow of positive charge.

17.2 A Microscopic View: Current and Drift Speed

The current in a conductor is related to the motion of the charge carriers by

$$I = nqv_dA \qquad [17.2]$$

where n is the number of mobile charge carriers per unit volume, q is the charge on each carrier, v_d is the drift speed

of the charges, and A is the cross-sectional area of the conductor.

17.4 Resistance and Ohm's Law

The **resistance** R of a conductor is defined as the ratio of the potential difference across the conductor to the current in it:

$$R \equiv \frac{\Delta V}{I} \qquad [17.3]$$

The SI units of resistance are volts per ampere, or **ohms** (Ω); 1 Ω = 1 V/A.

Ohm's law describes many conductors, for which the applied voltage is directly proportional to the current it causes. The proportionality constant is the resistance:

$$\Delta V = IR \qquad [17.4]$$

17.5 Resistivity

If a conductor has length l and cross-sectional area A, its **resistance** is

$$R = \rho \frac{l}{A} \qquad [17.5]$$

where ρ, is an intrinsic property of the conductor called the **electrical resistivity**. The SI unit of resistivity is the **ohm-meter** ($\Omega \cdot m$).

17.6 Temperature Variation of Resistance

Over a limited temperature range, the resistivity of a conductor varies with temperature according to the expression

$$\rho = \rho_0[1 + \alpha(T - T_0)] \qquad [17.6]$$

where α is the **temperature coefficient of resistivity** and ρ_0 is the resistivity at some reference temperature T_0 (usually taken to be 20°C).

The resistance of a conductor varies with temperature according to the expression

$$R = R_0[1 + \alpha(T - T_0)] \qquad [17.7]$$

17.8 Electrical Energy and Power

If a potential difference ΔV is maintained across an electrical device, the **power**, or rate at which energy is supplied to the device, is

$$\mathcal{P} = I\,\Delta V \qquad [17.8]$$

Because the potential difference across a resistor is $\Delta V = IR$, the **power delivered to a resistor** can be expressed as

$$\mathcal{P} = I^2R = \frac{\Delta V^2}{R} \qquad [17.9]$$

A **kilowatt-hour** is the amount of energy converted or consumed in one hour by a device supplied with power at the rate of 1 kW. This is equivalent to

$$1\ kWh = 3.60 \times 10^6\ J \qquad [17.10]$$

CONCEPTUAL QUESTIONS

1. Car batteries are often rated in ampere-hours. Does this unit designate the amount of current, power, energy, or charge that can be drawn from the battery?

2. We have seen that an electric field must exist inside a conductor that carries a current. How is that possible in view of the fact that in electrostatics we concluded that the electric field must be zero inside a conductor?

3. Why don't the free electrons in a metal fall to the bottom of the metal due to gravity? And charges in a conductor are supposed to reside on the surface—why don't the free electrons all go to the surface?

4. In an analogy between traffic flow and electrical current, what would correspond to the charge Q? What would correspond to the current I?

5. Newspaper articles often have statements such as "10 000 volts of electricity surged *through* the victim's body." What is wrong with this statement?

6. Two lightbulbs are each connected to a voltage of 120 V. One has a power of 25 W, the other 100 W. Which bulb has the higher resistance? Which bulb carries more current?

7. When the voltage across a certain conductor is doubled, the current is observed to triple. What can you conclude about the conductor?

8. There is an old admonition given to experimenters to "keep one hand in the pocket" when working around high voltages. Why is this warning a good idea?

9. What factors affect the resistance of a conductor?

10. Some homes have light dimmers that are operated by rotating a knob. What is being changed in the electric circuit when the knob is rotated?

11. Two wires A and B with circular cross section are made of the same metal and have equal lengths, but the resistance of wire A is three times greater than that of wire B. What is the ratio of their cross-sectional areas? How do the radii compare?

12. What single experimental requirement makes superconducting devices expensive to operate? In principle, can this limitation be overcome?

13. What could happen to the drift velocity of the electrons in a wire and to the current in the wire if the electrons could move through it freely without resistance?

14. Use the atomic theory of matter to explain why the resistance of a material should increase as its temperature increases.

15. When is more power delivered to a lightbulb, just after it is turned on and the glow of the filament is increasing or after it has been on for a few seconds and the glow is steady?

PROBLEMS

1, 2, 3 = straightforward, intermediate, challenging   ☐ = full solution available in *Student Solutions Manual/Study Guide*

Physics⊗Now™ = coached problem with hints available at **www.cp7e.com**   = biomedical application

Section 17.1 Electric Current
Section 17.2 A Microscopic View: Current and Drift Speed

1. If a current of 80.0 mA exists in a metal wire, how many electrons flow past a given cross section of the wire in 10.0 min? Sketch the direction of the current and the direction of the electrons' motion.

2. A certain conductor has 7.50×10^{28} free electrons per cubic meter, a cross-sectional area of 4.00×10^{-6} m^2, and carries a current of 2.50 A. Find the drift speed of the electrons in the conductor.

3. A 1.00-V potential difference is maintained across a 10.0-Ω resistor for a period of 20.0 s. What total charge passes through the wire in this time interval?

4. In a particular television picture tube, the measured beam current is 60.0 μA. How many electrons strike the screen every second?

5. In the Bohr model of the hydrogen atom, an electron in the lowest energy state moves at a speed of 2.19×10^6 m/s in a circular path having a radius of 5.29×10^{-11} m. What is the effective current associated with this orbiting electron?

6. If 3.25×10^{-3} kg of gold is deposited on the negative electrode of an electrolytic cell in a period of 2.78 h, what is the current in the cell during that period? Assume that the gold ions carry one elementary unit of positive charge.

7. A 200-km-long high-voltage transmission line 2.0 cm in diameter carries a steady current of 1 000 A. If the conductor is copper with a free charge density of 8.5×10^{28} electrons per cubic meter, how many years does it take one electron to travel the full length of the cable?

8. **Physics⊗Now™** An aluminum wire carrying a current of 5.0 A has a cross-sectional area of 4.0×10^{-6} m². Find the drift speed of the electrons in the wire. The density of aluminum is 2.7 g/cm³. (Assume that one electron is supplied by each atom.)

9. If the current carried by a conductor is doubled, what happens to (a) the charge carrier density? (b) the electron drift velocity?

Section 17.4 Resistance and Ohm's Law
Section 17.5 Resistivity

10. A lightbulb has a resistance of 240 Ω when operating at a voltage of 120 V. What is the current in the bulb?

11. A person notices a mild shock if the current along a path through the thumb and index finger exceeds 80 µA. Compare the maximum possible voltage without shock across the thumb and index finger with a dry-skin resistance of 4.0×10^5 Ω and a wet-skin resistance of 2 000 Ω.

12. Suppose that you wish to fabricate a uniform wire out of 1.00 g of copper. If the wire is to have a resistance $R = 0.500$ Ω, and if all of the copper is to be used, what will be (a) the length and (b) the diameter of the wire?

13. Calculate the diameter of a 2.0-cm length of tungsten filament in a small lightbulb if its resistance is 0.050 Ω.

14. Eighteen-gauge wire has a diameter of 1.024 mm. Calculate the resistance of 15 m of 18-gauge copper wire at 20°C.

15. A potential difference of 12 V is found to produce a current of 0.40 A in a 3.2-m length of wire with a uniform radius of 0.40 cm. What is (a) the resistance of the wire? (b) the resistivity of the wire?

16. Make an order-of-magnitude estimate of the cost of one person's routine use of a hair dryer for 1 yr. If you do not use a blow dryer yourself, observe or interview someone who does. State the quantities you estimate and their values.

17. A wire 50.0 m long and 2.00 mm in diameter is connected to a source with a potential difference of 9.11 V, and the current is found to be 36.0 A. Assume a temperature of 20°C, and, using Table 17.1, identify the metal out of which the wire is made.

18. A rectangular block of copper has sides of length 10 cm, 20 cm, and 40 cm. If the block is connected to a 6.0-V source across two of its opposite faces, what are (a) the

maximum current and (b) the minimum current that the block can carry?

19. A wire of initial length L_0 and radius r_0 has a measured resistance of 1.0 Ω. The wire is drawn under tensile stress to a new uniform radius of $r = 0.25r_0$. What is the new resistance of the wire?

Section 17.6 Temperature Variation of Resistance

20. A certain lightbulb has a tungsten filament with a resistance of 19 Ω when cold and 140 Ω when hot. Assume that Equation 17.7 can be used over the large temperature range involved here, and find the temperature of the filament when it is hot. Assume an initial temperature of 20°C.

21. While taking photographs in Death Valley on a day when the temperature is 58.0°C, Bill Hiker finds that a certain voltage applied to a copper wire produces a current of 1.000 A. Bill then travels to Antarctica and applies the same voltage to the same wire. What current does he register there if the temperature is -88.0°C? Assume that no change occurs in the wire's shape and size.

22. A metal wire has a resistance of 10.00 Ω at a temperature of 20°C. If the same wire has a resistance of 10.55 Ω at 90°C, what is the resistance of the wire when its temperature is -20°C?

23. At 20°C, the carbon resistor in an electric circuit connected to a 5.0-V battery has a resistance of 200 Ω. What is the current in the circuit when the temperature of the carbon rises to 80°C?

24. A wire 3.00 m long and 0.450 mm² in cross-sectional area has a resistance of 41.0 Ω at 20°C. If its resistance increases to 41.4 Ω at 29.0°C, what is the temperature coefficient of resistivity?

25. The copper wire used in a house has a cross-sectional area of 3.00 mm². If 10.0 m of this wire is used to wire a circuit in the house at 20.0°C, find the resistance of the wire at temperatures of (a) 30.0°C and (b) 10.0°C.

26. A 100-cm-long copper wire of radius 0.50 cm has a potential difference across it sufficient to produce a current of 3.0 A at 20°C. (a) What is the potential difference? (b) If the temperature of the wire is increased to 200°C, what potential difference is now required to produce a current of 3.0 A?

27. **Physics⊗Now™** (a) A 34.5-m length of copper wire at 20.0°C has a radius of 0.25 mm. If a potential difference of 9.0 V is applied across the length of the wire, determine the current in the wire. (b) If the wire is heated to 30.0°C while the 9.0-V potential difference is maintained, what is the resulting current in the wire?

28. A toaster rated at 1 050 W operates on a 120-V household circuit and a 4.00-m length of nichrome wire as its heating element. The operating temperature of this element is 320°C. What is the cross-sectional area of the wire?

29. In one form of plethysmograph (a device for measuring volume), a rubber capillary tube with an inside diameter of 1.00 mm is filled with mercury at 20°C. The resistance of the mercury is measured with the aid of electrodes sealed into the ends of the tube. If 100.00 cm of the tube is wound in a spiral around a patient's upper arm, the blood flow during a heartbeat causes the arm to expand, stretching the tube to a length of 100.04 cm. From this observation, and

assuming cylindrical symmetry, you can find the change in volume of the arm, which gives an indication of blood flow. (a) Calculate the resistance of the mercury. (b) Calculate the fractional change in resistance during the heartbeat. [*Hint:* The fraction by which the cross-sectional area of the mercury thread decreases is the fraction by which the length increases, since the volume of mercury is constant.] Take $\rho_{Hg} = 9.4 \times 10^{-7}\ \Omega \cdot m$.

30. A platinum resistance thermometer has resistances of 200.0 Ω when placed in a 0°C ice bath and 253.8 Ω when immersed in a crucible containing melting potassium. What is the melting point of potassium? [*Hint:* First determine the resistance of the platinum resistance thermometer at room temperature, 20°C.]

Section 17.8 Electrical Energy and Power

31. A toaster is rated at 600 W when connected to a 120-V source. What current does the toaster carry, and what is its resistance?

32. If electrical energy costs 12 cents, or $0.12, per kilowatt-hour, how much does it cost to (a) burn a 100-W lightbulb for 24 h? (b) operate an electric oven for 5.0 h if it carries a current of 20.0 A at 220 V?

33. How many 100-W lightbulbs can you use in a 120-V circuit without tripping a 15-A circuit breaker? (The bulbs are connected in parallel, which means that the potential difference across each lightbulb is 120 V.)

34. A high-voltage transmission line with a resistance of 0.31 Ω/km carries a current of 1 000 A. The line is at a potential of 700 kV at the power station and carries the current to a city located 160 km from the station. (a) What is the power loss due to resistance in the line? (b) What fraction of the transmitted power does this loss represent?

35. The heating element of a coffeemaker operates at 120 V and carries a current of 2.00 A. Assuming that the water absorbs all of the energy converted by the resistor, calculate how long it takes to heat 0.500 kg of water from room temperature (23.0°C) to the boiling point.

36. The power supplied to a typical black-and-white television set is 90 W when the set is connected to 120 V. (a) How much electrical energy does this set consume in 1 hour? (b) A color television set draws about 2.5 A when connected to 120 V. How much time is required for it to consume the same energy as the black-and-white model consumes in 1 hour?

37. What is the required resistance of an immersion heater that will increase the temperature of 1.50 kg of water from 10.0°C to 50.0°C in 10.0 min while operating at 120 V?

38. A certain toaster has a heating element made of Nichrome resistance wire. When the toaster is first connected to a 120-V source of potential difference (and the wire is at a temperature of 20.0°C), the initial current is 1.80 A. However, the current begins to decrease as the resistive element warms up. When the toaster reaches its final operating temperature, the current has dropped to 1.53 A. (a) Find the power the toaster converts when it is at its operating temperature. (b) What is the final temperature of the heating element?

39. A copper cable is designed to carry a current of 300 A with a power loss of 2.00 W/m. What is the required radius of this cable?

40. A small motor draws a current of 1.75 A from a 120-V line. The output power of the motor is 0.20 hp. (a) At a rate of $0.060/kWh, what is the cost of operating the motor for 4.0 h? (b) What is the efficiency of the motor?

41. It has been estimated that there are 270 million plug-in electric clocks in the United States, approximately one clock for each person. The clocks convert energy at the average rate of 2.50 W. To supply this energy, how many metric tons of coal are burned per hour in coal-fired electric generating plants that are, on average, 25.0% efficient? The heat of combustion for coal is 33.0 MJ/kg.

42. The cost of electricity varies widely throughout the United States; $0.120/kWh is a typical value. At this unit price, calculate the cost of (a) leaving a 40.0-W porch light on for 2 weeks while you are on vacation, (b) making a piece of dark toast in 3.00 min with a 970-W toaster, and (c) drying a load of clothes in 40.0 min in a 5 200-W dryer.

43. How much does it cost to watch a complete 21-hour-long World Series on a 180-W television set? Assume that electricity costs $0.070/kWh.

44. A house is heated by a 24.0-kW electric furnace that uses resistance heating. The rate for electrical energy is $0.080/kWh. If the heating bill for January is $200, how long must the furnace have been running on an average January day?

45. An 11-W energy-efficient fluorescent lamp is designed to produce the same illumination as a conventional 40-W lamp. How much does the energy-efficient lamp save during 100 hours of use? Assume a cost of $0.080/kWh for electrical energy.

46. An office worker uses an immersion heater to warm 250 g of water in a light, covered, insulated cup from 20°C to 100°C in 4.00 minutes. The heater is a Nichrome resistance wire connected to a 120-V power supply. Assume that the wire is at 100°C throughout the 4.00-min time interval. Specify a diameter and a length that the wire can have. Can it be made from less than 0.5 cm³ of Nichrome?

47. Physics⊗Now™ The heating coil of a hot-water heater has a resistance of 20 Ω and operates at 210 V. If electrical energy costs $0.080/kWh, what does it cost to raise the 200 kg of water in the tank from 15°C to 80°C? (See Chapter 11.)

ADDITIONAL PROBLEMS

48. One lightbulb is marked "25 W 120 V," and another "100 W 120 V"; this means that each converts its respective power when plugged into a constant 120-V potential difference. (a) Find the resistance of each bulb. (b) How long does it take for 1.00 C to pass through the dim bulb? How is this charge different upon its exit from, versus its entry into, the bulb? (c) How long does it take for 1.00 J to pass through the dim bulb? How is this energy different upon its exit from, versus its entry into, the bulb? (d) Find the cost of running the dim bulb continuously for 30.0 days if the electric company sells its product at $0.070 0 per kWh. What physical quantity *does* the electric company sell? What is its price for one SI unit of this quantity?

49. A particular wire has a resistivity of $3.0 \times 10^{-8}\ \Omega \cdot m$ and a cross-sectional area of $4.0 \times 10^{-6}\ m^2$. A length of this wire is to be used as a resistor that will develop 48 W of power when connected across a 20-V battery. What length of wire is required?

50. A steam iron draws 6.0 A from a 120-V line. (a) How many joules of internal energy are produced in 20 min? (b) How much does it cost, at $0.080/kWh, to run the steam iron for 20 min?

51. An experiment is conducted to measure the electrical resistivity of Nichrome in the form of wires with different lengths and cross-sectional areas. For one set of measurements, a student uses 30-gauge wire, which has a cross-sectional area of 7.30×10^{-8} m². The student measures the potential difference across the wire and the current in the wire with a voltmeter and an ammeter, respectively. For each of the measurements given in the following table taken on wires of three different lengths, calculate the resistance of the wires and the corresponding value of the resistivity:

L (m)	ΔV (V)	I (A)	R (Ω)	ρ (Ω·m)
0.540	5.22	0.500		
1.028	5.82	0.276		
1.543	5.94	0.187		

What is the average value of the resistivity, and how does this value compare with the value given in Table 17.1?

52. Birds resting on high-voltage power lines are a common sight. The copper wire on which a bird stands is 2.2 cm in diameter and carries a current of 50 A. If the bird's feet are 4.0 cm apart, calculate the potential difference across its body.

53. You are cooking breakfast for yourself and a friend using a 1 200-W waffle iron and a 500-W coffeepot. Usually, you operate these appliances from a 110-V outlet for 0.500 h each day. (a) At 12 cents per kWh, how much do you spend to cook breakfast during a 30.0 day period? (b) You find yourself addicted to waffles and would like to upgrade to a 2 400-W waffle iron that will enable you to cook twice as many waffles during a half-hour period, but you know that the circuit breaker in your kitchen is a 20-A breaker. Can you do the upgrade?

54. The current in a conductor varies in time as shown in Figure P17.54. (a) How many coulombs of charge pass through a cross section of the conductor in the interval from $t = 0$ to $t = 5.0$ s? (b) What constant current would transport the same total charge during the 5.0-s interval as does the actual current?

Figure P17.54

55. An electric car is designed to run off a bank of 12.0-V batteries with a total energy storage of 2.00×10^7 J. (a) If the electric motor draws 8.00 kW, what is the current delivered to the motor? (b) If the electric motor draws 8.00 kW as the car moves at a steady speed of 20.0 m/s, how far will the car travel before it is "out of juice"?

56. (a) A 115-g mass of aluminum is formed into a right circular cylinder, shaped so that its diameter equals its height. Calculate the resistance between the top and bottom faces of the cylinder at 20°C. (b) Calculate the resistance between opposite faces if the same mass of aluminum is formed into a cube.

57. A length of metal wire has a radius of 5.00×10^{-3} m and a resistance of 0.100 Ω. When the potential difference across the wire is 15.0 V, the electron drift speed is found to be 3.17×10^{-4} m/s. On the basis of these data, calculate the density of free electrons in the wire.

58. A carbon wire and a Nichrome wire are connected one after the other. If the combination has a total resistance of 10.0 kΩ at 20°C, what is the resistance of each wire at 20°C so that the resistance of the combination does not change with temperature?

59. (a) Determine the resistance of a lightbulb marked 100 W @ 120 V. (b) Assuming that the filament is tungsten and has a cross-sectional area of 0.010 mm², determine the length of the wire inside the bulb when the bulb is operating. (c) Why do you think the wire inside the bulb is tightly coiled? (d) If the temperature of the tungsten wire is 2 600°C when the bulb is operating, what is the length of the wire after the bulb is turned off and has cooled to 20°C? (See Chapter 10, and use $4.5 \times 10^{-6}/°C$ as the coefficient of linear expansion for tungsten.)

60. In a certain stereo system, each speaker has a resistance of 4.00 Ω. The system is rated at 60.0 W in each channel. Each speaker circuit includes a fuse rated at a maximum current of 4.00 A. Is this system adequately protected against overload?

61. A resistor is constructed by forming a material of resistivity 3.5×10^5 Ω·m into the shape of a hollow cylinder of length 4.0 cm and inner and outer radii 0.50 cm and 1.2 cm, respectively. In use, a potential difference is applied between the ends of the cylinder, producing a current parallel to the length of the cylinder. Find the resistance of the cylinder.

62. The graph in Figure P17.62a shows the current I in a diode as a function of the potential difference ΔV across the diode. Figure P17.62b shows the circuit used to make the measurements. The symbol ▶— represents the diode. (a) Using Equation 17.3, make a table of the resistance of the diode for different values of ΔV in the range from −1.5 V to +1.0 V. (b) Based on your results, what amazing electrical property does a diode possess?

63. An x-ray tube used for cancer therapy operates at 4.0 MV, with a beam current of 25 mA striking the metal target. Nearly all the power in the beam is transferred to a stream of water flowing through holes drilled in the target. What rate of flow, in kilograms per second, is needed if the rise in temperature (ΔT) of the water is not to exceed 50°C?

64. A 50.0-g sample of a conducting material is all that is available. The resistivity of the material is measured to be 11×10^{-8} Ω·m, and the density is 7.86 g/cm³. The material is to be shaped into a solid cylindrical wire that has a total resistance of 1.5 Ω. (a) What length of wire is required? (b) What must be the diameter of the wire?

(a)

(b)

Figure P17.62

65. **Physics⊗Now™** (a) A sheet of copper ($\rho = 1.7 \times 10^{-8}$ $\Omega \cdot$ m) is 2.0 mm thick and has surface dimensions of 8.0 cm $\times$ 24 cm. If the long edges are joined to form a tube 24 cm in length, what is the resistance between the ends? (b) What mass of copper is required to manufacture a 1 500-m-long spool of copper cable with a total resistance of 4.5 Ω?

66. When a straight wire is heated, its resistance changes according to the equation

$$R = R_0[1 + \alpha(T - T_0)]$$

(Eq. 17.7), where α is the temperature coefficient of resistivity. (a) Show that a more precise result, which includes the fact that the length and area of a wire change when it is heated, is

$$R = \frac{R_0[1 + \alpha(T - T_0)][1 + \alpha'(T - T_0)]}{[1 + 2\alpha'(T - T_0)]}$$

where α' is the coefficient of linear expansion. (See Chapter 10.) (b) Compare the two results for a 2.00-m-long copper wire of radius 0.100 mm, starting at 20.0°C and heated to 100.0°C.

67. A man wishes to vacuum his car with a canister vacuum cleaner marked 535 W at 120 V. The car is parked far from the building, so he uses an extension cord 15.0 m long to plug the cleaner into a 120-V source. Assume that the cleaner has constant resistance. (a) If the resistance of each of the two conductors of the extension cord is 0.900 Ω, what is the actual power delivered to the cleaner? (b) If, instead, the power is to be at least 525 W, what must be the diameter of each of two identical copper conductors in the cord the young man buys? (c) Repeat part (b) if the power is to be at least 532 W. [*Suggestion*: A symbolic solution can simplify the calculations.]

ACTIVITIES

1. Connect one terminal of a D-cell battery to the base of a flashlight bulb with insulated wire, tape a second wire to the other battery terminal, and tape a third wire to the center conductor of the bulb, as in Figure A17.1. Make

Touch objects with these wires

Figure A17.1

sure to remove about 1 cm of insulation from the ends of all wires before making the connections. Now bridge the gap between the open wires with different objects, such as a plastic pen, an aluminum can, a penny, a rubber band, and a spoon. Which objects make the bulb light up? Explain your observations.

2. When the lightbulbs in your home are turned on, they are always connected across the same potential difference. Which do you believe has a filament with the highest resistance when cool, a 60-W bulb or a 100-W bulb? To check your prediction, ask your instructor to lend you a device called an ohmmeter and to instruct you in its use. A resistor must always be disconnected from a circuit when its resistance is measured with an ohmmeter.

3. Examine the labels on several appliances, such as a toaster, a television set, a lamp, a stereo system, an air conditioner, and a clock. From each label, determine the power rating of the device in watts. Check the billing statement from your electric utility company to find the cost of electrical energy per kilowatt-hour. (Prices usually range from about a nickel to 20 cents.) Calculate the cost of running each appliance for 1 h. Estimate how many hours per day each appliance is used. Then, on the basis of your daily estimate, calculate the monthly cost of using each appliance.

The complex circuits in modern electronic devices allow a highly sophisticated control of current, which in turn can be used to obtain, store, manipulate, and transmit data.

© Lester Lefkowitz/Corbis

Direct-Current Circuits

Batteries, resistors, and capacitors can be used in various combinations to construct electric circuits, which direct and control the flow of electricity and the energy it conveys. Such circuits make possible all the modern conveniences in a home—electric lights, electric stove tops and ovens, washing machines, and a host of other appliances and tools. Electric circuits are also found in our cars, in tractors that increase farming productivity, and in all types of medical equipment that saves so many lives every day.

In this chapter, we study and analyze a number of simple direct-current circuits. The analysis is simplified by the use of two rules known as Kirchhoff's rules, which follow from the principle of conservation of energy and the law of conservation of charge. Most of the circuits are assumed to be in *steady state*, which means that the currents are constant in magnitude and direction. We close the chapter with a discussion of circuits containing resistors and capacitors, in which current varies with time.

18.1 SOURCES OF EMF

A current is maintained in a closed circuit by a source of emf.[1] Among such sources are any devices (for example, batteries and generators) that increase the potential energy of the circulating charges. A source of emf can be thought of as a "charge pump" that forces electrons to move in a direction opposite the electrostatic field inside the source. The emf $\mathcal{E}$ of a source is the work done per unit charge; hence the SI unit of emf is the volt.

Consider the circuit in Active Figure 18.1a consisting of a battery connected to a resistor. We assume that the connecting wires have no resistance. If we neglect the internal resistance of the battery, the potential drop across the battery (the terminal voltage) equals the emf of the battery. Because a real battery always has some

---

[1]The term was originally an abbreviation for *electromotive force*, but emf is not really a force, so the long form is discouraged.

internal resistance r, however, the terminal voltage is not equal to the emf. The circuit of Active Figure 18.1a can be described schematically by the diagram in Active Figure 18.1b. The battery, represented by the dashed rectangle, consists of a source of emf $\mathcal{E}$ in series with an internal resistance r. Now imagine a positive charge moving through the battery from a to b in the figure. As the charge passes from the negative to the positive terminal of the battery, the potential of the charge increases by $\mathcal{E}$. As the charge moves through the resistance r, however, its potential decreases by the amount Ir, where I is the current in the circuit. The terminal voltage of the battery, $\Delta V = V_b - V_a$, is therefore given by

$$\Delta V = \mathcal{E} - Ir \qquad [18.1]$$

From this expression, we see that $\mathcal{E}$ **is equal to the terminal voltage when the current is zero**, called the **open-circuit voltage**. By inspecting Figure 18.1b, we find that the terminal voltage ΔV must also equal the potential difference across the external resistance R, often called the **load resistance**; that is, $\Delta V = IR$. Combining this relationship with Equation 18.1, we arrive at

$$\mathcal{E} = IR + Ir \qquad [18.2]$$

Solving for the current gives

$$I = \frac{\mathcal{E}}{R + r}$$

The preceding equation shows that the current in this simple circuit depends on both the resistance external to the battery and the internal resistance of the battery. If R is much greater than r, we can neglect r in our analysis (an option we usually select).

If we multiply Equation 18.2 by the current I, we get

$$I\mathcal{E} = I^2 R + I^2 r$$

This equation tells us that the total power output $I\mathcal{E}$ of the source of emf is converted at the rate $I^2 R$ at which energy is delivered to the load resistance, *plus* the rate $I^2 r$ at which energy is delivered to the internal resistance. Again, if $r << R$, most of the power delivered by the battery is transferred to the load resistance.

Unless otherwise stated, we will assume in our examples and end-of-chapter problems that the internal resistance of a battery in a circuit is negligible.

18.2   RESISTORS IN SERIES

When two or more resistors are connected end to end as in Active Figure 18.2, they are said to be in *series*. The resistors could be simple devices, such as lightbulbs or heating elements. When two resistors R_1 and R_2 are connected to a battery as in Active Figure 18.2, **the current is the same in the two resistors, because any charge that flows through R_1 must also flow through R_2**. This is analogous to water flowing through a pipe with two constrictions, corresponding to R_1 and R_2. Whatever volume of water flows in one end in a given time interval must exit the opposite end.

Because the potential difference between a and b in Active Figure 18.2b equals IR_1 and the potential difference between b and c equals IR_2, the potential difference between a and c is

$$\Delta V = IR_1 + IR_2 = I(R_1 + R_2)$$

Regardless of how many resistors we have in series, the sum of the potential differences across the resistors is equal to the total potential difference across the combination. As we will show later, this is a consequence of the conservation of energy. Active Figure 18.2c shows an equivalent resistor R_{eq} that can replace the two resistors of the original circuit. The equivalent resistor has the same effect on the circuit because it results in the same current in the circuit as the two resistors. Applying Ohm's law to this equivalent resistor, we have

$$\Delta V = IR_{eq}$$

Battery

Resistor

(a)

(b)

ACTIVE FIGURE 18.1

(a) A circuit consisting of a resistor connected to the terminals of a battery. (b) A circuit diagram of a source of emf $\mathcal{E}$ having internal resistance r connected to an external resistor R.

Physics⊗Now™

Log into PhysicsNow at **www.cp7e.com** and go to Active Figure 18.1, where you can adjust the emf and the resistances r and R, and see the effect on the current in part (b).

An assortment of batteries.

TIP 18.1   What's Constant in a Battery?

Equation 18.2 shows that the current in a circuit depends on the resistance of the battery, so a battery can't be considered a source of constant current. Even the terminal voltage of a battery given by Equation 18.1 can't be considered constant, because the internal resistance can change (due to warming, for example, during the operation of the battery). A battery is, however, a source of constant emf.

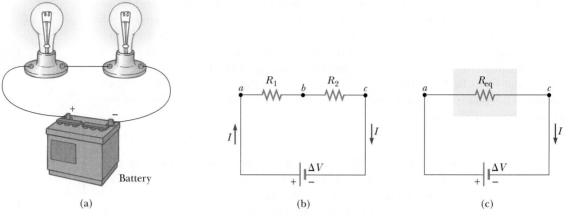

(a)   (b)   (c)

ACTIVE FIGURE 18.2
A series connection of two resistors, R_1 and R_2. The currents in the resistors are the same, and the equivalent resistance of the combination is given by $R_{eq} = R_1 + R_2$.

Physics⊗Now™
Log into PhysicsNow at **www.cp7e.com** and go to Active Figure 18.2, where you can adjust the battery voltage and resistances R_1 and R_2, observing the effect on the current and voltages of the individual resistors.

Equating the preceding two expressions, we have

$$IR_{eq} = I(R_1 + R_2)$$

or

$$R_{eq} = R_1 + R_2 \quad \text{(series combination)} \qquad [18.3]$$

An extension of the preceding analysis shows that the equivalent resistance of three or more resistors connected in series is

◀ Equivalent resistance of a series combination of resistors

$$R_{eq} = R_1 + R_2 + R_3 + \cdots \qquad [18.4]$$

Therefore, **the equivalent resistance of a series combination of resistors is the algebraic sum of the individual resistances and is always greater than any individual resistance**.

Note that if the filament of one lightbulb in Active Figure 18.2 were to fail, the circuit would no longer be complete (an open-circuit condition would exist) and the second bulb would also go out.

Applying Physics 18.1 Christmas Lights in Series

A new design for Christmas tree lights allows them to be connected in series. A failed bulb in such a string would result in an open circuit, and all of the bulbs would go out. How can the bulbs be redesigned to prevent this from happening?

Explanation If the string of lights contained the usual kind of bulbs, a failed bulb would be hard to locate. Each bulb would have to be replaced with a good bulb, one by one, until the failed bulb was found. If there happened to be two or more failed bulbs in the string of lights, finding them would be a lengthy and annoying task.

Filament

Jumper

Glass insulator

Figure 18.3 (Applying Physics 18.1) Diagram of a modern miniature holiday lightbulb, with a jumper connection to provide a current if the filament breaks.

Christmas lights use special bulbs that have an insulated loop of wire (a jumper) across the conducting supports to the bulb filaments (Fig. 18.3). If the filament breaks and the bulb fails, the bulb's resistance increases dramatically. As a result, most of the applied voltage appears across the loop of wire. This voltage causes the insulation around the loop of wire to burn, causing the metal wire to make electrical contact with the supports. This produces a conducting path through the bulb, so the other bulbs remain lit.

Quick Quiz 18.1

When a piece of wire is used to connect points b and c in Figure 18.2b, the brightness of bulb R_1 (a) increases, (b) decreases but remains lit, (c) stays the same, (d) goes out. The brightness of bulb R_2 (a) increases, (b) decreases but remains lit, (c) stays the same, (d) goes out. (Assume connecting wires have no resistance.)

Quick Quiz 18.2

In Figure 18.4a the current is measured with the ammeter at the right side of the circuit. When the switch is opened as in Figure 18.4b, the reading on the ammeter (a) increases (b) decreases (c) doesn't change.

(a) (b)

Figure 18.4 (Quick Quiz 18.2)

EXAMPLE 18.1 Four Resistors in Series

Goal Analyze several resistors connected in series.

Problem Four resistors are arranged as shown in Figure 18.5a. Find **(a)** the equivalent resistance of the circuit and **(b)** the current in the circuit if the emf of the battery is 6.0 V.

Strategy Because the resistors are connected in series, summing their resistances gives the equivalent resistance. Ohm's law can then be used to find the current.

Figure 18.5 (Example 18.1) (a) Four resistors connected in series. (b) The equivalent resistance of the circuit in (a).

Solution

(a) Find the equivalent resistance of the circuit.

Apply Equation 18.4, summing the resistances:

$$R_{eq} = R_1 + R_2 + R_3 + R_4 = 2.0\,\Omega + 4.0\,\Omega + 5.0\,\Omega + 7.0\,\Omega$$
$$= 18.0\,\Omega$$

(b) Find the current in the circuit.

Apply Ohm's law to the equivalent resistor in Figure 18.5b, solving for the current:

$$I = \frac{\Delta V}{R_{eq}} = \frac{6.0\text{ V}}{18.0\,\Omega} = \tfrac{1}{3}\text{ A}$$

Exercise 18.1

Because the current in the equivalent resistor is $\frac{1}{3}$ A, this must also be the current in each resistor of the original circuit. Find the voltage drop across each resistor.

Answers $\Delta V_{2\Omega} = \frac{2}{3}$ V; $\Delta V_{4\Omega} = \frac{4}{3}$ V; $\Delta V_{5\Omega} = \frac{5}{3}$ V; $\Delta V_{7\Omega} = \frac{7}{3}$ V.

18.3 RESISTORS IN PARALLEL

Now consider two resistors connected in parallel, as in Active Figure 18.6. In this case, **the potential differences across the resistors are the same because each is connected directly across the battery terminals**. The currents are generally not the same. When charges reach point a (called a junction) in Active Figure 18.6b, the current splits into two parts: I_1, flowing through R_1; and I_2, flowing through R_2. If R_1 is greater than R_2, then I_1 is less than I_2. In general, more charge travels through the path with less resistance. **Because charge is conserved, the current I that enters point a must equal the total current $I_1 + I_2$ leaving that point**. Mathematically, this is written

$$I = I_1 + I_2$$

The potential drop must be the same for the two resistors and must also equal the potential drop across the battery. Ohm's law applied to each resistor yields

$$I_1 = \frac{\Delta V}{R_1} \qquad\qquad I_2 = \frac{\Delta V}{R_2}$$

Ohm's law applied to the equivalent resistor in Active Figure 18.6c gives

$$I = \frac{\Delta V}{R_{eq}}$$

When these expressions for the currents are substituted into the equation $I = I_1 + I_2$ and the ΔV's are cancelled, we obtain

$$\frac{1}{R_{eq}} = \frac{1}{R_1} + \frac{1}{R_2} \qquad \text{(parallel combination)} \qquad\qquad \textbf{[18.5]}$$

ACTIVE FIGURE 18.6
(a) A parallel connection of two lightbulbs with resistances R_1 and R_2. (b) Circuit diagram for the two-resistor circuit. The potential differences across R_1 and R_2 are the same. (c) The equivalent resistance of the combination is given by the reciprocal relationship $1/R_{eq} = 1/R_1 + 1/R_2$.

Physics☒Now™
Log into PhysicsNow at **www.cp7e.com**, and go to Active Figure 18.6, where you can adjust the battery voltage and resistances R_1 and R_2 and see the effect on the currents and voltages in the individual resistors.

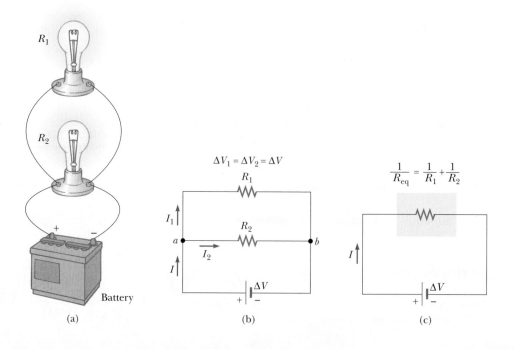

(a)   (b)   (c)

An extension of this analysis to three or more resistors in parallel produces the following general expression for the equivalent resistance:

$$\frac{1}{R_{eq}} = \frac{1}{R_1} + \frac{1}{R_2} + \frac{1}{R_3} + \cdots$$

[18.6]   ◀ Equivalent resistance of a parallel combination of resistors

From this expression, we see that **the inverse of the equivalent resistance of two or more resistors connected in parallel is the sum of the inverses of the individual resistances and is always less than the smallest resistance in the group.**

INTERACTIVE EXAMPLE 18.2   Three Resistors in Parallel

Goal   Analyze a circuit having resistors connected in parallel.

Problem   Three resistors are connected in parallel as in Figure 18.7. A potential difference of 18 V is maintained between points a and b. **(a)** Find the current in each resistor. **(b)** Calculate the power delivered to each resistor and the total power. **(c)** Find the equivalent resistance of the circuit. **(d)** Find the total power delivered to the equivalent resistance.

Strategy   We can use Ohm's law and the fact that the voltage drops across parallel resistors are all the same to get the current in each resistor. The rest of the problem just requires substitution into the equation for power delivered to a resistor, $\mathcal{P} = I^2 R$, and the reciprocal-sum law for parallel resistors.

Figure 18.7   (Example 18.2) Three resistors connected in parallel. The voltage across each resistor is 18 V.

Solution

(a) Find the current in each resistor.

Apply Ohm's law, solved for the current I delivered by the battery to find the current in each resistor:

$$I_1 = \frac{\Delta V}{R_1} = \frac{18\ \text{V}}{3.0\ \Omega} = \boxed{6.0\ \text{A}}$$

$$I_2 = \frac{\Delta V}{R_2} = \frac{18\ \text{V}}{6.0\ \Omega} = \boxed{3.0\ \text{A}}$$

$$I_3 = \frac{\Delta V}{R_3} = \frac{18\ \text{V}}{9.0\ \Omega} = \boxed{2.0\ \text{A}}$$

(b) Calculate the power delivered to each resistor and the total power.

Apply $\mathcal{P} = I^2 R$ to each resistor, substituting the results from part (a).

3 Ω: $\mathcal{P}_1 = I_1^2 R_1 = (6.0\ \text{A})^2(3.0\ \Omega) = \boxed{110\ \text{W}}$

6 Ω: $\mathcal{P}_2 = I_2^2 R_2 = (3.0\ \text{A})^2(6.0\ \Omega) = \boxed{54\ \text{W}}$

9 Ω: $\mathcal{P}_3 = I_3^2 R_3 = (2.0\ \text{A})^2(9.0\ \Omega) = \boxed{36\ \text{W}}$

Sum to get the total power:

$$\mathcal{P}_{tot} = 110\ \text{W} + 54\ \text{W} + 36\ \text{W} = \boxed{2.0 \times 10^2\ \text{W}}$$

(c) Find the equivalent resistance of the circuit.

Apply the reciprocal-sum rule, Equation 18.6:

$$\frac{1}{R_{eq}} = \frac{1}{R_1} + \frac{1}{R_2} + \frac{1}{R_3}$$

$$\frac{1}{R_{eq}} = \frac{1}{3.0\ \Omega} + \frac{1}{6.0\ \Omega} + \frac{1}{9.0\ \Omega} = \frac{11}{18\ \Omega}$$

$$R_{eq} = \frac{18}{11}\ \Omega = \boxed{1.6\ \Omega}$$

(d) Compute the power dissipated by the equivalent resistance.

Use the alternate power equation:

$$\mathcal{P} = \frac{(\Delta V)^2}{R_{eq}} = \frac{(18\ V)^2}{(1.6\ \Omega)} = \boxed{2.0 \times 10^2\ W}$$

Remarks   There's something important to notice in part (a): the smallest $3.0\ \Omega$ resistor carries the largest current, while the other, larger resistors of $6.0\ \Omega$ and $9.0\ \Omega$ carry smaller currents. The largest current is always found in the path of least resistance. In part (b), the power could also be found with $\mathcal{P} = (\Delta V)^2/R$. Note that $\mathcal{P}_1 = 108\ W$, but is rounded to 110 W because there are only two significant figures. Finally, notice that the total power dissipated in the equivalent resistor is the same as the sum of the power dissipated in the individual resistors, as it should be.

TIP 18.2   Don't Forget to Flip It!

The most common mistake in calculating the equivalent resistance for resistors in parallel is to forget to invert the answer after summing the reciprocals. Don't forget to flip it!

Exercise 18.2

Suppose the resistances in the example are $1.0\ \Omega$, $2.0\ \Omega$, and $3.0\ \Omega$, respectively, and a new voltage source is provided. If the current measured in the 3.0-Ω resistor is 2.0 A, find (a) the potential difference provided by the new battery, and the currents in each of the remaining resistors, (b) the power delivered to each resistor, and the total power, (c) the equivalent resistance, and (d) the total current, and the power dissipated by the equivalent resistor.

Answers   (a) $\mathcal{E} = 6.0\ V$, $I_1 = 6.0\ A$, $I_2 = 3.0\ A$ (b) $\mathcal{P}_1 = 36\ W$, $\mathcal{P}_2 = 18\ W$, $\mathcal{P}_3 = 12\ W$, $\mathcal{P}_{tot} = 66\ W$
(c) $\frac{6}{11}\ \Omega$ (d) $I = 11\ A$, $\mathcal{P}_{eq} = 66\ W$

Physics⊗Now™ Explore different configurations of the battery and resistors by logging into PhysicsNow at **www.cp7e.com** and going to Interactive Example 18.2.

Household circuits are always wired so that the electrical devices are connected in parallel, as in Active Figure 18.6a. In this way, each device operates independently of the others, so that if one is switched off, the others remain on. For example, if one of the lightbulbs in Active Figure 18.6 were removed from its socket, the other would continue to operate. Equally important, each device operates at the same voltage. If the devices were connected in series, the voltage across any one device would depend on how many devices were in the combination and on their individual resistances.

APPLICATION
Circuit Breakers

In many household circuits, circuit breakers are used in series with other circuit elements for safety purposes. A circuit breaker is designed to switch off and open the circuit at some maximum value of the current (typically 15 A or 20 A) that depends on the nature of the circuit. If a circuit breaker were not used, excessive currents caused by operating several devices simultaneously could result in excessive wire temperatures, perhaps causing a fire. In older home construction, fuses were used in place of circuit breakers. When the current in a circuit exceeded some value, the conductor in a fuse melted and opened the circuit. The disadvantage of fuses is that they are destroyed in the process of opening the circuit, whereas circuit breakers can be reset.

Applying Physics 18.2   Lightbulb Combinations

Compare the brightness of the four identical bulbs shown in Figure 18.8. What happens if bulb A fails, so that it cannot conduct current? What if C fails? What if D fails?

Explanation   Bulbs A and B are connected in series across the emf of the battery, whereas bulb C is connected by itself across the battery. This means the

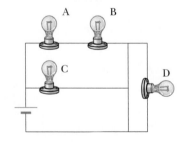

Figure 18.8   (Applying Physics 18.2)

voltage drop across C has the same magnitude as the battery emf, whereas this same emf is split between bulbs A and B. As a result, bulb C will glow more brightly than either of bulbs A and B, which will glow equally brightly. Bulb D has a wire connected across it—a short circuit—so the potential difference across bulb D is zero and it doesn't glow. If bulb A fails, B goes out, but C stays lit. If C fails, there is no effect on the other bulbs. If D fails, the event is undetectable, because D was not glowing initially.

Applying Physics 18.3 Three-Way Lightbulbs

Figure 18.9 illustrates how a three-way lightbulb is constructed to provide three levels of light intensity. The socket of the lamp is equipped with a three-way switch for selecting different light intensities. The bulb contains two filaments. Why are the filaments connected in parallel? Explain how the two filaments are used to provide the three different light intensities.

Explanation If the filaments were connected in series and one of them were to fail, there would be no current in the bulb, and the bulb would not glow, regardless of the position of the switch. However, when the filaments are connected in parallel and one of them (say, the 75-W filament) fails, the bulb will still operate in one of the switch positions because there is current in the other (100-W) filament. The three light intensities are made possible by selecting one of three values of filament resistance, using a single value of

120 V for the applied voltage. The 75-W filament offers one value of resistance, the 100-W filament offers a second value, and the third resistance is obtained by combining the two filaments in parallel. When switch S_1 is closed and switch S_2 is opened, only the 75-W filament carries current. When switch S_1 is open and switch S_2 is closed, only the 100-W filament carries current. When both switches are closed, both filaments carry current and a total illumination corresponding to 175 W is obtained.

Figure 18.9 (Applying Physics 18.3)

Quick Quiz 18.3

In Figure 18.10a the current is measured with the ammeter on the right side of the circuit diagram. When the switch is closed, the reading on the ammeter (a) increases, (b) decreases, or (c) remains the same.

(a)   (b)

Figure 18.10 (Quick Quiz 18.3)

Quick Quiz 18.4

Suppose you have three identical lightbulbs, some wire, and a battery. You connect one lightbulb to the battery and take note of its brightness. You add a second lightbulb, connecting it in parallel with the previous bulbs, again taking note of the brightness. Repeat the process with the third bulb, connecting it in parallel with the other two. As the lightbulbs are added, what happens to (a) the brightness of the bulbs? (b) the individual currents in the bulbs? (c) the power delivered by the battery? (d) the lifetime of the battery? (Neglect the battery's internal resistance)

If the lightbulbs in Quick Quiz 18.4 are connected, one by one, in series instead of in parallel, what happens to (a) the brightness of the bulbs? (b) the individual currents in the bulbs? (c) the power delivered by the battery? (d) the lifetime of the battery? (Again, neglect the battery's internal resistance)

Problem-Solving Strategy   Simplifying Circuits with Resistors

1. **Combine all resistors in series** by summing the individual resistances, and draw the new, simplified circuit diagram.
   Useful facts: $R_{eq} = R_1 + R_2 + R_3 + \cdots$
   The current in each resistor is the same.
2. **Combine all resistors in parallel** by summing the reciprocals of the resistances and then taking the reciprocal of the result. Draw the new, simplified circuit diagram.
   Useful facts: $\dfrac{1}{R_{eq}} = \dfrac{1}{R_1} + \dfrac{1}{R_2} + \dfrac{1}{R_3} + \cdots$
   The potential difference across each resistor is the same.
3. **Repeat** the first two steps as necessary, until no further combinations can be made. If there is only a single battery in the circuit, this will usually result in a single equivalent resistor in series with the battery.
4. **Use Ohm's Law**, $\Delta V = IR$, to determine the current in the equivalent resistor. Then work backwards through the diagrams, applying the useful facts listed in step 1 or step 2 to find the currents in the other resistors. (In more complex circuits, Kirchhoff's rules will be needed, as described in the next section).

EXAMPLE 18.3   Equivalent Resistance

Goal   Solve a problem involving both series and parallel resistors.

Problem   Four resistors are connected as shown in Figure 18.11a. **(a)** Find the equivalent resistance between points a and c. **(b)** What is the current in each resistor if a 42-V battery is connected between a and c?

Strategy   Reduce the circuit in steps, as shown in Figures 18.11b and 18.11c, using the sum rule for resistors in series and the reciprocal-sum rule for resistors in parallel. Finding the currents is a matter of applying Ohm's law while working backwards through the diagrams.

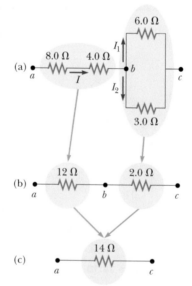

Figure 18.11   (Example 18.3) The four resistors shown in (a) can be reduced in steps to an equivalent 14-Ω resistor.

Solution

(a) Find the equivalent resistance of the circuit.

The 8.0-Ω and 4.0-Ω resistors are in series, so use the sum rule to find the equivalent resistance between a and b:

$$R_{eq} = R_1 + R_2 = 8.0 \ \Omega + 4.0 \ \Omega = 12 \ \Omega$$

The 6.0-Ω and 3.0-Ω resistors are in parallel, so use the reciprocal-sum rule to find the equivalent resistance between b and c (don't forget to invert!):

$$\frac{1}{R_{eq}} = \frac{1}{R_1} + \frac{1}{R_2} = \frac{1}{6.0 \ \Omega} + \frac{1}{3.0 \ \Omega} = \frac{1}{2.0 \ \Omega}$$

$$R_{eq} = 2.0 \ \Omega$$

In the new diagram, 18.11b, there are now two resistors in series. Combine them with the sum rule to find the equivalent resistance of the circuit:

$$R_{eq} = R_1 + R_2 = 12 \ \Omega + 2.0 \ \Omega = \boxed{14 \ \Omega}$$

(b) Find the current in each resistor if a 42-V battery is connected between points *a* and *c*.

Find the current in the equivalent resistor in Figure 18.11c, which is the total current. Resistors in series all carry the same current, so this is the current in the 12-Ω resistor in Figure 18.11b, and also in the 8.0-Ω and 4.0-Ω resistors in Figure 18.11a.

$$I = \frac{\Delta V_{ac}}{R_{eq}} = \frac{42 \ \text{V}}{14 \ \Omega} = \boxed{3.0 \ \text{A}}$$

Apply the junction rule to point *b*:

(1) $I = I_1 + I_2$

The 6.0-Ω and 3.0-Ω resistors are in parallel, so the voltage drops across them are the same:

$$\Delta V_{6\Omega} = \Delta V_{3\Omega} \quad \rightarrow \quad (6.0 \ \Omega)I_1 = (3.0 \ \Omega)I_2 \quad \rightarrow$$
$$2.0I_1 = I_2$$

Substitute this result into Equation (1), with $I = 3.0$ A:

$$3.0 \ \text{A} = I_1 + 2I_1 = 3I_1 \quad \rightarrow \quad I_1 = \boxed{1.0 \ \text{A}}$$

$$I_2 = \boxed{2.0 \ \text{A}}$$

Remarks   As a final check, note that $\Delta V_{bc} = (6.0 \ \Omega)I_1 = (3.0 \ \Omega)I_2 = 6.0$ V and $\Delta V_{ab} = (12 \ \Omega)I_1 = 36$ V; therefore, $\Delta V_{ac} = \Delta V_{ab} + \Delta V_{bc} = 42$ V, as expected.

Exercise 18.3
Suppose the series resistors in Example 18.3 are now 6.00 Ω and 3.00 Ω while the parallel resistors are 8.00 Ω (top) and 4.00 Ω (bottom), and the battery provides an emf of 27.0 V. Find (a) the equivalent resistance and (b) the currents I, I_1, and I_2.

Answers   (a) 11.7 Ω (b) $I = 2.31$ A, $I_1 = 0.770$ A, $I_2 = 1.54$ A

18.4   KIRCHHOFF'S RULES AND COMPLEX DC CIRCUITS

As demonstrated in the preceding section, we can analyze simple circuits using Ohm's law and the rules for series and parallel combinations of resistors. However, there are many ways in which resistors can be connected so that the circuits formed can't be reduced to a single equivalent resistor. The procedure for analyzing more complex circuits can be facilitated by the use of two simple rules called **Kirchhoff's rules**:

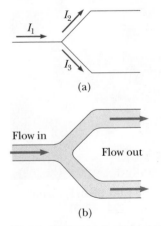

(a)

(b)

Figure 18.12   (a) A schematic diagram illustrating Kirchhoff's junction rule. Conservation of charge requires that whatever current enters a junction must leave that junction. In this case, therefore, $I_1 = I_2 + I_3$. (b) A mechanical analog of the junction rule: the net flow in must equal the net flow out.

1. The sum of the currents entering any junction must equal the sum of the currents leaving that junction. (This rule is often referred to as the **junction rule**.)
2. The sum of the potential differences across all the elements around any closed circuit loop must be zero. (This rule is usually called the **loop rule**.)

The junction rule is a statement of *conservation of charge*. Whatever current enters a given point in a circuit must leave that point because charge can't build up or disappear at a point. If we apply this rule to the junction in Figure 18.12a, we get

$$I_1 = I_2 + I_3$$

Figure 18.12b represents a mechanical analog of the circuit shown in Figure 18.12a. In this analog, water flows through a branched pipe with no leaks. The flow rate into the pipe equals the total flow rate out of the two branches.

Figure 18.13 Rules for determining the potential differences across a resistor and a battery, assuming the battery has no internal resistance.

GUSTAV KIRCHHOFF, German Physicist (1824–1887)

Together with German chemist Robert Bunsen, Kirchhoff, a professor at Heidelberg, invented the spectroscopy that we study in Chapter 28. He also formulated another rule that states, "A cool substance will absorb light of the same wavelengths that it emits when hot."

The loop rule is equivalent to the principle of *conservation of energy*. Any charge that moves around any closed loop in a circuit (starting and ending at the same point) must gain as much energy as it loses. It gains energy as it is pumped through a source of emf. Its energy may decrease in the form of a potential drop $-IR$ across a resistor or as a result of flowing backward through a source of emf, from the positive to the negative terminal inside the battery. In the latter case, electrical energy is converted to chemical energy as the battery is charged.

When applying Kirchhoff's rules, you must make two decisions at the beginning of the problem:

1. Assign symbols and directions to the currents in all branches of the circuit. Don't worry about guessing the direction of a current incorrectly; the resulting answer will be negative, but *its magnitude will be correct*. (This is because the equations are *linear* in the currents—all currents are to the first power.)
2. When applying the loop rule, you must choose a direction for traversing the loop, and be consistent in going either clockwise or counterclockwise. As you traverse the loop, record voltage drops and rises according to the following rules (summarized in Figure 18.13, where it is assumed that movement is from point *a* toward point *b*):

   (a) If a resistor is traversed in the direction of the current, the change in electric potential across the resistor is $-IR$ (Fig. 18.13a).
   (b) If a resistor is traversed in the direction opposite the current, the change in electric potential across the resistor is $+IR$ (Fig. 18.13b).
   (c) If a source of emf is traversed in the direction of the emf (from $-$ to $+$ on the terminals), the change in electric potential is $+\mathcal{E}$ (Fig. 18.13c).
   (d) If a source of emf is traversed in the direction opposite the emf (from $+$ to $-$ on the terminals), the change in electric potential is $-\mathcal{E}$ (Fig. 18.13d).

There are limits to the number of times the junction rule and the loop rule can be used. You can use the junction rule as often as needed as long as, each time you write an equation, you include in it a current that has not been used in a previous junction-rule equation. (If this procedure isn't followed, the new equation will just be a combination of two other equations that you already have.) In general, the number of times the junction rule can be used is one fewer than the number of junction points in the circuit. The loop rule can also be used as often as needed, so long as a new circuit element (resistor or battery) or a new current appears in each new equation. **To solve a particular circuit problem, you need as many independent equations as you have unknowns.**

Problem-Solving Strategy Applying Kirchhoff's Rules to a Circuit

1. **Assign labels and symbols** to all the known and unknown quantities.
2. **Assign *directions* to the currents** in each part of the circuit. Although the assignment of current directions is arbitrary, you must stick with your original choices throughout the problem as you apply Kirchhoff's rules.
3. **Apply the junction rule** to any junction in the circuit. The rule may be applied as many times as a new current (one not used in a previously found equation) appears in the resulting equation.
4. **Apply Kirchhoff's loop rule** to as many loops in the circuit as are needed to solve for the unknowns. In order to apply this rule, you must correctly identify the change in electric potential as you cross each element in traversing the closed loop. Watch out for signs!
5. **Solve the equations** simultaneously for the unknown quantities, using substitution or any other method familiar to the student.
6. **Check your answers** by substituting them into the original equations.

EXAMPLE 18.4   Applying Kirchhoff's Rules

Goal   Use Kirchhoff's rules to find currents in a circuit with three currents and one battery.

Problem   Find the currents in the circuit shown in Figure 18.14 by using Kirchhoff's rules.

Strategy   There are three unknown currents in this circuit, so we must obtain three independent equations, which then can be solved by substitution. We can find the equations with one application of the junction rule and two applications of the loop rule. We choose junction c. (Junction d gives the same equation.) For the loops, we choose the bottom loop and the top loop, both shown by blue arrows, which indicate the direction we are going to traverse the circuit mathematically (not necessarily the direction of the current). The third loop gives an equation that can be obtained by a linear combination of the other two, so it provides no additional information.

Figure 18.14   (Example 18.4)

Solution

Apply the junction rule to point c. I_1 is directed into the junction, I_2 and I_3 are directed out of the junction.

$$I_1 = I_2 + I_3$$

Select the bottom loop, and traverse it clockwise starting at point a, generating an equation with the loop rule:

$$\Sigma \Delta V = \Delta V_{\text{bat}} + \Delta V_{4.0\Omega} + \Delta V_{9.0\Omega} = 0$$
$$6.0 \text{ V} - (4.0 \ \Omega)I_1 - (9.0 \ \Omega)I_3 = 0$$

Select the top loop, and traverse it clockwise from point c. Notice the gain across the 9.0-Ω resistor, because it is traversed *against* the direction of the current!

$$\Sigma \Delta V = \Delta V_{5.0\Omega} + \Delta V_{9.0\Omega} = 0$$
$$-(5.0 \ \Omega)I_2 + (9.0 \ \Omega)I_3 = 0$$

Rewrite the three equations, rearranging terms and dropping units for the moment, for convenience:

(1) $I_1 = I_2 + I_3$

(2) $4.0I_1 + 9.0I_3 = 6.0$

(3) $-5.0I_2 + 9.0I_3 = 0$

Solve Equation 3 for I_2 and substitute into Equation 1:

$$I_2 = 1.8I_3$$
$$I_1 = I_2 + I_3 = 1.8I_3 + I_3 = 2.8I_3$$

Substitute the latter expression into Equation 2 and solve for I_3:

$$4.0(2.8I_3) + 9.0I_3 = 6.0 \quad \rightarrow \quad I_3 = \boxed{0.30 \text{ A}}$$

Substitute I_3 back into Equation 3 to get I_2:

$$-5.0I_2 + 9.0(0.30 \text{ A}) = 0 \quad \rightarrow \quad I_2 = \boxed{0.54 \text{ A}}$$

Substitute I_3 into Equation 2 to get I_1:

$$4.0I_1 + 9.0(0.30 \text{ A}) = 6.0 \quad \rightarrow \quad I_1 = \boxed{0.83 \text{ A}}$$

Remarks   Substituting these values back into the original equations verifies that they are correct, with any small discrepancies due to rounding. The problem can also be solved by first combining resistors.

Exercise 18.4

Suppose the 6.0-V battery is replaced by a battery of unknown emf, and an ammeter measures $I_1 = 1.5$ A. Find the other two currents and the emf of the battery.

Answers $I_2 = 0.96$ A, $I_3 = 0.54$ A, $\mathcal{E} = 11$ V

TIP 18.3   More Current Goes in the Path of Less Resistance

You may have heard the statement "Current takes the path of least resistance." For a parallel combination of resistors, this statement is inaccurate, because current actually follows all paths. The most current, however, travels in the path of least resistance.

INTERACTIVE EXAMPLE 18.5   Another Application of Kirchhoff's Rules

Goal   Find the currents in a circuit with three currents and two batteries when some current directions are chosen wrongly.

Problem   Find I_1, I_2, and I_3 in Figure 18.15a on page 604.

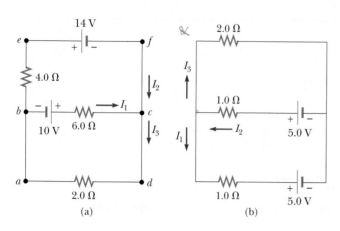

Figure 18.15   (a) (Example 18.5) (b) (Exercise 18.5)

Strategy   Use Kirchhoff's two rules, the junction rule once and the loop rule twice, to develop three equations for the three unknown currents. Solve the equations simultaneously.

Solution

Apply Kirchhoff's junction rule to junction c. Because of the chosen current directions, I_1 and I_2 are directed into the junction and I_3 is directed out of the junction.

(1) $I_3 = I_1 + I_2$

Apply Kirchhoff's loop rule to the loops $abcda$ and $befcb$. (Loop $aefda$ gives no new information.) In loop $befcb$, a positive sign is obtained when the 6.0-Ω resistor is traversed, because the direction of the path is opposite the direction of the current I_1.

(2) Loop $abcda$: $10\text{ V} - (6.0\ \Omega)I_1 - (2.0\ \Omega)I_3 = 0$

(3) Loop $befcb$:
$$-14\text{ V} + (6.0\ \Omega)I_1 - 10\text{ V} - (4.0\ \Omega)I_2 = 0$$

Using Equation (1), eliminate I_3 from Equation (2) (ignore units for the moment):

$$10 - 6.0I_1 - 2.0(I_1 + I_2) = 0$$

(4) $10 = 8.0I_1 + 2.0I_2$

Divide each term in Equation (3) by 2 and rearrange the equation so that the currents are on the right side:

(5) $-12 = -3.0I_1 + 2.0I_2$

Subtracting Equation (5) from Equation (4) eliminates I_2 and gives I_1:

$22 = 11I_1 \quad\rightarrow\quad I_1 = \boxed{2.0\text{ A}}$

Substituting this value of I_1 into Equation (5) gives I_2:

$2.0I_2 = 3.0I_1 - 12 = 3.0(2.0) - 12 = -6.0\text{ A}$

$I_2 = \boxed{-3.0\text{ A}}$

Finally, substitute the values found for I_1 and I_2 into Equation (1) to obtain I_3:

$I_3 = I_1 + I_2 = 2.0\text{ A} - 3.0\text{ A} = \boxed{-1.0\text{ A}}$

Remarks   The fact that I_2 and I_3 are both negative indicates that the wrong directions were chosen for these currents. Nonetheless, the magnitudes are correct. Choosing the right directions of the currents at the outset is unimportant because the equations are linear, and wrong choices result only in a minus sign in the answer.

Exercise 18.5   Find the three currents in Figure 18.15b. (Note that the direction of one current was deliberately chosen wrongly!)

Answers $I_1 = -1.0\text{ A}$, $I_2 = 1.0\text{ A}$, $I_3 = 2.0\text{ A}$

Physics⊗Now™   Practice applying Kirchhoff's rules for different values of resistance and voltage by logging into PhysicsNow at **www.cp7e.com** and going to Interactive Example 18.5.

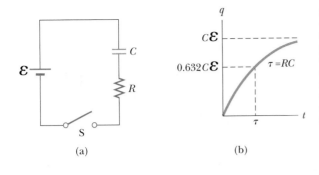

(a)   (b)

ACTIVE FIGURE 18.16
(a) A capacitor in series with a resistor, a battery, and a switch. (b) A plot of the charge on the capacitor versus time after the switch on the circuit is closed. After one time constant τ, the charge is 63% of the maximum value, $C\mathcal{E}$. The charge approaches its maximum value as t approaches infinity.

Physics⊗Now™
Log into PhysicsNow at **www.cp7e.com** and go to Active Figure 18.16, where you can adjust the values of R and C and observe the effect on the charging of the capacitor.

18.5   *RC* CIRCUITS

So far, we have been concerned with circuits with constant currents. We now consider direct-current circuits containing capacitors, in which the currents vary with time. Consider the series circuit in Active Figure 18.16. We assume that the capacitor is initially uncharged with the switch opened. After the switch is closed, the battery begins to charge the plates of the capacitor and the charge passes through the resistor. As the capacitor is being charged, the circuit carries a changing current. The charging process continues until the capacitor is charged to its maximum equilibrium value, $Q = C\mathcal{E}$, where $\mathcal{E}$ is the maximum voltage across the capacitor. Once the capacitor is fully charged, the current in the circuit is zero. If we assume that the capacitor is uncharged before the switch is closed, and if the switch is closed at $t = 0$, we find that the charge on the capacitor varies with time according to the equation

$$q = Q(1 - e^{-t/RC})   \text{[18.7]}$$

where $e = 2.718 \ldots$ is Euler's constant, the base of the natural logarithms. Active Figure 18.16b is a graph of this equation. The charge is zero at $t = 0$ and approaches its maximum value, Q, as t approaches infinity. The voltage ΔV across the capacitor at any time is obtained by dividing the charge by the capacitance: $\Delta V = q/C$.

As you can see from Equation 18.7, it would take an infinite amount of time, in this model, for the capacitor to become fully charged. The reason for this is mathematical: in obtaining that equation, charges are assumed to be infinitely small, whereas in reality the smallest charge is that of an electron, with a magnitude of 1.60×10^{-19} C. For all practical purposes, the capacitor is fully charged after a finite amount of time. The term RC that appears in Equation 18.7 is called the **time constant** τ (Greek letter tau), so

$$\tau = RC   \text{[18.8]}$$

◀ Time constant of an *RC* circuit

The time constant represents the time required for the charge to increase from zero to 63.2% of its maximum equilibrium value. This means that in a period of time equal to one time constant, the charge on the capacitor increases from zero to $0.632Q$. This can be seen by substituting $t = \tau = RC$ into Equation 18.7 and solving for q. (Note that $1/e = 0.632$.) It's important to note that a capacitor charges very slowly in a circuit with a long time constant, whereas it charges very rapidly in a circuit with a short time constant. After a time equal to ten time constants, the capacitor is over 99.99% charged.

Now consider the circuit in Active Figure 18.17a, consisting of a capacitor with an initial charge Q, a resistor, and a switch. Before the switch is closed, the potential difference across the charged capacitor is Q/C. Once the switch is closed, the charge begins to flow through the resistor from one capacitor plate to the other until the capacitor is fully discharged. If the switch is closed at $t = 0$, it can be shown that the charge q on the capacitor varies with time according to the equation

$$q = Qe^{-t/RC}   \text{[18.9]}$$

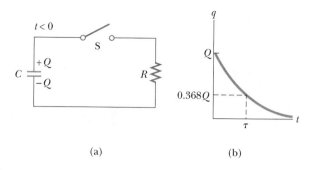

ACTIVE FIGURE 18.17
(a) A charged capacitor connected to a resistor and a switch. (b) A graph of the charge on the capacitor versus time after the switch is closed.

Physics ⊗ Now™
Log into PhysicsNow at **www.cp7e.com** and go to Active Figure 18.17, where you can adjust the values of R and C and see the effect on the discharging of the capacitor.

The charge decreases exponentially with time, as shown in Active Figure 18.17b. In the interval $t = \tau = RC$, the charge decreases from its initial value Q to $0.368Q$. In other words, in a time equal to one time constant, the capacitor loses 63.2% of its initial charge. Because $\Delta V = q/C$, the voltage across the capacitor also decreases exponentially with time according to the equation $\Delta V = \mathcal{E}e^{-t/RC}$, where $\mathcal{E}$ (which equals Q/C) is the initial voltage across the fully charged capacitor.

Applying Physics 18.4   Timed Windshield Wipers

Many automobiles are equipped with windshield wipers that can be used intermittently during a light rainfall. How does the operation of this feature depend on the charging and discharging of a capacitor?

Explanation   The wipers are part of an RC circuit with time constant that can be varied by selecting different values of R through a multiposition switch. The brief time that the wipers remain on and the time they are off are determined by the value of the time constant of the circuit.

Applying Physics 18.5   Bacterial Growth

In biological applications concerned with population growth, an equation is used that is similar to the exponential equations encountered in the analysis of RC circuits. Applied to a number of bacteria, this equation is

$$N_f = N_i 2^n$$

where N_f is the number of bacteria present after n doubling times, N_i is the number present initially, and n is the number of growth cycles or doubling times. Doubling times vary according to the organism. The doubling time for the bacteria responsible for leprosy is about 30 days, and that for the salmonella bacteria

responsible for food poisoning is about 20 minutes. Suppose only 10 salmonella bacteria find their way onto a turkey leg after your Thanksgiving meal. Four hours later you come back for a midnight snack. How many bacteria are present now?

Explanation   The number of doubling times is 240 min/20 min = 12. Thus, we have

$$N_f = N_i 2^n = (10\ \text{bacteria})(2^{12}) = 40\,960\ \text{bacteria}.$$

So your system will have to deal with an invading host of about 41 000 bacteria, which are going to continue to double in a very promising environment.

Applying Physics 18.6   Roadway Flashers

Many roadway construction sites have flashing yellow lights to warn motorists of possible dangers. What causes the lights to flash?

Explanation   A typical circuit for such a flasher is shown in Figure 18.18. The lamp L is a gas-filled lamp that acts as an open circuit until a large potential difference causes a discharge, which gives off a bright light. During this discharge, charge flows through the gas between the electrodes of the lamp. When the switch is closed, the battery charges the capacitor. At the beginning, the current is high and the charge on the capacitor is low, so that most of the potential difference appears across the resistance R. As the capacitor charges, more potential difference appears across it, reflecting the lower current and lower potential difference across the resistor. Eventually, the potential difference across the capacitor reaches a value at which the lamp will conduct, causing a flash. This discharges the capacitor through the lamp, and the process of charging begins again. The period between flashes can be adjusted by changing the time constant of the *RC* circuit.

Figure 18.18 (Applying Physics 18.6)

Quick Quiz 18.6

The switch is closed in Figure 18.19. After a long time compared to the time constant of the capacitor, what will the current be in the 2-Ω resistor? (a) 4 A (b) 3 A (c) 2 A (d) 1 A (e) more information is needed

Figure 18.19 (Quick Quiz 18.6)

EXAMPLE 18.6 Charging a Capacitor in an *RC* Circuit

Goal Calculate elementary properties of a simple *RC* circuit.

Problem An uncharged capacitor and a resistor are connected in series to a battery, as in Active Figure 18.16a. If $\mathcal{E} = 12.0$ V, $C = 5.00$ μF, and $R = 8.00 \times 10^5$ Ω, find (a) the time constant of the circuit, (b) the maximum charge on the capacitor, (c) the charge on the capacitor after 6.00 s, (d) the potential difference across the resistor after 6.00 s, and (e) the current in the resistor at that time.

Strategy Finding the time constant in part (a) requires substitution into Equation 18.8. For part (b), the maximum charge occurs after a long time, when the current has dropped to zero. By Ohm's law, $\Delta V = IR$, the potential difference across the resistor is also zero at that time, and Kirchhoff's loop rule then gives the maximum charge. Finding the charge at some particular time, as in part (c), is a matter of substituting into Equation 18.7. Kirchhoff's loop rule and the capacitance equation can be used to indirectly find the potential drop across the resistor in part (d), and then Ohm's law yields the current.

Solution
(a) Find the time constant of the circuit.

Use the definition of the time constant, Equation 18.8: $\tau = RC = (8.00 \times 10^5 \text{ Ω})(5.00 \times 10^{-6} \text{ F}) = \boxed{4.00 \text{ s}}$

(b) Calculate the maximum charge on the capacitor.

Apply Kirchhoff's loop rule to the RC circuit, going clockwise, which means that the voltage difference across the battery is positive and the differences across the capacitor and resistor are negative.

$$(1) \qquad \Delta V_{bat} + \Delta V_C + \Delta V_R = 0$$

From the definition of capacitance (Equation 16.8) and Ohm's law, we have $\Delta V_C = -q/C$ and $\Delta V_R = -IR$. These are voltage drops, so they're negative. Also, $\Delta V_{bat} = +\mathcal{E} = 12.0$ V.

$$(2) \qquad \mathcal{E} - \frac{q}{c} - IR = 0$$

When the maximum charge $q = Q$ is reached, $I = 0$. Solve Equation (2) for the maximum charge:

$$\mathcal{E} - \frac{Q}{C} = 0 \quad \rightarrow \quad Q = C\mathcal{E}$$

Substitute to find the maximum charge:

$$Q = (5.00 \times 10^{-6} \text{ F})(12.0 \text{ V}) = \boxed{60.0 \ \mu\text{C}}$$

(c) Find the charge on the capacitor after 6.00 s.

Substitute into Equation 18.7:

$$q = Q(1 - e^{-t/\tau}) = (60.0 \ \mu\text{C})(1 - e^{-6.00 \text{ s}/4.00 \text{ s}})$$
$$= \boxed{46.6 \ \mu\text{C}}$$

(d) Compute the potential difference across the resistor after 6.00 s.

Compute the voltage drop ΔV_C across the capacitor at that time:

$$\Delta V_C = -\frac{q}{C} = \frac{-46.6 \ \mu\text{C}}{5.00 \ \mu\text{F}} = -9.32 \text{ V}$$

Solve Equation 1 for ΔV_R, and substitute:

$$\Delta V_R = -\Delta V_{bat} - \Delta V_C = -12.0 - (-9.32 \text{ V})$$
$$= \boxed{-2.68 \text{ V}}$$

(e) Find the current in the resistor after 6.00 s.

Apply Ohm's law, using the results of part (d) (remember that $\Delta V_R = -IR$ here):

$$I = \frac{-\Delta V_R}{R} = -\frac{(-2.68 \text{ V})}{(8.0 \times 10^5 \ \Omega)}$$
$$= \boxed{3.4 \times 10^{-6} \text{ A}}$$

Remark   In solving this problem, we paid scrupulous attention to signs. These signs must always be chosen when applying Kirchhoff's loop rule, and must remain consistent throughout the problem. Alternately, magnitudes can be used and the signs chosen by physical intuition. For example, the magnitude of the potential difference across the resistor must equal the magnitude of the potential difference across the battery minus the magnitude of the potential difference across the capacitor.

Exercise 18.6
Find (a) the charge on the capacitor after 2.00 s have elapsed, (b) the magnitude of the potential difference across the capacitor after 2.00 s, and (c) the magnitude of the potential difference across the resistor at that same time.

Answers   (a) 23.6 μC   (b) 4.72 V   (c) 7.28 V

EXAMPLE 18.7   Discharging a Capacitor in an RC Circuit

Goal   Calculate some elementary properties of a discharging capacitor in an RC circuit.

Problem   Consider a capacitor C being discharged through a resistor R as in Figure 18.17a, page 606. **(a)** How long does it take for the charge on the capacitor to drop to one-fourth of its initial value? **(b)** Compute the initial charge and time constant. **(c)** How long does it take to discharge all but the last quantum of charge, 1.60×10^{-19} C, if the initial potential difference across the capacitor is 12.0 V, the capacitance is 3.50×10^{-6} F, and the resistance is 2.00 Ω? (Assume an exponential decrease during the entire discharge process.)

Strategy   This problem requires substituting given values into various equations, as well as a couple algebraic manipulations involving the natural logarithm. In part (a), set $q = \frac{1}{4}Q$ in Equation 18.9 for a discharging capacitor, where Q is the initial charge, and solve for time t. For part (b), substitute into Equations 16.8 and 18.8 to find the initial capacitor charge and time constant, respectively. In part (c), substitute the results of part (b) and $q = 1.60 \times 10^{-19}$ C into the discharging-capacitor equation, again solving for time.

Solution
(a) How long does it take for the capacitor to drain to one-fourth its initial value?

Apply Equation 18.9:

$$q(t) = Qe^{-t/RC}$$

Substitute $q(t) = Q/4$ into the preceding equation and cancel Q:

$$\tfrac{1}{4}Q = Qe^{-t/RC} \quad \rightarrow \quad \tfrac{1}{4} = e^{-t/RC}$$

Take natural logarithms of both sides and solve for the time t:

$$\ln\left(\tfrac{1}{4}\right) = -t/RC$$

$$t = -RC\,\ln\left(\tfrac{1}{4}\right) = 1.39RC = \boxed{1.39\tau}$$

(b) Compute the initial charge and time constant from the given data.

Use the capacitance equation to find the initial charge:

$$C = \frac{Q}{\Delta V} \quad \rightarrow \quad Q = C\Delta V = (3.50 \times 10^{-6}\,\text{F})(12.0\,\text{V})$$

$$Q = \boxed{4.20 \times 10^{-5}\,\text{C}}$$

Now calculate the time constant:

$$\tau = RC = (2.00\,\Omega)(3.50 \times 10^{-6}\,\text{F}) = \boxed{7.00 \times 10^{-6}\,\text{s}}$$

(c) How long does it take to drain all but the last quantum of charge?

Apply Equation 18.9, divide by Q, and take natural logarithms of both sides:

$$q(t) = Qe^{-t/\tau} \quad \rightarrow \quad e^{-t/\tau} = \frac{q}{Q}$$

Take the natural logs of both sides:

$$-t/\tau = \ln\left(\frac{q}{Q}\right) \quad \rightarrow \quad t = -\tau\ln\left(\frac{q}{Q}\right)$$

Substitute $q = 1.60 \times 10^{-19}$ C and the values for Q and τ found in part (b):

$$t = -(7.00 \times 10^{-6}\,\text{s})\ln\left(\frac{1.60 \times 10^{-19}\,\text{C}}{4.20 \times 10^{-5}\,\text{C}}\right)$$

$$= \boxed{2.32 \times 10^{-4}\,\text{s}}$$

Remarks   Part (a) shows how useful information can often be obtained even when no details concerning capacitances, resistances, or voltages are known. Part (c) demonstrates that capacitors can be rapidly discharged (or conversely, charged), despite the mathematical form of Equations 18.7 and 18.9, which indicate an infinite time would be required.

Exercise 18.7
Suppose the same type of series circuit has $R = 8.00 \times 10^4\,\Omega$, $C = 5.00\,\mu$F, and an initial voltage across the capacitor of 6.0 V. (a) How long does it take the capacitor to lose half its initial charge? (b) How long does it take to lose all but the last 10 electrons on the negative plate?

Answers   (a) 0.277 s   (b) 12.2 s

18.6   HOUSEHOLD CIRCUITS

Household circuits are a practical application of some of the ideas presented in this chapter. In a typical installation, the utility company distributes electric power to individual houses with a pair of wires, or power lines. Electrical devices in a

APPLICATION

Fuses and Circuit Breakers

Figure 18.20 A wiring diagram for a household circuit. The resistances R_1, R_2, and R_3 represent appliances or other electrical devices that operate at an applied voltage of 120 V.

Figure 18.21 A circuit breaker that uses a bimetallic strip for its operation.

house are then connected in parallel to these lines, as shown in Figure 18.20. The potential difference between the two wires is about 120 V. (These currents and voltages are actually alternating currents and voltages, but for the present discussion we will assume that they are direct currents and voltages.) One of the wires is connected to ground, and the other wire, sometimes called the "hot" wire, is at a potential of 120 V. A meter and a circuit breaker (or a fuse) are connected in series with the wire entering the house, as indicated in the figure.

In modern homes, circuit breakers are used in place of fuses. When the current in a circuit exceeds some value (typically 15 A), the circuit breaker acts as a switch and opens the circuit. Figure 18.21 shows one design for a circuit breaker. Current passes through a bimetallic strip, the top of which bends to the left when excessive current heats it. If the strip bends far enough to the left, it settles into a groove in the spring-loaded metal bar. When this occurs, the bar drops enough to open the circuit at the contact point. The bar also flips a switch which indicates that the circuit breaker is not operational. (After the overload is removed, the switch can be flipped back on.) Circuit breakers based on this design have the disadvantage that some time is required for the heating of the strip, so the circuit may not be opened rapidly enough when it is overloaded. Because of this, many circuit breakers are now designed to use electromagnets (discussed in Chapter 19).

The wire and circuit breaker are carefully selected to meet the current demands of a circuit. If the circuit is to carry currents as large as 30 A, a heavy-duty wire and an appropriate circuit breaker must be used. Household circuits that are normally used to power lamps and small appliances often require only 20 A. Each circuit has its own circuit breaker to accommodate its maximum safe load.

As an example, consider a circuit that powers a toaster, a microwave oven, and a heater (represented by R_1, R_2, and R_3 in Fig. 18.20). Using the equation $\mathcal{P} = I\,\Delta V$, we can calculate the current carried by each appliance. The toaster, rated at 1 000 W, draws a current of 1 000/120 = 8.33 A. The microwave oven, rated at 800 W, draws a current of 6.67 A, and the heater, rated at 1 300 W, draws a current of 10.8 A. If the three appliances are operated simultaneously, they draw a total current of 25.8 A. Therefore, the breaker should be able to handle at least this much current, or else it will be tripped. As an alternative, the toaster and microwave oven could operate on one 20-A circuit and the heater on a separate 20-A circuit.

Many heavy-duty appliances, such as electric ranges and clothes dryers, require 240 V to operate. The power company supplies this voltage by providing, in addition to a live wire that is 120 V above ground potential, another wire, also considered live, that is 120 V below ground potential (Fig. 18.22). Therefore, the potential drop across the two live wires is 240 V. An appliance operating from a 240-V line requires half the current of one operating from a 120-V line; consequently, smaller wires can be used in the higher-voltage circuit without becoming overheated.

Figure 18.22 Power connections for a 240-V appliance.

(a)   (b)

18.7 ELECTRICAL SAFETY

A person can be electrocuted by touching a live wire (which commonly is live because of a frayed cord and exposed conductors) while in contact with ground. The ground contact might be made by touching a water pipe (which is normally at ground potential) or by standing on the ground with wet feet, because impure water is a good conductor. Obviously, such situations should be avoided at all costs.

Electric shock can result in fatal burns, or it can cause the muscles of vital organs, such as the heart, to malfunction. The degree of damage to the body depends on the magnitude of the current, the length of time it acts, and the part of the body through which it passes. Currents of 5 mA or less can cause a sensation of shock, but ordinarily do little or no damage. If the current is larger than about 10 mA, the hand muscles contract and the person may be unable to let go of the live wire. If a current of about 100 mA passes through the body for just a few seconds, it can be fatal. Such large currents paralyze the respiratory muscles. In some cases, currents of about 1 A through the body produce serious (and sometimes fatal) burns.

As an additional safety feature for consumers, electrical equipment manufacturers now use electrical cords that have a third wire, called a *case ground*. To understand how this works, consider the drill being used in Figure 18.23. A two-wire device that has one wire, called the "hot" wire, is connected to the high-potential (120-V) side of the input power line, and the second wire is connected to ground (0 V). If the high-voltage wire comes in contact with the case of the drill (Fig 18.23a), a short circuit occurs. In this undesirable circumstance, the pathway for the current is from the high-voltage wire through the person holding the drill and to Earth—a pathway that can be fatal. Protection is provided by a third wire, connected to the case of the drill (Fig. 18.23b). In this case, if a short occurs, the path of least resistance for the current is from the high-voltage wire through the case

APPLICATION

Third Wire on Consumer Appliances

(a)

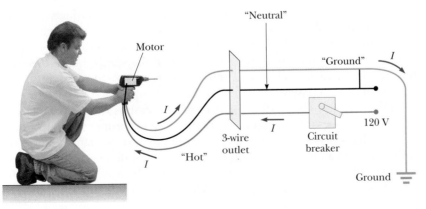

(b)

Figure 18.23 The "hot" (or "live") wire, at 120 V, always includes a circuit breaker for safety. (a) When the drill is operated with two wires, the normal current path is from the "hot" wire, through the motor connections, and back to ground through the "neutral" wire. However, here the high-voltage side has come in contact with the drill case, so that the person holding the drill receives an electrical shock. (b) Shock can be prevented by a third wire running from the drill case to the ground.

and back to ground through the third wire. The resulting high current produced will blow a fuse or trip a circuit breaker before the consumer is injured.

Special power outlets called ground-fault interrupters (GFIs) are now being used in kitchens, bathrooms, basements, and other hazardous areas of new homes. They are designed to protect people from electrical shock by sensing small currents—approximately 5 mA and greater—leaking to ground. When current above this level is detected, the device shuts off (interrupts) the current in less than a millisecond. (Ground-fault interrupters are discussed in Chapter 20.)

18.8 CONDUCTION OF ELECTRICAL SIGNALS BY NEURONS[2]

The most remarkable use of electrical phenomena in living organisms is found in the nervous system of animals. Specialized cells in the body called **neurons** form a complex network that receives, processes, and transmits information from one part of the body to another. The center of this network is located in the brain, which has the ability to store and analyze information. On the basis of this information, the nervous system controls parts of the body.

The nervous system is highly complex and consists of about 10^{10} interconnected neurons. Some aspects of the nervous system are well known. Over the past 45 years, the method of signal propagation through the nervous system has been established. The messages transmitted by neurons are voltage pulses called *action potentials*. When a neuron receives a strong enough stimulus, it produces identical voltage pulses that are actively propagated along its structure. The strength of the stimulus is conveyed by the number of pulses produced. When the pulses reach the end of the neuron, they activate either muscle cells or other neurons. There is a "firing threshold" for neurons: action potentials propagate along a neuron only if the stimulus is sufficiently strong.

Neurons can be divided into three classes: sensory neurons, motor neurons, and interneurons. The sensory neurons receive stimuli from sensory organs that monitor the external and internal environment of the body. Depending on their specialized functions, the sensory neurons convey messages about factors such as light, temperature, pressure, muscle tension, and odor to higher centers in the nervous system. The motor neurons carry messages that control the muscle cells. The messages are based on the information provided by the sensory neurons and by the brain. The interneurons transmit information from one neuron to another.

Each neuron consists of a cell body to which are attached input ends called **dendrites** and a long tail called the **axon**, which transmits the signal away from the cell (Fig. 18.24). The far end of the axon branches into nerve endings that transmit the signal across small gaps to other neurons or to muscle cells. A simple sensori-motor neuron circuit is shown in Figure 18.25. A stimulus from a muscle produces nerve impulses that travel to the spine. Here the signal is transmitted to a motor neuron, which in turn sends impulses to control the muscle. Figure 18.26 shows an electron microscope image of neurons in the brain.

The axon, which is an extension of the neuron cell, conducts electric impulses away from the cell body. Some axons are extremely long. In humans, for example, the axons connecting the spine with the fingers and toes are more than 1 m long. The neuron can transmit messages because of the special active electrical characteristics of the axon. (The axon acts as an **active** source of energy like a battery, rather than like a **passive** stretch of resistive wire.) Much of the information about the electrical and chemical properties of the axon is obtained by inserting small needlelike probes into it. Figure 18.27 shows an experimental setup.

Note that the outside of the axon is grounded, so that all measured voltages are with respect to a zero potential on the outside. With these probes, it is possible to inject current into the axon, measure the resulting action potential as a function of time at a fixed point, and sample the cell's chemical composition. Such experiments

[2]This section is based upon an essay by Paul Davidovits of Boston College.

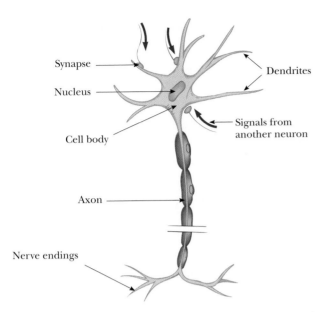

Figure 18.24 Diagram of a neuron.

Synapse

Nucleus

Cell body

Dendrites

Signals from another neuron

Axon

Nerve endings

are usually difficult to run because the diameter of most axons is very small. Even the largest axons in the human nervous system have a diameter of only about 20×10^{-4} cm. The giant squid, however, has an axon with a diameter of about 0.5 mm, which is large enough for the convenient insertion of probes. Much of the information about signal transmission in the nervous system has come from experiments with the squid axon.

In the aqueous environment of the body, salts and other molecules dissociate into positive and negative ions. As a result, body fluids are relatively good conductors of electricity. The inside of the axon is filled with an ionic fluid that is separated from the surrounding body fluid by a thin membrane that is only about 5 nm to 10 nm thick. The resistivities of the internal and external fluids are about the same, but their chemical compositions are substantially different. The external fluid is similar to seawater: Its ionic solutes are mostly positive sodium ions and negative chloride ions. Inside the axon, the positive ions are mostly potassium ions and the negative ions are mostly large organic ions.

Ordinarily, the concentrations of sodium and potassium ions inside and outside the axon would be equalized by diffusion. The axon, however, is a living cell with an energy supply and can change the permeability of its membranes on a time scale of milliseconds.

When the axon is not conducting an electric pulse, the axon membrane is highly permeable to potassium ions, slightly permeable to sodium ions, and impermeable to large organic ions. Consequently, although sodium ions cannot easily enter the axon, potassium ions can leave it. As the potassium ions leave the axon, however, they leave behind large negative organic ions, which cannot follow them through the membrane. As a result, a negative potential builds up inside the axon

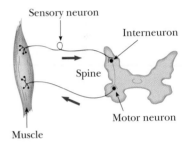

Sensory neuron

Interneuron

Spine

Motor neuron

Muscle

Figure 18.25 A simple neural circuit.

Figure 18.26 Stellate neuron from human cortex.

Juergen Berger, Max-Planck Institute/Science Photo Library/Photo Researchers, Inc.

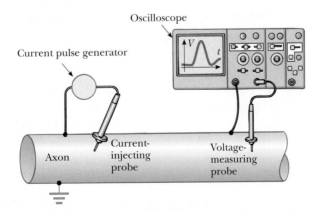

Oscilloscope

Current pulse generator

Axon

Current-injecting probe

Voltage-measuring probe

Figure 18.27 An axon stimulated electrically. The left probe injects a short pulse of current, and the right probe measures the resulting action potential as a function of time.

Figure 18.28 (a) Typical action potential as a function of time. (b) Current in the axon membrane wall as a function of time.

(a)

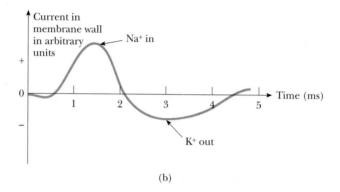

(b)

with respect to the outside. The final negative potential reached, which has been measured at about -70 mV, holds back the outflow of potassium ions so that at equilibrium, the concentration of ions is as we have stated.

The mechanism for the production of an electric signal by the neuron is conceptually simple, but was experimentally difficult to sort out. When a neuron changes its resting potential because of an appropriate stimulus, the properties of its membrane change locally. As a result, there is a sudden flow of sodium ions into the cell that lasts for about two milliseconds. This produces the $+30$ mV peak in the action potential shown in Figure 18.28a. Immediately after, there is an increase in potassium ion flow out of the cell which restores the resting action potential of -70 mV in an additional 3 ms. Both the Na$^+$ and K$^+$ ion flows have been measured by using radioactive Na and K tracers. The nerve signal has been measured to travel along the axon at speeds of 50 m/s to about 150 m/s. This flow of charged particles (or signal transmission) in a nerve axon is *unlike* signal transmission in a metal wire. In an axon, charges move in a direction perpendicular to the direction of travel of the nerve signal, and the nerve signal moves much more slowly than a voltage pulse traveling along a metallic wire.

Although the axon is a highly complex structure, and much of how Na$^+$ and K$^+$ ion channels open and close is not understood, standard electric circuit concepts of current and capacitance can be used to analyze axons. It is left as a problem (Problem 41) to show that the axon, having equal and opposite charges separated by a thin dielectric membrane, acts like a capacitor.

SUMMARY

Physics ⊗ Now™ Take a practice test by logging into PhysicsNow at **www.cp7e.com** and clicking on the Pre-Test link for this chapter.

18.1 Sources of emf

Any device, such as a battery or generator, that increases the electric potential energy of charges in an electric circuit is called a **source of emf**. Batteries convert chemical energy into electrical potential energy, and generators convert mechanical energy into electrical potential energy.

The terminal voltage ΔV of a battery is given by

$$\Delta V = \mathcal{E} - Ir \qquad \text{[18.1]}$$

where $\mathcal{E}$ is the emf of the battery, I is the current, and r is the internal resistance of the battery. Generally, the internal resistance is small enough to be neglected.

18.2   Resistors in Series

The **equivalent resistance** of a set of resistors connected in **series** is

$$R_{eq} = R_1 + R_2 + R_3 + \cdots \qquad [18.4]$$

The current remains at a constant value as it passes through a series of resistors. The potential difference across any two resistors in series is different, unless the resistors have the same resistance.

18.3   Resistors in Parallel

The **equivalent resistance** of a set of resistors connected in **parallel** is

$$\frac{1}{R_{eq}} = \frac{1}{R_1} + \frac{1}{R_2} + \frac{1}{R_3} + \cdots \qquad [18.6]$$

The potential difference across any two parallel resistors is the same; however, the current in each resistor will be different unless the two resistances are equal.

18.4   Kirchhoff's Rules and Complex DC Circuits

Complex circuits can be analyzed using **Kirchhoff's rules**:

1. The sum of the currents entering any junction must equal the sum of the currents leaving that junction.
2. The sum of the potential differences across all the elements around any closed circuit loop must be zero.

The first rule, called the junction rule, is a statement of **conservation of charge**. The second rule, called the loop rule, is a statement of **conservation of energy**. Solving problems involves using these rules to generate as many equations as there are unknown currents. The equations can then be solved simultaneously.

18.5   *RC* Circuits

In a simple *RC* circuit with a battery, a resistor, and a capacitor in series, the charge on the capacitor increases according to the equation

$$q = Q(1 - e^{-t/RC}) \qquad [18.7]$$

The term *RC* in Equation 18.7 is called the **time constant** τ (Greek letter tau), so

$$\tau = RC \qquad [18.8]$$

The time constant represents the time required for the charge to increase from zero to 63.2% of its maximum equilibrium value.

A simple *RC* circuit consisting of a charged capacitor in series with a resistor discharges according to the expression

$$q = Qe^{-t/RC} \qquad [18.9]$$

Problems can be solved by substituting into these equations. The voltage ΔV across the capacitor at any time is obtained by dividing the charge by the capacitance: $\Delta V = q/C$. Using Kirchhoff's loop rule yields the potential difference across the resistor. Ohm's law applied to the resistor then gives the current.

CONCEPTUAL QUESTIONS

1. Is the direction of current in a battery always from the negative terminal to the positive one? Explain.

2. Given three lightbulbs and a battery, sketch as many different circuits as you can.

3. Suppose the energy transferred to a dead battery during charging is *W*. The recharged battery is then used until fully discharged again. Is the total energy transferred out of the battery during use also *W*?

4. (a) A group of resistors connected in parallel have the same (i) current in them, (ii) potential difference across them, or (iii) neither of the above. (b) A group of resistors connected in series have the same (i) current in them, (ii) potential difference across them, or (iii) neither of the above. Justify your answers by considering a circuit consisting of a 3.0-Ω resistor and a 5.0-Ω resistor connected across a 12-V battery.

5. If you have your headlights on while you start your car, why do they dim while the car is starting?

6. (a) The equivalent resistance of a group of resistors connected in parallel is (i) greater than any of the resistors in the group, (ii) less than any of the resistors in the group, or (iii) neither of the above. (b) The equivalent resistance of a group of resistors connected in series is (i) greater than any of the resistors in the group, (ii) less than any of the resistors in the group, or (iii) neither of the above. Justify your answers by considering a circuit consisting of a 3.0-Ω resistor and a 5.0-Ω resistor connected across a 12-V battery.

7. Electrical devices are often rated with a voltage and a current—for example, 120 V, 5 A. Batteries, however, are rated only with a voltage—for example, 1.5 V. Why?

8. A short circuit is a circuit containing a path of very low resistance in parallel with some other part of the circuit. Discuss the effect of a short circuit on the portion of the circuit it parallels. Use a lamp with a frayed line cord as an example.

9. Connecting batteries in series increases the emf applied to a circuit. What advantage might there be to connecting them in parallel?

10. If electrical power is transmitted over long distances, the resistance of the wires becomes significant. Why? Which mode of transmission would result in less energy loss—high current and low voltage or low current and high voltage? Discuss.

11. Describe what happens to the lightbulb in Figure Q18.11 after the switch is closed. Assume the capacitor has a large capacitance and is initially uncharged. Assume also that the bulb lights up when connected directly across the battery terminals.

12. Two sets of Christmas tree lights are available. For set A, when one bulb is removed, the remaining bulbs remain illuminated. For set B, when one bulb is removed, the

Switch
+ | −
Battery

Figure Q18.11

remaining bulbs do not operate. Explain the difference in wiring for the two sets.

13. Why is it possible for a bird to sit on a high-voltage wire without being electrocuted? (See Fig. Q18.13.)

Figure Q18.13 Birds on a high-voltage wire.

14. (a) Two resistors are connected in series across a battery. The power delivered to each resistor is (i) the same or (ii) not necessarily the same. (b) Two resistors are connected in parallel across a battery. The power delivered to each resistor is (i) the same or (ii) not necessarily the same.

15. Embodied in Kirchhoff's rules are two conservation laws. What are they?

16. A ski resort consists of a few chairlifts and several interconnected downhill runs on the side of a mountain, with a lodge at the bottom. The lifts are analogous to batteries and the runs are analogous to resistors. Describe how two runs can be in series. Describe how three runs can be in parallel. Sketch a junction of one lift and two runs. One of the skiers is carrying an altimeter. State Kirchhoff's junction rule and Kirchhoff's loop rule for ski resorts.

17. Suppose you are flying a kite when it strikes a high-voltage wire (a very dangerous situation). What factors determine how great a shock you will receive?

18. Why is it dangerous to turn on a light when you are in a bathtub?

19. Suppose a parachutist lands on a high-voltage wire and grabs the wire as she prepares to be rescued. Will she be electrocuted? If the wire then breaks, should she continue to hold onto the wire as she falls to the ground?

20. Would a fuse or circuit breaker work successfully if it were placed in parallel with the device it was supposed to protect?

21. A series circuit consists of three identical lamps connected to a battery as in Figure Q18.21. When the switch S is closed, what happens (a) to the intensities of lamps A and B, (b) to the intensity of lamp C, (c) to the current in the circuit, and (d) to the voltage drop across the three lamps? (e) Does the power dissipated in the circuit increase, decrease, or remain the same?

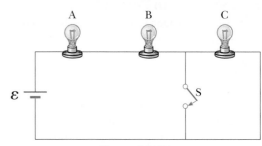

A   B   C

ε

S

Figure Q18.21

22. Figure Q18.22 shows a series connection of three lamps, all rated at 120 V, with power ratings of 60 W, 75 W, and 200 W, respectively. Why do the intensities of the lamps differ? Which lamp has the greatest resistance? How would their intensities differ if they were connected in parallel?

Figure Q18.22

23. A student claims that, of two lightbulbs connected, in series the second is less bright than the first, because the first bulb uses up some of the current. How would you respond to this statement?

24. Two identical, parallel copper wires are placed underground between two points 1.00 mile apart. These wires are usually not connected, but a construction accident shorts the wires together at some point. To try to isolate the spot so that repair can be initiated, a technician goes to end A of the lines and finds that a 12.0-V battery connected across the wires at that end produces a current of 1.00 A. Doing the same at the other end, B, produces a current of 0.20 A. Is the break closer to end A or to end B?

PROBLEMS

1, 2, 3 = straightforward, intermediate, challenging   ☐ = full solution available in *Student Solutions Manual/Study Guide*

Physics⊗Now™ = coached problem with hints available at **www.cp7e.com**   🔬 = biomedical application

Section 18.1 Sources of emf
Section 18.2 Resistors in Series
Section 18.3 Resistors in Parallel

1. A battery having an emf of 9.00 V delivers 117 mA when connected to a 72.0-Ω load. Determine the internal resistance of the battery.

2. A 4.0-Ω resistor, an 8.0-Ω resistor, and a 12-Ω resistor are connected in series with a 24-V battery. What are (a) the equivalent resistance and (b) the current in each resistor? (c) Repeat for the case in which all three resistors are connected in parallel across the battery.

3. A lightbulb marked "75 W [at] 120 V" is screwed into a socket at one end of a long extension cord in which each of the two conductors has a resistance of 0.800 Ω. The other end of the extension cord is plugged into a 120-V outlet. Draw a circuit diagram, and find the actual power of the bulb in the circuit described.

4. A 9.0-Ω resistor and a 6.0-Ω resistor are connected in series with a power supply. (a) The voltage drop across the 6.0-Ω resistor is measured to be 12 V. Find the voltage output of the power supply. (b) The two resistors are connected in parallel across a power supply, and the current through the 9.0-Ω resistor is found to be 0.25 A. Find the voltage setting of the power supply.

5. (a) Find the equivalent resistance between points *a* and *b* in Figure P18.5. (b) Calculate the current in each resistor if a potential difference of 34.0 V is applied between points *a* and *b*.

Figure P18.5

6. Find the equivalent resistance of the circuit in Figure P18.6.

Figure P18.6

7. What is the equivalent resistance of the combination of resistors between points *a* and *b* in Figure P18.7? Note that one end of the vertical resistor is left free.

Figure P18.7

8. (a) Find the equivalent resistance of the circuit in Figure P18.8. (b) If the total power supplied to the circuit is 4.00 W, find the emf of the battery.

Figure P18.8

9. Consider the circuit shown in Figure P18.9. Find (a) the current in the 20.0-Ω resistor and (b) the potential difference between points *a* and *b*.

Figure P18.9

10. Two resistors, *A* and *B*, are connected in parallel across a 6.0-V battery. The current through *B* is found to be 2.0 A. When the two resistors are connected in series to the 6.0-V battery, a voltmeter connected across resistor *A* measures a voltage of 4.0 V. Find the resistances of *A* and *B*.

11. The resistance between terminals *a* and *b* in Figure P18.11 is 75 Ω. If the resistors labeled *R* have the same value, determine *R*.

Figure P18.11

12. Three 100-Ω resistors are connected as shown in Figure P18.12. The maximum power that can safely be delivered to any one resistor is 25.0 W. (a) What is the maximum voltage that can be applied to the terminals a and b? (b) For the voltage determined in part (a), what is the power delivered to each resistor? What is the total power delivered?

Figure P18.12

13. **Physics⊗Now™** Find the current in the 12-Ω resistor in Figure P18.13.

Figure P18.13

14. Calculate the power delivered to each resistor in the circuit shown in Figure P18.14.

Figure P18.14

15. (a) You need a 45-Ω resistor, but the stockroom has only 20-Ω and 50-Ω resistors. How can the desired resistance be achieved under these circumstances? (b) What can you do if you need a 35-Ω resistor?

Section 18.4 Kirchhoff's Rules and Complex DC Circuits

Note: For some circuits, the currents are not necessarily in the direction shown.

16. The ammeter shown in Figure P18.16 reads 2.00 A. Find I_1, I_2, and $\mathcal{E}$.

Figure P18.16

17. Determine the current in each branch of the circuit shown in Figure P18.17.

Figure P18.17

18. Determine the potential difference ΔV_{ab} for the circuit in Figure P18.18.

Figure P18.18

19. Figure P18.19 shows a circuit diagram. Determine (a) the current, (b) the potential of wire A relative to ground, and (c) the voltage drop across the 1 500-Ω resistor.

Figure P18.19

20. In the circuit of Figure P18.20, the current I_1 is 3.0 A while the values of $\mathcal{E}$ and R are unknown. What are the currents I_2 and I_3?

Figure P18.20

21. What is the emf $\mathcal{E}$ of the battery in the circuit of Figure P18.21?

Figure P18.21

22. Four resistors are connected to a battery with a terminal voltage of 12 V, as shown in Figure P18.22. Determine the power delivered to the 50-Ω resistor.

Figure P18.22

23. Using Kirchhoff's rules, (a) find the current in each resistor shown in Figure P18.23 and (b) find the potential difference between points c and f.

Figure P18.23

24. Two 1.50-V batteries—with their positive terminals in the same direction—are inserted in series into the barrel of a flashlight. One battery has an internal resistance of 0.255 Ω, the other an internal resistance of 0.153 Ω. When the switch is closed, a current of 0.600 A passes through the lamp. (a) What is the lamp's resistance? (b) What fraction of the power dissipated is dissipated in the batteries?

25. Calculate each of the unknown currents I_1, I_2, and I_3 for the circuit of Figure P18.25.

Figure P18.25

26. A dead battery is charged by connecting it to the live battery of another car with jumper cables (Fig. P18.26). Determine the current in the starter and in the dead battery.

Figure P18.26

27. Find the current in each resistor in Figure P18.27.

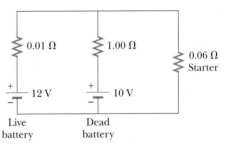

Figure P18.27

28. (a) Determine the potential difference ΔV_{ab} for the circuit in Figure P18.28. Note that each battery has an internal resistance as indicated in the figure. (b) If points a and b are connected by a 7.0-Ω resistor, what is the current in this resistor?

Figure P18.28

29. Find the potential difference across each resistor in Figure P18.29.

Figure P18.29

Section 18.5 *RC* Circuits

30. Show that $\tau = RC$ has units of time.

31. Consider a series *RC* circuit for which $C = 6.0\ \mu F$, $R = 2.0 \times 10^6\ \Omega$, and $\mathcal{E} = 20$ V. Find (a) the time constant of the circuit and (b) the maximum charge on the capacitor after a switch in the circuit is closed.

32. An uncharged capacitor and a resistor are connected in series to a source of emf. If $\mathcal{E} = 9.00$ V, $C = 20.0\ \mu F$, and $R = 100\ \Omega$, find (a) the time constant of the circuit, (b) the maximum charge on the capacitor, and (c) the charge on the capacitor after one time constant.

33. **Physics⊗Now™** Consider a series *RC* circuit for which $R = 1.0$ MΩ, $C = 5.0\ \mu F$, and $\mathcal{E} = 30$ V. Find the charge on the capacitor 10 s after the switch is closed.

34. A series combination of a 12-kΩ resistor and an unknown capacitor is connected to a 12-V battery. One second after the circuit is completed, the voltage across the capacitor is 10 V. Determine the capacitance of the capacitor.

35. A capacitor in an *RC* circuit is charged to 60.0% of its maximum value in 0.900 s. What is the time constant of the circuit.

36. A series *RC* circuit has a time constant of 0.960 s. The battery has an emf of 48.0 V, and the maximum current in the circuit is 0.500 mA. What are (a) the value of the capacitance and (b) the charge stored in the capacitor 1.92 s after the switch is closed?

Section 18.6 Household Circuits

37. An electric heater is rated at 1 300 W, a toaster at 1 000 W, and an electric grill at 1 500 W. The three appliances are connected in parallel to a common 120-V circuit. (a) How much current does each appliance draw? (b) Is a 30.0-A circuit breaker sufficient in this situation? Explain.

38. A lamp ($R = 150\ \Omega$), an electric heater ($R = 25\ \Omega$), and a fan ($R = 50\ \Omega$) are connected in parallel across a 120-V line. (a) What total current is supplied to the circuit? (b) What is the voltage across the fan? (c) What is the current in the lamp? (d) What power is expended in the heater?

39. A heating element in a stove is designed to dissipate 3 000 W when connected to 240 V. (a) Assuming that the resistance is constant, calculate the current in the heating element if it is connected to 120 V. (b) Calculate the power it dissipates at that voltage.

40. Your toaster oven and coffeemaker each dissipate 1 200 W of power. Can you operate them together if the 120-V line that feeds them has a circuit breaker rated at 15 A? Explain.

Section 18.8 Conduction of Electrical Signals by Neurons

41. Assume that a length of axon membrane of about 10 cm is excited by an action potential (length excited = nerve speed $\times$ pulse duration = 50 m/s $\times$ 2.0 ms = 10 cm). In the resting state, the outer surface of the axon wall is charged positively with K^+ ions and the inner wall has an equal and opposite charge of negative organic ions, as shown in Figure P18.41. Model the axon as a parallel-plate capacitor, and take $C = \kappa\varepsilon_0 A/d$ and $Q = C\Delta V$ to investigate the charge as follows: Use typical values for a cylindrical axon of cell wall thickness $d = 1.0 \times 10^{-8}$ m, axon radius $r = 10\ \mu m$, and cell-wall dielectric constant $\kappa = 3.0$. (a) Calculate the positive charge on the outside of a 10-cm piece of axon when it is not conducting an electric pulse. How many K^+ ions are on the outside of the axon? Is this a large charge per unit area? [*Hint*: Calculate the charge per unit area in terms of the number of angstroms ($Å^2$) per electronic charge. An atom has a cross section of about $1\ Å^2$ ($1\ Å = 10^{-10}$ m)] (b) How much positive charge must flow through the cell membrane to reach the excited state of $+30$ mV from the resting state of -70 mV? How many sodium ions (Na^+) is this? (c) If it takes 2.0 ms for the Na^+ ions to enter the axon, what is the average current in the axon wall in this process? (d) How much energy does it take to raise the potential of the inner axon wall to $+30$ mV, starting from the resting potential of -70 mV?

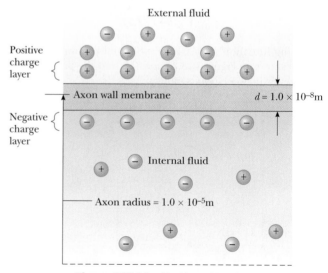

Figure P18.41   (Problems 41 and 42)

42. Consider the model of the axon as a capacitor from Problem 41 and Figure P18.41. (a) How much energy does it take to restore the inner wall of the axon to -70 mV, starting from $+30$ mV? (b) Find the average current in the axon wall during this process.

43. Using Figure 18.28b and the results of Problems 18.41d and 18.42a, find the power supplied by the axon per action potential.

ADDITIONAL PROBLEMS

44. Consider an *RC* circuit in which the capacitor is being charged by a battery connected in the circuit. After a time equal to two time constants, what percentage of the *final* charge is present on the capacitor?

45. Find the equivalent resistance between points *a* and *b* in Figure P18.45.

Figure P18.45

46. For the circuit in Figure P18.46, calculate (a) the equivalent resistance of the circuit and (b) the power dissipated by the entire circuit. (c) Find the current in the 5.0-Ω resistor.

Figure P18.46

47. Find (a) the equivalent resistance of the circuit in Figure P18.47, (b) each current in the circuit, (c) the potential difference across each resistor, and (d) the power dissipated by each resistor.

Figure P18.47

48. Three 60.0-W, 120-V lightbulbs are connected across a 120-V power source, as shown in Figure P18.48. Find (a) the total power delivered to the three bulbs and (b) the potential difference across each. Assume that the resistance of each bulb is constant (even though, in reality, the resistance increases markedly with current).

Figure P18.48

49. **Physics⊗Now™** An automobile battery has an emf of 12.6 V and an internal resistance of 0.080 Ω. The headlights have a total resistance of 5.00 Ω (assumed constant). What is the potential difference across the headlight bulbs (a) when they are the only load on the battery and (b) when the starter motor is operated, taking an additional 35.0 A from the battery?

50. In Figure P18.50, suppose that the switch has been closed for a length of time sufficiently long for the capacitor to become fully charged. Find (a) the steady-state current in each resistor and (b) the charge on the capacitor.

Figure P18.50

51. Find the values of I_1, I_2, and I_3 for the circuit in Figure P18.51.

Figure P18.51

52. The resistance between points *a* and *b* in Figure P18.52 drops to one-half its original value when switch S is closed. Determine the value of *R*.

Figure P18.52

53. A generator has a terminal voltage of 110 V when it delivers 10.0 A and 106 V when it delivers 30.0 A. Calculate the emf and the internal resistance of the generator.

54. An emf of 10 V is connected to a series RC circuit consisting of a resistor of $2.0 \times 10^6\ \Omega$ and a capacitor of 3.0 μF. Find the time required for the charge on the capacitor to reach 90% of its final value.

55. The student engineer of a campus radio station wishes to verify the effectiveness of the lightning rod on the antenna mast (Fig. P18.55). The unknown resistance R_x is between points C and E. Point E is a "true ground," but is inaccessible for direct measurement, since the stratum in which it is located is several meters below the Earth's surface. Two identical rods are driven into the ground at A and B, introducing an unknown resistance R_y. The procedure for finding the unknown resistance R_x is as follows: Measure resistance R_1 between points A and B. Then connect A and B with a heavy conducting wire, and measure resistance R_2 between points A and C. (a) Derive a formula for R_x in terms of the observable resistances R_1 and R_2. (b) A satisfactory ground resistance would be $R_x < 2.0\ \Omega$. Is the grounding of the station adequate if measurements give $R_1 = 13\ \Omega$ and $R_2 = 6.0\ \Omega$?

Figure P18.55

56. The resistor R in Figure P18.56 dissipates 20 W of power. Determine the value of R.

Figure P18.56

57. Physics⊗Now™ A voltage ΔV is applied to a series configuration of n resistors, each of resistance R. The circuit components are reconnected in a parallel configuration, and voltage ΔV is again applied. Show that the power consumed by the series configuration is $1/n^2$ times the power consumed by the parallel configuration.

58. For the network in Figure P18.58, show that the resistance between points a and b is $R_{ab} = \frac{27}{17}\ \Omega$. [*Hint*: Connect a battery with emf $\mathcal{E}$ across points a and b and determine $\mathcal{E}/I$, where I is the current in the battery.]

Figure P18.58

59. A battery with an internal resistance of 10.0 Ω produces an open-circuit voltage of 12.0 V. A variable load resistance with a range from 0 to 30.0 Ω is connected across the battery. (*Note*: A battery has a resistance that depends on the condition of its chemicals and that increases as the battery ages. This internal resistance can be represented in a simple circuit diagram as a resistor in series with the battery.) (a) Graph the power dissipated in the load resistor as a function of the load resistance. (b) With your graph, demonstrate the following important theorem: *The power delivered to a load is a maximum if the load resistance equals the internal resistance of the source.*

60. The circuit in Figure P18.60 contains two resistors, $R_1 = 2.0\ k\Omega$ and $R_2 = 3.0\ k\Omega$, and two capacitors, $C_1 = 2.0\ \mu$F and $C_2 = 3.0\ \mu$F, connected to a battery with emf $\mathcal{E} = 120$ V. If there are no charges on the capacitors before switch S is closed, determine the charges q_1 and q_2 on capacitors C_1 and C_2, respectively, as functions of time, after the switch is closed. [*Hint*: First reconstruct the circuit so that it becomes a simple RC circuit containing a single resistor and single capacitor in series, connected to the battery, and then determine the total charge q stored in the circuit.]

Figure P18.60

61. Consider the circuit shown in Figure P18.61. Find (a) the potential difference between points a and b and (b) the current in the 20.0-Ω resistor.

Figure P18.61

62. In Figure P18.62, $R_1 = 0.100\ \Omega$, $R_2 = 1.00\ \Omega$, and $R_3 = 10.0\ \Omega$. Find the equivalent resistance of the circuit and the current in each resistor when a 5.00-V power supply is connected between (a) points A and B, (b) points A and C, and (c) points A and D.

Figure P18.62

63. What are the expected readings of the ammeter and voltmeter for the circuit in Figure P18.63?

Figure P18.63

64. Consider the two arrangements of batteries and bulbs shown in Figure P18.64. The two bulbs are identical and have resistance R, and the two batteries are identical with output voltage ΔV. (a) In case 1, with the two bulbs in series, compare the brightness of each bulb, the current in each bulb, and the power delivered to each bulb? (b) In case 2, with the two bulbs in parallel, compare the brightness of each bulb, the current in each bulb, and the power supplied to each bulb. (c) Which bulbs are brighter, those in case 1 or those in case 2? (d) In each case, if one bulb fails, will the other go out as well? If the other bulb doesn't fail, will it get brighter or stay the same?

Case 1   Case 2

Figure P18.64

(Problem 64 is courtesy of E.F. Redish. For other problems of this type, visit http://www.physics.umd.edu/perg/.)

ACTIVITIES

1. Using insulated wire, connect one terminal of a D-cell battery to the base of a flashlight bulb, tape a second wire to the other battery terminal, and tape a third wire to the center conductor of the bulb, as shown in the Figure A18.1a. Be sure to remove about 1 cm of insulation from the ends of all wires before making connections. Connect the two open wires together to complete the circuit, and note the illumination of the bulb. Now add a second D-cell battery to the circuit as in Figure A18.1b to give a total voltage of 3.0 V, connect the two open wires together to complete the circuit, and note the illumination of the bulb again. Why does the bulb grow brighter in this case?

Figure A18.1

2. Use the basic equipment of activity 1 plus a few more items to test some additional features of circuits. First, note the brightness of a single bulb connected to the battery. Now connect two bulbs in series with each other and the battery. Predict whether the bulbs will be dimmer or brighter than when operated separately. Try it and see. Continue for three bulbs in series.

3. Repeat activity 2, but this time connect the bulbs in parallel with each other. Predict how the brightness will change as you add more bulbs in parallel. Explain.

4. Continue your experimentation with circuits by connecting the battery to one bulb, followed in the same circuit by two bulbs in parallel and then back to the battery. Predict the brightness of each bulb in this situation before you connect the circuit. Explain the results you obtain.

Aurora borealis, the Northern Lights. Displays such as this one are caused by cosmic ray particles trapped in the magnetic field of Earth. When the particles collide with atoms in the atmosphere, they cause the atoms to emit visible light.

Johnny Johnson/Stone/Getty

CHAPTER

19

Magnetism

In terms of applications, magnetism is one of the most important fields in physics. Large electromagnets are used to pick up heavy loads. Magnets are used in such devices as meters, motors, and loudspeakers. Magnetic tapes and disks are used routinely in sound- and video-recording equipment and to store computer data. Intense magnetic fields are used in magnetic resonance imaging (MRI) devices to explore the human body with better resolution and greater safety than x-rays can provide. Giant superconducting magnets are used in the cyclotrons that guide particles into targets at nearly the speed of light, and magnetic bottles hold antimatter, possibly the key to future space propulsion systems.

Magnetism is closely linked with electricity. Magnetic fields affect moving charges, and moving charges produce magnetic fields. Changing magnetic fields can even create electric fields. These phenomena signify an underlying unity of electricity and magnetism, which James Clerk Maxwell first described in the 19th century. The ultimate source of any magnetic field is electric current.

19.1 MAGNETS

Most people have had experience with some form of magnet. You are most likely familiar with the common iron horseshoe magnet that can pick up iron-containing objects such as paper clips and nails. Several commercially available magnets are shown in Figure 19.1. In the discussion that follows, we assume the magnet has the shape of a bar. Iron objects are most strongly attracted to either end of such a bar magnet, called its **poles**. One end is called the **north pole** and the other the **south pole**. The names come from the behavior of a magnet in the presence of Earth's magnetic field. If a bar magnet is suspended from its midpoint by a piece of string so that it can swing freely in a horizontal plane, it will rotate until its north pole points to the north and its south pole points to the south. The same idea is used to construct a simple compass. Magnetic poles also exert attractive or repulsive forces on each other similar to the electrical forces between charged objects. In fact, simple

Courtesy of Central Scientific Company

Figure 19.1   An assortment of commercially available magnets. The four red magnets and the large black magnet on the left are made of an alloy of iron, aluminum, and cobalt. The six horseshoe magnets on the right are made of different nickel–steel alloys. The rectangular magnets on the lower right are ceramics made of iron, nickel, and beryllium oxides.

experiments with two bar magnets show that **like poles repel each other and unlike poles attract each other**.

Although the force between opposite magnetic poles is similar to the force between positive and negative electric charges, there is an important difference: positive and negative electric charges can exist in isolation of each other; north and south poles don't. No matter how many times a permanent magnet is cut, each piece always has a north pole and a south pole. There is some theoretical basis for the speculation that magnetic monopoles (isolated north or south poles) exist in nature, and the attempt to detect them is currently an active experimental field of investigation.

An unmagnetized piece of iron can be magnetized by stroking it with a magnet. Magnetism can also be induced in iron (and other materials) by other means. For example, if a piece of unmagnetized iron is placed near a strong permanent magnet, the piece of iron eventually becomes magnetized. The process can be accelerated by heating and then cooling the iron.

Naturally occurring magnetic materials such as magnetite are magnetized in this way because they have been subjected to Earth's magnetic field for long periods of time. The extent to which a piece of material retains its magnetism depends on whether it is classified as magnetically hard or soft. **Soft** magnetic materials, such as iron, are easily magnetized, but also tend to lose their magnetism easily. In contrast, **hard** magnetic materials, such as cobalt and nickel, are difficult to magnetize, but tend to retain their magnetism.

In earlier chapters we described the interaction between charged objects in terms of electric fields. Recall that an electric field surrounds any stationary electric charge. The region of space surrounding a *moving* charge includes a magnetic field as well. A magnetic field also surrounds a properly magnetized magnetic material.

To describe any type of vector field, we must define its magnitude, or strength, and its direction. The direction of a magnetic field $\vec{\mathbf{B}}$ at any location is the direction in which the north pole of a compass needle points at that location. Active Figure 19.2a shows how the magnetic field of a bar magnet can be traced with the aid of a compass, defining a **magnetic field line**. Several magnetic field

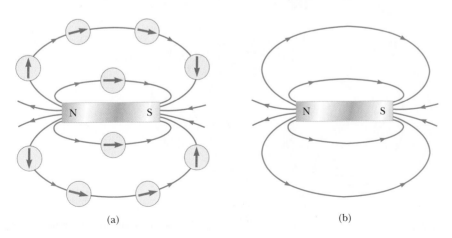

(a)   (b)

ACTIVE FIGURE 19.2
(a) Tracing the magnetic field of a bar magnet. (b) Several magnetic field lines of a bar magnet.

Physics⊗Now™

Log into PhysicsNow at **www.cp7e.com** and go to Active Figure 19.2, where you can move the compass around and trace the magnetic field for yourself.

(a) (b) (c)

Figure 19.3 (a) The magnetic field pattern of a bar magnet, as displayed with iron filings on a sheet of paper. (b) The magnetic field pattern between *unlike* poles of two bar magnets, as displayed with iron filings. (c) The magnetic field pattern between two *like* poles.

lines of a bar magnet traced out in this way appear in the two-dimensional representation in Active Figure19.2b. Magnetic field patterns can be displayed by placing small iron filings in the vicinity of a magnet, as in Figure 19.3.

Forensic scientists use a technique similar to that shown in Figure 19.3 to find fingerprints at a crime scene. One way to find latent, or invisible, prints is by sprinkling a powder of iron dust on a surface. The iron adheres to any perspiration or body oils that are present and can be spread around on the surface with a magnetic brush that never comes into contact with the powder or the surface.

APPLICATION

Dusting for Fingerprints

19.2 EARTH'S MAGNETIC FIELD

A small bar magnet is said to have north and south poles, but it's more accurate to say it has a "north-seeking" pole and a "south-seeking" pole. By these expressions, we mean that if such a magnet is used as a compass, one end will "seek," or point to, the geographic North Pole of Earth and the other end will "seek," or point to, the geographic South Pole of Earth. We conclude that **the geographic North Pole of Earth corresponds to a magnetic south pole, and the geographic South Pole of Earth corresponds to a magnetic north pole.** In fact, the configuration of Earth's magnetic field, pictured in Figure 19.4, very much resembles

TIP 19.1 The Geographic North Pole is the Magnetic South Pole

The north pole of a magnet in a compass points north because it's attracted to the Earth's *magnetic* south pole—located near the Earth's *geographic* north pole.

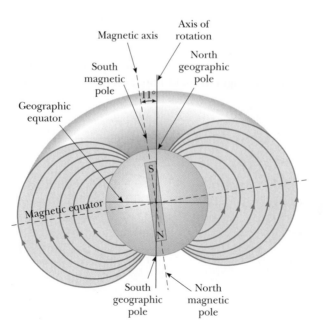

Figure 19.4 Earth's magnetic field lines. Note that magnetic south is at the north geographic pole and magnetic north is at the south geographic pole.

what would be observed if a huge bar magnet were buried deep in the Earth's interior.

If a compass needle is suspended in bearings that allow it to rotate in the vertical plane as well as in the horizontal plane, the needle is horizontal with respect to Earth's surface only near the equator. As the device is moved northward, the needle rotates so that it points more and more toward the surface of Earth. The angle between the direction of the magnetic field and the horizontal is called the **dip angle**. Finally, at a point just north of Hudson Bay in Canada, the north pole of the needle points directly downward, with a dip angle of 90°. This site, first found in 1832, is considered to be the location of the south magnetic pole of Earth. It is approximately 1 300 miles from Earth's geographic North Pole and varies with time. Similarly, Earth's magnetic north pole is about 1 200 miles from its geographic South Pole. This means that compass needles point only approximately north. The difference between true north, defined as the geographic North Pole, and north indicated by a compass varies from point to point on Earth, a difference referred to as *magnetic declination*. For example, along a line through South Carolina and the Great Lakes a compass indicates true north, whereas in Washington state it aligns 25° east of true north (Fig. 19.5).

Although the magnetic field pattern of Earth is similar to the pattern that would be set up by a bar magnet placed at its center the source of Earth's field can't consist of large masses of permanently magnetized material. Earth does have large deposits of iron ore deep beneath its surface, but the high temperatures in the core prevent the iron from retaining any permanent magnetization. It's considered more likely that the true source of Earth's magnetic field is electric current in the liquid part of its core. This current, which is not well understood, may be driven by an interaction between the planet's rotation and convection in the hot liquid core. There is some evidence that the strength of a planet's magnetic field is related to the planet's rate of rotation. For example, Jupiter rotates faster than Earth, and recent space probes indicate that Jupiter's magnetic field is stronger than Earth's, even though Jupiter lacks an iron core. Venus, on the other hand, rotates more slowly than Earth, and its magnetic field is weaker. Investigation into the cause of Earth's magnetism continues.

An interesting fact concerning Earth's magnetic field is that its direction reverses every few million years. Evidence for this phenomenon is provided by basalt (an iron-containing rock) that is sometimes spewed forth by volcanic activity on the ocean floor. As the lava cools, it solidifies and retains a picture of the direction of Earth's magnetic field. When the basalt deposits are dated, they provide evidence for periodic reversals of the magnetic field. The cause of these field reversals is still not understood.

It has long been speculated that some animals, such as birds, use the magnetic field of Earth to guide their migrations. Studies have shown that a type of anaerobic bacterium that lives in swamps has a magnetized chain of magnetite as part of its internal structure. (The term *anaerobic* means that these bacteria live and grow

APPLICATION

Magnetic Bacteria

Figure 19.5 A map of the continental United States showing the declination of a compass from true north.

without oxygen; in fact, oxygen is toxic to them.) The magnetized chain acts as a compass needle that enables the bacteria to align with Earth's magnetic field. When they find themselves out of the mud on the bottom of the swamp, they return to their oxygen-free environment by following the magnetic field lines of Earth. Further evidence for their magnetic sensing ability is the fact that bacteria found in the Northern Hemisphere have internal magnetite chains that are opposite in polarity to those of similar bacteria in the Southern Hemisphere. This is consistent with the fact that in the Northern Hemisphere Earth's field has a downward component, whereas in the Southern Hemisphere it has an upward component. Recently, a meteorite originating on Mars has been found to contain a chain of magnetite. NASA scientists believe it may be a fossil of ancient Martian bacterial life.

APPLICATION

Labeling Airport Runways

The magnetic field of Earth is used to label runways at airports according to their direction. A large number is painted on the end of the runway so that it can be read by the pilot of an incoming airplane. This number describes the direction in which the airplane is traveling, expressed as the magnetic heading, in degrees measured clockwise from magnetic north divided by 10. A runway marked 9 would be directed toward the east (90° divided by 10), while a runway marked 18 would be directed toward magnetic south.

Applying Physics 19.1   Compasses Down Under

On a business trip to Australia, you take along your American-made compass that you may have used on a camping trip. Does this compass work correctly in Australia?

Explanation   There's no problem with using the compass in Australia. The north pole of the magnet in the compass will be attracted to the south magnetic pole near the geographic North Pole, just as it was in the United States. The only difference in the magnetic field lines is that they have an upward component in Australia, whereas they have a downward component in the United States. Held in a horizontal plane, your compass can't detect this, however—it only displays the direction of the horizontal component of the magnetic field.

19.3   MAGNETIC FIELDS

Experiments show that a stationary charged particle doesn't interact with a static magnetic field. **When a charged particle is *moving* through a magnetic field, however, a magnetic force acts on it.** This force has its maximum value when the charge moves in a direction perpendicular to the magnetic field lines, decreases in value at other angles, and becomes zero when the particle moves along the field lines. This is quite different from the electric force, which exerts a force on a charged particle whether it's moving or at rest. Further, the electric force is directed parallel to the electric field while the magnetic force on a moving charge is directed perpendicular to the magnetic field.

In our discussion of electricity, the electric field at some point in space was defined as the electric force per unit charge acting on some test charge placed at that point. In a similar way, we can describe the properties of the magnetic field $\vec{\mathbf{B}}$ at some point in terms of the magnetic force exerted on a test charge at that point. Our test object is a charge q moving with velocity $\vec{\mathbf{v}}$. It is found experimentally that the strength of the magnetic force on the particle is proportional to the magnitude of the charge q, the magnitude of the velocity $\vec{\mathbf{v}}$, the strength of the external magnetic field $\vec{\mathbf{B}}$, and the sine of the angle θ between the direction of $\vec{\mathbf{v}}$ and the direction of $\vec{\mathbf{B}}$. These observations can be summarized by writing the magnitude of the magnetic force as

$$F = qvB \sin \theta \qquad\qquad [19.1]$$

This expression is used to define the magnitude of the magnetic field as

$$B \equiv \frac{F}{qv\sin\theta} \qquad \text{[19.2]}$$

If F is in newtons, q in coulombs, and v in meters per second, then the SI unit of magnetic field is the **tesla** (T), also called the **weber** (Wb) **per square meter** (1 T = 1 Wb/m^2). If a 1-C charge moves in a direction perpendicular to a magnetic field of magnitude 1 T with a speed of 1 m/s, the magnetic force exerted on the charge is 1 N. We can express the units of $\vec{\textbf{B}}$ as

$$[B] = T = \frac{Wb}{m^2} = \frac{N}{C \cdot m/s} = \frac{N}{A \cdot m} \qquad \text{[19.3]}$$

In practice, the cgs unit for magnetic field, the **gauss** (G), is often used. The gauss is related to the tesla through the conversion

$$1\ T = 10^4\ G$$

Conventional laboratory magnets can produce magnetic fields as large as about 25 000 G, or 2.5 T. Superconducting magnets that can generate magnetic fields as great as 3×10^5 G, or 30 T, have been constructed. These values can be compared with the value of Earth's magnetic field near its surface, which is about 0.5 G, or 0.5×10^{-4} T.

From Equation 19.1 we see that the force on a charged particle moving in a magnetic field has its maximum value when the particle's motion is *perpendicular* to the magnetic field, corresponding to $\theta = 90°$, so that $\sin\theta = 1$. The magnitude of this maximum force has the value

$$F_{\max} = qvB \qquad \text{[19.4]}$$

Also from Equation 19.1, F is zero when $\vec{\textbf{v}}$ is parallel to $\vec{\textbf{B}}$ (corresponding to $\theta = 0°$ or 180°), so no magnetic force is exerted on a charged particle when it moves in the direction of the magnetic field or opposite the field.

Experiments show that the direction of the magnetic force is always perpendicular to both $\vec{\textbf{v}}$ and $\vec{\textbf{B}}$, as shown in Figure 19.6 for a positively charged particle. To determine the direction of the force, we employ the **right-hand rule number 1**:

1. Point the fingers of your right hand in the direction of the velocity $\vec{\textbf{v}}$.
2. Curl the fingers in the direction of the magnetic field $\vec{\textbf{B}}$, moving through the smallest angle (as in Fig. 19.7).
3. Your thumb is now pointing in the direction of the magnetic force $\vec{\textbf{F}}$ exerted on a positive charge.

If the charge is negative rather than positive, the force $\vec{\textbf{F}}$ is directed *opposite* that shown in Figures 19.6 and 19.7. So if q is negative, simply use the right-hand rule to find the direction for positive q, and then reverse that direction for the negative charge.

A charged particle moves in a straight line through a region of space. Which of the following answers *must* be true? (Assume any other fields are negligible.) The magnetic field (a) has a magnitude of zero (b) has a zero component perpendicular to the particle's velocity (c) has a zero component parallel to the particle's velocity in that region.

The north-pole end of a bar magnet is held near a stationary positively charged piece of plastic. Is the plastic (a) attracted, (b) repelled, or (c) unaffected by the magnet?

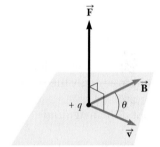

Figure 19.6 The direction of the magnetic force on a positively charged particle moving with a velocity $\vec{\textbf{v}}$ in the presence of a magnetic field. When $\vec{\textbf{v}}$ is at an angle θ with respect to $\vec{\textbf{B}}$, the magnetic force is perpendicular to both $\vec{\textbf{v}}$ and $\vec{\textbf{B}}$.

Figure 19.7 Right-hand rule number 1 for determining the direction of the magnetic force on a positive charge moving with a velocity $\vec{\textbf{v}}$ in a magnetic field $\vec{\textbf{B}}$. Point the fingers of your right hand in the direction of $\vec{\textbf{v}}$, and then curl them in the direction of $\vec{\textbf{B}}$. The magnetic force $\vec{\textbf{F}}$ points in the direction of your right thumb.

EXAMPLE 19.1   A Proton Traveling in Earth's Magnetic Field

Goal   Calculate the magnitude and direction of a magnetic force.

Problem   A proton moves with a speed of 1.00×10^5 m/s through Earth's magnetic field, which has a value of 55.0 μT at a particular location. When the proton moves eastward, the magnetic force acting on it is directed straight upward, and when it moves northward, no magnetic force acts on it. **(a)** What is the direction of the magnetic field, and **(b)** what is the strength of the magnetic force when the proton moves eastward? **(c)** Calculate the gravitational force on the proton and compare it with the magnetic force. Compare it also with the electric force if there were an electric field with magnitude $E = 1.50 \times 10^2$ N/C at that location, a common value at Earth's surface. Note that the mass of the proton is 1.67×10^{-27} kg.

Strategy   The direction of the magnetic field can be found from an application of the right-hand rule, together with the fact that no force is exerted on the proton when it's traveling north. Substituting into Equation 19.1 yields the magnitude of the magnetic field.

Solution
(a) Find the direction of the magnetic field.

No magnetic force acts on the proton when it's going north, so the angle such a proton makes with the magnetic field direction must be either 0° or 180°. Therefore, the magnetic field $\vec{\mathbf{B}}$ must point either north or south. Now apply the right-hand rule. When the particle travels east, the magnetic force is directed upward. Point your thumb in the direction of the force and your fingers in the direction of the velocity eastward. When you curl your fingers, they point north, which must therefore be the direction of the magnetic field.

(b) Find the magnitude of the magnetic force.

Substitute the given values and the charge of a proton into Equation 19.1. From part (a), the angle between the velocity $\vec{\mathbf{v}}$ of the proton and the magnetic field $\vec{\mathbf{B}}$ is 90.0°.

$$F = qvB \sin\theta = (1.60 \times 10^{-19} \text{ C})(1.00 \times 10^5 \text{ m/s})$$
$$\times (55.0 \times 10^{-6} \text{ T})\sin(90.0°)$$
$$= 8.80 \times 10^{-19} \text{ N}$$

(c) Calculate the gravitational force on the proton and compare it with the magnetic force, and also with the electric force if $E = 1.50 \times 10^2$ N/C:

$$F_{\text{grav}} = mg = (1.67 \times 10^{-27} \text{ kg})(9.80 \text{ m/s}^2)$$
$$= 1.64 \times 10^{-26} \text{ N}$$

$$F_{\text{elec}} = qE = (1.60 \times 10^{-19} \text{ C})(1.50 \times 10^2 \text{ N/C})$$
$$= 2.40 \times 10^{-17} \text{ N}$$

Remarks   The information regarding a proton moving north was necessary to fix the direction of the magnetic field. Otherwise, an upward magnetic force on an eastward-moving proton could be caused by a magnetic field pointing anywhere northeast or northwest. Notice in part (c) the relative strengths of the forces, with the electric force larger than the magnetic force and both much larger than the gravitational force, all for typical field values found in nature.

Exercise 19.1
Suppose an electron is moving due west in the same magnetic field as in Example 19.1 at a speed of 2.50×10^5 m/s. Find the magnitude and direction of the magnetic force on the electron.

Answer 2.20×10^{-18} N, straight up. (Don't forget, the electron is negatively charged!)

EXAMPLE 19.2   A Proton Moving in a Magnetic Field

Goal   Calculate the magnetic force and acceleration when a particle moves at an angle other than ninety degrees to the field.

Problem   A proton moves at 8.00×10^6 m/s along the x-axis. It enters a region in which there is a magnetic field of magnitude 2.50 T, directed at an angle of 60.0° with the x-axis and lying in the xy-plane (Fig. 19.8).

(a) Find the initial magnitude and direction of the magnetic force on the proton. **(b)** Calculate the proton's initial acceleration.

Strategy Finding the magnitude and direction of the magnetic force requires substituting values into the equation for magnetic force, Equation 19.1, and using the right-hand rule. Applying Newton's second law solves part (b).

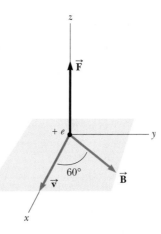

Figure 19.8 (Example 19.2) The magnetic force $\vec{\mathbf{F}}$ on a proton is in the positive z-direction when $\vec{\mathbf{v}}$ and $\vec{\mathbf{B}}$ lie in the xy-plane.

Solution
(a) Find the magnitude and direction of the magnetic force on the proton.

Substitute $v = 8.00 \times 10^6$ m/s, the magnetic field strength $B = 2.50$ T, the angle, and the charge of a proton into Equation 19.1:

$$F = qvB \sin \theta$$
$$= (1.60 \times 10^{-19} \text{ C})(8.00 \times 10^6 \text{ m/s})(2.50 \text{ T})(\sin 60°)$$
$$F = \boxed{2.77 \times 10^{-12} \text{ N}}$$

Apply the right-hand rule number 1 to find the initial direction of the magnetic force:

Point the fingers of the right hand in the x-direction (the direction of $\vec{\mathbf{v}}$), and then curl them towards $\vec{\mathbf{B}}$. The thumb points upwards, in the positive z-direction.

(b) Calculate the proton's initial acceleration.

Substitute the force and the mass of a proton into Newton's second law:

$$ma = F \quad \rightarrow \quad (1.67 \times 10^{-27} \text{ kg})a = 2.77 \times 10^{-12} \text{ N}$$
$$a = \boxed{1.66 \times 10^{15} \text{ m/s}^2}$$

Remarks The initial acceleration is also in the positive z-direction. Because the direction of $\vec{\mathbf{v}}$ changes, however, the subsequent direction of the magnetic force also changes. In applying right-hand rule number 1 to find the direction, it was important to take into consideration the charge. A negatively charged particle accelerates in the opposite direction.

Exercise 19.2
Calculate the acceleration of an electron that moves through the same magnetic field as in Example 19.2, at the same velocity as the proton. The mass of an electron is 9.11×10^{-31} kg.

Answer 3.04×10^{18} m/s^2 in the negative z-direction

19.4 MAGNETIC FORCE ON A CURRENT-CARRYING CONDUCTOR

If a magnetic field exerts a force on a single charged particle when it moves through a magnetic field, it should be no surprise that magnetic forces are exerted on a current-carrying wire, as well (see Fig. 19.9). This follows from the fact that the current is a collection of many charged particles in motion; hence, the resultant force on the wire is due to the sum of the individual forces on the charged particles. The force on the particles is transmitted to the "bulk" of the wire through collisions with the atoms making up the wire.

Some explanation is in order concerning notation in many of the figures. To indicate the direction of $\vec{\mathbf{B}}$, we use the following conventions:

If $\vec{\mathbf{B}}$ is directed into the page, as in Figure 19.10 (page 632), we use a series of blue crosses, representing the tails of arrows. If $\vec{\mathbf{B}}$ is directed out of the page, we use a series of blue dots, representing the tips of arrows. If $\vec{\mathbf{B}}$ lies in the plane of the page, we use a series of blue field lines with arrowheads.

Courtesy of Henry Leap and Jim Lehman

Figure 19.9 This apparatus demonstrates the force on a current-carrying conductor in an external magnetic field. Why does the bar swing *away* from the magnet after the switch is closed?

Figure 19.10 A segment of a flexible vertical wire partially stretched between the poles of a magnet, with the field (blue crosses) directed into the page. (a) When there is no current in the wire, it remains vertical. (b) When the current is upward, the wire deflects to the left. (c) When the current is downward, the wire deflects to the right.

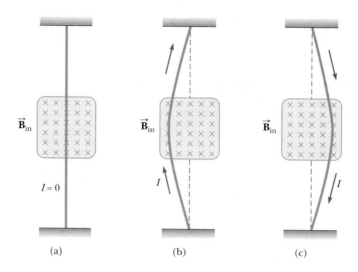

$\vec{B}_{in}$ $\vec{B}_{in}$ $\vec{B}_{in}$

$I = 0$ I I

(a)   (b)   (c)

Figure 19.11 A section of a wire containing moving charges in an external magnetic field $\vec{B}$.

TIP 19.2 The Origin of the Magnetic Force on a Wire

When a magnetic field is applied at some angle to a wire carrying a current, a magnetic force is exerted on each moving charge in the wire. The total magnetic force on the wire is the sum of all the magnetic forces on the individual charges producing the current.

Figure 19.12 A wire carrying a current I in the presence of an external magnetic field $\vec{B}$ that makes an angle θ with the wire.

Figure 19.13 A diagram of a loudspeaker.

The force on a current-carrying conductor can be demonstrated by hanging a wire between the poles of a magnet, as in Figure 19.10. In this figure, the magnetic field is directed into the page and covers the region within the shaded area. The wire deflects to the right or left when it carries a current.

We can quantify this discussion by considering a straight segment of wire of length ℓ and cross-sectional area A carrying current I in a uniform external magnetic field $\vec{B}$, as in Figure 19.11. We assume that the magnetic field is perpendicular to the wire and is directed into the page. A force of magnitude $F_{max} = qv_dB$ is exerted on each charge carrier in the wire, where v_d is the drift velocity of the charge. To find the total force on the wire, we multiply the force on one charge carrier by the number of carriers in the segment. Because the volume of the segment is $A\ell$, the number of carriers is $nA\ell$, where n is the number of carriers per unit volume. Hence, the magnitude of the total magnetic force on the wire of length ℓ is as follows:

Total force = force on each charge carrier × total number of carriers

$$F_{max} = (qv_dB)(nA\ell)$$

From Chapter 17, however, we know that the current in the wire is given by $I = nqv_dA$, so

$$F_{max} = BI\ell \qquad \text{[19.5]}$$

This equation can be used only when the current and the magnetic field are at right angles to each other.

If the wire is not perpendicular to the field, but is at some arbitrary angle, as in Figure 19.12, the magnitude of the magnetic force on the wire is

$$F = BI\ell \sin \theta \qquad \text{[19.6]}$$

where θ is the angle between $\vec{B}$ and the direction of the current. The direction of this force can be obtained by the use of right-hand rule number 1. However, in this case you must place your fingers in the direction of the positive current I, rather than in the direction of $\vec{v}$. The current, naturally, is made up of charges moving at some velocity, so this really isn't a separate rule. In Figure 19.12, the direction of the magnetic force on the wire is out of the page.

Finally, when the current is either in the direction of the field or opposite the direction of the field, the magnetic force on the wire is zero.

The fact that a magnetic force acts on a current-carrying wire in a magnetic field is the operating principle of most speakers in sound systems. One speaker design, shown in Figure 19.13, consists of a coil of wire called the voice coil, a flexible paper cone that acts as the speaker, and a permanent magnet. The coil of wire surrounding the north pole of the magnet is shaped so that the magnetic field lines are directed radially outward from the coil's axis. When an electrical signal is sent to the coil, producing a current in the coil as in Figure 19.13, a magnetic force to the left acts on the coil. (This can be seen by applying right-hand rule number 1 to each turn of wire.) When the current reverses direction, as it would for a current that varied sinusoidally, the magnetic force on the coil also reverses direction, and the cone, which is attached to the coil, accelerates to the right. An alternating current through the coil causes an alternating force on the coil, which results in vibrations of the cone. The vibrating cone creates sound waves as it pushes and pulls on the air in front of it. In this way, a 1-kHz electrical signal is converted to a 1-kHz sound wave.

An unusual application of the force on a current-carrying conductor is illustrated by the electromagnetic pump shown in Figure 19.14. Artificial hearts require a pump to keep the blood flowing, and kidney dialysis machines also require a pump to assist the heart in pumping blood that is to be cleansed. Ordinary mechanical pumps create problems because they damage the blood cells as they move through the pump. The mechanism shown in the figure has demonstrated some promise in such applications. A magnetic field is established across a segment of the tube containing the blood, flowing in the direction of the velocity $\vec{v}$. An electric current passing through the fluid in the direction shown has a magnetic force acting on it in the direction of $\vec{v}$, as applying the right-hand rule shows. This force helps to keep the blood in motion.

Figure 19.14   A simple electromagnetic pump has no moving parts to damage a conducting fluid, such as blood, passing through. Application of the right-hand rule #1 (right fingers in the direction of the current I, curl them in the direction of $\vec{B}$, thumb points in the direction of the force) shows that the force on the current-carrying segment of the fluid is in the direction of the velocity.

Applying Physics 19.2   Lightning Strikes

In a lightning strike there is a rapid movement of negative charge from a cloud to the ground. In what direction is a lightning strike deflected by Earth's magnetic field?

Explanation   The downward flow of negative charge in a lightning stroke is equivalent to a current moving upward. Consequently, we have an upward-moving current in a northward-directed magnetic field. According to right-hand rule number 1, the lightning strike would be deflected toward the west.

EXAMPLE 19.3   A Current-Carrying Wire in Earth's Magnetic Field

Goal   Compare the magnetic force on a current-carrying wire with the gravitational force exerted on the wire.

Problem   A wire carries a current of 22.0 A from west to east. Assume that at this location the magnetic field of Earth is horizontal and directed from south to north and that it has a magnitude of 0.500×10^{-4} T. **(a)** Find the magnitude and direction of the magnetic force on a 36.0-m length of wire. **(b)** Calculate the gravitational force on the same length of wire if it's made of copper and has a cross-sectional area of 2.50×10^{-6} m^2.

Solution
(a) Calculate the magnetic force on the wire.

Substitute into Equation 19.6, using the fact that the magnetic field and the current are at right angles to each other:

$$F = BI\ell \sin\theta = (0.500 \times 10^{-4}\,\text{T})(22.0\,\text{A})(36.0\,\text{m}) \sin 90.0°$$
$$= 3.96 \times 10^{-2}\,\text{N}$$

Apply right-hand rule number 1 to find the direction of the magnetic force:

With the fingers of your right hand pointing west to east in the direction of the current, curl them north in the direction of the magnetic field. Your thumb points upward.

(b) Calculate the gravitational force on the wire segment.

First, obtain the mass of the wire from the density of copper, the length, and cross-sectional area of the wire:

$$m = \rho V = \rho(A\ell)$$
$$= (8.92 \times 10^3\,\text{kg/m}^3)(2.50 \times 10^{-6}\,\text{m}^2 \cdot 36.0\,\text{m})$$
$$= 0.803\,\text{kg}$$

To get the gravitational force, multiply the mass by the acceleration of gravity:

$$F_{grav} = mg = \boxed{7.87 \text{ N}}$$

Remarks   This calculation demonstrates that under normal circumstances, the gravitational force on a current-carrying conductor is much greater than the magnetic force due to the Earth's magnetic field.

Exercise 19.3

What current would make the magnetic force in the example equal in magnitude to the gravitational force?

Answer 4.37×10^3 A, a large current that would very rapidly melt the wire.

19.5   TORQUE ON A CURRENT LOOP AND ELECTRIC MOTORS

In the preceding section we showed how a magnetic force is exerted on a current-carrying conductor when the conductor is placed in an external magnetic field. With this as a starting point, we now show that a torque is exerted on a current loop placed in a magnetic field. The results of this analysis will be of great practical value when we discuss generators and motors in Chapter 20.

Consider a rectangular loop carrying current I in the presence of an external uniform magnetic field in the plane of the loop, as shown in Figure 19.15a. The forces on the sides of length a are zero because these wires are parallel to the field. The magnitudes of the magnetic forces on the sides of length b, however, are

$$F_1 = F_2 = BIb$$

The direction of $\vec{F}_1$, the force on the left side of the loop, is out of the page, and that of $\vec{F}_2$, the force on the right side of the loop, is into the page. If we view the loop from the side, as in Figure 19.15b, the forces are directed as shown. If we assume that the loop is pivoted so that it can rotate about point O, we see that these two forces produce a torque about O that rotates the loop clockwise. The magnitude of this torque, τ_{max}, is

$$\tau_{max} = F_1 \frac{a}{2} + F_2 \frac{a}{2} = (BIb)\frac{a}{2} + (BIb)\frac{a}{2} = BIab$$

where the moment arm about O is $a/2$ for both forces. Because the area of the loop is $A = ab$, the torque can be expressed as

$$\tau_{max} = BIA \qquad \text{[19.7]}$$

This result is valid only when the magnetic field is *parallel* to the plane of the loop, as in Figure 19.15b. If the field makes an angle θ with a line perpendicular to the plane of the loop, as in Figure 19.15c, the moment arm for each force is given by $(a/2) \sin \theta$. An analysis similar to the previous gives, for the magnitude of the torque,

Figure 19.15   (a) Top view of a rectangular loop in a uniform magnetic field $\vec{B}$. No magnetic forces act on the sides of length a parallel to $\vec{B}$, but forces do act on the sides of length b. (b) A side view of the rectangular loop shows that the forces $\vec{F}_1$ and $\vec{F}_2$ on the sides of length b create a torque that tends to twist the loop clockwise. (c) If $\vec{B}$ is at an angle θ with a line perpendicular to the plane of the loop, the torque is given by $BIA \sin \theta$.

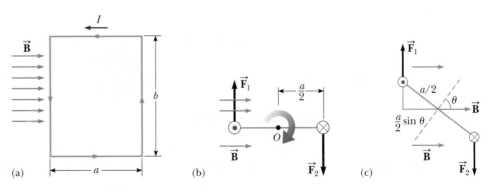

$$\tau = BIA \sin \theta \qquad \text{[19.8]}$$

This result shows that the torque has the *maximum* value *BIA* when the field is parallel to the plane of the loop ($\theta = 90°$) and is *zero* when the field is perpendicular to the plane of the loop ($\theta = 0$). As seen in Figure 19.15c, the loop tends to rotate to smaller values of θ (so that the normal to the plane of the loop rotates toward the direction of the magnetic field).

Although the foregoing analysis was for a rectangular loop, a more general derivation shows that Equation 19.8 applies regardless of the shape of the loop. Further, the torque on a coil with N turns is

$$\tau = BIAN \sin \theta \qquad \text{[19.9a]}$$

The quantity $\mu = IAN$ is defined as the magnitude of a vector $\vec{\mu}$ called the *magnetic moment* of the coil. The vector $\vec{\mu}$ always points perpendicular to the plane of the loop(s). The angle θ in Equations 19.8 and 19.9 lies between the directions of the magnetic moment $\vec{\mu}$ and the magnetic field $\vec{B}$. The magnetic torque can then be written

$$\tau = \mu B \sin \theta \qquad \text{[19.9b]}$$

A square and a circular loop with the same area lie in the *xy*-plane, where there is a uniform magnetic field $\vec{B}$ pointing at some angle θ with respect to the positive *z*-direction. Each loop carries the same current, in the same direction. Which magnetic torque is larger? (a) the torque on the square loop (b) the torque on the circular loop (c) the torques are the same (d) more information is needed

---

EXAMPLE 19.4   The Torque on a Circular Loop in a Magnetic Field

Goal   Calculate a magnetic torque on a loop of current.

Problem   A circular wire loop of radius 1.00 m is placed in a magnetic field of magnitude 0.500 T. The normal to the plane of the loop makes an angle of 30.0° with the magnetic field (Fig. 19.16a). The current in the loop is 2.00 A in the direction shown. **(a)** Find the magnetic moment of the loop and the magnitude of the torque at this instant. **(b)** The same current is carried by the rectangular 2.00-m by 3.00-m coil with three loops shown in Figure 19.16b. Find the magnetic moment of the coil and the magnitude of the torque acting on the coil at that instant.

Strategy   For each part, we just have to calculate the area, use it in the calculation of the magnetic moment, and multiply the result by $B \sin \theta$. Altogether, this amounts to substituting values into Equation 19.9b.

Solution
(a) Find the magnetic moment of the circular loop and the magnetic torque exerted on it.

Figure 19.16   (Example 19.4) (a) A circular current loop in an external magnetic field $\vec{B}$. (b) A rectangular current loop in the same field. (c) (Exercise 19.4)

First, calculate the enclosed area of the circular loop: $A = \pi r^2 = \pi (1.00 \text{ m})^2 = 3.14 \text{ m}^2$

Calculate the magnetic moment of the loop: $\mu = IAN = (2.00 \text{ A})(3.14 \text{ m}^2)(1) = \boxed{6.28 \text{ A} \cdot \text{m}^2}$

Now substitute values for the magnetic moment, magnetic field, and θ into Equation 19.9b:

$$\tau = \mu B \sin \theta = (6.28 \text{ A} \cdot \text{m}^2)(0.500 \text{ T})(\sin 30.0°)$$
$$= \boxed{1.57 \text{ N} \cdot \text{m}}$$

(b) Find the magnetic moment of the rectangular coil and the magnetic torque exerted on it.

Calculate the area of the coil:

$$A = L \times H = (2.00 \text{ m})(3.00 \text{ m}) = 6.00 \text{ m}^2$$

Calculate the magnetic moment of the coil:

$$\mu = IAN = (2.00 \text{ A})(6.00 \text{ m}^2)(3) = \boxed{36.0 \text{ A} \cdot \text{m}^2}$$

Substitute values into Equation 19.9b:

$$\tau = \mu B \sin \theta = (0.500 \text{ T})(36.0 \text{ A} \cdot \text{m}^2)(\sin 30.0°)$$
$$= \boxed{9.00 \text{ N} \cdot \text{m}}$$

Remarks   In calculating a magnetic torque, it's not strictly necessary to calculate the magnetic moment. Instead, Equation 19.9a can be used directly.

Exercise 19.4

Suppose a right triangular coil with base of 2.00 m and height 3.00 m having two loops carries a current of 2.00 A as shown in Figure 19.16c. Find the magnetic moment and the torque on the coil. The magnetic field is again 0.500 T and makes an angle of 30.0° with respect to the normal direction.

Answer $\mu = 12.0 \text{ A} \cdot \text{m}^2$, $\tau = 3.00 \text{ N} \cdot \text{m}$

Electric Motors

APPLICATION

Electric Motors

It's hard to imagine life in the 21st century without electric motors. Some appliances that contain motors include computer disk drives, CD players, VCR and DVD players, food processors and blenders, car starters, furnaces, and air conditioners. The motors convert electrical energy to kinetic energy of rotation, and consist basically of a rigid current-carrying loop that rotates when placed in the field of a magnet.

As we have just seen (Fig. 19.15), the torque on such a loop rotates the loop to smaller values of θ until the torque becomes zero, when the magnetic field is perpendicular to the plane of the loop and $\theta = 0$. If the loop turns past this angle and the current remains in the direction shown in the figure, the torque reverses direction and turns the loop in the opposite direction—that is, counterclockwise. To overcome this difficulty and provide continuous rotation in one direction, the current in the loop must periodically reverse direction. In alternating current (AC) motors, such a reversal occurs naturally 120 times each second. In direct current (DC) motors, the reversal is accomplished mechanically with split-ring contacts (commutators) and brushes, as shown in Active Figure 19.17.

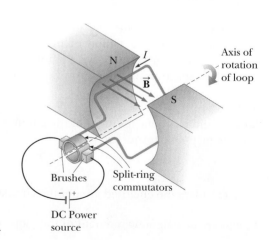

ACTIVE FIGURE 19.17
Simplified sketch of a DC electric motor.

Physics⚛Now™

Log into PhysicsNow at **www.cp7e.com** and go to Active Figure 19.17, where you can adjust the speed of rotation and the strength of the field and see the effects on the generated emf.

Figure 19.18   The Honda Insight combines a three-cylinder gasoline automobile engine with a thin electric motor for improved efficiency and added power when needed. The electric motor (circled) also acts as a generator during braking or coasting downhill to recharge the batteries, with the result that they never need to be recharged by the owner.

Although actual motors contain many current loops and commutators, for simplicity Active Figure 19.17 shows only a single loop and a single set of split-ring contacts rigidly attached to and rotating with the loop. Electrical stationary contacts called *brushes* are maintained in electrical contact with the rotating split ring. These brushes are usually made of graphite, because graphite is a good electrical conductor as well as a good lubricant. Just as the loop becomes perpendicular to the magnetic field and the torque becomes zero, inertia carries the loop forward in the clockwise direction and the brushes cross the gaps in the ring, causing the loop current to reverse its direction. This provides another pulse of torque in the clockwise direction for another 180°, the current reverses, and the process repeats itself. Figure 19.18 shows a modern motor used to power a hybrid gas–electric car.

19.6   MOTION OF A CHARGED PARTICLE IN A MAGNETIC FIELD

Consider the case of a positively charged particle moving in a uniform magnetic field so that the direction of the particle's velocity is *perpendicular to the field*, as in Active Figure 19.19. The label $\vec{\mathbf{B}}_{in}$ and the crosses indicate that $\vec{\mathbf{B}}$ is directed into the page. Application of the right-hand rule at point P shows that the direction of the magnetic force $\vec{\mathbf{F}}$ at that location is upward. The force causes the particle to alter its direction of travel and to follow a curved path. Application of the right-hand rule at any point shows that **the magnetic force is always directed toward the center of the circular path**; therefore, the magnetic force causes a centripetal acceleration, which changes only the direction of $\vec{\mathbf{v}}$ and not its magnitude. Because $\vec{\mathbf{F}}$ produces the centripetal acceleration, we can equate its magnitude, qvB in this case, to the mass of the particle multiplied by the centripetal acceleration v^2/r. From Newton's second law, we find that

$$F = qvB = \frac{mv^2}{r}$$

which gives

$$r = \frac{mv}{qB} \qquad [19.10]$$

This equation says that the radius of the path is proportional to the momentum mv of the particle and is inversely proportional to the charge and the magnetic field. Equation 19.10 is often called the *cyclotron equation*, because it's used in the design of these instruments (popularly known as atom smashers).

If the initial direction of the velocity of the charged particle is not perpendicular to the magnetic field, as shown in Active Figure 19.20, the path followed by the particle is a spiral (called a helix) along the magnetic field lines.

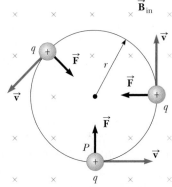

ACTIVE FIGURE 19.19
When the velocity of a charged particle is perpendicular to a uniform magnetic field, the particle moves in a circle whose plane is perpendicular to $\vec{\mathbf{B}}$, which is directed into the page. (The crosses represent the tails of the magnetic field vectors.) The magnetic force $\vec{\mathbf{F}}$ on the charge is always directed toward the center of the circle.

Physics⊗Now™
Log into PhysicsNow at **www.cp7e.com** and go to Active Figure 19.19, where you can adjust the mass, speed, charge of the particle, and the magnitude of the magnetic field, and observe the resulting circular motion.

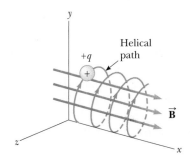

ACTIVE FIGURE 19.20
A charged particle having a velocity directed at an angle with a uniform magnetic field moves in a helical path.

Physics⊗Now™
Log into PhysicsNow at **www.cp7e.com** and go to Active Figure 19.20, where you can adjust the x-component of the velocity of the particle and observe the resulting helical motion.

Applying Physics 19.3   Trapping Charges

Storing charged particles is important for a variety of applications. Suppose a uniform magnetic field exists in a finite region of space. Can a charged particle be injected into this region from the outside and remain trapped in the region by magnetic force alone?

Explanation   It's best to consider separately the components of the particle velocity parallel and perpendicular to the field lines in the region. There is no magnetic force on the particle associated with the velocity component parallel to the field lines, so that velocity component remains unchanged.

Now consider the component of velocity perpendicular to the field lines. This component will result in a magnetic force that is perpendicular to both the field lines and the velocity component itself. The path of a particle for which the force is always perpendicular to the velocity is a circle. The particle therefore follows a circular arc and exits the field on the other side of the circle, as shown in Figure 19.21 for a particle with constant kinetic energy. On the other hand, a particle can become trapped if it loses some kinetic energy in a collision after entering the field, as in Active Figure 19.20.

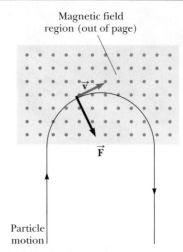

Figure 19.21   (Applying Physics 19.3)

Particles *can* be injected and contained if, in addition to the magnetic field, electrostatic fields are involved. These fields are used in the *Penning trap*. With these devices, it's possible to store charged particles for extended periods. Such traps are useful, for example, in the storage of antimatter, which disintegrates completely on contact with ordinary matter.

Quick Quiz 19.4

As a charged particle moves freely in a circular path in the presence of a constant magnetic field applied perpendicular to the particle's velocity, its kinetic energy (a) remains constant, (b) increases, or (c) decreases.

EXAMPLE 19.5   The Mass Spectrometer: Identifying Particles

Goal   Use the cyclotron equation to identify a particle.

Problem   A charged particle enters the magnetic field of a mass spectrometer at a speed of 1.79×10^6 m/s. It subsequently moves in a circular orbit with a radius of 16.0 cm in a uniform magnetic field of magnitude 0.350 T having a direction perpendicular to the particle's velocity. Find the particle's mass-to-charge ratio, and identify it based on the table on page 639.

Strategy   After finding the mass-to-charge ratio with Equation 19.10, compare it to the values in the table, identifying the particle.

Solution

Write the cyclotron equation:

$$r = \frac{mv}{qB}$$

Solve this equation for the mass divided by the charge, m/q and substitute values:

$$\frac{m}{q} = \frac{rB}{v} = \frac{(0.160 \text{ m})(0.350 \text{ T})}{1.79 \times 10^6 \text{ m/s}} = 3.13 \times 10^{-8} \frac{\text{kg}}{\text{C}}$$

Identify the particle from the table. All particles are completely ionized.

particle	m (kg)	q (C)	m/q (kg/C)
Hydrogen	1.67×10^{-27}	1.60×10^{-19}	1.04×10^{-8}
Deuterium	3.35×10^{-27}	1.60×10^{-19}	2.09×10^{-8}
Tritium	5.01×10^{-27}	1.60×10^{-19}	3.13×10^{-8}
Helium-3	5.01×10^{-27}	3.20×10^{-19}	1.57×10^{-8}

The particle is tritium.

Remarks   The mass spectrometer is an important tool in both chemistry and physics. Unknown chemicals can be heated and ionized, and the resulting particles passed through the mass spectrometer and subsequently identified.

Exercise 19.5
Suppose a second charged particle enters the mass spectrometer at the same speed as the particle in Example 19.5. If it travels in a circle with radius 10.7 cm, find the mass-to-charge ratio and identify the particle from the table above.

Answer 2.09×10^{-8} kg/C; deuterium

EXAMPLE 19.6   The Mass Spectrometer: Separating Isotopes

APPLICATION
Mass Spectrometers

Goal   Apply the cyclotron equation to the process of separating isotopes.

Problem   Two singly ionized atoms move out of a slit at point S in Figure 19.22 and into a magnetic field of magnitude 0.100 T pointing into the page. Each has a speed of 1.00×10^6 m/s. The nucleus of the first atom contains one proton and has a mass of 1.67×10^{-27} kg, while the nucleus of the second atom contains a proton and a neutron and has a mass of 3.34×10^{-27} kg. Atoms with the same number of protons in the nucleus but different masses are called isotopes. The two isotopes here are hydrogen and deuterium. Find their distance of separation when they strike a photographic plate at P.

Figure 19.22   (Example 19.6) Two isotopes leave the slit at point S and travel in different circular paths before striking a photographic plate at P.

Strategy   Apply the cyclotron equation to each atom, finding the radius of the path of each. Double the radii to find the path diameters, and then find their difference.

Solution
Use Equation 19.10 to find the radius of the circular path followed by the lighter isotope, hydrogen.

$$r_1 = \frac{m_1 v}{qB} = \frac{(1.67 \times 10^{-27} \text{ kg})(1.00 \times 10^6 \text{ m/s})}{(1.60 \times 10^{-19} \text{ C})(0.100 \text{ T})}$$
$$= 0.104 \text{ m}$$

Use the same equation to calculate the radius of the path of deuterium, the heavier isotope:

$$r_2 = \frac{m_2 v}{qB} = \frac{(3.34 \times 10^{-27} \text{ kg})(1.00 \times 10^6 \text{ m/s})}{(1.60 \times 10^{-19} \text{ C})(0.100 \text{ T})}$$
$$= 0.209 \text{ m}$$

Multiply the radii by 2 to find the diameters, and take the difference, getting the separation x between the isotopes:

$$x = 2r_2 - 2r_1 = \boxed{0.210 \text{ m}}$$

Remarks   During World War II, mass spectrometers were used to separate the highly radioactive uranium isotope U-235 from its far more common isotope, U-238.

Exercise 19.6
Use the same mass spectrometer as in Example 19.6 to find the separation between two isotopes of helium: normal helium-4, which has a nucleus consisting of two protons and two neutrons, and helium-3, which has two protons and a single neutron. Assume both nuclei, doubly ionized (having a charge of $2e = 3.20 \times 10^{-19}$ C), enter the field at 1.00×10^6 m/s. The helium-4 nucleus has a mass of 6.64×10^{-27} kg, and the helium-3 nucleus has a mass of 5.01×10^{-27} kg.

Answer   0.102 m

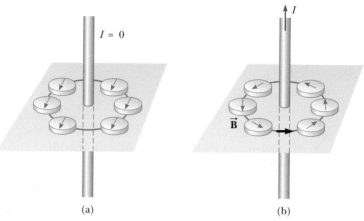

(a) (b)

ACTIVE FIGURE 19.23
(a) When there is no current in the vertical wire, all compass needles point in the same direction. (b) When the wire carries a strong current, the compass needles deflect in directions tangent to the circle, pointing in the direction of $\vec{\mathbf{B}}$, due to the current.

Physics Now™
Log into PhysicsNow at **www.cp7e.com** and go to Active Figure 19.23, where you can change the value of the current and see the effect on the compasses.

19.7 MAGNETIC FIELD OF A LONG, STRAIGHT WIRE AND AMPÈRE'S LAW

During a lecture demonstration in 1819, the Danish scientist Hans Oersted (1777–1851) found that an electric current in a wire deflected a nearby compass needle. This momentous discovery, linking a magnetic field with an electric current for the first time, was the beginning of our understanding of the origin of magnetism.

A simple experiment first carried out by Oersted in 1820 clearly demonstrates that a current-carrying conductor produces a magnetic field. In this experiment, several compass needles are placed in a horizontal plane near a long vertical wire, as in Active Figure 19.23a. When there is no current in the wire, all needles point in the same direction (that of Earth's field), as one would expect. However, when the wire carries a strong, steady current, the needles all deflect in directions tangent to the circle, as in Active Figure 19.23b. These observations show that the direction of $\vec{\mathbf{B}}$ is consistent with the following convenient rule, **right-hand rule number 2**:

> Point the thumb of your right hand along a wire in the direction of positive current, as in Figure 19.24a. Your fingers then naturally curl in the direction of the magnetic field $\vec{\mathbf{B}}$.

When the current is reversed, the filings in Figure 19.24b also reverse.

HANS CHRISTIAN OERSTED
(1777–1851), Danish Physicist and Chemist

Oersted is best known for observing that a compass needle deflects when placed near a wire carrying a current. This important discovery was the first evidence of the connection between electric and magnetic phenomena. Oersted was also the first to prepare pure aluminum.

TIP 19.3   Raise Your Right Hand!
We have introduced two right-hand rules in this chapter. Be sure to use *only* your right hand when applying these rules.

Figure 19.24 (a) Right-hand rule number 2 for determining the direction of the magnetic field due to a long, straight wire carrying a current. Note that the magnetic field lines form circles around the wire. (b) Circular magnetic field lines surrounding a current-carrying wire, displayed by iron filings.

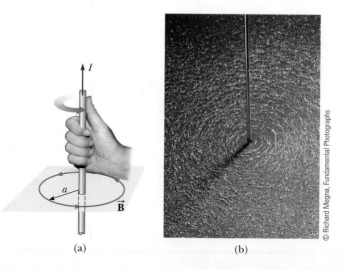

(a) (b)

Because the filings point in the direction of $\vec{\mathbf{B}}$, we conclude that the lines of $\vec{\mathbf{B}}$ form circles about the wire. By symmetry, the magnitude of $\vec{\mathbf{B}}$ is the same everywhere on a circular path centered on the wire and lying in a plane perpendicular to the wire. By varying the current and distance from the wire, it can be experimentally determined that $\vec{\mathbf{B}}$ is proportional to the current and inversely proportional to the distance from the wire. These observations lead to a mathematical expression for the strength of the magnetic field due to the current I in a long, straight wire:

$$B = \frac{\mu_0 I}{2\pi r} \qquad [19.11]$$

The proportionality constant μ_0, called the **permeability of free space**, has the value

$$\mu_0 \equiv 4\pi \times 10^{-7}\,\text{T·m/A} \qquad [19.12]$$

◄ Magnetic field due to a long, straight wire

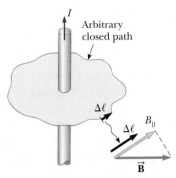

Figure 19.25 An arbitrary closed path around a current is used to calculate the magnetic field of the current by the use of Ampère's rule.

Ampère's Law and a Long, Straight Wire

Equation 19.11 enables us to calculate the magnetic field due to a long, straight wire carrying a current. A general procedure for deriving such equations was proposed by the French scientist André-Marie Ampère (1775–1836); it provides a relation between the current in an arbitrarily shaped wire and the magnetic field produced by the wire.

Consider an arbitrary closed path surrounding a current as in Figure 19.25. The path consists of many short segments, each of length $\Delta\ell$. Multiply one of these lengths by the component of the magnetic field parallel to that segment, where the product is labeled $B_\parallel \Delta\ell$. According to Ampère, the sum of all such products over the closed path is equal to μ_0 times the net current I that passes through the surface bounded by the closed path. This statement, known as **Ampère's circuital law**, can be written

$$\sum B_\parallel \Delta\ell = \mu_0 I \qquad [19.13]$$

where $B_\parallel$ is the component of $\vec{\mathbf{B}}$ parallel to the segment of length $\Delta\ell$ and $\sum B_\parallel \Delta\ell$ means that we take the sum over all the products $B_\parallel \Delta\ell$ around the closed path. Ampère's law is the fundamental law describing how electric currents create magnetic fields in the surrounding empty space.

We can use Ampère's circuital law to derive the magnetic field due to a long, straight wire carrying a current I. As discussed earlier, each of the magnetic field lines of this configuration forms a circle with the wire at its centers. The magnetic field is tangent to this circle at every point, and its magnitude has the same value B over the entire circumference of a circle of radius r, so that $B_\parallel = B$, as shown in Figure 19.26 (page 642). In calculating the sum $\sum B_\parallel \Delta\ell$ over the circular path, notice that $B_\parallel$ can be removed from the sum (because it has the same value B for each element on the circle). Equation 19.13 then gives

$$\sum B_\parallel \Delta\ell = B_\parallel \sum \Delta\ell = B(2\pi r) = \mu_0 I$$

Dividing both sides by the circumference $2\pi r$, we obtain

$$B = \frac{\mu_0 I}{2\pi r}$$

This is identical to Equation 19.11, which is the magnetic field due to the current I in a long, straight wire.

Leonard de Selva/CORBIS

ANDRÉ-MARIE AMPÈRE
(1775–1836)

Ampère, a Frenchman, is credited with the discovery of electromagnetism—the relationship between electric currents and magnetic fields. Ampère's genius, particularly in mathematics, became evident by the age of 12, but his personal life was filled with tragedy. His father, a wealthy city official, was guillotined during the French Revolution, and his wife died young, in 1803. Ampère died of pneumonia at the age of 61. His judgment of his life is clear from the epitaph he chose for his gravestone: *Tandem felix* (Happy at last).

Figure 19.26 A closed circular path of radius r around a long, straight current-carrying wire is used to calculate the magnetic field set up by the wire.

Ampère's circuital law provides an elegant and simple method for calculating the magnetic fields of highly symmetric current configurations. However, it can't easily be used to calculate magnetic fields for complex current configurations that lack symmetry. In addition, Ampère's circuital law in this form is valid only when the currents and fields don't change with time.

EXAMPLE 19.7   The Magnetic Field of a Long Wire

Goal   Calculate the magnetic field of a long, straight wire and the force that the field exerts on a particle.

Problem   A long, straight wire carries a current of 5.00 A. At one instant, a proton, 4.00 mm from the wire, travels at 1.50×10^3 m/s parallel to the wire and in the same direction as the current (Fig. 19.27). **(a)** Find the magnitude and direction of the magnetic field created by the wire. **(b)** Find the magnitude and direction of the magnetic force the wire's magnetic field exerts on the proton.

Figure 19.27 (Example 19.7) The magnetic field due to the current is into the page at the location of the proton, and the magnetic force on the proton is to the left.

Strategy   First use Equation 19.11 to find the magnitude of the magnetic field at the given point. Use right-hand rule number 2 to find the direction of the field. Finally, substitute into Equation 19.1, computing the magnetic force on the proton.

Solution

(a) Find the magnitude and direction of the wire's magnetic field.

Use Equation 19.11 to calculate the magnitude of the magnetic field 4.00 mm from the wire:

$$B = \frac{\mu_0 I}{2\pi r} = \frac{(4\pi \times 10^{-7}\,\text{T}\cdot\text{m/A})(5.00\,\text{A})}{2\pi(4.00 \times 10^{-3}\,\text{m})}$$

$$= 2.50 \times 10^{-4}\,\text{T}$$

Apply right-hand rule number 2 to find the direction of the magnetic field $\vec{\mathbf{B}}$:

With the right thumb pointing in the direction of the current in Figure 19.27, the fingers curl into the page at the location of the proton. The angle θ between $\vec{\mathbf{v}}$ and $\vec{\mathbf{B}}$ is therefore 90°.

(b) Compute the magnetic force exerted by the wire on the proton.

Substitute into Equation 19.1, which gives the magnitude of the magnetic force on a charged particle:

$$F = qvB \sin\theta = (1.60 \times 10^{-19}\,\text{C})(1.50 \times 10^3\,\text{m/s})$$
$$\times (2.50 \times 10^{-4}\,\text{T})(\sin 90°)$$

$$= 6.00 \times 10^{-20}\,\text{N}$$

Find the direction of the magnetic force with right-hand rule number 1:

Point your right fingers in the direction of $\vec{\mathbf{v}}$, curling them into the page toward $\vec{\mathbf{B}}$. Your thumb points to the left, which is the direction of the magnetic force.

Remarks   The location of the proton is important. On the left-hand side, the wire's magnetic field points outward, and the magnetic force on the proton is to the right.

Exercise 19.7

Find (a) the magnetic field created by the wire and (b) the magnetic force on a helium-3 nucleus located 7.50 mm to the left of the wire in Figure 19.27, traveling 2.50×10^3 m/s opposite the direction of the current. (See the data table presented in Example 19.5 on page 639).

Answers   (a) 1.33×10^{-4} T   (b) 1.07×10^{-19} N, directed to the left in Figure 19.27.

19.8 MAGNETIC FORCE BETWEEN TWO PARALLEL CONDUCTORS

As we have seen, a magnetic force acts on a current-carrying conductor when the conductor is placed in an external magnetic field. Because a conductor carrying a current creates a magnetic field around itself, it is easy to understand that two current-carrying wires placed close together exert magnetic forces on each other. Consider two long, straight, parallel wires separated by the distance d and carrying currents I_1 and I_2 in the same direction, as shown in Active Figure 19.28. Wire 1 is directly above wire 2. What's the magnetic force on one wire due to a magnetic field set up by the other wire?

In this calculation, we are finding the force on wire 1 due to the magnetic field of wire 2. The current I_2, sets up magnetic field $\vec{B}_2$ at wire 1. The direction of $\vec{B}_2$ is perpendicular to the wire, as shown in the figure. Using Equation 19.11, we find that the magnitude of this magnetic field is

$$B_2 = \frac{\mu_0 I_2}{2\pi d}$$

According to Equation 19.5, the magnitude of the magnetic force on wire 1 in the presence of field $\vec{B}_2$ due to I_2 is

$$F_1 = B_2 I_1 \ell = \left(\frac{\mu_0 I_2}{2\pi d}\right) I_1 \ell = \frac{\mu_0 I_1 I_2 \ell}{2\pi d}$$

We can rewrite this relationship in terms of the force per unit length:

$$\frac{F_1}{\ell} = \frac{\mu_0 I_1 I_2}{2\pi d} \qquad \text{[19.14]}$$

The direction of $\vec{F}_1$ is downward, toward wire 2, as indicated by right-hand rule number 1. This calculation is completely symmetric, which means that the force $\vec{F}_2$ on wire 2 is equal to and opposite $\vec{F}_1$, as expected from Newton's third law of action–reaction.

We have shown that parallel conductors carrying currents in the same direction *attract* each other. You should use the approach indicated by Figure 19.28 and the steps leading to Equation 19.14 to show that parallel conductors carrying currents in opposite directions *repel* each other.

The force between two parallel wires carrying a current is used to define the SI unit of current, the **ampere** (A), as follows:

> If two long, parallel wires 1 m apart carry the same current, and the magnetic force per unit length on each wire is 2×10^{-7} N/m, then the current is defined to be 1 A.

◀ Definition of the ampere

The SI unit of charge, the **coulomb** (C), can now be defined in terms of the ampere as follows:

> If a conductor carries a steady current of 1 A, then the quantity of charge that flows through any cross section in 1 s is 1 C.

◀ Definition of the coulomb

ACTIVE FIGURE 19.28
Two parallel wires, oriented vertically, carry steady currents and exert forces on each other. The field $\vec{B}_2$ at wire 1 due to wire 2 produces a force on wire 1 given by $F_1 = B_2 I_1 \ell$. The force is attractive if the currents have the same direction, as shown, and repulsive if the two currents have opposite directions.

Physics Now™
Log into PhysicsNow at **www.cp7e.com** and go to Active Figure 19.28, where you can adjust the currents in the wires and the distance between them, and see the effect on the force.

If, in Figure 19.28, $I_1 = 2$ A and $I_2 = 6$ A, which of the following is true? (a) $F_1 = 3F_2$ (b) $F_1 = F_2$ or (c) $F_1 = F_2/3$

EXAMPLE 19.8   Levitating a Wire

Goal   Calculate the magnetic force of one current-carrying wire on a parallel current-carrying wire.

Problem   Two wires, each having a weight per unit length of 1.00×10^{-4} N/m, are parallel with one directly above the other. Assume that the wires carry currents that are equal in magnitude and opposite in direction. The wires are 0.10 m apart, and the sum of the magnetic force and gravitational force on the upper wire is zero. Find the current in the wires. (Neglect Earth's magnetic field.)

Strategy   The upper wire must be in equilibrium under the forces of magnetic repulsion and gravity. Set the sum of the forces equal to zero and solve for the unknown current, I.

Solution

Set the sum of the forces equal to zero, and substitute the appropriate expressions. Notice that the magnetic force between the wires is repulsive.

$$\vec{F}_{grav} + \vec{F}_{mag} = 0$$

$$-mg + \frac{\mu_0 I_1 I_2}{2\pi d}\ell = 0$$

The currents are equal, so $I_1 = I_2 = I$. Make these substitutions and solve for I^2.

$$\frac{\mu_0 I^2}{2\pi d}\ell = mg \quad \rightarrow \quad I^2 = \frac{(2\pi d)(mg/\ell)}{\mu_0}$$

Substitute given values, finding I^2, then take the square root. Notice that the weight per unit length, mg/ℓ, is given.

$$I^2 = \frac{(2\pi \cdot 0.100 \text{ m})(1.00 \times 10^{-4} \text{ N/m})}{(4\pi \times 10^{-7} \text{ T} \cdot \text{m})} = 50.0 \text{ A}^2$$

$$I = \boxed{7.07 \text{ A}}$$

Remark   Exercise 19.3 showed that using the Earth's magnetic field to levitate a wire required extremely large currents. Currents in wires can create much stronger magnetic fields than Earth's in regions near the wire.

Exercise 19.8

If the current in each wire is doubled, how far apart should the wires be placed if the magnitudes of the gravitational and magnetic forces on the upper wire are to be equal?

Answer   0.400 m

Figure 19.29 All segments of the current loop produce a magnetic field at the center of the loop, directed *out of the page.*

19.9 MAGNETIC FIELDS OF CURRENT LOOPS AND SOLENOIDS

The strength of the magnetic field set up by a piece of wire carrying a current can be enhanced at a specific location if the wire is formed into a loop. You can understand this by considering the effect of several small segments of the current loop, as in Figure 19.29. The small segment at the top of the loop, labeled Δx_1, produces a magnetic field of magnitude B_1 at the loop's center, directed out of the page. The direction of $\vec{B}$ can be verified using right-hand rule number 2 for a long, straight wire. Imagine holding the wire with your right hand, with your thumb pointing in the direction of the current. Your fingers then curl around in the direction of $\vec{B}$.

A segment of length Δx_2 at the bottom of the loop also contributes to the field at the center, increasing its strength. The field produced at the center by the segment Δx_2 has the same magnitude as B_1 and is also directed out of the page. Similarly, all other such segments of the current loop contribute to the field. The net effect is a magnetic field for the current loop as pictured in Figure 19.30a.

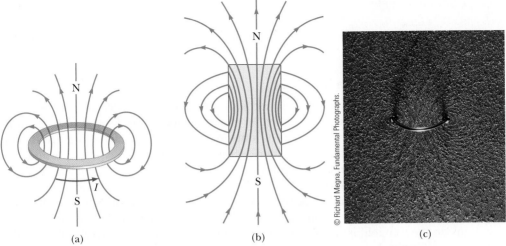

(a)   (b)   (c)

Figure 19.30   (a) Magnetic field lines for a current loop. Note that the lines resemble those of a bar magnet. (b) The magnetic field of a bar magnet is similar to that of a current loop. (c) Field lines of a current loop, displayed by iron filings.

Notice in Figure 19.30a that the magnetic field lines enter at the bottom of the current loop and exit at the top. Compare this to Figure 19.30b, illustrating the field of a bar magnet. The two fields are similar. One side of the loop acts as though it were the north pole of a magnet, and the other acts as a south pole. The similarity of these two fields will be used to discuss magnetism in matter in an upcoming section.

Applying Physics 19.4   Twisted Wires

In electrical circuits, it is often the case that insulated wires carrying currents in opposite directions are twisted together. What is the advantage of doing this?

Explanation   If the wires are not twisted together, the combination of the two wires forms a current loop,

which produces a relatively strong magnetic field. This magnetic field generated by the loop could be strong enough to affect adjacent circuits or components. When the wires are twisted together, their magnetic fields tend to cancel.

The magnitude of the magnetic field at the center of a circular loop carrying current I as in Figure 19.30a is given by

$$B = \frac{\mu_0 I}{2R}$$

This must be derived with calculus. However, it can be shown to be reasonable by calculating the field at the center of four long wires, each carrying current I and forming a square, as in Figure 19.31, with a circle of radius R inscribed within it. Intuitively, this arrangement should give a magnetic field at the center that is similar in magnitude to the field produced by the circular loop. The current in the circular wire is closer to the center, so that wire would have a magnetic field somewhat stronger than just the four legs of the rectangle, but the lengths of the straight wires beyond the rectangle compensate for this. Each wire contributes the same magnetic field at the exact center, so the total field is given by

$$B = 4 \times \frac{\mu_0 I}{2\pi R} = \frac{4}{\pi}\left(\frac{\mu_0 I}{2R}\right) = (1.27)\left(\frac{\mu_0 I}{2R}\right)$$

This is *approximately* the same as the field produced by the circular loop of current. When the coil has N loops, each carrying current I, the magnetic field at the

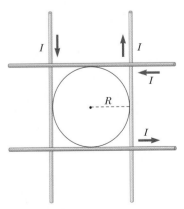

Figure 19.31   The field of a circular loop carrying current I can be approximated by the field due to four straight wires, each carrying current I.

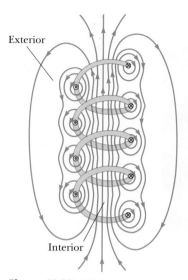

Figure 19.32 The magnetic field lines for a loosely wound solenoid.

The magnetic field inside a solenoid ▶

center is given by

$$B = N \frac{\mu_0 I}{2R} \qquad [19.15]$$

Magnetic Field of a Solenoid

If a long, straight wire is bent into a coil of several closely spaced loops, the resulting device is a **solenoid**, often called an **electromagnet**. This device is important in many applications because it acts as a magnet only when it carries a current. The magnetic field inside a solenoid increases with the current and is proportional to the number of coils per unit length.

Figure 19.32 shows the magnetic field lines of a loosely wound solenoid of length ℓ and total number of turns N. Notice that the field lines inside the solenoid are nearly parallel, uniformly spaced, and close together. As a result, the field inside the solenoid is strong and approximately uniform. The exterior field at the sides of the solenoid is nonuniform, much weaker than the interior field, and *opposite in direction* to the field inside the solenoid.

If the turns are closely spaced, the field lines are as shown in Figure 19.33a, entering at one end of the solenoid and emerging at the other. This means that one end of the solenoid acts as a north pole and the other end acts as a south pole. If the length of the solenoid is much greater than its radius, the lines that leave the north end of the solenoid spread out over a wide region before returning to enter the south end. The more widely separated the field lines are, the weaker the field. This is in contrast to a much stronger field *inside* the solenoid, where the lines are close together. Also, the field inside the solenoid has a constant magnitude at all points far from its ends. As will be shown subsequently, these considerations allow the application of Ampere's law to the solenoid, giving a result of

$$B = \mu_0 n I \qquad [19.16]$$

for the field inside the solenoid, where $n = N/\ell$ is the number of turns per unit length of the solenoid.

So-called steering magnets placed along the neck of the picture tube in a televi-

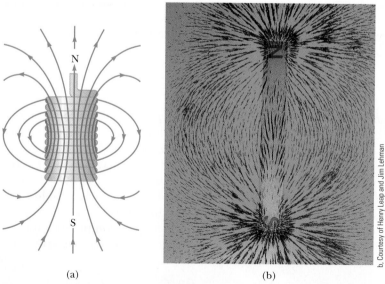

(a)   (b)

b. Courtesy of Henry Leap and Jim Lehman

Figure 19.33 (a) Magnetic field lines for a tightly wound solenoid of finite length carrying a steady current. The field inside the solenoid is nearly uniform and strong. Note that the field lines resemble those of a bar magnet, so the solenoid effectively has north and south poles. (b) The magnetic field pattern of a bar magnet, displayed by small iron filings on a sheet of paper.

EXAMPLE 19.9 The Magnetic Field inside a Solenoid

Goal Calculate the magnetic field of a solenoid from given data and the momentum of a charged particle in this field.

Problem A certain solenoid consists of 100 turns of wire and has a length of 10.0 cm. **(a)** Find the magnitude of the magnetic field inside the solenoid when it carries a current of 0.500 A. **(b)** What is the momentum of a proton orbiting inside the solenoid in a circle with a radius of 0.020 m? The axis of the solenoid is perpendicular to the plane of the orbit. **(c)** Approximately how much wire would be needed to build this solenoid? Assume the solenoid's radius is 5.00 cm.

Strategy In part (a), calculate the number of turns per meter and substitute that and given information into Equation 19.16, getting the magnitude of the magnetic field. Part (b) is an application of Newton's second law.

Solution

(a) Find the magnitude of the magnetic field inside the solenoid when it carries a current of 0.500 A.

Calculate the number of turns per unit length:

$$n = \frac{N}{\ell} = \frac{100 \text{ turns}}{0.100 \text{ m}} = 1.00 \times 10^3 \text{ turns/m}$$

Substitute n and I into Equation 19.16 to find the magnitude of the magnetic field:

$$B = \mu_0 n I$$
$$= (4\pi \times 10^{-7} \text{ T} \cdot \text{m/A})(1.00 \times 10^3 \text{ turns/m})(0.500 \text{ A})$$
$$= 6.28 \times 10^{-4} \text{ T}$$

(b) Find the momentum of a proton orbiting in a circle of radius 0.020 m near the center of the solenoid.

Write Newton's second law for the proton:

$$ma = F = qvB$$

Substitute the centripetal acceleration $a = v^2/r$:

$$m\frac{v^2}{r} = qvB$$

Cancel one factor of v on both sides and multiply by r, getting the momentum mv:

$$mv = rqB = (0.020 \text{ m})(1.60 \times 10^{-19} \text{ C})(6.28 \times 10^{-4} \text{ T})$$
$$p = mv = 2.01 \times 10^{-24} \text{ kg} \cdot \text{m/s}$$

(c) Approximately how much wire would be needed to build this solenoid?

Multiply the number of turns by the circumference of one loop:

$$\text{length of wire} \approx (\text{number of turns})(2\pi r)$$
$$= (1.00 \times 10^2 \text{ turns})(2\pi \cdot 0.0500 \text{ m})$$
$$= 31.4 \text{ m}$$

Remarks An electron in part (b) would have the same momentum as the proton, but a much higher speed. It would also orbit in the opposite direction. The length of wire in part (c) is only an estimate, because the wire has a certain thickness, slightly increasing the size of each loop. In addition the wire loops aren't perfect circles, because they wind slowly up along the solenoid.

Exercise 19.9

Suppose you have a 32.0-m length of copper wire. If the wire is wrapped into a solenoid 0.240 m long and having a radius of 0.0400 m, how strong is the resulting magnetic field in its center when the current is 12.0 A?

Answer $8.00 \times 10^{-3} \text{ T}$

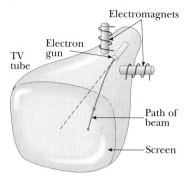

Figure 19.34 Electromagnets are used to deflect electrons to desired positions on the screen of a television tube.

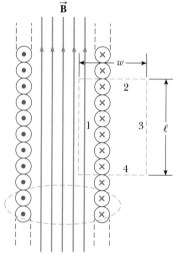

Figure 19.35 A cross-sectional view of a tightly wound solenoid. If the solenoid is long relative to its radius, we can assume that the magnetic field inside is uniform and the field outside is zero. Ampère's law applied to the blue dashed rectangular path can then be used to calculate the field inside the solenoid.

Figure 19.36 Classical model of a spinning electron.

sion set, as in Figure 19.34, are used to make the electron beam move to the desired locations on the screen, tracing out the images. The rate at which the electron beam sweeps over the screen is so fast that to the eye it looks like a picture rather than a sequence of dots.

Ampère's Law Applied to a Solenoid

We can use Ampère's law to obtain the expression for the magnetic field inside a solenoid carrying a current I. A cross section taken along the length of part of our solenoid is shown in Figure 19.35. $\vec{B}$ inside the solenoid is uniform and parallel to the axis, and $\vec{B}$ outside is approximately zero. Consider a rectangular path of length L and width w, as shown in the figure. We can apply Ampère's law to this path by evaluating the sum of $B_{\parallel} \Delta \ell$ over each side of the rectangle. The contribution along side 3 is clearly zero, because $\vec{B} = 0$ in this region. The contributions from sides 2 and 4 are both zero, because $\vec{B}$ is perpendicular to $\Delta \ell$ along these paths. Side 1 of length L gives a contribution BL to the sum, because $\vec{B}$ is uniform along this path, and parallel to $\Delta \ell$. Therefore, the sum over the closed rectangular path has the value

$$\Sigma \, B_{\parallel} \, \Delta \ell = BL$$

The right side of Ampère's law involves the total current that passes through the area bounded by the path chosen. In this case, the total current through the rectangular path equals the current through each turn of the solenoid, multiplied by the number of turns. If N is the number of turns in the length L, then the total current through the rectangular path equals NI. Ampère's law applied to this path therefore gives

$$\Sigma \, B_{\parallel} \, \Delta \ell = BL = \mu_0 NI$$

or

$$B = \mu_0 \frac{N}{L} I = \mu_0 nI$$

where $n = N/L$ is the number of turns per unit length.

19.10 MAGNETIC DOMAINS

The magnetic field produced by a current in a coil of wire gives us a hint as to what might cause certain materials to exhibit strong magnetic properties. A single coil like that in Figure 19.30a has a north pole and a south pole, but if this is true for a coil of wire, it should also be true for any current confined to a circular path. In particular, *an individual atom should act as a magnet because of the motion of the electrons about the nucleus.* Each electron, with its charge of 1.6×10^{-19} C, circles the atom once in about 10^{-16} s. If we divide the electric charge by this time interval, we see that the orbiting electron is equivalent to a current of 1.6×10^{-3} A. Such a current produces a magnetic field on the order of 20 T at the center of the circular path. From this we see that a very strong magnetic field would be produced if several of these atomic magnets could be aligned inside a material. This doesn't occur, however, because the simple model we have described is not the complete story. A thorough analysis of atomic structure shows that the magnetic field produced by one electron in an atom is often canceled by an oppositely revolving electron in the same atom. The net result is that **the magnetic effect produced by the electrons orbiting the nucleus is either zero or very small for most materials**.

The magnetic properties of many materials can be explained by the fact that an electron not only circles in an orbit, but also spins on its axis like a top, with spin magnetic moment as shown (Fig. 19.36). (This classical description should not be taken too literally. The property of electron *spin* can be understood only in the context of quantum mechanics, which we will not discuss here.) The spinning elec-

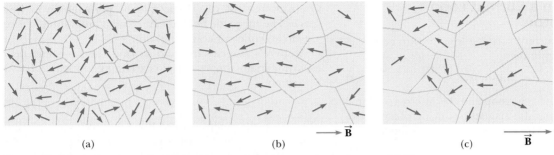

Figure 19.37 (a) Random orientation of domains in an unmagnetized substance. (b) When an external magnetic field $\vec{\mathbf{B}}$ is applied, the domains tend to align with the magnetic field. (c) As the field is made even stronger, the domains not aligned with the external field become very small.

tron represents a charge in motion that produces a magnetic field. The field due to the spinning is generally stronger than the field due to the orbital motion. In atoms containing many electrons, the electrons usually pair up with their spins opposite each other, so that their fields cancel each other. That is why most substances are not magnets. However, in certain strongly magnetic materials, such as iron, cobalt, and nickel, the magnetic fields produced by the electron spins don't cancel completely. Such materials are said to be **ferromagnetic**. In ferromagnetic materials, strong coupling occurs between neighboring atoms, forming large groups of atoms with spins that are aligned. Called **domains**, the sizes of these groups typically range from about 10^{-4} cm to 0.1 cm. In an unmagnetized substance the domains are randomly oriented, as shown in Figure 19.37a. When an external field is applied, as in Figure 19.37b, the magnetic field of each domain tends to come nearer to alignment with the external field, resulting in magnetization.

In what are called hard magnetic materials, domains remain aligned even after the external field is removed; the result is a **permanent magnet**. In soft magnetic materials, such as iron, once the external field is removed, thermal agitation produces motion of the domains and the material quickly returns to an unmagnetized state.

The alignment of domains explains why the strength of an electromagnet is increased dramatically by the insertion of an iron core into the magnet's center. The magnetic field produced by the current in the loops causes the domains to align, thus producing a large net external field. The use of iron as a core is also advantageous because it is a soft magnetic material that loses its magnetism almost instantaneously after the current in the coils is turned off.

TIP 19.4   The Electron Spins—but Doesn't!

Even though we use the word *spin*, the electron, unlike a child's top, isn't physically spinning in this sense. The electron has an intrinsic angular momentum that causes it to act *as if it were spinning*, but the concept of spin angular momentum is actually a relativistic quantum effect.

SUMMARY

Physics⊗Now™ Take a practice test by logging into Physics-Now at **www.cp7e.com** and clicking on the Pre-Test link for this chapter.

19.3   Magnetic Fields

The **magnetic force** that acts on a charge q moving with velocity $\vec{\mathbf{v}}$ in a magnetic field $\vec{\mathbf{B}}$ has magnitude

$$F = qvB \sin \theta \qquad [19.1]$$

where θ is the angle between $\vec{\mathbf{v}}$ and $\vec{\mathbf{B}}$.

To find the direction of this force, use **right-hand rule number 1**: point the fingers of your open right hand in the direction of $\vec{\mathbf{v}}$ and then curl them in the direction of $\vec{\mathbf{B}}$. Your thumb then points in the direction of the magnetic force $\vec{\mathbf{F}}$.

If the charge is *negative* rather than positive, the force is directed opposite the force given by the right-hand rule.

The SI unit of the magnetic field is the **tesla** (T), or weber per square meter (Wb/m^2). An additional commonly used unit for the magnetic field is the **gauss** (G); $1 \text{ T} = 10^4 \text{ G}$.

19.4   Magnetic Force on a Current-Carrying Conductor

If a straight conductor of length ℓ carries current I, the magnetic force on that conductor when it is placed in a uniform external magnetic field $\vec{\mathbf{B}}$ is

$$F = BI\ell \sin \theta \qquad [19.6]$$

where θ is the angle between the direction of the current and the direction of the magnetic field.

Right-hand rule number 1 also gives the direction of the magnetic force on the conductor. In this case, however, you must point your fingers in the direction of the current rather than in the direction of $\vec{\mathbf{v}}$.

19.5 Torque on a Current Loop and Electric Motors

The torque τ on a current-carrying loop of wire in a magnetic field $\vec{B}$ has magnitude

$$\tau = BIA \sin \theta \qquad \text{[19.8]}$$

where I is the current in the loop and A is its cross-sectional area. The magnitude of the magnetic moment of a current-carrying coil is defined by $\mu = IAN$, where N is the number of loops. The magnetic moment is considered a vector, $\vec{\mu}$, that is perpendicular to the plane of the loop. The angle between $\vec{B}$ and $\vec{\mu}$ is θ.

19.6 Motion of a Charged Particle in a Magnetic Field

If a charged particle moves in a uniform magnetic field so that its initial velocity is perpendicular to the field, it will move in a circular path in a plane perpendicular to the magnetic field. The radius r of the circular path can be found from Newton's second law and centripetal acceleration, and is given by

$$r = \frac{mv}{qB} \qquad \text{[19.10]}$$

where m is the mass of the particle and q is its charge.

19.7 Magnetic Field of a Long, Straight Wire and Ampère's Law

The magnetic field at distance r from a **long, straight wire** carrying current I has the magnitude

$$B = \frac{\mu_0 I}{2\pi r} \qquad \text{[19.11]}$$

where $\mu_0 = 4\pi \times 10^{-7} \text{ T·m/A}$ is the **permeability of free space**. The magnetic field lines around a long, straight wire are circles concentric with the wire.

Ampère's law can be used to find the magnetic field around certain simple current-carrying conductors. It can be written

$$\Sigma \, B_\parallel \, \Delta\ell = \mu_0 I \qquad \text{[19.13]}$$

where $B_\parallel$ is the component of $\vec{B}$ tangent to a small current element of length $\Delta\ell$ that is part of a closed path and I is the total current that penetrates the closed path.

19.8 Magnetic Force between Two Parallel Conductors

The force per unit length on each of two parallel wires separated by the distance d and carrying currents I_1 and I_2 has the magnitude

$$\frac{F}{\ell} = \frac{\mu_0 I_1 I_2}{2\pi d} \qquad \text{[19.14]}$$

The forces are attractive if the currents are in the same direction and repulsive if they are in opposite directions.

19.9 Magnetic Field of Current Loops and Solenoids

The magnetic field at the center of a coil of N circular loops of radius R, each carrying current I, is given by

$$B = N\frac{\mu_0 I}{2R} \qquad \text{[19.15]}$$

The magnetic field inside a solenoid has the magnitude

$$B = \mu_0 n I \qquad \text{[19.16]}$$

where $n = N/\ell$ is the number of turns of wire per unit length.

CONCEPTUAL QUESTIONS

1. In your home television set, a beam of electrons moves from the back of the picture tube to the screen, where it strikes a fluorescent dot that glows with a particular color when hit. The Earth's magnetic field at the location of the television set is horizontal and toward the north. In which direction(s) should the set be oriented so that the beam undergoes the largest deflection?

2. Can a constant magnetic field set a proton at rest into motion? Explain your answer.

3. A proton moving horizontally enters a region where a uniform magnetic field is directed perpendicular to the proton's velocity, as shown in Figure Q19.3. Describe the subsequent motion of the proton. How would an electron behave under the same circumstances.

4. No magnetic force acts upon a current-carrying conductor when it is placed in a certain manner in a uniform magnetic field. Explain.

5. How can the motion of a charged particle be used to distinguish between a magnetic field and an electric field in a certain region?

6. Which way would a compass point if you were at Earth's north magnetic pole?

7. Why does the picture on a television screen become distorted when a magnet is brought near the screen as in Figure Q19.7? [*Caution*: You should not do this at home on a color television set, because it may permanently affect the picture quality.]

Figure Q19.3

Figure Q19.7

8. A magnet attracts a piece of iron. The iron can then attract another piece of iron. On the basis of domain alignment, explain what happens in each piece of iron.

9. A Hindu ruler once suggested that he be entombed in a magnetic coffin with the polarity arranged so that he could be forever suspended between heaven and Earth. Is such magnetic levitation possible? Discuss.

10. Will a nail be attracted to either pole of a magnet? Explain what is happening inside the nail when it is placed near the magnet.

11. Suppose you move along a wire at the same speed as the drift speed of the electrons in the wire. Do you now measure a magnetic field of zero?

12. Describe the change in the magnetic field in the space enclosed by a solenoid carrying a steady current I if (a) the length of the solenoid is doubled, but the number of turns remains the same, and (b) the number of turns is doubled, but the length remains the same.

13. Can you use a compass to detect the currents in wires in the walls near light switches in your home?

14. Why do charged particles from outer space, called cosmic rays, strike Earth more frequently at the poles than at the equator?

15. Two wires carry currents in opposite directions and are oriented parallel, with one above the other. The wires repel each other. Is the upper wire in a stable levitation over the lower wire? Suppose the current in one wire is reversed, so that the wires now attract. Is the lower wire hanging in a stable attraction to the upper wire?

16. How can a current loop be used to determine the presence of a magnetic field in a given region of space?

17. A hanging Slinky® toy is attached to a powerful battery and a switch. When the switch is closed so that the toy now carries current, does the Slinky® compress or expand?

18. Is it possible to orient a current loop in a uniform magnetic field such that the loop will not tend to rotate?

19. Parallel wires exert magnetic forces on each other. What about perpendicular wires? Imagine two wires oriented perpendicular to each other and almost touching. Each wire carries a current. Is there a force between the wires?

20. Is the magnetic field created by a current loop uniform? Explain.

21. The electron beam in Figure Q19.21 is projected to the right. The beam deflects downward in the presence of a magnetic field produced by a pair of current-carrying coils. (a) What is the direction of the magnetic field? (b) What would happen to the beam if the magnetic field were reversed in direction?

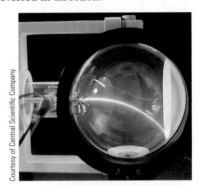

Figure Q19.21

22. Figure Q19.22 shows two permanent magnets, each having a hole through its center. Note that the upper magnet is levitated above the lower one. (a) How does this occur? (b) What purpose does the pencil serve? (c) What can you say about the poles of the magnets from this observation? (d) If the upper magnet were inverted, what do you suppose would happen?

Figure Q19.22

PROBLEMS

1, 2, 3 = straightforward, intermediate, challenging   □ = full solution available in *Student Solutions Manual/Study Guide*

Physics⊗Now™ = coached problem with hints available at **www.cp7e.com**   = biomedical application

Section 19.3 Magnetic Fields

1. An electron gun fires electrons into a magnetic field directed straight downward. Find the direction of the force exerted by the field on an electron for each of the following directions of the electron's velocity: (a) horizontal and due north; (b) horizontal and 30° west of north; (c) due north, but at 30° below the horizontal; (d) straight upward. (Remember that an electron has a negative charge.)

2. (a) Find the direction of the force on a proton (a positively charged particle) moving through the magnetic fields in Figure P19.2 (page 652), as shown. (b) Repeat part (a), assuming the moving particle is an electron.

3. Find the direction of the magnetic field acting on the positively charged particle moving in the various situations

shown in Figure P19.3 (page 652) if the direction of the magnetic force acting on it is as indicated.

4. Determine the initial direction of the deflection of charged particles as they enter the magnetic fields, as shown in Figure P19.4 (page 652).

5. At the equator, near the surface of Earth, the magnetic field is approximately 50.0 μT northward, and the electric field is about 100 N/C downward in fair weather. Find the gravitational, electric, and magnetic forces on an electron with an instantaneous velocity of 6.00×10^6 m/s directed to the east in this environment.

6. The magnetic field of the Earth at a certain location is directed vertically downward and has a magnitude of 50.0 μT. A proton is moving horizontally toward the west

Figure P19.2   (Problems 2 and 12) For Problem 12, replace the velocity vector with a current in that direction.

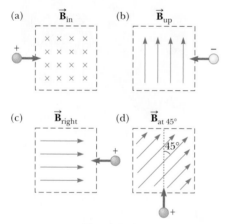

Figure P19.3   (Problems 3 and 13) For Problem 13, replace the velocity vector with a current in that direction.

(a) $\vec{B}_{in}$ (b) $\vec{B}_{up}$

(c) $\vec{B}_{right}$ (d) $\vec{B}_{at\ 45°}$

Figure P19.4

in this field with a speed of 6.20×10^6 m/s. What are the direction and magnitude of the magnetic force the field exerts on the proton?

7. **Physics⊗Now™** What velocity would a proton need to circle Earth 1 000 km above the magnetic equator, where Earth's magnetic field is directed horizontally north and has a magnitude of 4.00×10^{-8} T?

8. An electron is accelerated through 2 400 V from rest and then enters a region where there is a uniform 1.70-T magnetic field. What are (a) the maximum and (b) the minimum magnitudes of the magnetic force acting on this electron?

9. A proton moves perpendicularly to a uniform magnetic field $\vec{B}$ at 1.0×10^7 m/s and exhibits an acceleration of 2.0×10^{13} m/s² in the +x-direction when its velocity is in the +z-direction. Determine the magnitude and direction of the field.

10. Sodium ions (Na⁺) move at 0.851 m/s through a bloodstream in the arm of a person standing near a large magnet. The magnetic field has a strength of 0.254 T and makes an angle of 51.0° with the motion of the sodium ions. The arm contains 100 cm³ of blood with 3.00×10^{20} Na⁺ ions per cubic centimeter. If no other ions were present in the arm, what would be the magnetic force on the arm?

Section 19.4 Magnetic Force on a Current-Carrying Conductor

11. A current $I = 15$ A is directed along the positive x-axis and perpendicular to a magnetic field. A magnetic force per unit length of 0.12 N/m acts on the conductor in the negative y-direction. Calculate the magnitude and direction of the magnetic field in the region through which the current passes.

12. In Figure P19.2, assume that in each case the velocity vector shown is replaced with a wire carrying a current in the direction of the velocity vector. For each case, find the direction of the magnetic force acting on the wire.

13. In Figure P19.3, assume that in each case the velocity vector shown is replaced with a wire carrying a current in the direction of the velocity vector. For each case, find the direction of the magnetic field that will produce the magnetic force shown.

14. A wire having a mass per unit length of 0.500 g/cm carries a 2.00-A current horizontally to the south. What are the direction and magnitude of the minimum magnetic field needed to lift this wire vertically upward?

15. A wire carries a current of 10.0 A in a direction that makes an angle of 30.0° with the direction of a magnetic field of strength 0.300 T. Find the magnetic force on a 5.00-m length of the wire.

16. At a certain location, Earth has a magnetic field of 0.60×10^{-4} T, pointing 75° below the horizontal in a north–south plane. A 10.0-m-long straight wire carries a 15-A current. (a) If the current is directed horizontally toward the east, what are the magnitude and direction of the magnetic force on the wire? (b) What are the magnitude and direction of the force if the current is directed vertically upward?

17. A wire with a mass of 1.00 g/cm is placed on a horizontal surface with a coefficient of friction of 0.200. The wire carries a current of 1.50 A eastward and moves horizontally to the north. What are the magnitude and the direction of the *smallest* vertical magnetic field that enables the wire to move in this fashion?

18. A conductor suspended by two flexible wires as shown in Figure P19.18 has a mass per unit length of 0.040 0 kg/m. What current must exist in the conductor for the tension in the supporting wires to be zero when the magnetic field is 3.60 T into the page? What is the required direction for the current?

19. **Physics⊗Now™** An unusual message delivery system is pictured in Figure P19.19. A 15-cm length of conductor that is free to move is held in place between two thin conduc-

Figure P19.18

tors. When a 5.0-A current is directed as shown in the figure, the wire segment moves upward at a constant velocity. If the mass of the wire is 15 g, find the magnitude and direction of the minimum magnetic field that is required to move the wire. (The wire slides without friction on the two vertical conductors.)

Figure P19.19

20. A wire 2.80 m in length carries a current of 5.00 A in a region where a uniform magnetic field has a magnitude of 0.390 T. Calculate the magnitude of the magnetic force on the wire, assuming the angle between the magnetic field and the current is (a) 60.0°, (b) 90.0°, (c) 120°.

21. In Figure P19.21, the cube is 40.0 cm on each edge. Four straight segments of wire — ab, bc, cd, and da — form a closed loop that carries a current $I = 5.00$ A in the direction shown. A uniform magnetic field of magnitude $B = 0.020\ 0$ T is in the positive y-direction. Determine the magnitude and direction of the magnetic force on each segment.

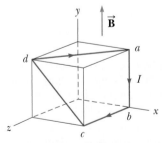

Figure P19.21

Section 19.5 Torque on a Current Loop and Electric Motors

22. A current of 17.0 mA is maintained in a single circular loop with a circumference of 2.00 m. A magnetic field of 0.800 T is directed parallel to the plane of the loop. What is the magnitude of the torque exerted by the magnetic field on the loop?

23. An eight-turn coil encloses an elliptical area having a major axis of 40.0 cm and a minor axis of 30.0 cm (Fig. P19.23). The coil lies in the plane of the page and has a 6.00-A current flowing clockwise around it. If the coil is in a uniform magnetic field of 2.00×10^{-4} T directed toward the left of the page, what is the magnitude of the torque on the coil? [*Hint*: The area of an ellipse is $A = \pi ab$, where a and b are, respectively, the semimajor and semiminor axes of the ellipse.]

Figure P19.23

24. A rectangular loop consists of 100 closely wrapped turns and has dimensions 0.40 m by 0.30 m. The loop is hinged along the y-axis, and the plane of the coil makes an angle of 30.0° with the x-axis (Fig. P19.24). What is the magnitude of the torque exerted on the loop by a uniform magnetic field of 0.80 T directed along the x-axis when the current in the windings has a value of 1.2 A in the direction shown? What is the expected direction of rotation of the loop?

Figure P19.24

25. A long piece of wire with a mass of 0.100 kg and a total length of 4.00 m is used to make a square coil with a side of 0.100 m. The coil is hinged along a horizontal side, carries a 3.40-A current, and is placed in a vertical magnetic field with a magnitude of 0.010 0 T. (a) Determine the angle that the plane of the coil makes with the vertical when the coil is in equilibrium. (b) Find the torque acting on the coil due to the magnetic force at equilibrium.

26. A copper wire is 8.00 m long and has a cross-sectional area of 1.00×10^{-4} m². The wire forms a one-turn loop in the shape of square and is then connected to a battery that applies a potential difference of 0.100 V. If the loop is placed in a uniform magnetic field of magnitude 0.400 T, what is the maximum torque that can act on it? The resistivity of copper is $1.70 \times 10^{-8}\ \Omega \cdot$m.

Section 19.6 Motion of a Charged Particle in a Magnetic Field

27. A proton moving freely in a circular path perpendicular to a constant magnetic field takes 1.00 μs to complete one revolution. Determine the magnitude of the magnetic field.

28. A cosmic-ray proton in interstellar space has an energy of 10.0 MeV and executes a circular orbit having a radius equal to that of Mercury's orbit around the Sun, which is 5.80×10^{10} m. What is the magnetic field in that region of space?

29. Figure P19.29a is a diagram of a device called a velocity selector, in which particles of a specific velocity pass through undeflected while those with greater or lesser velocities are deflected either upwards or downwards. An electric field is directed perpendicular to a magnetic field, producing an electric force and a magnetic force on the charged particle that can be equal in magnitude and opposite in direction (Fig. P19.29b) and hence cancel. Show that particles with a speed of $v = E/B$ will pass through the velocity selector undeflected.

(a)   (b)

Figure P19.29

30. Consider the mass spectrometer shown schematically in Figure P19.30. The electric field between the plates of the velocity selector is 950 V/m, and the magnetic fields in both the velocity selector and the deflection chamber have magnitudes of 0.930 T. Calculate the radius of the path in the system for a singly charged ion with mass $m = 2.18 \times 10^{-26}$ kg. [*Hint*: See Problem 29.]

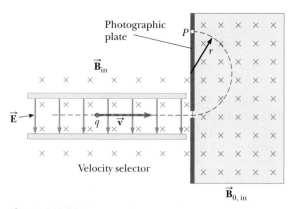

Figure P19.30 A mass spectrometer. Charged particles are first sent through a velocity selector. They then enter a region where a magnetic field $\vec{B}_0$ (directed inward) causes positive ions to move in a semicircular path and strike a photographic film at P.

31. A singly charged positive ion has a mass of 2.50×10^{-26} kg. After being accelerated through a potential difference of 250 V, the ion enters a magnetic field of 0.500 T, in a direction perpendicular to the field. Calculate the radius of the path of the ion in the field.

32. A mass spectrometer is used to examine the isotopes of uranium. Ions in the beam emerge from the velocity selector at a speed of 3.00×10^5 m/s and enter a uniform

magnetic field of 0.600 T directed perpendicularly to the velocity of the ions. What is the distance between the impact points formed on the photographic plate by singly charged ions of ^{235}U and ^{238}U?

33. A proton is at rest at the plane vertical boundary of a region containing a uniform vertical magnetic field B. An alpha particle moving horizontally makes a head-on elastic collision with the proton. Immediately after the collision, both particles enter the magnetic field, moving perpendicular to the direction of the field. The radius of the proton's trajectory is R. Find the radius of the alpha particle's trajectory. The mass of the alpha particle is four times that of the proton, and its charge is twice that of the proton.

Section 19.7 Magnetic Field of a Long, Straight Wire and Ampère's Law

34. In each of parts (a), (b), and (c) of Figure P19.34, find the direction of the current in the wire that would produce a magnetic field directed as shown.

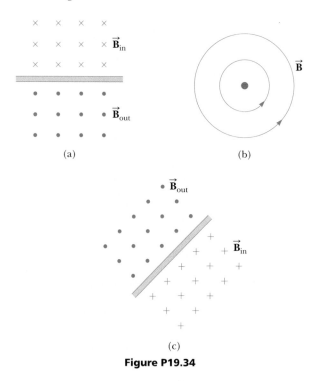

(a)   (b)

(c)

Figure P19.34

35. A lightning bolt may carry a current of 1.00×10^4 A for a short time. What is the resulting magnetic field 100 m from the bolt? Suppose that the bolt extends far above and below the point of observation.

36. In 1962, measurements of the magnetic field of a large tornado were made at the Geophysical Observatory in Tulsa, Oklahoma. If the magnitude of the tornado's field was $B = 1.50 \times 10^{-8}$ T pointing north when the tornado was 9.00 km east of the observatory, what current was carried up or down the funnel of the tornado? Model the vortex as a long, straight wire carrying a current.

37. A cardiac pacemaker can be affected by a static magnetic field as small as 1.7 mT. How close can a pacemaker wearer come to a long, straight wire carrying 20 A?

38. The two wires shown in Figure P19.38 carry currents of 5.00 A in opposite directions and are separated by 10.0 cm. Find the direction and magnitude of the net magnetic

field (a) at a point midway between the wires; (b) at point P_1, 10.0 cm to the right of the wire on the right, and (c) at point P_2, 20.0 cm to the left of the wire on the left.

Figure P19.38

39. Four long, parallel conductors carry equal currents of $I = 5.00$ A. Figure P19.39 is an end view of the conductors. The direction of the current is into the page at points A and B (indicated by the crosses) and out of the page at C and D (indicated by the dots). Calculate the magnitude and direction of the magnetic field at point P, located at the center of the square with edge of length 0.200 m.

Figure P19.39

40. The two wires in Figure P19.40 carry currents of 3.00 A and 5.00 A in the direction indicated. (a) Find the direction and magnitude of the magnetic field at a point midway between the wires. (b) Find the magnitude and direction of the magnetic field at point P, located 20.0 cm above the wire carrying the 5.00-A current.

Figure P19.40

41. A wire carries a 7.00-A current along the x-axis, and another wire carries a 6.00-A current along the y-axis, as shown in Figure P19.41. What is the magnetic field at point P, located at $x = 4.00$ m, $y = 3.00$ m?

42. A long, straight wire lies on a horizontal table and carries a current of 1.20 μA. In a vacuum, a proton moves parallel to the wire (opposite the direction of the current) with a constant velocity of 2.30×10^4 m/s at a constant distance d above the wire. Determine the value of d. (You may ignore the magnetic field due to Earth.)

Figure P19.41

43. The magnetic field 40.0 cm away from a long, straight wire carrying current 2.00 A is 1.00 μT. (a) At what distance is it 0.100 μT? (b) At one instant, the two conductors in a long household extension cord carry equal 2.00-A currents in opposite directions. The two wires are 3.00 mm apart. Find the magnetic field 40.0 cm away from the middle of the straight cord, in the plane of the two wires. (c) At what distance is it one-tenth as large? (d) The center wire in a coaxial cable carries current 2.00 A in one direction, and the sheath around it carries current 2.00 A in the opposite direction. What magnetic field does the cable create at points outside?

Section 19.8 Magnetic Force between Two Parallel Conductors

44. Two parallel wires are 10.0 cm apart, and each carries a current of 10.0 A. (a) If the currents are in the same direction, find the force per unit length exerted on one of the wires by the other. Are the wires attracted to or repelled by each other? (b) Repeat the problem with the currents in opposite directions.

45. **Physics⊗Now**™ A wire with a weight per unit length of 0.080 N/m is suspended directly above a second wire. The top wire carries a current of 30.0 A and the bottom wire carries a current of 60.0 A. Find the distance of separation between the wires so that the top wire will be held in place by magnetic repulsion.

46. In Figure P19.46, the current in the long, straight wire is $I_1 = 5.00$ A, and the wire lies in the plane of the rectangular loop, which carries 10.0 A. The dimensions shown are $c = 0.100$ m, $a = 0.150$ m, and $\ell = 0.450$ m. Find the magnitude and direction of the net force exerted by the magnetic field due to the straight wire on the loop.

Figure P19.46

Section 19.9 Magnetic Fields of Current Loops and Solenoids

47. What current is required in the windings of a long solenoid that has 1 000 turns uniformly distributed over a length of 0.400 m in order to produce a magnetic field of magnitude 1.00×10^{-4} T at the center of the solenoid?

48. It is desired to construct a solenoid that will have a resistance of $5.00 \ \Omega$ (at 20°C) and produce a magnetic field of 4.00×10^{-2} T at its center when it carries a current of 4.00 A. The solenoid is to be constructed from copper wire having a diameter of 0.500 mm. If the radius of the solenoid is to be 1.00 cm, determine (a) the number of turns of wire needed and (b) the length the solenoid should have.

49. A single-turn square loop of wire 2.00 cm on a side carries a counterclockwise current of 0.200 A. The loop is inside a solenoid, with the plane of the loop perpendicular to the magnetic field of the solenoid. The solenoid has 30 turns per centimeter and carries a counterclockwise current of 15.0 A. Find the force on each side of the loop and the torque acting on the loop.

50. An electron is moving at a speed of 1.0×10^4 m/s in a circular path of radius of 2.0 cm inside a solenoid. The magnetic field of the solenoid is perpendicular to the plane of the electron's path. Find (a) the strength of the magnetic field inside the solenoid and (b) the current in the solenoid if it has 25 turns per centimeter.

ADDITIONAL PROBLEMS

51. A circular coil consisting of a single loop of wire has a radius of 30.0 cm and carries a current of 25 A. It is placed in an external magnetic field of 0.30 T. Find the torque on the wire when the plane of the coil makes an angle of 35° with the direction of the field.

52. An electron enters a region of magnetic field of magnitude 0.010 0 T, traveling perpendicular to the linear boundary of the region. The direction of the field is perpendicular to the velocity of the electron. (a) Determine the time it takes for the electron to leave the "field-filled" region, noting that its path is a semicircle. (b) Find the kinetic energy of the electron if the radius of its semicircular path is 2.00 cm.

53. Two long, straight wires cross each other at right angles, as shown in Figure P19.53. (a) Find the direction and magnitude of the magnetic field at point P, which is in the same plane as the two wires. (b) Find the magnetic field at a point 30.0 cm above the point of intersection (30.0 cm out of the page, toward you).

Figure P19.53

54. A 0.200-kg metal rod carrying a current of 10.0 A glides on two horizontal rails 0.500 m apart. What vertical magnetic field is required to keep the rod moving at a constant speed if the coefficient of kinetic friction between the rod and rails is 0.100?

55. **Physics⊗Now**™ Two species of singly charged positive ions of masses 20.0×10^{-27} kg and 23.4×10^{-27} kg enter a magnetic field at the same location with a speed of 1.00×10^5 m/s. If the strength of the field is 0.200 T, and the ions move perpendicularly to the field, find their distance of separation after they complete one-half of their circular path.

56. Two parallel conductors carry currents in opposite directions, as shown in Figure P19.56. One conductor carries a current of 10.0 A. Point A is the midpoint between the wires, and point C is 5.00 cm to the right of the 10.0-A current. I is adjusted so that the magnetic field at C is zero. Find (a) the value of the current I and (b) the value of the magnetic field at A.

Figure P19.56

57. Using an electromagnetic flowmeter (Fig. P19.57), a heart surgeon monitors the flow rate of blood through an artery. Electrodes A and B make contact with the outer surface of the blood vessel, which has interior diameter 3.00 mm. (a) For a magnetic field magnitude of 0.040 0 T, a potential difference of 160 μV appears between the electrodes. Calculate the speed of the blood. (b) Verify that electrode A is positive, as shown. Does the sign of the emf depend on whether the mobile ions in the blood are predominantly positively or negatively charged? Explain.

Figure P19.57

58. Two circular loops are parallel, coaxial, and almost in contact 1.00 mm apart (Fig. P19.58). Each loop is 10.0 cm in radius. The top loop carries a clockwise current of 140 A. The bottom loop carries a counterclockwise current of 140 A. (a) Calculate the magnetic force that the bottom loop exerts on the top loop. (b) The upper loop has a mass of 0.021 0 kg. Calculate its acceleration, assuming that the only forces acting on it are the force in part (a) and its weight. [*Hint*: The distance between the loops is small in comparison to their radius of curvature, so the loops may be treated as long, straight parallel wires.]

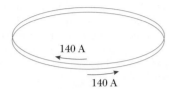

140 A

140 A

Figure P19.58

59. A 1.00-kg ball having net charge $Q = 5.00 \ \mu C$ is thrown out of a window horizontally at a speed $v = 20.0$ m/s. The window is at a height $h = 20.0$ m above the ground. A uniform horizontal magnetic field of magnitude $B = 0.010 \ 0$ T is perpendicular to the plane of the ball's trajectory. Find the magnitude of the magnetic force acting on the ball just before it hits the ground. [*Hint*: Ignore magnetic forces in finding the ball's final velocity.]

60. The idea that static magnetic fields might have a therapeutic value has been around for centuries. A currently available rare-Earth magnet that is advertised to relieve joint pain is shown in Figure P19.60. It is 1.0 mm thick and has a field strength of 5.0×10^{-2} T at the center of the flat surface. The magnetic field strength at points away from the center of the disk is inversely proportional to h^3, where h is the distance from the midplane of the disk. How far from the surface of the disk will the field strength be reduced to that of Earth (5.0×10^{-5} T)?

Figure P19.60

61. Two long parallel conductors carry currents $I_1 = 3.00$ A and $I_2 = 3.00$ A, both directed into the page in Figure P19.61. Determine the magnitude and direction of the resultant magnetic field at P.

$\times \ I_1$

5.00 cm

$\bullet P$

13.0 cm

12.0 cm

$\times \ I_2$

Figure P19.61

62. A uniform horizontal wire with a linear mass density of 0.50 g/m carries a 2.0-A current. It is placed in a constant magnetic field with a strength of 4.0×10^{-3} T. The field is horizontal and perpendicular to the wire. As the wire moves upward starting from rest, (a) what is its acceleration and (b) how long does it take to rise 50 cm? Neglect the magnetic field of Earth.

63. Three long parallel conductors carry currents of $I = 2.0$ A. Figure P19.63 is an end view of the conductors, with each current coming out of the page. Given that $a = 1.0$ cm, determine the magnitude and direction of the magnetic field at points A, B, and C.

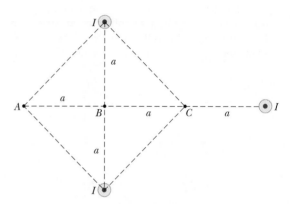

Figure P19.63

64. Two long parallel wires, each with a mass per unit length of 40 g/m, are supported in a horizontal plane by 6.0-cm-long strings, as shown in Figure P19.64. Each wire carries the same current I, causing the wires to repel each other so that the angle θ between the supporting strings is 16°. (a) Are the currents in the same or opposite directions? (b) Determine the magnitude of each current.

Figure P19.64

65. Protons having a kinetic energy of 5.00 MeV are moving in the positive x-direction and enter a magnetic field of 0.050 0 T in the z-direction, out of the plane of the page, and extending from $x = 0$ to $x = 1.00$ m as in Figure P19.65 (page 658). (a) Calculate the y-component of the protons' momentum as they leave the magnetic field. (b) Find the angle α between the initial velocity vector of the proton beam and the velocity vector after the beam emerges from the field. [*Hint*: Neglect relativistic effects and note that 1 eV = 1.60×10^{-19} J.]

Figure P19.65

66. A straight wire of mass 10.0 g and length 5.0 cm is suspended from two identical springs that, in turn, form a closed circuit (Fig. P19.66). The springs stretch a distance of 0.50 cm under the weight of the wire. The circuit has a total resistance of 12 Ω. When a magnetic field directed out of the page (indicated by the dots in the figure is turned on, the springs are observed to stretch an additional 0.30 cm. What is the strength of the magnetic field? (The upper portion of the circuit is fixed.)

24 V

←—5.0 cm—→

Figure P19.66

67. A solenoid 10.0 cm in diameter and 75.0 cm long is made from copper wire of diameter 0.100 cm with very thin insulation. The wire is wound onto a cardboard tube in a single layer, with adjacent turns touching each other. To produce a field of magnitude 20.0 mT at the center of the solenoid, what power must be delivered to the solenoid?

68. Assume that the region to the right of a certain vertical plane contains a vertical magnetic field of magnitude 1.00 mT and that the field is zero in the region to the left of the plane. An electron, originally traveling perpendicular to the boundary plane, passes into the region of the field. (a) Noting that the path of the electron is a semicircle, determine the time interval required for the electron to leave the "field-filled" region. (b) Find the kinetic energy of the electron if the maximum depth of penetration into the field is 2.00 cm.

69. Three long wires (wire 1, wire 2, and wire 3) are coplanar and hang vertically. The distance between wire 1 and wire 2 is 20.0 cm. On the left, wire 1 carries an upward current of 1.50 A. To the right, wire 2 carries a downward current of 4.00 A. Wire 3 is located such that when it carries a certain current, no net force acts upon any of the wires. Find (a) the position of wire 3 and (b) the magnitude and direction of the current in wire 3.

70. Two long parallel conductors separated by 10.0 cm carry currents in the same direction. The first wire carries a current $I_1 = 5.00$ A and the second carries $I_2 = 8.00$ A. (a) What is the magnitude of the magnetic field created by I_1 at the location of I_2? (b) What is the force per unit length exerted by I_1 on I_2? (c) What is the magnitude of the magnetic field created by I_2 at the

location of I_1? (d) What is the force per length exerted by I_2 on I_1?

ACTIVITIES

1. For this activity, you will need a small bar magnet, a small plastic container, and a bowl of water. Tape the magnet to the bottom of the container, and float the container and magnet on the surface of the bowl as in Figure A19.1. The magnet and the container should rotate and come to equilibrium, with the magnet pointing along a north–south line. The compass you have constructed is similar to the type used by early sailing vessels. How can you determine which direction is north and which is south?

Magnet

Water

Figure A19.01

2. In the Northern Hemisphere, the direction of Earth's magnetic field becomes more and more nearly vertical the farther north one goes. To find the variation from the horizontal of the magnetic field in your locale, try the following: press an unmagnetized needle through a Ping-Pong® ball, and balance the structure between two drinking glasses that are lined up along an east–west line. Next, press a magnetized needle through the ball at right angles to the unmagnetized needle so that the needle points north. The magnetized needle can now rotate in the vertical direction and will point in the direction of Earth's magnetic field, which is at some angle below the horizontal. Take several measurements of this dip angle and obtain an average value.

3. Construct an electromagnet by wrapping about 1 meter of small-diameter insulated wire around a steel nail. Tape the ends of the wires to a D-cell battery as in Figure A19.3. How many staples or paper clips can you pick up with your electromagnet? How would you increase the magnetic field set up by the nail? Disconnect the wires from

+   −

Figure A19.03

the battery and test how much magnetism is retained by the nail by seeing how many staples it can pick up. A convenient way to test the strength of a magnet is to attach a paper clip to a rubber band. Note how far the rubber band is stretched before the clip comes free of the magnet. Test your electromagnet in this way. Where is the magnetic field of the electromagnet strongest, at the ends of the nail or near its center? When you have your nail magnetized, bang it against a table or the floor and then check its magnetism. Why does the nail lose its magnetism by this procedure?

4. You can trace out the field pattern of a magnet with iron filings. Any machine shop will supply the filings, which should be soaked in a soap solution to remove grit and oil and then dried. Scatter them lightly over the surface of a paper covering the magnet, and then tap the paper gently to jar the filings into alignment. Explain why the filings form their pattern. Examine the field pattern set up in the following situations: (a) Arrange two bar magnets about 4 cm apart, aligned with opposite poles facing each other. (b) Use two bar magnets about 4 cm apart, aligned with like poles facing each other. (c) Use a horseshoe magnet.

The vibrating strings induce a voltage in pickup coils that detect and amplify the musical sounds being produced. The details of how this phenomenon works are discussed in this chapter.

PhotoDisc/Getty Images

Induced Voltages and Inductance

In 1819, Hans Christian Oersted discovered that an electric current exerted a force on a magnetic compass. Although there had long been speculation that such a relationship existed, Oersted's finding was the first evidence of a link between electricity and magnetism. Because nature is often symmetric, the discovery that electric currents produce magnetic fields led scientists to suspect that magnetic fields could produce electric currents. Indeed, experiments conducted by Michael Faraday in England and independently by Joseph Henry in the United States in 1831 showed that a changing magnetic field could induce an electric current in a circuit. The results of these experiments led to a basic and important law known as Faraday's law. In this chapter we discuss Faraday's law and several practical applications, one of which is the production of electrical energy in power generation plants throughout the world.

20.1 INDUCED EMF AND MAGNETIC FLUX

An experiment first conducted by Faraday demonstrated that a current can be produced by a changing magnetic field. The apparatus shown in Figure 20.1 (page 661) consists of a coil connected to a switch and a battery. We will refer to this coil as the *primary coil* and to the corresponding circuit as the primary circuit. The coil is wrapped around an iron ring to intensify the magnetic field produced by the current in the coil. A second coil, at the right, is wrapped around the iron ring and is connected to an ammeter. This is called the *secondary coil*, and the corresponding circuit is called the secondary circuit. It's important to notice that **there is no battery in the secondary circuit**.

At first glance, you might guess that no current would ever be detected in the secondary circuit. However, when the switch in the primary circuit in Figure 20.1 is suddenly closed, something amazing happens: the ammeter measures a current

Ammeter

Switch

Iron

Primary
coil

Secondary
coil

Battery

FIGURE 20.1 Faraday's experiment. When the switch in the primary circuit at the left is closed, the ammeter in the secondary circuit at the right measures a momentary current. The emf in the secondary circuit is induced by the changing magnetic field through the coil in that circuit.

in the secondary circuit and then returns to zero! When the switch is opened again, the ammeter reads a current in the opposite direction and again returns to zero. Finally, whenever there is a steady current in the primary circuit, the ammeter reads zero.

From observations such as these, Faraday concluded that an electric current could be produced by a *changing* magnetic field. (A steady magnetic field doesn't produce a current, unless the coil is moving, as explained below.) The current produced in the secondary circuit occurs only for an instant while the magnetic field through the secondary coil is changing. In effect, the secondary circuit behaves as though a source of emf were connected to it for a short time. It's customary to say that **an induced emf is produced in the secondary circuit by the changing magnetic field**.

Magnetic Flux

In order to evaluate induced emfs quantitatively, we need to understand what factors affect the phenomenon. While changing magnetic fields always induce electric fields, there are also situations in which the magnetic field remains constant, yet an induced electric field is still produced. The best example of this is an electric generator: A loop of conductor rotating in a constant magnetic field creates an electric current.

The physical quantity associated with magnetism that creates an electric field is a **changing magnetic flux**. Magnetic flux is defined in the same way as electric flux (Section 15.9) and is proportional to both the strength of the magnetic field passing through the plane of a loop of wire and the area of the loop.

The **magnetic flux Φ_B** through a loop of wire with area A is defined by

$$\Phi_B \equiv B_\perp A = BA \cos \theta \qquad \text{[20.1]}$$

where $B_\perp$ is the component of $\vec{B}$ perpendicular to the plane of the loop, as in Figure 20.2a, and θ is the angle between $\vec{B}$ and the normal (perpendicular) to the plane of the loop.
SI unit: weber (Wb)

MICHAEL FARADAY, British physicist and chemist (1791–1867)

Faraday is often regarded as the greatest experimental scientist of the 1800s. His many contributions to the study of electricity include the invention of the electric motor, electric generator, and transformer, as well as the discovery of electromagnetic induction and the laws of electrolysis. Greatly influenced by religion, he refused to work on military poison gas for the British government.

◀ Magnetic flux

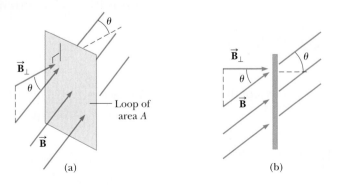

Loop of
area A

Figure 20.2 (a) A uniform magnetic field $\vec{B}$ making an angle θ with a direction normal to the plane of a wire loop of area A. (b) An edge view of the loop.

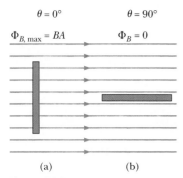

$\theta = 0°$ $\theta = 90°$

$\Phi_{B, max} = BA$ $\Phi_B = 0$

(a)   (b)

Figure 20.3 An edge view of a loop in a uniform magnetic field. (a) When the field lines are perpendicular to the plane of the loop, the magnetic flux through the loop is a maximum and equal to $\Phi_B = BA$. (b) When the field lines are parallel to the plane of the loop, the magnetic flux through the loop is zero.

From Equation 20.1, it follows that $B_\perp = B \cos \theta$. The magnetic flux, in other words, is the magnitude of the part of $\vec{B}$ that is perpendicular to the plane of the loop times the area of the loop. Figure 20.2b is an edge view of the loop and the penetrating magnetic field lines. When the field is perpendicular to the plane of the loop as in Figure 20.3a, $\theta = 0$ and Φ_B has a maximum value, $\Phi_{B, max} = BA$. When the plane of the loop is parallel to $\vec{B}$ as in Figure 20.3b, $\theta = 90°$ and $\Phi_B = 0$. The flux can also be negative. For example, when $\theta = 180°$, the flux is equal to $- BA$. Because the SI unit of B is the tesla, or weber per square meter, the unit of flux is $T \cdot m^2$, or weber (Wb).

We can emphasize the qualitative meaning of Equation 20.1 by first drawing magnetic field lines, as in Figure 20.3. The number of lines per unit area increases as the field strength increases. **The value of the magnetic flux is proportional to the total number of lines passing through the loop.** We see that the most lines pass through the loop when its plane is perpendicular to the field, as in Figure 20.3a, so the flux has its maximum value at that time. As Figure 20.3b shows, no lines pass through the loop when its plane is parallel to the field, so in that case $\Phi_B = 0$.

Applying Physics 20.1   Flux Compared

Argentina has more land area (2.8×10^6 km^2) than Greenland (2.2×10^6 km^2). Why is the magnetic flux of the Earth's magnetic field larger through Greenland than through Argentina?

Explanation Greenland (latitude 60° north to 80° north) is closer to a magnetic pole than Argentina (latitude 20° south to 50° south), so the magnetic field is stronger there. That in itself isn't sufficient to conclude that the magnetic flux is greater, but Greenland's proximity to a pole also means the angle magnetic field lines make with the vertical is smaller than in Argentina. As a result, more field lines penetrate the surface in Greenland, despite Argentina's slightly larger area.

EXAMPLE 20.1   Magnetic Flux

Goal   Calculate magnetic flux and a change in flux.

Problem   A conducting circular loop of radius 0.250 m is placed in the xy-plane in a uniform magnetic field of 0.360 T that points in the positive z-direction, the same direction as the normal to the plane. **(a)** Calculate the magnetic flux through the loop. **(b)** Suppose the loop is rotated clockwise around the x-axis, so the normal direction now points at a 45.0° angle with respect to the z-axis. Recalculate the magnetic flux through the loop. **(c)** What is the change in flux due to the rotation of the loop?

Strategy   After finding the area, substitute values into the equation for magnetic flux for each part.

Solution
(a) Calculate the initial magnetic flux through the loop.

First, calculate the area of the loop:

$$A = \pi r^2 = \pi (0.250 \text{ m})^2 = 0.196 \text{ m}^2$$

Substitute A, B, and $\theta = 0°$ into Equation 20.1 to find the initial magnetic flux:

$$\Phi_B = AB \cos \theta = (0.196 \text{ m}^2)(0.360 \text{ T}) \cos (0°)$$
$$= 0.070\ 6 \text{ T} \cdot m^2 = \boxed{0.070\ 6 \text{ Wb}}$$

(b) Calculate the magnetic flux through the loop after it has rotated 45.0° around the x-axis.

Make the same substitutions as in part (a), except the angle between $\vec{B}$ and the normal is now $\theta = 45.0°$:

$$\Phi_B = AB \cos \theta = (0.196 \text{ m}^2)(0.360 \text{ T}) \cos (45.0°)$$
$$= 0.049\ 9 \text{ T} \cdot m^2 = \boxed{0.049\ 9 \text{ Wb}}$$

(c) Find the change in the magnetic flux due to the rotation of the loop.

Subtract the result of part (a) from the result of part (b): $\Delta\Phi_B = 0.049\ 9\ \text{Wb} - 0.070\ 6\ \text{Wb} = \boxed{-0.020\ 7\ \text{Wb}}$

Remarks   Notice that the rotation of the loop, not any change in the magnetic field, is responsible for the change in flux. This changing magnetic flux is essential in the functioning of electric motors and generators.

Exercise 20.1
The loop, having rotated by 45°, rotates clockwise another 30°, so the normal to the plane points at an angle of 75° with respect to the direction of the magnetic field. Find (a) the magnetic flux through the loop when $\theta = 75°$ and (b) the change in magnetic flux during the rotation from 45° to 75°.

Answers   (a) 0.018 3 Wb   (b) −0.031 6 Wb

20.2   FARADAY'S LAW OF INDUCTION

The usefulness of the concept of magnetic flux can be made obvious by another simple experiment that demonstrates the basic idea of electromagnetic induction. Consider a wire loop connected to an ammeter as in Active Figure 20.4. If a magnet is moved toward the loop, the ammeter reads a current in one direction, as in Active Figure 20.4a. When the magnet is held stationary, as in Active Figure 20.4b, the ammeter reads zero current. If the magnet is moved away from the loop, the ammeter reads a current in the opposite direction, as in Active Figure 20.4c. If the magnet is held stationary and the loop is moved either toward or away from the magnet, the ammeter also reads a current. From these observations, it can be concluded that **a current is set up in the circuit as long as there is relative motion between the magnet and the loop**. The same experimental results are found whether the loop moves or the magnet moves. We call such a current an **induced current**, because it is produced by an **induced emf**.

This experiment is similar to the Faraday experiment discussed in Section 20.1. In each case, an emf is induced in a circuit when the magnetic flux through the circuit changes with time. It turns out that the instantaneous emf induced in a circuit equals the negative of the rate of change of magnetic flux with respect to time through the circuit. This is **Faraday's law of magnetic induction**.

TIP 20.1   Induced Current Requires a Change in Magnetic Flux

The existence of magnetic flux through an area is not sufficient to create an induced emf. A *change* in the magnetic flux over some time interval Δt must occur for an emf to be induced.

◀ Faraday's law

> If a circuit contains N tightly wound loops and the magnetic flux through each loop changes by the amount $\Delta\Phi_B$ during the interval Δt, the average emf induced in the circuit during time Δt is
>
> $$\varepsilon = -N\frac{\Delta\Phi_B}{\Delta t} \qquad\qquad \textbf{[20.2]}$$

Ammeter

(a)

Ammeter

(b)

Ammeter

(c)

ACTIVE FIGURE 20.4
(a) When a magnet is moved toward a wire loop connected to an ammeter, the ammeter reads a current as shown, indicating that a current I is induced in the loop. (b) When the magnet is held stationary, no current is induced in the loop, even when the magnet is inside the loop. (c) When the magnet is moved away from the loop, the ammeter reads a current in the opposite direction, indicating an induced current going opposite the direction of the current in part (a).

Physics⊗Now™
Log into PhysicsNow at **www.cp7e.com** and go to Active Figure 20.4, where you can move the magnet and observe the current in the ammeter.

Because $\Phi_B = BA \cos \theta$, a change of any of the factors B, A, or θ with time produces an emf. We explore the effect of a change in each of these factors in the following sections. The minus sign in Equation 20.2 is included to indicate the polarity of the induced emf. This polarity simply determines which of two different directions current will flow in a loop, a direction given by **Lenz's law**:

Lenz's law ▶

The current caused by the induced emf travels in the direction that creates a magnetic field with flux opposing the change in the original flux through the circuit.

Lenz's law says that if the magnetic flux through a loop is becoming more positive, say, then the induced emf creates a current and associated magnetic field that produces negative magnetic flux. Some mistakenly think this "counter magnetic field" created by the induced current, called $\vec{B}_{ind}$ ("ind" for induced) will always point in a direction opposite the applied magnetic field $\vec{B}$, but this is only true half the time! Figure 20.5 shows a field penetrating a loop. The graph in Figure 20.5b shows that the magnitude of the magnetic field $\vec{B}$ shrinks with time. This means the flux of $\vec{B}$ is shrinking with time, so the induced field $\vec{B}_{ind}$ will actually be in the same direction as $\vec{B}$. In effect, $\vec{B}_{ind}$ "shores up" the field $\vec{B}$, slowing the loss of flux through the loop.

The direction of the current in Figure 20.5 can be determined by right-hand rule number 2: Point your right thumb in the direction that will cause the fingers on your right hand to curl in the direction of the induced field $\vec{B}_{ind}$. In this case, that direction is counterclockwise: with the right thumb pointed in the direction of the current, your fingers curl down outside the loop and around and *up through the inside of the loop*. Remember, inside the loop is where it's important for the induced magnetic field to be pointing up.

Figure 20.5 (a) The magnetic field $\vec{B}$ becomes smaller with time, reducing the flux, so current is induced in a direction that creates an induced magnetic field $\vec{B}_{ind}$ opposing the change in magnetic flux. (b) Graph of the magnitude of the magnetic field as a function of time.

Quick Quiz 20.1

Figure 20.6 is a graph of the magnitude B versus time for a magnetic field that passes through a fixed loop and is oriented perpendicular to the plane of the loop. Rank the magnitudes of the emf generated in the loop at the three instants indicated, from largest to smallest.

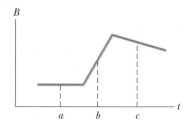

Figure 20.6 (Quick Quiz 20.1)

EXAMPLE 20.2   Faraday and Lenz to the Rescue

Goal Calculate an induced emf and current with Faraday's law, and apply Lenz's law, when the magnetic field changes with time.

Problem A coil with 25 turns of wire is wrapped on a frame with a square cross-section 1.80 cm on a side. Each turn has the same area, equal to that of the frame, and the total resistance of the coil is 0.350 Ω. An applied uniform magnetic field is perpendicular to the plane of the coil, as in Figure 20.7. **(a)** If the field changes uniformly from 0.00 T to 0.500 T in 0.800 s, find the induced emf in the coil while the field is changing. Find **(b)** the magnitude and **(c)** the direction of the induced current in the coil while the field is changing.

Figure 20.7 (Example 20.2)

Strategy Part (a) requires substituting into Faraday's law, Equation 20.2. The necessary information is given, except for $\Delta\Phi_B$, the change in the magnetic flux during the elapsed time. Compute the initial and final magnetic fluxes with Equation 20.1, find the difference, and assemble all terms in Faraday's law. The current can then be found with Ohm's law, and its direction with Lenz's law.

Solution

(a) Find the induced emf in the coil.

To compute the flux, the area of the coil is needed:

$$A = L^2 = (0.018\ 0\ \text{m})^2 = 3.24 \times 10^{-4}\ \text{m}^2$$

The magnetic flux $\Phi_{B,i}$ through the coil at $t = 0$ is zero because $B = 0$. Calculate the flux at $t = 0.800$ s:

$$\Phi_{B,f} = BA \cos \theta = (0.500\ \text{T})(3.24 \times 10^{-4}\ \text{m}^2) \cos (0°)$$
$$= 1.62 \times 10^{-4}\ \text{Wb}$$

Compute the change in the magnetic flux through the cross-section of the coil over the 0.800-s interval:

$$\Delta\Phi_B = \Phi_{B,f} - \Phi_{B,i} = 1.62 \times 10^{-4}\ \text{Wb}$$

Substitute into Faraday's law of induction to find the induced emf in the coil:

$$\mathcal{E} = -N\frac{\Delta\Phi_B}{\Delta t} = (25\ \text{turns})\left(\frac{1.62 \times 10^{-4}\ \text{Wb}}{0.800\ \text{s}}\right)$$
$$= -5.06 \times 10^{-3}\ \text{V}$$

(b) Find the magnitude of the induced current in the coil.

Substitute the voltage difference and the resistance into Ohm's law:

$$I = \frac{\Delta V}{R} = \frac{5.06 \times 10^{-3}\ \text{V}}{0.350\ \Omega} = 1.45 \times 10^{-2}\ \text{A}$$

(c) Find the direction of the induced current in the coil.

The magnetic field is increasing up through the loop, in the same direction as the normal to the plane; hence the flux is positive and increasing, also. A downward-pointing induced magnetic field will create negative flux, opposing the change. If you point your right thumb in the clockwise direction along the loop, your fingers curl down through the loop—the correct direction for the counter magnetic field.

Remark   Lenz's law can best be handled by sketching a diagram, first.

Exercise 20.2

Suppose the magnetic field changes uniformly from 0.500 T to 0.200 T in the next 0.600 s. Compute (a) the induced emf in the coil and (b) the magnitude and direction of the induced current.

Answers   (a) 4.05×10^{-3} V   (b) 1.16×10^{-2} A, counterclockwise

---

The ground fault interrupter (GFI) is an interesting safety device that protects people against electric shock when they touch appliances and power tools. Its operation makes use of Faraday's law. Figure 20.8 shows the essential parts of a ground fault interrupter. Wire 1 leads from the wall outlet to the appliance to be protected, and wire 2 leads from the appliance back to the wall outlet. An iron ring surrounds the two wires to confine the magnetic field set up by each wire. A sensing coil, which can activate a circuit breaker when changes in magnetic flux occur, is wrapped around part of the iron ring. Because the currents in the wires are in opposite directions, the net magnetic field through the sensing coil due to the currents is zero. However, if a short circuit occurs in the appliance so that there is no returning current, the net magnetic field through the sensing coil is no longer zero. This can happen if, for example, one of the wires loses its insulation, providing a path through you to ground if you happen to be touching the appliance and are grounded as in Figure 18.23a. Because the current is alternating, the magnetic flux through the sensing coil changes with time, producing an induced voltage in the coil. This induced voltage is used to trigger a circuit breaker, stopping the

Figure 20.8   Essential components of a ground fault interrupter (contents of the gray box in Fig. 20.9a). In newer homes, such devices are built directly into wall outlets. The purpose of the sensing coil and circuit breaker is to cut off the current before damage is done.

APPLICATION

Ground Fault Interrupters

b, George Semple

Figure 20.9   (a) This hair dryer has been plugged into a ground fault interrupter that is in turn plugged into an unprotected wall outlet. (b) You likely have seen this kind of ground fault interrupter in a hotel bathroom, where hair dryers and electric shavers are often used by people just out of the shower or who might touch a water pipe, providing a ready path to ground in the event of a short circuit.

current quickly in about a millisecond before it reaches a level that might be harmful to the person using the appliance. A ground fault interrupter provides faster and more complete protection than even the case-ground-and-circuit-breaker combination shown in Figure 18.23b. For this reason, ground fault interrupters are commonly found in bathrooms, where electricity poses a hazard to people. (See Fig. 20.9.)

Another interesting application of Faraday's law is the production of sound in an electric guitar. A vibrating string induces an emf in a coil (Fig. 20.10). The pickup coil is placed near the vibrating guitar string, which is made of a metal that can be magnetized. The permanent magnet inside the coil magnetizes the portion of the string nearest the coil. When the guitar string vibrates at some frequency, its magnetized segment produces a changing magnetic flux through the pickup coil. The changing flux induces a voltage in the coil; the voltage is fed to an amplifier. The output of the amplifier is sent to the loudspeakers, producing the sound waves that we hear.

Sudden infant death syndrome (SIDS) is a devastating affliction in which a baby suddenly stops breathing during sleep without an apparent cause. One type of monitoring device, called an apnea monitor, is sometimes used to alert caregivers of the cessation of breathing. The device uses induced currents, as shown in Figure 20.11. A coil of wire attached to one side of the chest carries an alternating current. The varying magnetic flux produced by this current passes through a pickup coil attached to the opposite side of the chest. Expansion and contraction of the chest caused by breathing or movement changes the strength of the voltage induced in the pickup coil. However, if breathing stops, the pattern of the induced voltage stabilizes, and external circuits monitoring the voltage sound an alarm to the caregivers after a momentary pause to ensure that a problem actually does exist.

APPLICATION

Electric Guitar Pickups

APPLICATION

Apnea Monitors

b, Charles D. Winters

Figure 20.10   (a) In an electric guitar, a vibrating string induces a voltage in the pickup coil. (b) Several pickups allow the vibration to be detected from different portions of the string.

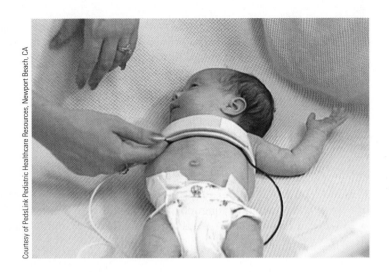

Courtesy of PedsLink Pediatric Healthcare Resources, Newport Beach, CA

20.3 MOTIONAL emf

In Section 20.2, we considered emfs induced in a circuit when the magnetic field changes with time. In this section we describe a particular application of Faraday's law in which a so-called **motional emf** is produced. This is the emf induced in a conductor moving through a magnetic field.

First consider a straight conductor of length ℓ moving with constant velocity through a uniform magnetic field directed into the paper, as in Figure 20.12. For simplicity, we assume that the conductor moves in a direction perpendicular to the field. A magnetic force of magnitude $F_m = qvB$, directed downward, acts on the electrons in the conductor. Because of this magnetic force, the free electrons move to the lower end of the conductor and accumulate there, leaving a net positive charge at the upper end. As a result of this charge separation, an electric field is produced in the conductor. The charge at the ends builds up until the downward magnetic force qvB is balanced by the upward electric force qE. At this point, charge stops flowing and the condition for equilibrium requires that

$$qE = qvB \quad \text{or} \quad E = vB$$

Because the electric field is uniform, the field produced in the conductor is related to the potential difference across the ends by $\Delta V = E\ell$, giving

$$\Delta V = E\ell = B\ell v \qquad \text{[20.3]}$$

Because there is an excess of positive charge at the upper end of the conductor and an excess of negative charge at the lower end, the upper end is at a higher potential than the lower end. There is a potential difference across a conductor as long as it moves through a field. If the motion is reversed, the polarity of the potential difference is also reversed.

A more interesting situation occurs if the moving conductor is part of a closed conducting path. This situation is particularly useful for illustrating how a changing loop area induces a current in a closed circuit described by Faraday's law. Consider a circuit consisting of a conducting bar of length ℓ, sliding along two fixed parallel conducting rails, as in Active Figure 20.13a. For simplicity, assume that the moving bar has zero resistance and that the stationary part of the circuit has constant resistance R. A uniform and constant magnetic field $\vec{B}$ is applied perpendicular to the plane of the circuit. As the bar is pulled to the right with velocity $\vec{v}$ under the influence of an applied force $\vec{F}_{app}$, a magnetic force along the length of the bar acts on the free charges in the bar. This force in turn sets up an induced current because the charges are free to move in a closed conducting path. In this case, the changing magnetic flux through the loop and the corresponding induced emf across the moving bar arise from the *change in area of the loop* as the bar moves through the magnetic field.

Figure 20.11 This infant is wearing a monitor designed to alert caregivers if breathing stops. Note the two wires attached to opposite sides of the chest.

Figure 20.12 A straight conductor of length ℓ moving with velocity $\vec{v}$ through a uniform magnetic field $\vec{B}$ directed perpendicular to $\vec{v}$. The vector $\vec{F}_m$ is the magnetic force on an electron in the conductor. An emf of $B\ell v$ is induced between the ends of the bar.

(a)

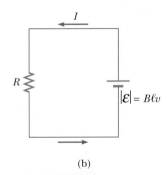

(b)

ACTIVE FIGURE 20.13
(a) A conducting bar sliding with velocity $\vec{v}$ along two conducting rails under the action of an applied force $\vec{F}_{app}$. The magnetic force $\vec{F}_m$ opposes the motion, and a counterclockwise current is induced in the loop. (b) The equivalent circuit of that in (a).

Physics⊗Now™
Log into PhysicsNow at **www.cp7e.com** and go to Active Figure 20.13, where you can adjust the applied force, the magnetic field, and the resistance, and observe the effects on the motion of the bar.

Figure 20.14 As the bar moves to the right, the area of the loop increases by the amount $\ell \Delta x$ and the magnetic flux through the loop increases by $B\ell \Delta x$.

Assume that the bar moves a distance Δx in time Δt, as shown in Figure 20.14. The increase in flux $\Delta \Phi_B$ through the loop in that time is the amount of flux that now passes through the portion of the circuit that has area $\ell \Delta x$:

$$\Delta \Phi_B = BA = B\ell \, \Delta x$$

Using Faraday's law and noting that there is one loop ($N = 1$), we find that the magnitude of the induced emf is

$$|\mathcal{E}| = \frac{\Delta \Phi_B}{\Delta t} = B\ell \frac{\Delta x}{\Delta t} = B\ell v \qquad \text{[20.4]}$$

This induced emf is often called a **motional emf** because it arises from the motion of a conductor through a magnetic field.

Further, if the resistance of the circuit is R, the magnitude of the induced current in the circuit is

$$I = \frac{|\mathcal{E}|}{R} = \frac{B\ell v}{R} \qquad \text{[20.5]}$$

Active Figure 20.13b shows the equivalent circuit diagram for this example.

Applying Physics 20.2   Space Catapult

Applying a force on the bar will result in an induced emf in the circuit shown in Active Figure 20.13. Suppose we remove the external magnetic field in the diagram and replace the resistor with a high-voltage source and a switch, as in Figure 20.15. What will happen when the switch is closed? Will the bar move, and does it matter which way we connect the high-voltage source?

Explanation Suppose the source is capable of establishing high current. Then the two horizontal conducting rods will create a strong magnetic field in the area between them, directed into the page. (The

movable bar also creates a magnetic field, but this field can't exert force on the bar itself.) Because the moving bar carries a downward current, a magnetic force is exerted on the bar, directed to the right. Hence, the bar accelerates along the rails away from the power supply. If the polarity of the power were reversed, the magnetic field would be out of the page, the current in the bar would be upward, and the force on the bar would still be directed to the right. The $BI\ell$ force exerted by a magnetic field according to Equation 19.6 causes the bar to accelerate away from the voltage source. Studies have shown that it's possible to launch payloads into space with this technology. (This is the working principle of a rail gun.) Very large accelerations can be obtained with currently available technology, with payloads being accelerated to a speed of several kilometers per second in a fraction of a second. This is a larger acceleration than humans can tolerate.

Rail guns have been proposed as propulsion systems for moving asteroids into more useful orbits. The material of the asteroid could be mined and launched off the surface by a rail gun, which would act like a rocket engine, modifying the velocity and hence the orbit of the asteroid. Some asteroids contain trillions of dollars worth of valuable metals.

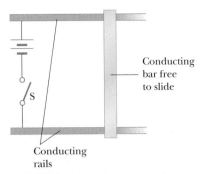

Figure 20.15 (Applying Physics 20.2)

Quick Quiz 20.2

A horizontal metal bar oriented east–west drops straight down in a location where the Earth's magnetic field is due north. As a result, an emf develops between the ends. Which end is positively charged? (a) the east end (b) the west end (c) neither end carries a charge

You intend to move a rectangular loop of wire into a region of uniform magnetic field at a given speed so as to induce an emf in the loop. The plane of the loop must remain perpendicular to the magnetic field lines. In which orientation should you hold the loop while you move it into the region with the magnetic field in order to generate the largest emf? (a) With the long dimension of the loop parallel to the velocity vector, (b) with the short dimension of the loop parallel to the velocity vector, or (c) either way—the emf is the same regardless of orientation.

EXAMPLE 20.3   The Electrified Airplane Wing

Goal   Find the emf induced by motion through a magnetic field.

Problem   An airplane with a wingspan of 30.0 m flies due north at a location where the downward component of the Earth's magnetic field is 0.600×10^{-4} T. There is also a component pointing due north which has a magnitude of 0.470×10^{-4} T. **(a)** Find the difference in potential between the wingtips when the speed of the plane is 2.50×10^2 m/s. **(b)** Which wingtip is positive?

Strategy   Because the plane is flying north, the northern component of the magnetic field won't have any effect on the induced emf. The induced emf across the wing is caused solely by the downward component of the Earth's magnetic field. Substitute the given quantities into Equation 20.4. Use right-hand rule number 1 to find the direction positive charges would be propelled by the magnetic force.

Solution
(a) Calculate the difference in potential across the wingtips.

Write the motional emf equation and substitute the given quantities:

$\mathcal{E} = B\ell v = (0.600 \times 10^{-4} \text{ T})(30.0 \text{ m})(2.50 \times 10^2 \text{ m/s})$
$= \boxed{0.450 \text{ V}}$

(b) Which wingtip is positive?

Apply right hand rule number 1:

Point your right fingers north, in the direction of the velocity, curl them down, in the direction of the magnetic field. Your thumb points west.

Remark   An induced emf such as this can cause problems on an aircraft.

Exercise 20.3
Suppose the magnetic field in a given region of space is parallel to the Earth's surface, points north, and has magnitude 1.80×10^{-4} T. A metal cable attached to a space station stretches radially outwards 2.50 km. (a) Estimate the potential difference that develops between the ends of the cable if it's traveling eastward around Earth at 7.70×10^3 m/s. (b) Which end of the cable is positive, the lower end or the upper end?

Answer   (a) 3.47×10^3 V   (b) The upper end is positive.

EXAMPLE 20.4   Where Is the Energy Source?

Goal   Use motional emf to find an induced emf and a current.

Problem   **(a)** The sliding bar in Figure 20.13a has a length of 0.500 m and moves at 2.00 m/s in a magnetic field of magnitude 0.250 T. Using the concept of motional emf, find the induced voltage in the moving rod. **(b)** If the resistance in the circuit is 0.500 Ω, find the current in the circuit and the power delivered to the resistor. (*Note*: The current, in this case, goes counterclockwise around the loop.) **(c)** Calculate the magnetic force on the bar. **(d)** Use the concepts of work and power to calculate the applied force.

Strategy   For part (a), substitute into Equation 20.4 for the motional emf. Once the emf is found, substitution into Ohm's law gives the current. In part (c), use Equation 19.6 for the magnetic force on a current-carrying conductor. In part (d), use the fact that the power dissipated by the resistor multiplied by the elapsed time must equal the work done by the applied force.

Solution

(a) Find the induced emf with the concept of motional emf.

Substitute into Equation 20.4 to find the induced emf:

$$\mathcal{E} = B\ell v = (0.250 \text{ T})(0.500 \text{ m})(2.00 \text{ m/s}) = \boxed{0.250 \text{ V}}$$

(b) Find the induced current in the circuit and the power dissipated by the resistor.

Substitute the emf and the resistance into Ohm's law to find the induced current:

$$I = \frac{\mathcal{E}}{R} = \frac{0.250 \text{ V}}{0.500 \ \Omega} = \boxed{0.500 \text{ A}}$$

Substitute I and $\mathcal{E} = 0.250$ V into Equation 17.8 to find the power dissipated by the 0.500-Ω resistor:

$$\mathcal{P} = I \Delta V = (0.500 \text{ A})(0.250 \text{ V}) = \boxed{0.125 \text{ W}}$$

(c) Calculate the magnitude and direction of the magnetic force on the bar.

Substitute values for I, B, and ℓ into Equation 19.6 (with $\sin \theta = \sin (90°) = 1$) to find the magnitude of the force:

$$F_m = IB\ell = (0.500 \text{ A})(0.250 \text{ T})(0.500 \text{ m}) = \boxed{6.25 \times 10^{-2} \text{ N}}$$

Apply right hand rule number 2 to find the direction of the force:

Point the fingers of your right hand in the direction of the positive current, then curl them in the direction of the magnetic field. Your thumb points in the negative x-direction.

(d) Find the value of F_{app}, the applied force.

Set the work done by the applied force equal to the dissipated power times the elapsed time:

$$W_{\text{app}} = F_{\text{app}} d = \mathcal{P} \Delta t$$

Solve for F_{app} and substitute $d = v \Delta t$:

$$F_{\text{app}} = \frac{\mathcal{P} \Delta t}{d} = \frac{\mathcal{P} \Delta t}{v \Delta t} = \frac{\mathcal{P}}{v} = \frac{0.125 \text{ W}}{2.00 \text{ m/s}} = \boxed{6.25 \times 10^{-2} \text{ N}}$$

Remarks   Part (d) could be solved by using Newton's second law for an object in equilibrium: two forces act horizontally on the bar, and the acceleration of the bar is zero, so the forces must be equal in magnitude and opposite in direction. Notice the agreement between the answers for F_m and F_{app}, despite the very different concepts used.

Exercise 20.4

Suppose the current suddenly increases to 1.25 A in the same direction as before, due to an increase in speed of the bar. Find (a) the emf induced in the rod, (b) the new speed of the rod.

Answers   (a) 0.625 V   (b) 5.00 m/s

20.4   LENZ'S LAW REVISITED (The Minus Sign in Faraday's Law)

To reach a better understanding of Lenz's law, consider the example of a bar moving to the right on two parallel rails in the presence of a uniform magnetic field directed into the paper (Fig. 20.16a). As the bar moves to the right, the magnetic flux through the circuit increases with time because the area of the loop increases. Lenz's law says that the induced current must be in a direction such that the flux *it* produces opposes the change in the external magnetic flux. Because the flux due to the external field is increasing *into* the paper, the induced current, to oppose the change, must produce a flux *out* of the paper. Hence, the induced current must be counterclockwise when the bar moves to the right. (Use right-hand rule number 2 from Chapter 19 to verify this direction.) On the other hand, if the bar is moving to the left, as in Figure 20.16b, the magnetic flux through the loop decreases with time. Because the flux is into the paper, the induced current has to be clockwise to produce its own flux into the paper (which opposes the decrease in the external flux). In either case, the induced current tends to maintain the original flux through the circuit.

Now we examine this situation from the viewpoint of energy conservation. Suppose that the bar is given a slight push to the right. In the preceding analysis, we found that this motion led to a counterclockwise current in the loop. Let's see what would happen if we assume that the current is clockwise, opposite the direction required by Lenz's law. For a clockwise current I, the direction of the magnetic force $BI\ell$ on the sliding bar is to the right. This force accelerates the rod and increases its velocity. This, in turn, causes the area of the loop to increase more rapidly, thereby increasing the induced current, which increases the force, which increases the current, and so forth. In effect, the system acquires energy with zero input energy. This is inconsistent with all experience and with the law of conservation of energy, so we're forced to conclude that the current must be counterclockwise.

Consider another situation. A bar magnet is moved to the right toward a stationary loop of wire, as in Figure 20.17a. As the magnet moves, the magnetic flux through the loop increases with time. To counteract this rise in flux, the induced current produces a flux to the left, as in Figure 20.17b; hence, the induced current is in the direction shown. Note that the magnetic field lines associated with the induced current oppose the motion of the magnet. The left face of the current loop is therefore a north pole and the right face is a south pole.

On the other hand, if the magnet were moving to the left, as in Figure 20.17c, its flux through the loop, which is toward the right, would decrease with time. Under these circumstances, the induced current in the loop would be in a direction to set up a field directed from left to right through the loop, in an effort to maintain a constant number of flux lines. Hence, the induced current in the loop would be as shown in Figure 20.17d. In this case, the left face of the loop would be a south pole and the right face would be a north pole.

As another example, consider a coil of wire placed near an electromagnet, as in Figure 20.18a (page 672). We wish to find the direction of the induced current in the coil at various times: at the instant the switch is closed, after the switch has been closed for several seconds, and when the switch is opened.

When the switch is closed, the situation changes from a condition in which no lines of flux pass through the coil to one in which lines of flux pass through in the

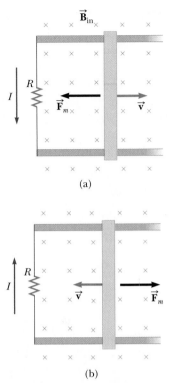

Figure 20.16   (a) As the conducting bar slides on the two fixed conducting rails, the magnetic flux through the loop increases with time. By Lenz's law, the induced current must be *counterclockwise* so as to produce a counteracting flux *out of the paper*. (b) When the bar moves to the left, the induced current must be *clockwise*. Why?

TIP 20.2   There are Two Magnetic Fields to Consider

When applying Lenz's law, there are *two* magnetic fields to consider. The first is the external changing magnetic field that induces the current in a conducting loop. The second is the magnetic field produced by the induced current in the loop.

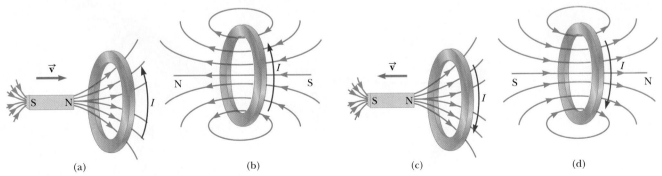

Figure 20.17   (a) When the magnet is moved toward the stationary conducting loop, a current is induced in the direction shown. (b) This induced current produces its own flux to the left to counteract the increasing external flux to the right. (c) When the magnet is moved away from the stationary conducting loop, a current is induced in the direction shown. (d) This induced current produces its own flux to the right to counteract the decreasing external flux to the right.

Figure 20.18 An example of Lenz's law.

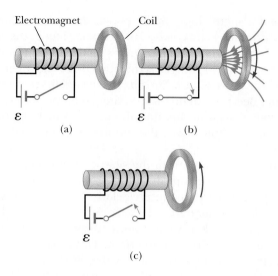

(a)   (b)

(c)

direction shown in Figure 20.18b. To counteract this change in the number of lines, the coil must set up a field from left to right in the figure. This requires a current directed as shown in Figure 20.18b.

After the switch has been closed for several seconds, there is no change in the number of lines through the loop; hence, the induced current is zero.

Opening the switch causes the magnetic field to change from a condition in which flux lines thread through the coil from right to left to a condition of zero flux. The induced current must then be as shown in Figure 20.18c, so as to set up its own field from right to left.

Quick Quiz 20.4

A bar magnet is falling through a loop of wire with constant velocity with the north pole entering first. Viewed from the same side of the loop as the magnet, as the north pole approaches the loop, what is the direction of the induced current? (a) clockwise (b) zero (c) counterclockwise (d) along the length of the magnet

Tape Recorders

APPLICATION

Magnetic Tape Recorders

One common practical use of induced currents and emfs is associated with the tape recorder. Many different types of tape recorders are made, but the basic principles are the same for all. A magnetic tape moves past a recording and playback head, as in Figure 20.19a. The tape is a plastic ribbon coated with iron oxide or

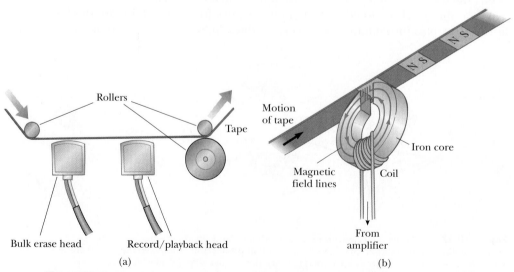

(a)   (b)

Figure 20.19 (a) Major parts of a magnetic tape recorder. If a new recording is to be made, the bulk erase head wipes the tape clean of signals before recording. (b) The fringing magnetic field magnetizes the tape during recording.

chromium oxide. The hard drives in computers work on the same principle, but use a coated disk instead of tape, allowing for faster access.

The recording process uses the fact that a current in an electromagnet produces a magnetic field. Figure 20.19b illustrates the steps in the process. A sound wave sent into a microphone is transformed into an electric current, amplified, and allowed to pass through a wire coiled around a doughnut-shaped piece of iron, which functions as the recording head. The iron ring and the wire constitute an electromagnet, in which the lines of the magnetic field are contained completely inside the iron except at the point where a slot is cut in the ring. Here the magnetic field fringes out of the iron and magnetizes the small pieces of iron oxide embedded in the tape. As the tape moves past the slot, it becomes magnetized in a pattern that reproduces both the frequency and the intensity of the sound signal entering the microphone.

To reconstruct the sound signal, the previously magnetized tape is allowed to pass through a recorder head operating in the playback mode. A second wire-wound doughnut-shaped piece of iron with a slot in it passes close to the tape, so that the varying magnetic fields on the tape produce changing field lines through the wire coil. The changing flux induces a current in the coil which corresponds to the current in the recording head that originally produced the tape. This changing electric current can be amplified and used to drive a speaker. Playback is thus an example of induction of a current by a moving magnet.

20.5 GENERATORS

Generators and motors are important practical devices that operate on the principle of electromagnetic induction. First, consider the **alternating-current** (AC) **generator**, a device that converts mechanical energy to electrical energy. In its simplest form, the AC generator consists of a wire loop rotated in a magnetic field by some external means (Active Fig. 20.20a). In commercial power plants, the energy required to rotate the loop can be derived from a variety of sources. For example, in a hydroelectric plant, falling water directed against the blades of a turbine produces the rotary motion; in a coal-fired plant, heat produced by burning coal is used to convert water to steam, and this steam is directed against the turbine blades. As the loop rotates, the magnetic flux through it changes with time, inducing an emf and a current in an external circuit. The ends of the loop are connected to slip rings that rotate with the loop.

APPLICATION

Alternating-Current Generators

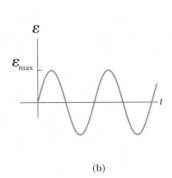

(a)

(b)

ACTIVE FIGURE 20.20

(a) A schematic diagram of an AC generator. An emf is induced in a coil, which rotates by some external means in a magnetic field. (b) A plot of the alternating emf induced in the loop versus time.

Physics⊗Now™

Log into PhysicsNow at **www.cp7e.com** and go to Active Figure 20.20, where you can adjust the speed of rotation and the strength of the field to see the effects on the generated emf.

Figure 20.21 (a) A loop rotating at a constant angular velocity in an external magnetic field. The emf induced in the loop varies sinusoidally with time. (b) An edge view of the rotating loop.

(a)

(b)

Turbines turn electric generators at a hydroelectric power plant.

Connections to the external circuit are made by stationary brushes in contact with the slip rings.

We can derive an expression for the emf generated in the rotating loop by making use of the equation for motional emf, $\mathcal{E} = B\ell v$. Figure 20.21a shows a loop of wire rotating clockwise in a uniform magnetic field directed to the right. The magnetic force (qvB) on the charges in wires AB and CD is not along the lengths of the wires. (The force on the electrons in these wires is perpendicular to the wires.) Hence, an emf is generated only in wires BC and AD. At any instant, wire BC has velocity $\vec{v}$ at an angle θ with the magnetic field, as shown in Figure 20.21b. (Note that the component of velocity parallel to the field has no effect on the charges in the wire, whereas the component of velocity perpendicular to the field produces a magnetic force on the charges that moves electrons from C to B.) The emf generated in wire BC equals $B\ell v_\perp$, where ℓ is the length of the wire and $v_\perp$ is the component of velocity perpendicular to the field. An emf of $B\ell v_\perp$ is also generated in wire DA, and the sense of this emf is the same as that in wire BC. Because $v_\perp = v \sin \theta$, the total induced emf is

$$\mathcal{E} = 2B\ell v_\perp = 2B\ell v \sin \theta \qquad [20.6]$$

If the loop rotates with a constant angular speed ω, we can use the relation $\theta = \omega t$ in Equation 20.6. Furthermore, because every point on the wires BC and DA rotates in a circle about the axis of rotation with the same angular speed ω, we have $v = r\omega = (a/2)\omega$, where a is the length of sides AB and CD. Equation 20.6 therefore reduces to

$$\mathcal{E} = 2B\ell \left(\frac{a}{2}\right) \omega \sin \omega t = B\ell a\omega \sin \omega t$$

If a coil has N turns, the emf is N times as large because each loop has the same emf induced in it. Further, because the area of the loop is $A = \ell a$, the total emf is

$$\mathcal{E} = NBA\omega \sin \omega t \qquad [20.7]$$

This result shows that the emf varies sinusoidally with time, as plotted in Active Figure 20.20b. Note that the maximum emf has the value

$$\mathcal{E}_{max} = NBA\omega \qquad [20.8]$$

which occurs when $\omega t = 90°$ or $270°$. In other words, $\mathcal{E} = \mathcal{E}_{max}$ when the plane of the loop is parallel to the magnetic field. Further, the emf is zero when $\omega t = 0$ or $180°$, which happens whenever the magnetic field is perpendicular to the plane of the loop. In the United States and Canada the frequency of rotation for commercial generators is 60 Hz, whereas in some European countries 50 Hz is used. (Recall that $\omega = 2\pi f$, where f is the frequency in hertz.)

APPLICATION

Direct Current Generators

The **direct current** (DC) **generator** is illustrated in Active Figure 20.22a. The components are essentially the same as those of the AC generator, except that the contacts to the rotating loop are made by a split ring, or commutator. In this design, the output voltage always has the same polarity and the current is a pulsating direct current, as in Active Figure 20.22b. This can be understood by noting

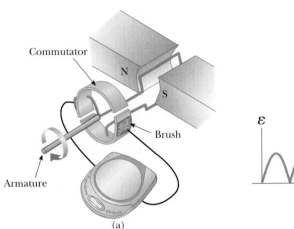

Commutator

N

S

Brush

Armature

(a)

(b)

ACTIVE FIGURE 20.22
(a) A schematic diagram of a DC generator. (b) The emf fluctuates in magnitude, but always has the same polarity.

Physics⊗Now™
Log into PhysicsNow at **www.cp7e.com** and go to Active Figure 20.22, where you can adjust the speed of rotation and the strength of the field, observing the effects on the generated emf.

that the contacts to the split ring reverse their roles every half cycle. At the same time, the polarity of the induced emf reverses. Hence, the polarity of the split ring remains the same.

A pulsating DC current is not suitable for most applications. To produce a steady DC current, commercial DC generators use many loops and commutators distributed around the axis of rotation so that the sinusoidal pulses from the loops overlap in phase. When these pulses are superimposed, the DC output is almost free of fluctuations.

EXAMPLE 20.5   Emf Induced in an AC Generator

Goal   Understand physical aspects of an AC generator.

Problem   An AC generator consists of eight turns of wire, each having area $A = 0.090\ 0\ \text{m}^2$, with a total resistance of $12.0\ \Omega$. The loop rotates in a magnetic field of $0.500\ \text{T}$ at a constant frequency of $60.0\ \text{Hz}$. **(a)** Find the maximum induced emf. **(b)** What is the maximum induced current? **(c)** Determine the induced emf and current as functions of time. **(d)** What maximum torque must be applied to keep the coil turning?

Strategy   From the given frequency, calculate the angular frequency ω and substitute it, together with given quantities, into Equation 20.8. As functions of time, the emf and current have the form $A \sin \omega t$, where A is the maximum emf or current, respectively. For part (d), calculate the magnetic torque on the coil when the current is at a maximum. (See Chapter 19.) The applied torque must do work against this magnetic torque to keep the coil turning.

Solution
(a) Find the maximum induced emf.

First, calculate the angular frequency of the rotational motion:

$$\omega = 2\pi f = 2\pi(60.0\ \text{Hz}) = 377\ \text{rad/s}.$$

Substitute the values for N, A, B, and ω into Equation 20.8, obtaining the maximum induced emf:

$$\mathcal{E}_{\text{max}} = NAB\omega = 8(0.090\ 0\ \text{m}^2)(0.500\ \text{T})(377\ \text{rad/s})$$

$$= \boxed{136\ \text{V}}$$

(b) What is the maximum induced current?

Substitute the maximum induced emf $\mathcal{E}_{\text{max}}$ and the resistance R into Ohm's law to find the maximum induced current:

$$I_{\text{max}} = \frac{\mathcal{E}_{\text{max}}}{R} = \frac{136\ \text{V}}{12.0\ \Omega} = \boxed{11.3\ \text{A}}$$

(c) Determine the induced emf and the current as functions of time.

Substitute $\mathcal{E}_{\text{max}}$ and ω into Equation 20.7 to obtain the variation of $\mathcal{E}$ with time t in seconds:

$$\mathcal{E} = \mathcal{E}_{\text{max}} \sin \omega t = \boxed{(136\ \text{V}) \sin 377t}$$

The time variation of the current looks just like this, except with the maximum current out in front:

$$I = \boxed{(11.3 \text{ A}) \sin 377t}$$

(d) Calculate the maximum applied torque necessary to keep the coil turning.

Write the equation for magnetic torque:

$$\tau = \mu B \sin \theta$$

Calculate the maximum magnetic moment of the coil, μ:

$$\mu = I_{max}AN = (11.3 \text{ A})(0.090 \text{ m}^2)(8) = 8.14 \text{ A} \cdot \text{m}^2$$

Substitute into the magnetic torque equation, with $\theta = 90°$ to find the maximum applied torque:

$$\tau_{max} = (8.14 \text{ A} \cdot \text{m}^2)(0.500 \text{ T}) \sin 90° = \boxed{4.07 \text{ N} \cdot \text{m}}$$

Remarks   The number of loops, N, can't be arbitrary, because there must be a force strong enough to turn the coil.

Exercise 20.5
An AC generator is to have a maximum output of 301 V. Each coil has an area of 0.100 m² and a resistance of 16.0 Ω and rotates in a magnetic field of 0.600 T with a frequency of 40.0 Hz. (a) How many turns of wire should the coil have to produce the desired emf? (b) Find the maximum current induced in the coil. (c) Determine the induced emf as a function of time.

Answers   (a) 20 turns   (b) 18.8 A   (c) $\mathcal{E} = (301 \text{ V}) \sin 251t$

Motors and Back emf

Motors are devices that convert electrical energy to mechanical energy. Essentially, **a motor is a generator run in reverse**: instead of a current being generated by a rotating loop, a current is supplied to the loop by a source of emf, and the magnetic torque on the current-carrying loop causes it to rotate.

A motor can perform useful mechanical work when a shaft connected to its rotating coil is attached to some external device. As the coil in the motor rotates, however, the changing magnetic flux through it induces an emf which acts to reduce the current in the coil. If it *increased* the current, Lenz's law would be violated. The phrase **back emf** is used for an emf that tends to reduce the applied current. The back emf increases in magnitude as the rotational speed of the coil increases. We can picture this state of affairs as the equivalent circuit in Figure 20.23. For illustrative purposes, assume that the external power source supplying current in the coil of the motor has a voltage of 120 V, that the coil has a resistance of 10 Ω, and that the back emf induced in the coil at this instant is 70 V. The voltage available to supply current equals the difference between the applied voltage and the back emf, 50 V in this case. The current is always reduced by the back emf.

When a motor is turned on, there is no back emf initially, and the current is very large because it's limited only by the resistance of the coil. As the coil begins to rotate, the induced back emf opposes the applied voltage and the current in the coil is reduced. If the mechanical load increases, the motor slows down, which decreases the back emf. This reduction in the back emf increases the current in the coil and therefore also increases the power needed from the external voltage source. As a result, the power requirements for starting a motor and for running it under heavy loads are greater than those for running the motor under average loads. If the motor is allowed to run under no mechanical load, the back emf reduces the current to a value just large enough to balance energy losses by heat and friction.

APPLICATION

Motors

120 V external source

Figure 20.23   A motor can be represented as a resistance plus a back emf.

EXAMPLE 20.6   Induced Current in a Motor

Goal   Apply the concept of a back emf in calculating the induced current in a motor.

Problem   A motor has coils with a resistance of 10.0 Ω and is supplied by a voltage of $\Delta V = 1.20 \times 10^2$ V. When the motor is running at its maximum speed, the back emf is 70.0 V. Find the current in the coils **(a)** when the motor is first turned on and **(b)** when the motor has reached its maximum rotation rate.

Strategy   For each part, find the net voltage, which is the applied voltage minus the induced emf. Divide the net voltage by the resistance to get the current.

Solution
(a) Find the initial current, when the motor is first turned on.

If the coil isn't rotating, the back emf is zero and the current has its maximum value. Calculate the difference between the emf and the initial back emf and divide by the resistance R, obtaining the initial current:

$$I = \frac{\mathcal{E} - \mathcal{E}_{back}}{R} = \frac{1.20 \times 10^2 \, \text{V} - 0}{10.0 \, \Omega} = \boxed{12.0 \, \text{A}}$$

(b) Find the current when the motor is rotating at its maximum rate.

Repeat the calculation, using the maximum value of the back emf:

$$I = \frac{\mathcal{E} - \mathcal{E}_{back}}{R} = \frac{1.20 \times 10^2 \, \text{V} - 70.0 \, \text{V}}{10.0 \, \Omega} = \frac{50.0 \, V}{10.0 \, \Omega}$$

$$= \boxed{5.00 \, \text{A}}$$

Remark   The phenomenon of back emf is one way in which the rotation rate of electric motors is limited.

Exercise 20.6
If the current in the motor is 8.00 A at some instant, what is the back emf at that time?

Answer   40.0 V

20.6 SELF-INDUCTANCE

Consider a circuit consisting of a switch, a resistor, and a source of emf, as in Figure 20.24. When the switch is closed, the current doesn't immediately change from zero to its maximum value, $\mathcal{E}/R$. The law of electromagnetic induction— Faraday's law—prevents this. What happens instead is the following: as the current increases with time, the magnetic flux through the loop due to this current also increases. The increasing flux induces an emf in the circuit that opposes the change in magnetic flux. By Lenz's law, the induced emf is in the direction indicated by the dashed battery in the figure. The net potential difference across the resistor is the emf of the battery minus the opposing induced emf. As the magnitude of the current increases, the *rate* of increase lessens and hence the induced emf decreases. This opposing emf results in a gradual increase in the current. For the same reason, when the switch is opened, the current doesn't immediately fall to zero. This effect is called **self-induction** because the changing flux through the circuit arises from the circuit itself. The emf that is set up in the circuit is called a **self-induced emf**.

As a second example of self-inductance, consider Figure 20.25 (page 678), which shows a coil wound on a cylindrical iron core. (A practical device would have several hundred turns.) Assume that the current changes with time. When the current is in the direction shown, a magnetic field is set up inside the coil, directed from right to left. As a result, some lines of magnetic flux pass through

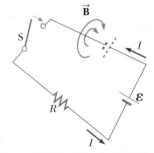

Figure 20.24   After the switch in the circuit is closed, the current produces its own magnetic flux through the loop. As the current increases towards its equilibrium value, the flux changes in time and induces an emf in the loop. The battery drawn with dashed lines is a symbol for the self-induced emf.

Figure 20.25   (a) A current in the coil produces a magnetic field directed to the left. (b) If the current increases, the coil acts as a source of emf directed as shown by the dashed battery. (c) The induced emf in the coil changes its polarity if the current decreases.

JOSEPH HENRY, American physicist (1797–1878)

Henry became the first director of the Smithsonian Institution and first president of the Academy of Natural Science. He was the first to produce an electric current with a magnetic field, but he failed to publish his results as early as Faraday because of his heavy teaching duties at the Albany Academy in New York State. He improved the design of the electromagnet and constructed one of the first motors. He also discovered the phenomenon of self-induction. The unit of inductance, the henry, is named in his honor.

the cross-sectional area of the coil. As the current changes with time, the flux through the coil changes and induces an emf in the coil. Lenz's law shows that this induced emf has a direction so as to oppose the change in the current. If the current is increasing, the induced emf is as pictured in Figure 20.25b, and if the current is decreasing, the induced emf is as shown in Figure 20.25c.

To evaluate self-inductance quantitatively, first note that, according to Faraday's law, the induced emf is given by Equation 20.2:

$$\mathcal{E} = -N\frac{\Delta\Phi_B}{\Delta t}$$

The magnetic flux is proportional to the magnetic field, which is proportional to the current in the coil. Therefore, **the self-induced emf must be proportional to the rate of change of the current with time**, or

$$\mathcal{E} \equiv -L\frac{\Delta I}{\Delta t}$$ [20.9]

where L is a proportionality constant called the **inductance** of the device. The negative sign indicates that a changing current induces an emf in opposition to the change. This means that if the current is increasing (ΔI positive), the induced emf is negative, indicating opposition to the increase in current. Likewise, if the current is decreasing (ΔI negative), the sign of the induced emf is positive, indicating that the emf is acting to oppose the decrease.

The inductance of a coil depends on the cross-sectional area of the coil and other quantities, all of which can be grouped under the general heading of geometric factors. The SI unit of inductance is the **henry** (H), which, from Equation 20.9, is equal to 1 volt-second per ampere:

$$1\,\text{H} = 1\,\text{V}\cdot\text{s/A}$$

In the process of calculating self-inductance, it is often convenient to equate Equations 20.2 and 20.9 to find an expression for L:

$$N\frac{\Delta\Phi_B}{\Delta t} = L\frac{\Delta I}{\Delta t}$$

Inductance ▶

$$L = N\frac{\Delta\Phi_B}{\Delta I} = \frac{N\Phi_B}{I}$$ [20.10]

Applying Physics 20.3   Making Sparks Fly

In some circuits, a spark occurs between the poles of a switch when the switch is opened. Why isn't there a spark when the switch for this circuit is closed?

Explanation According to Lenz's law, the direction of induced emfs is such that the induced magnetic field opposes change in the original magnetic flux. When

the switch is opened, the sudden drop in the magnetic field in the circuit induces an emf in a direction that opposes change in the original current. This induced emf can cause a spark as the current bridges the air gap between the poles of the switch. The spark doesn't occur when the switch is closed, because the original current is zero and the induced emf opposes any change in that current.

In general, determining the inductance of a given current element can be challenging. Finding an expression for the inductance of a common solenoid, however, is straightforward. Let the solenoid have N turns and length ℓ. Assume that ℓ is large compared with the radius and that the core of the solenoid is air. We take the interior magnetic field to be uniform and given by Equation 19.16,

$$B = \mu_0 n I = \mu_0 \frac{N}{\ell} I$$

where $n = N/\ell$ is the number of turns per unit length. The magnetic flux through each turn is therefore

$$\Phi_B = BA = \mu_0 \frac{N}{\ell} AI$$

where A is the cross-sectional area of the solenoid. From this expression and Equation 20.10, we find that

$$L = \frac{N\Phi_B}{I} = \frac{\mu_0 N^2 A}{\ell} \qquad \text{[20.11a]}$$

This shows that L depends on the geometric factors ℓ and A and on μ_0 and is proportional to the square of the number of turns. Because $N = n\ell$, we can also express the result in the form

$$L = \mu_0 \frac{(n\ell)^2}{\ell} A = \mu_0 n^2 A\ell = \mu_0 n^2 V \qquad \text{[20.11b]}$$

where $V = A\ell$ is the volume of the solenoid.

EXAMPLE 20.7   Inductance, Self-Induced emf, and Solenoids

Goal   Calculate the inductance and self-induced emf of a solenoid.

Problem   (a) Calculate the inductance of a solenoid containing 300 turns if the length of the solenoid is 25.0 cm and its cross-sectional area is 4.00×10^{-4} m^2. (b) Calculate the self-induced emf in the solenoid described in (a) if the current in the solenoid decreases at the rate of 50.0 A/s.

Strategy   Substituting given quantities into Equation 20.11a gives the inductance L. For part (b), substitute the result of part (a) and $\Delta I/\Delta t = -50.0$ A/s into Equation 20.9 to get the self-induced emf.

Solution
(a) Calculate the inductance of the solenoid.

Substitute the number N of turns, the area A, and the length ℓ into Equation 20.11a to find the inductance:

$$L = \frac{\mu_0 N^2 A}{\ell}$$
$$= (4\pi \times 10^{-7}\,\text{T·m/A}) \frac{(300)^2(4.00 \times 10^{-4}\,\text{m}^2)}{25.0 \times 10^{-2}\,\text{m}}$$
$$= 1.81 \times 10^{-4}\,\text{T·m}^2/\text{A} = \boxed{0.181\,\text{mH}}$$

(b) Calculate the self-induced emf in the solenoid.

Substitute L and $\Delta I/\Delta t = -50.0$ A/s into Equation 20.9, finding the self-induced emf:

$$\mathcal{E} = -L\frac{\Delta I}{\Delta t} = -(1.81 \times 10^{-4}\,\text{H})(-50.0\,\text{A/s})$$
$$= \boxed{9.05\,\text{mV}}$$

Remark   Notice that $\Delta I/\Delta t$ is negative because the current is decreasing with time.

Exercise 20.7

A solenoid is to have an inductance of 0.285 mH, a cross-sectional area of $6.00 \times 10^{-4}\ \text{m}^2$, and a length of 36.0 cm. (a) How many turns per unit length should it have? (b) If the self-induced emf is -12.5 mV at a given time, at what rate is the current changing at that instant?

Answers   (a) 1 025 turns/m   (b) 43.9 A/s

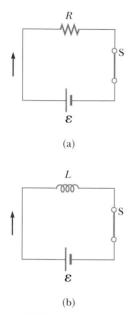

(a)

(b)

Figure 20.26 A comparison of the effect of a resistor with that of an inductor in a simple circuit.

ACTIVE FIGURE 20.27
A series *RL* circuit. As the current increases towards its maximum value, the inductor produces an emf that opposes the increasing current.

Physics⊗Now™

Log into PhysicsNow at **www.cp7e.com** and go to Active Figure 20.27, where you can adjust the values of *R* and *L* and observe the effect on current. A graphical display as in Active Figure 20.28 is available.

20.7   *RL* CIRCUITS

A circuit element that has a large inductance, such as a closely wrapped coil of many turns, is called an **inductor**. The circuit symbol for an inductor is ⎯⎯⎓⎯⎯ . We will always assume that the self-inductance of the remainder of the circuit is negligible compared with that of the inductor in the circuit.

To gain some insight into the effect of an inductor in a circuit, consider the two circuits in Figure 20.26. Figure 20.26a shows a resistor connected to the terminals of a battery. For this circuit, Kirchhoff's loop rule is $\boldsymbol{\mathcal{E}} - IR = 0$. The voltage drop across the resistor is

$$\Delta V_R = -IR \qquad\qquad [20.12]$$

In this case, **we interpret resistance as a measure of opposition to the current**. Now consider the circuit in Figure 20.26b, consisting of an inductor connected to the terminals of a battery. At the instant the switch in this circuit is closed, because $IR = 0$, the emf of the battery equals the back emf generated in the coil. Hence, we have

$$\mathcal{E}_L = -L\frac{\Delta I}{\Delta t} \qquad\qquad [20.13]$$

From this expression, **we can interpret *L* as a measure of opposition to the rate of change of current**.

Active Figure 20.27 shows a circuit consisting of a resistor, an inductor, and a battery. Suppose the switch is closed at $t = 0$. The current begins to increase, but the inductor produces an emf that opposes the increasing current. As a result, the current can't change from zero to its maximum value of $\boldsymbol{\mathcal{E}}/R$ instantaneously. Equation 20.13 shows that the induced emf is a maximum when the current is changing most rapidly, which occurs when the switch is first closed. As the current approaches its steady-state value, the back emf of the coil falls off because the current is changing more slowly. Finally, when the current reaches its steady-state value, the rate of change is zero and the back emf is also zero. Active Figure 20.28 plots current in the circuit as a function of time.[1] This plot is similar to that of the charge on a capacitor as a function of time, discussed in Chapter 18. In that case, we found it convenient to introduce a quantity called the *time constant of the circuit*, which told us something about the time required for the capacitor to approach its steady-state charge. In the same way, time constants are defined for circuits containing resistors and inductors. The **time constant τ for an *RL* circuit** is the time required for the current in the circuit to reach 63.2% of its final value $\boldsymbol{\mathcal{E}}/R$; the time constant of an *RL* circuit is given by

Time constant for an *RL* circuit ▶

$$\tau = \frac{L}{R} \qquad\qquad [20.14]$$

---

[1]The equation for the current in the circuit as a function of time may be obtained from calculus and is

$$I = \frac{\boldsymbol{\mathcal{E}}}{R}\,(1 - e^{-Rt/L})$$

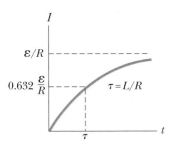

ACTIVE FIGURE 20.28
A plot of current versus time for the *RL* circuit shown in Figure 20.27. The switch is closed at $t = 0$, and the current increases towards its maximum value $\mathcal{E}/R$. The time constant τ is the time it takes the current to reach 63.2% of its maximum value.

Physics⊗Now™

Log into PhysicsNow at **www.cp7e.com** and go to Active Figure 20.27, where you can observe this graph develop after the switch in Active Figure 20.27 is closed.

Quick Quiz 20.5

The switch in the circuit shown in Figure 20.29 is closed and the lightbulb glows steadily. The inductor is a simple air-core solenoid. An iron rod is inserted into the interior of the solenoid, increasing the magnitude of the magnetic field in the solenoid. As the rod is inserted, the brightness of the lightbulb (a) increases, (b) decreases, or (c) remains the same.

Figure 20.29   (Quick Quiz 20.5)

EXAMPLE 20.8   An *RL* Circuit

Goal   Calculate a time constant and relate it to current in an *RL* circuit.

Problem   A 12.6-V battery is in a circuit with a 30.0-mH inductor and a 0.150-Ω resistor, as in Active Figure 20.27. The switch is closed at $t = 0$. **(a)** Find the time constant of the circuit. **(b)** Find the current after one time constant has elapsed. **(c)** Find the voltage drops across the resistance when $t = 0$ and $t =$ one time constant. **(d)** What's the rate of change of the current after one time constant?

Solution
(a) What's the time constant of the circuit?

Substitute the inductance L and resistance R into Equation 20.14, finding the time constant:

$$\tau = \frac{L}{R} = \frac{30.0 \times 10^{-3}\,\text{H}}{0.150\,\Omega} = \boxed{0.200\ \text{s}}$$

(b) Find the current after one time constant has elapsed.

First, use Ohm's law to compute the final value of the current after many time constants have elapsed:

$$I_{max} = \frac{\mathcal{E}}{R} = \frac{12.6\,\text{V}}{0.150\,\Omega} = 84.0\ \text{A}$$

After one time constant, the current rises to 63.2% of its final value:

$$I_{1\tau} = (0.632)I_{max} = (0.632)(84.0\ \text{A}) = \boxed{53.1\ \text{A}}$$

(c) Find the voltage drops across the resistance when $t = 0$ and $t =$ one time constant.

Initially, the current in the circuit is zero, so, from Ohm's law, the voltage across the resistor is zero:

$$\Delta V_R = IR$$
$$\Delta V_R\,(t = 0\ \text{s}) = (0\ \text{A})(0.150\ \Omega) = \boxed{0}$$

Next, using Ohm's law, find the magnitude of the voltage drop across the resistor after one time constant:

$$\Delta V_R\,(t = 0.200\ \text{s}) = (53.1\ \text{A})(0.150\ \Omega) = \boxed{7.97\ \text{V}}$$

(d) What's the rate of change of the current after one time constant?

Using Kirchhoff's voltage rule, calculate the voltage drop across the inductor at that time:

$$\mathcal{E} + \Delta V_R + \Delta V_L = 0$$

Solve for ΔV_L:

$$\Delta V_L = -\mathcal{E} - \Delta V_R = -12.6\text{ V} - (-7.97\text{ V}) = -4.6\text{ V}$$

Now solve Equation 20.13 for $\Delta I/\Delta t$ and substitute:

$$\Delta V_L = -L\frac{\Delta I}{\Delta t}$$

$$\frac{\Delta I}{\Delta t} = -\frac{\Delta V_L}{L} = -\frac{-4.6\text{ V}}{30.0 \times 10^{-3}\text{ H}} = \boxed{150\text{ A/s}}$$

Remarks   The values used in this problem were taken from actual components salvaged from the starter system of a car. Because the current in such an RL circuit is initially zero, inductors are sometimes referred to as "chokes," since they temporarily choke off the current. In solving part (d), we traversed the circuit in the direction of positive current, so the voltage difference across the battery was positive and the differences across the resistor and inductor were negative.

Exercise 20.8
A 12.6-V battery is in series with a resistance of 0.350 Ω and an inductor. (a) After a long time, what is the current in the circuit? (b) What is the current after one time constant? (c) What's the voltage drop across the inductor at this time? (d) Find the inductance if the time constant is 0.130 s.

Answers   (a) 36.0 A   (b) 22.8 A   (c) 4.62 V   (d) 4.55×10^{-2} H

20.8   ENERGY STORED IN A MAGNETIC FIELD

The emf induced by an inductor prevents a battery from establishing an instantaneous current in a circuit. The battery has to do work to produce a current. We can think of this needed work as energy stored in the inductor in its magnetic field. In a manner similar to that used in Section 16.9 to find the energy stored in a capacitor, we find that the energy stored by an inductor is

Energy stored in an inductor ▶

$$PE_L = \tfrac{1}{2}LI^2 \tag{20.15}$$

Note that the result is similar in form to the expression for the energy stored in a charged capacitor (Equation 16.18):

Energy stored in a capacitor ▶

$$PE_C = \tfrac{1}{2}C(\Delta V)^2$$

EXAMPLE 20.9   Magnetic Energy

Goal   Relate the storage of magnetic energy to currents in an RL circuit.

Problem   A 12.0-V battery is connected in series to a 25.0-Ω resistor and a 5.00-H inductor. (a) Find the maximum current in the circuit. (b) Find the energy stored in the inductor at this time. (c) How much energy is stored in the inductor when the current is changing at a rate of 1.50 A/s?

Strategy   In part (a), Ohm's law and Kirchhoff's voltage rule yield the maximum current, because the voltage across the inductor is zero when the current is maximal. Substituting the current into Equation 20.15 gives the energy stored in the inductor. In part (c), the given rate of change of the current can be used to calculate the voltage drop across the inductor at the specified time. Kirchhoff's voltage rule and Ohm's law then give the current I at that time, which can be used to find the energy stored in the inductor.

Solution
(a) Find the maximum current in the circuit.

Apply Kirchhoff's voltage rule to the circuit:

$$\Delta V_{\text{batt}} + \Delta V_R + \Delta V_L = 0$$

$$\mathcal{E} - IR - L\frac{\Delta I}{\Delta t} = 0$$

When the maximum current is reached, $\Delta I/\Delta t$ is zero, so the voltage drop across the inductor is zero. Solve for the maximum current I_{max}:

$$I_{max} = \frac{\mathcal{E}}{R} = \frac{12.0\,\text{V}}{25.0\,\Omega} = \boxed{0.480\,\text{A}}$$

(b) Find the energy stored in the inductor at this time.

Substitute known values into Equation 20.15:

$$PE_L = \tfrac{1}{2}LI_{max}^2 = \tfrac{1}{2}(5.00\,\text{H})(0.480\,\text{A})^2 = \boxed{0.576\,\text{J}}$$

(c) Find the energy in the inductor when the current changes at a rate of 1.50 A/s.

Apply Kirchhoff's voltage rule to the circuit, once again:

$$\mathcal{E} - IR - L\frac{\Delta I}{\Delta t} = 0$$

Solve this equation for the current I and substitute:

$$I = \frac{1}{R}\left(\mathcal{E} - L\frac{\Delta I}{\Delta t}\right)$$

$$= \frac{1}{25.0\,\Omega}[12.0\,\text{V} - (5.00\,\text{H})(1.50\,\text{A/s})] = 0.180\,\text{A}$$

Finally, substitute the value for the current into Equation 20.15, finding the energy stored in the inductor:

$$PE_L = \tfrac{1}{2}LI^2 = \tfrac{1}{2}(5.00\,\text{H})(0.180\,\text{A})^2 = \boxed{0.081\,0\,\text{J}}$$

Remark   Notice how important it is to combine concepts from previous chapters. Here, Ohm's law and Kirchhoff's loop rule were essential to the solution of the problem.

Exercise 20.9
For the same circuit, find the energy stored in the inductor when the rate of change of the current is 1.00 A/s.

Answer   0.196 J

SUMMARY

Physics⊗Now™ Take a practice test by logging into PhysicsNow at **www.cp7e.com** and clicking on the Pre-Test link for this chapter.

20.1   Induced emf and Magnetic Flux

The magnetic flux Φ_B through a closed loop is defined as

$$\Phi_B \equiv BA\cos\theta \qquad \text{[20.1]}$$

where B is the strength of the uniform magnetic field, A is the cross-sectional area of the loop, and θ is the angle between $\vec{B}$ and a direction perpendicular to the plane of the loop.

20.2   Faraday's Law of Induction

Faraday's law of induction states that the instantaneous emf induced in a circuit equals the negative of the rate of change of magnetic flux through the circuit,

$$\mathcal{E} = -N\frac{\Delta\Phi_B}{\Delta t} \qquad \text{[20.2]}$$

where N is the number of loops in the circuit. The magnetic flux Φ_B can change with time whenever the magnetic field $\vec{B}$, the area A, or the angle θ changes with time.

Lenz's law states that the current from the induced emf creates a magnetic field with flux opposing the *change* in magnetic flux through a circuit.

20.3   Motional emf

If a conducting bar of length $\vec{\ell}$ moves through a magnetic field with a speed v so that $\vec{B}$ is perpendicular to the bar, then the emf induced in the bar, often called a **motional emf**, is

$$|\mathcal{E}| = B\ell v \qquad \text{[20.4]}$$

20.5   Generators

When a coil of wire with N turns, each of area A, rotates with constant angular speed ω in a uniform magnetic field $\vec{B}$, the emf induced in the coil is

$$\mathcal{E} = NAB\omega\sin\omega t \qquad \text{[20.7]}$$

Such generators naturally produce alternating current (AC), which changes direction with frequency $\omega/2\pi$. The AC current can be transformed to direct current.

20.7   *RL* Circuits

When the current in a coil changes with time, an emf is induced in the coil according to Faraday's law. This

self-induced emf is defined by the expression

$$\mathcal{E} \equiv -L \frac{\Delta I}{\Delta t} \qquad \text{[20.9]}$$

where L is the inductance of the coil. The SI unit for inductance is the henry (H); 1 H = 1 V·s/A.

The **inductance** of a coil can be found from the expression

$$L = \frac{N\Phi_B}{I} \qquad \text{[20.10]}$$

where N is the number of turns on the coil, I is the current in the coil, and Φ_B is the magnetic flux through the coil produced by that current. For a solenoid, the inductance is given by

$$L = \frac{\mu_0 N^2 A}{\ell} \qquad \text{[20.11]}$$

If a resistor and inductor are connected in series to a battery and a switch is closed at $t = 0$, the current in the circuit doesn't rise instantly to its maximum value. After one **time constant** $\tau = L/R$, the current in the circuit is 63.2% of its final value $\mathcal{E}/R$. As the current approaches its final, maximum value, the voltage drop across the inductor approaches zero.

20.8 Energy Stored in a Magnetic Field

The **energy stored** in the magnetic field of an inductor carrying current I is

$$PE_L = \tfrac{1}{2}LI^2 \qquad \text{[20.15]}$$

As the current in an RL circuit approaches its maximum value, the stored energy also approaches a maximum value.

CONCEPTUAL QUESTIONS

1. A circular loop is located in a uniform and constant magnetic field. Describe how an emf can be induced in the loop in this situation.

2. Does dropping a magnet down a copper tube produce a current in the tube? Explain.

3. A spacecraft orbiting the Earth has a coil of wire in it. An astronaut measures a small current in the coil, although there is no battery connected to it and there are no magnets in the spacecraft. What is causing the current?

4. A loop of wire is placed in a uniform magnetic field. For what orientation of the loop is the magnetic flux a maximum? For what orientation is the flux zero?

5. As the conducting bar in Figure Q20.5 moves to the right, an electric field directed downward is set up. If the bar were moving to the left, explain why the electric field would be upward.

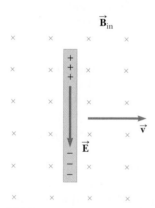

Figure Q20.5   (Conceptual Questions 5 and 6)

6. As the bar in Figure Q20.5 moves perpendicular to the field, is an external force required to keep it moving with constant speed?

7. Wearing a metal bracelet in a region of strong magnetic field could be hazardous. Discuss this statement.

8. How is electrical energy produced in dams? (That is, how is the energy of motion of the water converted to AC electricity?)

9. Eddy currents are induced currents set up in a piece of metal when it moves through a nonuniform magnetic field. For example, consider the flat metal plate swinging at the end of a bar as a pendulum, as shown in Figure Q20.9. At position 1, the pendulum is moving from a region where there is no magnetic field into a region where the field $\vec{\mathbf{B}}_{in}$ is directed into the paper. Show that at position 1 the direction of the eddy current is counterclockwise. Also, at position 2 the pendulum is moving out of the field into a region of zero field. Show that the direction of the eddy current is clockwise in this case. Use right-hand rule number 2 to show that these eddy currents lead to a magnetic force on the plate directed as shown in the figure. Because the induced eddy current always produces a retarding force when the plate enters or leaves the field, the swinging plate quickly comes to rest.

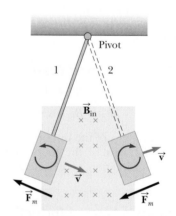

Figure Q20.9

10. Suppose you would like to steal power for your home from the electric company by placing a loop of wire near a transmission cable in order to induce an emf in the loop (Don't do this; it's illegal.) Should you locate the loop so that the transmission cable passes through your loop or simply place your loop near the transmission cable? Does the orientation of the loop matter?

11. A piece of aluminum is dropped vertically downward between the poles of an electromagnet. Does the magnetic

field affect the velocity of the aluminum? [*Hint:* See Conceptual Question 9.]

12. A bar magnet is dropped toward a conducting ring lying on the floor. As the magnet falls toward the ring, does it move as a freely falling object?

13. If the current in an inductor is doubled, by what factor does the stored energy change?

14. Is it possible to induce a constant emf for an infinite amount of time?

15. Why is the induced emf that appears in an inductor called a back (counter) emf?

16. A magneto is used to cause the spark in a spark plug in many lawn mowers today. A magneto consists of a permanent magnet mounted on a flywheel so that it spins past a fixed coil. Explain how this arrangement generates a large enough potential difference to cause the spark.

17. A ramp runs from the bed of a truck down to the level ground. The ramp holds two parallel conducting rails connected at its base. A metal bar slides on the rails without friction. A magnet supplies an external magnetic field directed toward the ground. It is found that the bar slides down the ramp at a constant speed. (a) What is the direction of the induced current in the bar as viewed from above? (b) What can you conclude about the forces exerted on the bar?

18. A bar magnet is held above the center of a wire loop in a horizontal plane, as shown in Figure Q20.18. The south end of the magnet is toward the loop. The magnet is dropped. Find the direction of the current in the resistor

as viewed from above (a) while the magnet is falling toward the loop and (b) after the magnet has passed through the loop and moved away from it.

Figure Q20.18

19. What is the direction of the current induced in the resistor when the current in the long, straight wire in Figure Q20.19 decreases rapidly to zero?

Figure Q20.19

PROBLEMS

1, 2, 3 = straightforward, intermediate, challenging   □ = full solution available in *Student Solutions Manual/Study Guide*
Physics⊗Now™ = coached problem with hints available at **www.cp7e.com**   🐚 = biomedical application

Section 20.1 Induced emf and Magnetic Flux

1. A magnetic field of strength 0.30 T is directed perpendicular to a plane circular loop of wire of radius 25 cm. Find the magnetic flux through the area enclosed by this loop.

2. Find the flux of the Earth's magnetic field of magnitude 5.00×10^{-5} T through a square loop of area 20.0 cm² (a) when the field is perpendicular to the plane of the loop, (b) when the field makes a 30.0° angle with the normal to the plane of the loop, and (c) when the field makes a 90.0° angle with the normal to the plane.

3. A square loop 2.00 m on a side is placed in a magnetic field of magnitude 0.300 T. If the field makes an angle of 50.0° with the normal to the plane of the loop, find the magnetic flux through the loop.

4. A long, straight wire carrying a current of 2.00 A is placed along the axis of a cylinder of radius 0.500 m and a length of 3.00 m. Determine the total magnetic flux through the cylinder.

5. A long, straight wire lies in the plane of a circular coil with a radius of 0.010 m. The wire carries a current of 2.0 A and is placed along a diameter of the coil. (a) What is the net flux through the coil? (b) If the wire passes through the center of the coil and is perpendicular to the plane of the coil, find the net flux through the coil.

6. A solenoid 4.00 cm in diameter and 20.0 cm long has 250 turns and carries a current of 15.0 A. Calculate the magnetic flux through the circular cross-sectional area of the solenoid.

7. A cube of edge length $\ell = 2.5$ cm is positioned as shown in Figure P20.7. There is a uniform magnetic field throughout the region with components $B_x = +5.0$ T, $B_y = +4.0$ T, and $B_z = +3.0$ T. (a) Calculate the flux through the shaded face of the cube. (b) What is the total flux

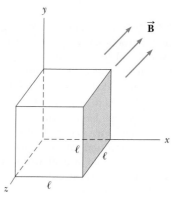

Figure P20.7

emerging from the volume enclosed by the cube (i.e., the total flux through all six faces)?

Section 20.2 Faraday's Law of Induction

8. Transcranial magnetic stimulation (TMS) is a noninvasive technique used to stimulate regions of the human brain. A small coil is placed on the scalp, and a brief burst of current in the coil produces a rapidly changing magnetic field inside the brain. The induced emf can be sufficient to stimulate neuronal activity. One such device generates a magnetic field within the brain that rises from zero to 1.5 T in 120 ms. Determine the induced emf within a circle of tissue of radius 1.6 mm and that is perpendicular to the direction of the field.

9. A square, single-turn coil 0.20 m on a side is placed with its plane perpendicular to a constant magnetic field. An emf of 18 mV is induced in the coil winding when the area of the coil decreases at the rate of $0.10 \text{ m}^2/\text{s}$. What is the magnitude of the magnetic field?

10. The flexible loop in Figure P20.10 has a radius of 12 cm and is in a magnetic field of strength 0.15 T. The loop is grasped at points A and B and stretched until its area is nearly zero. If it takes 0.20 s to close the loop, find the magnitude of the average induced emf in it during this time.

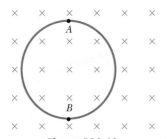

Figure P20.10

11. A wire loop of radius 0.30 m lies so that an external magnetic field of magnitude 0.30 T is perpendicular to the loop. The field reverses its direction, and its magnitude changes to 0.20 T in 1.5 s. Find the magnitude of the average induced emf in the loop during this time.

12. A 500-turn circular-loop coil 15.0 cm in diameter is initially aligned so that its axis is parallel to the Earth's magnetic field. In 2.77 ms, the coil is flipped so that its axis is perpendicular to the Earth's magnetic field. If an average voltage of 0.166 V is thereby induced in the coil, what is the value of the Earth's magnetic field at that location?

13. **Physics☒Now™** The plane of a rectangular coil, 5.0 cm by 8.0 cm, is perpendicular to the direction of a magnetic field $\vec{\textbf{B}}$. If the coil has 75 turns and a total resistance of 8.0 Ω, at what rate must the magnitude of $\vec{\textbf{B}}$ change to induce a current of 0.10 A in the windings of the coil?

14. A square, single-turn wire loop 1.00 cm on a side is placed inside a solenoid that has a circular cross section of radius 3.00 cm, as shown in Figure P20.14. The solenoid is 20.0 cm long and wound with 100 turns of wire. (a) If the current in the solenoid is 3.00 A, find the flux through the loop. (b) If the current in the solenoid is reduced to zero in 3.00 s, find the magnitude of the average induced emf in the loop.

Figure P20.14

15. A 300-turn solenoid with a length of 20 cm and a radius of 1.5 cm carries a current of 2.0 A. A second coil of four turns is wrapped tightly about this solenoid so that it can be considered to have the same radius as the solenoid. Find (a) the change in the magnetic flux through the coil and (b) the magnitude of the average induced emf in the coil when the current in the solenoid increases to 5.0 A in a period of 0.90 s.

16. A circular coil enclosing an area of 100 cm^2 is made of 200 turns of copper wire. The wire making up the coil has resistance of 5.0 Ω, and the ends of the wire are connected to form a closed circuit. Initially, a 1.1-T uniform magnetic field points perpendicularly upward through the plane of the coil. The direction of the field then reverses so that the final magnetic field has a magnitude of 1.1 T and points downward through the coil. If the time required for the field to reverse directions is 0.10 s, what average current flows through the coil during that time?

17. To monitor the breathing of a hospital patient, a thin belt is girded around the patient's chest as in Figure P20.17. The belt is a 200-turn coil. When the patient inhales, the area encircled by the coil increases by 39.0 cm^2. The magnitude of the Earth's magnetic field is 50.0 μT and makes an angle of 28.0° with the plane of the coil. Assuming a patient takes 1.80 s to inhale, find the magnitude of the average induced emf in the coil during that time.

Figure P20.17

Section 20.3 Motional emf

18. Consider the arrangement shown in Figure P20.18. Assume that $R = 6.00$ Ω and $\ell = 1.20$ m, and that a uniform 2.50-T magnetic field is directed *into* the page. At what speed should the bar be moved to produce a current of 0.500 A in the resistor?

19. A Boeing 747 jet with a wingspan of 60.0 m is flying horizontally at a speed of 300 m/s over Phoenix, Arizona, at a location where the Earth's magnetic field is 50.0 μT at

58.0° below the horizontal. What voltage is generated between the wingtips?

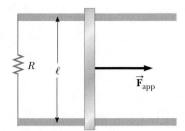

Figure P20.18   (Problems 18 and 57)

20. A 12.0-m-long steel beam is accidentally dropped by a construction crane from a height of 9.00 m. The horizontal component of the Earth's magnetic field over the region is 18.0 μT. What is the induced emf in the beam just before impact with the Earth? Assume the long dimension of the beam remains in a horizontal plane, oriented perpendicular to the horizontal component of the Earth's magnetic field.

21. An automobile has a vertical radio antenna 1.20 m long. The automobile travels at 65.0 km/h on a horizontal road where the Earth's magnetic field is 50.0 μT, directed toward the north and downwards at an angle of 65.0° below the horizontal. (a) Specify the direction the automobile should move in order to generate the maximum motional emf in the antenna, with the top of the antenna positive relative to the bottom. (b) Calculate the magnitude of this induced emf.

22. A helicopter has blades of length 3.0 m, rotating at 2.0 rev/s about a central hub. If the vertical component of Earth's magnetic field is 5.0×10^{-5} T, what is the emf induced between the blade tip and the central hub?

Section 20.4 Lenz's Law Revisited (the Minus Signin Faraday's Law)

23. A bar magnet is positioned near a coil of wire as shown in Figure P20.23. What is the direction of the current in the resistor when the magnet is moved (a) to the left? (b) to the right?

Figure P20.23

24. A conducting rectangular loop of mass M, resistance R, and dimensions w by ℓ falls from rest into a magnetic field $\vec{B}$ as shown in Figure P20.24. During the time interval before the top edge of the loop reaches the field, the loop approaches a terminal speed v_T. (a) Show that

$$v_T = \frac{MgR}{B^2w^2}$$

(b) Why is v_T proportional to R? (c) Why is it inversely proportional to B^2?

Figure P20.24

25. A rectangular coil with resistance R has N turns, each of length ℓ and width w as shown in Figure P20.25. The coil moves into a uniform magnetic field $\vec{B}$ with constant velocity $\vec{v}$. What are the magnitude and direction of the total magnetic force on the coil (a) as it enters the magnetic field, (b) as it moves within the field, and (c) as it leaves the field?

Figure P20.25

26. In Figure P20.26, what is the direction of the current induced in the resistor at the instant the switch is closed?

Figure P20.26

27. A copper bar is moved to the right while its axis is maintained in a direction perpendicular to a magnetic field, as shown in Figure P20.27. If the top of the bar becomes

Figure P20.27

positive relative to the bottom, what is the direction of the magnetic field?

28. Find the direction of the current in the resistor shown in Figure P20.28 (a) at the instant the switch is closed, (b) after the switch has been closed for several minutes, and (c) at the instant the switch is opened.

Figure P20.28

29. Find the direction of the current in the resistor R shown in Figure P20.29 after each of the following steps (taken in the order given): (a) The switch is closed. (b) The variable resistance in series with the battery is decreased. (c) The circuit containing resistor R is moved to the left. (d) The switch is opened.

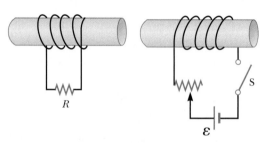

Figure P20.29

Section 20.5 Generators

30. A 100-turn square wire coil of area 0.040 m^2 rotates about a vertical axis at $1\,500$ rev/min, as indicated in Figure P20.30. The horizontal component of the Earth's magnetic field at the location of the loop is 2.0×10^{-5} T. Calculate the maximum emf induced in the coil by the Earth's field.

20 cm

20 cm

Figure P20.30

31. Considerable scientific work is currently underway to determine whether weak oscillating magnetic fields such as those found near outdoor electric power lines can effect human health. One study indicated that a magnetic field of magnitude 1.0×10^{-3} T, oscillating at 60 Hz, might stimulate red blood cells to become cancerous. If the diameter of a red blood cell is 8.0 μm, determine the maximum emf that can be generated around the perimeter of the cell.

32. A motor has coils with a resistance of 30 Ω and operates from a voltage of 240 V. When the motor is operating at its maximum speed, the back emf is 145 V. Find the current in the coils (a) when the motor is first turned on and (b) when the motor has reached maximum speed. (c) If the current in the motor is 6.0 A at some instant, what is the back emf at that time?

33. A coil of 10.0 turns is in the shape of an ellipse having a major axis of 10.0 cm and a minor axis of 4.00 cm. The coil rotates at 100 rpm in a region in which the magnitude of the Earth's magnetic field is 55 μT. What is the maximum voltage induced in the coil if the axis of rotation of the coil is along its major axis and is aligned (a) perpendicular to the Earth's magnetic field and (b) parallel to the Earth's magnetic field? (Note that the area of an ellipse is given by $A = \pi ab$, where a is the length of the *semi*major axis and b is the length of the *semi*minor axis.)

34. A flat coil enclosing an area of 0.10 m^2 is rotating at 60 rev/s, with its axis of rotation perpendicular to a 0.20-T magnetic field. (a) If there are 1 000 turns on the coil, what is the maximum voltage induced in the coil? (b) When the maximum induced voltage occurs, what is the orientation of the coil with respect to the magnetic field?

35. **Physics⊗Now™** In a model AC generator, a 500-turn rectangular coil 8.0 cm by 20 cm rotates at 120 rev/min in a uniform magnetic field of 0.60 T. (a) What is the maximum emf induced in the coil? (b) What is the instantaneous value of the emf in the coil at $t = (\pi/32)$ s? Assume that the emf is zero at $t = 0$. (c) What is the smallest value of t for which the emf will have its maximum value?

Section 20.6 Self-Inductance

36. A coiled telephone cord forms a spiral with 70.0 turns, a diameter of 1.30 cm, and an unstretched length of 60.0 cm. Determine the self-inductance of one conductor in the unstretched cord.

37. A coil has an inductance of 3.0 mH, and the current in it changes from 0.20 A to 1.5 A in 0.20 s. Find the magnitude of the average induced emf in the coil during this period.

38. Show that the two expressions for inductance given by

$$L = \frac{N\Phi_B}{I} \quad \text{and} \quad L = \frac{-\varepsilon}{\Delta I/\Delta t}$$

have the same units.

39. A solenoid of radius 2.5 cm has 400 turns and a length of 20 cm. Find (a) its inductance and (b) the rate at which current must change through it to produce an emf of 75 mV.

40. An emf of 24.0 mV is induced in a 500-turn coil when the current is changing at a rate of 10.0 A/s. What is the magnetic flux through each turn of the coil at an instant when the current is 4.00 A?

Section 20.7 *RL* Circuits

41. Show that the SI units for the inductive time constant $\tau = L/R$ are seconds.

42. An *RL* circuit with *L* = 3.00 H and an *RC* circuit with *C* = 3.00 μF have the same time constant. If the two circuits have the same resistance *R*, (a) what is the value of *R* and (b) what is this common time constant?

43. A 6.0-V battery is connected in series with a resistor and an inductor. The series circuit has a time constant of 600 μs, and the maximum current is 300 mA. What is the value of the inductance?

44. A 25-mH inductor, an 8.0-Ω resistor, and a 6.0-V battery are connected in series. The switch is closed at *t* = 0. Find the voltage drop across the resistor (a) at *t* = 0 and (b) after one time constant has passed. Also, find the voltage drop across the inductor (c) at *t* = 0 and (d) after one time constant has elapsed.

45. Physics⊗Now™ Calculate the resistance in an *RL* circuit in which *L* = 2.50 H and the current increases to 90.0% of its final value in 3.00 s.

46. Consider the circuit shown in Figure P20.46. Take ε = 6.00 V, *L* = 8.00 mH, and *R* = 4.00 Ω. (a) What is the inductive time constant of the circuit? (b) Calculate the current in the circuit 250 μs after the switch is closed. (c) What is the value of the final steady-state current? (d) How long does it take the current to reach 80.0% of its maximum value?

Figure P20.46

20.8 Energy Stored in a Magnetic Field

47. How much energy is stored in a 70.0-mH inductor at an instant when the current is 2.00 A?

48. A 300-turn solenoid has a radius of 5.00 cm and a length of 20.0 cm. Find (a) the inductance of the solenoid and (b) the energy stored in the solenoid when the current in its windings is 0.500 A.

49. A 24-V battery is connected in series with a resistor and an inductor, with *R* = 8.0 Ω and *L* = 4.0 H, respectively. Find the energy stored in the inductor (a) when the current reaches its maximum value and (b) one time constant after the switch is closed.

ADDITIONAL PROBLEMS

50. What is the time constant for (a) the circuit shown in Figure P20.50a and (b) the circuit shown in Figure P20.50b?

(a)   (b)

Figure P20.50

51. In Figure P20.51, the bar magnet is being moved toward the loop. Is (V_a − V_b) positive, negative, or zero during this motion? Explain.

Figure P20.51

52. Your physics teacher asks you to help her set up a demonstration of Faraday's law for the class. The apparatus consists of a strong permanent magnet that has a field of 0.10 T, a small 10-turn coil of radius 2.0 cm cemented on a wood frame with a handle, some flexible connecting wires, and an ammeter, as in Figure P20.52. The idea is to pull the coil out of the center of the magnetic field as quickly as possible and read the average current registered on the meter. The combined resistance of the coil, leads, and meter is 2.0 Ω, and you must flip the coil out of the field in about 0.20 s. The ammeter you must use has a full-scale sensitivity of 1 000 μA. Will this meter be sensitive enough to show the induced current clearly?

Figure P20.52

53. Physics⊗Now™ An 820-turn wire coil of resistance 24.0 Ω is placed on top of a 12 500-turn, 7.00-cm-long solenoid, as in Figure P20.53 (page 690). Both coil and solenoid have cross-sectional areas of 1.00×10^{-4} m². (a) How long does it take the solenoid current to reach 0.632 times its maximum value? (b) Determine the average back emf caused by the self-inductance of the solenoid during this interval. The magnetic field produced by the solenoid at the location of the coil is one-half as strong as the field at the center of the solenoid. (c) Determine the average rate of change in magnetic flux through each turn of the coil during the stated interval. (d) Find the magnitude of the average induced current in the coil.

Figure P20.53

54. Figure P20.54 is a graph of induced emf versus time for a coil of N turns rotating with angular speed ω in a uniform magnetic field directed perpendicular to the axis of rotation of the coil. Copy this sketch (increasing the scale), and, on the same set of axes, show the graph of emf versus t when (a) the number of turns in the coil is doubled, (b) the angular speed is doubled, and (c) the angular speed is doubled while the number of turns in the coil is halved.

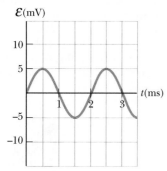

Figure P20.54

55. The plane of a square loop of wire with edge length $a = 0.200$ m is perpendicular to the Earth's magnetic field at a point where $B = 15.0 \ \mu$T, as in Figure P20.55. The total resistance of the loop and the wires connecting it to the ammeter is $0.500 \ \Omega$. If the loop is suddenly collapsed by horizontal forces as shown, what total charge passes through the ammeter?

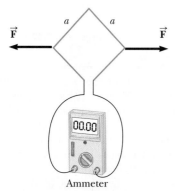

Ammeter

Figure P20.55

56. A novel method of storing electrical energy has been proposed. A huge underground superconducting coil

1.00 km in diameter would be fabricated. It would carry a maximum current of 50.0 kA through each winding of a 150-turn Nb_3Sn solenoid. (a) If the inductance of this huge coil were 50.0 H, what is the total stored energy? (b) What is the compressive force per meter acting between two adjacent windings 0.250 m apart? [*Hint*: Because the radius of the coil is so large, the magnetic field created by one winding and acting on an adjacent turn can be considered to be that of a long, straight wire.]

57. A conducting rod of length ℓ moves on two horizontal frictionless rails, as in Figure P20.18. A constant force of magnitude 1.00 N moves the bar at a uniform speed of 2.00 m/s through a magnetic field $\vec{\mathbf{B}}$ that is directed into the page. (a) What is the current in an 8.00-Ω resistor R? (b) What is the rate of energy dissipation in the resistor? (c) What is the mechanical power delivered by the constant force?

58. The square loop in Figure P20.58 is made of wires with a total series resistance of $10.0 \ \Omega$. It is placed in a uniform 0.100-T magnetic field directed perpendicular into the plane of the paper. The loop, which is hinged at each corner, is pulled as shown until the separation between points A and B is 3.00 m. If this process takes 0.100 s, what is the average current generated in the loop? What is the direction of the current?

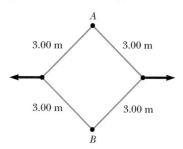

Figure P20.58

59. The bolt of lightning depicted in Figure P20.59 passes 200 m from a 100-turn coil oriented as shown. If the current in the lightning bolt falls from 6.02×10^6 A to zero in 10.5 μs, what is the average voltage induced in the coil? Assume that the distance to the center of the coil determines the average magnetic field at the coil's position. Treat the lightning bolt as a long, vertical wire.

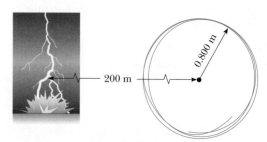

Figure P20.59

60. The wire shown in Figure P20.60 is bent in the shape of a "tent" with $\theta = 60°$ and $L = 1.5$ m, and is placed in a uniform magnetic field of 0.30 T directed perpendicular to the tabletop. The wire is "hinged" at points a and b. If the tent is flattened out on the table in 0.10 s,

what is the average induced emf in the wire during this time?

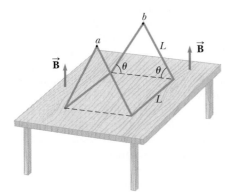

Figure P20.60

61. The magnetic field shown in Figure P20.61 has a uniform magnitude of 25.0 mT directed into the paper. The initial diameter of the kink is 2.00 cm. (a) The wire is quickly pulled taut, and the kink shrinks to a diameter of zero in 50.0 ms. Determine the average voltage induced between endpoints A and B. Include the polarity. (b) Suppose the kink is undisturbed, but the magnetic field increases to 100 mT in 4.00×10^{-3} s. Determine the average voltage across terminals A and B, including polarity, during this period.

Figure P20.61

62. An aluminum ring of radius 5.00 cm and resistance 3.00×10^{-4} Ω is placed around the top of a long air-core solenoid with 1 000 turns per meter and a smaller radius of 3.00 cm, as in Figure P20.62. If the current in the solenoid is increasing at a constant rate of 270 A/s, what is the induced current in the ring? Assume that the magnetic field produced by the solenoid over the area at the end of the solenoid is one-half as strong as the field at the center of the solenoid. Assume also that the solenoid produces a negligible field outside its cross-sectional area.

Figure P20.62

63. In Figure P20.63, the rolling axle, 1.50 m long, is pushed along horizontal rails at a constant speed $v = 3.00$ m/s. A resistor $R = 0.400$ Ω is connected to the rails at points a and b, directly opposite each other. (The wheels make good electrical contact with the rails, so the axle, rails, and R form a closed-loop circuit. The only significant resistance in the circuit is R.) A uniform magnetic field $B = 0.800$ T is directed vertically downwards. (a) Find the induced current I in the resistor. (b) What horizontal force $\vec{F}$ is required to keep the axle rolling at constant speed? (c) Which end of the resistor, a or b, is at the higher electric potential? (d) After the axle rolls past the resistor, does the current in R reverse direction?

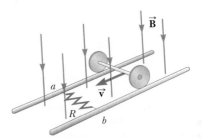

Figure P20.63

64. In 1832, Faraday proposed that the apparatus shown in Figure P20.64 could be used to generate electric current from the water flowing in the Thames River.[2] Two conducting plates of length a and width b are placed facing one another on opposite sides of the river, a distance w apart and immersed entirely. The flow velocity of the river is $\vec{v}$, and the vertical component of Earth's magnetic field is B. Show that the current in the load resistor R is

$$I = \frac{abvB}{\rho + abR/w}$$

where ρ is the resistivity of the water. (b) Calculate the short-circuit current $(R = 0)$ if $a = 100$ m, $b = 5.00$ m, $v = 3.00$ m/s, $B = 50.0$ μT, and $\rho = 100$ Ω·m.

Figure P20.64

65. A horizontal wire is free to slide on the vertical rails of a conducting frame, as in Figure P20.65 (page 692). The wire has mass m and length ℓ, and the resistance of the circuit is R. If a uniform magnetic field is directed perpendicular to

[2]The idea for this problem and Figure P20.64 is from Oleg D. Jefimenko, *Electricity and Magnetism: An Introduction to the Theory of Electric and Magnetic Fields* (Star City, WV, Electret Scientific Co., 1989).

the frame, what is the terminal speed of the wire as it falls under the force of gravity? (Neglect friction.)

Figure P20.65

66. A one-turn coil of wire of area 0.20 m^2 and resistance $0.25 \ \Omega$ is in a magnetic field that varies with time as shown in Figure P20.66a. The magnetic flux through the coil at $t = 0$ is as shown in Figure P20.66b. (a) When is the induced current the largest? (b) When is it zero? (c) Is the induced current always in the same direction? (d) Find the direction (clockwise or counterclockwise) and magnitude of the current between times $t = 0$ and $t = 2.0$ s, between $t = 2.0$ s and $t = 4.0$ s, and between $t = 4.0$ s and $t = 6.0$ s.

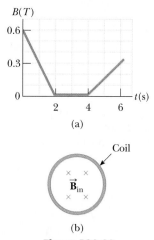

Figure P20.66

ACTIVITIES

1. Experimenting with induced currents is not easy. For small magnets and small coils of wire, the resulting induced currents are so small that they are difficult to detect. Thus, you may have to try this exercise several times before you are satisfied with your results. Wind a coil of wire on a cardboard mailing tube. Use insulated wire with as small a diameter as possible, because you need as many turns as possible on the coil. Connect the coil to a flashlight bulb, and see if you can get it to light by moving a bar magnet into and out of the coil in rapid succession. Why does the speed of movement make a difference? If you are unsuccessful, place two magnets side by side and repeat the experiment.

   After you have finished experimenting with the bulb, ask your instructor to let you use a galvanometer as a current detector. These devices are capable of measuring very small currents, and they have the added advantage of detecting the direction of the current in a circuit.

   Use your equipment to observe or test the following: (a) Does the magnitude of the induced current depend on the speed of movement of the magnet? (b) Can you induce a current by holding the magnet still and moving the coil over it? (c) Does the direction of the current depend on whether the magnet is pushed in or pulled out of the coil? (d) Does the direction of the current depend on whether the inserted pole of the magnet is the north pole or the south pole? (e) Can you predict the direction of the current by using Lenz's law? (f) Replace your bar magnet with the electromagnet you constructed in the last chapter, and repeat the preceding observations.

2. As explained in the text, a cassette tape is made up of tiny particles of metal oxide attached to a long plastic strip. Pull a tape out of a cassette that you do not mind destroying, and see if it is repelled or attracted by a refrigerator magnet. Also, try this with an expendable floppy computer disk.

3. This experiment takes steady hands, a dime, and a strong magnet. After verifying that a dime is not attracted to the magnet, carefully balance the coin on its edge. (This will not work with other coins, because they require too much force to topple them.) Hold one pole of the magnet within a millimeter of the face of the dime, but do not make contact with it. Now very rapidly pull the magnet straight back away from the coin. Which way does the dime tip? Does the coin fall the same way most of the time? Explain what is going on in terms of Lenz's law.

CHAPTER

21

Alternating Current Circuits and Electromagnetic Waves

Every time we turn on a television set, a stereo system, or any of a multitude of other electric appliances, we call on alternating currents (AC) to provide the power to operate them. We begin our study of AC circuits by examining the characteristics of a circuit containing a source of emf and one other circuit element: a resistor, a capacitor, or an inductor. Then we examine what happens when these elements are connected in combination with each other. Our discussion is limited to simple series configurations of the three kinds of elements.

We conclude this chapter with a discussion of **electromagnetic waves**, which are composed of fluctuating electric and magnetic fields. Electromagnetic waves in the form of visible light enable us to view the world around us; infrared waves warm our environment; radio-frequency waves carry our television and radio programs, as well as information about processes in the core of our galaxy. X-rays allow us to perceive structures hidden inside our bodies, and study properties of distant, collapsed stars. Light is key to our understanding of the universe.

21.1   RESISTORS IN AN AC CIRCUIT

An AC circuit consists of combinations of circuit elements and an AC generator or an AC source, which provides the alternating current. We have seen that the output of an AC generator is sinusoidal and varies with time according to

$$\Delta v = \Delta V_{\max} \sin 2\pi f t \qquad \textbf{[21.1]}$$

where Δv is the instantaneous voltage, $\Delta V_{\max}$ is the maximum voltage of the AC generator, and f is the frequency at which the voltage changes, measured in hertz (Hz). (Compare Equations 20.7 and 20.8 with Equation 21.1.) We first consider a simple

$$\Delta v = \Delta V_{\max} \sin 2\pi ft$$

ACTIVE FIGURE 21.1

A series circuit consisting of a resistor R connected to an AC generator, designated by the symbol

.

Physics⊗Now™

Log into PhysicsNow at **www.cp7e.com** and go to Active Figure 21.1, where you can adjust the resistance, the frequency, and the maximum voltage of the circuit shown. The results can be studied with the graph and phasor diagram in Active Figure 21.2.

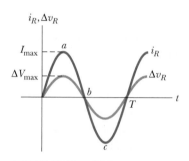

ACTIVE FIGURE 21.2

A plot of current and voltage across a resistor versus time.

Physics⊗Now™

Log into PhysicsNow at **www.cp7e.com** and go to Active Figure 21.1, where you can adjust the resistance, the frequency, and the maximum voltage of the circuit in Active Figure 21.1. The results can be studied with the graph and phasor diagram in Active Figure 21.2.

circuit consisting of a resistor and an AC source (designated by the symbol ──(∼)──), as in Active Figure 21.1. The current and the voltage across the resistor are shown in Active Figure 21.2.

To explain the concept of alternating current, we begin by discussing the current-versus-time curve in Active Figure 21.2. At point a on the curve, the current has a maximum value in one direction, arbitrarily called the positive direction. Between points a and b, the current is decreasing in magnitude but is still in the positive direction. At point b, the current is momentarily zero; it then begins to increase in the opposite (negative) direction between points b and c. At point c, the current has reached its maximum value in the negative direction.

The current and voltage are in step with each other because they vary identically with time. **Because the current and the voltage reach their maximum values at the same time, they are said to be in phase**. Notice that **the average value of the current over one cycle is zero**. This is because the current is maintained in one direction (the positive direction) for the same amount of time and at the same magnitude as it is in the opposite direction (the negative direction). However, the direction of the current has no effect on the behavior of the resistor in the circuit: the collisions between electrons and the fixed atoms of the resistor result in an increase in the resistor's temperature regardless of the direction of the current.

We can quantify this discussion by recalling that the rate at which electrical energy is dissipated in a resistor, the power $\mathcal{P}$, is

$$\mathcal{P} = i^2 R$$

where i is the *instantaneous* current in the resistor. Because the heating effect of a current is proportional to the *square* of the current, it makes no difference whether the sign associated with the current is positive or negative. However, the heating effect produced by an alternating current with a maximum value of $I_{\max}$ *is not the same* as that produced by a direct current of the same value. The reason is that the alternating current has this maximum value for only an instant of time during a cycle. The important quantity in an AC circuit is a special kind of average value of current, called the **rms current**—the direct current that dissipates the same amount of energy in a resistor that is dissipated by the actual alternating current. To find the rms current, we first square the current, Then find its average value, and finally take the square root of this average value. Hence, the rms current is the square *root* of the average (*mean*) of the *square* of the current. Because i^2 varies as $\sin^2 2\pi ft$, the average value of i^2 is $\frac{1}{2} I_{\max}^2$ (Fig. 21.3b).[1] Therefore, the rms current I_{rms} is related to the maximum value of the alternating current $I_{\max}$ by

$$I_{\text{rms}} = \frac{I_{\max}}{\sqrt{2}} = 0.707 I_{\max} \qquad [21.2]$$

This equation says that an alternating current with a maximum value of 3 A produces the same heating effect in a resistor as a direct current of $(3/\sqrt{2})$ A. We can therefore say that the average power dissipated in a resistor that carries alternating current I is

$$\mathcal{P}_{\text{av}} = I_{\text{rms}}^2 R.$$

---

[1]The fact that $(i^2)_{\text{av}} = I_{\max}^2/2$ can be shown as follows: The current in the circuit varies with time according to the expression $i = I_{\max} \sin 2\pi ft$, so $i^2 = I_{\max}^2 \sin^2 2\pi ft$. Therefore, we can find the average value of i^2 by calculating the average value of $\sin^2 2\pi ft$. Note that a graph of $\cos^2 2\pi ft$ versus time is identical to a graph of $\sin^2 2\pi ft$ versus time, except that the points are shifted on the time axis. Thus, the time average of $\sin^2 2\pi ft$ is equal to the time average of $\cos^2 2\pi ft$, taken over one or more cycles. That is,

$$(\sin^2 2\pi ft)_{\text{av}} = (\cos^2 2\pi ft)_{\text{av}}$$

With this fact and the trigonometric identity $\sin^2 \theta + \cos^2 \theta = 1$, we get

$$(\sin^2 2\pi ft)_{\text{av}} + (\cos^2 2\pi ft)_{\text{av}} = 2(\sin^2 2\pi ft)_{\text{av}} = 1$$

$$(\sin^2 2\pi ft)_{\text{av}} = \tfrac{1}{2}$$

When this result is substituted into the expression $i^2 = I_{\max}^2 \sin^2 2\pi ft$, we get $(i^2)_{\text{av}} = I_{\text{rms}}^2 = I_{\max}^2/2$, or $I_{\text{rms}} = I_{\max}/\sqrt{2}$, where I_{rms} is the rms current.

Figure 21.3   (a) Plot of the current in a resistor as a function of time. (b) Plot of the square of the current in a resistor as a function of time. Notice that the gray shaded regions *under* the curve and *above* the dashed line for $I_{max}^2/2$ have the same area as the gray shaded regions *above* the curve and *below* the dashed line for $I_{max}^2/2$. Thus, the average value of I^2 is $I_{max}^2/2$.

Alternating voltages are also best discussed in terms of rms voltages, with a relationship identical to the preceding one,

$$\Delta V_{rms} = \frac{\Delta V_{max}}{\sqrt{2}} = 0.707 \, \Delta V_{max} \qquad \qquad \text{[21.3]}$$

◀ rms voltage

where ΔV_{rms} is the rms voltage and ΔV_{max} is the maximum value of the alternating voltage.

When we speak of measuring an AC voltage of 120 V from an electric outlet, we really mean an rms voltage of 120 V. A quick calculation using Equation 21.3 shows that such an AC voltage actually has a peak value of about 170 V. In this chapter we use rms values when discussing alternating currents and voltages. One reason is that AC ammeters and voltmeters are designed to read rms values. Further, if we use rms values, many of the equations for alternating current will have the same form as those used in the study of direct-current (DC) circuits. Table 21.1 summarizes the notations used throughout the chapter.

Consider the series circuit in Figure 21.1, consisting of a resistor connected to an AC generator. A resistor impedes the current in an AC circuit, just as it does in a DC circuit. Ohm's law is therefore valid for an AC circuit, and we have

$$\Delta V_{R,rms} = I_{rms} R \qquad \qquad \text{[21.4a]}$$

The rms voltage across a resistor is equal to the rms current in the circuit times the resistance. This equation is also true if maximum values of current and voltage are used:

$$\Delta V_{R,max} = I_{max} R \qquad \qquad \text{[21.4b]}$$

TABLE 21.1

Notation Used in This Chapter

	Voltage	Current
Instantaneous value	Δv	i
Maximum value	ΔV_{max}	I_{max}
rms value	ΔV_{rms}	I_{rms}

Quick Quiz 21.1

Which of the following statements can be true for a resistor connected in a simple series circuit to an operating AC generator? (a) $\mathcal{P}_{av} = 0$ and $i_{av} = 0$ (b) $\mathcal{P}_{av} = 0$ and $i_{av} > 0$ (c) $\mathcal{P}_{av} > 0$ and $i_{av} = 0$ (d) $\mathcal{P}_{av} > 0$ and $i_{av} > 0$

EXAMPLE 21.1   What Is the rms Current?

Goal   Perform basic AC circuit calculations for a purely resistive circuit.

Problem   An AC voltage source has an output of $\Delta v = (2.00 \times 10^2 \text{ V}) \sin 2\pi ft$. This source is connected to a 1.00×10^2-Ω resistor as in Figure 21.1. Find the rms voltage and rms current in the resistor.

Strategy   Compare the expression for the voltage output just given with the general form, $\Delta v = \Delta V_{max} \sin 2\pi ft$, finding the maximum voltage. Substitute this result into the expression for the rms voltage.

Solution

Obtain the maximum voltage by comparison of the given expression for the output with the general expression:

$$\Delta v = (2.00 \times 10^2 \text{ V}) \sin 2\pi ft \qquad \Delta v = \Delta V_{max} \sin 2\pi ft$$
$$\rightarrow \quad \Delta V_{max} = 2.00 \times 10^2 \text{ V}$$

Next, substitute into Equation 21.3 to find the rms voltage of the source:

$$\Delta V_{rms} = \frac{\Delta V_{max}}{\sqrt{2}} = \frac{2.00 \times 10^2 \text{ V}}{\sqrt{2}} = \boxed{141 \text{ V}}$$

Substitute this result into Ohm's law to find the rms current:

$$I_{rms} = \frac{\Delta V_{rms}}{R} = \frac{141 \text{ V}}{1.00 \times 10^2 \, \Omega} = \boxed{1.41 \text{ A}}$$

Remarks   Notice how the concept of rms values allows the handling of an AC circuit quantitatively in much the same way as a DC circuit.

Exercise 21.1

Find the maximum current in the circuit and the average power delivered to the circuit.

Answer   2.00 A; 2.00×10^2 W

21.2   CAPACITORS IN AN AC CIRCUIT

To understand the effect of a capacitor on the behavior of a circuit containing an AC voltage source, we first review what happens when a capacitor is placed in a circuit containing a DC source, such as a battery. When the switch is closed in a series circuit containing a battery, a resistor, and a capacitor, the initial charge on the plates of the capacitor is zero. The motion of charge through the circuit is therefore relatively free, and there is a large current in the circuit. As more charge accumulates on the capacitor, the voltage across it increases, opposing the current. After some time interval, which depends on the time constant RC, the current approaches zero. Consequently, a capacitor in a DC circuit limits or impedes the current so that it approaches zero after a brief time.

Now consider the simple series circuit in Figure 21.4, consisting of a capacitor connected to an AC generator. We sketch curves of current versus time and voltage versus time, and then attempt to make the graphs seem reasonable. The curves are shown in Figure 21.5. First, note that the segment of the current curve from a to b indicates that the current starts out at a rather large value. This can be understood by recognizing that there is no charge on the capacitor at $t = 0$; as a consequence, there is nothing in the circuit except the resistance of the wires to hinder the flow of charge at this instant. However, the current decreases as the voltage across the capacitor increases from c to d on the voltage curve. When the voltage is at point d, the current reverses and begins to increase in the opposite direction (from b to e on the current curve). During this time, the voltage across the capacitor decreases from d to f because the plates are now losing the charge they accumulated earlier. The remainder of the cycle for both voltage and current is a repeat of what happened during the first half of the cycle. The current reaches a maximum value in the opposite direction at point e on the current curve and then decreases as the voltage across the capacitor builds up.

In a purely resistive circuit, the current and voltage are always in step with each other. This isn't the case when a capacitor is in the circuit. In Figure 21.5, when an alternating voltage is applied across a capacitor, the voltage reaches its maximum value one-quarter of a cycle after the current reaches its maximum value. We say that **the voltage across a capacitor always lags the current by 90°.**

The impeding effect of a capacitor on the current in an AC circuit is expressed in terms of a factor called the **capacitive reactance** X_C, defined as

$$\Delta v = \Delta V_{max} \sin 2\pi ft$$

Figure 21.4   A series circuit consisting of a capacitor C connected to an AC generator.

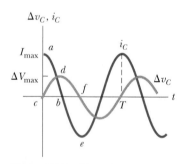

Figure 21.5   Plots of current and voltage across a capacitor versus time in an AC circuit. The voltage lags the current by 90°.

The voltage across a capacitor lags the current by 90° ▶

Capacitive reactance ▶

$$X_C \equiv \frac{1}{2\pi fC}$$

[21.5]

When C is in farads and f is in hertz, the unit of X_C is the ohm. Notice that $2\pi f = \omega$, the angular frequency.

From Equation 21.5, as the frequency f of the voltage source increases, the capacitive reactance X_C (the impeding effect of the capacitor) decreases, so the current increases. At high frequency, there is less time available to charge the capacitor, so less charge and voltage accumulate on the capacitor, which translates into less opposition to the flow of charge and, consequently, a higher current. The analogy between capacitive reactance and resistance means that we can write an equation of the same form as Ohm's law to describe AC circuits containing capacitors. This equation relates the rms voltage and rms current in the circuit to the capacitive reactance:

$$\Delta V_{C,\text{rms}} = I_{\text{rms}} X_C \qquad [21.6]$$

EXAMPLE 21.2   A Purely Capacitive AC Circuit

Goal   Perform basic AC circuit calculations for a capacitive circuit.

Problem   An 8.00-μF capacitor is connected to the terminals of an AC generator with an rms voltage of 1.50×10^2 V and a frequency of 60.0 Hz. Find the capacitive reactance and the rms current in the circuit.

Strategy   Substitute values into Equations 21.5 and 21.6.

Solution

Substitute the values of f and C into Equation 21.5:

$$X_C = \frac{1}{2\pi f C} = \frac{1}{2\pi\,(60.0\text{ Hz})\,(8.00 \times 10^{-6}\text{ F})} = \boxed{332\ \Omega}$$

Solve Equation 21.6 for the current, and substitute X_C and the rms voltage to find the rms current:

$$I_{\text{rms}} = \frac{\Delta V_{C,\text{rms}}}{X_C} = \frac{1.50 \times 10^2\text{ V}}{332\ \Omega} = \boxed{0.452\text{ A}}$$

Remark   Again, notice how similar the technique is to that of analyzing a DC circuit with a resistor.

Exercise 21.2

If the frequency is doubled, what happens to the capacitive reactance and the rms current?

Answer X_C is halved, and I_{rms} is doubled.

21.3   INDUCTORS IN AN AC CIRCUIT

Now consider an AC circuit consisting only of an inductor connected to the terminals of an AC source, as in Active Figure 21.6. (In any real circuit, there is some resistance in the wire forming the inductive coil, but we ignore this for now.) The changing current output of the generator produces a back emf that impedes the current in the circuit. The magnitude of this back emf is

$$\Delta v_L = L\,\frac{\Delta I}{\Delta t} \qquad [21.7]$$

The effective resistance of the coil in an AC circuit is measured by a quantity called the **inductive reactance**, X_L:

$$X_L \equiv 2\pi f L \qquad [21.8]$$

When f is in hertz and L is in henries, the unit of X_L is the ohm. The inductive reactance *increases* with increasing frequency and increasing inductance. Contrast these facts with capacitors, where increasing frequency or capacitance *decreases* the capacitive reactance.

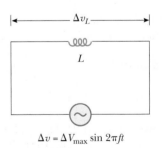

$\Delta v = \Delta V_{\text{max}} \sin 2\pi ft$

ACTIVE FIGURE 21.6
A series circuit consisting of an inductor L connected to an AC generator.

Physics⊗Now™
Log into PhysicsNow at **www.cp7e.com** and go to Active Figure 21.6, where you can adjust the inductance, the frequency, and the maximum voltage. The results can be studied with the graph and phasor diagram in Active Figure 21.7.

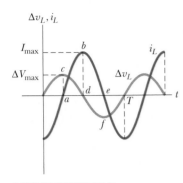

ACTIVE FIGURE 21.7

Plots of current and voltage across an inductor versus time in an AC circuit. The voltage leads the current by 90°.

Physics⊗Now™

Log into PhysicsNow at **www.cp7e.com** and go to Active Figure 21.6, where you can adjust the inductance, the frequency, and the maximum voltage. The results can be studied with the graph and phasor diagram in Active Figure 21.7.

To understand the meaning of inductive reactance, compare Equation 21.8 with Equation 21.7. First, note from Equation 21.8 that the inductive reactance depends on the inductance L. This is reasonable, because the back emf (Eq. 21.7) is large for large values of L. Second, note that the inductive reactance depends on the frequency f. This, too, is reasonable, because the back emf depends on $\Delta I/\Delta t$, a quantity that is large when the current changes rapidly, as it would for high frequencies.

With inductive reactance defined in this way, we can write an equation of the same form as Ohm's law for the voltage across the coil or inductor:

$$\Delta V_{L,\text{rms}} = I_{\text{rms}} X_L \qquad\qquad \textbf{[21.9]}$$

where $\Delta V_{L,\text{rms}}$ is the rms voltage across the coil and I_{rms} is the rms current in the coil.

Active Figure 21.7 shows the instantaneous voltage and instantaneous current across the coil as functions of time. When a sinusoidal voltage is applied across an inductor, the voltage reaches its maximum value one-quarter of an oscillation period before the current reaches its maximum value. In this situation, we say that **the voltage across an inductor always leads the current by 90°**.

To see why there is a phase relationship between voltage and current, we examine a few points on the curves of Active Figure 21.7. At point a on the current curve, the current is beginning to increase in the positive direction. At this instant, the rate of change of current, $\Delta I/\Delta t$ (the slope of the current curve), is at a maximum, and we see from Equation 21.7 that the voltage across the inductor is consequently also at a maximum. As the current rises between points a and b on the curve, $\Delta I/\Delta t$ gradually decreases until it reaches zero at point b. As a result, the voltage across the inductor is decreasing during this same time interval, as the segment between c and d on the voltage curve indicates. Immediately after point b, the current begins to decrease, although it still has the same direction it had during the previous quarter cycle. As the current decreases to zero (from b to e on the curve), a voltage is again induced in the coil (from d to f), but the polarity of this voltage is opposite the polarity of the voltage induced between c and d. This occurs because back emfs always oppose the change in the current.

We could continue to examine other segments of the curves, but no new information would be gained because the current and voltage variations are repetitive.

EXAMPLE 21.3   A Purely Inductive AC Circuit

Goal   Perform basic AC circuit calculations for an inductive circuit.

Problem   In a purely inductive AC circuit (see Active Fig. 21.6), $L = 25.0$ mH and the rms voltage is 1.50×10^2 V. Find the inductive reactance and rms current in the circuit if the frequency is 60.0 Hz.

Solution

Substitute L and f into Equation 21.8 to get the inductive reactance:

$$X_L = 2\pi f L = 2\pi (60.0 \text{ s}^{-1})(25.0 \times 10^{-3} \text{ H}) = \boxed{9.42 \ \Omega}$$

Solve Equation 21.9 for the rms current and substitute:

$$I_{\text{rms}} = \frac{\Delta V_{L,\text{rms}}}{X_L} = \frac{1.50 \times 10^2 \text{ V}}{9.42 \ \Omega} = \boxed{15.9 \text{ A}}$$

Remark   The analogy with DC circuits is even closer than in the capacitive case, because in the inductive equivalent of Ohm's law, the voltage across an inductor is *proportional* to the inductance L, just as the voltage across a resistor is proportional to R in Ohm's law.

Exercise 21.3

Calculate the inductive reactance and rms current in a similar circuit if the frequency is again 60.0 Hz, but the rms voltage is 85.0 V and the inductance is 47.0 mH.

Answers $X_L = 17.7 \ \Omega$; $I = 4.80$ A

21.4 THE *RLC* SERIES CIRCUIT

In the foregoing sections, we examined the effects of an inductor, a capacitor, and a resistor when they are connected separately across an AC voltage source. We now consider what happens when these devices are combined.

Active Figure 21.8 shows a circuit containing a resistor, an inductor, and a capacitor connected in series across an AC source that supplies a total voltage Δv at some instant. The current in the circuit is the same at all points in the circuit at any instant and varies sinusoidally with time, as indicated in Active Figure 21.9a. This fact can be expressed mathematically as

$$i = I_{max} \sin 2\pi ft$$

Earlier, we learned that the voltage across each element may or may not be in phase with the current. The instantaneous voltages across the three elements, shown in Active Figure 21.9, have the following phase relations to the instantaneous current:

1. The instantaneous voltage Δv_R across the resistor is *in phase* with the instantaneous current. (See Active Fig. 21.9b.)
2. The instantaneous voltage Δv_L across the inductor *leads* the current by 90°. (See Active Fig. 21.9c.)
3. The instantaneous voltage Δv_C across the capacitor *lags* the current by 90°. (See Active Fig. 21.9d.)

The net instantaneous voltage Δv supplied by the AC source equals the sum of the instantaneous voltages across the separate elements: $\Delta v = \Delta v_R + \Delta v_C + \Delta v_L$. This doesn't mean, however, that the voltages measured with an AC voltmeter across R, C, and L sum to the measured source voltage! In fact, the measured voltages *don't* sum to the measured source voltage, because the voltages across R, C, and L all have different phases.

To account for the different phases of the voltage drops, we use a technique involving vectors. We represent the voltage across each element with a rotating vector, as in Figure 21.10. The rotating vectors are referred to as **phasors**, and the diagram is called a **phasor diagram**. This particular diagram represents the circuit voltage given by the expression $\Delta v = \Delta V_{max} \sin(2\pi ft + \phi)$, where ΔV_{max} is the maximum voltage (the magnitude or length of the rotating vector or phasor) and ϕ is the angle between the phasor and the $+ x$-axis when $t = 0$. The phasor can be viewed as a vector of magnitude ΔV_{max} rotating at a constant frequency f so that its projection along the y-axis is the instantaneous voltage in the circuit. Because ϕ is the phase angle between the voltage and current in the circuit, the phasor for the current (not shown in Fig. 21.10) lies along the positive x-axis when $t = 0$ and is expressed by the relation $i = I_{max} \sin(2\pi ft)$.

The phasor diagrams in Figure 21.11 (page 700) are useful for analyzing the *series RLC* circuit. Voltages in phase with the current are represented by vectors along the positive x-axis, and voltages out of phase with the current lie along other directions. ΔV_R is horizontal and to the right because it's in phase with the current. Likewise, ΔV_L is represented by a phasor along the positive y-axis because it leads the current by 90°. Finally, ΔV_C is along the negative y-axis because it lags the current[2] by 90°. If the phasors are added as vector quantities in order to account for the different phases of the voltages across R, L, and C, Figure 21.11a shows that the only x-component for the voltages is ΔV_R and the net y-component is $\Delta V_L - \Delta V_C$. We now add the phasors vectorially to find the phasor ΔV_{max} (Fig. 21.11b), which represents the maximum voltage. The right triangle in Figure 21.11b gives the following equations for the maximum voltage and the phase angle ϕ between the maximum voltage and the current:

ACTIVE FIGURE 21.8
A series circuit consisting of a resistor, an inductor, and a capacitor connected to an AC generator.

Physics⊗Now™
Log into PhysicsNow at **www.cp7e.com** and go to Active Figure 21.8, where you can adjust the resistance, the inductance, and the capacitance. The results can be studied with the graph in Active Figure 21.9 and the phasor diagram in Figure 21.10.

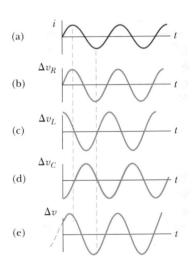

ACTIVE FIGURE 21.9
Phase relations in the series *RLC* circuit shown in Figure 21.8.

Physics⊗Now™
Log into PhysicsNow at **www.cp7e.com** and go to Active Figure 21.8, where you can adjust the resistance, the inductance, and the capacitance. The results can be studied with the graph in this figure and the phasor diagram in Figure 21.10.

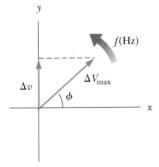

Figure 21.10 A phasor diagram for the voltage in an AC circuit, where ϕ is the phase angle between the voltage and the current and Δv is the instantaneous voltage.

---

[2]A mnemonic to help you remember the phase relationships in *RLC* circuits is " *ELI* the *ICE* man." *E* represents the voltage $\mathcal{E}$, *I* the current, *L* the inductance, and *C* the capacitance. Thus, the name *ELI* means that, in an inductive circuit, the voltage $\mathcal{E}$ leads the current *I*. In a capacitive circuit, *ICE* means that the current leads the voltage.

Figure 21.11 (a) A phasor diagram for the *RLC* circuit. (b) Addition of the phasors as vectors gives $\Delta V_{max} = \sqrt{\Delta V_R{}^2 + (\Delta V_L - \Delta V_C)^2}$. (c) The reactance triangle that gives the impedance relation $Z = \sqrt{R^2 + (X_L - X_C)^2}$.

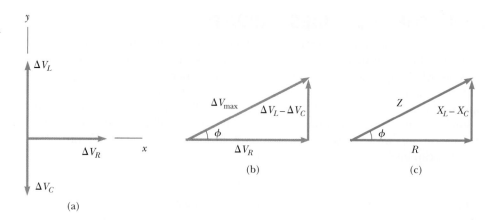

$$\Delta V_{max} = \sqrt{\Delta V_R{}^2 + (\Delta V_L - \Delta V_C)^2} \qquad [21.10]$$

$$\tan \phi = \frac{\Delta V_L - \Delta V_C}{\Delta V_R} \qquad [21.11]$$

In these equations, all voltages are maximum values. Although we choose to use maximum voltages in our analysis, the preceding equations apply equally well to rms voltages, because the two quantities are related to each other by the same factor for all circuit elements. The result for the maximum voltage ΔV_{max} given by Equation 21.10 reinforces the fact that **the voltages across the resistor, capacitor, and inductor are not in phase, so one cannot simply add them to get the voltage across the combination of element, or the source voltage**.

Quick Quiz 21.2

For the circuit of Figure 21.8, is the instantaneous voltage of the source equal to (a) the sum of the maximum voltages across the elements, (b) the sum of the instantaneous voltages across the elements, or (c) the sum of the rms voltages across the elements?

We can write Equation 21.10 in the form of Ohm's law, using the relations $\Delta V_R = I_{max}R$, $\Delta V_L = I_{max}X_L$, and $\Delta V_C = I_{max}X_C$, where I_{max} is the maximum current in the circuit:

$$\Delta V_{max} = I_{max}\sqrt{R^2 + (X_L - X_C)^2} \qquad [21.12]$$

It's convenient to define a parameter called the **impedance** Z of the circuit as

Impedance ▶

$$Z \equiv \sqrt{R^2 + (X_L - X_C)^2} \qquad [21.13]$$

so that Equation 21.12 becomes

$$\Delta V_{max} = I_{max}Z \qquad [21.14]$$

Equation 21.14 is in the form of Ohm's law, $\Delta V = IR$, with R replaced by the impedance in ohms. Indeed, Equation 21.14 can be regarded as a generalized form of Ohm's law applied to a series AC circuit. Both the impedance and, therefore, the current in an AC circuit depend on the resistance, the inductance, the capacitance, *and* the frequency (because the reactances are frequency dependent).

It's useful to represent the impedance Z with a vector diagram such as the one depicted in Figure 21.11c. A right triangle is constructed with right side $X_L - X_C$, base R, and hypotenuse Z. Applying the Pythagorean theorem to this triangle, we see that

$$Z = \sqrt{R^2 + (X_L - X_C)^2}$$

which is Equation 21.13. Furthermore, we see from the vector diagram in Figure 21.11c that the phase angle ϕ between the current and the voltage obeys the

TABLE 21.2

Impedance Values and Phase Angles for Various Combinations of Circuit Elements[a]

Circuit Elements	Impedance Z	Phase Angle ϕ
R ⟋⟍⟋⟍	R	$0°$
C ⊣⊢	X_C	$-90°$
L ⦾⦾⦾	X_L	$+90°$
R ⟋⟍⟋⟍ C ⊣⊢	$\sqrt{R^2 + X_C^2}$	Negative, between $-90°$ and $0°$
R ⟋⟍⟋⟍ L ⦾⦾⦾	$\sqrt{R^2 + X_L^2}$	Positive, between $0°$ and $90°$
R ⟋⟍⟋⟍ L ⦾⦾⦾ C ⊣⊢	$\sqrt{R^2 + (X_L - X_C)^2}$	Negative if $X_C > X_L$ Positive if $X_C < X_L$

[a] In each case, an AC voltage (not shown) is applied across the combination of elements (that is, across the dots).

relationship

$$\tan \phi = \frac{X_L - X_C}{R} \qquad \text{[21.15]}$$

◀ Phase angle ϕ

The physical significance of the phase angle will become apparent in Section 21.5.

Table 21.2 provides impedance values and phase angles for some series circuits containing different combinations of circuit elements.

Parallel alternating current circuits are also useful in everyday applications. We won't discuss them here, however, because their analysis is beyond the scope of this book.

Quick Quiz 21.3

The switch in the circuit shown in Figure 21.12 is closed and the lightbulb glows steadily. The inductor is a simple air-core solenoid. As an iron rod is being inserted into the interior of the solenoid, the brightness of the lightbulb (a) increases, (b) decreases, or (c) remains the same.

Figure 21.12 (Quick Quiz 21.3)

Bettmann/Corbis

NIKOLA TESLA (1856–1943)

Tesla was born in Croatia, but spent most of his professional life as an inventor in the United States. He was a key figure in the development of alternating-current electricity, high-voltage transformers, and the transport of electrical power via AC transmission lines. Tesla's viewpoint was at odds with the ideas of Edison, who committed himself to the use of direct current in power transmission. Tesla's AC approach won out.

Problem-Solving Strategy Alternating Current

The following procedure is recommended for solving alternating-current problems:
1. Calculate as many of the unknown quantities, such as X_L and X_C, as possible.
2. Apply the equation $\Delta V_{max} = I_{max} Z$ to the portion of the circuit of interest. For example, if you want to know the voltage drop across the combination of an inductor and a resistor, the equation for the voltage drop reduces to $\Delta V_{max} = I_{max}\sqrt{R^2 + X_L^2}$.

EXAMPLE 21.4 An *RLC* Circuit

Goal Analyze a series *RLC* AC circuit and find the phase angle.

Problem A series *RLC* AC circuit has resistance $R = 2.50 \times 10^2 \, \Omega$, inductance $L = 0.600$ H, capacitance $C = 3.50 \, \mu$F, frequency $f = 60.0$ Hz, and maximum voltage $\Delta V_{max} = 1.50 \times 10^2$ V. Find **(a)** the impedance, **(b)** the maximum current in the circuit, **(c)** the phase angle, and **(d)** the maximum voltages across the elements.

Strategy Calculate the inductive and capacitive reactances, then substitute them and given quantities into the appropriate equations.

Solution

(a) Find the impedance of the circuit.

First, calculate the inductive and capacitive reactances:

$$X_L = 2\pi f L = 226\ \Omega \qquad X_C = 1/2\pi f C = 758\ \Omega$$

Substitute these results and the resistance R into Equation 21.13 to obtain the impedance of the circuit:

$$Z = \sqrt{R^2 + (X_L - X_C)^2}$$
$$= \sqrt{(2.50 \times 10^2\ \Omega)^2 + (226\ \Omega - 758\ \Omega)^2} = \boxed{588\ \Omega}$$

(b) Find the maximum current.

Use Equation 21.12, the equivalent of Ohm's law, to find the maximum current:

$$I_{max} = \frac{\Delta V_{max}}{Z} = \frac{1.50 \times 10^2\ \text{V}}{588\ \Omega} = \boxed{0.255\ \text{A}}$$

(c) Find the phase angle.

Calculate the phase angle between the current and the voltage with Equation 21.15:

$$\phi = \tan^{-1}\frac{X_L - X_C}{R} = \tan^{-1}\left(\frac{226\ \Omega - 758\ \Omega}{2.50 \times 10^2\ \Omega}\right) = \boxed{-64.8°}$$

(d) Find the maximum voltages across the elements.

Substitute into the "Ohm's law" expressions for each individual type of current element:

$$\Delta V_{R,max} = I_{max}R = (0.255\ \text{A})(2.50 \times 10^2\ \Omega) = \boxed{63.8\ \text{V}}$$

$$\Delta V_{L,max} = I_{max}X_L = (0.255\ \text{A})(2.26 \times 10^2\ \Omega) = \boxed{57.6\ \text{V}}$$

$$\Delta V_{C,max} = I_{max}X_C = (0.255\ \text{A})(7.58 \times 10^2\ \Omega) = \boxed{193\ \text{V}}$$

Remarks Because the circuit is more capacitive than inductive ($X_C > X_L$), ϕ is negative. A negative phase angle means that the current leads the applied voltage. Notice also that the sum of the maximum voltages across the elements is $\Delta V_R + \Delta V_L + \Delta V_C = 314$ V, which is much greater than the maximum voltage of the generator, 150 V. As we saw in Quick Quiz 21.2, the sum of the maximum voltages is a meaningless quantity because when alternating voltages are added, *both their amplitudes and their phases* must be taken into account. We know that the maximum voltages across the various elements occur at different times, so it doesn't make sense to add all the maximum values. The correct way to "add" the voltages is through Equation 21.10.

Exercise 21.4

Analyze a series *RLC* AC circuit for which $R = 175\ \Omega$, $L = 0.500$ H, $C = 22.5\ \mu$F, $f = 60.0$ Hz, and $\Delta V_{max} = 325$ V. Find (a) the impedance, (b) the maximum current, (c) the phase angle, and (d) the maximum voltages across the elements.

Answers   (a) 189 Ω (b) 1.72 A   (c) 22.0°   (d) $\Delta V_{R,max} = 301$ V, $\Delta V_{L,max} = 324$ V, $\Delta V_{C,max} = 203$ V

21.5 POWER IN AN AC CIRCUIT

No power losses are associated with pure capacitors and pure inductors in an AC circuit. A pure capacitor, by definition, has no resistance or inductance, while a pure inductor has no resistance or capacitance. (These are idealizations: in a real capacitor, for example, inductive effects could become important at high frequencies.) We begin by analyzing the power dissipated in an AC circuit that contains only a generator and a capacitor.

When the current increases in one direction in an AC circuit, charge accumulates on the capacitor and a voltage drop appears across it. When the voltage reaches its maximum value, the energy stored in the capacitor is

$$PE_C = \tfrac{1}{2}C(\Delta V_{max})^2$$

However, this energy storage is only momentary: When the current reverses direction, the charge leaves the capacitor plates and returns to the voltage source. During one-half of each cycle the capacitor is being charged, and during the other half the charge is being returned to the voltage source. Therefore, the average power supplied by the source is zero. In other words, **no power losses occur in a capacitor in an AC circuit**.

Similarly, the source must do work against the back emf of an inductor that is carrying a current. When the current reaches its maximum value, the energy stored in the inductor is a maximum and is given by

$$PE_L = \tfrac{1}{2}LI_{max}^2$$

When the current begins to decrease in the circuit, this stored energy is returned to the source as the inductor attempts to maintain the current in the circuit. The average power delivered to a resistor in an *RLC* circuit is

$$\mathscr{P}_{av} = I_{rms}^2 R \qquad \qquad \textbf{[21.16]}$$

The average power delivered by the generator is converted to internal energy in the resistor. No power loss occurs in an ideal capacitor or inductor.

An alternate equation for the average power loss in an AC circuit can be found by substituting (from Ohm's law) $R = \Delta V_R / I_{rms}$ into Equation 21.16:

$$\mathscr{P}_{av} = I_{rms}\Delta V_R$$

It's convenient to refer to a voltage triangle that shows the relationship among ΔV_{rms}, ΔV_R, and $\Delta V_L - \Delta V_C$, such as Figure 21.11b. (Remember that Fig. 21.11 applies to *both* maximum and rms voltages.) From this figure, we see that the voltage drop across a resistor can be written in terms of the voltage of the source, ΔV_{rms}:

$$\Delta V_R = \Delta V_{rms} \cos \phi$$

Hence, the average power delivered by a generator in an AC circuit is

$$\mathscr{P}_{av} = I_{rms} \Delta V_{rms} \cos \phi \qquad \qquad \textbf{[21.17]}$$ ◀ Average power

where the quantity $\cos \phi$ is called the **power factor**.

Equation 21.17 shows that the power delivered by an AC source to any circuit depends on the phase difference between the source voltage and the resulting current. This fact has many interesting applications. For example, factories often use devices such as large motors in machines, generators, and transformers that have a large inductive load due to all the windings. To deliver greater power to such devices without using excessively high voltages, factory technicians introduce capacitance in the circuits to shift the phase.

APPLICATION

Shifting Phase to Deliver More Power

EXAMPLE 21.5   Average Power in an *RLC* Series Circuit

Goal   Understand power in *RLC* series circuits.

Problem   Calculate the average power delivered to the series *RLC* circuit described in Example 21.4.

Strategy   After finding the rms current and rms voltage with Equations 21.2 and 21.3, substitute into Equation 21.17, using the phase angle found in Example 21.4.

Solution

First, use Equations 21.2 and 21.3 to calculate the rms current and rms voltage:

$$I_{rms} = \frac{I_{max}}{\sqrt{2}} = \frac{0.255 \text{ A}}{\sqrt{2}} = 0.180 \text{ A}$$

$$\Delta V_{rms} = \frac{\Delta V_{max}}{\sqrt{2}} = \frac{1.50 \times 10^2 \text{ V}}{\sqrt{2}} = 106 \text{ V}$$

Substitute these results and the phase angle $\phi = -64.8°$ into Equation 21.17 to find the average power:

$$\mathscr{P}_{av} = I_{rms}\Delta V_{rms} \cos \phi = (0.180 \text{ A})(106 \text{ V}) \cos(-64.8°)$$

$$= \boxed{8.12 \text{ W}}$$

Remark The same result can be obtained from Equation 21.16, $\mathcal{P}_{av} = I_{rms}^2 R$.

Exercise 21.5
Repeat this problem, using the system described in Exercise 21.4.

Answer 259 W

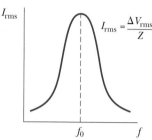

Figure 21.13 A plot of current amplitude in a series *RLC* circuit versus frequency of the generator voltage. Note that the current reaches its maximum value at the resonance frequency f_0.

Resonance frequency ▶

21.6 RESONANCE IN A SERIES *RLC* CIRCUIT

In general, the rms current in a series *RLC* circuit can be written

$$I_{rms} = \frac{\Delta V_{rms}}{Z} = \frac{\Delta V_{rms}}{\sqrt{R^2 + (X_L - X_C)^2}} \qquad \text{[21.18]}$$

From this equation, we see that if the frequency is varied, the current has its *maximum* value when the impedance has its *minimum* value. This occurs when $X_L = X_C$. In such a circumstance, the impedance of the circuit reduces to $Z = R$. The frequency f_0 at which this happens is called the **resonance frequency** of the circuit. To find f_0, we set $X_L = X_C$, which gives, from Equations 21.5 and 21.8,

$$2\pi f_0 L = \frac{1}{2\pi f_0 C}$$

$$f_0 = \frac{1}{2\pi\sqrt{LC}} \qquad \text{[21.19]}$$

Figure 21.13 is a plot of current as a function of frequency for a circuit containing a fixed value for both the capacitance and the inductance. From Equation 21.18, it must be concluded that the current would become infinite at resonance when $R = 0$. Although Equation 21.18 predicts this result, real circuits always have some resistance, which limits the value of the current.

The tuning circuit of a radio is an important application of a series resonance circuit. The radio is tuned to a particular station (which transmits a specific radio-frequency signal) by varying a capacitor, which changes the resonance frequency of the tuning circuit. When this resonance frequency matches that of the incoming radio wave, the current in the tuning circuit increases.

APPLICATION

Tuning Your Radio

Applying Physics 21.1 Metal Detectors in Airports

When you walk through the doorway of an airport metal detector, as the person in Figure 21.14 is doing, you are really walking through a coil of many turns. How might the metal detector work?

Explanation The metal detector is essentially a resonant circuit. The portal you step through is an inductor (a large loop of conducting wire) that is part of the circuit. The frequency of the circuit is tuned to the resonant frequency of the circuit when there is no metal in the inductor. When you walk through with metal in your pocket, you change the effective inductance of the resonance circuit, resulting in a change in the current in the circuit. This change in current is detected, and an electronic circuit causes a sound to be emitted as an alarm.

Figure 21.14 (Applying Physics 21.1) An airport metal detector.

EXAMPLE 21.6   A Circuit in Resonance

Goal   Understand resonance frequency and its relation to inductance, capacitance, and the rms current.

Problem   Consider a series RLC circuit for which $R = 1.50 \times 10^2\ \Omega$, $L = 20.0$ mH, $\Delta V_{rms} = 20.0$ V, and $f = 796$ s^{-1}. **(a)** Determine the value of the capacitance for which the rms current is a maximum. **(b)** Find the maximum rms current in the circuit.

Strategy   The current is a maximum at the resonance frequency f_0, which should be set equal to the driving frequency, 796 s^{-1}. The resulting equation can be solved for C. For part (b), substitute into Equation 21.18 to get the maximum rms current.

Solution

(a) Find the capacitance giving the maximum current in the circuit (the resonance condition).

Solve the resonance frequency for the capacitance:

$$f_0 = \frac{1}{2\pi\sqrt{LC}} \quad\rightarrow\quad \sqrt{LC} = \frac{1}{2\pi f_0} \quad\rightarrow\quad LC = \frac{1}{4\pi^2 f_0^2}$$

$$C = \frac{1}{4\pi^2 f_0^2 L}$$

Insert the given values, substituting the source frequency for the resonance frequency, f_0:

$$C = \frac{1}{4\pi^2 (796\ \text{Hz})^2 (20.0 \times 10^{-3}\ \text{H})} = \boxed{2.00 \times 10^{-6}\ \text{F}}$$

(b) Find the maximum rms current in the circuit.

The capacitive and inductive reactances are equal, so $Z = R = 1.50 \times 10^2\ \Omega$. Substitute into Equation 21.18 to find the rms current:

$$I_{rms} = \frac{\Delta V_{rms}}{Z} = \frac{20.0\ \text{V}}{1.50 \times 10^2\ \Omega} = \boxed{0.133\ \text{A}}$$

Remark   Because the impedance Z is in the denominator of Equation 21.18, the maximum current will always occur when $X_L = X_C$, since that yields the minimum value of Z.

Exercise 21.6

Consider a series RLC circuit for which $R = 1.20 \times 10^2\ \Omega$, $C = 3.10 \times 10^{-5}$ F, $\Delta V_{rms} = 35.0$ V, and $f = 60.0$ s^{-1}. **(a)** Determine the value of the inductance for which the rms current is a maximum. **(b)** Find the maximum rms current in the circuit.

Answers   (a) 0.227 H   (b) 0.292 A

21.7   THE TRANSFORMER

It's often necessary to change a small AC voltage to a larger one or vice versa. Such changes are effected with a device called a transformer.

In its simplest form, the **AC transformer** consists of two coils of wire wound around a core of soft iron, as shown in Figure 21.15. The coil on the left, which is connected to the input AC voltage source and has N_1 turns, is called the primary winding, or the *primary*. The coil on the right, which is connected to a resistor R and consists of N_2 turns, is the *secondary*. The purpose of the common iron core is to increase the magnetic flux and to provide a medium in which nearly all the flux through one coil passes through the other.

When an input AC voltage ΔV_1 is applied to the primary, the induced voltage across it is given by

$$\Delta V_1 = -N_1 \frac{\Delta \Phi_B}{\Delta t} \qquad \textbf{[21.20]}$$

Figure 21.15   An ideal transformer consists of two coils wound on the same soft iron core. An AC voltage ΔV_1 is applied to the primary coil, and the output voltage ΔV_2 is observed across the load resistance R after the switch is closed.

where Φ_B is the magnetic flux through each turn. If we assume that no flux leaks from the iron core, then the flux through each turn of the primary equals the flux through each turn of the secondary. Hence, the voltage across the secondary coil is

$$\Delta V_2 = -N_2 \frac{\Delta \Phi_B}{\Delta t} \qquad [21.21]$$

The term $\Delta \Phi_B / \Delta t$ is common to Equations 21.20 and 21.21 and can be algebraically eliminated, giving

$$\Delta V_2 = \frac{N_2}{N_1} \Delta V_1 \qquad [21.22]$$

When N_2 is greater than N_1, ΔV_2 exceeds ΔV_1 and the transformer is referred to as a *step-up transformer*. When N_2 is less than N_1, making ΔV_2 less than ΔV_1, we have a *step-down transformer*.

By Faraday's law, a voltage is generated across the secondary only when there is a *change* in the number of flux lines passing through the secondary. The input current in the primary must therefore change with time, which is what happens when an alternating current is used. When the input at the primary is a direct current, however, a voltage output occurs at the secondary only at the instant a switch in the primary circuit is opened or closed. Once the current in the primary reaches a steady value, the output voltage at the secondary is zero.

It may seem that a transformer is a device in which it is possible to get something for nothing. For example, a step-up transformer can change an input voltage from, say, 10 V to 100 V. This means that each coulomb of charge leaving the secondary has 100 J of energy, whereas each coulomb of charge entering the primary has only 10 J of energy. That is not the case, however, because **the power input to the primary equals the power output at the secondary**:

> In an ideal transformer, the input power equals the output power. ▶

$$I_1 \, \Delta V_1 = I_2 \, \Delta V_2 \qquad [21.23]$$

While the *voltage* at the secondary may be, say, ten times greater than the voltage at the primary, the *current* in the secondary will be smaller than the primary's current by a factor of ten. Equation 21.23 assumes an **ideal transformer**, in which there are no power losses between the primary and the secondary. Real transformers typically have power efficiencies ranging from 90% to 99%. Power losses occur because of such factors as eddy currents induced in the iron core of the transformer, which dissipate energy in the form of $I^2 R$ losses.

APPLICATION

Long-Distance Electric Power Transmission

When electric power is transmitted over large distances, it's economical to use a high voltage and a low current because the power lost via resistive heating in the transmission lines varies as $I^2 R$. This means that if a utility company can reduce the current by a factor of ten, for example, the power loss is reduced by a factor of one hundred. In practice, the voltage is stepped up to around 230 000 V at the generating station, then stepped down to around 20 000 V at a distribution station, and finally stepped down to 120 V at the customer's utility pole.

EXAMPLE 21.7 Distributing Power to a City

Goal Understand transformers and their role in reducing power loss.

Problem A generator at a utility company produces 1.00×10^2 A of current at 4.00×10^3 V. The voltage is stepped up to 2.40×10^5 V by a transformer before being sent on a high-voltage transmission line across a rural area to a city. Assume that the effective resistance of the power line is 30.0 Ω and that the transformers are ideal. **(a)** Determine the percentage of power lost in the transmission line. **(b)** What percentage of the original power would be lost in the transmission line if the voltage were not stepped up?

Strategy Solving this problem is just a matter of substitution into the equation for transformers and the equation for power loss. To obtain the fraction of power lost, it's also necessary to compute the power output of the generator—the current times the potential difference created by the generator.

Solution

(a) Determine the percentage of power lost in the line.

Substitute into Equation 21.23 to find the current in the transmission line:

$$I_2 = \frac{I_1 \Delta V_1}{\Delta V_2} = \frac{(1.00 \times 10^2 \, \text{A})(4.00 \times 10^3 \, \text{V})}{2.40 \times 10^5 \, \text{V}} = 1.67 \, \text{A}$$

Now use Equation 21.16 to find the power lost in the transmission line:

$$(1) \quad \mathscr{P}_{\text{lost}} = I_2{}^2 R = (1.67 \, \text{A})^2 (30.0 \, \Omega) = 83.7 \, \text{W}$$

Calculate the power output of the generator:

$$\mathscr{P} = I_1 \Delta V_1 = (1.00 \times 10^2 \, \text{A})(4.00 \times 10^3 \, \text{V}) = 4.00 \times 10^5 \, \text{W}$$

Finally, divide $\mathscr{P}_{\text{lost}}$ by the power output and multiply by 100 to find the percentage of power lost:

$$\% \text{ power lost} = \left(\frac{83.7 \, \text{W}}{4.00 \times 10^5 \, \text{W}} \right) \times 100 = \boxed{0.020 \, 9\%}$$

(b) What percentage of the original power would be lost in the transmission line if the voltage were not stepped up?

Replace the stepped-up current in equation (1) by the original current of 1.00×10^2 A.

$$\mathscr{P}_{\text{lost}} = I^2 R = (1.00 \times 10^2 \, \text{A})^2 (30.0 \, \Omega) = 3.00 \times 10^5 \, \text{W}$$

Calculate the percentage loss, as before:

$$\% \text{ power lost} = \left(\frac{3.00 \times 10^5 \, \text{W}}{4.00 \times 10^5 \, \text{W}} \right) \times 100 = \boxed{75\%}$$

Remarks This example illustrates the advantage of high-voltage transmission lines. At the city, a transformer at a substation steps the voltage back down to about 4 000 V, and this voltage is maintained across utility lines throughout the city. When the power is to be used at a home or business, a transformer on a utility pole near the establishment reduces the voltage to 240 V or 120 V.

Exercise 21.7
Suppose the same generator has the voltage stepped up to only 7.50×10^4 V and the resistance of the line is 85.0 Ω. Find the percentage of power lost in this case.

Answer 0.604%

This cylindrical step-down transformer drops the voltage from 4 000 V to 220 V for delivery to a group of residences.

George Semple

21.8 MAXWELL'S PREDICTIONS

During the early stages of their study and development, electric and magnetic phenomena were thought to be unrelated. In 1865, however, James Clerk Maxwell (1831–1879) provided a mathematical theory that showed a close relationship between all electric and magnetic phenomena. In addition to unifying the formerly separate fields of electricity and magnetism, his brilliant theory predicted that electric and magnetic fields can move through space as waves. The theory he developed is based on the following four pieces of information:

1. Electric field lines originate on positive charges and terminate on negative charges.
2. Magnetic field lines always form closed loops—they don't begin or end anywhere.
3. A varying magnetic field induces an emf and hence an electric field. This is a statement of Faraday's law (Chapter 20).
4. Magnetic fields are generated by moving charges (or currents), as summarized in Ampère's law (Chapter 19).

JAMES CLERK MAXWELL,
Scottish Theoretical Physicist
(1831–1879)

Maxwell developed the electromagnetic theory of light, the kinetic theory of gases, and explained the nature of Saturn's rings and color vision. Maxwell's successful interpretation of the electromagnetic field resulted in the equations that bear his name. Formidable mathematical ability combined with great insight enabled him to lead the way in the study of electromagnetism and kinetic theory.

The first statement is a consequence of the nature of the electrostatic force between charged particles, given by Coulomb's law. It embodies the fact that **free charges (electric monopoles) exist in nature**.

The second statement—that magnetic fields form continuous loops—is exemplified by the magnetic field lines around a long, straight wire, which are closed circles, and the magnetic field lines of a bar magnet, which form closed loops. It says, in contrast to the first statement, that **free magnetic charges (magnetic monopoles) don't exist in nature**.

The third statement is equivalent to Faraday's law of induction, and the fourth is equivalent to Ampère's law.

In one of the greatest theoretical developments of the 19th century, Maxwell used these four statements within a corresponding mathematical framework to prove that electric and magnetic fields play symmetric roles in nature. It was already known from experiments that a changing magnetic field produced an electric field according to Faraday's law. Maxwell believed that nature was symmetric, and he therefore hypothesized that a changing electric field should produce a magnetic field. This hypothesis could not be proven experimentally at the time it was developed, because the magnetic fields generated by changing electric fields are generally very weak and therefore difficult to detect.

To justify his hypothesis, Maxwell searched for other phenomena that might be explained by it. He turned his attention to the motion of rapidly oscillating (accelerating) charges, such as those in a conducting rod connected to an alternating voltage. Such charges are accelerated and, according to Maxwell's predictions, generate changing electric and magnetic fields. The changing fields cause electromagnetic disturbances that travel through space as waves, similar to the spreading water waves created by a pebble thrown into a pool. The waves sent out by the oscillating charges are fluctuating electric and magnetic fields, so they are called *electromagnetic waves*. From Faraday's law and from Maxwell's own generalization of Ampère's law, Maxwell calculated the speed of the waves to be equal to the speed of light, $c = 3 \times 10^8$ m/s. He concluded that visible light and other electromagnetic waves consist of fluctuating electric and magnetic fields traveling through empty space, with each varying field inducing the other! This was truly one of the greatest discoveries of science, on a par with Newton's discovery of the laws of motion. Like Newton's laws, it had a profound influence on later scientific developments.

21.9 HERTZ'S CONFIRMATION OF MAXWELL'S PREDICTIONS

HEINRICH RUDOLF HERTZ,
German Physicist (1857–1894)

Hertz made his most important discovery of radio waves in 1887. After finding that the speed of a radio wave was the same as that of light, Hertz showed that radio waves, like light waves, could be reflected, refracted, and diffracted. Hertz died of blood poisoning at the age of 36. During his short life, he made many contributions to science. The hertz, equal to one complete vibration or cycle per second, is named after him.

In 1887, after Maxwell's death, Heinrich Hertz (1857–1894) was the first to generate and detect electromagnetic waves in a laboratory setting, using *LC* circuits. In such a circuit, a charged capacitor is connected to an inductor, as in Figure 21.16. When the switch is closed, oscillations occur in the current in the circuit and in the charge on the capacitor. If the resistance of the circuit is neglected, no energy is dissipated and the oscillations continue.

In the following analysis, we neglect the resistance in the circuit. We assume that the capacitor has an initial charge of Q_{max} and that the switch is closed at $t = 0$. When the capacitor is fully charged, the total energy in the circuit is stored in the electric field of the capacitor and is equal to $Q_{max}^2/2C$. At this time, the current is zero, so no energy is stored in the inductor. As the capacitor begins to discharge, the energy stored in its electric field decreases. At the same time, the current increases and energy equal to $LI^2/2$ is now stored in the magnetic field of the inductor. Thus, energy is transferred from the electric field of the capacitor to the magnetic field of the inductor. When the capacitor is fully discharged, it stores no energy. At this time, the current reaches its maximum value and all of the energy is stored in the inductor. The process then repeats in the reverse direction. The energy continues to transfer between the inductor and the capacitor, corresponding to oscillations in the current and charge.

As we saw in Section 21.6, the frequency of oscillation of an *LC* circuit is called the *resonance frequency* of the circuit and is given by

$$f_0 = \frac{1}{2\pi\sqrt{LC}}$$

The circuit Hertz used in his investigations of electromagnetic waves is similar to that just discussed and is shown schematically in Figure 21.17. An induction coil (a large coil of wire) is connected to two metal spheres with a narrow gap between them to form a capacitor. Oscillations are initiated in the circuit by short voltage pulses sent via the coil to the spheres, charging one positive, the other negative. Because *L* and *C* are quite small in this circuit, the frequency of oscillation is quite high, $f \approx 100$ MHz. This circuit is called a transmitter because it produces electromagnetic waves.

Several meters from the transmitter circuit, Hertz placed a second circuit, the receiver, which consisted of a single loop of wire connected to two spheres. It had its own effective inductance, capacitance, and natural frequency of oscillation. Hertz found that energy was being sent from the transmitter to the receiver when the resonance frequency of the receiver was adjusted to match that of the transmitter. The energy transfer was detected when the voltage across the spheres in the receiver circuit became high enough to produce ionization in the air, which caused sparks to appear in the air gap separating the spheres. Hertz's experiment is analogous to the mechanical phenomenon in which a tuning fork picks up the vibrations from another, identical tuning fork.

Hertz hypothesized that the energy transferred from the transmitter to the receiver is carried in the form of waves, now recognized as electromagnetic waves. In a series of experiments, he also showed that the radiation generated by the transmitter exhibits wave properties: interference, diffraction, reflection, refraction, and polarization. As you will see shortly, all of these properties are exhibited by light. It became evident that Hertz's electromagnetic waves had the same known properties of light waves and differed only in frequency and wavelength. Hertz effectively confirmed Maxwell's theory by showing that Maxwell's mysterious electromagnetic waves existed and had all the properties of light waves.

Perhaps the most convincing experiment Hertz performed was the measurement of the speed of waves from the transmitter, accomplished as follows: waves of known frequency from the transmitter were reflected from a metal sheet so that an interference pattern was set up, much like the standing-wave pattern on a stretched string. As we learned in our discussion of standing waves, the distance between nodes is $\lambda/2$, so Hertz was able to determine the wavelength λ. Using the relationship $v = \lambda f$, he found that v was close to 3×10^8 m/s, the known speed of visible light. Hertz's experiments thus provided the first evidence in support of Maxwell's theory.

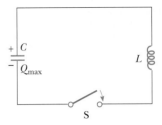

Figure 21.16   A simple *LC* circuit. The capacitor has an initial charge of Q_{max} and the switch is closed at $t = 0$.

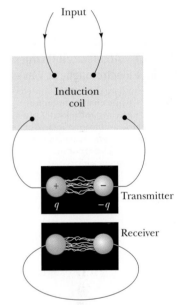

Figure 21.17   A schematic diagram of Hertz's apparatus for generating and detecting electromagnetic waves. The transmitter consists of two spherical electrodes connected to an induction coil, which provides short voltage surges to the spheres, setting up oscillations in the discharge. The receiver is a nearby single loop of wire containing a second spark gap.

21.10   PRODUCTION OF ELECTROMAGNETIC WAVES BY AN ANTENNA

In the previous section, we found that the energy stored in an *LC* circuit is continually transferred between the electric field of the capacitor and the magnetic field of the inductor. However, this energy transfer continues for prolonged periods of time only when the changes occur slowly. If the current alternates rapidly, the circuit loses some of its energy in the form of electromagnetic waves. In fact, electromagnetic waves are radiated by *any* circuit carrying an alternating current. The fundamental mechanism responsible for this radiation is the acceleration of a charged particle. **Whenever a charged particle accelerates it radiates energy**.

An alternating voltage applied to the wires of an antenna forces electric charges in the antenna to oscillate. This is a common technique for accelerating charged particles and is the source of the radio waves emitted by the broadcast antenna of a radio station.

APPLICATION

Radio-Wave Transmission

Figure 21.18 An electric field set up by oscillating charges in an antenna. The field moves away from the antenna at the speed of light.

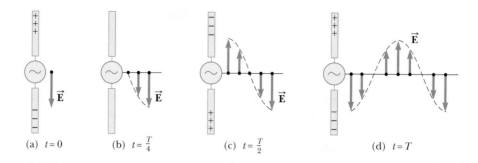

(a) $t = 0$ (b) $t = \frac{T}{4}$ (c) $t = \frac{T}{2}$ (d) $t = T$

TIP 21.1 Accelerated Charges Produce Electromagnetic Waves

Stationary charges produce only electric fields, while charges in uniform motion (i.e., constant velocity) produce electric and magnetic fields, but no electromagnetic waves. In contrast, accelerated charges produce electromagnetic waves as well as electric and magnetic fields. An accelerating charge also radiates energy.

Figure 21.19 Magnetic field lines around an antenna carrying a changing current.

Figure 21.18 illustrates the production of an electromagnetic wave by oscillating electric charges in an antenna. Two metal rods are connected to an AC source, which causes charges to oscillate between the rods. The output voltage of the generator is sinusoidal. At $t = 0$, the upper rod is given a maximum positive charge and the bottom rod an equal negative charge, as in Figure 21.18a. The electric field near the antenna at this instant is also shown in the figure. As the charges oscillate, the rods become less charged, the field near the rods decreases in strength, and the downward-directed maximum electric field produced at $t = 0$ moves away from the rod. When the charges are neutralized, as in Figure 21.18b, the electric field has dropped to zero, after an interval equal to one-quarter of the period of oscillation. Continuing in this fashion, the upper rod soon obtains a maximum negative charge and the lower rod becomes positive, as in Figure 21.18c, resulting in an electric field directed upward. This occurs after an interval equal to one-half the period of oscillation. The oscillations continue as indicated in Figure 21.18d. Note that the electric field near the antenna oscillates in phase with the charge distribution: the field points down when the upper rod is positive and up when the upper rod is negative. Further, the magnitude of the field at any instant depends on the amount of charge on the rods at that instant.

As the charges continue to oscillate (and accelerate) between the rods, the electric field set up by the charges moves away from the antenna in all directions at the speed of light. Figure 21.18 shows the electric field pattern on one side of the antenna at certain times during the oscillation cycle. As you can see, one cycle of charge oscillation produces one full wavelength in the electric field pattern.

Because the oscillating charges create a current in the rods, a magnetic field is also generated when the current in the rods is upward, as shown in Figure 21.19. The magnetic field lines circle the antenna (recall right-hand rule number 2) and are perpendicular to the electric field at all points. As the current changes with time, the magnetic field lines spread out from the antenna. At great distances from the antenna, the strengths of the electric and magnetic fields become very weak. At these distances, however, it is necessary to take into account the facts that (1) a changing magnetic field produces an electric field and (2) a changing electric field produces a magnetic field, as predicted by Maxwell. These induced electric and magnetic fields are in phase: at any point, the two fields reach their maximum values at the same instant. This synchrony is illustrated at one instant of time in Active Figure 21.20. Note that (1) the $\vec{E}$ and $\vec{B}$ fields are perpendicular to each other, and (2) both fields are perpendicular to the direction of motion of the wave. This second property is characteristic of transverse waves. Hence, we see that **an electromagnetic wave is a transverse wave.**

21.11 PROPERTIES OF ELECTROMAGNETIC WAVES

We have seen that Maxwell's detailed analysis predicted the existence and properties of electromagnetic waves. In this section we summarize what we know about electromagnetic waves thus far and consider some additional properties. In our discussion here and in future sections, we will often make reference to a type of wave called a **plane wave.** A plane electromagnetic wave is a wave traveling from a very distant source. Active Figure 21.20 pictures such a wave at a given instant of time. In

ACTIVE FIGURE 21.20
An electromagnetic wave sent out by oscillating charges in an antenna, represented at one instant of time and far from the antenna, moving in the positive x-direction with speed c. Note that the electric field is perpendicular to the magnetic field, and both are perpendicular to the direction of wave propagation. The variations of E and B with time are sinusoidal.

Physics⊗Now™
Log into PhysicsNow at **www.cp7e.com** and go to Active Figure 21.20, where you can observe the wave and the variations of the fields. In addition, you can take a "snapshot" of the wave at an instant of time and investigate the electric and magnetic fields at that instant.

this case, the oscillations of the electric and magnetic fields take place in planes perpendicular to the x-axis and are therefore perpendicular to the direction of travel of the wave. Because of the latter property, electromagnetic waves are transverse waves. In the figure, the electric field $\vec{E}$ is in the y-direction and the magnetic field $\vec{B}$ is in the z-direction. Light propagates in a direction perpendicular to these two fields. That direction is determined by yet another right-hand rule: (1) point the fingers of your right hand in the direction of $\vec{E}$, (2) curl them in the direction of $\vec{B}$, (3) the right thumb then points in the direction of propagation of the wave.

Electromagnetic waves travel with the speed of light. In fact, it can be shown that the speed of an electromagnetic wave is related to the permeability and permittivity of the medium through which it travels. Maxwell found this relationship for free space to be

$$c = \frac{1}{\sqrt{\mu_0 \epsilon_0}} \qquad\qquad \textbf{[21.24]} \qquad \blacktriangleleft \text{Speed of light.}$$

where c is the speed of light, $\mu_0 = 4\pi \times 10^{-7} \, \text{N} \cdot \text{s}^2/\text{C}^2$ is the permeability constant of vacuum, and $\epsilon_0 = 8.854\ 19 \times 10^{-12} \, \text{C}^2/\text{N} \cdot \text{m}^2$ is the permittivity of free space. Substituting these values into Equation 21.24, we find that

$$c = 2.997\ 92 \times 10^8 \text{ m/s} \qquad\qquad \textbf{[21.25]}$$

The fact that electromagnetic waves travel at the same speed as light in vacuum led scientists to conclude (correctly) that **light is an electromagnetic wave**.

Maxwell also proved the following relationship for electromagnetic waves:

$$\frac{E}{B} = c \qquad\qquad \textbf{[21.26]}$$

TIP 21.2   *E* Stronger Than *B*?

The relationship $E = Bc$ makes it appear that the electric fields associated with light are much larger than the magnetic fields. This is not the case: The units are different, so the quantities can't be directly compared. The two fields contribute equally to the energy of a light wave.

Light is an electromagnetic wave and transports energy and momentum. ▶

Figure 21.21 An apparatus for measuring the radiation pressure of light. In practice, the system is contained in a high vacuum.

Mirror

Black disk

Light

which states that the ratio of the magnitude of the electric field to the magnitude of the magnetic field equals the speed of light.

Electromagnetic waves carry energy as they travel through space, and this energy can be transferred to objects placed in their paths. The average rate at which energy passes through an area perpendicular to the direction of travel of a wave, or the average power per unit area, is called the **intensity *I*** of the wave, and is given by

$$I = \frac{E_{max} B_{max}}{2\mu_0} \qquad [21.27]$$

where E_{max} and B_{max} are the *maximum* values of E and B. The quantity I is analogous to the intensity of sound waves introduced in Chapter 14. From Equation 21.26, we see that $E_{max} = cB_{max} = B_{max}/\sqrt{\mu_0\epsilon_0}$. Equation 21.27 can therefore also be expressed as

$$I = \frac{E_{max}^2}{2\mu_0 c} = \frac{c}{2\mu_0} B_{max}^2 \qquad [21.28]$$

Note that in these expressions we use the *average* power per unit area. A detailed analysis would show that the energy carried by an electromagnetic wave is shared equally by the electric and magnetic fields.

Electromagnetic waves have an average intensity given by Equation 21.28. When the waves strike an area A of an object's surface for a given time Δt, energy $U = IA\Delta t$ is transferred to the surface. Momentum is transferred, as well. Hence, pressure is exerted on a surface when an electromagnetic wave impinges on it. In what follows, we assume that the electromagnetic wave transports a total energy U to a surface in a time Δt. If the surface absorbs all the incident energy U in this time, Maxwell showed that the total momentum $\vec{p}$ delivered to this surface has a magnitude

$$p = \frac{U}{c} \qquad \text{(complete absorption)} \qquad [21.29]$$

If the surface is a perfect reflector, then the momentum transferred in a time Δt for normal incidence is twice that given by Equation 21.29. This is analogous to a molecule of gas bouncing off the wall of a container in a perfectly elastic collision. If the molecule is initially traveling in the positive *x*-direction at velocity v, and after the collision is traveling in the negative *x*-direction at velocity $-v$, then its change in momentum is given by $\Delta p = mv - (-mv) = 2mv$. Light bouncing off a perfect reflector is a similar process, so for complete reflection,

$$p = \frac{2U}{c} \qquad \text{(complete reflection)} \qquad [21.30]$$

Although radiation pressures are very small (about 5×10^{-6} N/m² for direct sunlight), they have been measured with a device such as the one shown in Figure 21.21. Light is allowed to strike a mirror and a black disk that are connected to each other by a horizontal bar suspended from a fine fiber. Light striking the black disk is completely absorbed, so *all* of the momentum of the light is transferred to the disk. Light striking the mirror head-on is totally reflected; hence, the momentum transfer to the mirror is twice that transmitted to the disk. As a result, the horizontal bar supporting the disks twists counterclockwise as seen from above. The bar comes to equilibrium at some angle under the action of the torques caused by radiation pressure and the twisting of the fiber. The radiation pressure can be determined by measuring the angle at which equilibrium occurs. The apparatus must be placed in a high vacuum to eliminate the effects of air currents. It's interesting that similar experiments demonstrate that electromagnetic waves carry angular momentum, as well.

In summary, electromagnetic waves traveling through free space have the following properties:

1. Electromagnetic waves travel at the speed of light.
2. Electromagnetic waves are transverse waves, because the electric and magnetic fields are perpendicular to the direction of propagation of the wave and to each other.
3. The ratio of the electric field to the magnetic field in an electromagnetic wave equals the speed of light.
4. Electromagnetic waves carry both energy and momentum, which can be delivered to a surface.

◀ Some properties of electromagnetic waves

Applying Physics 21.2   Solar System Dust

In the interplanetary space in the Solar System, there is a large amount of dust. Although interplanetary dust can in theory have a variety of sizes—from molecular size upward—why are there very few dust particles smaller than about 0.2 μm in the Solar System? [*Hint:* The Solar System originally contained dust particles of all sizes.]

Explanation Dust particles in the Solar System are subject to two forces: the gravitational force toward the Sun and the force from radiation pressure, which is directed away from the Sun. The gravitational force is proportional to the cube of the radius of a spherical dust particle, because it is proportional to the mass (ρV) of the particle. The radiation pressure is proportional to the square of the radius, because it depends on the cross-sectional area of the particle. For large particles, the gravitational force is larger than the force of radiation pressure, and the weak attraction to the Sun causes such particles to move slowly towards it. For small particles, less than about 0.2 μm, the larger force from radiation pressure sweeps them out of the Solar System.

Quick Quiz 21.4

In an apparatus such as that in Figure 21.21, suppose the black disk is replaced by one with half the radius. Which of the following are different after the disk is replaced? (a) radiation pressure on the disk; (b) radiation force on the disk; (c) radiation momentum delivered to the disk in a given time interval.

EXAMPLE 21.8   A Hot Tin Roof (Solar-Powered Homes)

Goal   Calculate some basic properties of light and relate them to thermal radiation.

Problem   Assume that the Sun delivers an average power per unit area of about 1.00×10^3 W/m^2 to Earth's surface. (a) Calculate the total power incident on a flat tin roof 8.00 m by 20.0 m. Assume that the radiation is incident *normal* (perpendicular) to the roof. (b) Calculate the peak electric field of the light. (c) Compute the peak magnetic field of the light. (d) The tin roof reflects some light, and convection, conduction, and radiation transport the rest of the thermal energy away, until some equilibrium temperature is established. If the roof is a perfect blackbody and rids itself of one-half of the incident radiation through thermal radiation, what's its equilibrium temperature? Assume the ambient temperature is 298 K.

(Example 21.8) A solar home in Oregon.

John Neal/Photo Researchers, Inc.

Solution

(a) Calculate the power delivered to the roof.

Multiply the intensity by the area to get the power:

$$\mathscr{P} = IA = (1.00 \times 10^3 \text{ W/m}^2)(8.00 \text{ m} \times 20.0 \text{ m})$$
$$= 1.60 \times 10^5 \text{ W}$$

(b) Calculate the peak electric field of the light.

Solve Equation 21.28 for E_{max}:

$$I = \frac{E_{max}^2}{2\mu_0 c} \quad \rightarrow \quad E_{max} = \sqrt{2\mu_0 c I}$$

$$E_{max} = \sqrt{2(4\pi \times 10^{-7}\,\text{N}\cdot\text{s}^2/\text{C}^2)(3.00 \times 10^8\,\text{m/s})(1.00 \times 10^3\,\text{W/m}^2)}$$

$$= \boxed{868\,\text{V/m}}$$

(c) Compute the peak magnetic field of the light.

Obtain B_{max} using Equation 21.26:

$$B_{max} = \frac{E_{max}}{c} = \frac{868\,\text{V/m}}{3.00 \times 10^8\,\text{m/s}} = \boxed{2.89 \times 10^{-6}\,\text{T}}$$

(d) Find the equilibrium temperature of the roof.

Substitute into Stefan's law. Only one-half the incident power should be substituted, and twice the area of the roof (both the top and the underside of the roof count).

$$\mathcal{P} = \sigma e A (T^4 - T_0{}^4)$$

$$T^4 = T_0{}^4 + \frac{\mathcal{P}}{\sigma e A}$$

$$= (298\,\text{K})^4 + \frac{(0.500)(1.60 \times 10^5\,\text{W/m}^2)}{(5.67 \times 10^{-8}\,\text{W/m}^2\cdot\text{K}^4)(1)(3.20 \times 10^2\,\text{m}^2)}$$

$$T = 333\,\text{K} = \boxed{60.0°\,\text{C}}$$

Remarks   If the incident power could *all* be converted to electric power, it would be more than enough for the average home. Unfortunately, solar energy isn't easily harnessed, and the prospects for large-scale conversion are not as bright as they may appear from this simple calculation. For example, the conversion efficiency from solar to electrical energy is far less than 100%; 10% is typical for photovoltaic cells. Roof systems for using solar energy to raise the temperature of water with efficiencies of around 50% have been built. Other practical problems must be considered, however, such as overcast days, geographic location, and energy storage.

Exercise 21.8
A spherical satellite orbiting Earth is lighted on one side by the Sun, with intensity 1 340 W/m². (a) If the radius of the satellite is 1.00 m, what power is incident upon it? [*Note:* The satellite effectively intercepts radiation only over a cross section—an area equal to that of a disk, πr^2.) (b) Calculate the peak electric field. (c) Calculate the peak magnetic field.

Answer   (a) 4.21×10^3 W   (b) 1.01×10^3 V/m   (c) 3.35×10^{-6} T

EXAMPLE 21.9   Clipper Ships of Space

Goal   Relate the intensity of light to its mechanical effect on matter.

Problem   Aluminized mylar film is a highly reflective, lightweight material that could be used to make sails for spacecraft driven by the light of the sun. Suppose a sail with area 1.00 km² is orbiting the Sun at a distance of 1.50×10^{11} m. The sail has a mass of 5.00×10^3 kg and is tethered to a payload of mass 2.00×10^4 kg. (a) If the intensity of sunlight is 1.34×10^3 W and the sail is oriented perpendicular to the incident light, what radial force is exerted on the sail? (b) About how long would it take to change the radial speed of the sail by 1.00 km/s? Assume that the sail is perfectly reflecting.

Strategy   Equation 21.30 gives the momentum imparted when light strikes an object and is totally reflected. The change in this momentum with time is a force. For part (b), use Newton's second law to obtain the acceleration. The velocity kinematics equation then yields the necessary time to achieve the desired change in speed.

Solution
(a) Find the force exerted on the sail.

Write Equation 21.30, and substitute $U = \mathcal{P}\,\Delta t = IA\,\Delta t$ for the energy delivered to the sail:

$$\Delta p = \frac{2U}{c} = \frac{2\mathcal{P}\,\Delta t}{c} = \frac{2IA\,\Delta t}{c}$$

Divide both sides by Δt, obtaining the force $\Delta p/\Delta t$ exerted by the light on the sail:

$$F = \frac{\Delta p}{\Delta t} = \frac{2IA}{c} = \frac{2(1340 \text{ W/m}^2)(1.00 \times 10^6 \text{ m}^2)}{3.00 \times 10^8 \text{ m/s}}$$

$$= 8.93 \text{ N}$$

(b) Find the time it takes to change the radial speed by 1.00 km/s.

Substitute the force into Newton's second law and solve for the acceleration of the sail:

$$a = \frac{F}{m} = \frac{8.93 \text{ N}}{2.50 \times 10^4 \text{ kg}} = 3.57 \times 10^{-4} \text{ m/s}^2$$

Apply the kinematics velocity equation:

$$v = at + v_0$$

Solve for t:

$$t = \frac{v - v_0}{a} = \frac{1.00 \times 10^3 \text{ m/s}}{3.57 \times 10^{-4} \text{ m/s}^2} = 2.80 \times 10^6 \text{ s}$$

Remarks   The answer is a little over a month. While the acceleration is very low, there are no fuel costs, and within a few months the velocity can change sufficiently to allow the spacecraft to reach any planet in the solar system. Such spacecraft may be useful for certain purposes and are highly economical, but require a considerable amount of patience.

Exercise 21.9
A laser has a power of 22.0 W and a beam radius of 0.500 mm. (a) Find the intensity of the laser. (b) Suppose you were floating in space and pointed the laser beam away from you. What would your acceleration be? Assume your total mass, including equipment is 72.0 kg and that the force is directed through your center of mass. (*Hint:* The change in momentum is the same as in the nonreflective case.) (c) Compare the acceleration found in part (b) with the acceleration of gravity of a space station of mass 1.00×10^6 kg, if the station's center of mass is 100.0 m away.

Answers   (a) 2.80×10^7 W/m^2 (b) 1.02×10^{-9} m/s^2 (c) 6.67×10^{-9} m/s^2. If you were planning to use your laser welding torch as a thruster to get you back to the station, don't bother—the force of gravity is stronger. Better yet, get somebody to toss you a line.

21.12   THE SPECTRUM OF ELECTROMAGNETIC WAVES

All electromagnetic waves travel in a vacuum with the speed of light, c. These waves transport energy and momentum from some source to a receiver. In 1887, Hertz successfully generated and detected the radio-frequency electromagnetic waves predicted by Maxwell. Maxwell himself had recognized as electromagnetic waves both visible light and the infrared radiation discovered in 1800 by William Herschel. It is now known that other forms of electromagnetic waves exist that are distinguished by their frequencies and wavelengths.

Because all electromagnetic waves travel through free space with a speed c, their frequency f and wavelength λ are related by the important expression

$$c = f\lambda \qquad \textbf{[21.31]}$$

The various types of electromagnetic waves are presented in Figure 21.22 (page 716). Note the wide and overlapping range of frequencies and wavelengths. For instance, an AM radio wave with a frequency of 5.00 MHz (a typical value) has a wavelength of

$$\lambda = \frac{c}{f} = \frac{3.00 \times 10^8 \text{ m/s}}{5.00 \times 10^6 \text{ s}^{-1}} = 60.0 \text{ m}$$

The following abbreviations are often used to designate short wavelengths and distances:

Ron Chapple/Getty Images

Wearing sunglasses lacking ultraviolet (UV) protection is worse for your eyes than wearing no sunglasses at all. Sunglasses without protection absorb some visible light, causing the pupils to dilate. This allows more UV light to enter the eye, increasing the damage to the lens of the eye over time. Without the sunglasses, the pupils constrict, reducing both visible and dangerous UV radiation. Be cool: wear sunglasses with UV protection.

Figure 21.22 The electromagnetic spectrum. Note the overlap between adjacent types of waves. The expanded view to the right shows details of the visible spectrum.

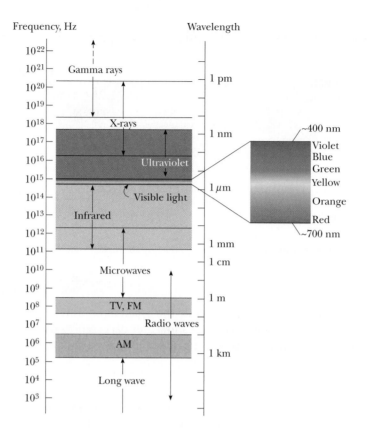

$$1 \text{ micrometer } (\mu m) = 10^{-6} \text{ m}$$

$$1 \text{ nanometer } (nm) = 10^{-9} \text{ m}$$

$$1 \text{ angstrom } (\text{Å}) = 10^{-10} \text{ m}$$

The wavelengths of visible light, for example, range from 0.4 μm to 0.7 μm, or 400 nm to 700 nm, or 4 000 Å to 7 000 Å.

Quick Quiz 21.5

Which of the following statements are true about light waves? (a) The higher the frequency, the longer the wavelength. (b) The lower the frequency, the longer the wavelength. (c) Higher frequency light travels faster than lower frequency light. (d) The shorter the wavelength, the higher the frequency. (e) The lower the frequency, the shorter the wavelength.

Brief descriptions of the wave types follow, in order of decreasing wavelength. There is no sharp division between one kind of electromagnetic wave and the next. All forms of electromagnetic radiation are produced by accelerating charges.

Radio waves, which were discussed in Section 21.10, are the result of charges accelerating through conducting wires. They are, of course, used in radio and television communication systems.

Microwaves (short-wavelength radio waves) have wavelengths ranging between about 1 mm and 30 cm and are generated by electronic devices. Their short wavelengths make them well suited for the radar systems used in aircraft navigation and for the study of atomic and molecular properties of matter. Microwave ovens are an interesting domestic application of these waves. It has been suggested that solar energy might be harnessed by beaming microwaves to Earth from a solar collector in space.

Infrared waves (sometimes incorrectly called "heat waves"), produced by hot objects and molecules, have wavelengths ranging from about 1 mm to the longest wavelength of visible light, 7×10^{-7} m. They are readily absorbed by most materials. The infrared energy absorbed by a substance causes it to get warmer because the energy agitates the atoms of the object, increasing their vibrational or translational

(a)   (b)   (c)   (d)

Figure 21.23 Observations in different parts of the electromagnetic spectrum show different features of the Crab Nebula. (a) X-ray image. (b) Optical image. (c) Polarized visible light. (d) Radio image.

motion. The result is a rise in temperature. Infrared radiation has many practical and scientific applications, including physical therapy, infrared photography, and the study of the vibrations of atoms.

Visible light, the most familiar form of electromagnetic waves, may be defined as the part of the spectrum that is detected by the human eye. Light is produced by the rearrangement of electrons in atoms and molecules. The wavelengths of visible light are classified as colors ranging from violet ($\lambda \approx 4 \times 10^{-7}$ m) to red ($\lambda \approx 7 \times 10^{-7}$ m). The eye's sensitivity is a function of wavelength and is greatest at a wavelength of about 5.6×10^{-7} m (yellow green).

Ultraviolet (UV) light covers wavelengths ranging from about 4×10^{-7} m (400 nm) down to 6×10^{-10} m (0.6 nm). The Sun is an important source of ultraviolet light (which is the main cause of suntans). Most of the ultraviolet light from the Sun is absorbed by atoms in the upper atmosphere, or stratosphere. This is fortunate, because UV light in large quantities has harmful effects on humans. One important constituent of the stratosphere is ozone (O_3), produced from reactions of oxygen with ultraviolet radiation. The resulting ozone shield causes lethal high-energy ultraviolet radiation to warm the stratosphere.

X-rays are electromagnetic waves with wavelengths from about 10^{-8} m (10 nm) down to 10^{-13} m (10^{-4} nm). The most common source of x-rays is the acceleration of high-energy electrons bombarding a metal target. X-rays are used as a diagnostic tool in medicine and as a treatment for certain forms of cancer. Because x-rays easily penetrate and damage or destroy living tissues and organisms, care must be taken to avoid unnecessary exposure and overexposure.

Gamma rays—electromagnetic waves emitted by radioactive nuclei—have wavelengths ranging from about 10^{-10} m to less than 10^{-14} m. They are highly penetrating and cause serious damage when absorbed by living tissues. Accordingly, those working near such radiation must be protected by garments containing heavily absorbing materials, such as layers of lead.

When astronomers observe the same celestial object using detectors sensitive to different regions of the electromagnetic spectrum, striking variations in the object's features can be seen. Figure 21.23 shows images of the Crab Nebula made in four different wavelength ranges. The Crab Nebula is the remnant of a supernova explosion that was seen on the Earth in 1054 A.D. (Compare with Fig. 8.28).

Applying Physics 21.3   The Sun and the Evolution of the Eye

The center of sensitivity of our eyes coincides with the center of the wavelength distribution of the Sun. Is this an amazing coincidence?

Explanation This is not a coincidence; rather it's the result of biological evolution. Humans have evolved with vision most sensitive to wavelengths that are strongest from the Sun. If aliens from another planet ever arrived at Earth, their eyes would have the center of sensitivity at wavelengths different from ours. If their sun were a red dwarf, for example, they'd be most sensitive to red light.

21.13 THE DOPPLER EFFECT FOR ELECTROMAGNETIC WAVES

As we saw in Section 14.6, sound waves exhibit the Doppler effect when the observer, the source, or both are moving relative to the medium of propagation. Recall that in the Doppler effect, the observed frequency of the wave is larger or smaller than the frequency emitted by the source of the wave.

A Doppler effect also occurs for electromagnetic waves, but it differs from the Doppler effect for sound waves in two ways. First, in the Doppler effect for sound waves, motion relative to the medium is most important, because sound waves require a medium in which to propagate. In contrast, the medium of propagation plays no role in the Doppler effect for electromagnetic waves, because the waves require no medium in which to propagate. Second, the speed of sound that appears in the equation for the Doppler effect for sound depends on the reference frame in which it is measured. In contrast, as we shall see in Chapter 26, the speed of electromagnetic waves has the same value in all coordinate systems that are either at rest or moving at constant velocity with respect to one another.

The single equation that describes the Doppler effect for electromagnetic waves is given by the approximate expression

$$f_O \approx f_S \left(1 \pm \frac{u}{c} \right) \qquad \text{if } u << c \qquad \text{[21.32]}$$

where f_O is the observed frequency, f_S is the frequency emitted by the source, c is the speed of light in a vacuum, and u is the *relative* speed of the observer and source. Note that Equation 21.32 is valid only if u is much smaller than c. Further, it can also be used for sound as long as the relative velocity of the source and observer is much less than the velocity of sound. The positive sign in the equation must be used when the source and observer are moving toward one another, while the negative sign must be used when they are moving away from each other. Thus, we anticipate an increase in the observed frequency if the source and observer are approaching each other and a decrease if the source and observer recede from each other.

Astronomers have made important discoveries using Doppler observations on light reaching Earth from distant stars and galaxies. Such measurements have shown that most distant galaxies are moving away from the Earth. Thus, the Universe is expanding. This Doppler shift is called a *red shift* because the observed wavelengths are shifted towards the red portion (longer wavelengths) of the visible spectrum. Further, measurements show that the speed of a galaxy increases with increasing distance from the Earth. More recent Doppler effect measurements made with the Hubble Space Telescope have shown that a galaxy labeled M87 is rotating, with one edge moving toward us and the other moving away. Its measured speed of rotation was used to identify a supermassive black hole located at its center.

SUMMARY

Physics⊗Now™ Take a practice test by logging into Physics-Now at **www.cp7e.com** and clicking on the Pre-Test link for this chapter.

21.1 Resistors in an AC Circuit

If an AC circuit consists of a generator and a resistor, the current in the circuit is in phase with the voltage, which means the current and voltage reach their maximum values at the same time.

In discussions of voltages and currents in AC circuits, **rms values** of voltages are usually used. One reason is that

AC ammeters and voltmeters are designed to read rms values. The rms values of currents and voltage (I_{rms} and ΔV_{rms}), are related to the maximum values of these quantities (I_{max} and ΔV_{max}) as follows:

$$I_{rms} = \frac{I_{max}}{\sqrt{2}} \qquad \text{and} \qquad \Delta V_{rms} = \frac{\Delta V_{max}}{\sqrt{2}} \qquad \text{[21.2, 21.3]}$$

The rms voltage across a resistor is related to the rms current in the resistor by **Ohm's law**:

$$\Delta V_{R,rms} = I_{rms}R \qquad \text{[21.4]}$$

21.2 Capacitors in an AC Circuit

If an AC circuit consists of a generator and a capacitor, the voltage lags behind the current by 90°. This means that the voltage reaches its maximum value one-quarter of a period after the current reaches its maximum value.

The impeding effect of a capacitor on current in an AC circuit is given by the **capacitive reactance** X_C, defined as

$$X_C \equiv \frac{1}{2\pi f C} \qquad \textbf{[21.5]}$$

where f is the frequency of the AC voltage source.

The rms voltage across and the rms current in a capacitor are related by

$$\Delta V_{C,\mathrm{rms}} = I_{\mathrm{rms}} X_C \qquad \textbf{[21.6]}$$

21.3 Inductors in an AC Circuit

If an AC circuit consists of a generator and an inductor, the voltage leads the current by 90°. This means the voltage reaches its maximum value one-quarter of a period before the current reaches its maximum value.

The effective impedance of a coil in an AC circuit is measured by a quantity called the **inductive reactance** X_L, defined as

$$X_L \equiv 2\pi f L \qquad \textbf{[21.8]}$$

The rms voltage across a coil is related to the rms current in the coil by

$$\Delta V_{L,\mathrm{rms}} = I_{\mathrm{rms}} X_L \qquad \textbf{[21.9]}$$

21.4 The *RLC* Series Circuit

In an *RLC* series AC circuit, the maximum applied voltage ΔV is related to the maximum voltages across the resistor (ΔV_R), capacitor (ΔV_C), and inductor (ΔV_L) by

$$\Delta V_{\mathrm{max}} = \sqrt{\Delta V_R^2 + (\Delta V_L - \Delta V_C)^2} \qquad \textbf{[21.10]}$$

If an AC circuit contains a resistor, an inductor, and a capacitor connected in series, the limit they place on the current is given by the **impedance Z** of the circuit, defined as

$$Z \equiv \sqrt{R^2 + (X_L - X_C)^2} \qquad \textbf{[21.13]}$$

The relationship between the maximum voltage supplied to an *RLC* series AC circuit and the maximum current in the circuit, which is the same in every element, is

$$\Delta V_{\mathrm{max}} = I_{\mathrm{max}} Z \qquad \textbf{[21.14]}$$

In an *RLC* series AC circuit, the applied rms voltage and current are out of phase. The **phase angle** ϕ between the current and voltage is given by

$$\tan \phi = \frac{X_L - X_C}{R} \qquad \textbf{[21.15]}$$

21.5 Power in an AC Circuit

The **average power** delivered by the voltage source in an *RLC* series AC circuit is

$$\mathscr{P}_{\mathrm{av}} = I_{\mathrm{rms}} \Delta V_{\mathrm{rms}} \cos \phi \qquad \textbf{[21.17]}$$

where the constant $\cos \phi$ is called the **power factor**.

21.6 Resonance in a Series *RLC* Circuit

In general, the rms current in a series *RLC* circuit can be written

$$I_{\mathrm{rms}} = \frac{\Delta V_{\mathrm{rms}}}{Z} = \frac{\Delta V_{\mathrm{rms}}}{\sqrt{R^2 + (X_L - X_C)^2}} \qquad \textbf{[21.18]}$$

The current has its *maximum* value when the impedance has its *minimum* value, corresponding to $X_L = X_C$ and $Z = R$. The frequency f_0 at which this happens is called the **resonance frequency** of the circuit, given by

$$f_0 = \frac{1}{2\pi\sqrt{LC}} \qquad \textbf{[21.19]}$$

21.7 The Transformer

If the primary winding of a transformer has N_1 turns and the secondary winding consists of N_2 turns, then if an input AC voltage ΔV_1 is applied to the primary, the induced voltage in the secondary winding is given by

$$\Delta V_2 = \frac{N_2}{N_1} \Delta V_1 \qquad \textbf{[21.22]}$$

When N_2 is greater than N_1, ΔV_2 exceeds ΔV_1 and the transformer is referred to as a *step-up transformer*. When N_2 is less than N_1, making ΔV_2 less than ΔV_1, we have a *step-down transformer*. In an ideal transformer, the power output equals the power input.

21.8–21.13 Electromagnetic Waves and their Properties

Electromagnetic waves were predicted by James Clerk Maxwell and experimentally confirmed by Heinrich Hertz. These waves are created by accelerating electric charges, and have the following properties:

1. Electromagnetic waves are transverse waves, because the electric and magnetic fields are perpendicular to the direction of propagation of the waves.
2. Electromagnetic waves travel at the speed of light.
3. The ratio of the electric field to the magnetic field at a given point in an electromagnetic wave equals the speed of light:

$$\frac{E}{B} = c \qquad \textbf{[21.26]}$$

4. Electromagnetic waves carry energy as they travel through space. The average power per unit area is the intensity I, given by

$$I = \frac{E_{\mathrm{max}} B_{\mathrm{max}}}{2\mu_0} = \frac{E_{\mathrm{max}}^2}{2\mu_0 c} = \frac{c}{2\mu_0} B_{\mathrm{max}}^2 \qquad \textbf{[21.27, 21.28]}$$

where E_{max} and B_{max} are the maximum values of the electric and magnetic fields.

5. Electromagnetic waves transport linear and angular momentum as well as energy. The momentum p delivered in time Δt at normal incidence to an object that completely absorbs light energy U is given by

$$p = \frac{U}{c} \qquad \text{(complete absorption)} \qquad \textbf{[21.29]}$$

If the surface is a perfect reflector, then the momentum delivered in time Δt at normal incidence is twice that given by Equation 21.29:

$$p = \frac{2U}{c} \qquad \text{(complete reflection)} \qquad \textbf{[21.30]}$$

6. The speed c, frequency f, and wavelength λ of an electromagnetic wave are related by

$$c = f\lambda \qquad \textbf{[21.31]}$$

The **electromagnetic spectrum** includes waves covering a broad range of frequencies and wavelengths. These waves have a variety of applications and characteristics, depend-ing on their frequencies or wavelengths. The frequency of a given wave can be shifted by the relative velocity of observer and source, with the observed frequency f_O given by

$$f_O \approx f_S \left(1 \pm \frac{u}{c}\right) \qquad \text{if } u << c \qquad \textbf{[21.32]}$$

where f_S is the frequency of the source, c is the speed of light in a vacuum, and u is the *relative* speed of the observer and source. The positive sign is used when the source and observer approach each other, the negative sign when they recede from each other.

CONCEPTUAL QUESTIONS

1. Before the advent of cable television and satellite dishes, homeowners either mounted a television antenna on the roof or used "rabbit ears" atop their sets. (See Fig. Q21.1.) Certain orientations of the receiving antenna on a television set gave better reception than others. Furthermore, the best orientation varied from station to station. Explain.

Figure Q21.1

2. What is the impedance of an *RLC* circuit at the resonance frequency?

3. When a DC voltage is applied to a transformer, the primary coil sometimes will overheat and burn. Why?

4. Why are the primary and secondary coils of a transformer wrapped on an iron core that passes through both coils?

5. Receiving radio antennas can be in the form of conducting lines or loops. What should the orientation of each of these antennas be relative to a broadcasting antenna that is vertical?

6. If the fundamental source of a sound wave is a vibrating object, what is the fundamental source of an electromagnetic wave?

7. In radio transmission, a radio wave serves as a carrier wave, and the sound signal is superimposed on the carrier wave. In amplitude modulation (AM) radio, the amplitude of the carrier wave varies according to the sound wave. The Navy sometimes uses flashing lights to send Morse code between neighboring ships, a process that has similarities to radio broadcasting. Is this process AM or FM? What is the carrier frequency? What is the signal frequency? What is the broadcasting antenna? What is the receiving antenna?

8. When light (or other electromagnetic radiation) travels across a given region, what is it that oscillates? What is it that is transported?

9. In space sailing, which is a proposed alternative for transport to the planets, a spacecraft carries a very large sail. Sunlight striking the sail exerts a force, accelerating the spacecraft. Should the sail be absorptive or reflective to be most effective?

10. How can the average value of an alternating current be zero, yet the square root of the average squared value not be zero?

11. Suppose a creature from another planet had eyes that were sensitive to infrared radiation. Describe what it would see if it looked around the room that you are now in. That is, what would be bright and what would be dim?

12. Why should an infrared photograph of a person look different from a photograph taken using visible light?

13. Radio stations often advertise "instant news." If what they mean is that you hear the news at the instant they speak it, is their claim true? About how long would it take for a message to travel across the United States by radio waves, assuming that the waves could travel that great distance and still be detected?

14. Would an inductor and a capacitor used together in an AC circuit dissipate any energy?

15. Does a wire connected to a battery emit an electromagnetic wave?

16. If a high-frequency current is passed through a solenoid containing a metallic core, the core becomes warm due to induction. Explain why the temperature of the material rises in this situation.

17. If the resistance in an *RLC* circuit remains the same, but the capacitance and inductance are each doubled, how will the resonance frequency change?

18. Why is the sum of the maximum voltages across each of the elements in a series *RLC* circuit usually greater than the maximum applied voltage? Doesn't this violate Kirchhoff's loop rule?

19. What is the advantage of transmitting power at high voltages?

20. What determines the maximum voltage that can be used on a transmission line?

21. Will a transformer operate if a battery is used for the input voltage across the primary? Explain.

PROBLEMS

1, 2, 3 = straightforward, intermediate, challenging   ☐ = full solution available in *Student Solutions Manual/Study Guide*

Physics⊗Now™ = coached problem with hints available at **www.cp7e.com**   🔬 = biomedical application

Section 21.1 Resistors in an AC Circuit

1. An rms voltage of 100 V is applied to a purely resistive load of 5.00 Ω. Find (a) the maximum voltage applied, (b) the rms current supplied, (c) the maximum current supplied, and (d) the power dissipated.

2. (a) What is the resistance of a lightbulb that uses an average power of 75.0 W when connected to a 60-Hz power source with an peak voltage of 170 V? (b) What is the resistance of a 100-W bulb?

3. An AC power supply that produces a maximum voltage of $\Delta V_{max} = 100$ V is connected to a 24.0-Ω resistor. The current and the resistor voltage are respectively measured with an ideal AC ammeter and an ideal AC voltmeter, as shown in Figure P21.3. What does each meter read? Note that an ideal ammeter has zero resistance and an ideal voltmeter has infinite resistance.

$$\Delta V_{max} = 100 \text{ V}$$

$$R = 24 \ \Omega$$

Figure P21.3

4. Figure P21.4 shows three lamps connected to a 120-V AC (rms) household supply voltage. Lamps 1 and 2 have 150-W bulbs; lamp 3 has a 100-W bulb. Find the rms current and the resistance of each bulb.

Figure P21.4

5. An audio amplifier, represented by the AC source and the resistor R in Figure P21.5, delivers alternating voltages at audio frequencies to the speaker. If the source puts out an alternating voltage of 15.0 V (rms), the resistance R is 8.20 Ω, and the speaker is equivalent to a resistance of 10.4 Ω, what is the time-averaged power delivered to the speaker?

Figure P21.5

6. An AC voltage source has an output voltage given by $\Delta v = (150 \text{ V}) \sin 377t$. Find (a) the rms voltage output, (b) the frequency of the source, and (c) the voltage at $t = 1/120$ s. (d) Find the maximum current in the circuit when the voltage source is connected to a 50.0-Ω resistor.

Section 21.2 Capacitors in an AC Circuit

7. Show that the SI unit of capacitive reactance X_c is the ohm.

8. What is the maximum current delivered to a circuit containing a 2.20-μF capacitor when it is connected across (a) a North American outlet having $\Delta V_{rms} = 120$ V and $f = 60.0$ Hz and (b) a European outlet having $\Delta V_{rms} = 240$ V and $f = 50.0$ Hz?

9. When a 4.0-μF capacitor is connected to a generator whose rms output is 30 V, the current in the circuit is observed to be 0.30 A. What is the frequency of the source?

10. What maximum current is delivered by an AC generator with a maximum voltage of $\Delta V_{max} = 48.0$ V and a frequency $f = 90.0$ Hz when it is connected across a 3.70-μF capacitor?

11. Physics⊗Now™ What must be the capacitance of a capacitor inserted in a 60-Hz circuit in series with a generator of 170 V maximum output voltage to produce an rms current output of 0.75 A?

12. The generator in a purely capacitive AC circuit has an angular frequency of 120π rad/s. If $\Delta V_{max} = 140$ V and $C = 6.00$ μF, what is the rms current in the circuit?

Section 21.3 Inductors in an AC Circuit

13. Show that the inductive reactance X_L has SI units of ohms.

14. The generator in a purely inductive AC circuit has an angular frequency of 120π rad/s. If $V_{max} = 140$ V and $L = 0.100$ H, what is the rms current in the circuit?

15. An inductor has a 54.0-Ω reactance at 60.0 Hz. What will be the *maximum* current if this inductor is connected to a 50.0-Hz source that produces a 100-V rms voltage?

16. An inductor is connected to a 20.0-Hz power supply that produces a 50.0-V rms voltage. What inductance is needed to keep the maximum current in the circuit below 80.0 mA?

17. Determine the maximum magnetic flux through an inductor connected to a standard outlet ($\Delta V_{rms} = 120$ V, $f = 60.0$ Hz).

Section 21.4 The *RLC* Series Circuit

18. An inductor ($L = 400$ mH), a capacitor ($C = 4.43$ μF), and a resistor ($R = 500$ Ω) are connected in series. A 50.0-Hz AC generator connected in series to these elements produces a maximum current of 250 mA in the circuit. (a) Calculate the required maximum voltage ΔV_{max}. (b) Determine the phase angle by which the current leads or lags the applied voltage.

19. A 40.0-μF capacitor is connected to a 50.0-Ω resistor and a generator whose rms output is 30.0 V at 60.0 Hz. Find

(a) the rms current in the circuit, (b) the rms voltage drop across the resistor, (c) the rms voltage drop across the capacitor, and (d) the phase angle for the circuit.

20. A 50.0-Ω resistor, a 0.100-H inductor, and a 10.0-μF capacitor are connected in series to a 60.0-Hz source. The rms current in the circuit is 2.75 A. Find the rms voltages across (a) the resistor, (b) the inductor, (c) the capacitor, and (d) the *RLC* combination. (e) Sketch the phasor diagram for this circuit.

21. A resistor (R = 900 Ω), a capacitor (C = 0.25 μF), and an inductor (L = 2.5 H) are connected in series across a 240-Hz AC source for which ΔV_{max} = 140 V. Calculate (a) the impedance of the circuit, (b) the maximum current delivered by the source, and (c) the phase angle between the current and voltage. (d) Is the current leading or lagging the voltage?

22. An AC source operating at 60 Hz with a maximum voltage of 170 V is connected in series with a resistor (R = 1.2 kΩ) and a capacitor (C = 2.5 μF). (a) What is the maximum value of the current in the circuit? (b) What are the maximum values of the potential difference across the resistor and the capacitor? (c) When the current is zero, what are the magnitudes of the potential difference across the resistor, the capacitor, and the AC source? How much charge is on the capacitor at this instant? (d) When the current is at a maximum, what are the magnitudes of the potential differences across the resistor, the capacitor, and the AC source? How much charge is on the capacitor at this instant?

23. A 60.0-Ω resistor, a 3.00-μF capacitor, and a 0.400-H inductor are connected in series to a 90.0-V (rms), 60.0-Hz source. Find (a) the voltage drop across the *LC* combination and (b) the voltage drop across the *RC* combination.

24. An AC source operating at 60 Hz with a maximum voltage of 170 V is connected in series with a resistor (R = 1.2 kΩ) and an inductor (L = 2.8 H). (a) What is the maximum value of the current in the circuit? (b) What are the maximum values of the potential difference across the resistor and the inductor? (c) When the current is at a maximum, what are the magnitudes of the potential differences across the resistor, the inductor, and the AC source? (d) When the current is zero, what are the magnitudes of the potential difference across the resistor, the inductor, and the AC source?

25. A person is working near the secondary of a transformer, as shown in Figure P21.25. The primary voltage is 120 V (rms) at 60.0 Hz. The capacitance C_s, which is the stray capacitance between the hand and the secondary winding, is 20.0 pF. Assuming that the person has a body resistance to ground of R_b = 50.0 kΩ, determine the rms voltage across the body. (*Hint:* Redraw the circuit with the secondary of the transformer as a simple AC source.)

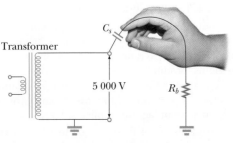

Figure P21.25

26. A coil of resistance 35.0 Ω and inductance 20.5 H is in series with a capacitor and a 200-V (rms), 100-Hz source. The rms current in the circuit is 4.00 A. (a) Calculate the capacitance in the circuit. (b) What is ΔV_{rms} across the coil?

27. **Physics⊗Now™** An AC source with a maximum voltage of 150 V and f = 50.0 Hz is connected between points a and d in Figure P21.27. Calculate the rms voltages between points (a) a and b, (b) b and c, (c) c and d, and (d) b and d.

Figure P21.27

Section 21.5 Power in an AC Circuit

28. A 50.0-Ω resistor is connected to a 30.0-μF capacitor and to a 60.0-Hz, 100-V (rms) source. (a) Find the power factor and the average power delivered to the circuit. (b) Repeat part (a) when the capacitor is replaced with a 0.300-H inductor.

29. A multimeter in an *RL* circuit records an rms current of 0.500 A and a 60.0-Hz rms generator voltage of 104 V. A wattmeter shows that the average power delivered to the resistor is 10.0 W. Determine (a) the impedance in the circuit, (b) the resistance R, and (c) the inductance L.

30. An AC voltage with an amplitude of 100 V is applied to a series combination of a 200-μF capacitor, a 100-mH inductor, and a 20.0-Ω resistor. Calculate the power dissipation and the power factor for frequencies of (a) 60.0 Hz and (b) 50.0 Hz.

31. An inductor and a resistor are connected in series. When connected to a 60-Hz, 90-V (rms) source, the voltage drop across the resistor is found to be 50 V (rms) and the power delivered to the circuit is 14 W. Find (a) the value of the resistance and (b) the value of the inductance.

32. Consider a series *RLC* circuit with R = 25 Ω, L = 6.0 mH, and C = 25 μF. The circuit is connected to a 10-V (rms), 600-Hz AC source. (a) Is the sum of the voltage drops across R, L, and C equal to 10 V (rms)? (b) Which is greatest, the power delivered to the resistor, to the capacitor, or to the inductor? (c) Find the average power delivered to the circuit.

Section 21.6 Resonance in a Series *RLC* circuit

33. An *RLC* circuit is used to tune a radio to an FM station broadcasting at 88.9 MHz. The resistance in the circuit is 12.0 Ω and the capacitance is 1.40 pF. What inductance should be present in the circuit?

34. Consider a series *RLC* circuit with R = 15 Ω, L = 200 mH, C = 75 μF, and a maximum voltage of 150 V. (a) What is the impedance of the circuit at resonance? (b) What is the resonance frequency of the circuit? (c) When will the current be greatest—at resonance, at ten percent below the resonant frequency, or at ten percent above the resonant frequency? (d) What is the rms current in the circuit at a frequency of 60 Hz?

35. The AM band extends from approximately 500 kHz to 1 600 kHz. If a 2.0-μH inductor is used in a tuning circuit for a radio, what are the extremes that a capacitor must reach in order to cover the complete band of frequencies?

36. A series circuit contains a 3.00-H inductor, a 3.00-μF capacitor, and a 30.0-Ω resistor connected to a 120-V (rms) source of variable frequency. Find the power delivered to the circuit when the frequency of the source is (a) the resonance frequency, (b) one-half the resonance frequency, (c) one-fourth the resonance frequency, (d) two times the resonance frequency, and (e) four times the resonance frequency. From your calculations, can you draw a conclusion about the frequency at which the maximum power is delivered to the circuit?

37. A 10.0-Ω resistor, a 10.0-mH inductor, and a 100-μF capacitor are connected in series to a 50.0-V (rms) source having variable frequency. Find the energy delivered to the circuit during one period if the operating frequency is twice the resonance frequency.

Section 21.7 The Transformer

38. An AC adapter for a telephone-answering unit uses a transformer to reduce the line voltage of 120 V (rms) to a voltage of 9.0 V. The rms current delivered to the answering system is 400 mA. (a) If the primary (input) coil in the transformer in the adapter has 240 turns, how many turns are there on the secondary (output) coil? (b) What is the rms power delivered to the transformer? Assume an ideal transformer.

39. An AC power generator produces 50 A (rms) at 3 600 V. The voltage is stepped up to 100 000 V by an ideal transformer, and the energy is transmitted through a long-distance power line that has a resistance of 100 Ω. What percentage of the power delivered by the generator is dissipated as heat in the power line?

40. A transformer is to be used to provide power for a computer disk drive that needs 6.0 V (rms) instead of the 120 V (rms) from the wall outlet. The number of turns in the primary is 400, and it delivers 500 mA (the secondary current) at an output voltage of 6.0 V (rms). (a) Should the transformer have more turns in the secondary compared with the primary, or fewer turns? (b) Find the current in the primary. (c) Find the number of turns in the secondary.

41. A transformer on a pole near a factory steps the voltage down from 3 600 V (rms) to 120 V (rms). The transformer is to deliver 1 000 kW to the factory at 90% efficiency. Find (a) the power delivered to the primary, (b) the current in the primary, and (c) the current in the secondary.

42. A transmission line that has a resistance per unit length of 4.50×10^{-4} Ω/m is to be used to transmit 5.00 MW over 400 miles (6.44×10^5 m). The output voltage of the generator is 4.50 kV (rms). (a) What is the line loss if a transformer is used to step up the voltage to 500 kV (rms)? (b) What fraction of the input power is lost to the line under these circumstances? (c) What difficulties would be encountered on attempting to transmit the 5.00 MW at the generator voltage of 4.50 kV (rms)?

Section 21.10 Production of Electromagnetic Waves by an Antenna

Section 21.11 Properties of Electromagnetic Waves

43. The U.S. Navy has long proposed the construction of extremely low frequency (ELF waves) communications

systems; such waves could penetrate the oceans to reach distant submarines. Calculate the length of a quarter-wavelength antenna for a transmitter generating ELF waves of frequency 75 Hz. How practical is this antenna?

44. Experimenters at the National Institute of Standards and Technology have made precise measurements of the speed of light using the fact that, in vacuum, the speed of electromagnetic waves is $c = 1/\sqrt{\mu_0\epsilon_0}$, where the constants $\mu_0 = 4\pi \times 10^{-7}$ N·s^2/C^2 and $\epsilon_0 = 8.854 \times 10^{-12}$ C^2/N·m^2. What value (to four significant figures) does this formula give for the speed of light in vacuum?

45. Oxygenated hemoglobin absorbs weakly in the red (hence its red color) and strongly in the near infrared, while deoxygenated hemoglobin has the opposite absorption. This fact is used in a "pulse oximeter" to measure oxygen saturation in arterial blood. The device clips onto the end of a person's finger and has two light-emitting diodes [a red (660 nm) and an infrared (940 nm)] and a photocell that detects the amount of light transmitted through the finger at each wavelength. (a) Determine the frequency of each of these light sources. (b) If 67% of the energy of the red source is absorbed in the blood, by what factor does the amplitude of the electromagnetic wave change? [Hint: The intensity of the wave is equal to the average power per unit area as given by Equation 21.28.]

46. **Operation of the pulse oximeter (see previous problem).** The transmission of light energy as it passes through a solution of light-absorbing molecules is described by the Beer–Lambert law

$$I = I_0 10^{-\epsilon CL} \quad \text{or} \quad \log_{10}\left(\frac{I}{I_0}\right) = -\epsilon CL$$

which gives the decrease in intensity I in terms of the distance L the light has traveled through a fluid with a concentration C of the light-absorbing molecule. The quantity ϵ is called the extinction coefficient, and its value depends on the frequency of the light. (It has units of m^2/mol.) Assume that the extinction coefficient for 660-nm light passing through a solution of oxygenated hemoglobin is identical to the coefficient for 940-nm light passing through deoxygenated hemoglobin. Assume also that 940-nm light has zero absorption ($\epsilon = 0$) in oxygenated hemoglobin and 660-nm light has zero absorption in deoxygenated hemoglobin. If 33% of the energy of the red source and 76% of the infrared energy is transmitted through the blood, determine the fraction of hemoglobin that is oxygenated.

47. A microwave oven is powered by an electron tube called a magnetron that generates electromagnetic waves of frequency 2.45 GHz. The microwaves enter the oven and are reflected by the walls. The standing-wave pattern produced in the oven can cook food unevenly, with hot spots in the food at antinodes and cool spots at nodes, so a turntable is often used to rotate the food and distribute the energy. If a microwave oven is used with a cooking dish in a fixed position, the antinodes can appear as burn marks on foods such as carrot strips or cheese. The separation distance between the burns is measured to be 6.00 cm. Calculate the speed of the microwaves from these data.

48. Assume that the solar radiation incident on Earth is 1 340 W/m^2 (at the top of Earth's atmosphere). Calculate

the total power radiated by the Sun, taking the average separation between Earth and the Sun to be 1.49×10^{11} m.

49. Physics⊗Now™ The Sun delivers an average power of $1\,340$ W/m^2 to the top of Earth's atmosphere. Find the magnitudes of $\vec{E}_{max}$ and $\vec{B}_{max}$ for the electromagnetic waves at the top of the atmosphere.

Section 21.12 The Spectrum of Electromagnetic Waves

50. A diathermy machine, used in physiotherapy, generates electromagnetic radiation that gives the effect of "deep heat" when absorbed in tissue. One assigned frequency for diathermy is 27.33 MHz. What is the wavelength of this radiation?

51. What are the wavelength ranges in (a) the AM radio band (540–1 600 kHz) and (b) the FM radio band (88–108 MHz)?

52. An important news announcement is transmitted by radio waves to people who are 100 km away, sitting next to their radios, and by sound waves to people sitting across the newsroom, 3.0 m from the newscaster. Who receives the news first? Explain. Take the speed of sound in air to be 343 m/s.

53. Infrared spectra are used by chemists to help identify an unknown substance. Atoms in a molecule that are bound together by a particular bond vibrate at a predictable frequency, and light at that frequency is absorbed strongly by the atom. In the case of the C=O double bond, for example, the oxygen atom is bound to the carbon by a bond that has an effective spring constant of 2 800 N/m. If we assume that the carbon atom remains stationary (it is attached to other atoms in the molecule), determine the resonant frequency of this bond and the wavelength of light that matches that frequency. Verify that this wavelength lies in the infrared region of the spectrum. (The mass of an oxygen atom is 2.66×10^{-26} kg.)

21.13 The Doppler Effect for Electromagnetic Waves

54. A spaceship is approaching a space station at a speed of 1.8×10^5 m/s. The space station has a beacon that emits green light with a frequency of 6.0×10^{14} Hz. What is the frequency of the beacon observed on the spaceship? What is the change in frequency? (Carry five digits in these calculations.)

55. While driving at a constant speed of 80 km/h, you are passed by a car traveling at 120 km/h. If the frequency of light emitted by the taillights of the car that passes you is 4.3×10^{14} Hz, what frequency will you observe? What is the change in frequency?

56. A speeder tries to explain to the police that the yellow warning lights on the side of the road looked green to her because of the Doppler shift. How fast would she have been traveling if yellow light of wavelength 580 nm had been shifted to green with a wavelength of 560 nm? (Note that, for speeds less than 0.03c, Equation 21.32 will lead to a value for the change of frequency accurate to approximately two significant digits.)

ADDITIONAL PROBLEMS

57. As a way of determining the inductance of a coil used in a research project, a student first connects the coil to a 12.0-V battery and measures a current of 0.630 A. The student

then connects the coil to a 24.0-V (rms), 60.0-Hz generator and measures an rms current of 0.570 A. What is the inductance?

58. The intensity of solar radiation at the top of Earth's atmosphere is $1\,340$ W/m^3. Assuming that 60% of the incoming solar energy reaches Earth's surface, and assuming that you absorb 50% of the incident energy, make an order-of-magnitude estimate of the amount of solar energy you absorb in a 60-minute sunbath.

59. A 200-Ω resistor is connected in series with a 5.0-μF capacitor and a 60-Hz, 120-V rms line. If electrical energy costs \$0.080/kWh, how much does it cost to leave this circuit connected for 24 h?

60. A series *RLC* circuit has a resonance frequency of $2\,000/\pi$ Hz. When it is operating at a frequency of $\omega > \omega_0$, $X_L = 12\ \Omega$ and $X_C = 8.0\ \Omega$. Calculate the values of *L* and *C* for the circuit.

61. Physics⊗Now™ Two connections allow contact with two circuit elements in series inside a box, but it is not known whether the circuit elements are *R*, *L*, or *C*. In an attempt to find what is inside the box, you make some measurements, with the following results: when a 3.0-V DC power supply is connected across the terminals, a maximum direct current of 300 mA is measured in the circuit after a suitably long time. When a 60-Hz source with maximum voltage of 3.0 V is connected instead, the maximum current is measured as 200 mA. (a) What are the two elements in the box? (b) What are their values of *R*, *L*, or *C*?

62. (a) What capacitance will resonate with a one-turn loop of inductance 400 pH to give a radar wave of wavelength 3.0 cm? (b) If the capacitor has square parallel plates separated by 1.0 mm of air, what should the edge length of the plates be? (c) What is the common reactance of the loop and capacitor at resonance?

63. A dish antenna with a diameter of 20.0 m receives (at normal incidence) a radio signal from a distant source, as shown in Figure P21.63. The radio signal is a continuous sinusoidal wave with amplitude $E_{max} = 0.20\ \mu$V/m. Assume that the antenna absorbs all the radiation that falls on the dish. (a) What is the amplitude of the magnetic field in this wave? (b) What is the intensity of the radiation received by the antenna? (c) What is the power received by the antenna?

Figure P21.63

64. A particular inductor has appreciable resistance. When the inductor is connected to a 12-V battery, the current in the inductor is 3.0 A. When it is connected to an AC source with an rms output of 12 V and a frequency of 60 Hz, the current drops to 2.0 A. What are (a) the impedance at 60 Hz and (b) the inductance of the inductor?

65. One possible means of achieving space flight is to place a perfectly reflecting aluminized sheet into Earth's orbit and to use the light from the Sun to push this solar sail. Suppose such a sail, of area $6.00 \times 10^4 \text{ m}^2$ and mass 6 000 kg, is placed in orbit facing the Sun. (a) What force is exerted on the sail? (b) What is the sail's acceleration? (c) How long does it take for this sail to reach the Moon, 3.84×10^8 m away? Ignore all gravitational effects, and assume a solar intensity of 1 340 W/m². [*Hint:* The radiation pressure by a reflected wave is given by 2(average power per unit area)/c.]

66. Suppose you wish to use a transformer as an impedance-matching device between an audio amplifier that has an output impedance of 8.0 kΩ and a speaker that has an input impedance of 8.0 Ω. What should be the ratio of primary to secondary turns on the transformer?

67. Compute the average energy content of a liter of sunlight as it reaches the top of Earth's atmosphere, where its intensity is 1 340 W/m².

68. In an *RLC* series circuit that includes a source of alternating current operating at fixed frequency and voltage, the resistance R is equal to the inductive reactance. If the plate separation of the capacitor is reduced to one-half of its original value, the current in the circuit doubles. Find the initial capacitive reactance in terms of R.

ACTIVITIES

1. For this observation, you will need some items that can be found at many electronics stores. You will need a bicolored light-emitting diode (LED), a resistor of about 100 Ω, 2 m of flexible wire, and a step-down transformer with an output of 3 to 6 V. Use the wire to connect the LED and the resistor in series with the transformer. A bicolored LED is designed such that it emits a red color when the current in the LED is in one direction and green when the current reverses. When connected to an AC source, the LED is yellow. Why?

Hold the wires and whirl the LED in a circular path. In a darkened room, you will see red and green bars at equally spaced intervals along the path of the LED. Why?

As you continue to whirl the LED in a circular path, have your partner count the number of green bars in the circle, then measure the time it takes for the LED to travel ten times around the circular path. Based on this information, determine the time it takes for the color of the LED to change from green to red to green. You should obtain an answer of (1/60) s. Why?

2. Rotate a portable radio (with a telescoping antenna) about a horizontal axis while it is tuned to a weak station. Such an antenna detects the varying electric field produced by the station. What can you determine about the direction of the electric field produced by the transmitter?

Now turn on your radio to a nearby station and experiment with shielding the radio from incoming waves. Is the reception affected by surrounding the radio by aluminum foil? By plastic wrap? Use any other material you have available. What kinds of material block the signal? Why?

Light is bent (refracted) as it passes through water, with different wavelengths bending by different amounts (which is called dispersion). Together with reflection, these physical phenomena lead to the creation of a rainbow when light passes through small, suspended droplets of water.

Peter Bowater/Photo Researchers, Inc.

Reflection and Refraction of Light

Light has a dual nature. In some experiments it acts like a particle, while in others it acts like a wave. In this part of the book, we concentrate on the aspects of light that are best understood through the wave model. First we discuss the reflection of light at the boundary between two media and the refraction (bending) of light as it travels from one medium into another. We use these ideas to study the refraction of light as it passes through lenses and the reflection of light from mirrored surfaces. Finally, we describe how lenses and mirrors can be used to view objects with telescopes and microscopes and how lenses are used in photography. The ability to manipulate light has greatly enhanced our capacity to investigate and understand the nature of the universe.

22.1 THE NATURE OF LIGHT

Until the beginning of the 19th century, light was modeled as a stream of particles emitted by a source that stimulated the sense of sight on entering the eye. The chief architect of the particle theory of light was Newton. With this theory, he provided simple explanations of some known experimental facts concerning the nature of light—namely, the laws of reflection and refraction.

Most scientists accepted Newton's particle theory of light. During Newton's lifetime, however, another theory was proposed. In 1678, the Dutch physicist and astronomer Christian Huygens (1629–1695) showed that a wave theory of light could also explain the laws of reflection and refraction.

The wave theory didn't receive immediate acceptance, for several reasons. First, all the waves known at the time (sound, water, and so on) traveled through some sort of medium, but light from the Sun could travel to Earth through empty space.

Further, it was argued that if light were some form of wave, it would bend around obstacles; hence, we should be able to see around corners. It is now known that light does indeed bend around the edges of objects. This phenomenon, known as *diffraction*, is difficult to observe because light waves have such short wavelengths. Even though experimental evidence for the diffraction of light was discovered by Francesco Grimaldi (1618–1663) around 1660, for more than a century most scientists rejected the wave theory and adhered to Newton's particle theory, probably due to Newton's great reputation as a scientist.

The first clear demonstration of the wave nature of light was provided in 1801 by Thomas Young (1773–1829), who showed that under appropriate conditions, light exhibits interference behavior. Light waves emitted by a single source and traveling along two different paths can arrive at some point and combine and cancel each other by destructive interference. Such behavior couldn't be explained at that time by a particle model, because scientists couldn't imagine how two or more particles could come together and cancel each other.

The most important development in the theory of light was the work of Maxwell, who predicted in 1865 that light was a form of high-frequency electromagnetic wave (Chapter 21). His theory also predicted that these waves should have a speed of 3×10^8 m/s, in agreement with the measured value.

Although the classical theory of electricity and magnetism explained most known properties of light, some subsequent experiments couldn't be explained by the assumption that light was a wave. The most striking of these was the *photoelectric effect* (which we will examine more closely in Chapter 27), discovered by Hertz. Hertz found that clean metal surfaces emit charges when exposed to ultraviolet light.

In 1905, Einstein published a paper that formulated the theory of light quanta ("particles") and explained the photoelectric effect. He reached the conclusion that light was composed of corpuscles, or discontinuous quanta of energy. These corpuscles or quanta are now called *photons* to emphasize their particlelike nature. According to Einstein's theory, the energy of a photon is proportional to the frequency of the electromagnetic wave associated with it, or:

$$E = hf \qquad\qquad \textbf{[22.1]}$$

◀ Energy of a photon

where $h = 6.63 \times 10^{-34}$ J·s is *Planck's constant*. This theory retains some features of both the wave and particle theories of light. As we will discuss later, the photoelectric effect is the result of energy transfer from a single photon to an electron in the metal. This means the electron interacts with one photon of light as if the electron had been struck by a particle. Yet the photon has wavelike characteristics, as implied by the fact that a frequency is used in its definition.

In view of these developments, light must be regarded as having a *dual nature*: **In some experiments light acts as a wave and in others it acts as a particle.** Classical electromagnetic wave theory provides adequate explanations of light propagation and of the effects of interference, whereas the photoelectric effect and other experiments involving the interaction of light with matter are best explained by assuming that light is a particle.

So in the final analysis, is light a wave or a particle? The answer is neither and both: light has a number of physical properties, some associated with waves and others with particles.

22.2 REFLECTION AND REFRACTION

When light traveling in one medium encounters a boundary leading into a second medium, the processes of reflection and refraction can occur. In **reflection**, part of the light encountering the second medium bounces off that medium. In **refraction**, the light passing into the second medium bends through an angle with respect to the normal to the boundary. Often, both processes occur at the same time, with part of the light being reflected and part refracted. To study reflection and refraction we need a way of thinking about beams of light, and this is given by the ray approximation.

CHRISTIAN HUYGENS
(1629–1695), Dutch Physicist and Astronomer

Huygens is best known for his contributions to the fields of optics and dynamics. To Huygens, light was a vibratory motion in the ether, spreading out and producing the sensation of light when impinging on the eye. On the basis of this theory, he deduced the laws of reflection and refraction and explained the phenomenon of double refraction.

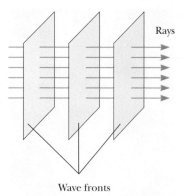

Figure 22.1 A plane wave traveling to the right. Note that the rays, corresponding to the direction of wave motion, are straight lines perpendicular to the wave fronts.

The Ray Approximation in Geometric Optics

An important property of light that can be understood based on common experience is the following: **light travels in a straight-line path in a homogeneous medium, until it encounters a boundary between two different materials**. When light strikes a boundary, it is either reflected from that boundary, passes into the material on the other side of the boundary, or partially does both.

The preceding observation leads us to use what is called the **ray approximation** to represent beams of light. As shown in Figure 22.1, a ray of light is an imaginary line drawn along the direction of travel of the light beam. For example, a beam of sunlight passing through a darkened room traces out the path of a light ray. We will also make use of the concept of wave fronts of light. A **wave front** is a surface passing through the points of a wave that have the same phase and amplitude. For instance, the wave fronts in Figure 22.1 could be surfaces passing through the crests of waves. The rays, corresponding to the direction of wave motion, are straight lines perpendicular to the wave fronts. When light rays travel in parallel paths, the wave fronts are planes perpendicular to the rays.

Reflection of Light

When a light ray traveling in a transparent medium encounters a boundary leading into a second medium, part of the incident ray is reflected back into the first medium. Figure 22.2a shows several rays of a beam of light incident on a smooth, mirrorlike reflecting surface. The reflected rays are parallel to each other, as indicated in the figure. The reflection of light from such a smooth surface is called **specular reflection**. On the other hand, if the reflecting surface is rough, as in Figure 22.2b, the surface reflects the rays in a variety of directions. Reflection from any rough surface is known as **diffuse reflection**. A surface behaves as a smooth surface as long as its variations are small compared with the wavelength of the incident light. Figures 22.2c and 22.2d are photographs of specular and diffuse reflection of laser light, respectively.

As an example, consider the two types of reflection from a road surface that a driver might observe while driving at night. When the road is dry, light from oncoming vehicles is scattered off the road in different directions (diffuse reflection) and the road is clearly visible. On a rainy night when the road is wet,

APPLICATION

Seeing the Road on a Rainy Night

Figure 22.2 A schematic representation of (a) specular reflection, where the reflected rays are all parallel to each other, and (b) diffuse reflection, where the reflected rays travel in random directions. (c, d) Photographs of specular and diffuse reflection, made with laser light.

Photographs Courtesy of Henry Leap and Jim Lehman

road's irregularities are filled with water. Because the wet surface is smooth, the light undergoes specular reflection. This means that the light is reflected straight ahead, and the driver of a car sees only what is directly in front of him. Light from the side never reaches his eye. In this book we concern ourselves only with specular reflection, and we use the term *reflection* to mean specular reflection.

Quick Quiz 22.1

Which part of Figure 22.3, (a) or (b), better shows specular reflection of light from the roadway?

(a)   (b)

Figure 22.3 (Quick Quiz 22.1)

Consider a light ray traveling in air and incident at some angle on a flat, smooth surface, as in Active Figure 22.4. The incident and reflected rays make angles θ_1 and θ_1', respectively, **with a line perpendicular to the surface** at the point where the incident ray strikes the surface. We call this line the *normal* to the surface. Experiments show that **the angle of reflection equals the angle of incidence**;

$$\theta_1' = \theta_1 \qquad \text{[22.2]}$$

You may have noticed a common occurrence in photographs of individuals: their eyes appear to be glowing red. This occurs when a photographic flash device is used and the flash unit is close to the camera lens. Light from the flash unit enters the eye and is reflected back along its original path from the retina. This type of reflection back along the original direction is called *retroreflection*. If the flash unit and lens are close together, retroreflected light can enter the lens. Most of the light reflected from the retina is red, due to the blood vessels at the back of the eye, giving the red-eye effect in the photograph.

ACTIVE FIGURE 22.4
According to the law of reflection, $\theta_1 = \theta_1'$.

Physics⊗Now™

Log into PhysicsNow at **www.cp7e.com** and go to Active Figure 22.4, where you can vary the incident angle and see the effect on the reflected ray.

APPLICATION

Red Eyes in Flash Photographs

Applying Physics 22.1   The Colors of Water Ripples at Sunset

An observer on the west-facing beach of a large lake is watching the beginning of a sunset. The water is very smooth except for some areas with small ripples. The observer notices that some areas of the water are blue and some are pink. Why does the water appear to be different colors in different areas?

Explanation The different colors arise from specular and diffuse reflection. The smooth areas of the water

will specularly reflect the light from the west, which is the pink light from the sunset. The areas with small ripples will reflect the light diffusely, so light from all parts of the sky will be reflected into the observer's eyes. Because most of the sky is still blue at the beginning of the sunset, these areas will appear to be blue.

Applying Physics 22.2 Double Images

When looking outdoors through a glass window at night, why do you sometimes see a double image of yourself?

Explanation Reflection occurs whenever there is an interface between two different media. For the glass in the window, there are two such surfaces. The first is the inner surface of the glass, and the second is the outer surface. Each of these interfaces results in an image.

INTERACTIVE EXAMPLE 22.1   The Double-Reflecting Light Ray

Goal Calculate a resultant angle from two reflections.

Problem Two mirrors make an angle of 120° with each other, as in Figure 22.5. A ray is incident on mirror M_1 at an angle of 65° to the normal. Find the angle the ray makes with the normal to M_2 after it is reflected from both mirrors.

Figure 22.5 (Example 22.1) Mirrors M_1 and M_2 make an angle of 120° with each other.

Strategy Apply the law of reflection twice. Given the incident ray at angle θ_{inc}, find the final resultant angle, β_{ref}.

Solution

Apply the law of reflection to M_1 to find the angle of reflection, θ_{ref}:	$\theta_{ref} = \theta_{inc} = 65°$
Find the angle ϕ that is the complement of the angle θ_{ref}:	$\phi = 90° - \theta_{ref} = 90° - 65° = 25°$
Find the unknown angle α in the triangle of M_1, M_2, and the ray traveling from M_1 to M_2, using the fact that the three angles sum to 180°:	$180° = 25° + 120° + \alpha \quad \rightarrow \quad \alpha = 35°$
The angle α is complementary to the angle of incidence, β_{inc}, for M_2:	$\alpha + \beta_{inc} = 90° \quad \rightarrow \quad \beta_{inc} = 90° - 35° = 55°$
Apply the law of reflection a second time, obtaining β_{ref}:	$\beta_{ref} = \beta_{inc} = \boxed{55°}$

Remarks Notice the heavy reliance on elementary geometry and trigonometry in these reflection problems.

Exercise 22.1

Repeat the problem if the angle of incidence is 55° and the second mirror makes an angle of 100° with the first mirror.

Answer 45°

Physics⊗Now™ Investigate reflection for various mirror angles by logging into PhysicsNow at **www.cp7e.com** and going to Interactive Example 22.1.

Refraction of Light

When a ray of light traveling through a transparent medium encounters a boundary leading into another transparent medium, as in Active Figure 22.6a, part of the ray is reflected and part enters the second medium. The ray that enters the second medium is bent at the boundary and is said to be *refracted*. The incident ray, the reflected ray, the refracted ray, and the normal at the point of incidence all lie

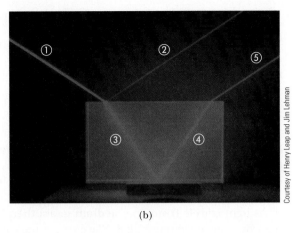

(a)   (b)

Courtesy of Henry Leap and Jim Lehman

ACTIVE FIGURE 22.6
(a) A ray obliquely incident on an
air–glass interface. The refracted ray
is bent toward the normal because
$v_2 < v_1$. (b) Light incident on the
Lucite block bends both when it
enters the block and when it leaves
the block.

Physics⊗Now™
Log into PhysicsNow at **www.cp7e.com**
and go to Active Figure 22.6, where
you can vary the incident angle and
see the effect on the reflected and
refracted rays.

in the same plane. The **angle of refraction**, θ_2, in Active Figure 22.6a depends on
the properties of the two media and on the angle of incidence, through the
relationship

$$\frac{\sin \theta_2}{\sin \theta_1} = \frac{v_2}{v_1} = \text{constant} \qquad [22.3]$$

where v_1 is the speed of light in medium 1 and v_2 is the speed of light in medium 2.
Note that the angle of refraction is also measured with respect to the normal. In
Section 22.7 we will derive the laws of reflection and refraction using Huygens'
principle.

Experiment shows that **the path of a light ray through a refracting surface is
reversible**. For example, the ray in Active Figure 22.6a travels from point A to
point B. If the ray originated at B, it would follow the same path to reach point A,
but the reflected ray would be in the glass.

Quick Quiz 22.2

If beam 1 is the incoming beam in Active Figure 22.6b, which of the other four
beams are due to reflection? Which are due to refraction?

When light moves from a material in which its speed is high to a material in
which its speed is lower, the angle of refraction θ_2 is less than the angle of incidence. The refracted ray therefore bends toward the normal, as shown in Active
Figure 22.7a. If the ray moves from a material in which it travels slowly to a material in which it travels more rapidly, θ_2 is greater than θ_1, so the ray bends away
from the normal, as shown in Active Figure 22.7b.

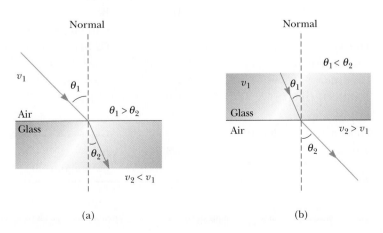

(a)   (b)

ACTIVE FIGURE 22.7
(a) When the light beam moves from
air into glass, its path is bent toward
the normal. (b) When the beam
moves from glass into air, its path is
bent away from the normal.

Physics⊗Now™
Log into PhysicsNow at **www.cp7e.com**
and go to Active Figure 22.7. Light
passes through three layers of material.
You can vary the incident angle and
see the effect on the refracted rays for
a variety of values of the index of refraction (Table 22.1, page 732) of the
three materials.

22.3 THE LAW OF REFRACTION

When light passes from one transparent medium to another, it's refracted because the speed of light is different in the two media[1]. The **index of refraction**, n, of a medium is defined as the ratio c/v;

◀ Index of refraction

$$n \equiv \frac{\text{speed of light in vacuum}}{\text{speed of light in a medium}} = \frac{c}{v} \qquad [22.4]$$

From this definition, we see that the index of refraction is a dimensionless number that is greater than or equal to one because v is always less than c. Further, n is equal to one for vacuum. Table 22.1 lists the indices of refraction for various substances.

As light travels from one medium to another, its frequency doesn't change. To see why, consider Figure 22.8. Wave fronts pass an observer at point A in medium 1 with a certain frequency and are incident on the boundary between medium 1 and medium 2. The frequency at which the wave fronts pass an observer at point B in medium 2 must equal the frequency at which they arrive at point A. If this were not the case, the wave fronts would either pile up at the boundary or be destroyed or created at the boundary. Because neither of these events occurs, the frequency must remain the same as a light ray passes from one medium into another.

Therefore, because the relation $v = f\lambda$ must be valid in both media, and because $f_1 = f_2 = f$, we see that

$$v_1 = f\lambda_1 \qquad \text{and} \qquad v_2 = f\lambda_2$$

Because $v_1 \neq v_2$, it follows that $\lambda_1 \neq \lambda_2$. A relationship between the index of refraction and the wavelength can be obtained by dividing these two equations and making use of the definition of the index of refraction given by Equation 22.4:

$$\frac{\lambda_1}{\lambda_2} = \frac{v_1}{v_2} = \frac{c/n_1}{c/n_2} = \frac{n_2}{n_1} \qquad [22.5]$$

which gives

$$\lambda_1 n_1 = \lambda_2 n_2 \qquad [22.6]$$

TIP 22.1 An Inverse Relationship

The index of refraction is *inversely* proportional to the wave speed. Therefore, as the wave speed v decreases, the index of refraction n *increases*.

TIP 22.2 The Frequency Remains the Same

The *frequency* of a wave does *not* change as the wave passes from one medium to another. Both the wave speed and the wavelength *do* change, but the frequency remains the same.

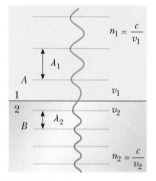

Figure 22.8 As the wave moves from medium 1 to medium 2, its wavelength changes, but its frequency remains constant.

TABLE 22.1

Indices of Refraction for Various Substances, Measured with Light of Vacuum Wavelength $\lambda_0 = 589$ mn

Substance	Index of Refraction	Substance	Index of Refraction
Solids at 20°C		**Liquids at 20°C**	
Diamond (C)	2.419	Benzene	1.501
Fluorite (CaF_2)	1.434	Carbon disulfide	1.628
Fused quartz (SiO_2)	1.458	Carbon tetrachloride	1.461
Glass, crown	1.52	Ethyl alcohol	1.361
Glass, flint	1.66	Glycerine	1.473
Ice (H_2O) (at 0°C)	1.309	Water	1.333
Polystyrene	1.49		
Sodium chloride (NaCl)	1.544	**Gases at 0°C, 1 atm**	
Zircon	1.923	Air	1.000 293
		Carbon dioxide	1.000 45

[1]The speed of light varies between media because the time lags caused by the absorption and reemission of light as it travels from atom to atom depend on the particular electronic structure of the atoms constituting each material.

Let medium 1 be the vacuum, so that $n_1 = 1$. It follows from Equation 22.6 that the index of refraction of any medium can be expressed as the ratio

$$n = \frac{\lambda_0}{\lambda_n}$$ [22.7]

where λ_0 is the wavelength of light in vacuum and λ_n is the wavelength in the medium having index of refraction n. Figure 22.9 is a schematic representation of this reduction in wavelength when light passes from a vacuum into a transparent medium.

Figure 22.9 A schematic diagram of the *reduction* in wavelength when light travels from a medium with a low index of refraction to one with a higher index of refraction.

We are now in a position to express Equation 22.3 in an alternate form. If we substitute Equation 22.5 into Equation 22.3, we get

$$n_1 \sin \theta_1 = n_2 \sin \theta_2$$ [22.8]

◀ Snell's law of refraction

The experimental discovery of this relationship is usually credited to Willebord Snell (1591–1627) and is therefore known as **Snell's law of refraction**.

Quick Quiz 22.3

A material has an index of refraction that increases continuously from top to bottom. Of the three paths shown in Figure 22.10, which path will a light ray follow as it passes through the material?

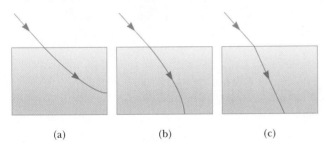

(a)   (b)   (c)   **Figure 22.10**   (Quick Quiz 22.3)

Quick Quiz 22.4

As light travels from a vacuum ($n = 1$) to a medium such as glass ($n > 1$), which of the following properties remains the same? (a) wavelength, (b) wave speed, or (c) frequency?

EXAMPLE 22.2   Angle of Refraction for Glass

Goal   Apply Snell's law to a slab of glass.

Problem   A light ray of wavelength 589 nm (produced by a sodium lamp) traveling through air is incident on a smooth, flat slab of crown glass at an angle of 30.0° to the normal, as sketched in Figure 22.11. Find the angle of refraction, θ_2.

Strategy   Substitute quantities into Snell's law and solve for the unknown angle of refraction, θ_2.

Figure 22.11   (Example 22.2) Refraction of light by glass.

Solution
Solve Snell's law (Eq. 22.8) for $\sin \theta_2$:

(1) $\sin \theta_2 = \dfrac{n_1}{n_2} \sin \theta_1$

From Table 22.1, find $n_1 = 1.00$ for air and $n_2 = 1.52$ for crown glass. Substitute these values into (1) and take the inverse sine of both sides:

$\sin \theta_2 = \left(\dfrac{1.00}{1.52}\right)(\sin 30.0°) = 0.329$

$\theta_2 = \sin^{-1}(0.329) = \boxed{19.2°}$

Remarks   Notice the light ray bends toward the normal when it enters a material of a higher index of refraction. If the ray left the material following the same path in reverse, it would bend away from the normal.

Exercise 22.2

If the light ray moves from inside the glass toward the glass–air interface at an angle of 30.0° to the normal, determine the angle of refraction.

Answer   The ray bends 49.5° *away* from the normal, as expected.

EXAMPLE 22.3   Light in Fused Quartz

Goal   Use the index of refraction to determine the effect of a medium on light's speed and wavelength.

Problem   Light of wavelength 589 nm in vacuum passes through a piece of fused quartz of index of refraction $n = 1.458$. **(a)** Find the speed of light in fused quartz. **(b)** What is the wavelength of this light in fused quartz? **(c)** What is the frequency of the light in fused quartz?

Strategy   Substitute values into Equations 22.4 and 22.7.

Solution
(a) Find the speed of light in fused quartz.

Obtain the speed from Equation 22.4:

$$v = \frac{c}{n} = \frac{3.00 \times 10^8 \text{ m/s}}{1.458} = \boxed{2.06 \times 10^8 \text{ m/s}}$$

(b) What is the wavelength of this light in fused quartz?

Use Eq. 22.7 to calculate the wavelength:

$$\lambda_n = \frac{\lambda_0}{n} = \frac{589 \text{ nm}}{1.458} = \boxed{404 \text{ nm}}$$

(c) What is the frequency of the light in fused quartz?

The frequency in quartz is the same as in vacuum. Solve $c = f\lambda$ for the frequency:

$$f = \frac{c}{\lambda} = \frac{3.00 \times 10^8 \text{ m/s}}{589 \times 10^{-9} \text{ m}} = \boxed{5.09 \times 10^{14} \text{ Hz}}$$

Remarks   It's interesting to note that the speed of light in vacuum, 3.00×10^8 m/s, is an upper limit for the speed of material objects. In our treatment of relativity in Chapter 26, we will find that this upper limit is consistent with experimental observations. However, it's possible for a particle moving in a medium to have a speed that exceeds the speed of light in that medium. For example, it's theoretically possible for a particle to travel through fused quartz at a speed greater than 2.06×10^8 m/s, but it must still have a speed less than 3.00×10^8 m/s.

Exercise 22.3

Light with wavelength 589 nm passes through crystalline sodium chloride. Find **(a)** the speed of light in this medium, **(b)** the wavelength, and **(c)** the frequency of the light.

Answer   **(a)** 1.94×10^8 m/s   **(b)** 381 nm   **(c)** 5.09×10^{14} Hz

---

INTERACTIVE EXAMPLE 22.4   Light Passing through a Slab

Goal   Apply Snell's law when a ray passes into and out of another medium.

Problem   A light beam traveling through a transparent medium of index of refraction n_1 passes through a thick transparent slab with parallel faces and index of refraction n_2 (Fig. 22.12). Show that the emerging beam is parallel to the incident beam.

Figure 22.12   (Example 22.4) When light passes through a flat slab of material, the emerging beam is parallel to the incident beam, and therefore $\theta_1 = \theta_3$.

Strategy   Apply Snell's law twice, once at the upper surface and once at the lower surface. The two equations will be related because the angle of refraction at the upper surface equals the angle of incidence at the lower surface. The ray passing through the slab makes equal angles with the normal at the entry and exit points. This procedure will enable us to compare angles θ_1 and θ_3.

Solution

Apply Snell's law to the upper surface:

$$(1) \quad \sin\theta_2 = \frac{n_1}{n_2}\sin\theta_1$$

Apply Snell's law to the lower surface:

$$(2) \quad \sin\theta_3 = \frac{n_2}{n_1}\sin\theta_2$$

Substitute Equation 1 into Equation 2:

$$\sin\theta_3 = \frac{n_2}{n_1}\left(\frac{n_1}{n_2}\sin\theta_1\right) = \sin\theta_1$$

Take the inverse sine of both sides, noting that the angles are positive and less than 90°:

$$\theta_3 = \theta_1$$

Remarks   The preceding result proves that the slab doesn't alter the direction of the beam. It does, however, produce a lateral displacement of the beam, as shown in Figure 22.12.

Exercise 22.4

Suppose the ray, in air with $n = 1.00$, enters a slab with $n = 2.50$ at a 45.0° angle with respect to the normal, then exits the bottom of the slab into water, with $n = 1.33$. At what angle to the normal does the ray leave the slab?

Answer   32.1°

Physics⊗Now™   Explore refraction through slabs of various thickness by logging into PhysicsNow at **www.cp7e.com** and going to Interactive Example 22.4.

EXAMPLE 22.5   Refraction of Laser Light in a Digital Video Disk (DVD)

Goal   Apply Snell's law together with geometric constraints.

Problem   A DVD is a video recording consisting of a spiral track about 1.0 μm wide with digital information. (See Fig. 22.13a.) The digital information consists of a series of pits that are "read" by a laser beam sharply focused on a track in the information layer. If the width a of the beam at the information layer must equal 1.0 μm to distinguish individual tracks, and the width w of the beam as it enters the plastic is 0.7000 mm, find the angle θ_1 at which the conical beam should enter the plastic. (See Fig. 22.13b.) Assume the plastic has a thickness $t = 1.20$ mm and an index of refraction $n = 1.55$. Note that this system is relatively immune to small dust particles degrading the video

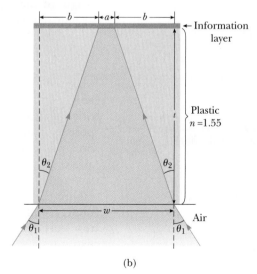

(a)   (b)

Figure 22.13   (Example 22.5)A micrograph of a DVD surface showing tracks and pits along each track. (b) Cross section of a cone-shaped laser beam used to read a DVD.

quality, because particles would have to be as large as 0.700 mm to obscure the beam at the point where it enters the plastic.

Strategy   Use right-triangle trigonometry to determine the angle θ_2, and then apply Snell's law to obtain the angle θ_1.

Solution

From the top and bottom of Figure 22.13b, obtain an equation relating w, b, and a:

$$w = 2b + a$$

Solve this equation for b and substitute given values:

$$b = \frac{w - a}{2} = \frac{700.0 \times 10^{-6}\,\text{m} - 1.0 \times 10^{-6}\,\text{m}}{2} = 349.5\,\mu\text{m}$$

Now use the tangent function to find θ_2:

$$\tan\theta_2 = \frac{b}{t} = \frac{349.5\,\mu\text{m}}{1.20 \times 10^3\,\mu\text{m}} \quad \rightarrow \quad \theta_2 = 16.2°$$

Finally, use Snell's law to find θ_1:

$$n_1 \sin\theta_1 = n_2 \sin\theta_2$$

$$\sin\theta_1 = \frac{n_2 \sin\theta_2}{n_1} = \frac{1.55 \sin 16.2°}{1.00} = 0.433$$

$$\theta_1 = \sin^{-1}(0.433) = \boxed{25.7°}$$

Remarks   Despite its apparent complexity, the problem isn't that different from Example 22.2.

Exercise 22.5
Suppose you wish to redesign the system to decrease the initial width of the beam from 0.700 0 mm to 0.600 0 mm, but leave the incident angle θ_1 and all other parameters the same as before, except the index of refraction for the plastic material (n_2) and the angle θ_2. What index of refraction should the plastic have?

Answer   1.79

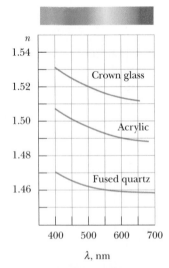

Figure 22.14   Variations of index of refraction in the visible spectrum with respect to vacuum wavelength for three materials.

22.4   DISPERSION AND PRISMS

In Table 22.1, we presented values for the index of refraction of various materials. If we make careful measurements, however, we find that the index of refraction in anything but vacuum depends on the wavelength of light. The dependence of the index of refraction on wavelength is called **dispersion**. Figure 22.14 is a graphical representation of this variation in the index of refraction with wavelength. Because n is a function of wavelength, Snell's law indicates that **the angle of refraction made when light enters a material depends on the wavelength of the light**. As seen in the figure, the index of refraction for a material usually decreases with increasing wavelength. This means that violet light ($\lambda \cong$ 400 nm) refracts more than red light ($\lambda \cong$ 650 nm) when passing from air into a material.

To understand the effects of dispersion on light, consider what happens when light strikes a prism, as in Figure 22.15a. A ray of light of a single wavelength that is incident on the prism from the left emerges bent away from its original direction of travel by an angle δ, called the **angle of deviation**. Now suppose a beam of white light (a combination of all visible wavelengths) is incident on a prism. Because of dispersion, the different colors refract through different angles of deviation, and the rays that emerge from the second face of the prism spread out in a series of colors known as a visible **spectrum**, as shown in Figure 22.16. These colors, in order of decreasing wavelength, are red, orange, yellow, green, blue, and violet. Violet light deviates the most, red light the least, and the remaining colors in the visible spectrum fall between these extremes.

Prisms are often used in an instrument known as a **prism spectrometer**, the essential elements of which are shown in Figure 22.17a (page 738). This instrument

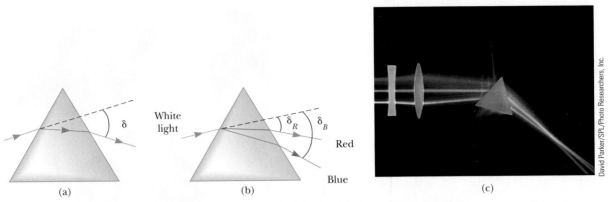

Figure 22.15 (a) A prism refracts a light ray and deviates the light through the angle δ. (b) When light is incident on a prism, the blue light is bent more than the red. (c) Light of different colors passes through a prism and two lenses. Note that as the light passes through the prism, different wavelengths are refracted at different angles.

is commonly used to study the wavelengths emitted by a light source, such as a sodium vapor lamp. Light from the source is sent through a narrow, adjustable slit and lens to produce a parallel, or collimated, beam. The light then passes through the prism and is dispersed into a spectrum. The refracted light is observed through a telescope. The experimenter sees different colored images of the slit through the eyepiece of the telescope. The telescope can be moved or the prism can be rotated in order to view the various wavelengths, which have different angles of deviation. Figure 22.17b (page 738) shows one type of prism spectrometer used in undergraduate laboratories.

All hot, low-pressure gases emit their own characteristic spectra. Thus, one use of a prism spectrometer is to identify gases. For example, sodium emits only two wavelengths in the visible spectrum: two closely spaced yellow lines. (The bright linelike images of the slit seen in a spectroscope are called *spectral lines*.) A gas emitting these, and only these, colors can thus be identified as sodium. Likewise, mercury vapor has its own characteristic spectrum, consisting of four prominent wavelengths—orange, green, blue, and violet lines—along with some wavelengths of lower intensity. The particular wavelengths emitted by a gas serve as "fingerprints" of that gas. Spectral analysis, which is the measurement of the wavelengths emitted or absorbed by a substance, is a powerful general tool in many scientific areas. As examples, chemists and biologists use infrared spectroscopy to identify molecules, astronomers use visible-light spectroscopy to identify elements on distant stars, and geologists use spectral analysis to identify minerals.

APPLICATION

Identifying Gases with
a Spectrometer

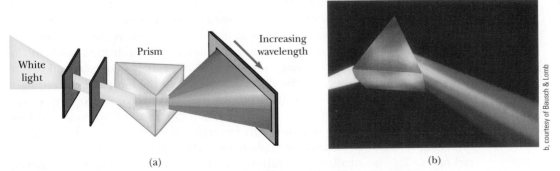

Figure 22.16 (a) Dispersion of white light by a prism. Since *n* varies with wavelength, the prism disperses the white light into its various spectral components. (b) Different colors of light that pass through a prism are refracted at different angles because the index of refraction of the glass depends on wavelength. Violet light bends the most, red light the least.

Slit

Source

Telescope

(a)

(b)

Figure 22.17   (a) A diagram of a prism spectrometer. The colors in the spectrum are viewed through a telescope. (b) A prism spectrometer with interchangeable components.

Applying Physics 22.3   Dispersion

When a beam of light enters a glass prism, which has nonparallel sides, the rainbow of color exiting the prism is a testimonial to the dispersion occurring in the glass. Suppose a beam of light enters a slab of material with parallel sides. When the beam exits the other side, traveling in the same direction as the original beam, is there any evidence of dispersion?

Explanation Due to dispersion, light at the violet end of the spectrum exhibits a larger angle of refraction on entering the glass than light at the red end. All colors of light return to their original direction of propagation as they refract back out into the air. As a result, the outgoing beam is white. But the net shift in the position of the violet light along the edge of the slab is larger than the shift of the red light, so one edge of the outgoing beam has a bluish tinge to it (it appears blue rather than violet, because the eye is not very sensitive to violet light), whereas the other edge has a reddish tinge. This effect is indicated in

Figure 22.18. The colored edges of the outgoing beam of white light are evidence of dispersion.

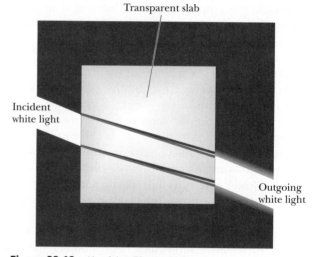

Transparent slab

Incident
white light

Outgoing
white light

Figure 22.18   (Applying Physics 22.3)

22.5   THE RAINBOW

The dispersion of light into a spectrum is demonstrated most vividly in nature through the formation of a rainbow, often seen by an observer positioned between the Sun and a rain shower. To understand how a rainbow is formed, consider Active Figure 22.19. A ray of light passing overhead strikes a drop of water in the atmosphere and is refracted and reflected as follows: It is first refracted at the front surface of the drop, with the violet light deviating the most and the red light the least. At the back surface of the drop, the light is reflected and returns to the front surface, where it again undergoes refraction as it moves from water into air. The rays leave the drop so that the angle between the incident white light and the returning violet ray is 40° and the angle between the white light and the returning red ray is 42°. This small angular difference between the returning rays causes us to see the bow as explained in the next paragraph.

Now consider an observer viewing a rainbow, as in Figure 22.20a. If a raindrop high in the sky is being observed, the red light returning from the drop can reach the observer because it is deviated the most, but the violet light passes over the observer because it is deviated the least. Hence, the observer sees this drop as being red. Similarly, a drop lower in the sky would direct violet light toward the observer and appear to be violet. (The red light from this drop would strike the ground and not be seen.) The remaining colors of the spectrum would reach the observer from raindrops lying between these two extreme positions. Figure 22.20b shows a beautiful rainbow and a secondary rainbow with its colors reversed.

22.6 HUYGENS' PRINCIPLE

The laws of reflection and refraction can be deduced using a geometric method proposed by Huygens in 1678. Huygens assumed that light is a form of wave motion rather than a stream of particles. He had no knowledge of the nature of light or of its electromagnetic character. Nevertheless, his simplified wave model is adequate for understanding many practical aspects of the propagation of light.

Huygens' principle is a geometric construction for determining at some instant the position of a new wave front from knowledge of the wave front that preceded it. (A wave front is a surface passing through those points of a wave which have the same phase and amplitude. For instance, a wave front could be a surface passing

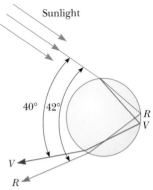

ACTIVE FIGURE 22.19
Refraction of sunlight by a spherical raindrop.

Physics⊗Now™
Log into PhysicsNow at **www.cp7e.com** and go to Active Figure 22.19, where you can vary the point at which the sunlight enters the raindrop and verify that the angles shown are maximum angles.

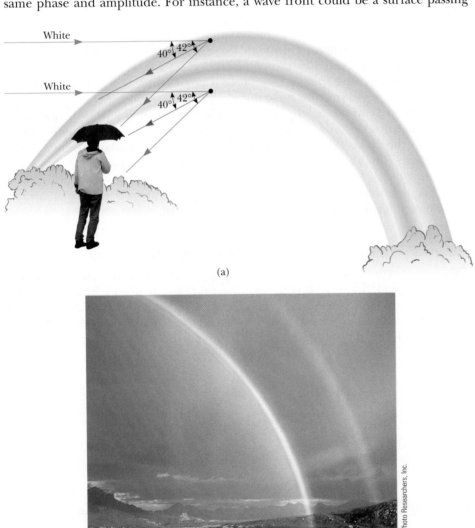

(a)

(b)

Figure 22.20   (a) The formation of a rainbow. (b) This photograph of a rainbow shows a distinct secondary rainbow with the colors reversed.

Figure 22.21 Huygens' constructions for (a) a plane wave propagating to the right and (b) a spherical wave.

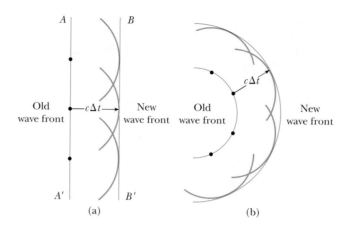

Old wave front

New wave front

Old wave front

New wave front

(a)   (b)

Huygens' principle ▶

through the crests of waves.) In Huygens' construction, **all points on a given wave front are taken as point sources for the production of spherical secondary waves, called wavelets, which propagate in the forward direction with speeds characteristic of waves in that medium. After some time has elapsed, the new position of the wave front is the surface tangent to the wavelets**.

Figure 22.21 illustrates two simple examples of Huygens' construction. First, consider a plane wave moving through free space, as in Figure 22.21a. At $t = 0$, the wave front is indicated by the plane labeled AA'. In Huygens' construction, each point on this wave front is considered a point source. For clarity, only a few points on AA' are shown. With these points as sources for the wavelets, we draw circles of radius $c\Delta t$, where c is the speed of light in vacuum and Δt is the period of propagation from one wave front to the next. The surface drawn tangent to these wavelets is the plane BB', which is parallel to AA'. In a similar manner, Figure 22.21b shows Huygens' construction for an outgoing spherical wave.

Figure 22.22 shows a convincing demonstration of Huygens' principle. Plane waves coming from far off shore emerge from the openings between the barriers as two-dimensional circular waves propagating outward.

Huygens' Principle Applied to Reflection and Refraction

The laws of reflection and refraction were stated earlier in the chapter without proof. We now derive these laws using Huygens' principle. Figure 22.23a illustrates the law of reflection. The line AA' represents a wave front of the incident light. As ray 3 travels from A' to C, ray 1 reflects from A and produces a spherical wavelet of radius AD. (Recall that the radius of a Huygens wavelet is $v\Delta t$.) Because the two

Figure 22.22 This photograph of the beach at Tel Aviv, Israel, shows Huygens wavelets radiating from each opening between breakwalls. Note how the beach has been shaped by the wave action.

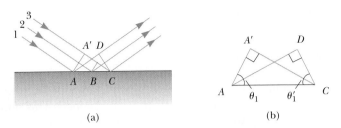

Figure 22.23 (a) Huygens' construction for proving the law of reflection. (b) Triangle ADC is congruent to triangle $AA'C$.

wavelets having radii $A'C$ and AD are in the same medium, they have the same speed v, so $AD = A'C$. Meanwhile, the spherical wavelet centered at B has spread only half as far as the one centered at A, because ray 2 strikes the surface later than ray 1.

From Huygens' principle, we find that the reflected wave front is CD, a line tangent to all the outgoing spherical wavelets. The remainder of our analysis depends on geometry, as summarized in Figure 22.23b. Note that the right triangles ADC and $AA'C$ are congruent because they have the same hypotenuse, AC, and because $AD = A'C$. From the figure, we have

$$\sin\theta_1 = \frac{A'C}{AC} \quad \text{and} \quad \sin\theta_1' = \frac{AD}{AC}$$

The right-hand sides are equal, so $\sin\theta = \sin\theta_1'$, and it follows that $\theta_1 = \theta_1'$, which is the law of reflection.

Huygens' principle and Figure 22.24a can be used to derive Snell's law of refraction. In the time interval Δt, ray 1 moves from A to B and ray 2 moves from A' to C. The radius of the outgoing spherical wavelet centered at A is equal to $v_2 \Delta t$. The distance $A'C$ is equal to $v_1 \Delta t$. Geometric considerations show that angle $A'AC$ equals θ_1 and angle ACB equals θ_2. From triangles $AA'C$ and ACB, we find that

$$\sin\theta_1 = \frac{v_1\Delta t}{AC} \quad \text{and} \quad \sin\theta_2 = \frac{v_2\Delta t}{AC}$$

If we divide the first equation by the second, we get

$$\frac{\sin\theta_1}{\sin\theta_2} = \frac{v_1}{v_2}$$

But from Equation 22.4 we know that $v_1 = c/n_1$ and $v_2 = c/n_2$. Therefore,

$$\frac{\sin\theta_1}{\sin\theta_2} = \frac{c/n_1}{c/n_2} = \frac{n_2}{n_1}$$

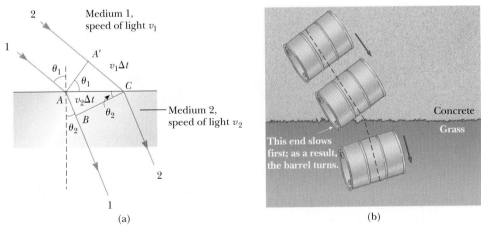

Figure 22.24 (a) Huygens' construction for proving the law of refraction. (b) Overhead view of a barrel rolling from concrete onto grass.

and it follows that

$$n_1 \sin \theta_1 = n_2 \sin \theta_2$$

which is the law of refraction.

A mechanical analog of refraction is shown in Figure 22.24b. When the left end of the rolling barrel reaches the grass, it slows down, while the right end remains on the concrete and moves at its original speed. This difference in speeds causes the barrel to pivot, changing its direction of its motion.

22.7 TOTAL INTERNAL REFLECTION

An interesting effect called *total internal reflection* can occur when light encounters the boundary between a medium with a *higher* index of refraction and one with a *lower* index of refraction. Consider a light beam traveling in medium 1 and meeting the boundary between medium 1 and medium 2, where n_1 is greater than n_2 (Active Fig. 22.25). Possible directions of the beam are indicated by rays 1 through 5. Note that the refracted rays are bent away from the normal because n_1 is greater than n_2. At some particular angle of incidence θ_c, called the **critical angle**, the refracted light ray moves parallel to the boundary, so that $\theta_2 = 90°$ (Active Fig. 22.25b). *For angles of incidence greater than θ_c*, the beam is entirely reflected at the boundary, as is ray 5 in Active Figure 22.25a. This ray is reflected as though it had struck a perfectly reflecting surface. It and all rays like it obey the law of reflection: the angle of incidence equals the angle of reflection.

We can use Snell's law to find the critical angle. When $\theta_1 = \theta_c$, $\theta_2 = 90°$, Snell's law (Eq. 22.8) gives

$$n_1 \sin \theta_c = n_2 \sin 90° = n_2$$

$$\sin \theta_c = \frac{n_2}{n_1} \qquad \text{for} \qquad n_1 > n_2 \qquad [22.9]$$

Equation 22.9 can be used only when n_1 is greater than n_2, because **total internal reflection occurs only when light attempts to move from a medium of higher index of refraction to a medium of lower index of refraction**. If n_1 were less than n_2, Equation 22.9 would give $\sin \theta_c > 1$, which is an absurd result because the sine of an angle can never be greater than one.

When medium 2 is air, the critical angle is small for substances with large indices of refraction, such as diamond, where $n = 2.42$ and $\theta_c = 24.0°$. By comparison, for crown glass, $n = 1.52$ and $\theta_c = 41.0°$. This property, combined with proper faceting, causes a diamond to sparkle brilliantly.

This photograph shows nonparallel light rays entering a glass prism. The bottom two rays undergo total internal reflection at the longest side of the prism. The top three rays are refracted at the longest side as they leave the prism.

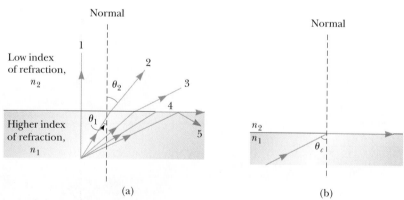

ACTIVE FIGURE 22.25
(a) Rays from a medium with index of refraction n_1 travel to a medium with index of refraction n_2, where $n_1 > n_2$. As the angle of incidence increases, the angle of refraction θ_2 increases until θ_2 is 90° (ray 4). For even larger angles of incidence, total internal reflection occurs (ray 5). (b) The angle of incidence producing a 90° angle of refraction is often called the *critical angle θ_c*.

Physics Now™
Log into PhysicsNow at **www.cp7e.com** and go to Active Figure 22.25, where you can vary the incident angle and see how the refracted ray undergoes total internal reflection when the incident angle exceeds the critical angle.

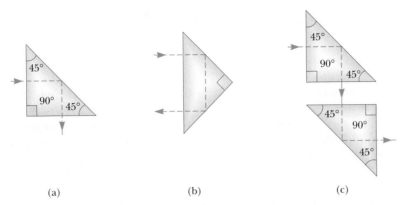

Figure 22.26 Internal reflection in a prism. (a) The ray is deviated by 90°. (b) The direction of the ray is reversed. (c) Two prisms used as a periscope.

(a)　　　　　　(b)　　　　　　(c)

A prism and the phenomenon of total internal reflection can alter the direction of travel of a light beam. Figure 22.26 illustrates two such possibilities. In one case the light beam is deflected by 90.0° (Fig. 22.26a), and in the second case the path of the beam is reversed (Fig. 22.26b). A common application of total internal reflection is a submarine periscope. In this device, two prisms are arranged as in Figure 22.26c, so that an incident beam of light follows the path shown and the user can "see around corners."

APPLICATION

Submarine Periscopes

Applying Physics 22.4 Total Internal Reflection and Dispersion

A beam of white light is incident on the curved edge of a semicircular piece of glass, as shown in Figure 22.27. The light enters the curved surface along the normal, so it shows no refraction. It encounters the straight side of the glass at the center of curvature of the curved side and refracts into the air. The incoming beam is moved clockwise (so that the angle θ increases) such that the beam always enters along the normal to the curved side and encounters the straight side at the center of curvature of the curved side. Why does the refracted beam become redder as it approaches a direction parallel to the straight side?

Explanation When the outgoing beam approaches the direction parallel to the straight side, the incident angle is approaching the critical angle for total internal reflection. Dispersion occurs as the light passes out of the glass. The index of refraction for light at the violet end of the visible spectrum is larger than at the red end. As a result, as the outgoing beam approaches the straight side, the violet light undergoes total internal reflection, followed by the other colors. The red light is the last to undergo total internal reflection, so just before the outgoing light disappears, it's composed of light from the red end of the visible spectrum.

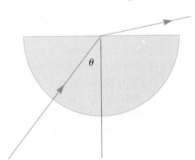

Figure 22.27 (Applying Physics 22.4)

EXAMPLE 22.6 A View from the Fish's Eye

Goal Apply the concept of total internal reflection. **(a)** Find the critical angle for a water–air boundary if the index of refraction of water is 1.33. **(b)** Use the result of part (a) to predict what a fish will see (Fig. 22.28) if it looks up toward the water surface at angles of 40.0°, 48.8°, and 60.0°.

Strategy After finding the critical angle by substitution, use the fact that the path of a light ray is reversible: at a given angle, wherever a light beam can go is also where a beam of light can come from, along the same path.

Figure 22.28 (Example 22.6) A fish looks upward toward the water's surface.

Solution

(a) Find the critical angle for a water–air boundary.

Substitute into Equation 22.9 to find the critical angle:
$$\sin\theta_c = \frac{n_2}{n_1} = \frac{1.00}{1.33} = 0.752$$
$$\theta_c = \boxed{48.8°}$$

(b) Predict what a fish will see if it looks up toward the water surface at angles of 40.0°, 48.8°, and 60.0°.

At an angle of 40.0°, a beam of light from underwater will be refracted at the surface and enter the air above. Because the path of a light ray is reversible (Snell's law works both going and coming), light from above can follow the same path and be perceived by the fish. At an angle of 48.8°, the critical angle for water, light from underwater is bent so that it travels along the surface. This means that light following the same path in reverse can reach the fish only by skimming along the water surface before being refracted towards the fish's eye. At angles greater than the critical angle of 48.8°, a beam of light shot toward the surface will be completely reflected down toward the bottom of the pool. Reversing the path, the fish sees a reflection of some object on the bottom.

Exercise 22.6

Suppose a layer of oil with $n = 1.50$ coats the surface of the water. What is the critical angle for total internal reflection for light traveling in the oil layer and encountering the oil-water boundary?

Answer   62.7°

Figure 22.29 Light travels in a curved transparent rod by multiple internal reflections.

APPLICATION

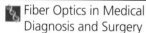

Fiber Optics in Medical Diagnosis and Surgery

Fiber Optics

Another interesting application of total internal reflection is the use of solid glass or transparent plastic rods to "pipe" light from one place to another. As indicated in Figure 22.29, light is confined to traveling within the rods, even around gentle curves, as a result of successive internal reflections. Such a light pipe can be quite flexible if thin fibers are used rather than thick rods. If a bundle of parallel fibers is used to construct an optical transmission line, images can be transferred from one point to another.

Very little light intensity is lost in these fibers as a result of reflections on the sides. Any loss of intensity is due essentially to reflections from the two ends and absorption by the fiber material. Fiber-optic devices are particularly useful for viewing images produced at inaccessible locations. Physicians often use fiber-optic cables to aid in the diagnosis and correction of certain medical problems without the intrusion of major surgery. For example, a fiber-optic cable can be threaded through the esophagus and into the stomach to look for ulcers. In this application, the cable consists of two fiber-optic lines: one to transmit a beam of light into the stomach for illumination and the other to allow the light to be transmitted out of the stomach. The resulting image can, in some cases, be viewed directly by the physician, but more often is displayed on a television monitor or captured on film.

(*Left*) Strands of glass optical fibers are used to carry voice, video, and data signals in telecommunication networks. (*Right*) A bundle of optical fibers is illuminated by a laser.

Dennis O'Clair/Tony Stone Images/Getty Images

Hank Morgan/Photo Researchers, Inc.

In a similar way, fiber-optic cables can be used to examine the colon or to help physicians perform surgery without the need for large incisions.

The field of fiber optics has revolutionized the entire communications industry. Billions of kilometers of optical fiber have been installed in the United States to carry high-speed internet traffic, radio and television signals, and telephone calls. The fibers can carry much higher volumes of telephone calls and other forms of communication than electrical wires because of the higher frequency of the infrared light used to carry the information on optical fibers. Optical fibers are also preferable to copper wires because they are insulators and don't pick up stray electric and magnetic fields or electronic "noise."

APPLICATION

Fiber Optics in
Telecommunications

Applying Physics 22.5   Design of an Optical Fiber

An optical fiber consists of a transparent core surrounded by cladding, which is a material with a lower index of refraction than the core (Fig. 22.30). A cone of angles, called the acceptance cone, is at the entrance to the fiber. Incoming light at angles within this cone will be transmitted through the fiber, whereas light entering the core from angles outside the cone will not be transmitted. The figure shows a light ray entering the fiber just within the acceptance cone and undergoing total internal reflection at the interface between the core and the cladding. If it is technologically difficult to produce light so that it enters the fiber from a small range of angles, how could you adjust the indices of refraction of the core and cladding to increase the size of the acceptance cone—would you design the indices to be farther apart or closer together?

Explanation The acceptance cone would become larger if the critical angle (θ_c in the figure) could be

made smaller. This can be done by making the index of refraction of the cladding material smaller, so that the indices of refraction of the core and cladding material would be farther apart.

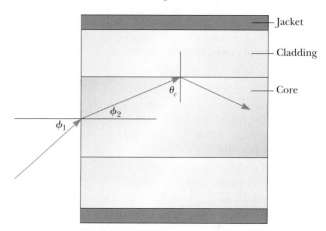

Figure 22.30 (Applying Physics 22.5)

SUMMARY

Physics⊗Now™ Take a practice test by logging into Physics-Now at **www.cp7e.com** and clicking on the Pre-Test link for this chapter.

22.1   The Nature of Light

Light has a dual nature. In some experiments it acts like a wave, in others like a particle, called a photon by Einstein. The energy of a photon is proportional to its frequency,

$$E = hf \qquad [22.1]$$

where $h = 6.63 \times 10^{-34}$ J · s is *Planck's constant*.

22.2   Reflection and Refraction

In the reflection of light off a flat, smooth surface, the angle of incidence, θ_1, with respect to a line perpendicular to the surface is equal to the angle of reflection, $\theta_1{}'$:

$$\theta_1{}' = \theta_1 \qquad [22.2]$$

Light that passes into a transparent medium is bent at the boundary and is said to be *refracted*. The angle of refraction is the angle the ray makes with respect to a line

perpendicular to the surface after it has entered the new medium.

22.3   The Law of Refraction

The **index of refraction** of a material, n, is defined as

$$n \equiv \frac{c}{v} \qquad [22.4]$$

where c is the speed of light in a vacuum and v is the speed of light in the material. The index of refraction of a material is also

$$n = \frac{\lambda_0}{\lambda_n} \qquad [22.7]$$

where λ_0 is the wavelength of the light in vacuum and λ_n is its wavelength in the material.

The **law of refraction**, or **Snell's law**, states that

$$n_1 \sin \theta_1 = n_2 \sin \theta_2 \qquad [22.8]$$

where n_1 and n_2 are the indices of refraction in the two media. The incident ray, the reflected ray, the refracted ray, and the normal to the surface all lie in the same plane.

22.4 Dispersion and Prisms & 22.5 The Rainbow

The index of refraction of a material depends on the wavelength of the incident light, an effect called *dispersion*. Light at the violet end of the spectrum exhibits a larger angle of refraction on entering glass than light at the red end. Rainbows are a consequence of dispersion.

22.6 Huygens' Principle

Huygens' principle states that all points on a wave front are point sources for the production of spherical secondary waves called wavelets. These wavelets propagate forward at a speed characteristic of waves in a particular medium. After some time has elapsed, the new position of the wave front is the surface tangent to the wavelets. This principle can be used to deduce the laws of reflection and refraction.

22.7 Total Internal Reflection

Total internal reflection can occur when light, traveling in a medium with higher index of refraction, is incident on the boundary of a material with a lower index of refraction. The *maximum angle of incidence* θ_c for which light can move from a medium with index n_1 into a medium with index n_2, where n_1 is greater than n_2, is called the **critical angle** and is given by

$$\sin \theta_c = \frac{n_2}{n_1} \qquad \text{for} \qquad n_1 > n_2 \qquad \textbf{[22.9]}$$

Total internal reflection is used in the optical fibers that carry data at high speed around the world.

CONCEPTUAL QUESTIONS

1. Under certain circumstances, sound can be heard from extremely far away. This frequently happens over a body of water, where the air near the water surface is cooler than the air at higher altitudes. Explain how the refraction of sound waves could increase the distance over which sound can be heard.

2. What are some reasons that most ceilings are made of white textured material?

3. The color of an object is said to depend on the wavelengths the object reflects. So, if you view colored objects under water, in which the wavelength of the light will be different, does the color change?

4. How is it possible that a complete circle of a rainbow can sometimes be seen from an airplane?

5. A ray of light is moving from a material having a high index of refraction into a material with a lower index of refraction. (a) Is the ray bent toward the normal or away from it? (b) If the wavelength is 600 nm in the material with the high index of refraction, is it greater, smaller, or the same in the material with the lower index of refraction? (c) How does the frequency change as the light moves between the two materials? Does it increase, decrease, or remain the same?

6. Why does the arc of a rainbow appear with red on top and violet on the bottom?

7. A scientific supply catalog advertises a material having an index of refraction of 0.85. Is this a good product to buy? Why or why not?

8. Under what conditions is a mirage formed? On a hot day, what are we seeing when we observe a mirage of a water puddle on the road?

9. In dispersive materials, the angle of refraction for a light ray depends on the wavelength of the light. Does the angle of reflection from the surface of the material depend on the wavelength? Why or why not?

10. A type of mirage called a *pingo* is often observed in Alaska. Pingos occur when the light from a small hill passes to an observer by a path that takes the light over a body of water warmer than the air. What is seen is the hill and an inverted image directly below it. Explain how these mirages are formed.

11. Explain why a diamond loses most of its sparkle when submerged in carbon disulfide.

12. Suppose you are told that only two colors of light (*X* and *Y*) are sent through a glass prism and that *X* is bent more than *Y*. Which color travels more slowly in the prism?

13. The level of water in a clear, colorless glass can easily be observed with the naked eye. The level of liquid helium in a clear glass vessel is extremely difficult to see with the naked eye. Explain. [*Hint*: The index of refraction of liquid helium is close to that of air.]

14. Is it possible to have total internal reflection for light incident from air on water? Explain.

15. Why does a diamond show flashes of color when observed under white light?

16. Explain why an oar partially submearged in water appears to be bent.

17. Why do astronomers looking at distant galaxies talk about looking backward in time?

18. If a beam of light with a given cross section enters a new medium, the cross section of the refracted beam is different from that of the incident beam. Is it larger or smaller, or is there no definite direction to the change?

PROBLEMS

1, 2, 3 = straightforward, intermediate, challenging ☐ = full solution available in *Student Solutions Manual/Study Guide*

Physics⊗Now™ = coached solution with hints available at **www.cp7e.com** 🎷 = biomedical application

Section 22.1 The Nature of Light

1. During the Apollo XI Moon landing, a retroreflecting panel was erected on the Moon's surface. The speed of light can be found by measuring the time it takes a laser beam to travel from Earth, reflect from the panel, and return to Earth. If this interval is found to be 2.51 s, what is the measured speed of light? Take the center-to-center distance from Earth to Moon to be 3.84×10^8 m. Assume that the Moon is directly overhead and do not neglect the sizes of the Earth and Moon.

2. Figure P22.2 shows the apparatus used by Armand H. L. Fizeau (1819–1896) to measure the speed of light. The basic idea is to measure the total time it takes light to travel from some point to a distant mirror and back. If d is the distance between the light source and the mirror, and if the transit time for one round-trip is t, then the speed of light is $c = 2d/t$. To measure the transit time, Fizeau used a rotating toothed wheel, which converts an otherwise continuous beam of light to a series of light pulses. The rotation of the wheel controls what an observer at the light source sees. For example, assume that the toothed wheel of the Fizeau experiment has 360 teeth and is rotating at a speed of 27.5 rev/s when the light from the source is extinguished—that is, when a burst of light passing through opening A in Figure P22.2 is blocked by tooth B on return. If the distance to the mirror is 7 500 m, find the speed of light.

Figure P22.2   (Problems 2 and 3)

3. In an experiment designed to measure the speed of light using the apparatus of Fizeau described in the preceding problem, the distance between light source and mirror was 11.45 km and the wheel had 720 notches. The experimentally determined value of c was 2.998×10^8 m/s. Calculate the minimum angular speed of the wheel for this experiment.

4. Albert A. Michelson very carefully measured the speed of light using an alternative version of the technique developed by Fizeau. (See Problem 22.2.) Figure P22.4 shows the approach Michelson used. Light was reflected from one face of a rotating eight-sided mirror towards a stationary mirror 35.0 km away. At certain rates of rotation, the returning beam of light was directed toward the eye of an observer as shown. (a) What minimum angular speed must the rotating mirror have in order that side A will

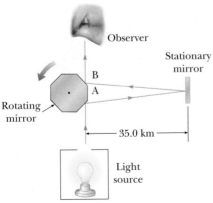

Figure P22.4

have rotated to position B, causing the light to be reflected to the eye? (b) What is the next-higher angular velocity that will enable the source of light to be seen?

5. Figure P22.5 shows an apparatus used to measure the distribution of the speeds of gas molecules. The device consists of two slotted rotating disks separated by a distance d, with the slots displaced by the angle θ. Suppose the speed of light is measured by sending a light beam toward the left disk of this apparatus. (a) Show that a light beam will be seen in the detector (that is, will make it through both slots) only if its speed is given by $c = \omega d/\theta$, where ω is the angular speed of the disks and θ is measured in radians. (b) What is the measured speed of light if the distance between the two slotted rotating disks is 2.500 m, the slot in the second disk is displaced 1/60 of 1 degree from the slot in the first disk, and the disks are rotating at 5 555 rev/s?

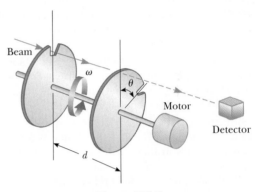

Figure P22.5

Section 22.2 Reflection and Refraction
Section 22.3 The Law of Refraction

6. The two mirrors in Figure P22.6 meet at a right angle. The beam of light in the vertical plane P strikes mirror 1 as shown. (a) Determine the distance the reflected light beam travels before striking mirror 2. (b) In what direction does the light beam travel after being reflected from mirror 2?

Figure P22.6

7. An underwater scuba diver sees the Sun at an apparent angle of 45.0° from the vertical. What is actual direction of the Sun?

8. Light is incident normal to a 1.00-cm layer of water that lies on top of a flat Lucite® plate with a thickness of 0.500 cm. How much more time is required for light to pass through this double layer than is required to traverse the same distance in air ($n_{Lucite} = 1.59$)?

9. A laser beam is incident at an angle of 30.0° to the vertical onto a solution of corn syrup in water. If the beam is refracted to 19.24° to the vertical, (a) what is the index of refraction of the syrup solution? Suppose the light is red, with wavelength 632.8 nm in a vacuum. Find its (b) wavelength, (c) frequency, and (d) speed in the solution.

10. Light containing wavelengths of 400 nm, 500 nm, and 650 nm is incident from air on a block of crown glass at an angle of 25.0°. (a) Are all colors refracted alike, or is one color bent more than the others? (b) Calculate the angle of refraction in each case to verify your answer.

11. Light of wavelength λ_0 in a vacuum has a wavelength of 438 nm in water and a wavelength of 390 nm in benzene. (a) What is the wavelength λ_0? (b) Using only the given wavelengths, determine the ratio of the index of refraction of benzene to that of water.

12. Light of wavelength 436 nm in air enters a fishbowl filled with water, then exits through the crown-glass wall of the container. Find the wavelengths of the light (a) in the water and (b) in the glass.

13. A ray of light is incident on the surface of a block of clear ice at an angle of 40.0° with the normal. Part of the light is reflected and part is refracted. Find the angle between the reflected and refracted light.

14. The laws of refraction and reflection are the same for sound as for light. The speed of sound is 340 m/s in air and 1 510 m/s in water. If a sound wave traveling in air approaches a plane water surface at an angle of incidence of 12.0°, what is the angle of refraction?

15. The light emitted by a helium–neon laser has a wavelength of 632.8 nm in air. As the light travels from air into zircon, find (a) its speed in zircon, (b) its wavelength in zircon, and (c) its frequency.

16. A flashlight on the bottom of a 4.00-m-deep swimming pool sends a ray upward and at an angle so that the ray strikes the surface of the water 2.00 m from the point directly above the flashlight. What angle (in air) does the emerging ray make with the water's surface?

17. How many times will the incident beam shown in Figure P22.17 be reflected by each of the parallel mirrors?

Figure P22.17

18. A ray of light strikes a flat 2.00-cm-thick block of glass ($n = 1.50$) at an angle of 30.0° with the normal (Fig. P22.18). Trace the light beam through the glass, and find the angles of incidence and refraction at each surface.

19. PhysicsNow™ When the light ray in Problem 18 passes through the glass block, it is shifted laterally by a distance d (Fig. P22.18). Find the value of d.

Figure P22.18   (Problems 18, 19, and 20)

20. Find the time required for the light to pass through the glass block described in Problem 19.

21. The light beam shown in Figure P22.21 makes an angle of 20.0° with the normal line NN' in the linseed oil. Determine the angles θ and θ'. (The refractive index for linseed oil is 1.48.)

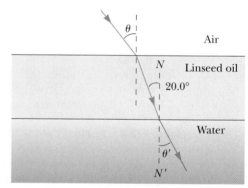

Figure P22.21

22. A submarine is 300 m horizontally out from the shore and 100 m beneath the surface of the water. A laser beam is sent from the sub so that it strikes the surface of the water at a point 210 m from the shore. If the beam just strikes the top of a building standing directly at the water's edge, find the height of the building.

23. Two light pulses are emitted simultaneously from a source. The pulses take parallel paths to a detector 6.20 m away, but one moves through air and the other through a block of ice. Determine the difference in the pulses' times of arrival at the detector.

24. A narrow beam of ultrasonic waves reflects off the liver tumor in Figure P22.24. If the speed of the wave is 10.0% less in the liver than in the surrounding medium, determine the depth of the tumor.

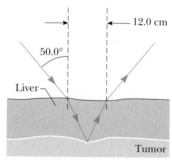

Figure P22.24

25. A beam of light both reflects and refracts at the surface between air and glass, as shown in Figure P22.25. If the index of refraction of the glass is n_g, find the angle of incidence, θ_1, in the air that would result in the reflected ray and the refracted ray being perpendicular to each other. [*Hint*: Remember the identity $\sin(90° - \theta) = \cos\theta$.]

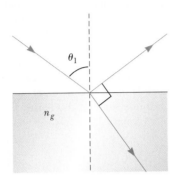

Figure P22.25

26. Three sheets of plastic have unknown indices of refraction. Sheet 1 is placed on top of sheet 2, and a laser beam is directed onto the sheets from above so that it strikes the interface at an angle of 26.5° with the normal. The refracted beam in sheet 2 makes an angle of 31.7° with the normal. The experiment is repeated with sheet 3 on top of sheet 2, and with the same angle of incidence, the refracted beam makes an angle of 36.7° with the normal. If the experiment is repeated again with sheet 1 on top of sheet 3, what is the expected angle of refraction in sheet 3? Assume the same angle of incidence.

27. An opaque cylindrical tank with an open top has a diameter of 3.00 m and is completely filled with water. When the afternoon Sun reaches an angle of 28.0° above the horizon, sunlight ceases to illuminate the bottom of the tank. How deep is the tank?

28. A cylindrical cistern, constructed below ground level, is 3.0 m in diameter and 2.0 m deep and is filled to the brim with a liquid whose index of refraction is 1.5. A small object rests on the bottom of the cistern at its center. How far from the edge of the cistern can a girl whose eyes are 1.2 m from the ground stand and still see the object?

Section 22.4 Dispersion and Prisms

29. The index of refraction for red light in water is 1.331, and that for blue light is 1.340. If a ray of white light enters the water at an angle of incidence of 83.00°, what are the underwater angles of refraction for the blue and red components of the light?

30. A certain kind of glass has an index of refraction of 1.650 for blue light of wavelength 430 nm and an index of 1.615 for red light of wavelength 680 nm. If a beam containing these two colors is incident at an angle of 30.00° on a piece of this glass, what is the angle between the two beams inside the glass?

31. A ray of light strikes the midpoint of one face of an equiangular (60°–60°–60°) glass prism ($n = 1.5$) at an angle of incidence of 30°. (a) Trace the path of the light ray through the glass, and find the angles of incidence and refraction at each surface. (b) If a small fraction of light is also reflected at each surface, find the angles of reflection at the surfaces.

32. The index of refraction for violet light in silica flint glass is 1.66, and that for red light is 1.62. What is the angular dispersion of visible light passing through an equilateral prism of apex angle 60.0° if the angle of incidence is 50.0°? (See Fig. P22.32.)

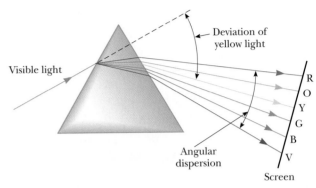

Figure P22.32

Section 22.7 Total Internal Reflection

33. Calculate the critical angles for the following materials when surrounded by air: (a) zircon, (b) fluorite, and (c) ice. Assume that $\lambda = 589$ nm.

34. For light of wavelength 589 nm, calculate the critical angle for the following materials surrounded by air: (a) diamond and (b) flint glass.

35. Repeat Problem 34, but this time suppose that the materials are surrounded by water.

36. **Physics⊗Now™** A beam of light is incident from air on the surface of a liquid. If the angle of incidence is 30.0° and the angle of refraction is 22.0°, find the critical angle for the liquid when surrounded by air.

37. A plastic light pipe has an index of refraction of 1.53. For total internal reflection, what is the minimum angle of incidence if the pipe is in (a) air? (b) water?

38. Determine the maximum angle θ for which the light rays incident on the end of the light pipe in Figure P22.38 are subject to total internal reflection along the walls of the pipe. Assume that the light pipe has an index of refraction of 1.36 and that the outside medium is air.

2.00 μm

θ

Figure P22.38

39. Consider a common mirage formed by superheated air just above a roadway. A truck driver whose eyes are 2.00 m above the road, where $n = 1.000\,3$, looks forward. She has the illusion of seeing a patch of water ahead on the road, where her line of sight makes an angle of 1.20° below the horizontal. Find the index of refraction of the air just above the road surface. [*Hint*: Treat this as a problem one involving total internal reflection.]

40. A jewel thief hides a diamond by placing it on the bottom of a public swimming pool. He places a circular raft on the surface of the water directly above and centered over the diamond, as shown in Figure P22.40. If the surface of the water is calm and the pool is 2.00 m deep, find the minimum diameter of the raft that would prevent the diamond from being seen.

Figure P22.40

41. A room contains air in which the speed of sound is 343 m/s. The walls of the room are made of concrete, in which the speed of sound is 1 850 m/s. (a) Find the critical angle for total internal reflection of sound at the concrete–air boundary. (b) In which medium must the sound be traveling in order to undergo total internal reflection? (c) "A bare concrete wall is a highly efficient mirror for sound." Give evidence for or against this statement.

42. Three adjacent faces (that all share a corner) of a plastic cube of index of refraction n are painted black. A clear spot at the painted corner serves as a source of diverging rays when light comes through it. Show that a ray from this corner to the center of a clear face is totally reflected if $n \geq \sqrt{3}$.

43. The light beam in Figure P22.43 strikes surface 2 at the critical angle. Determine the angle of incidence, θ_i.

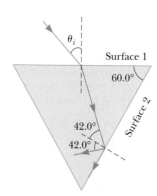

Figure P22.43

ADDITIONAL PROBLEMS

44. (a) Consider a horizontal interface between air above and glass with an index of 1.55 below. Draw a light ray incident from the air at an angle of incidence of 30.0°. Determine the angles of the reflected and refracted rays, and show them on the diagram. (b) Suppose instead that the light ray is incident from the glass at an angle of incidence of 30.0°. Determine the angles of the reflected and refracted rays, and show all three rays on a new diagram. (c) For rays incident from the air onto the air–glass surface, determine and tabulate the angles of reflection and refraction for all the angles of incidence at

10.0° intervals from 0° to 90.0°. (d) Do the same for light rays traveling up to the interface through the glass.

45. **Physics⊗Now™** A layer of ice having parallel sides floats on water. If light is incident on the upper surface of the ice at an angle of incidence of 30.0°, what is the angle of refraction in the water?

46. A light ray of wavelength 589 nm is incident at an angle θ on the top surface of a block of polystyrene surrounded by air, as shown in Figure P22.46. (a) Find the maximum value of θ for which the refracted ray will undergo total internal reflection at the left vertical face of the block. (b) Repeat the calculation for the case in which the polystyrene block is immersed in water. (c) What happens if the block is immersed in carbon disulfide?

Figure P22.46

47. Figure P22.47 shows the path of a beam of light through several layers with different indices of refraction. (a) If $\theta_1 = 30.0°$, what is the angle θ_2 of the emerging beam? (b) What must the incident angle θ_1 be in order to have total internal reflection at the surface between the medium with $n = 1.20$ and the medium with $n = 1.00$?

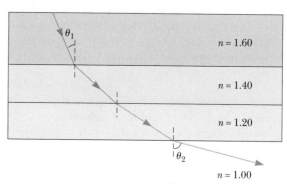

Figure P22.47

48. The walls of a prison cell are perpendicular to the four cardinal compass directions. On the first day of spring, light from the rising Sun enters a rectangular window in the eastern wall. The light traverses 2.37 m horizontally to shine perpendicularly on the wall opposite the window. A prisoner observes the patch of light moving across this western wall and for the first time forms his own understanding of the rotation of the Earth. (a) With what speed does the illuminated rectangle move? (b) The prisoner holds a small square mirror flat against the wall at one corner of the rectangle of light. The mirror reflects light back to a spot on the eastern wall close beside the window. How fast does the smaller square of light move across that wall? (c) Seen from a latitude of 40.0° north, the rising Sun moves through the sky along a line making a 50.0° angle with the southeastern horizon. In what direction does the rectangular patch of light on the western wall of the prisoner's cell move? (d) In what direction does the smaller square of light on the eastern wall move?

49. As shown in Figure P22.49, a light ray is incident normal to on one face of a 30°–60°–90° block of dense flint glass (a prism) that is immersed in water. (a) Determine the exit angle θ_4 of the ray. (b) A substance is dissolved in the water to increase the index of refraction. At what value of n_2 does total internal reflection cease at point P?

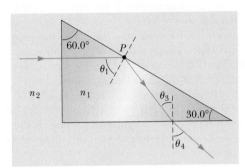

Figure P22.49

50. A narrow beam of light is incident from air onto a glass surface with index of refraction 1.56. Find the angle of incidence for which the corresponding angle of refraction is one-half the angle of incidence. [*Hint:* You might want to use the trigonometric identity $\sin 2\theta = 2 \sin \theta \cos \theta$.]

51. One technique for measuring the angle of a prism is shown in Figure P22.51. A parallel beam of light is directed onto the apex of the prism so that the beam reflects from opposite faces of the prism. Show that the angular separation of the two reflected beams is given by $B = 2A$.

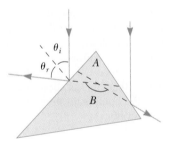

Figure P22.51

52. An optical fiber with index of refraction n and diameter d is surrounded by air. Light is sent into the fiber along its axis, as shown in Figure P22.52. (a) Find the smallest outside radius R permitted for a bend in the fiber if no light is to escape. (b) Does the result for part (a) predict reasonable behavior as d approaches zero? As n increases? As n approaches unity? (c) Evaluate R, assuming that the diameter of the fiber is 100 μm and its index of refraction is 1.40.

Figure P22.52

53. A piece of wire is bent through an angle θ. The bent wire is partially submerged in benzene (index of refraction = 1.50), so that, to a person looking along the dry part, the wire appears to be straight and makes an angle of 30.0° with the horizontal. Determine the value of θ.

54. A light ray traveling in air is incident on one face of a right-angle prism with index of refraction $n = 1.50$, as shown in Figure P22.54, and the ray follows the path shown in the figure. Assuming that $\theta = 60.0°$ and the base of the prism is mirrored, determine the angle ϕ made by the outgoing ray with the normal to the right face of the prism.

Figure P22.54

55. **Physics⊗Now™** A transparent cylinder of radius $R = 2.00$ m has a mirrored surface on its right half, as shown in Figure P22.55. A light ray traveling in air is incident on the left side of the cylinder. The incident light ray and the exiting light ray are parallel, and $d = 2.00$ m. Determine the index of refraction of the material.

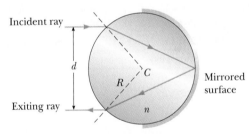

Figure P22.55

56. A laser beam strikes one end of a slab of material, as in Figure P22.56. The index of refraction of the slab is 1.48. Determine the number of internal reflections of the beam before it emerges from the opposite end of the slab.

Figure P22.56

57. For this problem, refer to Figure 22.15. For various angles of incidence, it can be shown that the deviation angle δ is a minimum when the ray passes through the glass so that the interior ray is parallel to the base of the prism. A measurement of this minimum angle of deviation enables one to find the index of refraction of the prism material.

Show that n is given by the expression

$$n = \frac{\sin\left[\frac{1}{2}(A + \delta_{min})\right]}{\sin\left(\frac{A}{2}\right)}$$

where A is the apex angle of the prism.

58. A hiker stands on a mountain peak near sunset and observes a rainbow caused by water droplets in the air about 8.00 km away. The valley is 2.00 km below the mountain peak and entirely flat. What fraction of the complete circular arc of the rainbow is visible to the hiker?

59. A light ray incident on a prism is refracted at the first surface, as shown in Figure P22.59. Let ϕ represent the apex angle of the prism and n its index of refraction. Find, in terms of n and ϕ, the smallest allowed value of the angle of incidence at the first surface for which the refracted ray will not undergo total internal reflection at the second surface.

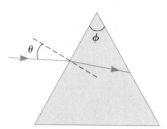

Figure P22.59

60. Students allow a narrow beam of laser light to strike a water surface. They arrange to measure the angle of refraction for selected angles of incidence and record the data shown in the following table:

Angle of Incidence	Angle of Refraction
(degrees)	(degrees)
10.0	7.5
20.0	15.1
30.0	22.3
40.0	28.7
50.0	35.2
60.0	40.3
70.0	45.3
80.0	47.7

Use the data to verify Snell's law of refraction by plotting the sine of the angle of incidence versus the sine of the angle of refraction. From the resulting plot, deduce the index of refraction of water.

61. A light ray enters a rectangular block of plastic at an angle $\theta_1 = 45.0°$ and emerges at an angle $\theta_2 = 76.0°$, as shown in Figure P22.61. (a) Determine the index of refraction of

the plastic. (b) If the light ray enters the plastic at a point $L = 50.0$ cm from the bottom edge, how long does it take the light ray to travel through the plastic?

62. A. H. Pfund's method for measuring the index of refraction of glass is illustrated in Figure P22.62. One face of a slab of thickness t is painted white, and a small hole scraped clear at point P serves as a source of diverging rays when the slab is illuminated from below. Ray PBB' strikes the clear surface at the critical angle and is totally reflected, as are rays such as PCC'. Rays such as PAA' emerge from the clear surface. On the painted surface there appears a dark circle of diameter d, surrounded by an illuminated region, or halo. (a) Derive an equation for n in terms of the measured quantities d and t. (b) What is the diameter of the dark circle if $n = 1.52$ for a slab 0.600 cm thick? (c) If white light is used, the critical angle depends on color caused by dispersion. Is the inner edge of the white halo tinged with red light or violet light? Explain.

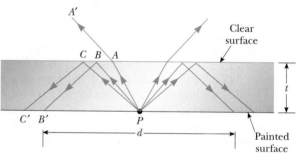

Figure P22.62

ACTIVITIES

1. Tape a coin to the bottom of a large opaque bowl, as shown in Figure A22.1a. Stand over the bowl so that you are looking at the coin, and then move backwards away from the bowl until you can no longer see the coin over the bowl's rim. Remain at that position, and have a friend fill the bowl with water, as shown in Figure A22.1b. You can now see the coin again because the light is refracted at the water–air interface.

Figure A22.1

2. Tape a piece of black paper to the end of a flashlight and cut a narrow slit in the middle of the paper, as shown in Figure A22.2. Lean a flat mirror against one end of a tray partially filled with water. Shine your flashlight on that

Figure P22.61

part of the mirror which is under water, and hold a sheet of white paper such that the reflected light shines on the paper. You should observe a spectrum of colors on the paper as the light is dispersed when it travels from air into water and then from water into air. According to your observations, which color is bent the most? Which is bent the least?

Figure A22.2

3. Create an artificial rainbow by standing with your back to the sun and spraying water into the air with a hose. You should cover the end of the hose slightly with a finger or use a nozzle so that the water is broken up into tiny droplets. Which color is on the outside of the arc? The inside?

If the droplets are close to you in the experiment above, you may be able to see two rainbows, one for each eye. Close one eye, and only one bow is seen.

With the Sun high in the sky, stand on a ladder and spray the hose toward the ground. In this case, you should be able to form a complete circle rainbow.

4. By observing the shadows formed by large and small light sources, you can demonstrate that light rays travel in straight-line paths. Figure A22.4 shows the shadows formed on a screen by a baseball when the light from a lightbulb (a large source) falls on it. Use the figure to explain why the shadow is dark at location A and less dark at B. Replace the light source with a small source, such as a high-intensity lamp with a small, clear bulb. (A substitute for a high-intensity lamp is an ordinary lightbulb with a piece of cardboard close in front of it with a hole about the size of a penny punched in the cardboard. The hole serves as the small source of light.) What kind of shadow is formed in this case? Why?

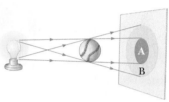

Figure A22.4

This beautiful photograph shows a raindrop suspended from a leaf. The raindrop acts as a lens. It refracts light twice to produce a real, inverted image of the foliage beyond.

Don Hammond/CORBIS

CHAPTER

23

Mirrors and Lenses

The development of the technology of mirrors and lenses led to a revolution in the progress of science. These devices, relatively simple to construct from cheap materials, led to microscopes and telescopes, extending human sight and opening up new pathways to knowledge, from microbes to distant planets.

This chapter covers the formation of images when plane and spherical light waves fall on plane and spherical surfaces. Images can be formed by reflection from mirrors or by refraction through lenses. In our study of mirrors and lenses, we continue to assume that light travels in straight lines (the ray approximation), ignoring diffraction.

23.1 FLAT MIRRORS

We begin by examining the flat mirror. Consider a point source of light placed at O in Figure 23.1, a distance p in front of a flat mirror. The distance p is called the **object distance**. Light rays leave the source and are reflected from the mirror. After reflection, the rays diverge (spread apart), but they appear to the viewer to come from a point I behind the mirror. Point I is called the **image** of the object at O. Regardless of the system under study, **images are formed at the point where rays of light actually intersect or where they appear to originate**. Because the rays in the figure appear to originate at I, which is a distance q behind the mirror, that is the location of the image. The distance q is called the **image distance**.

Images are classified as real or virtual. In the formation of a *real image*, light actually passes through the image point. For a *virtual image*, the light doesn't pass through the image point, but appears to come (diverge) from there. The image formed by the flat mirror in Figure 23.1 is a virtual image. In fact, the images seen in flat mirrors are always virtual (for real objects). Real images can be displayed on a screen (as at a movie), but virtual images cannot.

We will examine some of the properties of the images formed by flat mirrors by using the simple geometric techniques shown in Active Figure 23.2. To find out *dim*). This dim reflected light is responsible for the image observed when the mirror is in the night setting, as in Figure 23.5b. Now the wedge is rotated so that the path followed by the bright light (ray *B*) doesn't lead to the eye. Instead, the dim light reflected from the front surface travels to the eye, and the brightness of trailing headlights doesn't become a hazard.

Applying Physics 23.1   Illusionist's Trick

The professor in the box shown in Figure 23.6 appears to be balancing himself on a few fingers

Figure 23.6   (Applying Physics 23.1)

with both of his feet elevated from the floor. He can maintain this position for a long time, and appears to defy gravity. How do you suppose this illusion was created?

Explanation This is an example of an optical illusion, used by magicians, that makes use of a mirror. The box that the professor is standing in is a cubical open frame that contains a flat vertical mirror through a diagonal plane. The professor straddles the mirror so that one leg is in front of the mirror and the other leg is behind it, out of view. When he raises his front leg, that leg's reflection rises also, making it appear both his feet are off the ground, creating the illusion that he's floating in the air. In fact, he supports himself with the leg behind the mirror, which remains in contact with the ground.

23.2   IMAGES FORMED BY SPHERICAL MIRRORS

Concave Mirrors

A **spherical mirror**, as its name implies, has the shape of a segment of a sphere. Figure 23.7 shows a spherical mirror with a silvered inner, concave surface; this type of mirror is called a **concave mirror**. The mirror has radius of curvature *R*, and its center of curvature is at point *C*. Point *V* is the center of the spherical segment, and a line drawn from *C* to *V* is called the **principal axis** of the mirror.

Now consider a point source of light placed at point *O* in Figure 23.7b, on the principal axis and outside point *C*. Several diverging rays originating at *O* are shown. After reflecting from the mirror, these rays converge to meet at *I*, called the **image point**. The rays then continue and diverge from *I* as if there were an object there. As a result, a real image is formed. **Whenever reflected light actually passes through a point, the image formed there is real**.

We often assume that all rays that diverge from the object make small angles with the principal axis. All such rays reflect through the image point, as in Figure 23.7b.

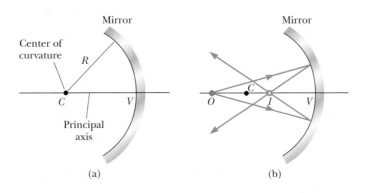

(a)   (b)

Figure 23.7   (a) A concave mirror of radius *R*. The center of curvature, *C*, is located on the principal axis. (b) A point object placed at *O* in front of a concave spherical mirror of radius *R*, where *O* is any point on the principal axis farther than *R* from the surface of the mirror, forms a real image at *I*. If the rays diverge from *O* at small angles, they all reflect through the same image point.

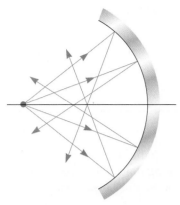

Figure 23.8 Rays at large angles from the horizontal axis reflect from a spherical, concave mirror to intersect the principal axis at different points, resulting in a blurred image. This phenomenon is called *spherical aberration*.

Rays that make a large angle with the principal axis, as in Figure 23.8, converge to other points on the principal axis, producing a blurred image. This effect, called **spherical aberration**, is present to some extent with any spherical mirror and will be discussed in Section 23.7.

We can use the geometry shown in Figure 23.9 to calculate the image distance q from the object distance p and radius of curvature, R. By convention, these distances are measured from point V. The figure shows two rays of light leaving the tip of the object. One ray passes through the center of curvature, C, of the mirror, hitting the mirror head-on (perpendicular to the mirror surface) and reflecting back on itself. The second ray strikes the mirror at point V and reflects as shown, obeying the law of reflection. The image of the tip of the arrow is at the point at which the two rays intersect. From the largest triangle in Figure 23.9 we see that $\tan \theta = h/p$; the light-blue triangle gives $\tan \theta = -h'/q$. The negative sign has been introduced to satisfy our convention that h' is negative when the image is inverted with respect to the object, as it is here. From Equation 23.1 and these results, we find that the magnification of the mirror is

$$M = \frac{h'}{h} = -\frac{q}{p} \qquad \text{[23.2]}$$

From two other triangles in the figure, we get

$$\tan \alpha = \frac{h}{p - R} \qquad \text{and} \qquad \tan \alpha = -\frac{h'}{R - q}$$

from which we find that

$$\frac{h'}{h} = -\frac{R - q}{p - R} \qquad \text{[23.3]}$$

If we compare Equation 23.2 to Equation 23.3, we see that

$$\frac{R - q}{p - R} = \frac{q}{p}$$

Simple algebra reduces this to

Mirror equation ▶

$$\frac{1}{p} + \frac{1}{q} = \frac{2}{R} \qquad \text{[23.4]}$$

This expression is called the **mirror equation**.

If the object is very far from the mirror—if the object distance p is great enough compared with R that p can be said to approach infinity—then $1/p \approx 0$, and we see from Equation 23.4 that $q \approx R/2$. In other words, when the object is very far from the mirror, **the image point is halfway between the center of curvature and the center of the mirror**, as in Figure 23.10a. The incoming rays are essentially parallel in that figure because the source is assumed to be very far from the mirror. In this special case, we call the image point the **focal point** F and the

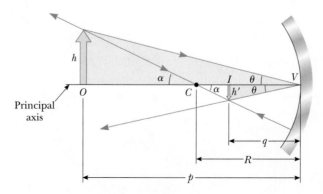

Figure 23.9 The image formed by a spherical concave mirror, where the object at O lies outside the center of curvature, C.

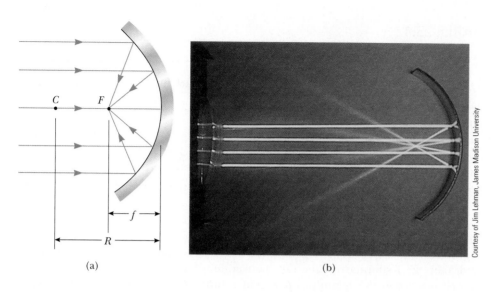

(a)   (b)

Courtesy of Jim Lehman, James Madison University

Figure 23.10 (a) Light rays from a distant object ($p = \infty$) reflect from a concave mirror through the focal point F. In this case, the image distance $q = R/2 = f$, where f is the focal length of the mirror. (b) A photograph of the reflection of parallel rays from a concave mirror.

image distance the **focal length** f, where

$$f = \frac{R}{2}$$ [23.5] ◀ Focal length

The mirror equation can therefore be expressed in terms of the focal length:

$$\frac{1}{p} + \frac{1}{q} = \frac{1}{f}$$ [23.6]

Note that rays from objects at infinity are always focused at the focal point.

23.3 CONVEX MIRRORS AND SIGN CONVENTIONS

Figure 23.11 shows the formation of an image by a **convex mirror**, which is silvered so that light is reflected from the outer, convex surface. This is sometimes called a **diverging mirror** because the rays from any point on the object diverge after reflection, as though they were coming from some point behind the mirror. The image in Figure 23.11 is virtual rather than real because it lies behind the mirror at the point at which the reflected rays appear to originate. In general, the image formed by a convex mirror is upright, virtual, and smaller than the object.

We won't derive any equations for convex spherical mirrors. If we did, we would find that the equations developed for concave mirrors can be used with convex mirrors if particular sign conventions are used. We call the region in which light rays move the *front side* of the mirror, and the other side, where virtual images are formed, the *back side*. For example, in Figures 23.9 and 23.11, the side to the left of the mirror is the front side and the side to the right is the back side.

TIP 23.2   Focal Point ≠ *Focus Point*

The focal point is *not* the point at which light rays focus to form an image. The focal point of a mirror is determined *solely* by its curvature—it doesn't depend on the location of any object.

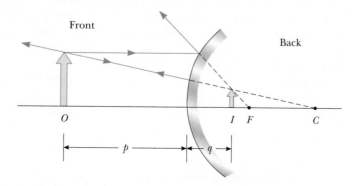

Front   Back

O   I F   C

p   q

Figure 23.11 Formation of an image by a spherical, convex mirror. Note that the image is virtual and upright.

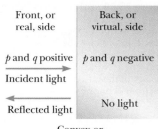

Front, or real, side | Back, or virtual, side

p and q positive | p and q negative

Incident light

Reflected light | No light

Convex or concave mirror

Figure 23.12 A diagram describing the signs of p and q for convex and concave mirrors.

TABLE 23.1

Sign Conventions for Mirrors

Quantity	Symbol	In Front	In Back	Upright Image	Inverted Image
Object location	p	+	−		
Image location	q	+	−		
Focal Length	f	+	−		
Image height	h'			+	−
Magnification	M			+	−

Figure 23.12 is helpful for understanding the rules for object and image distances, and Table 23.1 summarizes the sign conventions for all the necessary quantities. Notice that when the quantities p, q, and f (and R) are located where the light is—in front of the mirror—they are positive, whereas when they are located behind the mirror (where the light isn't), they are negative.

Ray Diagrams for Mirrors

We can conveniently determine the positions and sizes of images formed by mirrors by constructing *ray diagrams* similar to the ones we have been using. This kind of graphical construction tells us the overall nature of the image and can be used to check parameters calculated from the mirror and magnification equations. Making a ray diagram requires knowing the position of the object and the location of the center of curvature. To locate the image, three rays are constructed (rather than just the two we have been constructing so far), as shown by the examples in Active Figure 23.13. All three rays start from the same object point; for these examples, the tip of the arrow was chosen. For the concave mirrors in Active Figure 23.13a and b, the rays are drawn as follows:

1. Ray 1 is drawn parallel to the principal axis and is reflected back through the focal point F.
2. Ray 2 is drawn through the focal point and is reflected parallel to the principal axis.
3. Ray 3 is drawn through the center of curvature, C, and is reflected back on itself.

Note that rays actually go in all directions from the object; we choose to follow those moving in a direction that simplifies our drawing.

The intersection of any *two* of these rays at a point locates the image. The third ray serves as a check of our construction. The image point obtained in this fashion must always agree with the value of q calculated from the mirror formula.

In the case of a concave mirror, note what happens as the object is moved closer to the mirror. The real, inverted image in Active Figure 23.13a moves to the left as the object approaches the focal point. When the object is at the focal point, the image is infinitely far to the left. However, when the object lies between the focal point and the mirror surface, as in Active Figure 23.13b, the image is virtual and upright.

With the convex mirror shown in Active Figure 23.13c, the image of a real object is always virtual and upright. As the object distance increases, the virtual image shrinks and approaches the focal point as p approaches infinity. You should construct a ray diagram to verify this.

The image-forming characteristics of curved mirrors obviously determine their uses. For example, suppose you want to design a mirror that will help

(a)

(b)

(c)

ACTIVE FIGURE 23.13
Ray diagrams for spherical mirrors and corresponding photographs of the images of candles. (a) When an object is outside the center of curvature of a concave mirror, the image is real, inverted, and reduced in size. (b) When an object is between a concave mirror and the focal point, the image is virtual, upright, and magnified. (c) When an object is in front of a convex mirror, the image is virtual, upright, and reduced in size.

Photos courtesy of David Rogers

Physics⊗Now™
Log into PhysicsNow at **www.cp7e.com** and go to Active Figure 23.13, where you can move the objects, change the focal lengths of the mirrors, and see the effect on the images.

people shave or apply cosmetics. For this, you need a concave mirror that puts the user inside the focal point, such as the mirror in Active Figure 23.13b. With that mirror, the image is upright and greatly enlarged. In contrast, suppose that the primary purpose of a mirror is to observe a large field of view. In that case,

Figure 23.14 A convex side-view mirror on a vehicle produces an upright image that is smaller than the object. The smaller image means the object is closer than its apparent distance as observed in the mirror.

you need a convex mirror such as the one in Active Figure 23.13c. The diminished size of the image means that a fairly large field of view is seen in the mirror. Mirrors like this one are often placed in stores to help employees watch for shoplifters. A second use of such a mirror is as a side-view mirror on a car (Fig. 23.14). This kind of mirror is usually placed on the passenger side of the car and carries the warning "Objects are closer than they appear." Without such warning, a driver might think she is looking into a flat mirror, which doesn't alter the size of the image. She could be fooled into believing that a truck is far away because it looks small, when it's actually a large semi very close behind her, but diminished in size because of the image formation characteristics of the convex mirror.

Applying Physics 23.2 Concave versus Convex

A virtual image can be anywhere behind a concave mirror. Why is there a maximum distance at which the image can exist behind a *convex* mirror?

Explanation Consider the concave mirror first, and imagine two different light rays leaving a tiny object and striking the mirror. If the object is at the focal point, the light rays reflecting from the mirror will be parallel to the mirror axis. They can be interpreted as forming a virtual image infinitely far away behind the mirror. As the object is brought closer to the mirror, the reflected rays will diverge through larger and larger angles, resulting in their extensions converging closer

and closer to the back of the mirror. When the object is brought right up to the mirror, the image is right behind the mirror. When the object is much closer to the mirror than the focal length, the mirrors acts like a flat mirror, and the image is just as far behind the mirror as the object is in front of it. The image can therefore be anywhere from infinitely far away to right at the surface of the mirror. For the convex mirror, an object at infinity produces a virtual image at the focal point. As the object is brought closer, the reflected rays diverge more sharply and the image moves closer to the mirror. As a result, the virtual image is restricted to the region between the mirror and the focal point.

Applying Physics 23.3 Reversible Waves

Large trucks often have a sign on the back saying, "If you can't see my mirror, I can't see you." Explain this sign.

Explanation The trucking companies are making use of the principle of the reversibility of light rays. In or-

der for an image of you to be formed in the driver's mirror, there must be a pathway for rays of light to reach the mirror, allowing the driver to see your image. If you can't see the mirror, then this pathway doesn't exist.

INTERACTIVE EXAMPLE 23.2 Images Formed by a Concave Mirror

Goal Calculate properties of a concave mirror.

Problem Assume that a certain concave spherical mirror has a focal length of 10.0 cm. **(a)** Locate the image and find the magnification for an object distance of 25.0 cm. Determine whether the image is real or virtual, inverted or upright, and larger or smaller. Do the same for object distances of **(b)** 10.0 cm and **(c)** 5.00 cm.

Strategy For each part, substitute into the mirror and magnification equations. Part (b) involves a limiting process, because the answers are infinite.

Solution
(a) Find the image position for an object distance of 25.0 cm.
Calculate the magnification and describe the image.

Use the mirror equation to find the image distance:

$$(1) \quad \frac{1}{p} + \frac{1}{q} = \frac{1}{f}$$

Substitute and solve for q. According to Table 23.1, p and f are positive.

$$\frac{1}{25.0 \text{ cm}} + \frac{1}{q} = \frac{1}{10.0 \text{ cm}}$$

$$q = \boxed{16.7 \text{ cm}}$$

Because q is positive, the image is in front of the mirror and is real. The magnification is given by substituting into Equation 23.2:

$$M = -\frac{q}{p} = -\frac{16.7 \text{ cm}}{25.0 \text{ cm}} = \boxed{-0.668}$$

The image is smaller than the object because $|M| < 1$, and inverted because M is negative. (See Fig. 23.13a.)

(b) Locate the image distance when the object distance is 10.0 cm. Calculate the magnification and describe the image.

The object is at the focal point. Substitute $p = 10.0$ cm and $f = 10.0$ cm into the mirror equation:

$$\frac{1}{10.0 \text{ cm}} + \frac{1}{q} = \frac{1}{10.0 \text{ cm}}$$

$$\frac{1}{q} = 0 \quad \rightarrow \quad \boxed{q = \infty}$$

Since $M = -q/p$, the magnification is infinite, also.

(c) Locate the image distance when the object distance is 5.00 cm. Calculate the magnification and describe the image.

Once again, substitute into the mirror equation:

$$\frac{1}{5.00 \text{ cm}} + \frac{1}{q} = \frac{1}{10.0 \text{ cm}}$$

$$\frac{1}{q} = \frac{1}{10.0 \text{ cm}} - \frac{1}{5.00 \text{ cm}} = -\frac{1}{10.0 \text{ cm}}$$

$$q = \boxed{-10.0 \text{ cm}}$$

The image is virtual (behind the mirror) because q is negative. Use Equation 23.2 to calculate the magnification:

$$M = -\frac{q}{p} = -\left(\frac{-10.0 \text{ cm}}{5.00 \text{ cm}} \right) = \boxed{2.00}$$

The image is larger (magnified by a factor of 2) because $|M| > 1$, and upright because M is positive. (See Fig. 23.13b.)

Remarks   Note the characteristics of an image formed by a concave, spherical mirror. When the object is outside the focal point, the image is inverted and real; at the focal point, the image is formed at infinity; inside the focal point, the image is upright and virtual.

Exercise 23.2

If the object distance is 20.0 cm, find the image distance and the magnification of the mirror.

Answer $q = 20.0$ cm, $M = -1.00$

Physics⊗Now™ Investigate the image formed for various object positions and mirror focal lengths by logging into PhysicsNow at **www.cp7e.com** and going to Interactive Example 23.2.

EXAMPLE 23.3   Images Formed by a Convex Mirror

Goal   Calculate properties of a convex mirror.

Problem   An object 3.00 cm high is placed 20.0 cm from a convex mirror with a focal length of 8.00 cm. Find **(a)** the position of the image, **(b)** the magnification of the mirror, and **(c)** the height of the image.

Strategy   This problem again requires only substitution into the mirror and magnification equations. Multiplying the object height by the magnification gives the image height.

Solution
(a) Find the position of the image.

Because the mirror is convex, its focal length is negative. Substitute into the mirror equation:

$$\frac{1}{p} + \frac{1}{q} = \frac{1}{f}$$

$$\frac{1}{20.0 \text{ cm}} + \frac{1}{q} = \frac{1}{-8.00 \text{ cm}}$$

Solve for q:

$$q = \boxed{-5.71 \text{ cm}}$$

(b) Find the magnification of the mirror.

Substitute into Equation 23.2:

$$M = -\frac{q}{p} = -\left(\frac{-5.71 \text{ cm}}{20.0 \text{ cm}}\right) = \boxed{0.286}$$

(c) Find the height of the image.

Multiply the object height by the magnification:

$$h' = hM = (3.00 \text{ cm})(0.286) = \boxed{0.858 \text{ cm}}$$

Remarks   The negative value of q indicates the image is virtual, or behind the mirror, as in Figure 23.13c. The image is upright because M is positive.

Exercise 23.3
Suppose the object is moved so it is 4.00 cm from the same mirror. Repeat parts (a)–(c).

Answer   (a) −2.67 cm   (b) 0.668   (c) 2.00 cm; the image is upright and virtual.

EXAMPLE 23.4   The Face in the Mirror

Goal   Find a focal length from a magnification and an object distance.

Problem   When a woman stands with her face 40.0 cm from a cosmetic mirror, the upright image is twice as tall as her face. What is the focal length of the mirror?

Strategy   To find f in this example, we must first find q, the image distance. Because the problem states that the image is upright, the magnification must be positive (in this case, $M = +2$), and because $M = -q/p$, we can determine q.

Solution
Obtain q from the magnification equation:

$$M = -\frac{q}{p} = 2$$

$$q = -2p = -2(40.0 \text{ cm}) = -80.0 \text{ cm}$$

Because q is negative, the image is on the opposite side of the mirror and hence is virtual. Substitute q and p into the mirror equation and solve for f:

$$\frac{1}{40.0 \text{ cm}} - \frac{1}{80.0 \text{ cm}} = \frac{1}{f}$$

$$f = \boxed{80.0 \text{ cm}}$$

Remarks   The positive sign for the focal length tells us the mirror is concave, a fact we already knew because the mirror magnified the object. (A convex mirror would have produced a smaller image.)

Exercise 23.4
Suppose a fun-house mirror makes you appear to have one-third your normal height. If you are 1.20 m away from the mirror, find its focal length. Is the mirror concave or convex?

Answer   − 0.600 m, convex

23.4   IMAGES FORMED BY REFRACTION

In this section we describe how images are formed by refraction at a spherical surface. Consider two transparent media with indices of refraction n_1 and n_2, where the boundary between the two media is a spherical surface of radius R (Fig. 23.15). We assume that the medium to the right has a higher index of refraction than the one to the left: $n_2 > n_1$. This would be the case for light entering a curved piece of glass from air or for light entering the water in a fishbowl from air. The rays originating at the object location O are refracted at the spherical surface and then converge to the image point I. We can begin with Snell's law of refraction and use simple geometric techniques to show that the object distance, image distance, and radius of curvature are related by the equation

$$\frac{n_1}{p} + \frac{n_2}{q} = \frac{n_2 - n_1}{R} \tag{23.7}$$

Further, the magnification of a refracting surface is

$$M = \frac{h'}{h} = -\frac{n_1 q}{n_2 p} \tag{23.8}$$

As with mirrors, certain sign conventions hold, depending on circumstances. First note that real images are formed by refraction on the side of the surface *opposite* the side from which the light comes, in contrast to mirrors, where real images are formed on the *same* side of the reflecting surface. This makes sense, because light reflects off mirrors, so any real images must form on the same side the light comes from. With a transparent medium, the rays pass through and naturally form real images on the opposite side. We define the side of the surface in which light rays originate as the front side. The other side is called the back side. Because of the difference in location of real images, the refraction sign conventions for q and R are the opposite of those for reflection. For example, p, q, and R are all positive in Figure 23.15. The sign conventions for spherical refracting surfaces are summarized in Table 23.2 (page 766).

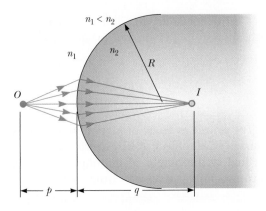

Figure 23.15   An image formed by refraction at a spherical surface. Rays making small angles with the principal axis diverge from a point object at O and pass through the image point I.

TABLE 23.2

Sign Conventions for Refracting Surfaces

Quantity	Symbol	In Front	In Back	Upright Image	Inverted Image
Object location	p	+	−		
Image location	q	−	+		
Radius	R	−	+		
Image height	h'			+	−

Applying Physics 23.4   Underwater Vision

Why does a person with normal vision see a blurry image if the eyes are opened underwater with no goggles or diving mask in use?

Explanation The eye presents a spherical refraction surface. The eye normally functions so that light entering from the air is refracted to form an image in the retina located at the back of the eyeball. The difference in the index of refraction between water and the eye is smaller than the difference in the index of refraction between air and the eye. Consequently, light entering the eye from the water doesn't undergo as much refraction as does light entering from the air, and the image is formed behind the retina. A diving mask or swimming goggles have no optical action of their own; they are simply flat pieces of glass or plastic in a rubber mount. However, they provide a region of air adjacent to the eyes, so that the correct refraction relationship is established and images will be in focus.

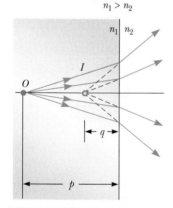

ACTIVE FIGURE 23.16
The image formed by a flat refracting surface is virtual and on the same side of the surface as the object. Note that if the light rays are reversed in direction, we have the situation described in Example 22.6.

Physics⊗Now™
Log into PhysicsNow at **www.cp7e.com** and go to Active Figure 23.16, where you can move the object and observe the effect on the location of the image.

Flat Refracting Surfaces

If the refracting surface is flat, then R approaches infinity and Equation 23.7 reduces to

$$\frac{n_1}{p} = -\frac{n_2}{q}$$

$$q = -\frac{n_2}{n_1}p \qquad \text{[23.9]}$$

From Equation 23.9 we see that the sign of q is opposite that of p. Consequently, **the image formed by a flat refracting surface is on the same side of the surface as the object.** This is illustrated in Active Figure 23.16 for the situation in which n_1 is greater than n_2, where a virtual image is formed between the object and the surface. Note that the refracted ray bends *away* from the normal in this case, because $n_1 > n_2$.

Quick Quiz 23.2

A person spearfishing from a boat sees a fish located 3 m from the boat at an apparent depth of 1 m. To spear the fish, should the person aim (a) at, (b) above, or (c) below the image of the fish?

Quick Quiz 23.3

True or false? (a) The image of an object placed in front of a concave mirror is always upright. (b) The height of the image of an object placed in front of a concave mirror must be smaller than or equal to the height of the object. (c) The image of an object placed in front of a convex mirror is always upright and smaller than the object.

EXAMPLE 23.5   Gaze into the Crystal Ball

Goal   Calculate the properties of an image created by a spherical lens.

Problem   A coin 2.00 cm in diameter is embedded in a solid glass ball of radius 30.0 cm (Fig. 23.17). The index of refraction of the ball is 1.50, and the coin is 20.0 cm from the surface. **(a)** Find the position of the image of the coin, and **(b)** the height of the coin's image.

Strategy   Because the rays are moving from a medium of high index of refraction (the glass ball) to a medium of lower index of refraction (air), those originating at the coin are refracted away from the normal at the surface and diverge outward. The image is formed in the glass and is virtual. Substitute into Equations 23.7 and 23.8 for the image position and magnification, respectively.

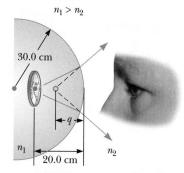

$n_1 > n_2$

Figure 23.17   (Example 23.5) A coin embedded in a glass ball forms a virtual image between the coin and the surface of the glass.

Solution

Apply Equation 23.7, and take $n_1 = 1.50$, $n_2 = 1.00$, $p = 20.0$ cm, and $R = -30.0$ cm:

$$\frac{n_1}{p} + \frac{n_2}{q} = \frac{n_2 - n_1}{R}$$

$$\frac{1.50}{20.0 \text{ cm}} + \frac{1.00}{q} = \frac{1.00 - 1.50}{-30.0 \text{ cm}}$$

Solve for q:

$$q = \boxed{-17.1 \text{ cm}}$$

To find the image height, we use Equation 23.8 for the magnification:

$$M = -\frac{n_1 q}{n_2 p} = -\frac{1.50(-17.1 \text{ cm})}{1.00(20.0 \text{ cm})} = \frac{h'}{h}$$

$$h' = 1.28h = (1.28)(2.00 \text{ cm}) = \boxed{2.56 \text{ cm}}$$

Remarks   The negative sign on q indicates that the image is in the same medium as the object (the side of incident light), in agreement with our ray diagram, and therefore must be virtual. The positive value for M means the image is upright.

Exercise 23.5

A coin is embedded 20.0 cm from the surface of a similar ball of transparent substance having radius 30.0 cm and unknown composition. If the coin's image is virtual and located 15.0 cm from the surface, find **(a)** the index of refraction of the substance and **(b)** the magnification.

Answers   (a) 2.00   (b) 1.50

EXAMPLE 23.6   The One That Got Away

Goal   Calculate the properties of an image created by a flat refractive surface.

Problem   A small fish is swimming at a depth d below the surface of a pond (Fig. 23.18). **(a)** What is the *apparent depth* of the fish as viewed from directly overhead? **(b)** If the fish is 12 cm long, how long is its image?

Strategy   In this example the refracting surface is flat, so R is infinite. Hence, we can use Equation 23.9 to determine the location of the image, which is the apparent location of the fish.

$n_2 = 1.00$
$n_1 = 1.33$

Figure 23.18   (Example 23.6) The apparent depth q of the fish is less than the true depth d.

Solution
(a) Find the apparent depth of the fish.

Substitute $n_1 = 1.33$ for water and $p = d$ into Equation 23.9:

$$q = -\frac{n_2}{n_1}p = -\frac{1}{1.33}d = \boxed{-0.752d}$$

(b) What is the size of the fish's image?

Use Equation 23.9 to eliminate q from the Equation 23.8, the magnification equation:

$$M = \frac{h'}{h} = -\frac{n_1 q}{n_2 p} = -\frac{n_1\left(-\dfrac{n_2}{n_1}p\right)}{n_2 p} = 1$$

$$h' = h = \boxed{12\ \text{cm}}$$

Remarks Again, because q is negative, the image is virtual, as indicated in Figure 23.18. The apparent depth is three-fourths the actual depth. For instance, if $d = 4.0$ m, then $q = -3.0$ m.

Exercise 23.6

A spear fisherman estimates that a trout is 1.5 m below the water's surface. What is the actual depth of the fish?

Answer   2.0 m

23.5 ATMOSPHERIC REFRACTION

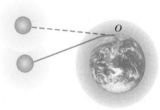

Figure 23.19 Because light is refracted by the Earth's atmosphere, an observer at O sees the Sun even though it has fallen below the horizon.

Images formed by refraction in our atmosphere lead to some interesting phenomena. One such phenomenon that occurs daily is the visibility of the Sun at dusk even though it has passed below the horizon. Figure 23.19 shows why this occurs. Rays of light from the Sun strike the Earth's atmosphere (represented by the shaded area around the planet) and are bent as they pass into a medium that has an index of refraction different from that of the almost empty space in which they have been traveling. The bending in this situation differs somewhat from the bending we have considered previously in that it is gradual and continuous as the light moves through the atmosphere toward an observer at point O. This is because the light moves through layers of air that have a continuously changing index of refraction. When the rays reach the observer, the eye follows them back along the direction from which they appear to have come (indicated by the dashed path in the figure). The end result is that the Sun appears to be above the horizon even after it has fallen below it.

(a)

(b)

John M. Dunay IV, Fundamental Photographs, NYC

Figure 23.20 (a) A mirage is produced by the bending of light rays in the atmosphere when there are large temperature differences between the ground and the air. (b) Notice the reflection of the cars in this photograph of a mirage. The road looks like it's flooded with water, but is actually dry.

The **mirage** is another phenomenon of nature produced by refraction in the atmosphere. A mirage can be observed when the ground is so hot that the air directly above it is warmer than the air at higher elevations. The desert is a region in which such circumstances prevail, but mirages are also seen on heated roadways during the summer. The layers of air at different heights above the Earth have different densities and different refractive indices. The effect this can have is pictured in Figure 23.20a. The observer sees the sky and a tree in two different ways. One group of light rays reaches the observer by the straight-line path A, and the eye traces these rays back to see the tree in the normal fashion. In addition, a second group of rays travels along the curved path B. These rays are directed toward the ground and are then bent as a result of refraction. As a consequence, the observer also sees an inverted image of the tree and the background of the sky as he traces the rays back to the point at which they appear to have originated. Because an upright image and an inverted image are seen when the image of a tree is observed in a reflecting pool of water, the observer unconsciously calls on this past experience and concludes that the sky is reflected by a pool of water in front of the tree.

23.6   THIN LENSES

A typical **thin lens** consists of a piece of glass or plastic, ground so that each of its two refracting surfaces is a segment of either a sphere or a plane. Lenses are commonly used to form images by refraction in optical instruments, such as cameras, telescopes, and microscopes. The equation that relates object and image distances for a lens is virtually identical to the mirror equation derived earlier, and the method used to derive it is also similar.

Figure 23.21 shows some representative shapes of lenses. Notice that we have placed these lenses in two groups. Those in Figure 23.21a are thicker at the center than at the rim, and those in Figure 23.21b are thinner at the center than at the rim. The lenses in the first group are examples of **converging lenses**, and those in the second group are **diverging lenses**. The reason for these names will become apparent shortly.

As we did for mirrors, it is convenient to define a point called the **focal point** for a lens. For example, in Figure 23.22a (page 770), a group of rays parallel to the axis passes through the focal point F after being converged by the lens. The distance from the focal point to the lens is called the **focal length** f. **The focal length is the image distance that corresponds to an infinite object distance**. Recall that we are considering the lens to be very thin. As a result, it makes no difference whether we take the focal length to be the distance from the focal point to the surface of the lens or the distance from the focal point to the center of the lens, because the difference between these two lengths is negligible. A thin lens has *two* focal points, as illustrated in Figure 23.22, one on each side of the lens. One focal point corresponds to parallel rays traveling from the left and the other corresponds to parallel rays traveling from the right.

Rays parallel to the axis diverge after passing through a lens of biconcave shape, shown in Figure 23.22b. In this case, the focal point is defined to be the point at which the diverged rays appear to originate, labeled F in the figure. Figures 23.22a

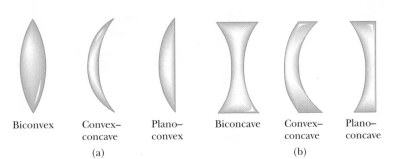

Biconvex   Convex–concave   Plano–convex   Biconcave   Convex–concave   Plano–concave

(a)   (b)

Figure 23.21 Various lens shapes. (a) Converging lenses have positive focal lengths and are thickest at the middle. (b) Diverging lenses have negative focal lengths and are thickest at the edges.

(a)

(b)

Figure 23.22 (*Left*) Photographs of the effects of converging and diverging lenses on parallel rays. (*Right*) The focal points of (a) the biconvex lens and (b) the biconcave lens.

and 23.22b indicate why the names *converging* and *diverging* are applied to these lenses.

Now consider a ray of light passing through the center of a lens. Such a ray is labeled ray 1 in Figure 23.23. For a thin lens, a ray passing through the center is undeflected. Ray 2 in the same figure is parallel to the principal axis of the lens (the horizontal axis passing through O), and as a result it passes through the focal point F after refraction. Rays 1 and 2 intersect at the point that is the tip of the image arrow.

We first note that the tangent of the angle α can be found by using the blue and gold shaded triangles in Figure 23.23:

$$\tan \alpha = \frac{h}{p} \quad \text{or} \quad \tan \alpha = -\frac{h'}{q}$$

From this we find that

$$M = \frac{h'}{h} = -\frac{q}{p} \qquad \text{[23.10]}$$

The equation for magnification by a lens is the same as the equation for magnification by a mirror. We also note from Figure 23.23 that

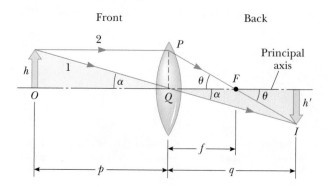

Figure 23.23 A geometric construction for developing the thin-lens equation.

TABLE 23.3

Sign Conventions for Thin Lenses

Quantity	Symbol	In Front	In Back	Convergent	Divergent
Object location	p	+	−		
Image location	q	−	+		
Lens Radii	R_1, R_2	−	+		
Focal Length	f			+	−

$$\tan \theta = \frac{PQ}{f} \quad \text{or} \quad \tan \theta = -\frac{h'}{q - f}$$

However, the height PQ used in the first of these equations is the same as h, the height of the object. Therefore,

$$\frac{h}{f} = -\frac{h'}{q - f}$$

$$\frac{h'}{h} = -\frac{q - f}{f}$$

Using the latter equation in combination with Equation 23.10 gives

$$\frac{q}{p} = \frac{q - f}{f}$$

which reduces to

$$\frac{1}{p} + \frac{1}{q} = \frac{1}{f} \qquad \text{[23.11]}$$

◀ Thin-lens equation

This equation, called the **thin-lens equation**, can be used with both converging and diverging lenses if we adhere to a set of sign conventions. Figure 23.24 is useful for obtaining the signs of p and q, and Table 23.3 gives the complete sign conventions for lenses. Note that **a converging lens has a positive focal length** under this convention and **a diverging lens has a negative focal length**. Hence the names *positive* and *negative* are often given to these lenses.

The focal length for a lens in air is related to the curvatures of its front and back surfaces and to the index of refraction n of the lens material by

$$\frac{1}{f} = (n - 1)\left(\frac{1}{R_1} - \frac{1}{R_2}\right) \qquad \text{[23.12]}$$

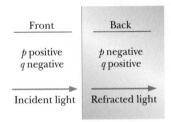

Figure 23.24 A diagram for obtaining the signs of p and q for a thin lens or a refracting surface.

where R_1 is the radius of curvature of the front surface of the lens and R_2 is the radius of curvature of the back surface. (As with mirrors, we arbitrarily call the side from which the light approaches the *front* of the lens.) Table 23.3 gives the sign conventions for R_1 and R_2. Equation 23.12, called the **lens maker's equation**, enables us to calculate the focal length from the known properties of the lens.

Ray Diagrams for Thin Lenses

Ray diagrams are essential for understanding the overall image formation by a thin lens or a system of lenses. They should also help clarify the sign conventions we have already discussed. Active Figure 23.25 (page 772) illustrates this method for three single-lens situations. To locate the image formed by a converging lens (Active Fig. 23.25a and b), the following three rays are drawn from the top of the object:

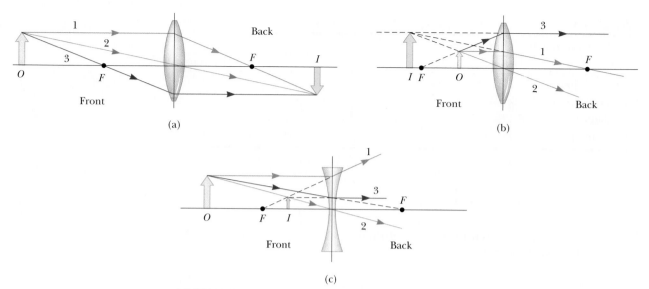

(a)

(b)

(c)

ACTIVE FIGURE 23.25
Ray diagrams for locating the image of an object. (a) The object is outside the focal point of a converging lens. (b) The object is inside the focal point of a converging lens. (c) The object is outside the focal point of a diverging lens.

Physics⊗Now™
Log into PhysicsNow at **www.cp7e.com** and go to Active Figure 23.25, where you can move the objects and change the focal lengths of the lenses, observing the effect on the images.

1. The first ray is drawn parallel to the principal axis. After being refracted by the lens, this ray passes through (or appears to come from) one of the focal points.
2. The second ray is drawn through the center of the lens. This ray continues in a straight line.
3. The third ray is drawn through the other focal point and emerges from the lens parallel to the principal axis.

> **TIP 23.3 We Choose Only a Few Rays**
>
> Although our ray diagrams in Figure 23.25 only show three rays leaving an object, an infinite number of rays can be drawn between the object and its image.

A similar construction is used to locate the image formed by a diverging lens, as shown in Active Figure 23.25c. The point of intersection of *any two* of the rays in these diagrams can be used to locate the image. The third ray serves as a check on construction.

For the converging lens in Active Figure 23.25a, where the object is *outside* the front focal point ($p > f$), the ray diagram shows the image is real and inverted. When the real object is *inside* the front focal point ($p < f$), as in Active Figure 23.25b, the image is virtual and upright. For the diverging lens of Active Figure 23.25c, the image is virtual and upright.

Quick Quiz 23.4

A clear plastic sandwich bag filled with water can act as a crude converging lens in air. If the bag is filled with air and placed under water, is the effective lens (a) converging or (b) diverging?

Quick Quiz 23.5

In Active Figure 23.25a, the blue object arrow is replaced by one that is much taller than the lens. How many rays from the object will strike the lens?

Quick Quiz 23.6

An object is placed to the left of a converging lens. Which of the following statements are true and which are false? (a) The image is always to the right of the lens. (b) The image can be upright or inverted. (c) The image is always smaller or the same size as the object. Justify your answers with ray diagrams.

Your success in working lens or mirror problems will be determined largely by whether you make sign errors when substituting into the lens or mirror equations. The only way to ensure that you don't make sign errors is to become adept at using the sign conventions. The best way to do this is to work a multitude of problems on your own and to construct confirming ray diagrams. Watching an instructor or reading the example problems is no substitute for practice.

Applying Physics 23.5   Vision and Diving Masks

Diving masks often have a lens built into the glass faceplate for divers who don't have perfect vision. This allows the individual to dive without the necessity of glasses, because the faceplate performs the necessary refraction to produce clear vision. Normal glasses have lenses that are curved on both the front and rear surfaces. The lenses in a diving-mask faceplate often have curved surfaces only on the inside of the glass. Why is this design desirable?

Solution The main reason for curving only the inner surface of the lenses in the diving-mask faceplate is to enable the diver to see clearly while underwater and in the air. If there were curved surfaces on both the front and the back of the diving lens, there would be two refractions. The lens could be designed so that these two refractions would give clear vision while the diver is in air. When the diver went underwater, however, the refraction between the water and the glass at the first interface would differ, because the index of refraction of water is different from that of air. Consequently, the diver's vision wouldn't be clear underwater.

INTERACTIVE EXAMPLE 23.7   Images Formed by a Converging Lens

Goal   Calculate geometric quantities associated with a converging lens.

Problem   A converging lens of focal length 10.0 cm forms images of an object situated at various distances. **(a)** If the object is placed 30.0 cm from the lens, locate the image, state whether it's real or virtual, and find its magnification. **(b)** Repeat the problem when the object is at 10.0 cm and **(c)** again when the object is 5.00 cm from the lens.

Strategy   All three problems require only substitution into the thin-lens equation and the associated magnification equation—Equations 23.10 and 23.11, respectively. The conventions of Table 23.3 must be followed.

Solution
(a) Find the image distance and describe the image when the object is placed at 30.0 cm.

The ray diagram is shown in Figure 23.26a. Substitute into the thin-lens equation to locate the image:

$$\frac{1}{p} + \frac{1}{q} = \frac{1}{f}$$

$$\frac{1}{30.0 \text{ cm}} + \frac{1}{q} = \frac{1}{10.0 \text{ cm}}$$

Solve for q, the image distance. It's positive, so the image is real and on the far side of the lens.

$$q = \;+15.0 \text{ cm}$$

The magnification of the lens is obtained from Equation 23.10. M is negative and less than one in absolute value, so the image is inverted and smaller than the object:

$$M = -\frac{q}{p} = -\frac{15.0 \text{ cm}}{30.0 \text{ cm}} = -0.500$$

(b) Repeat the problem, when the object is placed at 10.0 cm.

Locate the image by substituting into the thin-lens equation:

$$\frac{1}{10.0 \text{ cm}} + \frac{1}{q} = \frac{1}{10.0 \text{ cm}} \;\rightarrow\; \frac{1}{q} = 0$$

This equation is satisfied only in the limit as q becomes infinite.

$$q \;\rightarrow\; \infty$$

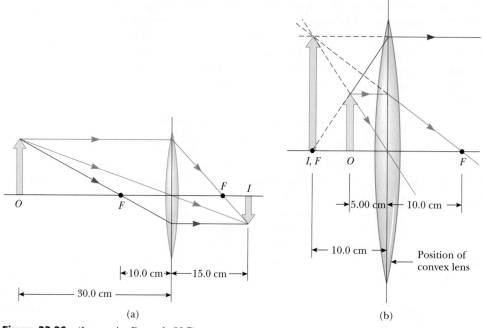

Figure 23.26   (Interactive Example 23.7)

(c) Repeat the problem when the object is placed 5.00 cm from the lens.

See the ray diagram in Figure 23.26b. Substitute into the thin-lens equation to locate the image:

$$\frac{1}{5.00\ cm} + \frac{1}{q} = \frac{1}{10.0\ cm}$$

Solve for q, which is negative, meaning the image is on the same side as the object and is virtual:

$$q = \boxed{-10.0\ cm}$$

Substitute the values of p and q into the magnification equation. M is positive and larger than one, so the image is upright and double the object size:

$$M = -\frac{q}{p} = -\left(\frac{-10.0\ cm}{5.00\ cm}\right) = \boxed{+2.00}$$

Remarks   The ability of a lens to magnify objects led to the inventions of reading glasses, microscopes, and telescopes.

Exercise 23.7
Suppose the image of an object is upright and magnified 1.75 times when the object is placed 15.0 cm from a lens. Find the location of the image and the focal length of the lens.

Answers   (a) −26.3 cm (virtual, on the same side as the object)   (b) 34.9 cm

Physics⊠Now™   Investigate the image formed for various object positions and mirror focal lengths by logging into PhysicsNow at **www.cp7e.com** and going to Interactive Example 23.7.

INTERACTIVE EXAMPLE 23.8   The Case of a Diverging Lens

Goal   Calculate geometric quantities associated with a diverging lens.

Problem   Repeat the problem of Example 23.7 for a *diverging* lens of focal length 10.0 cm.

Strategy   Once again, substitution into the thin-lens equation and the associated magnification equation, together with the conventions in Table 23.3, solve the various parts. The only difference is the negative focal length.

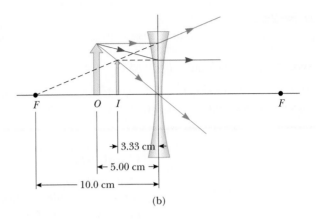

Figure 23.27   (Interactive Example 23.8)

Solution

(a) Locate the image and its magnification if the object is at 30.0 cm.

The ray diagram is given in Figure 23.27a. Apply the thin-lens equation with $p = 30.0$ cm to locate the image:

$$\frac{1}{p} + \frac{1}{q} = \frac{1}{f}$$

$$\frac{1}{30.0 \text{ cm}} + \frac{1}{q} = -\frac{1}{10.0 \text{ cm}}$$

Solve for q, which is negative, hence virtual:

$$q = \boxed{-7.50 \text{ cm}}$$

Substitute into Equation 23.10 to get the magnification. Because M is positive and has absolute value less than one, the image is upright and smaller than the object.

$$M = -\frac{q}{p} = -\left(\frac{-7.50 \text{ cm}}{30.0 \text{ cm}}\right) = \boxed{+0.250}$$

(b) Locate the image and find its magnification if the object is 10.0 cm from the lens.

Apply the thin-lens equation, taking $p = 10.0$ cm:

$$\frac{1}{10.0 \text{ cm}} + \frac{1}{q} = -\frac{1}{10.0 \text{ cm}}$$

Solve for q (once again, the result is negative, so the image is virtual):

$$q = \boxed{-5.00 \text{ cm}}$$

Calculate the magnification. Because M is positive and has absolute value less than 1, the image is upright and smaller than the object.

$$M = -\frac{q}{p} = -\left(\frac{-5.00 \text{ cm}}{10.0 \text{ cm}}\right) = \boxed{+0.500}$$

(c) Locate the image and find its magnification when the object is at 5.00 cm.

The ray diagram is given in Figure 23.27b. Substitute $p = 5.00$ cm into the thin-lens equation to locate the image:

$$\frac{1}{5.00 \text{ cm}} + \frac{1}{q} = -\frac{1}{10.0 \text{ cm}}$$

Solve for q. The answer is negative, so once again the image is virtual.

$$q = \boxed{-3.33 \text{ cm}}$$

Calculate the magnification. Because M is positive and less than one, the image is upright and smaller than the object.

$$M = -\left(\frac{-3.33 \text{ cm}}{5.00 \text{ cm}}\right) = \boxed{+0.666}$$

Remarks   Notice that in every case the image is virtual, hence on the same side of the lens as the object. Further, the image is smaller than the object. For a diverging lens and a real object, this is *always* the case, as can be proven mathematically.

Exercise 23.8

Repeat the calculation, finding the position of the image and the magnification if the object is 20.0 cm from the lens.

Answers $q = -6.67$ cm, $M = 0.334$

Physics⊗Now™ Investigate the image formed for various object positions and mirror focal lengths by logging into PhysicsNow at **www.cp7e.com** and going to Interactive Example 23.7.

Combinations of Thin Lenses

Many useful optical devices require two lenses. Handling problems involving two lenses is not much different from dealing with a single-lens problem twice. First, the image produced by the first lens is calculated as though the second lens were not present. The light then approaches the second lens *as if* it had come from the image formed by the first lens. Hence, **the image formed by the first lens is treated as the object for the second lens**. The image formed by the second lens is the final image of the system. If the image formed by the first lens lies on the back side of the second lens, then the image is treated as a virtual object for the second lens, so p is negative. The same procedure can be extended to a system of three or more lenses. The overall magnification of a system of thin lenses is the *product* of the magnifications of the separate lenses.

INTERACTIVE EXAMPLE 23.9   Two Lenses in a Row

Goal Calculate geometric quantities for a sequential pair of lenses.

Problem   Two converging lenses are placed 20.0 cm apart, as shown in Figure 23.28a, with an object 30.0 cm in front of lens 1 on the left. **(a)** If lens 1 has a focal length of 10.0 cm, locate the image formed by this lens and determine its

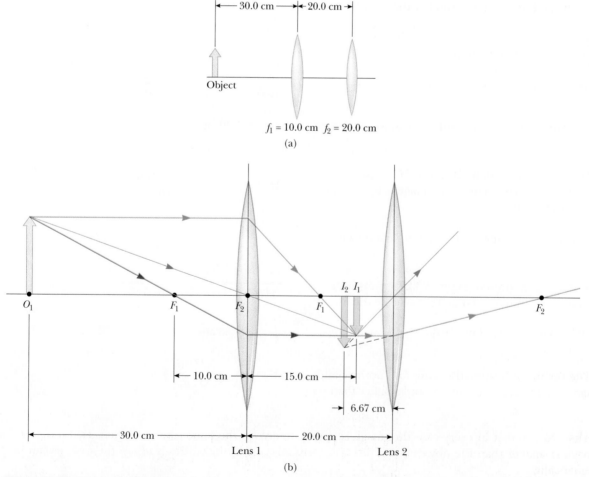

Figure 23.28   (Interactive Example 23.9)

magnification. **(b)** If lens 2 on the right has a focal length of 20.0 cm, locate the final image formed and find the total magnification of the system.

Strategy   We apply the thin-lens equation to each lens. The image formed by lens 1 is treated as the object for lens 2. Also, we use the fact that the total magnification of the system is the product of the magnifications produced by the separate lenses.

Solution
(a) Locate the image and determine the magnification of lens 1.

See the ray diagram, Figure 23.28b. Apply the thin-lens equation to lens 1:

$$\frac{1}{30.0 \text{ cm}} + \frac{1}{q} = \frac{1}{10.0 \text{ cm}}$$

Solve for q, which is positive, hence to the right of the first lens:

$$q = \boxed{+15.0 \text{ cm}}$$

Compute the magnification of lens 1:

$$M_1 = -\frac{q}{p} = -\frac{15.0 \text{ cm}}{30.0 \text{ cm}} = \boxed{-0.500}$$

(b) Locate the final image, and total magnification.

The image formed by lens 1 becomes the object for lens 2. Compute the object distance for lens 2:

$$p = 20.0 \text{ cm} - 15.0 \text{ cm} = 5.00 \text{ cm}$$

Once again apply the thin-lens equation to lens 2 to locate the final image:

$$\frac{1}{5.00 \text{ cm}} + \frac{1}{q} = \frac{1}{20.0 \text{ cm}}$$

$$q = \boxed{-6.67 \text{ cm}}$$

Calculate the magnification of lens 2:

$$M_2 = -\frac{q}{p} = -\frac{(-6.67 \text{ cm})}{5.00 \text{ cm}} = +1.33$$

Multiply the two magnifications to get the overall magnification of the system:

$$M = M_1 M_2 = (-0.500)(1.33) = \boxed{-0.665}$$

Remarks   The negative sign for M indicates that the final image is inverted, and smaller than the object because the absolute value of M is less than one. Because q is negative, the final image is virtual.

Exercise 23.9
If the two lenses in Figure 23.28 are separated by 10.0 cm, locate the final image and find the magnification of the system. [*Hint:* The object for the second lens is virtual!]

Answer   4.00 cm behind the second lens; $M = -0.400$

Physics Now™   Investigate the image formed by a combination of two lenses by logging into PhysicsNow at **www.cp7e.com** and going to Interactive Example 23.9.

23.7   LENS AND MIRROR ABERRATIONS

One of the basic problems of systems containing mirrors and lenses is the imperfect quality of the images, which is largely the result of defects in shape and form. The simple theory of mirrors and lenses assumes that rays make small angles with the principal axis and that all rays reaching the lens or mirror from a point source are focused at a single point, producing a sharp image. This is not always true in the real world. Where the approximations used in this theory do not hold, imperfect images are formed.

   If one wishes to analyze image formation precisely, it is necessary to trace each ray, using Snell's law, at each refracting surface. This procedure shows that there is no single point image; instead, the image is blurred. The departures of real (imperfect)

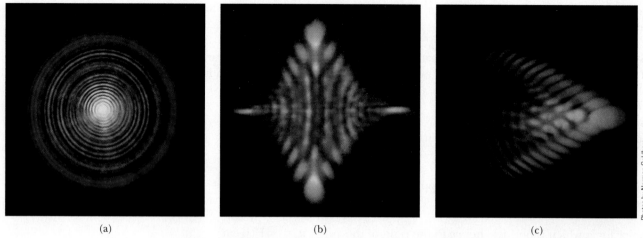

(a)   (b)   (c)

Figure 23.29 Lenses can produce various forms of aberrations, as shown by these blurred photographic images of a point source. (a) Spherical aberration occurs when light passing through the lens at different distances from the principal axis is focused at different points. (b) Astigmatism is an aberration that occurs when the object is not on the principal axis of the lens. (c) Coma. This aberration occurs when light passing through the lens far from the principal axis focuses at a different part of the focal plane than light passing near the center of the lens.

images from the ideal predicted by the simple theory are called **aberrations**. Two common types of aberrations are spherical aberration and chromatic aberration. Photographs of three forms of lens aberrations are shown in Figure 23.29.

Spherical Aberration

Spherical aberration results from the fact that the focal points of light rays passing far from the principal axis of a spherical lens (or mirror) are different from the focal points of rays with the same wavelength passing near the axis. Figure 23.30 illustrates spherical aberration for parallel rays passing through a converging lens. Rays near the middle of the lens are imaged farther from the lens than rays at the edges. Hence, there is no single focal length for a spherical lens.

Most cameras are equipped with an adjustable aperture to control the light intensity and, when possible, reduce spherical aberration. (An aperture is an opening that controls the amount of light transmitted through the lens.) As the aperture size is reduced, sharper images are produced, because only the central portion of the lens is exposed to the incident light when the aperture is very small. At the same time, however, progressively less light is imaged. To compensate for this loss, a longer exposure time is used. An example of the results obtained with small apertures is the sharp image produced by a pinhole camera, with an aperture size of approximately 0.1 mm.

In the case of mirrors used for very distant objects, one can eliminate, or at least minimize, spherical aberration by employing a parabolic rather than spherical surface. Parabolic surfaces are not used in many applications, however, because they are very expensive to make with high-quality optics. Parallel light rays incident on such a surface focus at a common point. Parabolic reflecting surfaces are used in many astronomical telescopes to enhance the image quality. They are also used in flashlights, in which a nearly parallel light beam is produced from a small lamp placed at the focus of the reflecting surface.

Chromatic Aberration

The fact that different wavelengths of light refracted by a lens focus at different points gives rise to chromatic aberration. In Chapter 22 we described how the index of refraction of a material varies with wavelength. When white light passes through a lens, for example, violet light rays are refracted more than red light rays (see Fig. 23.31), so the focal length for red light is greater than for violet light.

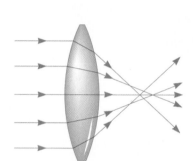

Figure 23.30 Spherical aberration produced by a converging lens. Does a diverging lens produce spherical aberration? (Angles are greatly exaggerated for clarity.)

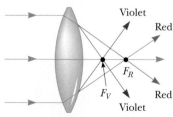

Figure 23.31 Chromatic aberration produced by a converging lens. Rays of different wavelengths focus at different points. (Angles are greatly exaggerated for clarity.)

Other wavelengths (not shown in the figure) would have intermediate focal points. Chromatic aberration for a diverging lens is opposite that for a converging lens. Chromatic aberration can be greatly reduced by a combination of converging and diverging lenses.

SUMMARY

Physics⊗Now™ Take a practice test by logging into Physics-Now at **www.cp7e.com** and clicking on the Pre-Test link for this chapter.

23.1 Flat Mirrors

Images are formed where rays of light intersect or where they appear to originate. A **real image** is formed when light intersects, or passes through, an image point. In a **virtual image** the light doesn't pass through the image point, but appears to diverge from it.

The image formed by a flat mirror has the following properties:

1. The image is as far behind the mirror as the object is in front.
2. The image is unmagnified, virtual, and upright.

23.2 Images Formed by Spherical Mirrors &
23.3 Convex Mirrors and Sign Conventions

The **magnification** M of a spherical mirror is defined as the ratio of the **image height** h' to the **object height** h, which is the negative of the ratio of the image distance q to the object distance p:

$$M = \frac{h'}{h} = -\frac{q}{p} \qquad [23.2]$$

The **object distance** and **image distance** for a spherical mirror of radius R are related by the **mirror equation**:

$$\frac{1}{p} + \frac{1}{q} = \frac{1}{f} \qquad [23.6]$$

where $f = R/2$ is the **focal length** of the mirror.

Equations 23.2 and 23.6 hold for both concave and convex mirrors, subject to the sign conventions given in Table 23.1.

23.4 Images Formed by Refraction

An image can be formed by refraction at a spherical surface of radius R. The object and image distances for refraction from such a surface are related by

$$\frac{n_1}{p} + \frac{n_2}{q} = \frac{n_2 - n_1}{R} \qquad [23.7]$$

The **magnification of a refracting surface** is

$$M = \frac{h'}{h} = -\frac{n_1 q}{n_2 p} \qquad [23.8]$$

where the object is located in the medium with index of refraction n_1 and the image is formed in the medium with index of refraction n_2. Equations 23.7 and 23.8 are subject to the sign conventions of Table 23.2.

23.6 Thin Lenses

The **magnification of a thin lens** is

$$M = \frac{h'}{h} = -\frac{q}{p} \qquad [23.10]$$

The object and image distances of a thin lens are related by the **thin-lens equation**:

$$\frac{1}{p} + \frac{1}{q} = \frac{1}{f} \qquad [23.11]$$

Equations 23.10 and 23.11 are subject to the sign conventions of Table 23.3.

23.7 Lens and Mirror Aberrations

Aberrations are responsible for the formation of imperfect images by lenses and mirrors. **Spherical aberration** results from the fact that the focal points of light rays far from the principal axis of a spherical lens or mirror are different from those of rays passing through the center. **Chromatic aberration** arises from the fact that light rays of different wavelengths focus at different points when refracted by a lens.

CONCEPTUAL QUESTIONS

1. Tape a picture of yourself on a bathroom mirror. Stand several centimeters away from the mirror. Can you focus your eyes on *both* the picture taped to the mirror *and* your image in the mirror *at the same time*? So where is the image of yourself?

2. One method for determining the position of an image, either real or virtual, is by means of *parallax*. If a finger or another object is placed at the position of the image, as shown in Figure Q23.2, and the finger and the image are viewed simultaneously (the image is viewed through the lens if it is virtual), the finger and image have the same parallax; that is, if the image is viewed from different

positions, it will appear to move along with the finger. Use this method to locate the image formed by a lens. Explain why the method works.

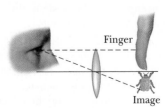

Figure Q23.02

3. A flat mirror creates a virtual image of your face. Suppose the flat mirror is combined with another optical element. Can the mirror form a real image in such a combination?

4. Explain why a mirror cannot give rise to chromatic aberration.

5. You are taking a picture of yourself with a camera that uses an ultrasonic range finder to measure the distance to the object. When you take a picture of yourself in a mirror with this camera, your image is out of focus. Why?

6. A solar furnace can be constructed by using a concave mirror to reflect and focus sunlight into the enclosure of a furnace. What factors in the design of the reflecting mirror will guarantee that very high temperatures can be reached?

7. A virtual image is often described as an image through which light rays don't actually travel, as they do for a real image. Can a virtual image be photographed?

8. What is wrong with the caption of the cartoon shown in Figure Q23.8?

Figure Q23.08 "Most mirrors reverse left and right. This one reverses top and bottom."

9. Suppose you want to use a converging lens to project the image of two trees onto a screen. One tree is a distance x from the lens; the other is at $2x$, as in Figure Q23.9. You adjust the screen so that the near tree is in focus. If you now want the far tree to be in focus, do you move the screen towards or away from the lens?

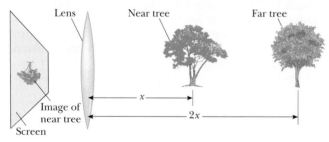

Figure Q23.09

10. Why does a clear stream always appear to be shallower than it actually is?

11. Can a converging lens be made to diverge light if placed in a liquid? How about a converging mirror?

12. A common mirage is formed when the air gets gradually cooler as the height above the ground increases. What might happen if the air grows gradually warmer as the height increases? This often happens over bodies of water or snow-covered ground; the effect is called *looming*.

13. In a Jules Verne novel, a piece of ice is shaped into a magnifying lens to focus sunlight to start a fire. Is this possible?

14. Lenses used in eyeglasses, whether converging or diverging, are always designed such that the middle of the lens curves away from the eye. Why?

15. Why does the focal length of a mirror not depend on the mirror material when the focal length of a lens does depend on the lens material?

16. If a cylinder of solid glass or clear plastic is placed above the words LEAD OXIDE and viewed from the side, as shown in Figure Q23.16, the word LEAD appears inverted, but the word OXIDE does not. Explain.

Figure Q23.16

17. A concave makeup mirror has a focal length of 15 cm. (a) If an object is placed 25 cm in front of the mirror, determine the signs of the focal length, the object distance, and the image distance. (b) Repeat part (a) if the object is placed 5 cm in front of the mirror.

18. Light from an object passes through a lens and forms a visible image on a screen. If the screen is removed, would you be able to see the image if (a) you remained in your present position? (b) you could look at the lens along the its axis, beyond the original position of the screen?

19. An object placed to the left of a converging lens forms a sharp image on a screen to the right of the lens. If the screen is moved towards the lens, the image on the screen (a) gets larger, but remains sharp, (b) gets smaller, but remains sharp, (c) becomes fuzzy and disappears, or (d) remains unchanged.

20. An inverted image of an object is viewed on a screen from the side facing a converging lens. An opaque card is then introduced covering only the upper half of the lens. What happens to the image on the screen? (a) Half the image would disappear. (b) The entire image would appear and remain unchanged. (c) Half the image would disappear and be dimmer. (d) The entire image would appear, but would be dimmer.

PROBLEMS

1, 2, 3 = straightforward, intermediate, challenging   □ = full solution available in *Student Solutions Manual/Study Guide*
Physics⊗Now™ = coached solution with hints available at **www.cp7e.com**   🔥 = biomedical application

Section 23.1 Flat Mirrors

1. Does your bathroom mirror show you older or younger than your actual age? Compute an order-of-magnitude estimate for the age difference, based on data that you specify.

2. Use Active Figure 23.2 to give a geometric proof that the virtual image formed by a plane mirror is the same distance behind the mirror as the object is in front of it.

3. A person walks into a room that has, on opposite walls, two plane mirrors producing multiple images. Find the distances from the person to the first three images seen in the left-hand mirror when the person is 5.00 ft from the mirror on the left wall and 10.0 ft from the mirror on the right wall.

4. In a church choir loft, two parallel walls are 5.30 m apart. The singers stand against the north wall. The organist faces the south wall, sitting 0.800 m away from it. So that she can see the choir, a flat mirror 0.600 m wide is mounted on the south wall, straight in front of the organist. What width of the north wall can she see? [*Hint:* Draw a top-view diagram to justify your answer.]

Section 23.2 Images Formed by Spherical Mirrors
Section 23.3 Convex Mirrors and Sign Conventions

In the following problems, algebraic signs are not given. We leave it to you to determine the correct sign to use with each quantity, based on an analysis of the problem and the sign conventions in Table 23.1.

5. At an intersection of hospital hallways, a convex mirror is mounted high on a wall to help people avoid collisions. The mirror has a radius of curvature of 0.550 m. Locate and describe the image of a patient 10.0 m from the mirror. Determine the magnification of the image.

6. To fit a contact lens to a patient's eye, a *keratometer* can be used to measure the curvature of the cornea—the front surface of the eye. This instrument places an illuminated object of known size at a known distance p from the cornea, which then reflects some light from the object, forming an image of it. The magnification M of the image is measured by using a small viewing telescope that allows a comparison of the image formed by the cornea with a second calibrated image projected into the field of view by a prism arrangement. Determine the radius of curvature of the cornea when $p = 30.0$ cm and $M = 0.013\ 0$.

7. A concave spherical mirror has a radius of curvature of 20.0 cm. Locate the images for object distances of (a) 40.0 cm, (b) 20.0 cm, and (c) 10.0 cm. In each case, state whether the image is real or virtual and upright or inverted, and find the magnification.

8. A dentist uses a mirror to examine a tooth that is 1.00 cm in front of the mirror. The image of the tooth is formed 10.0 cm behind the mirror. Determine (a) the mirror's radius of curvature and (b) the magnification of the image.

9. A large church has a niche in one wall. On the floor plan it appears as a semicircular indentation of radius 2.50 m. A worshiper stands on the centerline of the niche, 2.00 m out from its deepest point, and whispers a prayer. Where is the sound concentrated after it reflects from the back wall of the niche?

10. While looking at her image in a cosmetic mirror, Dina notes that her face is highly magnified when she is close to the mirror, but as she backs away from the mirror, her image first becomes blurry, then disappears when she is about 30 cm from the mirror, and then inverts when she is beyond 30 cm. Based on these observations, what can she conclude about the properties of the mirror?

11. A 2.00-cm-high object is placed 3.00 cm in front of a concave mirror. If the image is 5.00 cm high and virtual, what is the focal length of the mirror?

12. A dedicated sports car enthusiast polishes the inside and outside surfaces of a hubcap that is a section of a sphere. When he looks into one side of the hubcap, he sees an image of his face 30.0 cm in back of it. He then turns the hubcap over, keeping it the same distance from his face. He now sees an image of his face 10.0 cm in back of the hubcap. (a) How far is his face from the hubcap? (b) What is the radius of curvature of the hubcap?

13. Physics⊗Now™ A concave makeup mirror is designed so that a person 25 cm in front of it sees an upright image magnified by a factor of two. What is the radius of curvature of the mirror?

14. A certain Christmas tree ornament is a silver sphere having a diameter of 8.50 cm. Determine an object location for which the size of the reflected image is three-fourths the size of the object. Use a principal-ray diagram to arrive at a description of the image.

15. A man standing 1.52 m in front of a shaving mirror produces an inverted image 18.0 cm in front of it. How close to the mirror should he stand if he wants to form an upright image of his chin that is twice the chin's actual size?

16. A convex spherical mirror with a radius of curvature of 10.0 cm produces a virtual image one-third the size of the real object. Where is the object?

17. A child holds a candy bar 10.0 cm in front of a convex mirror and notices that the image is only one-half the size of the candy bar. What is the radius of curvature of the mirror?

18. It is observed that the size of a *real* image formed by a concave mirror is four times the size of the object when the object is 30.0 cm in front of the mirror. What is the radius of curvature of this mirror?

19. A spherical mirror is to be used to form an image, five times as tall as an object, on a screen positioned 5.0 m from the mirror. (a) Describe the type of mirror required. (b) Where should the mirror be positioned relative to the object?

20. A ball is dropped from rest 3.00 m directly above the vertex of a concave mirror having a radius of 1.00 m and lying in a horizontal plane. (a) Describe the motion of the ball's image in the mirror. (b) At what time do the ball and its image coincide?

Section 23.4 Images Formed by Refraction

21. A cubical block of ice 50.0 cm on an edge is placed on a level floor over a speck of dust. Locate the image of the speck, when viewed from directly above, if the index of refraction of ice is 1.309.

22. The top of a swimming pool is at ground level. If the pool is 2 m deep, how far below ground level does the bottom of the pool appear to be located when (a) the pool is completely filled with water? (b) the pool is filled halfway with water?

23. Physics⊗Now™ A paperweight is made of a solid glass hemisphere with index of refraction 1.50. The radius of the circular cross section is 4.0 cm. The hemisphere is placed on its flat surface, with the center directly over a 2.5-mm-long line drawn on a sheet of paper. What length of line is seen by someone looking vertically down on the hemisphere?

24. A flint glass plate ($n = 1.66$) rests on the bottom of an aquarium tank. The plate is 8.00 cm thick (vertical dimension) and covered with water ($n = 1.33$) to a depth of 12.0 cm. Calculate the apparent thickness of the plate as viewed from above the water. (Assume nearly normal incidence of light rays.)

25. A transparent sphere of unknown composition is observed to form an image of the Sun on its surface opposite the Sun. What is the refractive index of the sphere material?

26. A goldfish is swimming at 2.00 cm/s toward the front wall of a rectangular aquarium. What is the apparent speed of the fish as measured by an observer looking in from outside the front wall of the tank? The index of refraction of water is 1.333.

Section 23.6 Thin Lenses

27. A contact lens is made of plastic with an index of refraction of 1.50. The lens has an outer radius of curvature of $+2.00$ cm and an inner radius of curvature of $+2.50$ cm. What is the focal length of the lens?

28. The left face of a biconvex lens has a radius of curvature of 12.0 cm, and the right face has a radius of curvature of 18.0 cm. The index of refraction of the glass is 1.44. (a) Calculate the focal length of the lens. (b) Calculate the focal length if the radii of curvature of the two faces are interchanged.

29. A converging lens has a focal length of 20.0 cm. Locate the images for object distances of (a) 40.0 cm, (b) 20.0 cm, and (c) 10.0 cm. For each case, state whether the image is real or virtual and upright or inverted, and find the magnification.

30. Where must an object be placed to have unit magnification ($|M| = 1.00$) (a) for a converging lens of focal length 12.0 cm? (b) for a diverging lens of focal length 12.0 cm?

31. A diverging lens has a focal length of 20.0 cm. Locate the images for object distances of (a) 40.0 cm, (b) 20.0 cm, and (c) 10.0 cm. For each case, state whether the image is real or virtual and upright or inverted, and find the magnification.

32. The use of a lens in a certain situation is described by the equation

$$\frac{1}{p} + \frac{1}{-3.50p} = \frac{1}{7.50 \text{ cm}}$$

Determine (a) the object distance and (b) the image distance. (c) Use a ray diagram to obtain a description of the image. (d) Identify a practical device described by the given equation, and write the statement of a problem having a solution that contains the equation.

33. A transparent photographic slide is placed in front of a converging lens with a focal length of 2.44 cm. The lens forms an image of the slide 12.9 cm from it. How far is the lens from the slide if the image is (a) real? (b) virtual?

34. The nickel's image in Figure P23.34 has twice the diameter of the nickel when the lens is 2.84 cm from the nickel. Determine the focal length of the lens.

Figure P23.34

35. A certain LCD projector contains a single thin lens. An object 24.0 mm high is to be projected so that its image fills a screen 1.80 m high. The object-to-screen distance is 3.00 m. (a) Determine the focal length of the projection lens. (b) How far from the object should the lens of the projector be placed in order to form the image on the screen?

36. A person uses a converging lens that has a focal length of 12.5 cm to inspect a gem. The lens forms a virtual image 30.0 cm away. Determine the magnification. Is the image upright or inverted?

37. Physics⊗Now™ A diverging lens is to be used to produce a virtual image one-third as tall as the object. Where should the object be placed?

38. An object is 5.00 m to the left of a flat screen. A converging lens for which the focal length is $f = 0.800$ m is placed between object and screen. (a) Show that there are two lens positions that form an image on the screen, and determine how far these positions are from the object. (b) How do the two images differ from each other?

39. A converging lens is placed 30.0 cm to the right of a diverging lens of focal length 10.0 cm. A beam of parallel light enters the diverging lens from the left, and the beam is again parallel when it emerges from the converging lens. Calculate the focal length of the converging lens.

40. An object is placed 20.0 cm to the left of a converging lens of focal length 25.0 cm. A diverging lens of focal length 10.0 cm is 25.0 cm to the right of the converging lens. Find the position and magnification of the final image.

41. Two converging lenses, each of focal length 15.0 cm, are placed 40.0 cm apart, and an object is placed 30.0 cm in front of the first lens. Where is the final image formed, and what is the magnification of the system?

42. Object O_1 is 15.0 cm to the left of a converging lens with a 10.0-cm focal length. A second lens is positioned 10.0 cm to the right of the first lens and is observed to form a final image at the position of the original object O_1. (a) What is the focal length of the second lens? (b) What is the overall magnification of this system? (c) What is the nature (i.e., real or virtual, upright or inverted) of the final image?

43. A 1.00-cm-high object is placed 4.00 cm to the left of a converging lens of focal length 8.00 cm. A diverging lens of focal length −16.00 cm is 6.00 cm to the right of the converging lens. Find the position and height of the final image. Is the image inverted or upright? Real or virtual?

44. Two converging lenses having focal lengths of 10.0 cm and 20.0 cm are placed 50.0 cm apart, as shown in Figure P23.44. The final image is to be located between the lenses, at the position indicated. (a) How far to the left of the first lens should the object be positioned? (b) What is the overall magnification of the system? (c) Is the final image upright or inverted?

Figure P23.44

45. Lens L_1 in Figure P23.45 has a focal length of 15.0 cm and is located a fixed distance in front of the film plane of a camera. Lens L_2 has a focal length of 13.0 cm, and its distance d from the film plane can be varied from 5.00 cm to 10.0 cm. Determine the range of distances for which objects can be focused on the film.

Figure P23.45

46. Consider two thin lenses, one of focal length f_1 and the other of focal length f_2, placed in contact with each other as shown in Figure P23.46. Apply the thin-lens equation to each of these lenses and combine the results to show that this combination of lenses behaves like a thin lens having a focal length f given by $1/f = 1/f_1 + 1/f_2$. Assume that the thicknesses of the lenses can be ignored in comparison to the other distances involved.

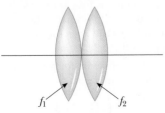

Figure P23.46

ADDITIONAL PROBLEMS

47. An object placed 10.0 cm from a concave spherical mirror produces a real image 8.00 cm from the mirror. If the object is moved to a new position 20.0 cm from the mirror, what is the position of the image? Is the final image real or virtual?

48. An object is placed 12 cm to the left of a diverging lens of focal length −6.0 cm. A converging lens of focal length 12 cm is placed a distance d to the right of the diverging lens. Find the distance d that places the final image at infinity.

49. A convergent lens with a 50.0-mm focal length is used to focus an image of a very distant scene onto a flat screen 35.0 mm wide. What is the angular width α of the scene included in the image on the screen?

50. The object in Figure P23.50 is midway between the lens and the mirror. The mirror's radius of curvature is 20.0 cm, and the lens has a focal length of −16.7 cm. Considering only the light that leaves the object and travels first towards the mirror, locate the final image formed by this system. Is the image real or virtual? Is it upright or inverted? What is the overall magnification of the image?

Figure P23.50

51. **Physics⊗Now™** The lens and mirror in Figure P23.51 are separated by 1.00 m and have focal lengths of +80.0 cm and −50.0 cm, respectively. If an object is placed 1.00 m to the left of the lens, locate the final image. State whether the image is upright or inverted, and determine the overall magnification.

Figure P23.51

52. A diverging lens ($n = 1.50$) is shaped like that in Active Figure 23.25c. The radius of the first surface is 15.0 cm, and that of the second surface is 10.0 cm. (a) Find the focal length of the lens. Determine the positions of the images for object distances of (b) infinity, (c) $3|f|$, (d) $|f|$, and (e) $|f|/2$.

53. A parallel beam of light enters a glass hemisphere perpendicular to the flat face, as shown in Figure P23.53. The radius of the hemisphere is $R = 6.00$ cm, and the index of refraction is $n = 1.560$. Determine the point at which the beam is focused. (Assume paraxial rays—that is, assume that all rays are located close to the principal axis.)

Figure P23.53

54. A converging lens of focal length 20.0 cm is separated by 50.0 cm from a converging lens of focal length 5.00 cm. (a) Find the position of the final image of an object placed 40.0 cm in front of the first lens. (b) If the height of the object is 2.00 cm, what is the height of the final image? Is the image real or virtual? (c) If the two lenses are now placed in contact with each other and the object is 5.00 cm in front of this combination, where will the image be located? (See Problem 46.)

55. To work this problem, use the fact that the image formed by the first surface becomes the object for the second surface. Figure P23.55 shows a piece of glass with index of refraction 1.50. The ends are hemispheres with radii 2.00 cm and 4.00 cm, and the centers of the hemispherical ends are separated by a distance of 8.00 cm. A point object is in air, 1.00 cm from the left end of the glass. Locate the image of the object due to refraction at the two spherical surfaces.

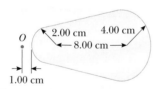

Figure P23.55

56. In a darkened room, a burning candle is placed 1.50 m from a white wall. A lens is placed between candle and wall at a location that causes a larger, inverted image to form on the wall. When the lens is moved 90.0 cm toward the wall, another image of the candle is formed. Find (a) the two object distances that produce the images just described and (b) the focal length of the lens. (c) Characterize the second image.

57. An object 2.00 cm high is placed 40.0 cm to the left of a converging lens having a focal length of 30.0 cm. A diverging lens having a focal length of -20.0 cm is placed 110 cm to the right of the converging lens. (a) Determine the final position and magnification of the final image. (b) Is the image upright or inverted? (c) Repeat parts (a) and (b) for the case where the second lens is a converging lens having a focal length of $+20.0$ cm.

58. A "floating strawberry" illusion can be produced by two parabolic mirrors, each with a focal length of 7.5 cm, facing each other so that their centers are 7.5 cm apart

(Fig. P23.58). If a strawberry is placed on the bottom mirror, an image of the strawberry forms at the small opening at the center of the top mirror. Show that the final image forms at that location, and describe its characteristics. [*Note:* A flashlight beam shone on these *images* has a very startling effect: Even at a glancing angle, the incoming light beam is seemingly reflected off the *images* of the strawberry! Do you understand why?]

Figure P23.58

59. Figure P23.59 shows a converging lens with radii $R_1 = 9.00$ cm and $R_2 = -11.00$ cm, in front of a concave spherical mirror of radius $R = 8.00$ cm. The focal points (F_1 and F_2) for the thin lens and the center of curvature (C) of the mirror are also shown. (a) If the focal points F_1 and F_2 are 5.00 cm from the vertex of the thin lens, determine the index of refraction of the lens. (b) If the lens and mirror are 20.0 cm apart, and an object is placed 8.00 cm to the left of the lens, determine the position of the final image and its magnification as seen by the eye in the figure. (c) Is the final image inverted or upright? Explain.

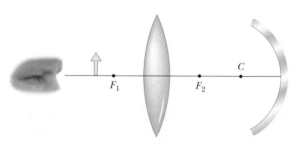

Figure P23.59

60. Find the object distances (in terms of f) for a thin converging lens of focal length f if (a) the image is real and the image distance is four times the focal length; (b) the image is virtual and the image distance is three times the focal length. (c) Calculate the magnification of the lens for cases (a) and (b).

61. The lens maker's equation for a lens with index n_1 immersed in a medium with index n_2 takes the form

$$\frac{1}{f} = \left(\frac{n_1}{n_2} - 1\right)\left(\frac{1}{R_1} - \frac{1}{R_2}\right)$$

A thin diverging glass (index = 1.50) lens with $R_1 = -3.00$ m and $R_2 = -6.00$ m is surrounded by air. An arrow is placed 10.0 m to the left of the lens. (a) Determine the position of the image. Repeat part (a) with the arrow and lens immersed in (b) water (index = 1.33); (c) a medium with an index of refraction of 2.00. (d) How can a lens that is diverging in air be changed into a converging lens.

62. An observer to the right of the mirror–lens combination shown in Figure P23.62 sees two real images that are the same size and in the same location. One image is upright and the other is inverted. Both images are 1.50 times larger than the object. The lens has a focal length of 10.0 cm. The lens and mirror are separated by 40.0 cm. Determine the focal length of the mirror. (Don't assume that the figure is drawn to scale.)

Figure P23.62

63. The lens maker's equation applies to a lens immersed in a liquid if n in the equation is replaced by n_1/n_2. Here n_1 refers to the refractive index of the lens material and n_2 is that of the medium surrounding the lens. (a) A certain lens has focal length of 79.0 cm in air and a refractive index of 1.55. Find its focal length in water. (b) A certain mirror has focal length of 79.0 cm in air. Find its focal length in water.

ACTIVITIES

1. This experiment will enable you to examine some of the properties of images formed by flat mirrors. As shown in Figure A23.1, place a clear plastic sheet between two small candles of the same height. Light one candle placed about 6 inches from the sheet and observe the reflected flame from the front side. Move the unlit candle until it also appears to be lit. The candles should be approximately equidistant from the sheet at this point. Why?

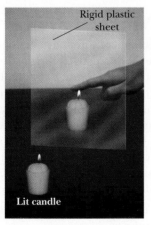

Figure A23.1

When you place your finger on the wick of the unlit candle, you will get the illusion that your finger is burning. Explain your observation.

A similar experiment can be performed with a flat mirror and two pencils. Hold one of the pencils in front of the mirror at a distance of about 10 inches from the mirror and at a position such that about half of the image appears in the mirror. Move the second pencil back and forth behind the mirror until it appears to align perfectly with the image in the mirror. When the two pencils are aligned, measure the distance from the mirror to the location of the second pencil. This pencil is situated at the apparent position of the virtual image. Why? The two pencils should be at equal distances from the mirror. Why?

2. Fill a clear glass tumbler with water and place a pencil or straw into the tumbler as in Figure A23.2. Now observe the pencil from the side at an angle of about 45° to the surface, and note that the line of the portion of the pencil under water is not parallel with the line of the portion in air. That is, the pencil appears to be bent at the point where it enters the water. Use the techniques of Section 23.4 to explain this observation.

Figure A23.2

3. View yourself in a full-length mirror. Stand close to the mirror, and place one piece of tape at the top of the image of your head and another piece at the very bottom of the image of your feet. Now step back a few meters and observe your image. How big is it relative to the original size? How does the distance between the pieces of tape compare with your actual height?

Move to a position in front of the mirror such that you can see a full image of yourself with the top of your head just level with the top of the mirror. Have a friend gradually block off the lower portion of the mirror with a sheet or newspaper page until you can see your complete image—but no more! Measure the length of the mirror and compare this measurement with your height. How do the two compare?

4. Compare the images formed of your face when you look first at the front side and then at the back side of a shiny soupspoon. For each side, observe the change in the image of your face as you move closer to and farther away from the spoon.

5. Draw a sketch of you and your image in a plane mirror when you are walking away from it with velocity $\vec{v}$. Indicate on your diagram the velocity vectors for you and your image. Keep the relative lengths of the two velocities as you believe they will be. Explain your answer. (b) Repeat the procedure in part (a), but now do it as you are walking towards the mirror. Explain why you have the vectors drawn as you do.

The colors in many of a humming-bird's feathers are not due to pigment. The *iridescence* which makes the brilliant colors that often appear on the bird's throat and belly is due to an interference effect caused by structures in the feathers. The colors vary with the viewing angle.

RO-MA/Index Stock Imagery

Wave Optics

Colors swirl on a soap bubble as it drifts through the air on a summer day, and vivid rainbows reflect from the filth of oil films in the puddles of a dirty city street. Beachgoers, covered with thin layers of oil, wear their coated sunglasses that absorb half the incoming light. In laboratories, scientists determine the precise composition of materials by analyzing the light they give off when hot, and in observatories around the world, telescopes gather light from distant galaxies, filtering out individual wavelengths in bands and thereby determining the speed of expansion of the universe.

Understanding how these rainbows are made and how certain scientific instruments can determine wavelengths is the domain of *wave optics*. Light can be viewed as either a particle or a wave. Geometric optics, the subject of the previous chapter, depends on the particle nature of light. Wave optics depends on the wave nature of light. The three primary topics we examine in this chapter are interference, diffraction, and polarization. These phenomena can't be adequately explained with ray optics, but can be understood if light is viewed as a wave.

24.1 CONDITIONS FOR INTERFERENCE

In our discussion of interference of mechanical waves in Chapter 13, we found that two waves could add together either constructively or destructively. In constructive interference, the amplitude of the resultant wave is greater than that of either of the individual waves, whereas in destructive interference, the resultant amplitude is less than that of either individual wave. Light waves also interfere with each other. Fundamentally, all interference associated with light waves arises when the electromagnetic fields that constitute the individual waves combine.

Interference effects in light waves aren't easy to observe because of the short wavelengths involved (about 4×10^{-7} m to about 7×10^{-7} m). For sustained interference between two sources of light to be observed, the following conditions must be met:

Conditions for interference ▶

1. The sources must be **coherent**, which means the waves they emit must maintain a constant phase with respect to each other.
2. The waves must have identical wavelengths.

Two sources (producing two traveling waves) are needed to create interference. To produce a stable interference pattern, the individual waves must maintain a constant phase with one another. When this situation prevails, the sources are said to be coherent. The sound waves emitted by two side-by-side loudspeakers driven by a single amplifier can produce interference because the two speakers respond to the amplifier in the same way at the same time—they are in phase.

If two light sources are placed side by side, however, no interference effects are observed, because the light waves from one source are emitted independently of the waves from the other source; hence, the emissions from the two sources don't maintain a constant phase relationship with each other during the time of observation. An ordinary light source undergoes random changes about once every 10^{-8} s. Therefore, the conditions for constructive interference, destructive interference, and intermediate states have durations on the order of 10^{-8} s. The result is that no interference effects are observed, because the eye can't follow such short-term changes. Ordinary light sources are said to be **incoherent**.

An older method for producing two coherent light sources is to pass light from a single wavelength (monochromatic) source through a narrow slit and then allow the light to fall on a screen containing two other narrow slits. The first slit is needed to create a single wave-front that illuminates both slits coherently. The light emerging from the two slits is coherent because a single source produces the original light beam and the slits serve only to separate the original beam into two parts. Any random change in the light emitted by the source will occur in the two separate beams at the same time, and interference effects can be observed.

Currently it's much more common to use a laser as a coherent source to demonstrate interference. A laser produces an intense, coherent, monochromatic beam over a width of several millimeters. This means that the laser may be used to illuminate multiple slits directly and that interference effects can be easily observed in a fully lighted room. The principles of operation of a laser are explained in Chapter 28.

24.2 YOUNG'S DOUBLE-SLIT EXPERIMENT

Thomas Young first demonstrated interference in light waves from two sources in 1801. Active Figure 24.1a is a schematic diagram of the apparatus used in this experiment. (Young used pinholes rather than slits in his original experiments.)

(a)

b. M. Cagnet, M. Francon, J. C. Thier

(b)

ACTIVE FIGURE 24.1
(a) A diagram of Young's double-slit experiment. The narrow slits act as sources of waves. Slits S_1 and S_2 behave as coherent sources that produce an interference pattern on screen C. (The drawing is not to scale.) (b) The fringe pattern formed on screen C could look like this.

Physics⊗Now™

Log into PhysicsNow at **www.cp7e.com** and go to Active Figure 24.1, where you can adjust the slit separation and the wavelength of the light, observing the effect on the interference pattern.

Figure 24.2 An interference pattern involving water waves is produced by two vibrating sources at the water's surface. The pattern is analogous to that observed in Young's double-slit experiment. Note the regions of constructive and destructive interference.

Light is incident on a screen containing a narrow slit S_0. The light waves emerging from this slit arrive at a second screen that contains two narrow, parallel slits S_1 and S_2. These slits serve as a pair of coherent light sources because waves emerging from them originate from the same wave front and therefore are always in phase. The light from the two slits produces a visible pattern on screen C consisting of a series of bright and dark parallel bands called **fringes** (Active Fig. 24.1b). When the light from slits S_1 and S_2 arrives at a point on the screen so that constructive interference occurs at that location, a bright fringe appears. When the light from the two slits combines destructively at any location on the screen, a dark fringe results. Figure 24.2 is a photograph of an interference pattern produced by two coherent vibrating sources in a water tank.

Figure 24.3 is a schematic diagram of some of the ways in which the two waves can combine at screen C of Figure 24.1. In Figure 24.3a, two waves, which leave the two slits in phase, strike the screen at the central point P. Because these waves travel equal distances, they arrive in phase at P, and as a result, constructive interference occurs there and a bright fringe is observed. In Figure 24.3b, the two light waves again start in phase, but the upper wave has to travel one wavelength farther to reach point Q on the screen. Because the upper wave falls behind the lower one by exactly one wavelength, the two waves still arrive in phase at Q, so a second bright fringe appears at that location. Now consider point R, midway between P and Q, in Figure 24.3c. At R, the upper wave has fallen half a wavelength behind the lower wave. This means that the trough of the bottom wave overlaps the crest of the upper wave, giving rise to destructive interference. As a result, a dark fringe can be observed at R.

We can describe Young's experiment quantitatively with the help of Figure 24.4. Consider point P on the viewing screen; the screen is positioned a perpendicular distance L from the screen containing slits S_1 and S_2, which are separated by distance d, and r_1 and r_2 are the distances the secondary waves travel from slit to screen. We assume the waves emerging from S_1 and S_2 have the same constant frequency, have the same amplitude, and start out in phase. The light intensity on the screen at P is the result of light from both slits. A wave from the lower slit, however, travels farther than a wave from the upper slit by the amount $d \sin \theta$. This distance is called the **path difference** δ (lower case Greek delta), where

Path difference ▶

$$\delta = r_2 - r_1 = d \sin \theta \qquad [24.1]$$

Equation 24.1 assumes that the two waves travel in parallel lines, which is approximately true, because L is much greater than d. As noted earlier, the value of this path difference determines whether the two waves are in phase when they arrive at P. If the path difference is either zero or some integral multiple of the wavelength, the two waves are in phase at P and constructive interference results. Therefore, the condition for bright fringes, or **constructive interference**, at P is

Condition for constructive interference (two slits) ▶

$$\delta = d \sin \theta_{\text{bright}} = m\lambda \qquad m = 0, \pm 1, \pm 2, \ldots \qquad [24.2]$$

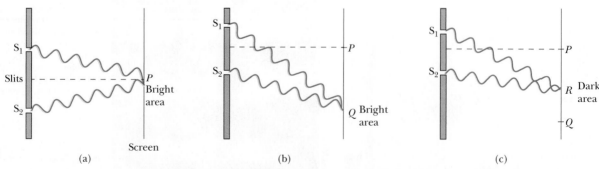

Figure 24.3 (a) Constructive interference occurs at P when the waves combine. (b) Constructive interference also occurs at Q. (c) Destructive interference occurs at R when the wave from the upper slit falls half a wavelength behind the wave from the lower slit. (These figures are not drawn to scale.)

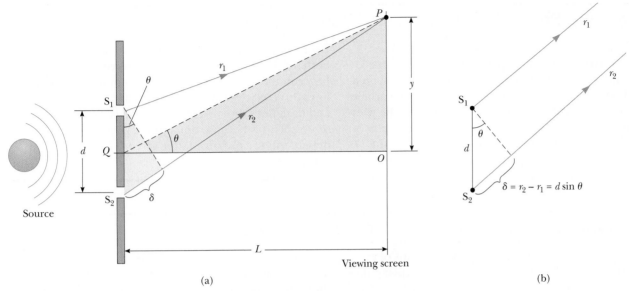

Figure 24.4 A geometric construction that describes Young's double-slit experiment. The path difference between the two rays is $\delta = r_2 - r_1 = d \sin \theta$. (This figure is not drawn to scale.)

The number m is called the **order number**. The central bright fringe at $\theta_{bright} = 0$ ($m = 0$) is called the *zeroth-order maximum*. The first maximum on either side, where $m = \pm 1$, is called the *first-order maximum*, and so forth.

When δ is an odd multiple of $\lambda/2$, the two waves arriving at P are 180° out of phase and give rise to destructive interference. Therefore, the condition for dark fringes, or **destructive interference**, at P is

$$\delta = d \sin \theta_{dark} = (m + \tfrac{1}{2})\lambda \quad m = 0, \pm 1, \pm 2, \ldots \qquad \textbf{[24.3]}$$

◀ Condition for destructive interference (two slits)

If $m = 0$ in this equation, the path difference is $\delta = \lambda/2$, which is the condition for the location of the first dark fringe on either side of the central (bright) maximum. Likewise, if $m = 1$, the path difference is $\delta = 3\lambda/2$, which is the condition for the second dark fringe on each side, and so forth.

It's useful to obtain expressions for the positions of the bright and dark fringes measured vertically from O to P. In addition to our assumption that $L \gg d$, we assume that $d \gg \lambda$. These can be valid assumptions because, in practice, L is often on the order of 1 m, d is a fraction of a millimeter, and λ is a fraction of a micrometer for visible light. Under these conditions θ is small, so we can use the approximation $\sin \theta \cong \tan \theta$. Then, from triangle OPQ in Figure 24.4, we see that

$$y = L \tan \theta \approx L \sin \theta \qquad \textbf{[24.4]}$$

TIP 24.1 Small-Angle Approximation: Size Matters!

The small-angle approximation $\sin \theta \cong \tan \theta$ is true to three-digit precision only for angles less than about 4°.

Solving Equation 24.2 for $\sin \theta$ and substituting the result into Equation 24.4, we find that the positions of the *bright fringes*, measured from O, are

$$y_{bright} = \frac{\lambda L}{d} m \qquad m = 0, \pm 1, \pm 2, \ldots \qquad \textbf{[24.5]}$$

Using Equations 24.3 and 24.4, we find that the *dark fringes* are located at

$$y_{dark} = \frac{\lambda L}{d} (m + \tfrac{1}{2}) \qquad m = 0, \pm 1, \pm 2, \ldots \qquad \textbf{[24.6]}$$

As we will show in Example 24.1, Young's double-slit experiment provides a method for measuring the wavelength of light. In fact, Young used this technique to do just that. In addition, his experiment gave the wave model of light a great deal of credibility. It was inconceivable that particles of light coming through the slits could cancel each other in a way that would explain the dark fringes.

Reflection, interference, and diffraction can be seen in this aerial photograph of waves in the sea.

Applying Physics 24.1   A Smoky Young's Experiment

Consider a double-slit experiment in which a laser beam is passed through a pair of very closely spaced slits and a clear interference pattern is displayed on a distant screen. Now suppose you place smoke particles between the double slit and the screen. With the presence of the smoke particles, will you see the effects of interference in the space between the slits and the screen, or will you see only the effects on the screen?

Explanation You will see the interference pattern both on the screen and in the area filled with smoke between the slits and the screen. There will be bright lines directed toward the bright areas on the screen and dark lines directed toward the dark areas on the screen. This is because Equations 24.5 and 24.6 depend on the distance to the screen, L, which can take any value.

Applying Physics 24.2   Television Signal Interference

Suppose you are watching television by means of an antenna rather than a cable system. If an airplane flies near your location, you may notice wavering ghost images in the television picture. What might cause this phenomenon?

Explanation Your television antenna receives two signals: the direct signal from the transmitting antenna and a signal reflected from the surface of the airplane. As the airplane changes position, there are some times when these two signals are in phase and other times when they are out of phase. As a result, the intensity of the combined signal received at your antenna will vary. The wavering of the ghost images of the picture is evidence of this variation.

Quick Quiz 24.1

In a two-slit interference pattern projected on a screen, the fringes are equally spaced on the screen (a) everywhere (b) only for large angles (c) only for small angles.

INTERACTIVE EXAMPLE 24.1   Measuring the Wavelength of a Light Source

Goal   Show how Young's experiment can be used to measure the wavelength of coherent light.

Problem   A screen is separated from a double-slit source by 1.20 m. The distance between the two slits is 0.030 0 mm. The second-order bright fringe ($m = 2$) is measured to be 4.50 cm from the centerline. Determine (a) the wavelength of the light and (b) the distance between adjacent bright fringes.

Strategy   Equation 24.5 relates the positions of the bright fringes to the other variables, including the wavelength of the light. Substitute into this equation and solve for λ. Taking the difference between y_{m+1} and y_m results in a general expression for the distance between bright fringes.

Solution

(a) Determine the wavelength of the light.

Solve Equation 24.5 for the wavelength and substitute $m = 2$, $y_2 = 4.50 \times 10^{-2}$ m, $L = 1.20$ m, and $d = 3.00 \times 10^{-5}$ m:

$$\lambda = \frac{y_2 d}{mL} = \frac{(4.50 \times 10^{-2}\,\text{m})(3.00 \times 10^{-5}\,\text{m})}{2(1.20\,\text{m})}$$

$$= 5.63 \times 10^{-7}\,\text{m} = \boxed{563\,\text{nm}}$$

(b) Determine the distance between adjacent bright fringes.

Use Equation 24.5 to find the distance between *any* adjacent bright fringes (here, those characterized by m and $m + 1$):

$$\Delta y = y_{m+1} - y_m = \frac{\lambda L}{d}(m + 1) - \frac{\lambda L}{d}m = \frac{\lambda L}{d}$$

$$= \frac{(5.63 \times 10^{-7}\,\text{m})(1.20\,\text{m})}{3.00 \times 10^{-5}\,\text{m}} = \boxed{2.25\,\text{cm}}$$

Remarks   This calculation depends on the angle θ being small, because the small-angle approximation was implicitly used. The measurement of the position of the bright fringes yields the wavelength of light, which in turn is a signature of atomic processes, as will be discussed in the chapters on modern physics. This kind of measurement, therefore, helped open the world of the atom.

Exercise 24.1
Suppose the same experiment is run with a different light source. If the first-order maximum is found at 1.85 cm from the centerline, what is the wavelength of the light?

Answer   463 nm

Physics ⊗ Now™ Investigate the double-slit interference pattern by logging into PhysicsNow at **www.cp7e.com** and going to Interactive Example 24.1.

24.3 CHANGE OF PHASE DUE TO REFLECTION

Young's method of producing two coherent light sources involves illuminating a pair of slits with a single source. Another simple, yet ingenious, arrangement for producing an interference pattern with a single light source is known as *Lloyd's mirror*. A point source of light is placed at point S, close to a mirror, as illustrated in Figure 24.5. Light waves can reach the viewing point P either by the direct path SP or by the path involving reflection from the mirror. The reflected ray can be treated as a ray originating at the source S' behind the mirror. Source S', which is the image of S, can be considered a virtual source.

At points far from the source, an interference pattern due to waves from S and S' is observed, just as for two real coherent sources. However, the positions of the dark and bright fringes are *reversed* relative to the pattern obtained from two real coherent sources (Young's experiment). This is because the coherent sources S and S' differ in phase by 180°, a phase change produced by reflection.

To illustrate the point further, consider P', the point where the mirror intersects the screen. This point is equidistant from S and S'. If path difference alone were responsible for the phase difference, a bright fringe would be observed at P' (because the path difference is zero for this point), corresponding to the central fringe of the two-slit interference pattern. Instead, we observe a *dark* fringe at P, from which we conclude that a 180° phase change must be produced by reflection from the mirror. In general, **an electromagnetic wave undergoes a phase change of 180° upon reflection from a medium that has an index of refraction higher than the one in which the wave was traveling**.

An analogy can be drawn between reflected light waves and the reflections of a transverse wave on a stretched string when the wave meets a boundary, as in Figure 24.6. The reflected pulse on a string undergoes a phase change of 180°

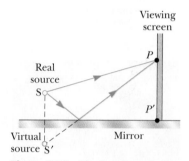

Figure 24.5 Lloyd's mirror. An interference pattern is produced on a screen at P as a result of the combination of the direct ray (blue) and the reflected ray (brown). The reflected ray undergoes a phase change of 180°.

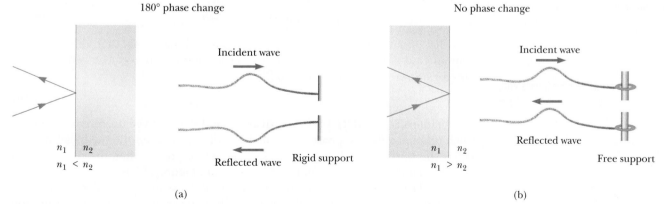

(a)   (b)

Figure 24.6 (a) A ray reflecting from a medium of higher refractive index undergoes a 180° phase change. The right side shows the analogy with a reflected pulse on a string. (b) A ray reflecting from a medium of lower refractive index undergoes no phase change.

Figure 24.7 Interference observed in light reflected from a thin film is due to a combination of rays reflected from the upper and lower surfaces.

Soap bubbles on water. The colors are due to interference between light rays reflected from the front and back of the thin film of soap making up the bubble. The color depends on the thickness of the film, ranging from black where the film is at its thinnest to magenta where it is thickest.

when it is reflected from the boundary of a denser string or from a rigid barrier and undergoes no phase change when it is reflected from the boundary of a less dense string. Similarly, an electromagnetic wave undergoes a 180° phase change when reflected from the boundary of a medium with index of refraction higher than the one in which it has been traveling. There is no phase change when the wave is reflected from a boundary leading to a medium of lower index of refraction. The transmitted wave that crosses the boundary also undergoes no phase change.

24.4 INTERFERENCE IN THIN FILMS

Interference effects are commonly observed in thin films, such as the thin surface of a soap bubble or thin layers of oil on water. The varied colors observed when incoherent white light is incident on such films result from the interference of waves reflected from the two surfaces of the film.

Consider a film of uniform thickness t and index of refraction n, as in Figure 24.7. Assume that the light rays traveling in air are nearly normal to the two surfaces of the film. To determine whether the reflected rays interfere constructively or destructively, we first note the following facts:

1. An electromagnetic wave traveling from a medium of index of refraction n_1 toward a medium of index of refraction n_2 undergoes a 180° phase change on reflection when $n_2 > n_1$. There is no phase change in the reflected wave if $n_2 < n_1$.
2. The wavelength of light λ_n in a medium with index of refraction n is

$$\lambda_n = \frac{\lambda}{n} \qquad [24.7]$$

where λ is the wavelength of light in vacuum.

We apply these rules to the film of Figure 24.7. According to the first rule, ray 1, which is reflected from the upper surface A, undergoes a phase change of 180° with respect to the incident wave. Ray 2, which is reflected from the lower surface B, undergoes no phase change with respect to the incident wave. Therefore, ray 1 is 180° out of phase with respect to ray 2, which is equivalent to a path difference of $\lambda_n/2$. However, we must also consider the fact that ray 2 travels an extra distance of $2t$ before the waves recombine in the air above the surface. For example, if $2t = \lambda_n/2$, then rays 1 and 2 recombine in phase, and constructive interference results. In general, the condition for *constructive interference* in thin films is

$$2t = (m + \tfrac{1}{2})\lambda_n \qquad m = 0, 1, 2, \ldots \qquad [24.8]$$

This condition takes into account two factors: (1) the difference in path length for the two rays (the term $m\lambda_n$) and (2) the 180° phase change upon reflection (the term $\lambda_n/2$). Because $\lambda_n = \lambda/n$, we can write Equation 24.8 in the form

Condition for constructive interference (thin film) ▶

$$2nt = (m + \tfrac{1}{2})\lambda \qquad m = 0, 1, 2, \ldots \qquad [24.9]$$

If the extra distance $2t$ traveled by ray 2 is a multiple of λ_n, then the two waves combine out of phase and the result is destructive interference. The general equation for *destructive interference* in thin films is

Condition for destructive interference (thin film) ▶

$$2nt = m\lambda \qquad m = 0, 1, 2, \ldots \qquad [24.10]$$

Equations 24.9 and 24.10 for constructive and destructive interference are valid when there is only one phase reversal. This will occur when the media above and below the thin film both have indices of refraction greater than the film or when both have indices of refraction less than the film. Figure 24.7 is a case in point: the air ($n = 1$) that is both above and below the film has an index of refraction less than that of the film. As a result, there is a phase reversal on reflection off the top layer of the film, but not the bottom, and Equations 24.9 and 24.10 apply. **If the film is placed between two different media, one of lower refractive index than the**

film and one of higher refractive index, Equations 24.9 and 24.10 are reversed: Equation 24.9 is used for destructive interference, and Equation 24.10 for obstructive interference. In this case, either there is a phase change of 180° for both ray 1 reflecting from surface A and ray 2 reflecting from surface B, as in Figure 24.9 of Example 24.3 (page 794), or there is no phase change for either ray, which would be the case if the incident ray came from underneath the film. Hence, the net change in relative phase due to the reflections is *zero*.

A thin film of oil on water displays interference, evidenced by the pattern of colors when white light is incident on the film. Variations in the film's thickness produce the intersecting color pattern. The razor blade gives you an idea of the size of the colored bands.

> ## Quick Quiz 24.2
>
> Suppose Young's experiment is carried out in air, and then, in a second experiment, the apparatus is immersed in water. In what way does the distance between bright fringes change? (a) They move further apart. (b) They move closer together. (c) There is no change.

Newton's Rings

Another method for observing interference in light waves is to place a planoconvex lens on top of a flat glass surface, as in Figure 24.8a. With this arrangement, the air film between the glass surfaces varies in thickness from zero at the point of contact to some value t at P. If the radius of curvature R of the lens is much greater than the distance r, and if the system is viewed from above light of wavelength λ, a pattern of light and dark rings is observed (Fig. 24.8b). These circular fringes, discovered by Newton, are called **Newton's rings**. The interference is due to the combination of ray 1, reflected from the plate, with ray 2, reflected from the lower surface of the lens. Ray 1 undergoes a phase change of 180° on reflection, because it is reflected from a boundary leading into a medium of higher refractive index, whereas ray 2 undergoes no phase change, because it is reflected from a medium of lower refractive index. Hence, the conditions for constructive and destructive interference are given by Equations 24.9 and 24.10, respectively, with $n = 1$ because the "film" is air. The contact point at O is dark, as seen in Figure 24.8b, because there is no path difference and the total phase change is due only to the 180° phase change upon reflection. Using the geometry shown in Figure 24.8a, we can obtain expressions for the radii of the bright and dark bands in terms of the radius of curvature R, and vacuum wavelength λ. For example, the dark rings have radii of $r \approx \sqrt{m\lambda R/n}$.

One of the important uses of Newton's rings is in the testing of optical lenses. A circular pattern like that in Figure 24.8b is achieved only when the lens is ground to a perfectly spherical curvature. Variations from such symmetry might produce a

TIP 24.2 The Two Tricks of Thin Films

Be sure to include *both* effects—path length and phase change—when you analyze an interference pattern from a thin film.

APPLICATION

Checking for Imperfections in Optical Lenses

Figure 24.8 (a) The combination of rays reflected from the glass plate and the curved surface of the lens gives rise to an interference pattern known as Newton's rings. (b) A photograph of Newton's rings. (c) This asymmetric interference pattern indicates imperfections in the lens.

Interference in a vertical film of variable thickness. The top of the film appears darkest where the film is thinnest.

Richard Megna, Fundamental Photographs

pattern like that in Figure 24.8c. These variations give an indication of how the lens must be reground and repolished to remove imperfections.

Problem-Solving Strategy   Thin-Film Interference

The following steps are recommended in addressing thin-film interference problems:
1. Identify the thin film causing the interference, and the indices of refraction in the film and in the media on either side of it.
2. Determine the number of phase reversals: zero, one, or two.
3. Consult the following table, which contains Equations 24.9 and 24.10, and select the correct column for the problem in question:

Equation ($m = 0, 1, \ldots$)	1 phase reversal	0 or 2 phase reversals
$2nt = (m + \frac{1}{2})\lambda$ [24.9]	constructive	destructive
$2nt = m\lambda$ [24.10]	destructive	constructive

4. Substitute values in the appropriate equations, as selected in the previous step.

EXAMPLE 24.2   Interference in a Soap-Film

Goal   Calculate interference effects in a thin film when there is one phase reversal.

Problem   Calculate the minimum thickness of a soap-bubble film ($n = 1.33$) that will result in constructive interference in the reflected light if the film is illuminated by light with wavelength 602 nm in free space.

Strategy   There is only one inversion, so the condition for constructive interference is $2nt = (m + \frac{1}{2})\lambda$. The minimum film thickness for constructive interference corresponds to $m = 0$ in this equation.

Solution
Solve $2nt = \lambda/2$ for the thickness t, and substitute:
$$t = \frac{\lambda}{4n} = \frac{602 \text{ nm}}{4(1.33)} = \boxed{113 \text{ nm}}$$

Remark   The swirling colors in a soap bubble are due to the fact that the thickness of the soap layer varies from one place to another.

Exercise 24.2
What other film thicknesses will produce constructive interference?

Answer   339 nm, 566 nm, 792 nm, and so on

INTERACTIVE EXAMPLE 24.3   Nonreflective Coatings for Solar Cells and Optical Lenses

Goal   Calculate interference effects in a thin film when there are two inversions.

Problem   Semiconductors such as silicon are used to fabricate solar cells—devices that generate electric energy when exposed to sunlight. Solar cells are often coated with a transparent thin film, such as silicon monoxide (SiO; $n = 1.45$) to minimize reflective losses (Fig. 24.9). A silicon solar cell ($n = 3.50$) is coated with a thin film of silicon monoxide for this purpose. Assuming normal incidence, determine the minimum thickness of the film that will produce the least reflection at a wavelength of 552 nm.

Figure 24.9   (Example 24.3) Reflective losses from a silicon solar cell are minimized by coating it with a thin film of silicon monoxide (SiO).

Strategy   Reflection is least when rays 1 and 2 in Figure 24.9 meet the condition for destructive interference. Note that *both* rays undergo 180° phase changes on reflection. The condition for a reflection *minimum* is therefore $2nt = \lambda/2$.

Solution
Solve $2nt = \lambda/2$ for t, the required thickness:

$$t = \frac{\lambda}{4n} = \frac{552\text{ nm}}{4(1.45)} = \boxed{95.2\text{ nm}}$$

Remarks   Typically, such coatings reduce the reflective loss from 30% (with no coating) to 10% (with a coating), thereby increasing the cell's efficiency because more light is available to create charge carriers in the cell. In reality, the coating is never perfectly nonreflecting, because the required thickness is wavelength dependent and the incident light covers a wide range of wavelengths.

Exercise 24.3
Glass lenses used in cameras and other optical instruments are usually coated with one or more transparent thin films, such as magnesium fluoride (MgF_2), to reduce or eliminate unwanted reflection. Carl Zeiss developed this method; his first coating was 1.00×10^2 nm thick. Using $n = 1.38$ for MgF_2, what visible wavelength would be eliminated by destructive interference in the reflected light?

Answer   552 nm

Physics⊗Now™   Investigate the interference for various film properties by logging into PhysicsNow at **www.cp7e.com** and going to Interactive Example 24.3.

EXAMPLE 24.4   Interference in a Wedge-Shaped Film

Goal   Calculate interference effects when the film has variable thickness.

Problem   A pair of glass slides 10.0 cm long and with $n = 1.52$ are separated on one end by a hair, forming a triangular wedge of air as illustrated in Figure 24.10. When coherent light from a helium–neon laser with wavelength 633 nm is incident on the film from above, 15.0 dark fringes per centimeter are observed. How thick is the hair?

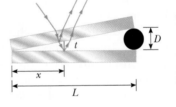

Figure 24.10   (Example 24.4) Interference bands in reflected light can be observed by illuminating a wedge-shaped film with monochromatic light. The dark areas correspond to positions of destructive interference.

Strategy   The interference pattern is created by the thin film of air having variable thickness. The pattern is a series of alternating bright and dark parallel bands. A dark band corresponds to destructive interference, and there is one phase reversal, so $2nt = m\lambda$ should be used. We can also use the similar triangles in Figure 24.10 to obtain the relation $t/x = D/L$. We can find the thickness for any m, and if the position x can also be found, this last equation gives the diameter of the hair, D.

Solution
Solve the destructive-interference equation for the thickness of the film, t, with $n = 1$ for air:

$$t = \frac{m\lambda}{2}$$

If d is the distance from one dark band to the next, then the x-coordinate of the mth band is a multiple of d:

$$x = md$$

By dimensional analysis, d is just the inverse of the number of bands per centimeter.

$$d = \left(15.0\ \frac{\text{bands}}{\text{cm}}\right)^{-1} = 6.67 \times 10^{-2}\ \frac{\text{cm}}{\text{band}}$$

Now use similar triangles, and substitute all the information:

$$\frac{t}{x} = \frac{m\lambda/2}{md} = \frac{\lambda}{2d} = \frac{D}{L}$$

Solve for D and substitute given values:

$$D = \frac{\lambda L}{2d} = \frac{(633 \times 10^{-9}\text{ m})(0.100\text{ m})}{2(6.67 \times 10^{-4}\text{ m})} = \boxed{4.75 \times 10^{-5}\text{ m}}$$

Remarks   Some may be concerned about interference caused by light bouncing off the top and bottom of, say, the upper glass slide. It's unlikely, however, that the thickness of the slide will be half an integer multiple of the wavelength of the helium-neon laser (for some very large value of m). In addition, in contrast to the air wedge, the thickness of the glass doesn't vary.

Exercise 24.4

The air wedge is replaced with water, with $n = 1.33$. Find the distance between dark bands when the helium–neon laser light hits the glass slides.

Answer 5.01×10^{-4} m

Figure 24.11   A photomicrograph of adjacent tracks on a compact disc (CD). The information encoded in these pits and smooth areas is read by a laser beam.

Courtesy of Sony Disc Manufacturing

24.5   USING INTERFERENCE TO READ CD'S AND DVD'S

Compact disks (CD's) and digital video disks (DVD's) have revolutionized the computer and entertainment industries by providing fast access; high-density storage of text, graphics, and movies; and high-quality sound recordings. The data on these disks are stored digitally as a series of zeros and ones, and these zeros and ones are read by laser light reflected from the disk. Strong reflections (constructive interference) from the disk are chosen to represent zeros and weak reflections (destructive interference) represent ones.

To see in more detail how thin-film interference plays a crucial role in reading CD's, consider Figure 24.11. This shows a photomicrograph of several CD tracks which consist of a sequence of pits (when viewed from the top or label side of the disk) of varying length formed in a reflecting-metal information layer. A cross-sectional view of a CD as shown in Figure 24.12 reveals that the pits appear as bumps to the laser beam, which shines on the metallic layer through a clear plastic coating from below.

As the disk rotates, the laser beam reflects off the sequence of bumps and lower areas into a photodetector, which converts the fluctuating reflected light intensity into an electrical string of zeros and ones. To make the light fluctuations more pronounced and easier to detect, the pit depth t is made equal to one-quarter of a wavelength of the laser light in the plastic. When the beam hits a rising or falling bump edge, part of the beam reflects from the top of the bump and part from the lower adjacent area, ensuring destructive interference and very low intensity when the reflected beams combine at the detector. Bump edges are read as ones, and flat bump tops and intervening flat plains are read as zeros.

In Example 24.5 the pit depth for a standard CD, using an infrared laser of wavelength 780 nm, is calculated. DVDs use shorter wavelength lasers of 635 nm,

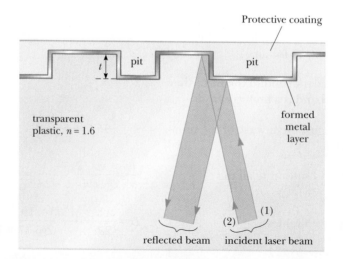

Figure 24.12   Cross section of a CD showing metallic pits of depth t and a laser beam detecting the edge of a pit.

and the track separation, pit depth, and minimum pit length are all smaller. This allows a DVD to store about 30 times more information than a CD.

EXAMPLE 24.5   Pit Depth in a CD

Goal   Apply interference principles to a CD.

Problem   Find the pit depth in a CD that has a plastic transparent layer with index of refraction of 1.60 and is designed for use in a CD player using a laser with a wavelength of 7.80×10^2 nm in air.

Strategy   (See Fig. 24.12.) Rays (1) and (2) both reflect from the metal layer which acts like a mirror, so there is no phase difference due to reflection between those rays. There is, however, the usual phase difference caused by the extra distance $2t$ traveled by ray (2). The wavelength is λ/n, where n is the index of refraction in the substance.

Solution

Use the appropriate condition for destructive interference in a thin film:

$$2t = \frac{\lambda}{2n}$$

Solve for the thickness t and substitute:

$$t = \frac{\lambda}{4n} = \frac{7.80 \times 10^2 \text{ nm}}{(4)(1.60)} = \boxed{1.22 \times 10^2 \text{ nm}}$$

Remarks   Different CD systems have different tolerances for scratches. Anything that changes the reflective properties of the disk can affect the readability of the disk.

Exercise 24.5
Repeat the example for a laser with wavelength 635 nm.

Answer   99.2 nm

24.6   DIFFRACTION

Suppose a light beam is incident on two slits, as in Young's double-slit experiment. If the light truly traveled in straight-line paths after passing through the slits, as in Figure 24.13a, the waves wouldn't overlap and no interference pattern would be seen. Instead, Huygens's principle requires that the waves spread out from the slits, as shown in Figure 24.13b. In other words, the light bends from a straight-line path and enters the region that would otherwise be shadowed. This spreading out of light from its initial line of travel is called **diffraction**.

   In general, diffraction occurs when waves pass through small openings, around obstacles, or by sharp edges. For example, when a single narrow slit is placed between a distant light source (or a laser beam) and a screen, the light produces a diffraction pattern like that in Figure 24.14. The pattern consists of a broad, intense central band flanked by a series of narrower, less intense secondary bands (called **secondary maxima**) and a series of dark bands, or **minima**. This phenomenon

Figure 24.14   The diffraction pattern that appears on a screen when light passes through a narrow vertical slit. The pattern consists of a broad central band and a series of less intense and narrower side bands.

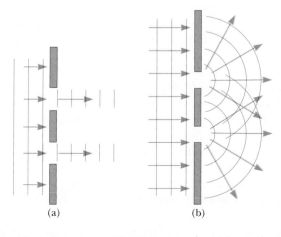

(a)   (b)

Figure 24.13   (a) If light did not spread out after passing through the slits, no interference would occur. (b) The light from the two slits overlaps as it spreads out, filling the expected shadowed regions with light and producing interference fringes.

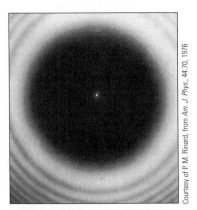

Courtesy of P. M. Rinard, from *Am. J. Phys.*, 44:70, 1976

Figure 24.15 The diffraction pattern of a penny placed midway between the screen and the source.

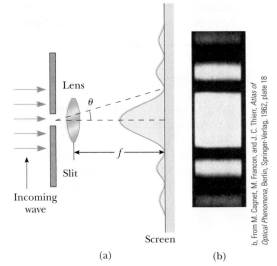

b. From M. Cagnet, M. Francon, and J. C. Thier, *Atlas of Optical Phenomena*, Berlin, Springer-Verlag, 1962, plate 18

(a)   (b)

ACTIVE FIGURE 24.16

(a) The Fraunhofer diffraction pattern of a single slit. The parallel rays are brought into focus on the screen with a converging lens. The pattern consists of a central bright region flanked by much weaker maxima. (This drawing is not to scale.) (b) A photograph of a single-slit Fraunhofer diffraction pattern.

Physics⊗Now™

Log into PhysicsNow at **www.cp7e.com** and go to Active Figure 24.16, where you can adjust the slit width and the wavelength of the light, observing the effect on the diffraction pattern.

can't be explained within the framework of geometric optics, which says that light rays traveling in straight lines should cast a sharp image of the slit on the screen.

Figure 24.15 shows the diffraction pattern and shadow of a penny. The pattern consists of the shadow, a bright spot at its center, and a series of bright and dark circular bands of light near the edge of the shadow. The bright spot at the center (called the *Fresnel bright spot*) is explained by Augustin Fresnel's wave theory of light, which predicts constructive interference at this point for certain locations of the penny. From the viewpoint of geometric optics, there shouldn't be any bright spot: the center of the pattern would be completely screened by the penny.

One type of diffraction, called **Fraunhofer diffraction**, occurs when the rays leave the diffracting object in parallel directions. Fraunhofer diffraction can be achieved experimentally either by placing the observing screen far from the slit or by using a converging lens to focus the parallel rays on a nearby screen, as in Active Figure 24.16a. A bright fringe is observed along the axis at $\theta = 0$, with alternating dark and bright fringes on each side of the central bright fringe. Active Figure 24.16b is a photograph of a single-slit Fraunhofer diffraction pattern.

24.7 SINGLE-SLIT DIFFRACTION

Until now we have assumed that slits have negligible width, acting as line sources of light. In this section we determine how their nonzero widths are the basis for understanding the nature of the Fraunhofer diffraction pattern produced by a single slit.

We can deduce some important features of this problem by examining waves coming from various portions of the slit, as shown in Figure 24.17. According to Huygens' principle, **each portion of the slit acts as a source of waves. Hence, light from one portion of the slit can interfere with light from another portion**, and the resultant intensity on the screen depends on the direction θ.

To analyze the diffraction pattern, it's convenient to divide the slit into halves, as in Figure 24.17. All the waves that originate at the slit are in phase. Consider waves 1 and 3, which originate at the bottom and center of the slit, respectively. Wave 1 travels farther than wave 3 by an amount equal to the path difference

Figure 24.17 Diffraction of light by a narrow slit of width a. Each portion of the slit acts as a point source of waves. The path difference between rays 1 and 3 or between rays 2 and 4 is equal to $(a/2)\sin\theta$. (This drawing is not to scale, and the rays are assumed to converge at a distant point.)

$(a/2) \sin \theta$, where a is the width of the slit. Similarly, the path difference between waves 3 and 5 is $(a/2) \sin \theta$. If this path difference is exactly half of a wavelength (corresponding to a phase difference of 180°), the two waves cancel each other and destructive interference results. This is true, in fact, for any two waves that originate at points separated by half the slit width, because the phase difference between two such points is 180°. Therefore, waves from the upper half of the slit interfere *destructively* with waves from the lower half of the slit when

$$\frac{a}{2} \sin \theta = \frac{\lambda}{2}$$

or when

$$\sin \theta = \frac{\lambda}{a}$$

If we divide the slit into four parts rather than two and use similar reasoning, we find that the screen is also dark when

$$\sin \theta = \frac{2\lambda}{a}$$

Continuing in this way, we can divide the slit into six parts and show that darkness occurs on the screen when

$$\sin \theta = \frac{3\lambda}{a}$$

Therefore, the general condition for **destructive interference** for a single slit of width a is

$$\sin \theta_{\text{dark}} = m \frac{\lambda}{a} \qquad m = \pm 1, \pm 2, \pm 3, \ldots \qquad \text{[24.11]}$$

◀ Condition for destructive interference (single slit)

Equation 24.11 gives the values of θ for which the diffraction pattern has zero intensity, where a dark fringe forms. However, the equation tells us nothing about the variation in intensity along the screen. The general features of the intensity distribution along the screen are shown in Figure 24.18. A broad central bright fringe is flanked by much weaker bright fringes alternating with dark fringes. The various dark fringes (points of zero intensity) occur at the values of θ that satisfy Equation 24.11. The points of constructive interference lie approximately halfway between the dark fringes. Note that the central bright fringe is twice as wide as the weaker maxima having $m > 1$.

Quick Quiz 24.3

In a single-slit diffraction experiment, as the width of the slit is made smaller, the width of the central maximum of the diffraction pattern (a) becomes smaller, (b) becomes larger, or (c) remains the same.

TIP 24.3 The Same, But Different
Although Equations 24.2 and 24.11 have the same form, they have different meanings. Equation 24.2 describes the *bright* regions in a two-slit interference pattern, while Equation 24.11 describes the *dark* regions in a single-slit interference pattern.

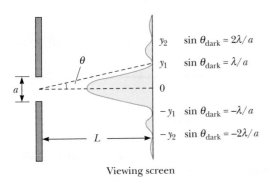

$y_2 \quad \sin \theta_{\text{dark}} = 2\lambda/a$

$y_1 \quad \sin \theta_{\text{dark}} = \lambda/a$

0

$-y_1 \quad \sin \theta_{\text{dark}} = -\lambda/a$

$-y_2 \quad \sin \theta_{\text{dark}} = -2\lambda/a$

Viewing screen

Figure 24.18 Positions of the minima for the Fraunhofer diffraction pattern of a single slit of width a. (This drawing is not to scale.)

Applying Physics 24.3   Diffraction of Sound Waves

If a classroom door is open even just a small amount, you can hear sounds coming from the hallway. Yet you can't see what is going on in the hallway. How can this difference be explained?

Explanation The space between the slightly open door and the wall is acting as a single slit for waves.

Sound waves have wavelengths larger than the width of the slit, so sound is effectively diffracted by the opening, and the central maximum spreads throughout the room. Light wavelengths are much smaller than the slit width, so there is virtually no diffraction for the light. You must have a direct line of sight to detect the light waves.

INTERACTIVE EXAMPLE 24.6   A Single-Slit Experiment

Goal   Find the positions of the dark fringes in single-slit diffraction.

Problem   Light of wavelength 5.80×10^2 nm is incident on a slit of width 0.300 mm. The observing screen is placed 2.00 m from the slit. Find the positions of the first dark fringes and the width of the central bright fringe.

Strategy   This problem requires substitution into Equation 24.11 to find the sines of the angles of the first dark fringes. The positions can then be found with the tangent function, since for small angles $\sin \theta \approx \tan \theta$. The extent of the central maximum is defined by these two dark fringes.

Solution

The first dark fringes that flank the central bright fringe correspond to $m = \pm 1$ in Equation 24.11:

$$\sin \theta = \pm \frac{\lambda}{a} = \pm \frac{5.80 \times 10^{-7} \text{ m}}{0.300 \times 10^{-3} \text{ m}} = \pm 1.93 \times 10^{-3}$$

Use the triangle in Figure 24.18 to relate the position of the fringe to the tangent function:

$$\tan \theta = \frac{y_1}{L}$$

Because θ is very small, we can use the approximation $\sin \theta \approx \tan \theta$ and then solve for y_1:

$$\sin \theta \approx \tan \theta \approx \frac{y_1}{L}$$

$$y_1 \approx L \sin \theta = \pm L \frac{\lambda}{a} = \boxed{\pm 3.86 \times 10^{-3} \text{ m}}$$

Compute the distance between the positive and negative first-order maxima, which is the width w of the central maximum:

$$w = + 3.86 \times 10^{-3} \text{ m} - (-3.86 \times 10^{-3} \text{ m}) = \boxed{7.72 \times 10^{-3} \text{ m}}$$

Remarks   Note that this value of w is much greater than the width of the slit. However, as the width of the slit is *increased*, the diffraction pattern *narrows*, corresponding to smaller values of θ. In fact, for large values of a, the maxima and minima are so closely spaced that the only observable pattern is a large central bright area resembling the geometric image of the slit. Because the width of the geometric image increases as the slit width increases, the narrowest image occurs when the geometric and diffraction widths are equal.

Exercise 24.6

Determine the width of the first-order bright fringe.

Answer   3.86 mm

Physics⊗Now™   Investigate the single-slit diffraction pattern by logging into PhysicsNow at **www.cp7e.com** and going to Interactive Example 24.6.

24.8   THE DIFFRACTION GRATING

The diffraction grating, a useful device for analyzing light sources, consists of a large number of equally spaced parallel slits. A grating can be made by scratching parallel lines on a glass plate with a precision machining technique. The clear panes between

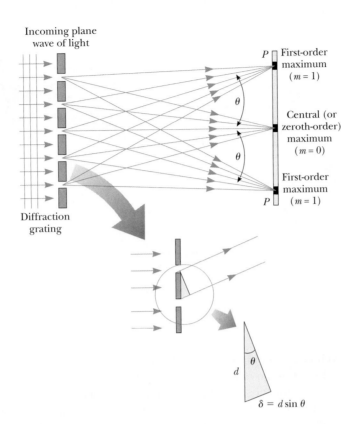

Incoming plane
wave of light

P First-order
maximum
($m = 1$)

θ

Central (or
zeroth-order)
maximum
($m = 0$)

θ

First-order
maximum
P ($m = 1$)

Diffraction
grating

d θ

$\delta = d \sin \theta$

Figure 24.19 A side view of a diffraction grating. The slit separation is d, and the path difference between adjacent slits is $d \sin \theta$.

scratches act like slits. A typical grating contains several thousand lines per centimeter. For example, a grating ruled with 5 000 lines/cm has a slit spacing d equal to the reciprocal of that number; hence, $d = (1/5\ 000)$ cm $= 2 \times 10^{-4}$ cm.

Figure 24.19 is a schematic diagram of a section of a plane diffraction grating. A plane wave is incident from the left, normal to the plane of the grating. The intensity of the pattern on the screen is the result of the combined effects of interference and diffraction. Each slit causes diffraction, and the diffracted beams in turn interfere with one another to produce the pattern. Moreover, each slit acts as a source of waves, and all waves start in phase at the slits. For some arbitrary direction θ measured from the horizontal, however, the waves must travel *different* path lengths before reaching a particular point P on the screen. From Figure 24.19, note that the path difference between waves from any two adjacent slits is $d \sin \theta$. If this path difference equals one wavelength or some integral multiple of a wavelength, waves from all slits will be in phase at P and a bright line will be observed at that point. Therefore, the condition for **maxima** in the interference pattern at the angle θ is

$$d \sin \theta_{\text{bright}} = m\lambda \qquad m = 0, 1, 2, \ldots \qquad \text{[24.12]}$$

Light emerging from a slit at an angle other than that for a maximum interferes nearly completely destructively with light from some other slit on the grating. All such pairs will result in little or no transmission in that direction, as illustrated in Active Figure 24.20.

Equation 24.12 can be used to calculate the wavelength from the grating spacing and the angle of deviation, θ. The integer m is the **order number** of the diffraction pattern. If the incident radiation contains several wavelengths, each wavelength deviates through a specific angle, which can be found from Equation 24.12. All wavelengths are focused at $\theta = 0$, corresponding to $m = 0$. This is called the *zeroth-order maximum*. The *first-order maximum*, corresponding to $m = 1$, is observed at an angle that satisfies the relationship $\sin \theta = \lambda/d$; the *second-order maximum*, corresponding to $m = 2$, is observed at a larger angle θ, and so on. Active Figure 24.20 is a sketch of the intensity distribution for some of the orders

m -2 -1 0 1 2

$-\dfrac{2\lambda}{d}$ $-\dfrac{\lambda}{d}$ 0 $\dfrac{\lambda}{d}$ $\dfrac{2\lambda}{d}$

$\sin \theta \longrightarrow$

ACTIVE FIGURE 24.20
Intensity versus $\sin \theta$ for the diffraction grating. The zeroth-, first-, and second-order principal maxima are shown.

Physics⊗Now™

Log into PhysicsNow at **www.cp7e.com** and go to Active Figure 24.20, where you can choose the number of slits to be illuminated and observe the effect on the diffraction pattern.

◀ Condition for maxima in the interference pattern of a diffraction grating

produced by a diffraction grating. Note the sharpness of the principal maxima and
the broad range of the dark areas, a pattern in direct contrast to the broad bright
fringes characteristic of the two-slit interference pattern.

A simple arrangement that can be used to measure the angles in a diffraction
pattern is shown in Active Figure 24.21. This is a form of diffraction-grating
spectrometer. The light to be analyzed passes through a slit and is formed into a
parallel beam by a lens. The light then strikes the grating at a 90° angle. The
diffracted light leaves the grating at angles that satisfy Equation 24.12. A telescope is used to view the image of the slit. The wavelength can be determined
by measuring the angles at which the images of the slit appear for the various
orders.

Quick Quiz 24.4

If laser light is reflected from a phonograph record or a compact disc, a diffraction pattern appears. The pattern arises because both devices contain parallel tracks of information that act as a reflection diffraction grating. Which device, record or compact disc, results in diffraction maxima that are farther
apart?

Applying Physics 24.4   Prism vs. Grating

When white light enters through an opening in an
opaque box and exits through an opening on the
other side of the box, a spectrum of colors appears on
the wall. From this observation, how would you be
able to determine whether the box contains a prism
or a diffraction grating?

Explanation The determination could be made by
noticing the order of the colors in the spectrum relative to the direction of the original beam of white
light. For a prism, in which the separation of light is a
result of dispersion, the violet light will be refracted
more than the red light. Hence, the order of the spectrum from a prism will be from red, closest to the original direction, to violet. For a diffraction grating, the
angle of diffraction increases with wavelength, so the
spectrum from the diffraction grating will have colors
in the order from violet, closest to the original direction, to red. Furthermore, the diffraction grating will
produce *two* first-order spectra on either side of the
grating, while the prism will produce only a single
spectrum.

Applying Physics 24.5   Rainbows from a Compact Disc

White light reflected from the surface of a compact disc has a multicolored appearance, as shown in Figure 24.22. The observation depends on the orientation of the disc relative to the eye and the position of the light source. Explain how all this works.

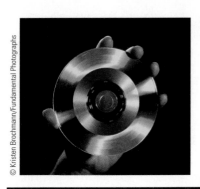

Figure 24.22   (Applying Physics 24.5) Compact discs act as diffraction gratings when observed under white light.

Explanation   The surface of a compact disc has a spiral-shaped track (with a spacing of approximately 1 μm) that acts as a reflection grating. The light scattered by these closely spaced parallel tracks interferes constructively in certain directions that depend on both the wavelength and the direction of the incident light. Any one section of the disc serves as a diffraction grating for white light, sending beams of constructive interference for different colors in different directions. The different colors you see when viewing one section of the disc change as the light source, the disc, or you move to change the angles of incidence or diffraction.

Use of a Diffraction Grating in CD Tracking

If a CD player is to reproduce sound faithfully, the laser beam must follow the spiral track of information perfectly. Sometimes the laser beam can drift off track, however, and without a feedback procedure to let the player know this is happening, the fidelity of the music can be greatly reduced.

Figure 24.23 shows how a diffraction grating is used in a three-beam method to keep the beam on track. The central maximum of the diffraction pattern reads the information on the CD track, and the two first-order maxima steer the beam. The grating is designed so that the first-order maxima fall on the smooth surfaces on either side of the information track. Both of these reflected beams have their own detectors, and because both beams are reflected from smooth surfaces, they should have the same strong intensity when they are detected. If the central beam wanders off the track, however, one of the steering beams will begin to strike bumps on the information track and the amount of light reflected will decrease. This information is then used by electronic circuits to drive the main beam back to its desired location.

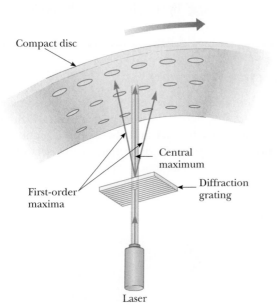

Figure 24.23   The laser beam in a CD player is able to follow the spiral track by using three beams produced with a diffraction grating.

INTERACTIVE EXAMPLE 24.7   A Diffraction Grating

Goal   Calculate different-order principal maxima for a diffraction grating.

Problem   Monochromatic light from a helium–neon laser ($\lambda = 632.8$ nm) is incident normally on a diffraction grating containing 6.00×10^3 lines/cm. Find the angles at which one would observe the first-order maximum, the second-order maximum, and so forth.

Strategy   Find the slit separation by inverting the number of lines per centimeter, then substitute values into Equation 24.12.

Solution

Invert the number of lines per centimeter to obtain the slit separation:

$$d = \frac{1}{6.00 \times 10^3 \, \text{cm}^{-1}} = 1.67 \times 10^{-4} \, \text{cm} = 1.67 \times 10^3 \, \text{nm}$$

Substitute $m = 1$ into Equation 24.12 to find the sine of the angle corresponding to the first-order maximum:

$$\sin \theta_1 = \frac{\lambda}{d} = \frac{632.8 \, \text{nm}}{1.67 \times 10^3 \, \text{nm}} = 0.379$$

Take the inverse sine of the preceding result to find θ_1:

$$\theta_1 = \sin^{-1} 0.379 = \boxed{22.3°}$$

Repeat the calculation for $m = 2$:

$$\sin \theta_2 = \frac{2\lambda}{d} = \frac{2(632.8 \, \text{nm})}{1.67 \times 10^3 \, \text{nm}} = 0.758$$

$$\theta_2 = \boxed{49.3°}$$

Repeat the calculation for $m = 3$:

$$\sin \theta_3 = \frac{3\lambda}{d} = \frac{3(632.8 \, \text{nm})}{1.67 \times 10^3 \, \text{nm}} = 1.14$$

Because $\sin \theta$ can't exceed one, there is no solution for θ_3.

Remarks   The foregoing calculation shows that there can only be a finite number of principal maxima. In this case, only zeroth-, first-, and second-order maxima would be observed.

Exercise 24.7

Suppose light with wavelength 7.80×10^2 nm is used instead and the diffraction grating has 3.30×10^3 lines per centimeter. Find the angles of all the principal maxima.

Answers   0°, 14.9°, 31.0°, 50.6°

Physics⊗Now™ Investigate the diffraction pattern from a diffraction grating by logging into PhysicsNow at **www.cp7e.com** and going to Interactive Example 24.7.

24.9   POLARIZATION OF LIGHT WAVES

In Chapter 21, we described the transverse nature of electromagnetic waves. Figure 24.24 shows that the electric and magnetic field vectors associated with an electromagnetic wave are at right angles to each other and also to the direction of wave propagation. The phenomenon of polarization, described in this section, is firm evidence of the transverse nature of electromagnetic waves.

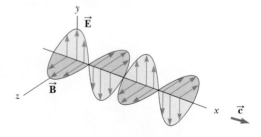

Figure 24.24   A schematic diagram of a polarized electromagnetic wave propagating in the x-direction. The electric field vector $\vec{\mathbf{E}}$ vibrates in the xy-plane, while the magnetic field vector $\vec{\mathbf{B}}$ vibrates in the xz-plane.

An ordinary beam of light consists of a large number of electromagnetic waves emitted by the atoms or molecules of the light source. The vibrating charges associated with the atoms act as tiny antennas. Each atom produces a wave with its own orientation of $\vec{\mathbf{E}}$, as in Figure 24.24, corresponding to the direction of atomic vibration. However, because all directions of vibration are possible, the resultant electromagnetic wave is a superposition of waves produced by the individual atomic sources. The result is an **unpolarized** light wave, represented schematically in Figure 24.25a. The direction of wave propagation shown in the figure is perpendicular to the page. Note that *all* directions of the electric field vector are equally probable and lie in a plane (such as the plane of this page) perpendicular to the direction of propagation.

A wave is said to be **linearly polarized** if the resultant electric field $\vec{\mathbf{E}}$ vibrates in the same direction *at all times* at a particular point, as in Figure 24.25b. (Sometimes such a wave is described as *plane polarized* or simply *polarized*.) The wave in Figure 24.24 is an example of a wave that is linearly polarized in the *y*-direction. As the wave propagates in the *x*-direction, $\vec{\mathbf{E}}$ is always in the *y*-direction. The plane formed by $\vec{\mathbf{E}}$ and the direction of propagation is called the *plane of polarization* of the wave. In Figure 24.24, the plane of polarization is the *xy*-plane.

It's possible to obtain a linearly polarized beam from an unpolarized beam by removing all waves from the beam except those with electric field vectors that oscillate in a single plane. We now discuss three processes for doing this: (1) selective absorption, (2) reflection, and (3) scattering.

Polarization by Selective Absorption

The most common technique for polarizing light is to use a material that transmits waves having electric field vectors that vibrate in a plane parallel to a certain direction and absorbs those waves with electric field vectors vibrating in directions perpendicular to that direction.

In 1932, E. H. Land discovered a material, which he called **Polaroid**, that polarizes light through selective absorption by oriented molecules. This material is fabricated in thin sheets of long-chain hydrocarbons, which are stretched during manufacture so that the molecules align. After a sheet is dipped into a solution containing iodine, the molecules become good electrical conductors. However, conduction takes place primarily along the hydrocarbon chains, because the valence electrons of the molecules can move easily only along those chains. (Recall that valence electrons are "free" electrons that can move easily through the conductor.) As a result, the molecules readily *absorb* light having an electric field vector parallel to their lengths and *transmit* light with an electric field vector perpendicular to their lengths. It's common to refer to the direction perpendicular to the molecular chains as the **transmission axis**. In an ideal polarizer, all light with $\vec{\mathbf{E}}$ parallel to the transmission axis is transmitted and all light with $\vec{\mathbf{E}}$ perpendicular to the transmission axis is absorbed.

Polarizing material reduces the intensity of light passing through it. In Active Figure 24.26, an unpolarized light beam is incident on the first polarizing sheet, called the **polarizer**; the transmission axis is as indicated. The light that passes

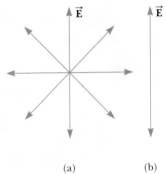

Figure 24.25 (a) An unpolarized light beam viewed along the direction of propagation (perpendicular to the page). The transverse electric field vector can vibrate in any direction with equal probability. (b) A linearly polarized light beam with the electric field vector vibrating in the vertical direction.

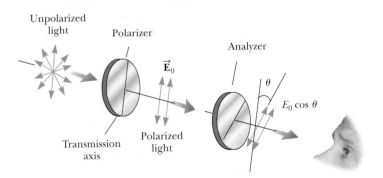

Unpolarized light
Polarizer
Analyzer
$\vec{\mathbf{E}}_0$
θ
$E_0 \cos \theta$
Transmission axis
Polarized light

ACTIVE FIGURE 24.26
Two polarizing sheets whose transmission axes make an angle θ with each other. Only a fraction of the polarized light incident on the analyzer is transmitted.

Physics ☉ Now™
Log into PhysicsNow at **www.cp7e.com** and go to Active Figure 24.26, where you can rotate the second polarizer and see the effect on the transmitted light.

(a)

(b)

(c)

Photos courtesy of Henry Leap

Figure 24.27   The intensity of light transmitted through two polarizers depends on the relative orientations of their transmission axes. (a) The transmitted light has *maximum* intensity when the transmission axes are *aligned* with each other. (b) The transmitted light intensity diminishes when the transmission axes are at an angle of 45° with each other. (c) The transmitted light intensity is a *minimum* when the transmission axes are at *right angles* to each other.

through this sheet is polarized vertically, and the transmitted electric field vector is $\vec{E}_0$. A second polarizing sheet, called the **analyzer**, intercepts this beam with its transmission axis at an angle of θ to the axis of the polarizer. The component of $\vec{E}_0$ that is perpendicular to the axis of the analyzer is completely absorbed. The component of $\vec{E}_0$ that is parallel to the analyzer axis, $E_0 \cos \theta$, is allowed to pass through the analyzer. Because the intensity of the transmitted beam varies as the *square* of its amplitude E, we conclude that the intensity of the (polarized) beam transmitted through the analyzer varies as

Malus's law ▶

$$I = I_0 \cos^2 \theta \qquad \text{[24.13]}$$

where I_0 is the intensity of the polarized wave incident on the analyzer. This expression, known as **Malus's law**, applies to any two polarizing materials having transmission axes at an angle of θ to each other. Note from Equation 24.13 that the transmitted intensity is a maximum when the transmission axes are parallel ($\theta = 0$ or 180°) and is zero (complete absorption by the analyzer) when the transmission axes are perpendicular to each other. This variation in transmitted intensity through a pair of polarizing sheets is illustrated in Figure 24.27.

When unpolarized light of intensity I_0 is sent through a single ideal polarizer, the transmitted linearly polarized light has intensity $I_0/2$. This fact follows from Malus's law, because the average value of $\cos^2 \theta$ is one-half.

Applying Physics 24.6   Polarizing Microwaves

A polarizer for microwaves can be made as a grid of parallel metal wires about a centimeter apart. Is the electric field vector for microwaves transmitted through this polarizer parallel or perpendicular to the metal wires?

Explanation Electric field vectors parallel to the metal wires cause electrons in the metal to oscillate parallel to the wires. Thus, the energy from the waves with these electric field vectors is transferred to the metal by accelerating the electrons and is eventually transformed to internal energy through the resistance of the metal. Waves with electric field vectors perpendicular to the metal wires are not able to accelerate electrons and pass through the wires. Consequently, the electric field polarization is perpendicular to the metal wires.

EXAMPLE 24.8   Polarizer

Goal   Understand how polarizing materials affect light intensity.

Problem   Unpolarized light is incident upon three polarizers. The first polarizer has a vertical transmission axis, the second has a transmission axis rotated 30.0° with respect to the first, and the third has a transmission axis rotated 75.0° relative to the first. If the initial light intensity of the beam is I_b, calculate the light intensity after the beam passes through (a) the second polarizer and (b) the third polarizer.

Strategy   After the beam passes through the first polarizer, it is polarized and its intensity is cut in half. Malus's law can then be applied to the second and third polarizers. The angle used in Malus's law must be relative to the immediately preceding transmission axis.

Solution
(a) Calculate the intensity of the beam after it passes through the second polarizer:

The incident intensity is $I_b/2$. Apply Malus's law to the second polarizer:

$$I_2 = I_0 \cos^2 \theta = \frac{I_b}{2} \cos^2 (30.0^\circ) = \frac{I_b}{2} \left(\frac{\sqrt{3}}{2}\right)^2 = \boxed{\frac{3}{8} I_b}$$

(b) Calculate the intensity of the beam after it passes through the third polarizer.

The incident intensity is now $3I_b/8$. Apply Malus's law to the third polarizer:

$$I_3 = I_2 \cos^2 \theta = \frac{3}{8} I_b \cos^2 (45.0^\circ) = \frac{3}{8} I_b \left(\frac{\sqrt{2}}{2}\right)^2 = \boxed{\frac{3}{16} I_b}$$

Remarks   Notice that the angle used in part (b) was not 75.0°, but $75.0^\circ - 30.0^\circ = 45.0^\circ$. The angle is always with respect to the previous polarizer's transmission axis, because the polarizing material physically determines what direction the transmitted electric fields can have.

Exercise 24.8
The polarizers are rotated, so that the second polarizer has a transmission axis of 40.0° with respect to the first polarizer and the third polarizer has an angle of 90.0° with respect to the first. If I_b is the intensity of the original unpolarized light, what is the intensity of the beam after it passes through (a) the second polarizer, and (b) the third polarizer? (c) What is the final transmitted intensity if the second polarizer is removed?

Answers   (a) $0.293I_b$ (b) $0.121I_b$ (c) 0

Polarization by Reflection

When an unpolarized light beam is reflected from a surface, the reflected light is completely polarized, partially polarized, or unpolarized, depending on the angle of incidence. If the angle of incidence is either 0° or 90° (a normal or grazing angle), the reflected beam is unpolarized. For angles of incidence between 0° and 90°, however, the reflected light is polarized to some extent. For one particular angle of incidence the reflected beam is completely polarized.

Suppose an unpolarized light beam is incident on a surface, as in Figure 24.28a (page 808). The beam can be described by two electric field components, one parallel to the surface (represented by dots) and the other perpendicular to the first component and to the direction of propagation (represented by brown arrows). It is found that the parallel component reflects more strongly than the other components, and this results in a partially polarized beam. In addition, the refracted beam is also partially polarized.

Now suppose that the angle of incidence, θ_1, is varied until the angle between the reflected and refracted beams is 90° (Fig. 24.28b). At this particular angle of incidence, called the **polarizing angle** θ_p, the reflected beam is completely polarized, with its electric field vector parallel to the surface, while the refracted beam is partially polarized.

An expression relating the polarizing angle to the index of refraction of the reflecting surface can be obtained by the use of Figure 24.28b. From this figure we see that at the polarizing angle, $\theta_p + 90^\circ + \theta_2 = 180^\circ$, so that $\theta_2 = 90^\circ - \theta_p$. Using Snell's law and taking $n_1 = n_{\text{air}} = 1.00$ and $n_2 = n$ yields

$$n = \frac{\sin \theta_1}{\sin \theta_2} = \frac{\sin \theta_p}{\sin \theta_2}$$

Because $\sin \theta_2 = \sin(90^\circ - \theta_p) = \cos \theta_p$, the expression for n can be written

Figure 24.28 (a) When unpolarized light is incident on a reflecting surface, the reflected and refracted beams are partially polarized. (b) The reflected beam is completely polarized when the angle of incidence equals the polarizing angle θ_p, satisfying the equation $n = \tan \theta_p$.

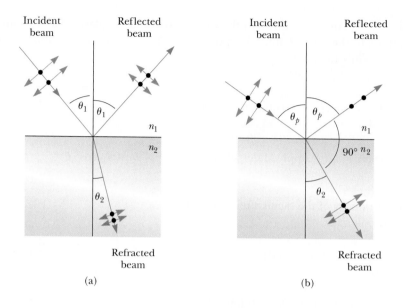

(a)   (b)

◀ Brewster's law

$$n = \frac{\sin \theta_p}{\cos \theta_p} = \tan \theta_p \qquad [24.14]$$

Equation 24.14 is called **Brewster's law**, and the polarizing angle θ_p is sometimes called **Brewster's angle** after its discoverer, Sir David Brewster (1781–1868). For example, Brewster's angle for crown glass (where $n = 1.52$) has the value $\theta_p = \tan^{-1}(1.52) = 56.7°$. Because n varies with wavelength for a given substance, Brewster's angle is also a function of wavelength.

Polarization by reflection is a common phenomenon. Sunlight reflected from water, glass, or snow is partially polarized. If the surface is horizontal, the electric field vector of the reflected light has a strong horizontal component. Sunglasses made of polarizing material reduce the glare, which *is* the reflected light. The transmission axes of the lenses are oriented vertically to absorb the strong horizontal component of the reflected light. Because the reflected light is mostly polarized, most of the glare can be eliminated without removing most of the normal light.

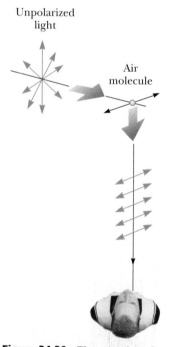

Figure 24.29 The scattering of unpolarized sunlight by air molecules. The light observed at right angles is linearly polarized because the vibrating molecule has a horizontal component of vibration.

Polarization by Scattering

When light is incident on a system of particles, such as a gas, the electrons in the medium can absorb and reradiate part of the light. The absorption and reradiation of light by the medium, called **scattering**, is what causes sunlight reaching an observer on Earth from straight overhead to be polarized. You can observe this effect by looking directly up through a pair of sunglasses made of polarizing glass. Less light passes through at certain orientations of the lenses than at others.

Figure 24.29 illustrates how the sunlight becomes polarized. The left side of the figure shows an incident unpolarized beam of sunlight on the verge of striking an air molecule. When the beam strikes the air molecule, it sets the electrons of the molecule into vibration. These vibrating charges act like those in an antenna except that they vibrate in a complicated pattern. The horizontal part of the electric field vector in the incident wave causes the charges to vibrate horizontally, and the vertical part of the vector simultaneously causes them to vibrate vertically. A horizontally polarized wave is emitted by the electrons as a result of their horizontal motion, and a vertically polarized wave is emitted parallel to the Earth as a result of their vertical motion.

Scientists have found that bees and homing pigeons use the polarization of sunlight as a navigational aid.

Optical Activity

Many important practical applications of polarized light involve the use of certain materials that display the property of **optical activity**. A substance is said to be optically active if it rotates the plane of polarization of transmitted light. Suppose

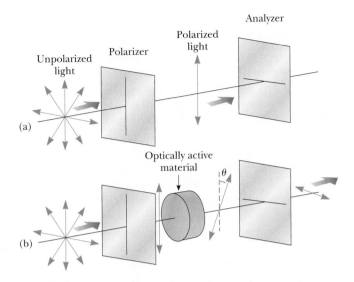

Figure 24.30 (a) When crossed polarizers are used, none of the polarized light can pass through the analyzer. (b) An optically active material rotates the direction of polarization through the angle θ, enabling some of the polarized light to pass through the analyzer.

unpolarized light is incident on a polarizer from the left, as in Figure 24.30a. The transmitted light is polarized vertically, as shown. If this light is then incident on an analyzer with its axis perpendicular to that of the polarizer, no light emerges from it. If an optically active material is placed between the polarizer and analyzer, as in Figure 24.30b, the material causes the direction of the polarized beam to rotate through the angle θ. As a result, some light is able to pass through the analyzer. The angle through which the light is rotated by the material can be found by rotating the polarizer until the light is again extinguished. It is found that the angle of rotation depends on the length of the sample and, if the substance is in solution, on the concentration. One optically active material is a solution of common sugar, dextrose. A standard method for determining the concentration of a sugar solution is to measure the rotation produced by a fixed length of the solution.

Optical activity occurs in a material because of an asymmetry in the shape of its constituent molecules. For example, some proteins are optically active because of their spiral shapes. Other materials, such as glass and plastic, become optically active when placed under stress. If polarized light is passed through an unstressed piece of plastic and then through an analyzer with an axis perpendicular to that of the polarizer, none of the polarized light is transmitted. If the plastic is placed under stress, however, the regions of greatest stress produce the largest angles of rotation of polarized light, and a series of light and dark bands are observed in the transmitted light. Engineers often use this property in the design of structures ranging from bridges to small tools. A plastic model is built and analyzed under different load conditions to determine positions of potential weakness and failure under stress. If the design is poor, patterns of light and dark bands will indicate the points of greatest weakness, and the design can be corrected at an early stage. Figure 24.31 shows examples of stress patterns in plastic.

APPLICATION

Finding the Concentrations of Solutions by Means of Their Optical Activity

(a)   (b)

Figure 24.31 (a) Strain distribution in a plastic model of a replacement hip used in a medical research laboratory. The pattern is produced when the model is placed between a polarizer and an analyzer oriented perpendicular to each other. (b) A plastic model of an arch structure under load conditions observed between perpendicular polarizers. Such patterns are useful in the optimum design of architectural components.

APPLICATION

Liquid Crystal Displays (LCD's)

Liquid Crystals

An effect similar to rotation of the plane of polarization is used to create the familiar displays on pocket calculators, wristwatches, notebook computers, and so forth. The properties of a unique substance called a liquid crystal make these displays (called LCD's, for *liquid crystal displays*) possible. As its name implies, a **liquid crystal** is a substance with properties intermediate between those of a crystalline solid and those of a liquid; that is, the molecules of the substance are more orderly than those in a liquid, but less orderly than those in a pure crystalline solid. The forces that hold the molecules together in such a state are just barely strong enough to enable the substance to maintain a definite shape, so it is reasonable to call it a solid. However, small inputs of mechanical or electrical energy can disrupt these weak bonds and make the substance flow, rotate, or twist.

To see how liquid crystals can be used to create a display, consider Figure 24.32a. The liquid crystal is placed between two glass plates in the pattern shown, and electrical contacts, indicated by the thin lines, are made. When a voltage is applied across any segment in the display, that segment turns dark. In this fashion, any number between 0 and 9 can be formed by the pattern, depending on the voltages applied to the seven segments.

To see why a segment can be changed from dark to light by the application of a voltage, consider Figure 24.32b, which shows the basic construction of a portion of

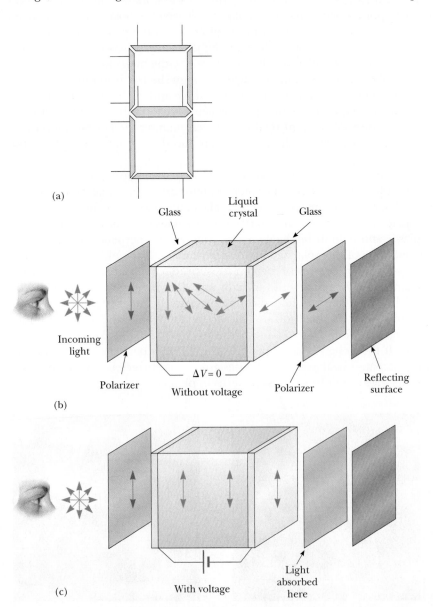

(a)

Glass   Liquid crystal   Glass

Incoming light

Polarizer $\Delta V = 0$ Polarizer   Reflecting surface

Without voltage

(b)

With voltage   Light absorbed here

(c)

Figure 24.32 (a) The light-segment pattern of a liquid crystal display. (b) Rotation of a polarized light beam by a liquid crystal when the applied voltage is zero. (c) Molecules of the liquid crystal align with the electric field when a voltage is applied.

the display. The liquid crystal is placed between two glass substrates that are packaged between two pieces of Polaroid material with their transmission axes perpendicular. A reflecting surface is placed behind one of the pieces of Polaroid. First consider what happens when light falls on this package and no voltages are applied to the liquid crystal, as shown in Figure 24.32b. Incoming light is polarized by the polarizer on the left and then falls on the liquid crystal. As the light passes through the crystal, its plane of polarization is rotated by 90°, allowing it to pass through the polarizer on the right. It reflects from the reflecting surface and retraces its path through the crystal. Thus, an observer to the left of the crystal sees the segment as being bright. When a voltage is applied as in Figure 24.32c, the molecules of the liquid crystal don't rotate the plane of polarization of the light. In this case, the light is absorbed by the polarizer on the right and none is reflected back to the observer to the left of the crystal. As a result, the observer sees this segment as black. Changing the applied voltage to the crystal in a precise pattern at precise times can make the pattern tick off the seconds on a watch, display a letter on a computer display, and so forth.

SUMMARY

Physics⊗Now™ Take a practice test by logging into Physics-Now at **www.cp7e.com** and clicking on the Pre-Test link for this chapter.

24.1 Conditions for Interference

Interference occurs when two or more light waves overlap at a given point. A sustained interference pattern is observed if (1) the sources are coherent (that is, they maintain a constant phase relationship with one another), (2) the sources have identical wavelengths, and (3) the superposition principle is applicable.

24.2 Young's Double-Slit Experiment

In **Young's double-slit experiment**, two slits separated by distance d are illuminated by a single-wavelength light source. An interference pattern consisting of bright and dark fringes is observed on a screen a distance L from the slits. The condition for **bright fringes** (constructive interference) is

$$d \sin \theta_{bright} = m\lambda \qquad m = 0, \pm 1, \pm 2, \ldots \qquad \textbf{[24.2]}$$

The number m is called the **order number** of the fringe.
The condition for **dark fringes** (destructive interference) is

$$d \sin \theta_{dark} = (m + \tfrac{1}{2})\lambda \qquad m = 0, \pm 1, \pm 2, \ldots \qquad \textbf{[24.3]}$$

The position y_m of the bright fringes on the screen can be determined by using the relation $\sin \theta \approx \tan \theta = y_m/L$, which is true for small angles. This can be substituted into Equations 24.2 and 24.3, yielding the location of the bright fringes:

$$y_{bright} = \frac{\lambda L}{d} m \qquad m = 0, \pm 1, \pm 2, \ldots \qquad \textbf{[24.5]}$$

A similar expression can be derived for the dark fringes. This equation can be used either to locate the maxima or to determine the wavelength of light by measuring y_m.

24.3 Change of Phase Due to Reflection &
24.4 Interference in Thin Films

An electromagnetic wave undergoes a phase change of 180° on reflection from a medium with an index of refraction higher than that of the medium in which the wave is traveling. There is no change when the wave, traveling in a medium with higher index of refraction, reflects from a medium with a lower index of refraction.

The wavelength λ_n of light in a medium with index of refraction n is

$$\lambda_n = \frac{\lambda}{n} \qquad \textbf{[24.7]}$$

where λ is the wavelength of the light in free space. Light encountering a thin film of thickness t will reflect off the top and bottom of the film, each ray undergoing a possible phase change as described above. The two rays recombine, and bright and dark fringes will be observed, with the conditions of interference given by the following table:

Equation ($m = 0, 1, \ldots$)	1 phase reversal	0 or 2 phase reversals
$2nt = (m + \tfrac{1}{2})\lambda$ **[24.9]**	constructive	destructive
$2nt = m\lambda$ **[24.10]**	destructive	constructive

24.6 Diffraction &
24.7 Single-Slit Diffraction

Diffraction occurs when waves pass through small openings, around obstacles, or by sharp edges. The **diffraction pattern** produced by a single slit on a distant screen consists of a central bright maximum flanked by less bright fringes alternating with dark regions. The angles θ at which the diffraction pattern has zero intensity (regions of destructive interference) are described by

$$\sin \theta_{dark} = m \frac{\lambda}{a} \qquad m = \pm 1, \pm 2, \pm 3, \ldots \qquad \textbf{[24.11]}$$

where a is the width of the slit and λ is the wavelength of the light incident on the slit.

24.8 The Diffraction Grating

A **diffraction grating** consists of many equally spaced, identical slits. The condition for **maximum intensity** in the interference pattern of a diffraction grating is

$$d \sin \theta_{bright} = m\lambda \qquad m = 0, 1, 2, \ldots \quad [24.12]$$

where d is the spacing between adjacent slits and m is the order number of the diffraction pattern. A diffraction grating can be made by putting a large number of evenly spaced scratches on a glass slide. The number of such lines per centimeter is the inverse of the spacing d.

24.9 Polarization of Light Waves

Unpolarized light can be polarized by selective absorption, reflection, or scattering. A material can polarize light if it transmits waves having electric field vectors that vibrate in a plane parallel to a certain direction and absorbs waves with electric field vectors vibrating in directions perpendicular to that direction. When unpolarized light pass through a polarizing sheet, its intensity is reduced by half, and the light becomes polarized. When this light passes through a second polarizing sheet with transmission axis at an angle of θ with respect to the transmission axis of the first sheet, the transmitted intensity is given by

$$I = I_0 \cos^2 \theta \qquad [24.13]$$

where I_0 is the intensity of the light after passing through the first polarizing sheet.

In general, light reflected from an amorphous material, such as glass, is partially polarized. Reflected light is completely polarized, with its electric field parallel to the surface, when the angle of incidence produces a 90° angle between the reflected and refracted beams. This angle of incidence, called the **polarizing angle** θ_p, satisfies **Brewster's law**, given by

$$n = \tan \theta_p \qquad [24.14]$$

where n is the index of refraction of the reflecting medium.

CONCEPTUAL QUESTIONS

1. Your automobile has two headlights. What sort of interference pattern do you expect to see from them? Why?

2. Holding your hand at arm's length, you can readily block sunlight from your eyes. Why can you not block sound from your ears this way?

3. Consider a dark fringe in an interference pattern, at which almost no light energy is arriving. Light from both slits is arriving at this point, but the waves cancel. Where does the energy go?

4. If Young's double-slit experiment were performed under water, how would the observed interference pattern be affected?

5. In a laboratory accident, you spill two liquids onto water, neither of which mixes with the water. They both form thin films on the water surface. As the films spread and become very thin, you notice that one film becomes bright and the other black in reflected light. Why might this be?

6. If white light is used in Young's double-slit experiment, rather than monochromatic light, how does the interference pattern change?

7. In our discussion of thin-film interference, we looked at light *reflecting* from a thin film. Consider one light ray, the direct ray, that transmits through the film without reflecting. Then consider a second ray, the reflected ray, that transmits through the first surface, reflects back to the second, reflects again from the first, and then transmits out into the air, parallel to the direct ray. For normal incidence, how thick must the film be, in terms of the wavelength of the light, for the outgoing rays to interfere destructively? Is it the same thickness as for reflected destructive interference?

8. What is the necessary condition on the difference in path length between two waves that interfere (a) constructively and (b) destructively? Assume that the wave sources are coherent.

9. A lens with outer radius of curvature R and index of refraction n rests on a flat glass plate, and the combination is illuminated from white light from above. Is there a dark spot or a light spot at the center of the lens? What does it mean if the observed rings are noncircular?

10. Often, fingerprints left on a piece of glass such as a windowpane show colored spectra like that from a diffraction grating. Why?

11. In everyday experience, why are radio waves polarized, while light is not?

12. Suppose reflected white light is used to observe a thin, transparent coating on glass as the coating material is gradually deposited by evaporation in a vacuum. Describe some color changes that might occur during the process of building up the thickness of the coating.

13. Would it be possible to place a nonreflective coating on an airplane to cancel radar waves of wavelength 3 cm?

14. Certain sunglasses use a polarizing material to reduce the intensity of light reflected from shiny surfaces, such as water or the hood of a car. What orientation of the transmission axis should the material have to be most effective?

15. Why is it so much easier to perform interference experiments with a laser than with an ordinary light source?

16. A simple way of observing an interference pattern is to look at a distant light source through a stretched handkerchief or an open umbrella. Explain how this works.

17. When you receive a chest x-ray at a hospital, the x-rays pass through a series of parallel ribs in your chest. Do the ribs act as a diffraction grating for x-rays?

18. Can a sound wave be polarized? Explain.

19. Astronomers often observe occulations, in which a star passes behind another object, such as the Moon. During an occultation, the intensity of light from the star doesn't suddenly drop to zero as the star passes behind the edge of the Moon. Instead, the intensity fluctuates for a short time before dropping to zero. Why should this happen?

20. In one experiment, light from a laser passes through a double slit and forms an interference pattern on a distant screen. The experiment is repeated after increasing the slit separation by 50%. In which experiment is the distance from the central maximum to the next maximum the greatest?

21. Light in air that is reflected from a water surface is found to be completely polarized at an angle θ. If the light is instead reflected from a glass coffee table, will the new angle for complete polarization be larger or smaller?

22. In one experiment, blue light passes through a diffraction grating and forms an interference pattern on a screen. In a second experiment, red light passes through the same diffraction grating and forms another interference pattern. How do the separations between bright lines in the two experiments compare with each other?

PROBLEMS

1, 2, 3 = straightforward, intermediate, challenging   □ = full solution available in *Student Solutions Manual/Study Guide*

Physics⊗Now™ = coached solution with hints available at **www.cp7e.com**   🎓 = biomedical application

Section 24.2 Young's Double-Slit Experiment

1. A laser beam ($\lambda = 632.8$ nm) is incident on two slits 0.200 mm apart. How far apart are the bright interference fringes on a screen 5.00 m away from the double slits?

2. In a Young's double-slit experiment, a set of parallel slits with a separation of 0.100 mm is illuminated by light having a wavelength of 589 nm, and the interference pattern is observed on a screen 4.00 m from the slits. (a) What is the difference in path lengths from each of the slits to the location of a third-order bright fringe on the screen? (b) What is the difference in path lengths from the two slits to the location of the third dark fringe on the screen, away from the center of the pattern?

3. A pair of narrow, parallel slits separated by 0.250 mm is illuminated by the green component from a mercury vapor lamp ($\lambda = 546.1$ nm). The interference pattern is observed on a screen 1.20 m from the plane of the parallel slits. Calculate the distance (a) from the central maximum to the first bright region on either side of the central maximum and (b) between the first and second dark bands in the interference pattern.

4. Light of wavelength 460 nm falls on two slits spaced 0.300 mm apart. What is the required distance from the slit to a screen if the spacing between the first and second dark fringes is to be 4.00 mm?

5. In a location where the speed of sound is 354 m/s, a 2 000-Hz sound wave impinges on two slits 30.0 cm apart. (a) At what angle is the first maximum located? (b) If the sound wave is replaced by 3.00-cm microwaves, what slit separation gives the same angle for the first maximum? (c) If the slit separation is 1.00 μm, what frequency of light gives the same first maximum angle?

6. White light spans the wavelength range between about 400 nm and 700 nm. If white light passes through two slits 0.30 mm apart and falls on a screen 1.5 m from the slits, find the distance between the first-order violet and the first-order red fringes.

7. Two radio antennas separated by 300 m, as shown in Figure P24.7, simultaneously transmit identical signals of the same wavelength. A radio in a car traveling due north receives the signals. (a) If the car is at the position of the second maximum, what is the wavelength of the signals? (b) How much farther must the car travel to encounter the next minimum in reception? [*Hint:* Determine the path difference between the two signals at the two locations of the car.]

Figure P24.7

8. If the distance between two slits is 0.050 mm and the distance to a screen is 2.50 m, find the spacing between the first- and second-order bright fringes for yellow light of 600-nm wavelength.

9. Waves from a radio station have a wavelength of 300 m. They travel by two paths to a home receiver 20.0 km from the transmitter. One path is a direct path, and the second is by reflection from a mountain directly behind the home receiver. What is the minimum distance from the mountain to the receiver that produces destructive interference at the receiver? (Assume that no phase change occurs on reflection from the mountain.)

10. A pair of slits, separated by 0.150 mm, is illuminated by light having a wavelength of $\lambda = 643$ nm. An interference pattern is observed on a screen 140 cm from the slits. Consider a point on the screen located at $y = 1.80$ cm from the central maximum of this pattern. (a) What is the path difference δ for the two slits at the location y? (b) Express this path difference in terms of the wavelength. (c) Will the interference correspond to a maximum, a minimum, or an intermediate condition?

11. A riverside warehouse has two open doors, as in Figure P24.11. Its interior is lined with a sound-absorbing material. A boat on the river sounds its horn. To person A, the

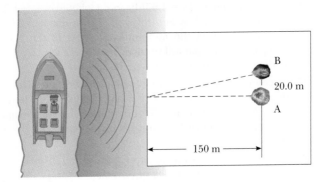

Figure P24.11

sound is loud and clear. To person B, the sound is barely audible. The principal wavelength of the sound waves is 3.00 m. Assuming person B is at the position of the first minimum, determine the distance between the doors, center to center.

12. **Physics⊗Now™** The waves from a radio station can reach a home receiver by two different paths. One is a straight-line path from the transmitter to the home, a distance of 30.0 km. The second path is by reflection from a storm cloud. Assume that this reflection takes place at a point midway between receiver and transmitter. If the wavelength broadcast by the radio station is 400 m, find the minimum height of the storm cloud that will produce destructive interference between the direct and reflected beams. (Assume no phase changes on reflection.)

13. Radio waves from a star, of wavelength 250 m, reach a radio telescope by two separate paths, as shown in Figure P24.13. One is a direct path to the receiver, which is situated on the edge of a cliff by the ocean. The second is by reflection off the water. The first minimum of destructive interference occurs when the star is 25.0° above the horizon. Find the height of the cliff. (Assume no phase change on reflection.)

Direct path

Reflected path

Figure P24.13

Section 24.3 Change of Phase Due to Reflection
Section 24.4 Interference in Thin Films

14. Determine the minimum thickness of a soap film ($n = 1.330$) that will result in constructive interference of (a) the red H_α line ($\lambda = 656.3$ nm); (b) the blue H_γ line ($\lambda = 434.0$ nm).

15. Suppose the film shown in Figure 24.7 has an index of refraction of 1.36 and is surrounded by air on both sides. Find the minimum thickness that will produce constructive interference in the reflected light when the film is illuminated by light of wavelength 500 nm.

16. A thin film of glass ($n = 1.50$) floats on a liquid of $n = 1.35$ and is illuminated by light of $\lambda = 580$ nm incident from air above it. Find the minimum thickness of the glass, other than zero, that will produce destructive interference in the reflected light.

17. A coating is applied to a lens to minimize reflections. The index of refraction of the coating is 1.55, and that of the lens is 1.48. If the coating is 177.4 nm thick, what wavelength is minimally reflected for normal incidence in the lowest order?

18. A transparent oil with index of refraction 1.29 spills on the surface of water (index of refraction 1.33), producing a maximum of reflection with normally incident orange light (wavelength 600 nm in air). Assuming the maximum occurs in the first order, determine the thickness of the oil slick.

19. A possible means for making an airplane invisible to radar is to coat the plane with an antireflective polymer. If radar waves have a wavelength of 3.00 cm and the index of refraction of the polymer is $n = 1.50$, how thick would you make the coating?

20. A beam of light of wavelength 580 nm passes through two closely spaced glass plates, as shown in Figure P24.20. For what minimum non-zero value of the plate separation d will the transmitted light be bright? This arrangement is often used to measure the wavelength of light and is called a Fabry–Perot interferometer.

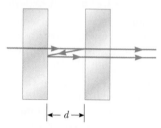

$\leftarrow d \rightarrow$

Figure P24.20

21. Astronomers observe the chromosphere of the sun with a filter that passes the red hydrogen spectral line of wavelength 656.3 nm, called the H_α line. The filter consists of a transparent dielectric of thickness d held between two partially aluminized glass plates. The filter is kept at a constant temperature. (a) Find the minimum value of d that will produce maximum transmission of perpendicular H_α light if the dielectric has an index of refraction of 1.378. (b) If the temperature of the filter increases above the normal value increasing its thickness, what happens to the transmitted wavelength? (c) The dielectric will also pass what near-visible wavelength? One of the glass plates is colored red to absorb this light.

22. Two rectangular optically flat plates ($n = 1.52$) are in contact along one end and are separated along the other end by a 2.00-μm-thick spacer (Fig. P24.22). The top plate is illuminated by monochromatic light of wavelength 546.1 nm. Calculate the number of dark parallel bands crossing the top plate (including the dark band at zero thickness along the edge of contact between the plates).

Figure P24.22   (Problems 22 and 23)

23. An air wedge is formed between two glass plates separated at one edge by a very fine wire, as in Figure P24.22. When the wedge is illuminated from above by 600-nm light, 30 dark fringes are observed. Calculate the radius of the wire.

24. A planoconvex lens with radius of curvature $R = 3.0$ m is in contact with a flat plate of glass. A light source and the observer's eye are both close to the normal, as shown in

Figure 24.8a. The radius of the 50th bright Newton's ring is found to be 9.8 mm. What is the wavelength of the light produced by the source?

25. A planoconvex lens rests with its curved side on a flat glass surface and is illuminated from above by light of wavelength 500 nm. (See Fig. 24.8.) A dark spot is observed at the center, surrounded by 19 concentric dark rings (with bright rings in between). How much thicker is the air wedge at the position of the 19th dark ring than at the center?

26. Nonreflective coatings on camera lenses reduce the loss of light at the surfaces of multilens systems and prevent internal reflections that might mar the image. Find the minimum thickness of a layer of magnesium fluoride ($n = 1.38$) on flint glass ($n = 1.66$) that will cause destructive interference of reflected light of wavelength 550 nm near the middle of the visible spectrum.

27. **Physics⊗Now™** A thin film of MgF$_2$ ($n = 1.38$) with thickness 1.00×10^{-5} cm is used to coat a camera lens. Are any wavelengths in the visible spectrum intensified in the reflected light?

28. A flat piece of glass is supported horizontally above the flat end of a 10.0-cm-long metal rod that has its lower end rigidly fixed. The thin film of air between the rod and the glass is observed to be bright when illuminated by light of wavelength 500 nm. As the temperature is slowly increased by 25.0°C, the film changes from bright to dark and back to bright 200 times. What is the coefficient of linear expansion of the metal?

Section 24.7 Single-Slit Diffraction

29. Helium–neon laser light ($\lambda = 632.8$ nm) is sent through a 0.300-mm-wide single slit. What is the width of the central maximum on a screen 1.00 m from the slit?

30. Light of wavelength 600 nm falls on a 0.40-mm-wide slit and forms a diffraction pattern on a screen 1.5 m away. (a) Find the position of the first dark band on each side of the central maximum. (b) Find the width of the central maximum.

31. Light of wavelength 587.5 nm illuminates a slit of width 0.75 mm. (a) At what distance from the slit should a screen be placed if the first minimum in the diffraction pattern is to be 0.85 mm from the central maximum? (b) Calculate the width of the central maximum.

32. Microwaves of wavelength 5.00 cm enter a long, narrow window in a building that is otherwise essentially opaque to the incoming waves. If the window is 36.0 cm wide, what is the distance from the central maximum to the first-order minimum along a wall 6.50 m from the window?

33. A slit of width 0.50 mm is illuminated with light of wavelength 500 nm, and a screen is placed 120 cm in front of the slit. Find the widths of the first and second maxima on each side of the central maximum.

34. A screen is placed 50.0 cm from a single slit, which is illuminated with light of wavelength 680 nm. If the distance between the first and third minima in the diffraction pattern is 3.00 mm, what is the width of the slit?

Section 24.8 The Diffraction Grating

35. Three discrete spectral lines occur at angles of 10.1°, 13.7°, and 14.8°, respectively, in the first-order spectrum of a diffraction-grating spectrometer. (a) If the grating has 3 660 slits/cm, what are the wavelengths of the light? (b) At what angles are these lines found in the second-order spectra?

36. Intense white light is incident on a diffraction grating that has 600 lines/mm. (a) What is the highest order in which the complete visible spectrum can be seen with this grating? (b) What is the angular separation between the violet edge (400 nm) and the red edge (700 nm) of the first-order spectrum produced by the grating?

37. The hydrogen spectrum has a red line at 656 nm and a violet line at 434 nm. What angular separation between these two spectral lines obtained with a diffraction grating that has 4 500 lines/cm?

38. A grating with 1 500 slits per centimeter is illuminated with light of wavelength 500 nm. (a) What is the highest-order number that can be observed with this grating? (b) Repeat for a grating of 15 000 slits per centimeter.

39. A light source emits two major spectral lines: an orange line of wavelength 610 nm and a blue-green line of wavelength 480 nm. If the spectrum is resolved by a diffraction grating having 5 000 lines/cm and viewed on a screen 2.00 m from the grating, what is the distance (in centimeters) between the two spectral lines in the second-order spectrum?

40. White light is spread out into its spectral components by a diffraction grating. If the grating has 2 000 lines per centimeter, at what angle does red light of wavelength 640 nm appear in the first-order spectrum?

41. Sunlight is incident on a diffraction grating that has 2 750 lines/cm. The second-order spectrum over the visible range (400–700 nm) is to be limited to 1.75 cm along a screen that is a distance L from the grating. What is the required value of L?

42. Light containing two different wavelengths passes through a diffraction grating with 1 200 slits/cm. On a screen 15.0 cm from the grating, the third-order maximum of the shorter wavelength falls midway between the central maximum and the first side maximum for the longer wavelength. If the neighboring maxima of the longer wavelength are 8.44 mm apart on the screen, what are the wavelengths in the light? [*Hint:* Use the small-angle approximation.]

43. **Physics⊗Now™** A beam of 541-nm light is incident on a diffraction grating that has 400 lines/mm. (a) Determine the angle of the second-order ray. (b) If the entire apparatus is immersed in water, determine the new second-order angle of diffraction. (c) Show that the two diffracted rays of parts (a) and (b) are related through the law of refraction.

44. Light from a helium–neon laser ($\lambda = 632.8$ nm) is incident on a single slit. What is the maximum width for which no diffraction minima are observed? [*Hint:* Values of $\sin \theta > 1$ are not possible.]

Section 24.9 Polarization of Light Waves

45. The angle of incidence of a light beam in air onto a reflecting surface is continuously variable. The reflected ray is found to be completely polarized when the angle of incidence is 48.0°. (a) What is the index of refraction of the reflecting material? (b) If some of the incident light

(at an angle of 48.0°) passes into the material below the surface, what is the angle of refraction?

46. Unpolarized light passes through two polaroid sheets. The axis of the first is vertical, and that of the second is at 30.0° to the vertical. What fraction of the initial light is transmitted?

47. The index of refraction of a glass plate is 1.52. What is the Brewster's angle when the plate is (a) in air? (b) in water? (See Problem 51.)

48. At what angle above the horizon is the Sun if light from it is completely polarized upon reflection from water?

49. A light beam is incident on heavy flint glass ($n = 1.65$) at the polarizing angle. Calculate the angle of refraction for the transmitted ray.

50. The critical angle for total internal reflection for sapphire surrounded by air is 34.4°. Calculate the Brewster angle for sapphire if the light is incident from the air.

51. Equation 24.14 assumes that the incident light is in air. If the light is incident from a medium of index n_1 onto a medium of index n_2, follow the procedure used to derive Equation 24.14 to show that $\tan \theta_p = n_2/n_1$.

52. Plane-polarized light is incident on a single polarizing disk, with the direction of E_0 parallel to the direction of the transmission axis. Through what angle should the disk be rotated so that the intensity in the transmitted beam is reduced by a factor of (a) 3.00, (b) 5.00, (c) 10.0?

53. Three polarizing plates whose planes are parallel are centered on a common axis. The directions of the transmission axes relative to the common vertical direction are shown in Figure P24.53. A linearly polarized beam of light with plane of polarization parallel to the vertical reference direction is incident from the left onto the first disk with intensity $I_i = 10.0$ units (arbitrary). Calculate the transmitted intensity I_f when $\theta_1 = 20.0°$, $\theta_2 = 40.0°$, and $\theta_3 = 60.0°$. [*Hint:* Make repeated use of Malus's law.]

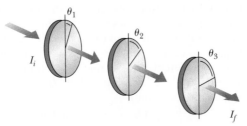

Figure P24.53   (Problems 53 and 62)

54. **Physics⊗Now**™ Light of intensity I_0 and polarized parallel to the transmission axis of a polarizer, is incident on an analyzer. (a) If the transmission axis of the analyzer makes an angle of 45° with the axis of the polarizer, what is the intensity of the transmitted light? (b) What should the angle between the transmission axes be to make $I/I_0 = 1/3$?

55. Light with a wavelength in vacuum of 546.1 nm falls perpendicularly on a biological specimen that is 1.000 μm thick. The light splits into two beams polarized at right angles, for which the indices of refraction are 1.320 and 1.333, respectively. (a) Calculate the wavelength of each component of the light while it is traversing the specimen. (b) Calculate the phase difference between the two beams when they emerge from the specimen.

ADDITIONAL PROBLEMS

56. A beam containing light of wavelengths λ_1 and λ_2 is incident on a set of parallel slits. In the interference pattern, the fourth bright line of the λ_1 light occurs at the same position as the fifth bright line of the λ_2 light. If λ_1 is known to be 540 nm, what is the value of λ_2?

57. Light of wavelength 546 nm (the intense green line from a mercury source) produces a Young's interference pattern in which the second minimum from the central maximum is along a direction that makes an angle of 18.0 min of arc with the axis through the central maximum. What is the distance between the parallel slits?

58. The two speakers are placed 35.0 cm apart. A single oscillator makes the speakers vibrate in phase at a frequency of 2.00 kHz. At what angles, measured from the perpendicular bisector of the line joining the speakers, would a distant observer hear maximum sound intensity? Minimum sound intensity? (Take the speed of sound to be 340 m/s.)

59. Interference effects are produced at point P on a screen as a result of direct rays from a 500-nm source and reflected rays off a mirror, as in Figure P24.59. If the source is 100 m to the left of the screen and 1.00 cm above the mirror, find the distance y (in millimeters) to the first dark band above the mirror.

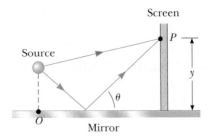

Figure P24.59

60. Many cells are transparent and colorless. Structures of great interest in biology and medicine can be practically invisible to ordinary microscopy. An *interference microscope* reveals a difference in refractive index as a shift in interference fringes, to indicate the size and shape of cell structures. The idea is exemplified in the following problem: An air wedge is formed between two glass plates in contact along one edge and slightly separated at the opposite edge. When the plates are illuminated with monochromatic light from above, the reflected light has 85 dark fringes. Calculate the number of dark fringes that appear if water ($n = 1.33$) replaces the air between the plates.

61. A thin layer of oil ($n = 1.25$) is floating on water. How thick is the oil in the region that strongly reflects green light ($\lambda = 525$ nm)?

62. Three polarizers, centered on a common axis and with their planes parallel to each other, have transmission axes oriented at angles of θ_1, θ_2, and θ_3 from the vertical, as shown in Figure P24.53. Light of intensity I_i, polarized with its plane of polarization oriented vertically, is incident from the left onto the first polarizer. What is the ratio I_f/I_i of the final transmitted intensity to the incident intensity if (a) $\theta_1 = 45°$, $\theta_2 = 90°$, and $\theta_3 = 0°$? (b) $\theta_1 = 0°$, $\theta_2 = 45°$, and $\theta_3 = 90°$?

63. Figure P24.63 shows a radio-wave transmitter and a receiver, both $h = 50.0$ m above the ground and $d = 600$ m apart. The receiver can receive signals directly from the transmitter, and indirectly, from signals that bounce off the ground. If the ground is level between the transmitter and receiver and a $\lambda/2$ phase shift occurs upon reflection, determine the longest wavelengths that interfere (a) constructively and (b) destructively.

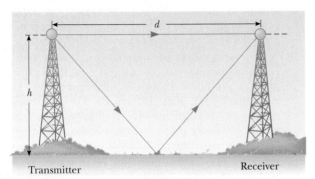

Figure P24.63

64. A planoconvex lens (flat on one side, convex on the other) with index of refraction n rests with its curved side (radius of curvature R) on a flat glass surface of the same index of refraction with a film of index n_{film} between them. The lens is illuminated from above by light of wavelength λ. Show that the dark Newton rings which appear have radii of

$$r \approx \sqrt{m\lambda R/n_{film}}$$

where m is an integer.

65. The transmitting antenna on a submarine is 5.00 m above the water when the ship surfaces. The captain wishes to transmit a message to a receiver on a 90.0-m-tall cliff at the ocean shore. If the signal is to be completely polarized by reflection off the ocean surface, how far must the ship be from the shore?

66. (a) If light is incident at an angle θ from a medium of index n_1 on a medium of index n_2 so that the angle between the reflected ray and refracted ray is β, show that

$$\tan \theta = \frac{n_2 \sin \beta}{n_1 - n_2 \cos \beta}$$

Hint: Use the trigonometric identity:

$$\sin(A + B) = \sin A \cos B + \cos A \sin B$$

(b) Show that the foregoing equation for $\tan \theta$ reduces to Brewster's law when $\beta = 90°$, $n_1 = 1$, and $n_2 = n$.

67. A diffraction pattern is produced on a screen 140 cm from a single slit, using monochromatic light of wavelength 500 nm, The distance from the center of the central maximum to the first-order maximum is 3.00 mm. Calculate the slit width. [*Hint:* Assume that the first-order maximum is halfway between the first- and second-order minima.]

68. A glass plate ($n = 1.61$) is covered with a thin, uniform layer of oil ($n = 1.20$). A light beam of variable wavelength is normally incident from air onto the oil surface. Observation of the reflected beam shows destructive interference at 500 nm and constructive interference at 750 nm. From this information, calculate the thickness of the oil film.

69. The condition for constructive interference by reflection from a thin film in air, as developed in Section 24.4, assumes nearly normal incidence. (a) Show that for large angles of incidence, the condition for constructive interference of light reflecting from a thin film of thickness t, with index of refraction n, and surrounded by air may be written as

$$2nt \cos \theta_2 = \left(m + \frac{1}{2} \right) \lambda$$

where θ_2 is the angle of refraction. (b) Calculate the minimum thickness for constructive interference if sodium light ($\lambda = 590$ nm) is incident at an angle of 30.0° on a film with an index of refraction of 1.38.

70. Figure P24.70 illustrates the formation of an interference pattern by the Lloyd's mirror method. Light from source S reaches the screen via two different pathways. One is a direct path, and the second is by reflection from a horizontal mirror. The effect is as if light from two different sources S and S' had interfered as in the Young's double-slit arrangement. Assume that the actual source S and the virtual source S' are in a plane 25 cm to the left of the mirror and the screen is a distance $L = 120$ cm to the right of that plane. Source S is a distance $h = 2.5$ mm above the top surface of the mirror, and the light is monochromatic with $\lambda = 620$ nm. Determine the distance of the first bright fringe above the surface of the mirror.

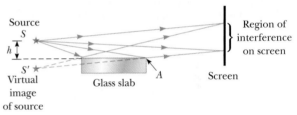

Figure P24.70

71. A piece of transparent material having an index of refraction n is cut into the shape of a wedge as shown in Figure P24.71. The angle of the wedge is small. Monochromatic light of wavelength λ is normally incident from above, and viewed from above. Let h represent the height of the wedge and ℓ its width. Show that bright fringes occur at the positions $x = \lambda \ell (m + \frac{1}{2})/2hn$ and dark fringes occur at the positions $x = \lambda \ell m/2hn$, where $m = 0, 1, 2, \ldots$ and x is measured as shown.

Figure P24.71

ACTIVITIES

1. Place a clear dish or plate on a black surface, such as a sheet of black construction paper. Now add a thin layer of

water to the glass, and place a few drops of kerosene or light machine oil on the water. Darken the room and shine a flashlight from an angle, as in Figure A24.1. Note the interference pattern of various colors you observe under the white light. How does the pattern change if you cover the flashlight with a sheet of red, blue, or green cellophane, which acts as a filter?

As an extension of the preceding experiment, observe the colors appearing to swirl on the surface of a soap bubble. What color do you see just before a bubble bursts?

Oil

Water

Figure A24.1

2. Stand a couple of meters from a lightbulb. Facing away from the light, hold a compact disc about 10 cm from your eye and tilt it until the reflection of the bulb is located in the hole at the disc's center. You should see spectra radiating out from the center, with violet on the inside and red on the outside. Now move the disc away from your eye until the violet band is at the outer edge. Carefully measure the distance from your eye to the center of the disc,

and also determine the radius of the disc. Use this information to find the angle θ to the first-order maximum for violet light. Now use the relationship $d \sin \theta = m\lambda$ to determine the spacing between the grooves of the disc. The industry standard is 1.6 μm. How close did you come? While you are observing the spectrum from a CD, note that the color of the light from a given point changes with the viewing angle. Explain this effect in terms of changes in the path length. It is of interest that the blues and blue-greens in hummingbird feathers and butterflies are caused by diffraction off finely aligned structures in feathers and wings. (See chapter opener photo.)

3. (a) Devise a way to use a protractor, a desk lamp, and polarizing sunglasses to measure Brewster's angle for the glass in a window. From this, determine the index of refraction of the glass. (b) Put on a pair of polarizing sunglasses and close one eye. Hold up a lens of a second pair of polarizing glasses in front of your open eye so that light must pass through a lens of each pair before entering your eye. Now rotate the second pair of glasses around. You will note that the light reaching your eye is considerably reduced at some orientations and will pass freely at others. (c) On a sunny day, rotate your polarizing sunglasses in front of your eye and observe how light reflects from a window or the surface of water. Note the change in the amount of light entering your eye for various orientations of the glasses. (d) For a final observation concerning polarized light, rotate a pair of polarizing sunglasses while looking at various areas of the sky. From what direction do you find the light to be most highly polarized?

© Denis Scott/CORBIS

The Hubble Space Telescope does its viewing above the atmosphere and doesn't suffer from the atmospheric blurring, caused by air turbulence, that plagues ground-based telescopes. Despite this advantage, it does have limitations due to diffraction effects. In this chapter, we show how the wave nature of light limits the ability of any optical system to distinguish between closely spaced objects.

CHAPTER

25

Optical Instruments

We use devices made from lenses, mirrors, or other optical components every time we put on a pair of eyeglasses or contact lenses, take a photograph, look at the sky through a telescope, and so on. In this chapter we examine how these and other optical instruments work. For the most part, our analyses will involve the laws of reflection and refraction and the procedures of geometric optics. To explain certain phenomena, however, we must use the wave nature of light.

25.1 THE CAMERA

The single-lens photographic **camera** is a simple optical instrument having the features shown in Figure 25.1 (page 820). It consists of an opaque box, a converging lens that produces a real image, and a film behind the lens to receive the image. Focusing is accomplished by varying the distance between lens and film—with an adjustable bellows in antique cameras and with some other mechanical arrangements in contemporary models. For proper focusing, which leads to sharp images, the lens-to-film distance depends on the object distance as well as on the focal length of the lens. The shutter, located behind the lens, is a mechanical device that is opened for selected time intervals. With this arrangement, moving objects can be photographed by using short exposure times, dark scenes (with low light levels) by using long exposure times. If this adjustment were not available, it would be impossible to take stop-action photographs. A rapidly moving vehicle, for example, could move far enough while the shutter was open to produce a blurred image. Another major cause of blurred images is movement of the *camera* while the shutter is open. To prevent such movement, you should mount the camera on a tripod or use short exposure times. Typical shutter speeds (that is, exposure times) are 1/30, 1/60, 1/125, and 1/250 s. Stationary objects are often shot with a shutter speed of 1/60 s.

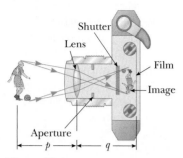

Figure 25.1 A cross-sectional view of a simple camera.

Most cameras also have an aperture of adjustable diameter to further control the intensity of the light reaching the film. When an aperture of small diameter is used, only light from the central portion of the lens reaches the film, so spherical aberration is reduced.

The intensity I of the light reaching the film is proportional to the area of the lens. Because this area in turn is proportional the square of the lens diameter D, the intensity is also proportional to D^2. Light intensity is a measure of the rate at which energy is received by the film per unit area of the image. Because the area of the image is proportional to q^2 in Figure 25.1, and $q \approx f$ (when $p \gg f$, so that p can be approximated as infinite), we conclude that the intensity is also proportional to $1/f^2$, so that $I \propto D^2/f^2$. The brightness of the image formed on the film depends on the light intensity, so we see that it ultimately depends on both the focal length f and diameter D of the lens. The ratio f/D is called the **f-number** (or focal ratio) of a lens:

$$f\text{-number} \equiv \frac{f}{D} \qquad \qquad [25.1]$$

The f-number is often given as a description of the lens "speed." A lens with a low f-number is a "fast" lens. Extremely fast lenses, which have an f-number as low as approximately 1.2, are expensive because of the difficulty of keeping aberrations acceptably small with light rays passing through a large area of the lens. Camera lenses are often marked with a range of f-numbers, such as 1.4, 2, 2.8, 4, 5.6, 8, 11, ... Any one of these settings can be selected by adjusting the aperture, which changes the value of D. Increasing the setting from one f-number to the next-higher value (for example, from 2.8 to 4) decreases the area of the aperture by a factor of two. The lowest f-number setting on a camera corresponds to a wide open aperture, and the use of the maximum possible lens area.

Simple cameras usually have a fixed focal length and fixed aperture size, with an f-number of about 11. This high value for the f-number allows for a large **depth of field**. This means that objects at a wide range of distances from the lens form reasonably sharp images on the film. In other words, the camera doesn't have to be focused. Most cameras with variable f-numbers adjust them automatically.

25.2 THE EYE

Like a camera, a normal eye focuses light and produces a sharp image. However, the mechanisms by which the eye controls the amount of light admitted and adjusts to produce correctly focused images are far more complex, intricate, and effective than those in even the most sophisticated camera. In all respects, the eye is a physiological wonder.

Figure 25.2a shows the essential parts of the eye. Light entering the eye passes through a transparent structure called the *cornea*, behind which are a clear liquid (the *aqueous humor*), a variable aperture (the *pupil*, which is an opening in the *iris*), and the *crystalline lens*. Most of the refraction occurs at the outer surface of the eye, at which the cornea is covered with a film of tears. Relatively little refraction occurs in the crystalline lens, because the aqueous humor in contact with the lens has an average index of refraction close to that of the lens. The iris, which is the colored portion of the eye, is a muscular diaphragm that controls pupil size. The iris regulates the amount of light entering the eye by dilating the pupil in low-light conditions and contracting the pupil under conditions of bright light. The f-number range of the eye is from about 2.8 to 16.

The cornea–lens system focuses light onto the back surface of the eye—the *retina*—which consists of millions of sensitive receptors called *rods* and *cones*. When stimulated by light, these structures send impulses to the brain via the optic nerve, converting them into our conscious view of the world. The process by which the brain performs this conversion is not well understood and is the subject of much speculation and research. Unlike film in a camera, the rods and cones chemically

Figure 25.2   (a) Essential parts of the eye. Can you correlate the essential parts of the eye with those
of the simple camera in Figure 25.1? (b) Close-up photograph of the human cornea.

adjust their sensitivity according to the prevailing light conditions. This adjustment, which takes about 15 minutes, is responsible for the experience of "getting
used to the dark" in such places as movie theaters. Iris aperture control, which
takes less than one second, helps protect the retina from overload in the adjustment process.

The eye focuses on an object by varying the shape of the pliable crystalline lens
through an amazing process called **accommodation**. An important component in
accommodation is the *ciliary muscle*, which is situated in a circle around the rim of
the lens. Thin filaments, called *zonules*, run from this muscle to the edge of the
lens. When the eye is focused on a distant object, the ciliary muscle is relaxed,
tightening the zonules that attach the ciliary muscle to the edge of the lens. The
force of the zonules causes the lens to flatten, increasing its focal length. For an
object distance of infinity, the focal length of the eye is equal to the fixed distance
between lens and retina, about 1.7 cm. The eye focuses on nearby objects by tens-
ing the ciliary muscle, which relaxes the zonules. This action allows the lens to
bulge a bit and its focal length decreases, resulting in the image being focused on
the retina. All these lens adjustments take place so swiftly that we are not even
aware of the change. In this respect, even the finest electronic camera is a toy compared with the eye.

There is a limit to accommodation because objects that are very close to the eye
produce blurred images. The **near point** is the closest distance for which the lens
can accommodate to focus light on the retina. This distance usually increases with
age and has an average value of 25 cm. Typically, at age 10 the near point of the
eye is about 18 cm. This increases to about 25 cm at age 20, 50 cm at age 40, and
500 cm or greater at age 60. The **far point** of the eye represents the farthest distance for which the lens of the relaxed eye can focus light on the retina. A person
with normal vision is able to see very distant objects, such as the Moon, and so has
a far point at infinity.

Conditions of the Eye

When the eye suffers a mismatch between the focusing power of the lens–cornea
system and the length of the eye so that light rays reach the retina before they converge to form an image, as in Figure 25.3a, the condition is known as **farsightedness** (or *hyperopia*). A farsighted person can usually see faraway objects clearly but

Figure 25.3 (a) A farsighted eye is
slightly shorter than normal; hence,
the image of a nearby object focuses
behind the retina. (b) The condition
can be corrected with a converging
lens. (The object is assumed to be
very small in these figures.)

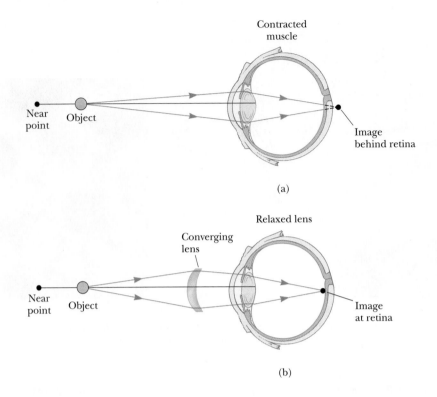

(a)

(b)

not nearby objects. Although the near point of a normal eye is approximately
25 cm, the near point of a farsighted person is much farther than that. The eye of
a farsighted person tries to focus by accommodation, by shortening its focal
length. This works for distant objects, but because the focal length of the farsighted eye is longer than normal, the light from nearby objects can't be brought
to a sharp focus before it reaches the retina, causing a blurred image. The condition can be corrected by placing a converging lens in front of the eye, as in Figure
25.3b. The lens refracts the incoming rays more toward the principal axis before
entering the eye, allowing them to converge and focus on the retina.

Nearsightedness (or *myopia*) is another mismatch condition in which a person is
able to focus on nearby objects, but not faraway objects. In the case of *axial myopia*,
nearsightedness is caused by the lens being too far from the retina. It is also possible
to have *refractive myopia*, in which the lens–cornea system is too powerful for the normal length of the eye. The far point of the nearsighted eye is not at infinity and may
be less than a meter. The maximum focal length of the nearsighted eye is insufficient to produce a sharp image on the retina, and rays from a distant object converge to a focus in front of the retina. They then continue past that point, diverging
before they finally reach the retina and produce a blurred image (Fig. 25.4a).

Nearsightedness can be corrected with a diverging lens, as shown in Figure
25.4b. The lens refracts the rays away from the principal axis before they enter the
eye, allowing them to focus on the retina.

Beginning with middle age, most people lose some of their accommodation ability as the ciliary muscle weakens and the lens hardens. Unlike farsightedness, which
is a mismatch of focusing power and eye length, **presbyopia** (literally, "old-age
vision") is due to a reduction in accommodation ability. The cornea and lens aren't
able to bring nearby objects into focus on the retina. The symptoms are the same as
with farsightedness, and the condition can be corrected with converging lenses.

In the eye defect known as **astigmatism**, light from a point source produces a
line image on the retina. This condition arises when either the cornea or the lens
(or both) are not perfectly symmetric. Astigmatism can be corrected with lenses
having different curvatures in two mutually perpendicular directions.

Optometrists and ophthalmologists usually prescribe lenses measured in
diopters:

APPLICATION

Using Optical Lenses to
Correct for Defects

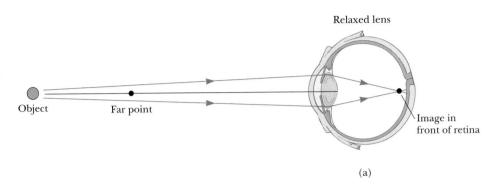

Relaxed lens

Object   Far point

Image in
front of retina

(a)

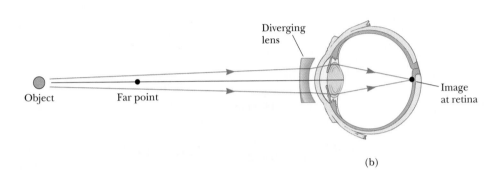

Diverging
lens

Object   Far point

Image
at retina

(b)

Figure 25.4 (a) A nearsighted eye is slightly longer than normal; hence, the image of a distant object focuses *in front* of the retina. (b) The condition can be corrected with a diverging lens. (The object is assumed to be very small in these figures.)

> The **power** $\mathcal{P}$ of a lens in diopters equals the inverse of the focal length in meters: $\mathcal{P} = 1/f$.

For example, a converging lens with a focal length of $+20$ cm has a power of $+5.0$ diopters, and a diverging lens with a focal length of -40 cm has a power of -2.5 diopters. (Although the symbol is the same as for mechanical power, there is no relationship between the two concepts.)

The position of the lens relative the eye causes differences in power, but this usually amounts to less than a quarter diopter, which isn't noticeable to most patients. As a result, practicing optometrists deal in increments of a quarter diopter. Neglecting the eye–lens distance is equivalent to doing the calculation for a contact lens, which rests directly on the eye.

EXAMPLE 25.1   Prescribing a Corrective Lens for a Farsighted Patient

Goal   Apply geometric optics to correct farsightedness.

Problem   The near point of a patient's eye is 50.0 cm. **(a)** What focal length must a corrective lens have to enable the eye to see clearly an object 25.0 cm away? Neglect the eye–lens distance. **(b)** What is the power of this lens? **(c)** Repeat the problem, taking into account the fact that, for typical eyeglasses, the corrective lens is 2.00 cm in front of the eye.

Strategy   This problem requires substitution into the thin-lens equation (Eq. 23.11) and then using the definition of lens power in terms of diopters. The object is at 25.0 cm, but the lens must form an image at the patient's near point, 50.0 cm, the closest point at which the patient's eye can see clearly. In part (c), 2.00 cm must be subtracted from both the object distance and the image distance to account for the position of the lens.

Solution

(a) Find the focal length of the corrective lens, neglecting its distance from the eye.

Apply the thin-lens equation:

$$\frac{1}{p} + \frac{1}{q} = \frac{1}{f}$$

Substitute $p = 25.0$ cm and $q = -50.0$ cm (the latter is negative because the image must be virtual) on the same side of the lens as the object:

$$\frac{1}{25.0 \text{ cm}} + \frac{1}{-50.0 \text{ cm}} = \frac{1}{f}$$

Solve for f. The focal length is positive, corresponding to a converging lens.

$$f = \boxed{50.0 \text{ cm}}$$

(b) What is the power of this lens?

The power is the reciprocal of the focal length in meters:

$$\mathcal{P} = \frac{1}{f} = \frac{1}{0.500 \text{ m}} = \boxed{+2.00 \text{ diopters}}$$

(c) Repeat the problem, noting that the corrective lens is actually 2.00 cm in front of the eye.

Substitute the corrected values of p and q into the thin-lens equation:

$$\frac{1}{p} + \frac{1}{q} = \frac{1}{23.0 \text{ cm}} + \frac{1}{(-48.0 \text{ cm})} = \frac{1}{f}$$
$$f = 44.2 \text{ cm}$$

Compute the power:

$$\mathcal{P} = \frac{1}{f} = \frac{1}{0.442 \text{ m}} = \boxed{+2.26 \text{ diopters}}$$

Remarks   Notice that the calculation in part (c), which doesn't neglect the eye–lens distance, results in a difference of 0.26 diopter.

Exercise 25.1
Suppose a lens is placed in a device that determines its power as 2.75 diopters. Find (a) the focal length of the lens and (b) the minimum distance at which a patient will be able to focus on an object if the patient's near point is 60.0 cm. Neglect the eye–lens distance.

Answers   (a) 36.4 cm   (b) 22.7 cm

EXAMPLE 25.2   A Corrective Lens for Nearsightedness

Goal   Apply geometric optics to correct nearsightedness.

Problem   A particular nearsighted patient can't see objects clearly when they are beyond 25 cm (the far point of the eye). **(a)** What focal length should the prescribed contact lens have to correct this problem? **(b)** Find the power of the lens, in diopters. Neglect the distance between the eye and the corrective lens.

Strategy   The purpose of the lens in this instance is to take objects at infinity and create an image of them at the patient's far point. Apply the thin-lens equation.

Solution
(a) Find the focal length of the corrective lens.

Apply the thin-lens equation for an object at infinity and image at 25.0 cm:

$$\frac{1}{p} + \frac{1}{q} = \frac{1}{\infty} + \frac{1}{(-25.0 \text{ cm})} = \frac{1}{f}$$
$$f = \boxed{-25.0 \text{ cm}}$$

(b) Find the power of the lens in diopters:

$$\mathcal{P} = \frac{1}{f} = \frac{1}{-0.250 \text{ m}} = -4.00 \text{ diopters}$$

Remarks   The focal length is negative, consistent with a diverging lens. Notice that the power is also negative and has the same numeric value as the sum on the left side of the thin-lens equation.

Exercise 25.2
(a) What power lens would you prescribe for a patient with a far point of 35.0 cm? Neglect the eye–lens distance.
(b) Repeat, assuming an eye-corrective lens distance of 2.00 cm.

Answer   (a) − 2.86 diopters   (b) − 3.03 diopters

Applying Physics 25.1   Vision of the Invisible Man

A classic science fiction story, *The Invisible Man* by H.G. Wells, tells of a person who becomes invisible by changing the index of refraction of his body to that of air. Students who know how the eye works have criticized this story; they claim the invisible man would be unable to see. On the basis of your knowledge of the eye, would he be able to see ?

Explanation He wouldn't be able to see. In order for the eye to see an object, incoming light must be refracted at the cornea and lens to form an image on the retina. If the cornea and lens have the same index of refraction as air, refraction can't occur, and an image wouldn't be formed.

Quick Quiz 25.1

Two campers wish to start a fire during the day. One camper is nearsighted and one is farsighted. Whose glasses should be used to focus the Sun's rays onto some paper to start the fire? (a) either camper (b) the nearsighted camper (c) the farsighted camper.

25.3   THE SIMPLE MAGNIFIER

The **simple magnifier** is one of the most basic of all optical instruments because it consists only of a single converging lens. As the name implies, this device is used to increase the apparent size of an object. Suppose an object is viewed at some distance p from the eye, as in Figure 25.5. Clearly, the size of the image formed at the retina depends on the angle θ subtended by the object at the eye. As the object moves closer to the eye, θ increases and a larger image is observed. However, a normal eye can't focus on an object closer than about 25 cm, the near point (Fig. 25.6a, page 826). (Try it!) Therefore, θ is a maximum at the near point.

   To further increase the apparent angular size of an object, a converging lens can be placed in front of the eye with the object positioned at point O, just inside the focal point of the lens, as in Figure 25.6b. At this location, the lens forms a virtual, upright, and enlarged image, as shown. The lens allows the object to be viewed closer to the eye than is possible without the lens. We define the **angular magnification** m as the ratio of the angle subtended by a small object when the lens is in use (angle θ in Fig. 25.6b) to the angle subtended by the object placed at the near point with no lens in use (angle θ_0 in Fig. 25.6a):

$$m \equiv \frac{\theta}{\theta_0} \qquad [25.2]$$

For the case where the lens is held close to the eye, the angular magnification is a maximum when the image formed by the lens is at the near point of the eye, which corresponds to $q = -25$ cm (see Fig. 25.6b). The object distance corresponding to this image distance can be calculated from the thin-lens equation:

$$\frac{1}{p} + \frac{1}{-25 \text{ cm}} = \frac{1}{f} \qquad [25.3]$$

$$p = \frac{25f}{25 + f}$$

Figure 25.5 The size of the image formed on the retina depends on the angle θ subtended at the eye.

◀ Angular magnification with the object at the near point

Figure 25.6 (a) An object placed at the near point ($p = 25$ cm) subtends an angle of $\theta_0 \approx h/25$ at the eye. (b) An object placed near the focal point of a converging lens produces a magnified image, which subtends an angle of $\theta \approx h'/25$ at the eye. Note that, in this situation, $q = -25$ cm.

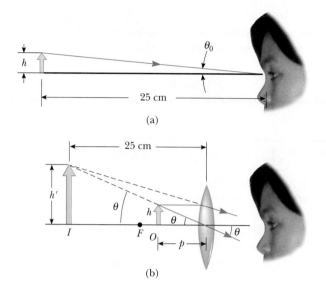

(a)

(b)

Here, f is the focal length of the magnifier in centimeters. From Figures 25.6a and 25.6b, the small-angle approximation gives

$$\tan \theta_0 \approx \theta_0 \approx \frac{h}{25} \quad \text{and} \quad \tan \theta \approx \theta \approx \frac{h}{p} \qquad \text{[25.4]}$$

Equation 25.2 therefore becomes

$$m_{max} = \frac{\theta}{\theta_0} = \frac{h/p}{h/25} = \frac{25}{p} = \frac{25}{25f/(25 + f)}$$

so that

$$m_{max} = 1 + \frac{25 \text{ cm}}{f} \qquad \text{[25.5]}$$

The maximum angular magnification given by Equation 25.5 is the ratio of the angular size seen with the lens to the angular size seen without the lens, with the object at the near point of the eye. Although the normal eye can focus on an image formed anywhere between the near point and infinity, it's most relaxed when the image is at infinity (Sec. 25.2). For the image formed by the magnifying lens to appear at infinity, the object must be placed at the focal point of the lens, so that $p = f$. In this case, Equation 25.4 becomes

$$\theta_0 \approx \frac{h}{25} \quad \text{and} \quad \theta \approx \frac{h}{f}$$

and the angular magnification is

$$m = \frac{\theta}{\theta_0} = \frac{25 \text{ cm}}{f} \qquad \text{[25.6]}$$

With a single lens, it's possible to achieve angular magnifications up to about 4 without serious aberrations. Magnifications up to about 20 can be achieved by using one or two additional lenses to correct for aberrations.

EXAMPLE 25.3 Magnification of a Lens

Goal Compute magnifications of a lens when the image is at the near point and when it's at infinity.

Problem (a) What is the maximum angular magnification of a lens with a focal length of 10.0 cm? (b) What is the angular magnification of this lens when the eye is relaxed? Assume an eye–lens distance of zero.

Strategy   The maximum angular magnification occurs when the image formed by the lens is at the near point of the eye. Under these circumstances, Equation 25.5 gives us the maximum angular magnification. In part (b), the eye is relaxed only if the image is at infinity, so Equation 25.6 applies.

Solution

(a) Find the maximum angular magnification of the lens.

Substitute into Equation 25.5:

$$m_{max} = 1 + \frac{25 \text{ cm}}{f} = 1 + \frac{25 \text{ cm}}{10.0 \text{ cm}} = \boxed{3.5}$$

(b) Find the magnification of the lens when the eye is relaxed.

When the eye is relaxed, the image is at infinity, so substitute into Equation 25.6:

$$m = \frac{25 \text{ cm}}{f} = \frac{25 \text{ cm}}{10.0 \text{ cm}} = \boxed{2.5}$$

Exercise 25.3

What focal length would be necessary if the lens were to have a maximum angular magnification of 4.00?

Answer   8.3 cm

25.4   THE COMPOUND MICROSCOPE

A simple magnifier provides only limited assistance with inspection of the minute details of an object. Greater magnification can be achieved by combining two lenses in a device called a compound microscope, a schematic diagram of which is shown in Active Figure 25.7a. The instrument consists of two lenses: an objective with a very short focal length f_o (where $f_o < 1$ cm), and an ocular lens, or eyepiece, with a focal length f_e of a few centimeters. The two lenses are separated by distance L that is much greater than either f_o or f_e.

The basic approach used to analyze the image formation properties of a microscope is that of two lenses in a row: the image formed by the first becomes the object for the second. The object O placed just outside the focal length of the objective forms a real, inverted image at I_1 that is at or just inside the focal point of the eyepiece. This image is much enlarged. (For clarity, the enlargement of I_1 is not shown in Active Fig. 25.7a.) The eyepiece, which serves as a simple magnifier, uses the image at I_1 as its object and produces an image at I_2. The image seen by the eye at I_2 is virtual, inverted, and very much enlarged.

The lateral magnification M_1 of the first image is $-q_1/p_1$. Note that q_1 is approximately equal to L, because the object is placed close to the focal point of the

(a)

b. © Tony Freeman/Photo Edit

(b)

ACTIVE FIGURE 25.7
(a) A diagram of a compound microscope, which consists of an objective and an eyepiece, or ocular lens. (b) A compound microscope. The three-objective turret allows the user to switch to several different powers of magnification. Combinations of eyepieces with different focal lengths and different objectives can produce a wide range of magnifications.

Physics⊗Now™

Log into PhysicsNow at **www.cp7e.com** and go to Active Figure 25.7, where you can adjust the focal lengths of the objective and eyepiece lenses, and see the effect on the final image.

objective lens, which ensures that the image formed will be far from the objective lens. Further, because the object is very close to the focal point of the objective lens, $p_1 \approx f_o$. Therefore, the lateral magnification of the objective is

$$M_1 = -\frac{q_1}{p_1} \approx -\frac{L}{f_o}$$

From Equation 25.6, the angular magnification of the eyepiece for an object (corresponding to the image at I_1) placed at the focal point is found to be

$$m_e = \frac{25 \text{ cm}}{f_e}$$

The overall magnification of the compound microscope is defined as the product of the lateral and angular magnifications:

Magnification of a microscope ▶

$$m = M_1 m_e = -\frac{L}{f_o}\left(\frac{25 \text{ cm}}{f_e}\right) \qquad [25.7]$$

The negative sign indicates that the image is inverted with respect to the object.

The microscope has extended our vision into the previously unknown realm of incredibly small objects, and the capabilities of this instrument have increased steadily with improved techniques in precision grinding of lenses. A natural question is whether there is any limit to how powerful a microscope could be. For example, could a microscope be made powerful enough to allow us to see an atom? The answer to this question is no, as long as visible light is used to illuminate the object. In order to be seen, the object under a microscope must be at least as large as a wavelength of light. An atom is many times smaller than the wavelength of visible light, so its mysteries must be probed via other techniques.

The wavelength dependence of the "seeing" ability of a wave can be illustrated by water waves set up in a bathtub in the following way: Imagine that you vibrate your hand in the water until waves with a wavelength of about 6 in. are moving along the surface. If you fix a small object, such as a toothpick, in the path of the waves, you will find that the waves are not appreciably disturbed by the toothpick, but continue along their path. Now suppose you fix a larger object, such as a toy sailboat, in the path of the waves. In this case, the waves are considerably disturbed by the object. The toothpick was much smaller than the wavelength of the waves, and as a result, the waves didn't "see" it. The toy sailboat, however, is about the same size as the wavelength of the waves and hence creates a disturbance. Light waves behave in this same general way. The ability of an optical microscope to view an object depends on the size of the object relative to the wavelength of the light used to observe it. Hence, it will never be possible to observe atoms or molecules with such a microscope, because their dimensions are so small (≈ 0.1 nm) relative to the wavelength of the light (≈ 500 nm).

EXAMPLE 25.4   Microscope Magnifications

Goal   Understand the critical factors involved in determining the magnifying power of a microscope.

Problem   A certain microscope has two interchangeable objectives. One has a focal length of 2.0 cm, and the other has a focal length of 0.20 cm. Also available are two eyepieces of focal lengths 2.5 cm and 5.0 cm. If the length of the microscope is 18 cm, compute the magnifications for the following combinations: the 2.0-cm objective and 5.0-cm eyepiece; the 2.0-cm objective and 2.5-cm eyepiece; the 0.20-cm objective and 5.0-cm eyepiece.

Strategy   The solution consists of substituting into Equation 25.7 for three different combinations of lenses.

Solution

Apply Equation 25.7 and combine the 2.0-cm objective with the 5.0-cm eyepiece:

$$m = -\frac{L}{f_o}\left(\frac{25 \text{ cm}}{f_e}\right) = -\frac{18 \text{ cm}}{2.0 \text{ cm}}\left(\frac{25 \text{ cm}}{5.0 \text{ cm}}\right) = \boxed{-45}$$

Combine the 2.0-cm objective with the 2.5-cm eyepiece: $m = -\dfrac{18\text{ cm}}{2.0\text{ cm}}\left(\dfrac{25\text{ cm}}{2.5\text{ cm}}\right) = \boxed{-9.0 \times 10^1}$

Combine the 0.20-cm objective with the 5.0-cm eyepiece: $m = -\dfrac{18\text{ cm}}{0.20\text{ cm}}\left(\dfrac{25\text{ cm}}{5.0\text{ cm}}\right) = \boxed{-450}$

Remarks   Much higher magnifications can be achieved, but the resolution starts to fall, resulting in fuzzy images that don't convey any details. (See Section 25.6 for further discussion of this point.)

Exercise 25.4
Combine the 0.20-cm objective with the 2.5-cm eyepiece.

Answer 9.0×10^2

25.5 THE TELESCOPE

There are two fundamentally different types of telescope, both designed to help us view distant objects such as the planets in our Solar System. These two types are (1) the **refracting telescope**, which uses a combination of lenses to form an image, and (2) the **reflecting telescope**, which uses a curved mirror and a lens to form an image. Once again, we will be able to analyze the telescope by considering it to be a system of two optical elements in a row. As before, the basic technique followed is that the image formed by the first element becomes the object for the second.

In the refracting telescope, two lenses are arranged so that the objective forms a real, inverted image of the distant object very near the focal point of the eyepiece (Active Fig. 25.8a, page 830). Further, the image at I_1 is formed at the focal point of the objective because the object is essentially at infinity. Hence, the two lenses are separated by the distance $f_o + f_e$, which corresponds to the length of the telescope's tube. Finally, at I_2, the eyepiece forms an enlarged, inverted image of the image at I_1.

The angular magnification of the telescope is given by θ/θ_o, where θ_o is the angle subtended by the object at the objective and θ is the angle subtended by the final image. From the triangles in Active Figure 25.8a, and for small angles, we have

$$\theta \approx \frac{h'}{f_e} \quad \text{and} \quad \theta_o \approx \frac{h'}{f_o}$$

Therefore, the angular magnification of the telescope can be expressed as

$$m = \frac{\theta}{\theta_o} = \frac{h'/f_e}{h'/f_o} = \frac{f_o}{f_e} \qquad \textbf{[25.8]}$$

This equation says that the angular magnification of a telescope equals the ratio of the objective focal length to the eyepiece focal length. Here again, the angular magnification is the ratio of the angular size seen with the telescope to the angular size seen with the unaided eye.

In some applications—for instance, the observation of relatively nearby objects such as the Sun, the Moon, or planets—angular magnification is important. Stars, however, are so far away that they always appear as small points of light regardless of how much angular magnification is used. The large research telescopes used to study very distant objects must have great diameters to gather as much light as possible. It's difficult and expensive to manufacture such large lenses for refracting telescopes. In addition, the heaviness of large lenses leads to sagging, which is another source of aberration.

These problems can be partially overcome by replacing the objective lens with a reflecting, concave mirror, usually having a parabolic shape so as to avoid spherical aberration. Figure 25.9 (page 830) shows the design of a typical reflecting telescope. Incoming light rays pass down the barrel of the telescope and are reflected

NASA

The Hubble Space Telescope enables us to see both further into space and further back in time than ever before.

◀ Angular magnification of a telescope

(a)   (b)

ACTIVE FIGURE 25.8

(a) A diagram of a refracting telescope, with the object at infinity. (b) A refracting telescope.

Physics⊗Now™

Log into PhysicsNow at **www.cp7e.com** and go to Active Figure 25.8, where you can adjust the focal lengths of the objective and eyepiece lenses, and observe the effect on the final image.

Figure 25.9 A reflecting telescope with a Newtonian focus.

by a parabolic mirror at the base. These rays converge toward point *A* in the figure, where an image would be formed on a photographic plate or another detector. However, before this image is formed, a small flat mirror at *M* reflects the light toward an opening in the side of the tube that passes into an eyepiece. This design is said to have a *Newtonian focus*, after its developer. Note that in the reflecting telescope the light never passes through glass (except for the small eyepiece). As a result, problems associated with chromatic aberration are virtually eliminated.

The largest optical telescopes in the world are the two 10-m-diameter Keck reflectors on Mauna Kea in Hawaii. The largest single-mirrored reflecting telescope in the United States is the 5-m-diameter instrument on Mount Palomar in California. (See Fig. 25.10.) In contrast, the largest refracting telescope in the world, at the Yerkes Observatory in Williams Bay, Wisconsin, has a diameter of only 1 m.

Figure 25.10 The Hale telescope at Mount Palomar Observatory. Just before taking the elevator up to the prime-focus cage, a first-time observer is always told, "Good viewing! And, if you should fall, try to miss the mirror."

EXAMPLE 25.5   Hubble Power

Goal   Understand magnification in telescopes.

Problem   The Hubble telescope is 13.2 m long, but has a secondary mirror that increases its effective focal length to 57.8 m. (See Fig. 25.11.) The telescope doesn't have an eyepiece, because various instruments, not a human eye, record the collected light. However, it can produce images several thousand times larger than they would appear with the unaided human eye. What focal length eyepiece used with the Hubble mirror system would produce a magnification of 8.00×10^3?

Strategy   Equation 25.8 for telescope magnification can be solved for the eyepiece focal length. The equation for finding the angular magnification of a reflector is the same as that for a refractor.

Figure 25.11   A schematic of the Hubble telescope.

Solution

Solve for f_e in Equation 25.8 and substitute values:

$$m = \frac{f_o}{f_e} \;\rightarrow\; f_e = \frac{f_o}{m} = \frac{57.8 \text{ m}}{8.00 \times 10^3} = \boxed{7.23 \times 10^{-3} \text{ m}}$$

Remarks   The result of this magnification is an image with "good" resolution. However, the light-gathering power of a telescope largely determines the resolution of the image, and is far more important than magnification. A high-resolution image can always be magnified so its details can be examined. Such details are often blurred when a low-resolution image is magnified.

Exercise 25.5

The Hale telescope on Mt. Palomar has a focal length of 16.8 m. Find the magnification of the telescope in conjunction with an eyepiece having a focal length of 5.00 mm.

Answer 3.36×10^3

25.6   RESOLUTION OF SINGLE-SLIT AND CIRCULAR APERTURES

The ability of an optical system such as the eye, a microscope, or a telescope to distinguish between closely spaced objects is limited because of the wave nature of light. To understand this difficulty, consider Figure 25.12, which shows two light sources far

(a)

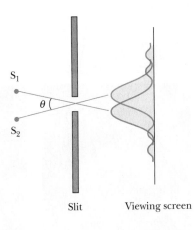

(b)

Figure 25.12   Two point sources far from a narrow slit each produce a diffraction pattern. (a) The angle subtended by the sources at the aperture is large enough so that the diffraction patterns are distinguishable. (b) The angle subtended by the sources is so small that the diffraction patterns are not distinguishable. (Note that the angles are greatly exaggerated. The drawing is not to scale.)

from a narrow slit of width a. The sources can be taken as two point sources S_1 and S_2 that are *not* coherent. For example, they could be two distant stars. If no diffraction occurred, two distinct bright spots (or images) would be observed on the screen at the right in the figure. However, because of diffraction, each source is imaged as a bright central region flanked by weaker bright and dark rings. What is observed on the screen is the sum of two diffraction patterns, one from S_1 and the other from S_2.

If the two sources are separated so that their central maxima don't overlap, as in Figure 25.12a, their images can be distinguished and are said to be *resolved*. If the sources are close together, however, as in Figure 25.12b, the two central maxima may overlap and the images are *not resolved*. To decide whether two images are resolved, the following condition is often applied to their diffraction patterns:

Rayleigh's criterion ▶

> When the central maximum of one image falls on the first minimum of another image, the images are said to be just resolved. This limiting condition of resolution is known as **Rayleigh's criterion**.

Figure 25.13 shows diffraction patterns in three situations. The images are just resolved when their angular separation satisfies Rayleigh's criterion (Fig. 25.13a). As the objects are brought closer together, their images are barely resolved (Fig. 25.13b). Finally, when the sources are very close to each other, their images are not resolved (Fig. 25.13c).

From Rayleigh's criterion, we can determine the minimum angular separation θ_{min} subtended by the source at the slit so that the images will be just resolved. In Chapter 24 we found that the first minimum in a single-slit diffraction pattern occurs at the angle that satisfies the relationship

$$\sin\theta = \frac{\lambda}{a}$$

where a is the width of the slit. According to Rayleigh's criterion, this expression gives the smallest angular separation for which the two images can be resolved. Because $\lambda \ll a$ in most situations, $\sin\theta$ is small and we can use the approximation $\sin\theta \approx \theta$. Therefore, the limiting angle of resolution for a slit of width a is

Limiting angle for a slit ▶

$$\theta_{min} \approx \frac{\lambda}{a} \qquad\qquad \textbf{[25.9]}$$

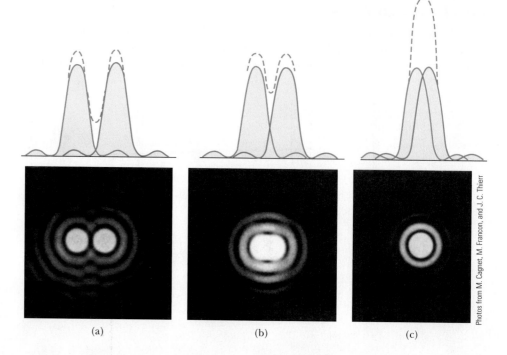

Figure 25.13 The diffraction patterns of two point sources (solid curves) and the resultant pattern (dashed curve) for three angular separations of the sources. (a) The sources are separated such that their patterns are just resolved. (b) The sources are closer together, and their patterns are barely resolved. (c) The sources are so close together that their patterns are not resolved.

(a)   (b)   (c)

where θ_{min} is in radians. Hence, the angle subtended by the two sources at the slit must be *greater* than λ/a if the images are to be resolved.

Many optical systems use circular apertures rather than slits. The diffraction pattern of a circular aperture (Fig. 25.14) consists of a central circular bright region surrounded by progressively fainter rings. Analysis shows that the limiting angle of resolution of the circular aperture is

$$\theta_{min} = 1.22\frac{\lambda}{D} \qquad [25.10]$$

where D is the diameter of the aperture. Note that Equation 25.10 is similar to Equation 25.9, except for the factor 1.22, which arises from a complex mathematical analysis of diffraction from a circular aperture.

Figure 25.14   The diffraction pattern of a circular aperture consists of a central bright disk surrounded by concentric bright and dark rings.

From M. Cagnet, M. Francon, and J. C. Thier, *Atlas of Optical Phenomena*, Berlin, Springer-Verlag, 1962, plate 34

Quick Quiz 25.2

Suppose you are observing a binary star with a telescope and are having difficulty resolving the two stars. Which color filter will better help resolve the stars? (a) blue (b) red (c) neither—colored filters have no affect on resolution

---

Applying Physics 25.2   Cat's Eyes

Cats' eyes have vertical pupils in dim light. Which would cats be most successful at resolving at night, headlights on a distant car or vertically separated running lights on a distant boat's mast having the same separation as the car's headlights?

Explanation The effective slit width in the vertical direction of the cat's eye is larger than that in the horizontal direction. Thus, it has more resolving power for lights separated in the vertical direction and would be more effective at resolving the mast lights on the boat.

EXAMPLE 25.6   Resolution of a Microscope

Goal   Study limitations on the resolution of a microscope.

Problem   Sodium light of wavelength 589 nm is used to view an object under a microscope. The aperture of the objective has a diameter of 0.90 cm. **(a)** Find the limiting angle of resolution for this microscope. **(b)** Using visible light of any wavelength you desire, find the maximum limit of resolution for this microscope. **(c)** Water of index of refraction 1.33 now fills the space between the object and the objective. What effect would this have on the resolving power of the microscope, using 589 nm light?

Strategy   Parts (a) and (b) require substitution into Equation 25.10. Because the wavelength appears in the numerator, violet light, with the shortest visible wavelength, gives the maximum resolution. In part (c), the only difference is that the wavelength changes to λ/n, where n is the index of refraction of water.

Solution
(a) Find the limiting angle of resolution for this microscope.

Substitute into Equation 25.10 to obtain the limiting angle of resolution:

$$\theta_{min} = 1.22\frac{\lambda}{D} = 1.22\left(\frac{589 \times 10^{-9}\ \text{m}}{0.90 \times 10^{-2}\ \text{m}}\right)$$
$$= 8.0 \times 10^{-5}\ \text{rad}$$

(b) Calculate the microscope's maximum limit of resolution.

To obtain the maximum resolution, substitute the shortest visible wavelength available—violet light, of wavelength 4.0×10^2 nm:

$$\theta_{min} = 1.22\frac{\lambda}{D} = 1.22\left(\frac{4.0 \times 10^{-7}\ \text{m}}{0.90 \times 10^{-2}\ \text{m}}\right)$$
$$= 5.4 \times 10^{-5}\ \text{rad}$$

(c) What effect does water between the object and the objective lens have on the resolution, with 589-nm light?

Calculate the wavelength of the sodium light in the water: $\lambda_w = \dfrac{\lambda_a}{n} = \dfrac{589 \text{ nm}}{1.33} = 443 \text{ nm}$

Substitute this wavelength into Equation 25.10 to get the resolution: $\theta_{min} = 1.22 \left(\dfrac{443 \times 10^{-9} \text{ m}}{0.90 \times 10^{-2} \text{ m}} \right) = \boxed{6.0 \times 10^{-5} \text{ rad}}$

Remarks   In each case, any two points on the object subtending an angle of less than the limiting angle θ_{min} at the objective cannot be distinguished in the image. Consequently, it may be possible to see a cell, but then be unable to clearly see smaller structures within the cell. Obtaining an increase in resolution is the motivation behind placing a drop of oil on the slide for certain objective lenses.

Exercise 25.6

Suppose oil with $n = 1.50$ fills the space between the object and the objective for this microscope. Calculate the limiting angle θ_{min} for sodium light of wavelength 589 nm in air.

Answer 5.3×10^{-5} rad

EXAMPLE 25.7   Resolving Craters on the Moon

Goal   Calculate the resolution of a telescope.

Problem   The Hubble Space Telescope has an aperture of diameter 2.40 m. **(a)** What is its limiting angle of resolution at a wavelength of 6.00×10^2 nm? **(b)** What's the smallest crater it could resolve on the Moon? (The Moon is 3.84×10^8 m from Earth.)

Strategy   After substituting into Equation 25.10 to find the limiting angle, use $s = r\theta$ to compute the minimum size of crater that can be resolved.

Solution

(a) What is the limiting angle of resolution at a wavelength of 6.00×10^2 nm?

Substitute $D = 2.40$ m and $\lambda = 6.00 \times 10^{-7}$ m into Equation 25.10:

$$\theta_{min} = 1.22 \frac{\lambda}{D} = 1.22 \left(\frac{6.00 \times 10^{-7} \text{ m}}{2.40 \text{ m}} \right)$$
$$= \boxed{3.05 \times 10^{-7} \text{ rad}}$$

(b) What's the smallest lunar crater the Hubble Space Telescope can resolve?

The two opposite sides of the crater must subtend the minimum angle. Use the arc length formula:

$$s = r\theta = (3.84 \times 10^8 \text{ m})(3.05 \times 10^{-7} \text{ rad}) = \boxed{117 \text{ m}}$$

Remarks   The distance is so great and the angle so small that using the arc length of a circle is justified—the circular arc is very nearly a straight line. The Hubble Space Telescope has produced several gigabytes of data every day for over 20 years.

Exercise 25.7

The Hale telescope on Mount Palomar has a diameter of 5.08 m (200 in.). **(a)** Find the limiting angle of resolution for a wavelength of 6.00×10^2 nm. **(b)** Calculate the smallest crater diameter the telescope can resolve on the Moon. **(c)** The answers appear better than what the Hubble can achieve. Why are the answers misleading?

Answers   (a) 1.44×10^{-7} rad (b) 55.3 m (c) While the numbers are better than Hubble's, the Hale telescope must contend with the effects of atmospheric turbulence, so the smaller space-based telescope actually obtains far better results.

It's interesting to compare the resolution of the Hale telescope with that of a large radio telescope, such as the system at Arecibo, Puerto Rico, which has a diameter of 1 000 ft (305 m). This telescope detects radio waves at a wavelength of 0.75 m. The corresponding minimum angle of resolution can be calculated as 3.0×10^{-3} rad (10 min 19 s of arc), which is more than 10 000 times larger than the calculated minimum angle for the Hale telescope.

With such relatively poor resolution, why is Arecibo considered a valuable astronomical instrument? Unlike its optical counterparts, Arecibo can see through clouds of dust. The center of our Milky Way galaxy is obscured by such dust clouds, which absorb and scatter visible light. Radio waves easily penetrate the clouds, so radio telescopes allow direct observations of the galactic core.

Resolving Power of the Diffraction Grating

The diffraction grating studied in Chapter 24 is most useful for making accurate wavelength measurements. Like the prism, it can be used to disperse a spectrum into its components. Of the two devices, the grating is better suited to distinguishing between two closely spaced wavelengths. We say that the grating spectrometer has a higher *resolution* than the prism spectrometer. If λ_1 and λ_2 are two nearly equal wavelengths between which the spectrometer can just barely distinguish, the **resolving power** of the grating is defined as

$$R \equiv \frac{\lambda}{\lambda_2 - \lambda_1} = \frac{\lambda}{\Delta\lambda} \qquad \textbf{[25.11]}$$

where $\lambda \approx \lambda_1 \approx \lambda_2$ and $\Delta\lambda = \lambda_2 - \lambda_1$. From this equation, it's clear that a grating with a high resolving power can distinguish small differences in wavelength. Further, if N lines of the grating are illuminated, it can be shown that the resolving power in the mth-order diffraction is given by

$$R = Nm \qquad \textbf{[25.12]} \qquad \blacktriangleleft \text{ Resolving power of a grating}$$

So the resolving power R increases with the order number m and is large for a grating with a great number of illuminated slits. Note that for $m = 0$, $R = 0$, which signifies that *all wavelengths are indistinguishable* for the zeroth-order maximum. (All wavelengths fall at the same point on the screen.) However, consider the second-order diffraction pattern of a grating that has 5 000 rulings illuminated by the light source. The resolving power of such a grating in second order is $R = 5\,000 \times 2 = 10\,000$. Therefore, the *minimum* wavelength separation between two spectral lines that can be just resolved, assuming a mean wavelength of 600 nm, is calculated from Equation 25.12 to be $\Delta\lambda = \lambda/R = 6 \times 10^{-2}$ nm. For the third-order principal maximum, $R = 15\,000$ and $\Delta\lambda = 4 \times 10^{-2}$ nm, and so on.

EXAMPLE 25.8 Light from Sodium Atoms

Goal Find the necessary resolving power to distinguish spectral lines.

Problem Two bright lines in the spectrum of sodium have wavelengths of 589.00 nm and 589.59 nm, respectively. **(a)** What must the resolving power of a grating be in order to distinguish these wavelengths? **(b)** To resolve these lines in the second-order spectrum, how many lines of the grating must be illuminated?

Strategy This problem requires little more than substituting into Equations 25.11 and 25.12.

Solution
(a) What must the resolving power of a grating be in order to distinguish the given wavelengths?

Substitute into Equation 25.11 to find R:

$$R = \frac{\lambda}{\Delta\lambda} = \frac{589.00 \text{ nm}}{589.59 \text{ nm} - 589.00 \text{ nm}} = \frac{589 \text{ nm}}{0.59 \text{ nm}}$$

$$= 1.0 \times 10^3$$

(b) To resolve these lines in the second-order spectrum, how many lines of the grating must be illuminated?

Solve Equation 25.12 for N and substitute:

$$N = \frac{R}{m} = \frac{1.0 \times 10^3}{2} = 5.0 \times 10^2 \text{ lines}$$

Remarks   The ability to resolve spectral lines is particularly important in experimental atomic physics.

Exercise 25.8

When the lines of a spectrum are examined at high resolution, each line is actually found to be two closely spaced lines called a doublet, due to a phenomenon called electron spin. An example is the doublet in the hydrogen spectrum having wavelengths of 656.272 nm and 656.285 nm. (a) What must be the resolving power of a grating in order to distinguish these wavelengths? (b) How many lines of the grating must be illuminated to resolve these lines in the third-order spectrum?

Answer   (a) 5.0×10^4 (b) 1.7×10^4 lines

25.7   THE MICHELSON INTERFEROMETER

The Michelson interferometer is an optical instrument having great scientific importance. Invented by the American physicist A. A. Michelson (1852–1931), it is an ingenious device that splits a light beam into two parts and then recombines them to form an interference pattern. The interferometer is used to make accurate length measurements.

Active Figure 25.15 is a schematic diagram of an interferometer. A beam of light provided by a monochromatic source is split into two rays by a partially silvered mirror M inclined at an angle of 45° relative to the incident light beam. One ray is reflected vertically upward to mirror M_1, and the other ray is transmitted horizontally through mirror M to mirror M_2. Hence, the two rays travel separate paths, L_1 and L_2. After reflecting from mirrors M_1 and M_2, the two rays eventually recombine to produce an interference pattern, which can be viewed through a telescope. The glass plate P, equal in thickness to mirror M, is placed in the path of the horizontal ray to ensure that the two rays travel the same distance through glass.

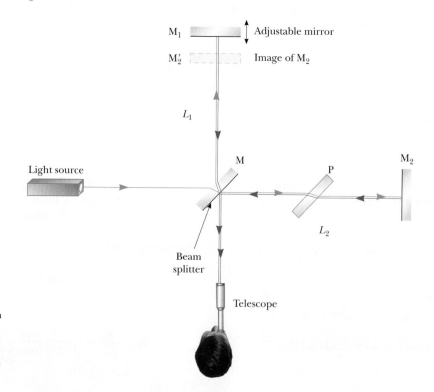

ACTIVE FIGURE 25.15
A diagram of the Michelson interferometer. A single beam is split into two rays by the half-silvered mirror M. The path difference between the two rays is varied with the adjustable mirror M_1.

Physics⊗Now™
Log into PhysicsNow at **www.cp7e.com** and go to Active Figure 25.15, where you can move one of the mirrors, observing the effect on the interference pattern, and use the interferometer to measure the speed of light.

The interference pattern for the two rays is determined by the difference in their path lengths. When the two rays are viewed as shown, the image of M_2 is at M_2', parallel to M_1. Hence, the space between M_2' and M_1 forms the equivalent of a parallel air film. The effective thickness of the air film is varied by using a finely threaded screw to move mirror M_1 in the direction indicated by the arrows in Active Figure 25.15. If one of the mirrors is tipped slightly with respect to the other, the thin film between the two is wedge shaped, and an interference pattern consisting of parallel fringes is set up, as described in Example 24.4. Now suppose we focus on one of the dark lines with the crosshairs of a telescope. As the mirror M_1 is moved to lengthen the path L_1, the thickness of the wedge increases. When the thickness increases by $\lambda/4$, the destructive interference that initially produced the dark fringe has changed to constructive interference, and we now observe a bright fringe at the location of the crosshairs. The term *fringe shift* is used to describe the change in a fringe from dark to light or light to dark. Thus, successive light and dark fringes are formed each time M_1 is moved a distance of $\lambda/4$. The wavelength of light can be measured by counting the number of fringe shifts for a measured displacement of M_1. Conversely, if the wavelength is accurately known (as with a laser beam), the mirror displacement can be determined to within a fraction of the wavelength. Because the interferometer can measure displacements precisely, it is often used to make highly accurate measurements of the dimensions of mechanical components.

If the mirrors are perfectly aligned, rather than tipped with respect to one another, the path difference differs slightly for different angles of view. This arrangement results in an interference pattern that resembles Newton's rings. The pattern can be used in a fashion similar to that for tipped mirrors. An observer pays attention to the center spot in the interference pattern. For example, suppose the spot is initially dark, indicating that destructive interference is occurring. If M_1 is now moved a distance of $\lambda/4$, this central spot changes to a light region, corresponding to a fringe shift.

SUMMARY

Physics⊗Now™ Take a practice test by logging into Physics-Now at **www.cp7e.com** and clicking on the Pre-Test link for this chapter.

25.1 The Camera

The light-concentrating power of a lens of focal length f and diameter D is determined by the **f-number**, defined as

$$f\text{-number} \equiv \frac{f}{D} \qquad [25.1]$$

The smaller the f-number of a lens, the brighter is the image formed.

25.2 The Eye

Hyperopia (farsightedness) is a defect of the eye that occurs either when the eyeball is too short or when the ciliary muscle cannot change the shape of the lens enough to form a properly focused image. **Myopia** (nearsightedness) occurs either when the eye is longer than normal or when the maximum focal length of the lens is insufficient to produce a clearly focused image on the retina.

The **power** of a lens in **diopters** is the inverse of the focal length in meters.

25.3 The Simple Magnifier

The **angular magnification of a lens** is defined as

$$m \equiv \frac{\theta}{\theta_0} \qquad [25.2]$$

where θ is the angle subtended by an object at the eye with a lens in use and θ_0 is the angle subtended by the object when it is placed at the near point of the eye and no lens is used. The **maximum angular magnification of a lens** is

$$m_{\max} = 1 + \frac{25 \text{ cm}}{f} \qquad [25.5]$$

When the eye is relaxed, the angular magnification is

$$m = \frac{25 \text{ cm}}{f} \qquad [25.6]$$

25.4 The Compound Microscope

The overall **magnification of a compound microscope** of length L is the product of the magnification produced by the objective, of focal length f_o, and the magnification produced by the eyepiece, of focal length f_e:

$$M = -\frac{L}{f_o}\left(\frac{25 \text{ cm}}{f_e}\right) \qquad \text{[25.7]}$$

25.5   The Telescope

The **angular magnification of a telescope** is

$$m = \frac{f_o}{f_e} \qquad \text{[25.8]}$$

where f_o is the focal length of the objective and f_e is the focal length of the eyepiece.

25.6   Resolution of Single-Slit and Circular Apertures

Two images are said to be **just resolved** when the central maximum of the diffraction pattern for one image falls on the first minimum of the other image. This limiting condition of resolution is known as **Rayleigh's criterion**. The limiting angle of resolution for a **slit** of width a is

$$\theta_{\min} \approx \frac{\lambda}{a} \qquad \text{[25.9]}$$

The limiting angle of resolution of a **circular aperture** is

$$\theta_{\min} = 1.22\frac{\lambda}{D} \qquad \text{[25.10]}$$

where D is the diameter of the aperture.

If λ_1 and λ_2 are two nearly equal wavelengths between which a grating spectrometer can just barely distinguish, the **resolving power** R of the grating is defined as

$$R \equiv \frac{\lambda}{\lambda_2 - \lambda_1} = \frac{\lambda}{\Delta\lambda} \qquad \text{[25.11]}$$

where $\lambda \approx \lambda_1 \approx \lambda_2$ and $\Delta\lambda = \lambda_2 - \lambda_1$. The **resolving power** of a diffraction grating in the mth order is

$$R = Nm \qquad \text{[25.12]}$$

where N is the number of illuminated rulings on the grating.

CONCEPTUAL QUESTIONS

1. A lens is used to examine an object across a room. Is the lens probably being used as a simple magnifier?

2. Why is it difficult or impossible to focus a microscope on an object across a room?

3. The optic nerve and the brain invert the image formed on the retina. Why don't we see everything upside down?

4. If you want to examine the fine detail of an object with a magnifying glass with a power of + 20.0 diopters, where should the object be placed in order to observe a magnified image of the object?

5. Suppose you are observing the interference pattern formed by a Michelson interferometer in a laboratory and a joking colleague holds a lit match in the light path of one arm of the interferometer. Will this have an effect on the interference pattern?

6. Compare and contrast the eye and a camera. What parts of the camera correspond to the iris, the retina, and the cornea of the eye?

7. Large telescopes are usually reflecting rather than refracting. List some reasons for this choice.

8. If you want to use a converging lens to set fire to a piece of paper, why should the light source be farther from the lens than its focal point?

9. Explain why it is theoretically impossible to see an object as small as an atom regardless of the quality of the light microscope being used.

10. Which is most important in the use of a camera photoflash unit, the intensity of the light (the energy per unit area per unit time) or the product of the intensity and the time of the flash, assuming the time is less than the shutter speed?

11. A patient has a near point of 1.25 m. Is she nearsighted or farsighted? Should the corrective lens be converging or diverging?

12. A lens with a certain power is used as a simple magnifier. If the power of the lens is doubled, does the angular magnification increase or decrease?

PROBLEMS

1, 2, 3 = straightforward, intermediate, challenging   □ = full solution available in *Student Solutions Manual/Study Guide*

Physics⊗Now™ = coached problem with hints available at **www.cp7e.com**   = biomedical application

Section 25.1   The Camera

1. A camera used by a professional photographer to shoot portraits has a focal length of 25.0 cm. The photographer takes a portrait of a person 1.50 m in front of the camera. Where is the image formed, and what is the lateral magnification?

2. The lens of a certain 35 mm camera (35 mm is the width of the film strip) has a focal length of 55 mm and a speed (an f-number) of $f/1.8$. Determine the diameter of the lens.

3. A photographic image of a building is 0.092 0 m high. The image was made with a lens with a focal length of 52.0 mm. If the lens was 100 m from the building when the photograph was made, determine the height of the building.

4. The full Moon is photographed using a camera with a 120-mm-focal-length lens. Determine the diameter of the Moon's image on the film. [*Note:* The radius of the Moon is 1.74×10^6 m, and the distance from the Earth to the Moon is 3.84×10^8 m.]

5. A camera is being used with the correct exposure at $f/4$ and a shutter speed of $1/32$ s. In order to "stop" a fast-moving subject, the shutter speed is changed to $1/256$ s. Find the new f-stop that should be used to maintain

satisfactory exposure, assuming no change in lighting conditions.

6. **Physics⊗Now™** (a) Use conceptual arguments to show that the intensity of light (energy per unit area per unit time) reaching the film in a camera is proportional to the square of the reciprocal of the *f*-number, as

$$I \propto \frac{1}{(f/D)^2}$$

(b) The correct exposure time for a camera set to $f/1.8$ is $(1/500)$ s. Calculate the correct exposure time if the *f*-number is changed to $f/4$ under the same lighting conditions.

7. A certain type of film requires an exposure time of 0.010 s with an $f/11$ lens setting. Another type of film requires twice the light energy to produce the same level of exposure. What *f*-stop does the second type of film need with the 0.010-s exposure time?

8. Assume that the camera in Figure 25.1 has a fixed focal length of 65.0 mm and is adjusted to properly focus the image of a distant object. How far and in what direction must the lens be moved to focus the image of an object that is 2.00 m away?

Section 25.2 The Eye

9. A retired bank president can easily read the fine print of the financial page when the newspaper is held no closer than arm's length, 60.0 cm from the eye. What should be the focal length of an eyeglass lens that will allow her to read at the more comfortable distance of 24.0 cm?

10. A person has far points 84.4 cm from the right eye and 122 cm from the left eye. Write a prescription for the powers of the corrective lenses.

11. The accommodation limits for Nearsighted Nick's eyes are 18.0 cm and 80.0 cm. When he wears his glasses, he is able to see faraway objects clearly. At what minimum distance is he able to see objects clearly?

12. The near point of an eye is 100 cm. A corrective lens is to be used to allow this eye to clearly focus on objects 25.0 in front of it. (a) What should be the focal length of the lens? (b) What is the power of the needed corrective lens?

13. An individual is nearsighted; his near point is 13.0 cm and his far point is 50.0 cm. (a) What lens power is needed to correct his nearsightedness? (b) When the lenses are in use, what is this person's near point?

14. A certain child's near point is 10.0 cm; her far point (with eyes relaxed) is 125 cm. Each eye lens is 2.00 cm from the retina. (a) Between what limits, measured in diopters, does the power of this lens–cornea combination vary? (b) Calculate the power of the eyeglass lens the child should use for relaxed distance vision. Is the lens converging or diverging?

15. An artificial lens is implanted in a person's eye to replace a diseased lens. The distance between the artificial lens and the retina is 2.80 cm. In the absence of the lens, an image of a distant object (formed by refraction at the cornea) falls 2.53 cm behind the retina. The lens is designed to put the image of the distant object on the retina. What is the power of the implanted lens? [*Hint:* Consider the image formed by the cornea to be a virtual object.]

16. A person is to be fitted with bifocals. She can see clearly when the object is between 30 cm and 1.5 m from the eye. (a) The upper portions of the bifocals (Fig. P25.16) should be designed to enable her to see distant objects clearly. What power should they have? (b) The lower portions of the bifocals should enable her to see objects comfortably at 25 cm. What power should they have?

Far vision
Near vision
Figure P25.16

Section 25.3 The Simple Magnifier

17. A stamp collector uses a lens with 7.5-cm focal length as a simple magnifier. The virtual image is produced at the normal near point (25 cm). (a) How far from the lens should the stamp be placed? (b) What is the expected angular magnification?

18. A lens having a focal length of 25 cm is used as a simple magnifier. (a) What is the angular magnification obtained when the image is formed at the normal near point ($q = -25$ cm)? (b) What is the angular magnification produced by this lens when the eye is relaxed?

19. A biology student uses a simple magnifier to examine the structural features of the wing of an insect. The wing is held 3.50 cm in front of the lens, and the image is formed 25.0 cm from the eye. (a) What is the focal length of the lens? (b) What angular magnification is achieved?

20. A lens that has a focal length of 5.00 cm is used as a magnifying glass. (a) To obtain maximum magnification, where should the object be placed? (b) What is the magnification?

21. A leaf of length h is positioned 71.0 cm in front of a converging lens with a focal length of 39.0 cm. An observer views the image of the leaf from a position 1.26 m behind the lens, as shown in Figure P25.21. (a) What is the magnitude of the lateral magnification (the ratio of the image size to the object size) produced by the lens? (b) What angular magnification is achieved by viewing the image of the leaf rather than viewing the leaf directly?

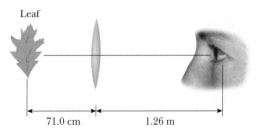
Leaf
71.0 cm   1.26 m
Figure P25.21

Section 25.4 The Compound Microscope &
Section 25.5 The Telescope

22. The objective lens in a microscope with a 20.0-cm-long tube has a magnification of 50.0, and the eyepiece has a magnification of 20.0. What are the focal lengths of (a) the objective and (b) the eyepiece? (c) What is the overall magnification of the microscope?

23. The desired overall magnification of a compound microscope is 140×. The objective alone produces a lateral magnification of 12×. Determine the required focal length of the eyepiece.

24. A microscope has an objective lens with a focal length of 16.22 mm and an eyepiece with a focal length of 9.50 mm. With the length of the barrel set at 29.0 cm, the diameter of a red blood cell's image subtends an angle of 1.43 mrad with the eye. If the final image distance is 29.0 cm from the eyepiece, what is the actual diameter of the red blood cell?

25. Physics⊗Now™ The length of a microscope tube is 15.0 cm. The focal length of the objective is 1.00 cm, and the focal length of the eyepiece is 2.50 cm. What is the magnification of the microscope, assuming it is adjusted so that the eye is relaxed? [*Hint:* To solve this question go back to basics and use the thin lens equation.]

26. A certain telescope has an objective of focal length 1 500 cm. If the Moon is used as an object, a 1.0-cm-long image formed by the objective corresponds to what distance, in miles, on the Moon? Assume 3.8×10^8 m for the Earth–Moon distance.

27. The lenses of an astronomical telescope are 92 cm apart when adjusted for viewing a distant object with minimum eyestrain. The angular magnification produced by the telescope is 45. Compute the focal length of each lens.

28. An elderly sailor is shipwrecked on a desert island, but manages to save his eyeglasses. The lens for one eye has a power of + 1.20 diopters, and the other lens has a power of + 9.00 diopters. (a) What is the magnifying power of the telescope he can construct with these lenses? (b) How far apart are the lenses when the telescope is adjusted for minimum eyestrain?

29. Astronomers often take photographs with the objective lens or mirror of a telescope alone, without an eyepiece. (a) Show that the image size h' for a telescope used in this manner is given by $h' = fh/(f - p)$, where h is the object size, f is the objective focal length, and p is the object distance. (b) Simplify the expression in part (a) if the object distance is much greater than the objective focal length. (c) The "wingspan" of the International Space Station is 108.6 m, the overall width of its solar panel configuration. When it is orbiting at an altitude of 407 km, find the width of the image formed by a telescope objective of focal length 4.00 m.

30. Galileo devised a simple terrestrial telescope that produces an upright image. It consists of a converging objective lens and a diverging eyepiece at opposite ends of the telescope tube. For distant objects, the tube length is the objective focal length less the absolute value of the eyepiece focal length. (a) Does the user of the telescope see a real or virtual image? (b) Where is the final image? (c) If a telescope is to be constructed with a tube of length 10.0 cm and a magnification of 3.00, what are the focal lengths of the objective and eyepiece?

31. A person decides to use an old pair of eyeglasses to make some optical instruments. He knows that the near point in his left eye is 50.0 cm and the near point in his right eye is 100 cm. (a) What is the maximum angular magnification he can produce in a telescope? (b) If he places the lenses 10.0 cm apart, what is the maximum overall magnification he can produce in a microscope? (Go back to basics and use the thin-lens equation to solve part (b).)

Section 25.6 Resolution of Single-Slit
and Circular Apertures

32. If the distance from the Earth to the Moon is 3.8×10^8 m, what diameter would be required for a telescope objective to resolve a Moon crater 300 m in diameter? Assume a wavelength of 500 nm.

33. A converging lens with a diameter of 30.0 cm forms an image of a satellite passing overhead. The satellite has two green lights (wavelength 500 nm) spaced 1.00 m apart. If the lights can just be resolved according to the Rayleigh criterion, what is the altitude of the satellite?

34. The pupil of a cat's eye narrows to a vertical slit of width 0.500 mm in daylight. What is the angular resolution for a pair of horizontally separated mice? (Use 500-nm light in your calculation.)

35. To increase the resolving power of a microscope, the object and the objective are immersed in oil ($n = 1.5$). If the limiting angle of resolution without the oil is 0.60 μrad, what is the limiting angle of resolution with the oil? [*Hint:* The oil changes the wavelength of the light.]

36. (a) Calculate the limiting angle of resolution for the eye, assuming a pupil diameter of 2.00 mm, a wavelength of 500 nm *in air*, and an index of refraction for the eye of 1.33. (b) What is the maximum distance from the eye at which two points separated by 1.00 cm could be resolved?

37. Two stars in a binary system are 8.0 lightyears away from the observer and can just be resolved by a 20-in. telescope equipped with a filter that allows only light of wavelength 500 nm to pass. What is the distance between the two stars?

38. A spy satellite circles the Earth at an altitude of 200 km and carries out surveillance with a special high-resolution telescopic camera having a lens diameter of 35 cm. If the angular resolution of this camera is limited by diffraction, estimate the separation of two small objects on the Earth's surface that are just resolved in yellow-green light ($\lambda = 550$ nm).

39. Suppose a 5.00-m-diameter telescope were constructed on the Moon, where the absence of atmospheric distortion would permit excellent viewing. If observations were made using 500-nm light, what minimum separation between two objects could just be resolved on Mars at closest approach (when Mars is 8.0×10^7 km from the Moon)?

40. The H_α line in hydrogen has a wavelength of 656.20 nm. This line differs in wavelength from the corresponding spectral line in deuterium (the heavy stable isotope of hydrogen) by 0.18 nm. (a) Determine the minimum number of lines a grating must have to resolve these two wavelengths in the first order. (b) Repeat part (a) for the second order.

41. A 15.0-cm-long grating has 6 000 slits per centimeter. Can two lines of wavelengths 600.000 nm and 600.003 nm be separated with this grating? Explain.

Section 25.7 The Michelson Interferometer

42. Light of wavelength 550 nm is used to calibrate a Michelson interferometer. With the use of a micrometer screw, the platform on which one mirror is mounted is moved 0.180 mm. How many fringe shifts are counted?

43. An interferometer is used to measure the length of a bacterium. The wavelength of the light used is 650 nm. As one arm of the interferometer is moved from one end of the cell to the other, 310 fringe shifts are counted. How long is the bacterium?

44. Mirror M_1 in Figure 25.15 is displaced a distance ΔL. During this displacement, 250 fringe shifts are counted. The light being used has a wavelength of 632.8 nm. Calculate the displacement ΔL.

45. A thin sheet of transparent material has an index of refraction of 1.40 and is 15.0 μm thick. When it is inserted in the light path along one arm of an interferometer, how many fringe shifts occur in the pattern? Assume that the wavelength (in a vacuum) of the light used is 600 nm. [*Hint:* The wavelength will change within the material.]

46. The Michelson interferometer can be used to measure the index of refraction of a gas by placing an evacuated transparent tube in the light path along one arm of the device. Fringe shifts occur as the gas is slowly added to the tube. Assume that 600-nm light is used, that the tube is 5.00 cm long, and that 160 fringe shifts occur as the pressure of the gas in the tube increases to atmospheric pressure. What is the index of refraction of the gas? [*Hint:* The fringe shifts occur because the wavelength of the light changes inside the gas-filled tube.]

47. **Physics⊗Now™** The light path in one arm of a Michelson interferometer includes a transparent cell that is 5.00 cm long. How many fringe shifts would be observed if all the air were evacuated from the cell? The wavelength of the light source is 590 nm and the refractive index of air is 1.000 29. (See the hint in Problem 46.)

ADDITIONAL PROBLEMS

48. A person with a nearsighted eye has near and far points of 16 cm and 25 cm, respectively. (a) Assuming a lens is placed 2.0 cm from the eye, what power must the lens have to correct this condition? (b) Suppose that contact lenses placed directly on the cornea are used to correct the person's eye. What is the power of the lens required in this case, and what is the new near point? [*Hint*: The contact lens and the eyeglass lens require slightly different powers because they are at different distances from the eye.]

49. The near point of an eye is 75.0 cm. (a) What should be the power of a corrective lens prescribed to enable the eye to see an object clearly at 25.0 cm? (b) If, using the corrective lens, the person can see an object clearly at 26.0 cm, but not at 25.0 cm, by how many diopters did the lens grinder miss the prescription?

50. If a typical eyeball is 2.00 cm long and has a pupil opening that can range from about 2.00 mm to 6.00 mm, what

are (a) the focal length of the eye when it is focused on objects 1.00 m away, (b) the smallest f-number of the eye when it is focused on objects 1.00 m away, and (c) the largest f-number of the eye when it is focused on objects 1.00 m away?

51. **Physics⊗Now™** A cataract-impaired lens in an eye may be surgically removed and replaced by a manufactured lens. The focal length required for the new lens is determined by the lens-to-retina distance, which is measured by a sonarlike device, and by the requirement that the implant provide for correct distance vision. (a) If the distance from lens to retina is 22.4 mm, calculate the power of the implanted lens in diopters. (b) Since there is no accommodation and the implant allows for correct distance vision, a corrective lens for close work or reading must be used. Assume a reading distance of 33.0 cm, and calculate the power of the lens in the reading glasses.

52. Estimate the minimum angle subtended at the eye of a hawk flying at an altitude of 50 m necessary to recognize a mouse on the ground.

53. The wavelengths of the sodium spectrum are $\lambda_1 = 589.00$ nm and $\lambda_2 = 589.59$ nm. Determine the minimum number of lines in a grating that will allow resolution of the sodium spectrum in (a) the first order and (b) the third order.

54. The text discusses the astronomical telescope. Another type is the Galilean telescope, in which an objective lens gathers light (Fig. P25.54) and tends to form an image at point A. An eyepiece consisting of a diverging lens intercepts the light before it comes to a focus and forms a virtual image at point B. When adjusted for minimum eyestrain, B is an infinite distance in front of the lens and parallel rays emerge from the lens, as in Figure P25.54b. An opera glass, which is a Galilean telescope, is used to view a 30.0-cm-tall singer's head that is 40.0 m from the objective lens. The focal length of the objective is $+8.00$ cm, and that of the eyepiece is -2.00 cm. The telescope is adjusted so parallel rays enter the eye. Compute (a) the size of the real image that would have been formed by the objective, (b) the virtual object distance for the diverging lens, (c) the distance between the lenses, and (d) the overall angular magnification.

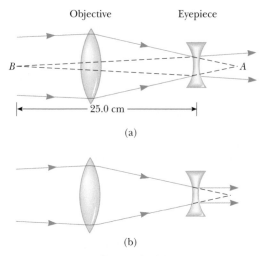

Figure P25.54

55. A laboratory (astronomical) telescope is used to view a scale that is 300 cm from the objective, which has a focal length of 20.0 cm; the eyepiece has a focal length of 2.00 cm. Calculate the angular magnification when the telescope is adjusted for minimum eyestrain. [*Note*: The object is not at infinity, so the simple expression $m = f_o/f_e$ is not sufficiently accurate for this problem. Also, assume small angles, so that $\tan \theta \approx \theta$.)

56. If the aqueous humor of the eye has an index of refraction of 1.34 and the distance from the vertex of the cornea to the retina is 2.00 cm, what is the radius of curvature of the cornea for which distant objects will be focused on the retina? (For simplicity, assume that all refraction occurs in the aqueous humor.)

57. A boy scout starts a fire by using a lens from his eyeglasses to focus sunlight on kindling 5.0 cm from the lens. The boy scout has a near point of 15 cm. When the lens is used as a simple magnifier, (a) what is the maximum magnification that can be achieved, and (b) what is the magnification when the eye is relaxed? [*Caution*: The equations derived in the text for a simple magnifier assume a "normal" eye.]

ACTIVITIES

1. (a) Move this book toward your face until the letters just begin to blur. The distance from the book to your eye is your near point. (b) On a sheet of paper, make a dot near the center. Then place an *x* about 3 inches to the left of the dot and another *x* about 3 inches to the right of the dot. With one eye shut and while looking at the dot, move the paper slowly toward your eye. You will notice that a certain distance from your eye, one of the **x**'s will disappear. This is the location of the blind spot of your eye—the point where the optic nerve enters the eye. (c) Stand before a mirror in a darkened room for a few minutes. Then turn on a light in the room and observe your pupils in the mirror as they change size. Such adaptation to the dark also takes place at the rods and cones as they chemically adjust their sensitivity. This adjustment takes 15–30 minutes, as you may have noted whenever you entered a darkened movie theater. Iris aperture control takes less than a second and helps protect the retina from overload.

2. On a sunny day, hold a magnifying glass above a nonflammable surface, such as a sidewalk, so the image of the Sun forms a round spot of light on the surface. Note where the spot formed by the lens is most distinct, or smallest. Use a ruler to measure the distance between the glass and the image. The distance is equal to the focal length of the lens.

3. Hold a pair of prescription glasses about 12 cm from your eye, and look at different objects through the lenses. Try this with different types of glasses, such as those for farsightedness and nearsightedness, and describe what effects the differences have on the image you see. If you have bifocals, how do the images produced by the top and bottom portions of the bifocal lens compare?

4. If you have never experimented with a 35-mm camera with adjustable *f*-numbers and shutter speeds, use up a couple of rolls of film to see what happens. Take several shots of the same object with different settings for these two variables. (You should record your *f*-numbers and shutter speeds for each photograph.) Explain any differences you see in the final images in terms of the settings used.

Courtesy of the Archives, California Institute of Technology

Albert Einstein revolutionized modern physics. He explained the random movements of pollen grains, which proved the existence of atoms, and the photoelectric effect, which showed that light was a particle as well as a wave. His theory of special relativity made clear the foundations of space and time, and his theory of gravitation—general relativity—is the most accurate theory in physics today. He was also deeply concerned with the social impact of scientific discovery.

Relativity

Most of our everyday experiences and observations have to do with objects that move at speeds much less than the speed of light. Newtonian mechanics was formulated to describe the motion of such objects, and its formalism is quite successful in describing a wide range of phenomena that occur at low speeds. It fails, however, when applied to particles having speeds approaching that of light.

This chapter introduces Einstein's theory of special relativity and includes a section on general relativity. The concepts of special relativity often violate our common sense. Moving clocks run slow, and the length of a moving meter stick is contracted. Nonetheless, the theory has been rigorously tested, correctly predicting the results of experiments involving speeds near the speed of light. The theory is verified daily in particle accelerators around the world.

26.1 INTRODUCTION

Experimentally, the predictions of Newtonian theory can be tested at high speeds by accelerating electrons or other charged particles through a large electric potential difference. For example, it's possible to accelerate an electron to a speed of $0.99c$ (where c is the speed of light) by using a potential difference of several million volts. According to Newtonian mechanics, if the potential difference is increased by a factor of 4, the electron's kinetic energy is four times greater and its speed should double to $1.98c$. However, experiments show that the speed of the electron—as well as the speed of any other particle that has mass—always remains *less* than the speed of light, regardless of the size of the accelerating voltage.

The existence of a universal speed limit has far-reaching consequences. It means that the usual concepts of force, momentum, and energy no longer apply for rapidly moving objects. Less obvious consequences include the fact that observers moving at different speeds will measure different time intervals and displacements between the same two events. Newtonian mechanics was contradicted by experimental observations, so it was necessary to replace it with another theory.

In 1905, at the age of 26, Einstein published his special theory of relativity. Regarding the theory, Einstein wrote:

> The relativity theory arose from necessity, from serious and deep contradictions in the old theory from which there seemed no escape. The strength of the new theory lies in the consistency and simplicity with which it solves all these difficulties, using only a few very convincing assumptions.[1]

Although Einstein made many other important contributions to science, his theory of relativity alone represents one of the greatest intellectual achievements of all time. With this theory, experimental observations can be correctly predicted over the range of speeds from $v = 0$ to speeds approaching the speed of light. Newtonian mechanics, which was accepted for more than 200 years, remains valid, but only for speeds much smaller than the speed of light.

At the foundation of special relativity is reconciling the measurements of two observers moving relative to each other. Normally, two such observers will measure different outcomes for the same event. If the measurement is the speed of a car, for example, an observer standing on the road will measure a different speed for the car than an observer in a truck traveling at speed v relative the stationary observer. Special relativity is all about relating two such measurements—and this rather innocuous relating of measurements leads to some of the most bizarre consequences in physics!

26.2 THE PRINCIPLE OF GALILEAN RELATIVITY

In order to describe a physical event, it's necessary to choose a *frame of reference*. For example, when you perform an experiment in a laboratory, you select a coordinate system, or frame of reference, that is at rest with respect to the laboratory. However, suppose an observer in a passing car moving at a constant velocity with respect to the lab were to observe your experiment. Would the observations made by the moving observer differ dramatically from yours? That is, if you found Newton's first law to be valid in your frame of reference, would the moving observer agree with you?

According to the principle of Galilean relativity, **the laws of mechanics must be the same in all inertial frames of reference**. Inertial frames of reference are those reference frames in which Newton's laws are valid. Practically, such frames are those in which objects subjected to no forces move in straight lines at constant speed—thus the name "inertial frame" because objects observed from these frames obey Newton's first law, the law of inertia. For the situation described in the previous paragraph, the laboratory coordinate system and the coordinate system of the moving car are both inertial frames of reference. Consequently, if the laws of mechanics are found to be true in the laboratory, then the person in the car must also observe the same laws.[2]

Consider an airplane in flight, moving with a constant velocity, as in Figure 26.1a. If a passenger in the airplane throws a ball straight up in the air, the passenger observes that the ball moves in a vertical path. The motion of the ball is precisely the same as it would be if the ball were thrown while at rest on Earth. The law of gravity and the equations of motion under constant acceleration are obeyed whether the airplane is at rest or in uniform motion.

Now consider the same experiment when viewed by another observer at rest on Earth. This stationary observer views the path of the ball in the plane to be a parabola, as in Figure 26.1b. Further, according to this observer, the ball has a velocity to the right equal to the velocity of the plane. Although the two observers disagree on the shape of the ball's path, both agree that the motion of the ball obeys the law of gravity and Newton's laws of motion, and even agree on how long

[1]A. Einstein and L. Infeld, *The Evolution of Physics* (New York: Simon and Schuster, 1961).
[2]What is an example of a *non*inertial frame? A frame undergoing translational acceleration or a frame rotating with respect to the two inertial frames just mentioned.

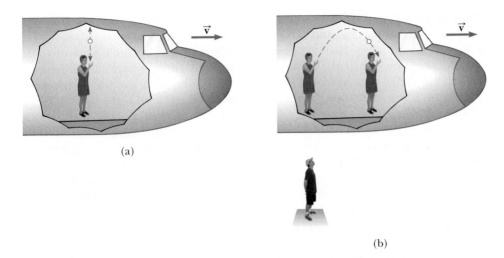

(a)

(b)

the ball is in the air. We draw the following important conclusion: **There is no preferred frame of reference for describing the laws of mechanics**.

26.3 THE SPEED OF LIGHT

It's natural to ask whether the concept of Galilean relativity in mechanics also applies to experiments in electricity, magnetism, optics, and other areas. Experiments indicate the answer is no. For example, if we assume that the laws of electricity and magnetism are the same in all inertial frames, a paradox concerning the speed of light immediately arises. This can be understood by recalling that, according to electromagnetic theory, the speed of light always has the fixed value of $2.997\,924\,58 \times 10^8$ m/s in free space. But this is in direct contradiction to common sense. For example, suppose a light pulse is sent out by an observer in a boxcar moving with a velocity $\vec{v}$ (Fig. 26.2). The light pulse has a velocity $\vec{c}$ relative to observer S' in the boxcar. According to Galilean relativity, the speed of the pulse relative to the stationary observer S outside the boxcar should be $c + v$. This obviously contradicts Einstein's theory, which postulates that the velocity of the light pulse is the same for all observers.

In order to resolve this paradox, we must conclude that either (1) the addition law for velocities is incorrect or (2) the laws of electricity and magnetism are not the same in all inertial frames. Assume that the second conclusion is true; then a preferred reference frame must exist in which the speed of light has the value c, but in any other reference frame the speed of light must have a value that is greater or less than c. It's useful to draw an analogy with sound waves, which propagate through a medium such as air. The speed of sound in air is about 330 m/s when measured in a reference frame in which the air is stationary. However, the speed of sound is greater or less than this value when measured from a reference frame that is moving with respect to the air.

In the case of light signals (electromagnetic waves), recall that electromagnetic theory predicted that such waves must propagate through free space with a speed

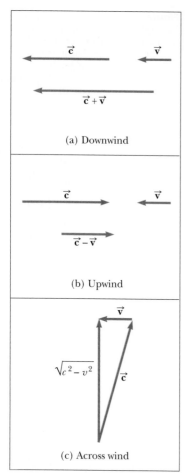

(a) Downwind

(b) Upwind

(c) Across wind

Figure 26.3 If the speed of the ether wind relative to Earth is v, and c is the speed of light relative to the ether, the speed of light relative to Earth is (a) $c + v$ in the downwind direction, (b) $c - v$ in the upwind direction, and (c) $\sqrt{c^2 - v^2}$ in the direction perpendicular to the wind.

equal to the speed of light. However, the theory doesn't require the presence of a medium for wave propagation. This is in contrast to other types of waves, such as water and sound waves, that do require a medium to support the disturbances. In the 19th century, physicists thought that electromagnetic waves also required a medium in order to propagate. They proposed that such a medium existed and gave it the name **luminiferous ether**. The ether was assumed to be present everywhere, even in empty space, and light waves were viewed as ether oscillations. Further, the ether would have to be a massless but rigid medium with no effect on the motion of planets or other objects. These are strange concepts indeed. In addition, it was found that the troublesome laws of electricity and magnetism would take on their simplest forms in a special frame of reference at *rest* with respect to the ether. This frame was called the *absolute frame*. The laws of electricity and magnetism would be valid in this absolute frame, but they would have to be modified in any reference frame moving with respect to the absolute frame.

As a result of the importance attached to the ether and the absolute frame, it became of considerable interest in physics to prove by experiment that they existed. Since it was considered likely that Earth was in motion through the ether, from the view of an experimenter on Earth, there was an "ether wind" blowing through his laboratory. A direct method for detecting the ether wind would use an apparatus fixed to Earth to measure the wind's influence on the speed of light. If v is the speed of the ether relative to Earth, then the speed of light should have its maximum value, $c + v$, when propagating downwind, as shown in Figure 26.3a. Likewise, the speed of light should have its minimum value, $c - v$, when propagating upwind, as in Figure 26.3b, and an intermediate value, $(c^2 - v^2)^{1/2}$, in the direction perpendicular to the ether wind, as in Figure 26.3c. If the Sun were assumed to be at rest in the ether, then the velocity of the ether wind would be equal to the orbital velocity of Earth around the Sun, which has a magnitude of approximately 3×10^4 m/s. Because $c = 3 \times 10^8$ m/s, it should be possible to detect a change in speed of about 1 part in 10^4 for measurements in the upwind or downwind directions. However, as we will see in the next section, all attempts to detect such changes and establish the existence of the ether (and hence the absolute frame) proved futile.

In conclusion, we see that the second hypothesis in our introduction to this section is false—and we now believe that **the laws of electricity and magnetism are the same in all inertial frames**. It is the simple classical addition laws for velocities that are incorrect and must be modified, as shown in Section 26.8.

26.4   THE MICHELSON–MORLEY EXPERIMENT

The most famous experiment designed to detect small changes in the speed of light was first performed in 1881 by Albert A. Michelson (1852–1931) and later repeated under various conditions by Michelson and Edward W. Morley (1838–1923). We state at the outset that the outcome of the experiment contradicted the ether hypothesis.

The experiment was designed to determine the velocity of Earth relative to the hypothetical ether. The experimental tool used was the Michelson interferometer, which was discussed in Section 25.7 and is shown again in Active Figure 26.4. Arm 2 is aligned along the direction of Earth's motion through space. Earth's moving through the ether at speed v is equivalent to the ether's flowing past Earth in the opposite direction with speed v. This ether wind blowing in the direction opposite the direction of Earth's motion should cause the speed of light measured in Earth frame to be $c - v$ as the light approaches mirror M_2 and $c + v$ after reflection, where c is the speed of light in the ether frame.

The two beams reflected from M_1 and M_2 recombine, and an interference pattern consisting of alternating dark and bright fringes is formed. The interference pattern was observed while the interferometer was rotated through an angle of 90°. This rotation supposedly would change the speed of the ether wind along the direction of arm 1. The effect of such rotation should have been to cause the fringe pattern to shift slightly but measurably; however, measurements failed to show any change in the interference pattern! The Michelson–Morley experiment was repeated at different times of the year when the ether wind was expected to change direction, but the results were always the same: **no fringe shift of the magnitude required was ever observed**.

The negative results of the Michelson–Morley experiment not only contradicted the ether hypothesis, but also showed that it was impossible to measure the absolute velocity of Earth with respect to the ether frame. However, as we will see in the next section, Einstein suggested a postulate in the special theory of relativity that places quite a different interpretation on these negative results. In later years, when more was known about the nature of light, the idea of an ether that permeates all of space was relegated to the theoretical graveyard. **Light is now understood to be an electromagnetic wave, which requires no medium for its propagation**. As a result, the idea of an ether in which these waves could travel became unnecessary.

Details of the Michelson–Morley Experiment

As we mentioned earlier, the Michelson–Morley experiment was designed to detect the motion of Earth with respect to the ether. Before we examine the details of this historical experiment, it is instructive to consider a race between two airplanes, as shown in Figure 26.5a. One airplane flies from point O to point A perpendicular to the direction of the wind, and the second airplane flies from point O to point B parallel to the wind. We will assume that they start at O at the same time, travel the same distance L with the same cruising speed c with respect to the wind, and return to O. Which airplane will win the race? In order to answer this question, we calculate the time of flight for both airplanes.

First, consider the airplane that moves along path I parallel to the wind. As it moves to the right, its speed is enhanced by the wind, and its speed with respect to Earth is $c + v$. As it moves to the left on its return journey, it must fly opposite the wind; hence, its speed with respect to Earth is $c - v$. The times of flight to the right and to the left are, respectively,

$$t_R = \frac{L}{c + v} \quad \text{and} \quad t_L = \frac{L}{c - v}$$

and the total time of flight for the airplane moving along path I is

$$t_1 = t_R + t_L = \frac{L}{c + v} + \frac{L}{c - v} = \frac{2Lc}{c^2 - v^2}$$

$$= \frac{2L}{c\left(1 - \dfrac{v^2}{c^2}\right)} \qquad\qquad \textbf{[26.1]}$$

Now consider the airplane flying along path II. If the pilot aims the airplane directly toward point A, it will be blown off course by the wind and won't reach its

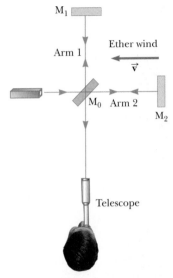

ACTIVE FIGURE 26.4
According to the ether wind theory, the speed of light should be $c - v$ as the beam approaches mirror M_2 and $c + v$ after reflection.

Physics⊗Now™
Log into PhysicsNow at **www.cp7e.com** and go to Active Figure 26.4, where you can adjust the speed of a fictitious ether wind and observe the effect on beams of light.

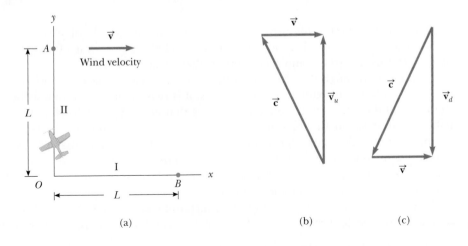

Figure 26.5 (a) If an airplane travels from O to A with a wind blowing to the right, it must head into the wind at some angle. (b) Vector diagram for determining the airplane's direction for the trip from O to A. (c) Vector diagram for determining its direction for the trip from A to O.

destination. To compensate for the wind, the pilot must point the airplane into the wind at some angle, as shown in Figure 26.5a. This angle must be selected so that the vector sum of $\vec{\mathbf{c}}$ and $\vec{\mathbf{v}}$ leads to a velocity vector pointed directly toward A. The resultant vector diagram is shown in Figure 26.5b, where $\vec{\mathbf{v}}_u$ is the velocity of the airplane with respect to the ground as it moves from O to A. From the Pythagorean theorem, the magnitude of the vector $\vec{\mathbf{v}}_u$ is

$$v_u = \sqrt{c^2 - v^2} = c\sqrt{1 - \frac{v^2}{c^2}}$$

Likewise, on the return trip from A to O, the pilot must again head into the wind so that the airplane's velocity $\vec{\mathbf{v}}_d$ with respect to Earth will be directed toward O, as shown in Figure 26.5c. From this figure, we see that

$$v_d = \sqrt{c^2 - v^2} = c\sqrt{1 - \frac{v^2}{c^2}}$$

The total time of flight for the trip along path II is therefore

$$t_2 = \frac{L}{v_u} + \frac{L}{v_d} = \frac{L}{c\sqrt{1 - \dfrac{v^2}{c^2}}} + \frac{L}{c\sqrt{1 - \dfrac{v^2}{c^2}}}$$

$$= \frac{2L}{c\sqrt{1 - \dfrac{v^2}{c^2}}} \qquad\qquad \text{[26.2]}$$

Comparing Equations 26.1 and 26.2, we see that the airplane flying along path II wins the race. The difference in flight times is given by

$$\Delta t = t_1 - t_2 = \frac{2L}{c}\left[\frac{1}{\left(1 - \dfrac{v^2}{c^2}\right)} - \frac{1}{\sqrt{1 - \dfrac{v^2}{c^2}}}\right]$$

This expression can be simplified by noting that the ratio of wind speed to plane speed, v/c, is usually much smaller than 1, and by using the following binomial expansions in v/c after dropping all terms higher than second order:

$$\left(1 - \frac{v^2}{c^2}\right)^{-1} \approx 1 + \frac{v^2}{c^2}$$

and

$$\left(1 - \frac{v^2}{c^2}\right)^{-1/2} \approx 1 + \frac{1}{2}\frac{v^2}{c^2}$$

The difference in flight times is therefore

$$\Delta t \approx \frac{Lv^2}{c^3} \qquad \text{for} \qquad v/c \ll 1 \qquad\qquad \text{[26.3]}$$

The analogy between this airplane race and the Michelson–Morley experiment is shown in Figure 26.6a. Two beams of light travel along two arms of an interferometer. In this case, the "wind" is the ether blowing across Earth from left to right as Earth moves through the ether from right to left. Because the speed of Earth in its orbital path is approximately 3×10^4 m/s, it is reasonable to use that value for the speed of the ether wind. Notice in this case that $v/c \approx 1 \times 10^{-4} \ll 1$. The two light beams start out in phase and return to form an interference pattern. We assume that the interferometer is adjusted for parallel fringes and that a telescope is focused on one of these fringes. The time difference between the two light beams gives rise to a phase difference between the beams, producing an interference pattern when they combine at the position of the telescope. The difference in the pattern is detected by rotating the interferometer through 90° in a horizontal plane, so that the two beams exchange roles (Fig. 26.6b). This results in a net

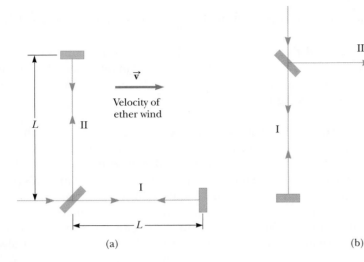

(a)   (b)

Figure 26.6 (a) Top view of the Michelson–Morley interferometer, where $\vec{v}$ is the velocity of the ether and L is the length of each arm. (b) When the interferometer is rotated by 90°, the role of each arm is reversed.

time shift of twice the time difference given by Equation 26.3. The net time difference is therefore

$$\Delta t_{\text{net}} = 2\,\Delta t = \frac{2Lv^2}{c^3} \qquad \textbf{[26.4]}$$

The corresponding path difference is

$$\Delta d = c\,\Delta t_{\text{net}} = \frac{2Lv^2}{c^2} \qquad \textbf{[26.5]}$$

In the first experiments by Michelson and Morley, each light beam was reflected by the mirrors many times to give an increased effective path length L of about 11 meters. Using this value and taking v to be equal to 3×10^4 m/s gives a path difference of

$$\Delta d = \frac{2(11\text{ m})(3.0 \times 10^4 \text{ m/s})^2}{(3.0 \times 10^8 \text{ m/s})^2} = 2.2 \times 10^{-7} \text{ m}$$

This extra travel distance should produce a noticeable shift in the fringe pattern. Specifically, calculations show that if the pattern is viewed while the interferometer is rotated through 90°, a shift of about 0.4 fringe should be observed. The instrument used by Michelson and Morley was capable of detecting a shift in the fringe pattern as small as 0.01 fringe. However, *it detected no shift whatsoever in the fringe pattern*. Since then, the experiment has been repeated many times by different scientists under a wide variety of conditions and no fringe shift has ever been detected. The inescapable conclusion is that motion of Earth with respect to the ether can't be detected.

Many efforts were made to explain the null results of the Michelson–Morley experiment and to save the ether frame concept and the Galilean addition law for the velocity of light. All proposals resulting from these efforts have been shown to be wrong. No experiment in the history of physics has received such valiant efforts to explain the absence of an expected result as was the Michelson–Morley experiment. The stage was set for Einstein, who, at the age of only 26, solved the problem in 1905 with his special theory of relativity.

26.5 EINSTEIN'S PRINCIPLE OF RELATIVITY

In the previous section we noted the serious contradiction between the Galilean addition law for velocities and the fact that the speed of light is the same for all observers. In 1905 Albert Einstein proposed a theory that resolved this contradiction but at the same time completely altered our notions of space and time. He based his special theory of relativity on two postulates:

ALBERT EINSTEIN,
German-American Physicist
(1879–1955)

One of the greatest physicists of all time, Einstein was born in Ulm, Germany. In 1905, at the age of 26, he published four scientific papers that revolutionized physics. Two of these papers were concerned with what is now considered his most important contribution: the special theory of relativity. In 1916, Einstein published his work on the general theory of relativity. The most dramatic prediction of this theory is the degree to which light is deflected by a gravitational field. Measurements made by astronomers on bright stars in the vicinity of the eclipsed Sun in 1919 confirmed Einstein's prediction, and as a result, Einstein became a world celebrity. Einstein was deeply disturbed by the development of quantum mechanics in the 1920s despite his own role as a scientific revolutionary. In particular, he could never accept the probabilistic view of events in nature that is a central feature of quantum theory. The last few decades of his life were devoted to an unsuccessful search for a unified theory that would combine gravitation and electromagnetism.

Postulates of relativity ▶

1. **The principle of relativity**: All the laws of physics are the same in all inertial frames.
2. **The constancy of the speed of light:** The speed of light in a vacuum has the same value, $c = 2.997\,924\,58 \times 10^8$ m/s, in all inertial reference frames, regardless of the velocity of the observer or the velocity of the source emitting the light.

The first postulate asserts that *all* the laws of physics are the same in all reference frames moving with constant velocity relative to each other. This postulate is a sweeping generalization of the principle of Galilean relativity, which refers only to the laws of mechanics. From an experimental point of view, Einstein's principle of relativity means that *any* kind of experiment—mechanical, thermal, optical, or electrical—performed in a laboratory at rest, must give the same result when performed in a laboratory moving at a constant speed past the first one. Hence, no preferred inertial reference frame exists, and it is impossible to detect absolute motion.

Although postulate 2 was a brilliant theoretical insight on Einstein's part in 1905, it has since been confirmed experimentally in many ways. Perhaps the most direct demonstration involves measuring the speed of photons emitted by particles traveling at 99.99% of the speed of light. The measured photon speed in this case agrees to five significant figures with the speed of light in empty space.

The null result of the Michelson–Morley experiment can be readily understood within the framework of Einstein's theory. According to his principle of relativity, the premises of the Michelson–Morley experiment were incorrect. In the process of trying to explain the expected results, we stated that when light traveled against the ether wind its speed was $c - v$. However, if the state of motion of the observer or of the source has no influence on the value found for the speed of light, the measured value must always be c. Likewise, the light makes the return trip after reflection from the mirror at a speed of c, not at a speed of $c + v$. Thus, the motion of Earth does not influence the fringe pattern observed in the Michelson–Morley experiment, and a null result should be expected.

If we accept Einstein's theory of relativity, we must conclude that uniform relative motion is unimportant when measuring the speed of light. At the same time, we have to adjust our commonsense notions of space and time and be prepared for some rather bizarre consequences.

26.6 CONSEQUENCES OF SPECIAL RELATIVITY

Almost everyone who has dabbled even superficially in science is aware of some of the startling predictions that arise because of Einstein's approach to relative motion. As we examine some of the consequences of relativity in this section, we'll find that they conflict with some of our basic notions of space and time. We will restrict our discussion to the concepts of length, time, and simultaneity, which are quite different in relativistic mechanics from what they are in Newtonian mechanics. For example, in relativistic mechanics, the distance between two points and the time interval between two events depend on the frame of reference in which they are measured. **In relativistic mechanics, there is no such thing as absolute length or absolute time.** Further, **events at different locations that are observed to occur simultaneously in one frame are not observed to be simultaneous in another frame moving uniformly past the first**.

◀ Absolute length and absolute time intervals are meaningless in relativity.

Simultaneity and the Relativity of Time

A basic premise of Newtonian mechanics is that a universal time scale exists that is the same for all observers. In fact, Newton wrote, "Absolute, true, and mathematical time, of itself, and from its own nature, flows equably without relation to anything external." Newton and his followers simply took simultaneity for granted. In his special theory of relativity, Einstein abandoned that assumption.

Figure 26.7   Two lightning bolts strike the ends of a moving boxcar. (a) The events appear to be simultaneous to the stationary observer at O, who is midway between A and B. (b) The events don't appear to be simultaneous to the observer at O', who claims that the front of the train is struck *before* the rear.

Einstein devised the following thought experiment to illustrate this point: a boxcar moves with uniform velocity, and two lightning bolts strike its ends, as in Figure 26.7a, leaving marks on the boxcar and the ground. The marks on the boxcar are labeled A' and B', and those on the ground are labeled A and B. An observer at O' moving with the boxcar is midway between A' and B', and an observer on the ground at O is midway between A and B. The events recorded by the observers are the striking of the boxcar by the two lightning bolts.

The light signals recording the instant at which the two bolts struck reach observer O at the same time, as indicated in Figure 26.7b. This observer realizes that the signals have traveled at the same speed over equal distances, and so rightly concludes that the events at A and B occurred simultaneously. Now consider the same events as viewed by observer O'. By the time the signals have reached observer O, observer O' has moved as indicated in Figure 26.7b. Thus, the signal from B' has already swept past O', but the signal from A' has not yet reached O'. In other words, O' sees the signal from B' before seeing the signal from A'. According to Einstein, *the two observers must find that light travels at the same speed.* Therefore, observer O' concludes that the lightning struck the front of the boxcar before it struck the back.

This thought experiment clearly demonstrates that the two events which appear to be simultaneous to observer O do not appear to be simultaneous to observer O'. In other words,

> Two events that are simultaneous in one reference frame are in general not simultaneous in a second frame moving relative to the first. Simultaneity depends on the state of motion of the observer, and is therefore not an absolute concept.

At this point, you might wonder which observer is right concerning the two events. The answer is that *both* are correct, because the principle of relativity states that **there is no preferred inertial frame of reference**. Although the two observers reach different conclusions, both are correct in their own reference frames because the concept of simultaneity is not absolute. In fact, this is the central point of relativity. Any inertial frame of reference can be used to describe events and do physics.

Time Dilation

We can illustrate the fact that observers in different inertial frames may measure different time intervals between a pair of events by considering a vehicle moving to the right with a speed v as in Active Figure 26.8a (page 852). A mirror is fixed to the ceiling of the vehicle, and an observer O' at rest in this system holds a laser a distance d below the mirror. At some instant, the laser emits a pulse of light

TIP 26.1   Who's Right?

Which person is correct concerning the simultaneity of the two events? Both are correct, because the principle of relativity states that no inertial frame of reference is preferred. Although the two observers may reach different conclusions, both are correct in their own reference frame. Any uniformly moving frame of reference can be used to describe events and do physics.

ACTIVE FIGURE 26.8
(a) A mirror is fixed to a moving vehicle, and a light pulse leaves O' at rest in the vehicle. (b) Relative to a stationary observer on Earth, the mirror and O' move with a speed v. Note that the distance the pulse travels is greater than $2d$ as measured by the stationary observer. (c) The right triangle for calculating the relationship between Δt and Δt_p.

Physics⊗Now™
Log into PhysicsNow at **www.cp7e.com** and go to Active Figure 26.8, where you can observe the bouncing of the light pulse for various speeds of the train.

directed toward the mirror (event 1), and at some later time after reflecting from the mirror, the pulse arrives back at the laser (event 2). Observer O' carries a clock and uses it to measure the time interval Δt_p between these two events which she views as occurring at the same place. (The subscript p stands for *proper*, as we'll see in a moment.) Because the light pulse has a speed c, the time it takes it to travel from point A to the mirror and back to point A is

$$\Delta t_p = \frac{\text{Distance traveled}}{\text{Speed}} = \frac{2d}{c} \qquad [26.6]$$

The time interval Δt_p measured by O' requires only a single clock located at the same place as the laser in this frame.

Now consider the same set of events as viewed by O in a second frame, as shown in Active Figure 26.8b. According to this observer, the mirror and laser are moving to the right with a speed v, and as a result, the sequence of events appears different. By the time the light from the laser reaches the mirror, the mirror has moved to the right a distance $v\,\Delta t/2$, where Δt is the time it takes the light pulse to travel from point A to the mirror and back to point A as measured by O. In other words, O concludes that, because of the motion of the vehicle, if the light is to hit the mirror, it must leave the laser at an angle with respect to the vertical direction. Comparing Active Figures 26.8a and 26.8b, we see that the light must travel farther in (b) than in (a). (Note that neither observer "knows" that he or she is moving. Each is at rest in his or her own inertial frame.)

According to the second postulate of the special theory of relativity, both observers must measure c for the speed of light. Because the light travels farther in the frame of O, it follows that the time interval Δt measured by O is longer than the time interval Δt_p measured by O'. To obtain a relationship between these two time intervals, it is convenient to examine the right triangle shown in Active Figure 26.8c. The Pythagorean theorem gives

$$\left(\frac{c\Delta t}{2}\right)^2 = \left(\frac{v\Delta t}{2}\right)^2 + d^2$$

Solving for Δt yields

$$\Delta t = \frac{2d}{\sqrt{c^2 - v^2}} = \frac{2d}{c\sqrt{1 - v^2/c^2}}$$

Because $\Delta t_p = 2d/c$, we can express this result as

$$\Delta t = \frac{\Delta t_p}{\sqrt{1 - v^2/c^2}} = \gamma \, \Delta t_p \qquad \text{[26.7]} \qquad \blacktriangleleft \text{Time dilation}$$

where

$$\gamma = \frac{1}{\sqrt{1 - v^2/c^2}} \qquad \text{[26.8]}$$

Because γ is always greater than one, Equation 26.7 says that **the time interval Δt between two events measured by an observer moving with respect to a clock[3] is longer than the time interval Δt_p between the same two events measured by an observer at rest with respect to the clock**. Consequently, $\Delta t > \Delta t_p$, and the proper time interval is expanded or dilated by the factor γ. Hence, this effect is known as **time dilation**.

For example, suppose the observer at rest with respect to the clock measures the time required for the light flash to leave the laser and return. We assume that the measured time interval in this frame of reference, Δt_p, is one second. (This would require a very tall vehicle.) Now we find the time interval as measured by observer O moving with respect to the same clock. If observer O is traveling at half the speed of light ($v = 0.500c$), then $\gamma = 1.15$, and according to Equation 26.7, $\Delta t = \gamma \, \Delta t_p = 1.15(1.00 \text{ s}) = 1.15 \text{ s}$. Therefore, when observer O' claims that 1.00 s has passed, observer O claims that 1.15 s has passed. Observer O considers the clock of O' to be reading too low a value for the elapsed time between the two events and says that the clock of O' is "running slow." From this phenomenon, we may conclude the following:

A clock moving past an observer at speed v runs more slowly than an identical clock at rest with respect to the observer by a factor of γ^{-1}.

$\blacktriangleleft$ A clock in motion runs more slowly than an identical stationary clock.

The time interval Δt_p in Equations 26.6 and 26.7 is called the **proper time**. In general, **proper time is the time interval between two events as measured by an observer who sees the events occur at the same position**.

Although you may have realized it by now, it's important to spell out that relativity is a scientific democracy: the view of O' that O is really the one moving with speed v to the left and that O's clock is running more slowly is just as valid as the view of O. The principle of relativity requires that the views of two observers in uniform relative motion be equally valid and capable of being checked experimentally.

We have seen that moving clocks run slow by a factor of γ^{-1}. This is true for ordinary mechanical clocks as well as for the light clock just described. In fact, we can generalize these results by stating that all physical processes, including chemical and biological ones, slow down relative to a clock when those processes occur in a frame moving with respect to the clock. For example, the heartbeat of an astronaut moving through space would keep time with a clock inside the spaceship. Both the astronaut's clock and heartbeat would be slowed down relative to a clock back on Earth (although the astronaut would have no sensation of life slowing down in the spaceship).

Time dilation is a very real phenomenon that has been verified by various experiments involving the ticking of natural clocks. An interesting example of time dilation involves the observation of *muons*—unstable elementary particles that are very similar to electrons, having the same charge, but 207 times the mass. Muons can be produced by the collision of cosmic radiation with atoms high in the atmosphere. These particles have a lifetime of 2.2 μs when measured in a reference frame at rest with respect to them. If we take 2.2 μs as the average lifetime of a muon and assume that their speed is close to the speed of light, we find that

TIP 26.2 Proper Time Interval
You must be able to correctly identify the observer who measures the proper time interval. The proper time interval between two events is the time interval measured by an observer for whom the two events take place at the same position.

---

[3]Actually, Figure 26.8 shows the clock moving and not the observer, but this is equivalent to observer O moving to the left with velocity $\vec{v}$ with respect to the clock.

Muon is created

6.6 × 10² m

Muon decays

(a)

Muon is created

4.8 × 10³ m

Muon decays

(b)

Figure 26.9 (a) If not for time dilation, a muon moving at 0.99c relative an observer on Earth would travel only 6.6 × 10² m, on average, before decaying. (b) Because of time dilation, the observed lifetime is longer, so such a muon travels about 4.80 × 10³ m, on average, as measured by an observer on Earth.

these particles can travel only about 600 m before they decay (Fig. 26.9a). Hence, they could never reach Earth from the upper atmosphere where they are produced. However, experiments show that a large number of muons *do* reach Earth, and the phenomenon of time dilation explains how. Relative to an observer on Earth, the muons have a lifetime equal to $\gamma\tau_p$, where $\tau_p = 2.2$ μs is the lifetime in a frame of reference traveling with the muons. For example, for $v = 0.99c$, $\gamma \approx 7.1$ and $\gamma\tau_p \approx 16$ μs. Hence, the average distance muons travel as measured by an observer on Earth is $\gamma v\tau_p \approx 4\,800$ m, as indicated in Figure 26.9b. Consequently, muons can reach Earth's surface.

In 1976 experiments with muons were conducted at the laboratory of the European Council for Nuclear Research (CERN) in Geneva. Muons were injected into a large storage ring, reaching speeds of about 0.9994c. Electrons produced by the decaying muons were detected by counters around the ring, enabling scientists to measure the decay rate, and hence the lifetime of the muons. The lifetime of the moving muons was measured to be about 30 times as long as that of stationary muons to within two parts in a thousand, in agreement with the prediction of relativity.

Quick Quiz 26.1

Suppose you're an astronaut being paid according to the time you spend traveling in space. You take a long voyage traveling at a speed near that of light. Upon your return to Earth, you're asked how you'd like to be paid: according to the time elapsed on a clock on Earth or according to your ship's clock. Which should you choose in order to maximize your paycheck? (a) the Earth clock (b) the ship's clock (c) Either clock, it doesn't make a difference.

EXAMPLE 26.1   Pendulum Periods

Goal   Apply the concept of time dilation.

Problem   The period of a pendulum is measured to be 3.00 s in the inertial frame of the pendulum. What is the period as measured by an observer moving at a speed of 0.950c with respect to the pendulum?

Strategy   Here, we're given the period of the clock as measured by an observer in the rest frame of the clock, so that's a proper time interval Δt_p. We want to know how much time passes as measured by an observer in a frame moving relative to the clock, which is Δt. Substitution into Equation 26.7 then solves the problem.

Solution

Substitute the proper time and relative speed into Equation 26.7:

$$\Delta t = \frac{\Delta t_p}{\sqrt{1 - v^2/c^2}} = \frac{3.00 \text{ s}}{\sqrt{1 - \frac{(0.950c)^2}{c^2}}} = 9.61 \text{ s}$$

Remarks   The moving observer considers the *pendulum* to be moving, and moving clocks are observed to run more slowly: while the pendulum oscillates once in 3 s for an observer in the rest frame of the clock, it takes nearly 10 s to oscillate once according the moving observer.

Exercise 26.1

What pendulum period does a third observer moving at 0.900c measure?

Answer   6.88 s

The confusion that arises in problems like Example 26.1 lies in the fact that movement is relative: from the point of view of someone in the pendulum's rest frame, the pendulum is standing still (except, of course, for the swinging motion), whereas to someone in a frame that is moving with respect to the pendulum, it's the pendulum that's doing the moving. To keep this straight, always focus on the observer who is doing the measurement, and ask yourself whether the clock being measured is moving with respect to that observer. If the answer is no, then the observer is in the rest frame of the clock and measures the clock's proper time. If the answer is yes, then the time measured by the observer will be dilated—larger than the clock's proper time.

This confusion of perspectives led to the famous "twin paradox."

The Twin Paradox

An intriguing consequence of time dilation is the so-called twin paradox (Fig. 26.10). Consider an experiment involving a set of twins named Speedo and Goslo. When they are 20 years old, Speedo, the more adventuresome of the two, sets out on an epic journey to Planet X, located 20 lightyears from Earth. Further, his spaceship is capable of reaching a speed of $0.95c$ relative to the inertial frame of his twin brother back home. After reaching Planet X, Speedo becomes homesick and immediately returns to Earth at the same speed of $0.95c$. Upon his return, Speedo is shocked to discover that Goslo has aged $2D/v = 2(20 \text{ ly})/(0.95 \text{ ly/y}) = 42$ years and is now 62 years old. Speedo, on the other hand, has aged only 13 years.

Some wrongly consider *this* the paradox; that twins could age at different rates and end up after a period of time having very different ages. While contrary to our common sense, this isn't the paradox at all. The paradox lies in the fact that from Speedo's point of view, *he* was at rest while Goslo (on Earth) sped away from *him* at $0.95c$ and returned later. So Goslo's clock was moving relative to Speedo and hence running slow compared with Speedo's clock. The conclusion: Speedo, not Goslo, should be the older of the twins!

To resolve this apparent paradox, consider a third observer moving at a constant speed of $0.5c$ relative to Goslo. To the third observer, Goslo never changes inertial frames: His speed relative to the third observer is always the same. The third observer notes, however, that Speedo accelerates during his journey, *changing reference frames in the process*. From the third observer's perspective, it's clear that there is something very different about the motion of Goslo when compared to Speedo. The roles played by Goslo and Speedo are not symmetric, so it isn't surprising that time flows differently for each. Further, because Speedo accelerates, he is in a noninertial frame of reference—technically outside the bounds of special relativity (though there are methods for dealing with accelerated motion in relativity). Only Goslo, who is in a single inertial frame, can apply the simple time-dilation formula to Speedo's trip. Goslo finds that instead of aging 42 years,

(a)   (b)

Figure 26.10 (a) As the twins depart, they're the same age. (b) When Speedo returns from his journey to Planet X, he's younger than his twin Goslo, who remained on Earth.

Speedo ages only $(1 - v^2/c^2)^{1/2}(42 \text{ years}) = 13$ years. Of these 13 years, Speedo spends 6.5 years traveling to Planet X and 6.5 years returning, for a total travel time of 13 years, in agreement with our earlier statement.

Length Contraction

The measured distance between two points depends on the frame of reference of the observer. The **proper length** L_p of an object is **the length of the object as measured by an observer at rest relative to the object**. The length of an object measured in a reference frame that is moving with respect to the object is always less than the proper length. This effect is known as **length contraction**.

To understand length contraction quantitatively, consider a spaceship traveling with a speed v from one star to another, as seen by two observers, one on Earth and the other in the spaceship. The observer at rest on Earth (and also assumed to be at rest with respect to the two stars) measures the distance between the stars to be L_p. According to this observer, the time it takes the spaceship to complete the voyage is $\Delta t = L_p/v$. Because of time dilation, the space traveler, using his spaceship clock, measures a smaller time of travel: $\Delta t_p = \Delta t/\gamma$. The space traveler claims to be at rest and sees the destination star moving toward the spaceship with speed v. Because the space traveler reaches the star in time Δt_p, he concludes that the distance L between the stars is shorter than L_p. The distance measured by the space traveler is

$$L = v\,\Delta t_p = v\,\frac{\Delta t}{\gamma}$$

Because $L_p = v\,\Delta t$, it follows that

Length contraction ▶

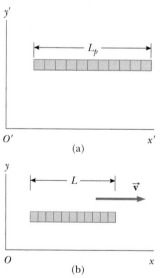

$$L = \frac{L_p}{\gamma} = L_p\sqrt{1 - v^2/c^2} \qquad [26.9]$$

According to this result, illustrated in Active Figure 26.11, if an observer at rest with respect to an object measures its length to be L_p, an observer moving at a speed v relative to the object will find it to be shorter than its proper length by the factor $\sqrt{1 - v^2/c^2}$. Note that **length contraction takes place only along the direction of motion**.

Time-dilation and length contraction effects have interesting applications for future space travel to distant stars. In order for the star to be reached in a fraction of a human lifetime, the trip must be taken at very high speeds. According to an Earth-bound observer, the time for a spacecraft to reach the destination star will be dilated compared with the time interval measured by travelers. As was discussed in the treatment of the twin paradox, the travelers will be younger than their twins when they return to Earth. Therefore, by the time the travelers reach the star, they will have aged by some number of years, while their partners back on Earth will have aged a larger number of years, the exact ratio depending on the speed of the spacecraft. At a spacecraft speed of $0.94c$, this ratio is about 3:1.

ACTIVE FIGURE 26.11
A meter stick moves to the right with a speed v. (a) The meter stick as viewed by an observer at rest with respect to the meter stick. (b) The meter stick as seen by an observer moving with a speed v with respect to it. The moving meter stick is always measured to be *shorter* than in its own rest frame by a factor of $\sqrt{1 - v^2/c^2}$.

Physics⊗Now™

Quick Quiz 26.2

You are packing for a trip to another star, and on your journey you will be traveling at a speed of $0.99c$. Can you sleep in a smaller cabin than usual, because you will be shorter when you lie down? Explain your answer.

Quick Quiz 26.3

You observe a rocket moving away from you. Compared to its length when it was at rest on the ground, you will measure its length to be (a) shorter, (b) longer, or (c) the same. Compared to the passage of time measured by the watch on your wrist, the passage of time on the rocket's clock is (d) faster, (e) slower, or (f) the same. Answer the same questions if the rocket turns around and comes toward you.

EXAMPLE 26.2   Starship Contraction

Goal   Apply the concept of length contraction to a moving object.

Problem   A starship is measured to be 125 m long while it is at rest with respect to an observer. If this starship now flies past the observer at a speed of $0.99c$, what length will the observer measure for the starship?

Strategy   Moving objects are observed to be contracted, or shorter. Substitute into Equation 26.9.

Solution
Substitute into Equation 26.9 to find the length as measured by the observer:

$$L = L_p \sqrt{1 - v^2/c^2} = (125 \text{ m})\sqrt{1 - (0.99\,c)^2/c^2} = \boxed{17.6 \text{ m}}$$

Exercise 26.2
If the ship moves past the observer with a speed of $0.80c$, what length will the observer measure?

Answer   75.0 m

EXAMPLE 26.3   Speedy Plunge

Goal   Apply the concept of length contraction to a distance.

Problem   **(a)** An observer on Earth sees a spaceship at an altitude of 4 350 km moving downward toward Earth with a speed of $0.970c$. What is the distance from the spaceship to Earth as measured by the spaceship's captain? **(b)** After firing his engines, the captain measures her ship's altitude as 267 km, while the observer on Earth measures it to be 625 km. What is the speed of the spaceship at this instant?

Strategy   To the captain, the Earth is rushing toward her ship at $0.970c$; hence the distance between her ship and the Earth is contracted. Substitution into Equation 26.9 yields the answer. In part (b) use the same equation, substituting the distances and solving for the speed.

Solution
(a) Find the distance from the ship to Earth as measured by the captain.

Substitute into Equation 26.9, getting the altitude as measured by the captain in the ship.

$$L = L_p\sqrt{1 - v^2/c^2} = (4\,350 \text{ km})\sqrt{1 - (0.970c)^2/c^2}$$
$$= \boxed{1.06 \times 10^3 \text{ km}}$$

(b) What is the subsequent speed of the spaceship if the Earth observer measures the distance from the ship to Earth as 625 km and the captain measures it as 267 km?

Apply the length-contraction equation:

$$L = L_p\sqrt{1 - v^2/c^2}$$

Square both sides of this equation and solve for v:

$$L^2 = L_p^2(1 - v^2/c^2) \quad \rightarrow \quad 1 - v^2/c^2 = \left(\frac{L}{L_p}\right)^2$$
$$v = c\sqrt{1 - (L/L_p)^2} = c\sqrt{1 - (267 \text{ km}/625 \text{ km})^2}$$
$$v = \boxed{0.904c}$$

Remarks   The proper length is always the length measured by an observer at rest with respect to that length.

Exercise 26.3
Suppose the observer on the ship measures the distance from Earth as 50.0 km, while the observer on Earth measures the distance as 125 km. At what speed is the spacecraft approaching Earth?

Answer $0.917c$

Length contraction occurs only in the direction of the observer's motion. No contraction occurs perpendicular to that direction. For example, a spaceship at rest relative to an observer may have the shape of an equilateral triangle, but if it passes the observer at relativistic speed in a direction parallel to its base, the base will shorten while the height remains the same. Hence, the craft will be observed to be isosceles. An observer traveling with the ship will still observe it to be equilateral.

26.7 RELATIVISTIC MOMENTUM

Properly describing the motion of particles within the framework of special relativity requires generalizing Newton's laws of motion and the definitions of momentum and energy. These generalized definitions reduce to the classical (nonrelativistic) definitions when v is much less than c.

First, recall that conservation of momentum states that when two objects collide, the total momentum of the system remains constant, assuming that the objects are isolated, reacting only with each other. However, analyzing such collisions from rapidly moving inertial frames, it is found that momentum is not conserved if the classical definition of momentum, $p = mv$, is used. In order to have momentum conservation in all inertial frames—even those moving at an appreciable fraction of c—the definition of momentum must be modified to read

Relativistic momentum ▶

$$p \equiv \frac{mv}{\sqrt{1 - v^2/c^2}} = \gamma\, mv \qquad \text{[26.10]}$$

where v is the speed of the particle and m is its mass as measured by an observer at rest with respect to the particle. Note that when v is much less than c, the denominator of Equation 26.10 approaches one, so that p approaches mv. Therefore, the relativistic equation for momentum reduces to the classical expression when v is small compared with c.

EXAMPLE 26.4   The Relativistic Momentum of an Electron

Goal   Contrast the classical and relativistic definitions of momentum.

Problem   An electron, which has a mass of 9.11×10^{-31} kg, moves with a speed of $0.750c$. Find the classical (nonrelativistic) momentum and compare it to its relativistic counterpart p_{rel}.

Strategy   Substitute into the classical definition to get the classical momentum, then multiply by the gamma factor to obtain the relativistic version.

Solution

First, compute the classical (nonrelativistic) momentum with $v = 0.750c$:

$p = mv = (9.11 \times 10^{-31}\text{ kg})(0.750 \times 3.00 \times 10^8\text{ m/s})$

$= 2.05 \times 10^{-22}\text{ kg·m/s}$

Multiply this result by γ to obtain the relativistic momentum:

$p_{\text{rel}} = \dfrac{mv}{\sqrt{1 - v^2/c^2}} = \dfrac{2.05 \times 10^{-22}\text{ kg·m/s}}{\sqrt{1 - (0.750c/c)^2}}$

$= 3.10 \times 10^{-22}\text{ kg·m/s}$

Remark   The (correct) relativistic result is 50% greater than the classical result. In subsequent calculations, no notational distinction will be made between classical and relativistic momentum. For problems involving relative speeds of $0.2c$, the answer using the classical expression is about 2% below the correct answer.

Exercise 26.4
Repeat the calculation for a proton traveling at $0.600c$.

Answers $p = 3.01 \times 10^{-19}$ kg·m/s, $p_{\text{rel}} = 3.76 \times 10^{-19}$ kg·m/s

26.8 RELATIVISTIC ADDITION OF VELOCITIES

Imagine a motorcycle rider moving with a speed of $0.80c$ past a stationary observer, as shown in Figure 26.12. If the rider tosses a ball in the forward direction with a speed of $0.70c$ relative to himself, what is the speed of the ball as seen by the stationary observer at the side of the road? Common sense and the ideas of Newtonian relativity say that the speed should be the sum of the two speeds, or $1.50c$. This answer must be incorrect because it contradicts the assertion that no material object can travel faster than the speed of light.

Einstein resolved this dilemma by deriving an equation for the relativistic addition of velocities. Here, only one dimension of motion will be considered. Let two frames or reference be labeled b and d, and suppose that frame d is moving at velocity v_{db} in the position x-direction relative frame b. If the velocity of an object a as measured in frame d is called v_{ad}, then the velocity of a as measured in frame b, v_{ab}, is given by

$$v_{ab} = \frac{v_{ad} + v_{db}}{1 + \dfrac{v_{ad}v_{db}}{c^2}}$$

[26.11]   ◀ Relativistic velocity addition

The left side of this equation and the numerator on the right are like the equations of Galilean relativity discussed in Chapter 3, and the evaluation of subscripts is applied in the same way as discussed in Section 3.6. The denominator of Equation 26.11 is a correction to Galilean relativity based on length contraction and time dilation.

We apply Equation 26.11 to Figure 26.13, which shows a motorcyclist, his ball, and a stationary observer. We are given

v_{bm} = the velocity of the ball with respect to the motorcycle = $0.70c$

v_{mo} = the velocity of the motorcycle with respect to the stationary
observer = $0.80c$,

and we want to find

v_{bo} = the velocity of the ball with respect to the stationary observer.

Thus,

$$v_{bo} = \frac{v_{bm} + v_{mo}}{1 + \dfrac{v_{bm}v_{mo}}{c^2}} = \frac{0.70c + 0.80c}{1 + \dfrac{(0.70c)(0.80c)}{c^2}} = 0.96c$$

The speed of light is the speed limit of the Universe.

Figure 26.12   A motorcycle moves past a stationary observer with a speed of $0.80c$; the motorcyclist throws a ball in the direction of motion with a speed of $0.70c$ relative to himself.

EXAMPLE 26.5   Urgent Course Correction Needed!

Goal   Apply the concept of the relativistic addition of velocities.

Problem   Suppose that Bob's spacecraft is traveling at $0.600c$ in the positive x-direction, as measured by a nearby observer, while Mike is traveling in his own vehicle directly toward Bob in the negative x-direction at $-0.800c$ relative the nearby observer. What's the velocity of Bob relative to Mike?

Strategy   This problem requires correctly identifying the quantities that go into Equation 26.11, followed by substitution. The measurement of Bob's velocity as determined in the observer's frame O is given, and the measurement of Bob's velocity in Mike's frame is desired.

Solution

Identify the velocity terms in Equation 26.11.

v_{BM} = the velocity of Bob with respect to Mike. This will be substituted for v_{ad} in Equation 26.11.

v_{MO} = the velocity of the Mike with respect to the stationary observer = $-0.800c$. This will be substituted for v_{db} in Equation 26.11.

v_{BO} = the velocity of the Bob with respect to the stationary observer = $0.600c$. This will be substituted for v_{ab} in Equation 26.11.

Substitute the velocity expressions into Equation 26.11. Examining the form of Equation 26.11, we can see intuitively that v_{BM} and v_{MO} belong on the right hand side (the letter M appears in both a first and a second position), so our previous choices are verified.

$$v_{BO} = \frac{v_{BM} + v_{MO}}{1 + \frac{v_{BM} v_{MO}}{c^2}}$$

Substitute given quantities and solve for v_{BM}:

$$0.600c = \frac{v_{BM} - 0.800c}{1 + \frac{v_{BM}(-0.800c)}{c^2}}$$

$$\left(1 - \frac{0.800 v_{BM}}{c}\right) 0.600c = v_{BM} - 0.800c$$

$$0.600c - 0.480\, v_{BM} = v_{BM} - 0.800c$$

$$v_{BM} = \boxed{0.946c}$$

Remarks   Notice how much care had to be taken in identifying quantities and their proper signs. Common sense might lead us to believe that Mike would measure Bob's velocity as $1.40c$, but as the calculation shows, Mike measures Bob's velocity as less than that of light.

Exercise 26.5

Suppose Bob shines a laser beam in the direction of his ship's motion. What speed would the nearby observer measure for the beam? Don't guess: do the calculation that proves the answer.

Answer c

26.9   RELATIVISTIC ENERGY AND THE EQUIVALENCE OF MASS AND ENERGY

We have seen that the definition of momentum required generalization to make it compatible with the principle of relativity. Likewise, the definition of kinetic energy requires modification in relativistic mechanics. Einstein found that the correct expression for the **kinetic energy** of an object is

Kinetic energy ▶

$$KE = \gamma mc^2 - mc^2$$

[26.12]

The constant term mc^2 in Equation 26.12, which is independent of the speed of the object, is called the **rest energy** of the object, E_R:

$$E_R = mc^2$$ [26.13]   ◀ Rest energy

The term γmc^2 in Equation 26.12 depends on the object's speed and is the sum of the kinetic and rest energies. We define γmc^2 to be the **total energy** E, so that

$$\text{total energy} = \text{kinetic energy} + \text{rest energy}$$

or, using Equation 26.12,

$$E = KE + mc^2 = \gamma mc^2$$ [26.14]

Because $\gamma = (1 - v^2/c^2)^{-1/2}$, we can also express E as

$$E = \frac{mc^2}{\sqrt{1 - v^2/c^2}}$$ [26.15]   ◀ Total energy

This is Einstein's famous mass–energy equivalence equation.[4]

The relation $E = \gamma mc^2 = KE + mc^2$ shows the amazing result that **a stationary particle with zero kinetic energy has an energy proportional to its mass**. Further, a small mass corresponds to an enormous amount of energy because the proportionality constant between mass and energy is large: $c^2 = 9 \times 10^{16}\ \text{m}^2/\text{s}^2$. The equation $E_R = mc^2$, as Einstein first suggested, indirectly implies that the mass of a particle may be completely convertible to energy and that pure energy—for example, electromagnetic energy—may be converted to particles having mass. This is indeed the case, as has been shown in the laboratory many times. For example, the coming together of a slowly moving electron and its antiparticle, the positron, a particle with the same mass m_e as the electron, but opposite charge, results in the disappearance of both particles and the appearance of a burst of electromagnetic energy in the amount $2m_ec^2$. The reverse process is also fairly easily observed in the laboratory: A high-energy pulse of electromagnetic energy, a gamma ray—disappears near an atom and an electron–positron pair is created with nearly 100% conversion of the gamma ray's energy into mass. Such a pair-production process is shown in the bubble chamber photo of Figure 26.13. We will discuss pair production and annihilation in more detail in Section 26.10.

On a larger scale, nuclear power plants produce energy by the fission of uranium, which involves the conversion of a small amount of the mass of the uranium into energy. The Sun, too, converts mass into energy, and continually loses mass in pouring out a tremendous amount of electromagnetic energy in all directions.

It's extremely interesting that while we have been talking about the interconversion of mass and energy for particles, the expression $E = mc^2$ is universal and applies to all objects, processes, and systems: a hot object has slightly more mass and is slightly more difficult to accelerate than an identical cold object because it has more thermal energy, and a stretched spring has more elastic potential energy and more mass than an identical unstretched spring. A key point, however, is that these changes in mass are often far too small to measure. Our best bet for measuring mass changes is in nuclear transformations, where a measurable fraction of the mass is converted into energy.

Figure 26.13 Bubble-chamber photograph of electron (green) and positron (red) tracks produced by energetic gamma rays. The highly curved tracks at the top are due to the electron and positron in an electron–positron pair bending in opposite directions in the magnetic field.

Lawrence Berkeley Laboratory/Science Photo Library/Photo Researchers, Inc.

EXAMPLE 26.6   Pool Heater

Goal   Combine the concepts of density, rest mass, and heat capacity.

Problem   Suppose some mechanism allowed the conversion of the rest mass of water completely into energy. **(a)** How much rest energy is contained in 0.500 mm³ of water? **(b)** If all this energy is used to heat an Olympic swim-

---

[4]Although this doesn't look exactly like the famous equation $E = mc^2$, it used to be common to write $m = \gamma m_0$ (Einstein himself wrote it that way), where m is the effective mass of an object moving at speed v and m_0 is the mass of that object as measured by an observer at rest with respect to the object. Then our $E = \gamma mc^2$ becomes the familiar $E = mc^2$. It is currently unfashionable to use $m = \gamma m_0$.

ming pool with dimensions 2.00 m deep, 25.0 m wide, and 50.0 m long, what is the change in temperature of the water?

Strategy   Use the density of water to find the mass in the given volume of water, and multiply by c^2 to get the energy. The heat capacity equation then yields the temperature change.

Solution
(a) How much rest energy is contained in 0.500 mm³ of water?

Use the density to find the mass of this volume of water:

$$\rho = \frac{m}{V} \rightarrow m = \rho V$$

$$m = (1.00 \times 10^3 \, \text{kg/m}^3)(0.500 \, \text{mm}^3)\left(\frac{1.00 \, \text{m}}{1.00 \times 10^3 \, \text{mm}}\right)^3$$

$$= 5.00 \times 10^{-7} \, \text{kg}$$

The energy equivalent of the water is found from Equation 26.13:

$$E_R = mc^2 = (5.00 \times 10^{-7} \, \text{kg})(3.00 \times 10^8 \, \text{m/s})^2$$

$$= 4.50 \times 10^{10} \, \text{J}$$

(b) Find the change in temperature of the pool water.

First find the volume of water in the pool:

$$V = L \times W \times H = (50.0 \, \text{m})(25.0 \, \text{m})(2.00 \, \text{m})$$

$$= 2.50 \times 10^3 \, \text{m}^3$$

Using the definition of density, calculate the mass of the water in the pool:

$$m = \rho V = (1.00 \times 10^3 \, \text{kg/m}^3)(2.50 \times 10^3 \, \text{m}^3)$$

$$= 2.50 \times 10^6 \, \text{kg}$$

Use the heat capacity equation and the result of part (a) to calculate the temperature change of the water in the pool:

$$Q = mc\Delta T$$

$$\Delta T = \frac{Q}{mc} = \frac{4.50 \times 10^{10} \, \text{J}}{(2.50 \times 10^6 \, \text{kg})(4.19 \times 10^3 \, \text{J/kg·K})}$$

$$= 4.30 \, \text{K}$$

Remarks   Only 12 mm³ of water, completely converted to energy, could raise the water temperature of an Olympic-sized pool by 100 K! However, it's generally impossible to achieve the complete conversion of mass to energy. Nuclear power plants convert only a tiny percentage of the mass of uranium. An exception is the interaction of matter with antimatter.

Exercise 26.6
(a) What mass, when completely converted into energy, would provide the annual energy needs of the entire world (about 4×10^{20} J)? **(b)** What volume of water contains that much energy?

Answers   (a) 4×10^3 kg   (b) 4 m³

Energy and Relativistic Momentum

Often the momentum or energy of a particle is measured rather than its speed, so it's useful to have an expression relating the total energy E to the relativistic momentum p. This is accomplished by using the expressions $E = \gamma mc^2$ and $p = \gamma mv$. By squaring these equations and subtracting, we can eliminate v. The result, after some algebra, is

$$E^2 = p^2c^2 + (mc^2)^2 \qquad \text{[26.16]}$$

When the particle is at rest, $p = 0$, so $E = E_R = mc^2$. In this special case, the total energy equals the rest energy. For the case of particles that have zero mass, such as photons (massless, chargeless particles of light), we set $m = 0$ in Equation 26.16 and find that

$$E = pc \qquad\qquad [26.17]$$

This equation is an exact expression relating energy and momentum for photons, which always travel at the speed of light.

In dealing with subatomic particles, it's convenient to express their energy in electron volts (eV), because the particles are given energy when accelerated through an electrostatic potential difference. The conversion factor is

$$1 \text{ eV} = 1.60 \times 10^{-19} \text{ J}$$

For example, the mass of an electron is 9.11×10^{-31} kg. Hence, the rest energy of the electron is

$$m_e c^2 = (9.11 \times 10^{-31} \text{ kg})(3.00 \times 10^8 \text{ m/s})^2 = 8.20 \times 10^{-14} \text{ J}$$

Converting to eV, we have

$$m_e c^2 = (8.20 \times 10^{-14} \text{ J})(1 \text{ eV}/1.60 \times 10^{-19} \text{ J}) = 0.511 \text{ MeV}$$

Because we frequently use the expression $E = \gamma mc^2$ in nuclear physics, and because m is usually in atomic mass units, u, it is useful to have the conversion factor $1 \text{ u} = 931.494 \text{ MeV}/c^2$. Using this factor makes it easy, for example, to find the rest energy in MeV of the nucleus of a uranium atom with a mass of 235.043 924 u:

$$E_R = mc^2 = (235.043\ 924 \text{ u})(931.494 \text{ MeV/u} \cdot c^2)(c^2) = 2.189\ 42 \times 10^5 \text{ MeV}$$

Quick Quiz 26.4

A photon is reflected from a mirror. **True or false**: (a) Because a photon has zero mass, it does not exert a force on the mirror. (b) Although the photon has energy, it can't transfer any energy to the surface because it has zero mass. (c) The photon carries momentum, and when it reflects off the mirror, it undergoes a change in momentum and exerts a force on the mirror. (d) Although the photon carries momentum, its change in momentum is zero when it reflects from the mirror, so it can't exert a force on the mirror.

EXAMPLE 26.7   A Speedy Electron

Goal   Compute a total energy and a relativistic kinetic energy.

Problem   An electron moves with a speed $v = 0.850c$. Find its total energy and kinetic energy in mega electron volts (MeV), and compare the latter to the classical kinetic energy (10^6 eV = 1 MeV).

Strategy   Substitute into Equation 26.15 to get the total energy, and subtract the rest mass energy to obtain the kinetic energy.

Solution

Substitute values into Equation 26.15 to obtain the total energy:

$$E = \frac{m_e c^2}{\sqrt{1 - v^2/c^2}} = \frac{(9.11 \times 10^{-31} \text{ kg})(3.00 \times 10^8 \text{ m/s})^2}{\sqrt{1 - (0.850c/c)^2}}$$

$$= 1.56 \times 10^{-13} \text{ J} = (1.56 \times 10^{-13} \text{ J})\left(\frac{1.00 \text{ eV}}{1.60 \times 10^{-19} \text{ J}}\right)$$

$$= \boxed{0.975 \text{ MeV}}$$

The kinetic energy is obtained by subtracting the rest energy from the total energy:

$$KE = E - m_e c^2 = 0.975 \text{ MeV} - 0.511 \text{ MeV} = \boxed{0.464 \text{ MeV}}$$

Calculate the classical kinetic energy:

$$KE_{\text{classical}} = \tfrac{1}{2} m_e v^2$$

$$= \tfrac{1}{2}(9.11 \times 10^{-31} \text{ kg})(0.850 \times 3.00 \times 10^8 \text{ m/s})^2$$

$$= 2.96 \times 10^{-14} \text{ J} = 0.185 \text{ MeV}$$

Remarks Notice the large discrepancy between the relativistic kinetic energy and the classical kinetic energy.

Exercise 26.7

Calculate the total energy and the kinetic energy in MeV of a proton traveling at $0.600c$. (The rest energy of a proton is approximately 938 MeV.)

Answers $E = 1.17 \times 10^3$ MeV, $KE = 232$ MeV

EXAMPLE 26.8 The Conversion of Mass to Kinetic Energy in Uranium Fission

Goal Understand the production of energy from nuclear sources.

Problem The fission, or splitting, of uranium was discovered in 1938 by Lise Meitner, who successfully interpreted some curious experimental results found by Otto Hahn as due to fission. (Hahn received the Nobel prize.) The fission of $^{235}_{92}U$ begins with the absorption of a slow-moving neutron that produces an unstable nucleus of ^{236}U. The ^{236}U nucleus then quickly decays into two heavy fragments moving at high speed, as well as several neutrons. Most of the kinetic energy released in such a fission is carried off by the two large fragments. **(a)** For the typical fission process

$$^{1}_{0}n + ^{235}_{92}U \rightarrow ^{141}_{56}Ba + ^{92}_{36}Kr + 3^{1}_{0}n$$

calculate the kinetic energy in MeV carried off by the fission fragments. **(b)** What percentage of the initial energy is converted into kinetic energy? The atomic masses involved are given below in atomic mass units.

$$^{1}_{0}n = 1.008\ 665\ u \qquad ^{235}_{92}U = 235.043\ 924\ u \qquad ^{141}_{56}Ba = 140.903\ 496\ u \qquad ^{92}_{36}Kr = 91.907\ 936\ u$$

Strategy This is an application of the conservation of relativistic energy. Write the conservation law as a sum of kinetic energy and rest energy, and solve for the final kinetic energy. Equation 26.15, solved for v, then yields the speeds.

Solution

(a) Calculate the final kinetic energy for the given process.

Apply the conservation of relativistic energy equation, assuming that $KE_{initial} = 0$:

$$(KE + mc^2)_{initial} = (KE + mc^2)_{final}$$

$$0 + m_n c^2 + m_U c^2 = m_{Ba} c^2 + m_{Kr} c^2 + 3m_n c^2 + KE_{final}$$

Solve for KE_{final} and substitute, converting to MeV in the last step:

$$KE_{final} = [(m_n + m_U) - (m_{Ba} + m_{Kr} + 3m_n)]c^2$$

$$KE_{final} = (1.008\ 665u + 235.043\ 924\ u)\,c^2$$

$$\qquad - [140.903\ 496\ u + 91.907\ 936\ u + 3(1.008\ 665\ u)]c^2$$

$$\qquad = (0.215\ 162\ u)(931.494\ MeV/u \cdot c^2)(c^2)$$

$$\qquad = \boxed{200.422\ MeV}$$

(b) What percentage of the initial energy is converted into kinetic energy?

Compute the total energy, which is the initial energy:

$$E_{initial} = 0 + m_n c^2 + m_U c^2$$

$$= (1.008\ 665u + 235.043\ 924\ u)\,c^2$$

$$= (236.052\ 59\ u)\,(931.494\ MeV/u \cdot c^2)\,(c^2)$$

$$= 2.198\ 82 \times 10^5\ MeV$$

Divide the kinetic energy by the total energy and multiply by 100%:

$$\frac{200.422\ MeV}{2.198\ 82 \times 10^5\ MeV} \times 100\% = \boxed{9.115 \times 10^{-2}\%}$$

Remarks This calculation shows that nuclear reactions liberate only about a tenth of one percent of the rest energy of the constituent particles. Some fusion reactions better that number by several times.

Exercise 26.8

In a fusion reaction, light elements combine to form a heavier element. Deuterium, which is also called heavy hydrogen, has an extra neutron in its nucleus. Two such particles can fuse into a heavier form of hydrogen, called tritium, plus an ordinary hydrogen atom. The reaction is

$$^2_1D + {}^2_1D \rightarrow {}^3_1T + {}^1_1H$$

(a) Calculate the energy released in the form of kinetic energy, assuming for simplicity that the initial kinetic energy is zero. (b) What percentage of the rest mass was converted to energy? The atomic masses involved are as follows:

$$^2_1D = 2.014\ 102\ u \qquad ^3_1T = 3.016\ 049\ u \qquad ^1_1H = 1.007\ 825\ u$$

Answers   (a) 4.033 37 MeV   (b) 0.1075%

26.10 PAIR PRODUCTION AND ANNIHILATION

In general, converting mass into energy is a low-yield process. Burning wood or coal, or even the fission or fusion processes presented in Example 26.8, convert only a very small percentage of the available energy. An exception is the reaction of matter with antimatter.

A common process in which a photon creates matter is called **pair production**, illustrated in Figure 26.14. In this process, an electron and a positron are simultaneously produced, while the photon disappears. (Note that the positron is a positively charged particle having the same mass as an electron. The positron is often called the *antiparticle* of the electron.) In order for pair production to occur, energy, momentum, and charge must all be conserved during the process. It's impossible for a photon to produce a single electron because the photon has zero charge and charge would not be conserved in the process.

As we explain in more detail in Chapter 27, the energy of a photon having a frequency f is given by $E = hf$, where h is Planck's constant. The *minimum* energy that a photon must have to produce an electron–positron pair can be found using conservation of energy by equating the photon energy hf_{min} to the total rest energy of the pair. That is,

$$hf_{min} = 2m_e c^2 \qquad [26.18]$$

Because the energy of an electron is $m_e c^2 = 0.51$ MeV, the minimum energy required for pair production is 1.02 MeV.

Pair production can't occur in a vacuum, but can only take place in the presence of a massive particle such as an atomic nucleus. The massive particle must participate in the interaction in order that energy and momentum be conserved simultaneously.

Pair annihilation is a process in which an electron–positron pair produces two photons—the inverse of pair production. Figure 26.15 is one example of pair annihilation in which an electron and positron initially at rest combine with each other, disappear, and create two photons. Because the initial momentum of the pair is zero, it's impossible to produce a single photon. Momentum can be conserved only if two photons moving in opposite directions, both with the same energy and magnitude of momentum, are produced. We will discuss particles and their antiparticles further in Chapter 30.

26.11 GENERAL RELATIVITY

Special relativity relates observations of inertial observers. Einstein sought a more general theory that would address accelerating systems. His search was motivated in part by the following curious fact: mass determines the inertia of an object and also the strength of the gravitational field. The mass involved in inertia is called inertial mass, m_i, whereas the mass responsible for the gravitational field is called the

Before

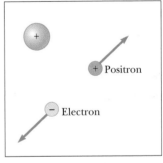

After

Figure 26.14 Representation of the process of pair production.

Before

After

Figure 26.15 Representation of the process of pair annihilation.

gravitational mass, m_g. They appear in Newton's law of gravitation and in the second law of motion:

$$\text{Gravitational property} \qquad F_g = G\,\frac{m_g m_g'}{r^2}$$

$$\text{Inertial property} \qquad F_i = m_i a$$

The value for the gravitational constant G was chosen to make the magnitudes of m_g and m_i numerically equal. Regardless of how G is chosen, however, the strict proportionality of m_g and m_i has been established experimentally to an extremely high degree: a few parts in 10^{12}. It appears that gravitational mass and inertial mass may indeed be exactly equal: $m_i = m_g$.

There is no reason a priori, however, why these two very different quantities should be equal. They seem to involve two entirely different concepts: a force of mutual gravitational attraction between two masses and the resistance of a single mass to being accelerated. This question puzzled Newton and many other physicists over the years and was finally incorporated as a fundamental principle of Einstein's remarkable theory of gravitation, known as *general relativity*, in 1916.

In Einstein's view, the remarkable coincidence that m_g and m_i were exactly equal was evidence for an intimate connection between the two concepts. He pointed out that no mechanical experiment (such as releasing a mass) could distinguish between the two situations illustrated in Figures 26.16a and 26.16b. In each case, a mass released by the observer undergoes a downward acceleration of g relative to the floor.

Einstein carried this idea further and proposed that *no* experiment, mechanical or otherwise, could distinguish between the two cases. This extension to include all phenomena (not just mechanical ones) has interesting consequences. For example, suppose that a light pulse is sent horizontally across the box, as in Figure 26.16c. The trajectory of the light pulse bends downward as the box accelerates upward to meet it. Einstein proposed that a beam of light should also be bent downward by a gravitational field (Fig. 26.16d).

The two postulates of Einstein's **general relativity** are as follows:

1. All the laws of nature have the same form for observers in any frame of reference, accelerated or not.
2. In the vicinity of any given point, a gravitational field is equivalent to an accelerated frame of reference without a gravitational field. (This is the *principle of equivalence*.)

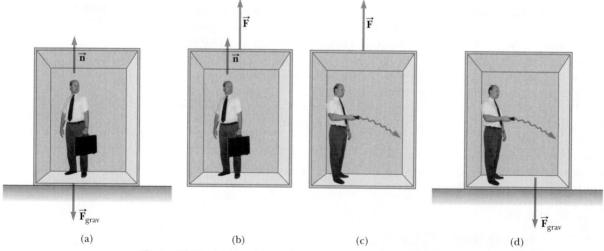

(a)   (b)   (c)   (d)

Figure 26.16   (a) The observer in the cubicle is at rest in a uniform gravitational field $\vec{\mathbf{g}}$. He experiences a normal force $\vec{\mathbf{n}}$. (b) Now the observer is in a region where gravity is negligible, but an external force $\vec{\mathbf{F}}$ acts on the frame of reference, producing an acceleration with magnitude g. Again, the man experiences a normal force $\vec{\mathbf{n}}$ that accelerates him along with the cubicle. According to Einstein, the frames of reference in parts (a) and (b) are equivalent in every way. No local experiment could distinguish between them. (c) The observer turns on his pocket flashlight. Because of the acceleration of the cubicle, the beam would appear to bend toward the floor, just as a tossed ball would. (d) Given the equivalence of the frames, the same phenomenon should be observed in the presence of a gravity field.

The second postulate implies that gravitational mass and inertial mass are completely equivalent, not just proportional. What were thought to be two different types of mass are actually identical.

One interesting effect predicted by general relativity is that time scales are altered by gravity. A clock in the presence of gravity runs more slowly than one in which gravity is negligible. As a consequence, light emitted from atoms in a strong gravity field, such as the Sun's, is observed to have a lower frequency than the same light emitted by atoms in the laboratory. This gravitational shift has been detected in spectral lines emitted by atoms in massive stars. It has also been verified on Earth by comparing the frequencies of gamma rays emitted from nuclei separated vertically by about 20 m.

Quick Quiz 26.5

Two identical clocks are in the same house, one upstairs in a bedroom and the other downstairs in the kitchen. Which statement is correct? (a) The clock in the kitchen runs more slowly than the clock in the bedroom. (b) The clock in the bedroom runs more slowly than the clock in the kitchen. (c) Both clocks keep the same time.

The second postulate suggests that a gravitational field may be "transformed away" at any point if we choose an appropriate accelerated frame of reference—a freely falling one. Einstein developed an ingenious method of describing the acceleration necessary to make the gravitational field "disappear." He specified a certain quantity, the *curvature of spacetime,* that describes the gravitational effect at every point. In fact, the curvature of spacetime completely replaces Newton's gravitational theory. According to Einstein, there is no such thing as a gravitational force. Rather, the presence of a mass causes a curvature of spacetime in the vicinity of the mass. Planets going around the Sun follow the natural contours of the spacetime, much as marbles roll around inside a bowl. In 1979, John Wheeler summarized Einstein's general theory of relativity in a single sentence: "Mass one tells spacetime how to curve; curved spacetime tells mass two how to move." The fundamental equation of general relativity can be roughly stated as a proportion as follows:

$$\text{Average curvature of spacetime} \propto \text{energy density}$$

The equation corresponding to this proportion is solved for a mathematical quantity called the *metric,* which can be used to measure the lengths of vectors and to compute trajectories of bodies through space. The metric looks something like a matrix, with different entries at each point of space and time. (There are a few important differences, beyond the level of this course.)

Einstein pursued a new theory of gravity in large part because of a discrepancy in the orbit of Mercury as calculated from Newton's second law. The closest approach of the Mercury to the Sun, called the perihelion, changes position slowly over time. Newton's theory accounted for all but 43 seconds of arc per century; Einstein's general relativity explained the discrepancy.

The most dramatic test of general relativity came shortly after the end of World War I. The theory predicts that a star would bend a light ray by a certain precise amount. Sir Arthur Eddington mounted an expedition to Africa and, during a solar eclipse, confirmed that starlight bent on passing the Sun in an amount matching the prediction of general relativity (Fig. 26.17). When this discovery was announced, Einstein became an international celebrity.

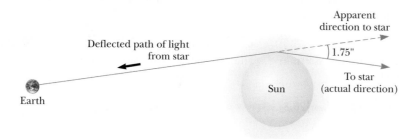

Figure 26.17 Deflection of starlight passing near the Sun. Because of this effect, the Sun and other remote objects can act as a *gravitational lens.* In his general theory of relativity, Einstein calculated that starlight just grazing the Sun's surface should be deflected by an angle of 1.75″.

Other tests were proposed and verified long after Einstein's death, including the time delay of radar bounced off Venus, and the gradual lengthening of the period of binary pulsars due to the emission of gravitational radiation. The latter has been verified with such precision that general relativity can lay claim to being the most accurate theory in physics.

General relativity also predicts extreme states of matter created by gravitational collapse. If the concentration of mass becomes very great, as is believed to occur when a large star exhausts its nuclear fuel and collapses to a very small volume, a **black hole** may form. Here the curvature of spacetime is so extreme that all matter and light within a certain radius becomes trapped. This radius, called the *Schwarzschild radius* or *event horizon*, is about 3 km for a black hole with the mass of our Sun. At the black hole's center may lurk a *singularity*—a point of infinite density and curvature where spacetime comes to an end.

There is strong evidence for the existence of a black hole having a mass of millions of Suns at the center of our galaxy.

Applying Physics 26.1   Faster Clocks in a "Mile High City"

Atomic clocks are extremely accurate; in fact, an error of 1 second in 3 million years is typical. This error can be described as about one part in 10^{14}. On the other hand, the atomic clock in Boulder, Colorado, is often 15 ns faster than the one in Washington after only one day. This is an error of about one part in 6×10^{12}, which is about 17 times larger than the typical error. If atomic clocks are so accurate, why does a clock in Boulder not remain synchronous with one in Washington?

Explanation According to the general theory of relativity, the passage of time depends on gravity—clocks run more slowly in strong gravitational fields. Washington is at an elevation very close to sea level, whereas Boulder is about a mile higher in altitude. Hence, the gravitational field at Boulder is weaker than at Washington. As a result, an atomic clock runs more rapidly in Boulder than in Washington. (This effect has been verified by experiment.)

SUMMARY

Physics ⊗ Now™ Take a practice test by logging into Physics-Now at **www.cp7e.com** and clicking on the Pre-Test link for this chapter.

26.5 Einstein's Principle of Relativity

The two basic postulates of the **special theory of relativity** are as follows:

1. The laws of physics are the same in all inertial frames of reference.
2. The speed of light is the same for all inertial observers, independently of their motion or of the motion of the source of light.

26.6 Consequences of Special Relativity

Some of the consequences of the special theory of relativity are as follows:

1. Clocks in motion relative to an observer slow down, a phenomenon known as **time dilation**. The relationship between time intervals in the moving and at-rest systems is

$$\Delta t = \gamma \Delta t_p \qquad [26.7]$$

where Δt is the time interval measured in the system in relative motion with respect to the clock,

$$\gamma = \frac{1}{\sqrt{1 - v^2/c^2}} \qquad [26.8]$$

and Δt_p is the proper time interval measured in the system moving with the clock.

2. The length of an object in motion is *contracted* in the direction of motion. The equation for **length contraction** is

$$L = L_p\sqrt{1 - v^2/c^2} \qquad [26.9]$$

where L is the length measured by an observer in motion relative to the object and L_p is the proper length measured by an observer for whom the object is at rest.

3. Events that are simultaneous for one observer are not simultaneous for another observer in motion relative to the first.

26.7 Relativistic Momentum

The relativistic expression for the **momentum** of a particle moving with velocity v is

$$p \equiv \frac{mv}{\sqrt{1 - v^2/c^2}} = \gamma mv \qquad [26.10]$$

26.8 Relativistic Addition of Velocities

The relativistic expression for the addition of velocities is

$$v_{ab} = \frac{v_{ad} + v_{db}}{1 + \dfrac{v_{ad}\,v_{db}}{c^2}} \qquad [26.11]$$

where v_{ab} is the velocity of object a with as measured in frame b, v_{ad} is the velocity of object a as measured in frame d, and v_{db} is the velocity of frame d as measured in frame b.

26.9 Relativistic Energy and the Equivalence of Mass and Energy

The relativistic expression for the **kinetic energy** of an object is

$$KE = \gamma mc^2 - mc^2 \qquad [26.12]$$

where mc^2 is the **rest energy** of the object, E_R.

The **total energy** of a particle is

$$E = \frac{mc^2}{\sqrt{1 - v^2/c^2}} \qquad [26.15]$$

This is Einstein's famous mass–energy equivalence equation.

The relativistic momentum is related to the total energy through the equation

$$E^2 = p^2c^2 + (mc^2)^2 \qquad [26.16]$$

26.10 Pair Production and Annihilation

Pair production is a process in which the energy of a photon is converted into mass. In this process, the photon disappears as an electron–positron pair is created. Likewise, the energy of an electron–positron pair can be converted into electromagnetic radiation by the process of **pair annihilation**.

CONCEPTUAL QUESTIONS

1. A spacecraft with the shape of a sphere of diameter D moves past an observer on Earth with a speed $0.5c$. What shape does the observer measure for the spacecraft as it moves past?

2. The equation $E = mc^2$ is often given in popular descriptions of Einstein's theory of relativity. Is this expression strictly correct? For example, does it accurately account for the kinetic energy of a moving mass?

3. You are in a speedboat on a lake. You see ahead of you a wave front, caused by the previous passage of another boat, moving away from you. You accelerate, catch up with, and pass the wave front. Is this scenario possible if you are in a rocket and you detect a wave front of light ahead of you?

4. What two speed measurements will two observers in relative motion always agree upon?

5. The speed of light in water is 2.30×10^8 m/s. Suppose an electron is moving through water at 2.50×10^8 m/s. Does this particle speed violate the principle of relativity?

6. With regard to reference frames, how does general relativity differ from special relativity?

7. Some distant starlike objects, called quasars, are receding from us at half the speed of light (or greater).

What is the speed of the light we receive from these quasars?

8. It is said that Einstein, in his teenage years, asked the question, "What would I see in a mirror if I carried it in my hands and ran at a speed near that of light?" How would you answer this question?

9. List some ways our day-to-day lives would change if the speed of light were only 50 m/s.

10. Two identically constructed clocks are synchronized. One is put into orbit around Earth while the other remains on Earth. Which clock runs more slowly? When the moving clock returns to Earth, will the two clocks still be synchronized.

11. Photons of light have zero mass. How is it possible that they have momentum?

12. Imagine an astronaut on a trip to Sirius, which lies 8 lightyears from Earth. Upon arrival at Sirius, the astronaut finds that the trip lasted 6 years. If the trip was made at a constant speed of $0.8c$, how can the 8-lightyear distance be reconciled with the 6-year duration?

13. Explain why it is necessary, when defining length, to specify that the positions of the ends of a rod are to be measured simultaneously.

PROBLEMS

1, 2, 3 = straightforward, intermediate, challenging   □ = full solution available in *Student Solutions Manual/Study Guide*

Physics⊗Now™ = coached problem with hints available at **www.cp7e.com**   🎵 = biomedical application

Section 26.4 The Michelson–Morley Experiment

1. Two airplanes fly paths I and II specified in Figure 26.5a. Both planes have air speeds of 100 m/s and fly a distance $L = 200$ km. The wind blows at 20.0 m/s in the direction shown in the figure. Find (a) the time of flight to each city, (b) the time to return, and (c) the difference in total flight times.

2. In one version of the Michelson–Morley experiment, the lengths L in Figure 26.6 were 28 m. Take v to be 3.0×10^4 m/s, and find the time difference caused by rotation of the interferometer and (b) the expected fringe shift, assuming that the light used has a wavelength of 550 nm.

Section 26.6 Consequences of Special Relativity

3. A deep-space probe moves away from Earth with a speed of $0.80c$. An antenna on the probe requires 3.0 s, in probe time, to rotate through 1.0 rev. How much time is required for 1.0 rev according to an observer on Earth?

4. If astronauts could travel at $v = 0.950c$, we on Earth would say it takes $(4.20/0.950) = 4.42$ years to reach Alpha Centauri, 4.20 lightyears away. The astronauts disagree. (a) How much time passes on the astronauts' clocks? (b) What is the distance to Alpha Centauri as measured by the astronauts?

5. A friend in a spaceship travels past you at a high speed. He tells you that his ship is 20 m long and that the identical ship you are sitting in is 19 m long. According to your observations, (a) how long is your ship, (b) how long is his ship, and (c) what is the speed of your friend's ship?

6. An astronaut at rest on Earth has a heartbeat rate of 70 beats/min. When the astronaut is traveling in a spaceship at $0.90c$, what will this rate be as measured by (a) an observer also in the ship and (b) an observer at rest on Earth?

7. The average lifetime of a pi meson in its own frame of reference (i.e., the proper lifetime) is 2.6×10^{-8} s. If the meson moves with a speed of $0.98c$, what is (a) its mean lifetime as measured by an observer on Earth, and (b) the average distance it travels before decaying, as measured by an observer on Earth? (c) What distance would it travel if time dilation did not occur?

8. An astronaut is traveling in a space vehicle that has a speed of $0.500c$ relative to Earth. The astronaut measures his pulse rate at 75.0 per minute. Signals generated by the astronaut's pulse are radioed to Earth when the vehicle is moving perpendicularly to a line that connects the vehicle with an Earth observer. What pulse rate does Earth observer measure? What would be the pulse rate if the speed of the space vehicle were increased to $0.990c$?

9. Physics⊗Now™ A muon formed high in Earth's atmosphere travels at speed $v = 0.99c$ for a distance of 4.6 km before it decays into an electron, a neutrino, and an antineutrino ($\mu^- \rightarrow e^- + \nu + \bar{\nu}$). (a) How long does the muon live, as measured in its reference frame? (b) How far does the muon travel, as measured in its frame?

10. A box is cubical with sides of proper lengths $L_1 = L_2 = L_3$, as shown in Figure P26.10, when viewed in its own rest frame. If this block moves parallel to one of its edges with a speed of $0.80c$ past an observer, (a) what shape does it appear to have to this observer, and (b) what is the length of each side as measured by the observer?

Figure P26.10

11. The proper length of one spaceship is three times that of another. The two spaceships are traveling in the same direction and, while both are passing overhead, an Earth observer measures the two spaceships to have the same length. If the slower spaceship is moving with a speed of $0.350c$, determine the speed of the faster spaceship.

12. Observer A measures the length of two rods, one stationary, the other moving with a speed of $0.955c$. She finds that the rods have the same length, L_0. A second observer B travels along with the moving rod. (a) What is the length observer B measures for the rod in observer A's frame? (b) What is the ratio of the length of A's rod to the length of B's rod, according to observer B?

13. A supertrain of proper length 100 m travels at a speed of $0.95c$ as it passes through a tunnel having proper length 50 m. As seen by a trackside observer, is the train ever completely within the tunnel? If so, by how much?

14. An observer moving at a speed of $0.995c$ relative to a rod (Fig. P26.14) measures its length to be 2.00 m and sees its length to be oriented at $30.0°$ with respect to its direction of motion. What is the proper length of the rod? (b) What is the orientation angle in a reference frame moving with the rod?

2.00 m

30.0°

Motion of observer

Figure P26.14 View of rod as seen by an observer moving to the right.

15. In 1963, when Mercury astronaut Gordon Cooper orbited Earth 22 times, the press stated that for each orbit, he aged 2 millionths of a second less than that if he remained on Earth. (a) Assuming that he was 160 km above Earth in a circular orbit, determine the time difference between someone on Earth and the orbiting astronaut for the 22 orbits. You will need to use the approximation

$$\frac{1}{\sqrt{1-x}} \approx 1 + \frac{x}{2} \qquad \text{for } x \ll 1$$

(b) Did the press report accurate information? Explain.

16. An interstellar space probe is launched from Earth. After a brief period of acceleration it moves with a constant velocity, 70.0% of the speed of light. Its nuclear-powered batteries supply the energy to keep its data transmitter active continuously. The batteries have a lifetime of 15.0 years as measured in a rest frame. (a) How long do the batteries on the space probe last as measured by mission control on Earth? (b) How far is the probe from Earth when its batteries fail, as measured by mission control? (c) How far is the probe from Earth, as measured by its built-in trip odometer, when its batteries fail? (d) For what total time after launch are data received from the probe by mission control? Note that radio waves travel at the speed of light and fill the space between the probe and Earth at the time the battery fails.

Section 26.7 Relativistic Momentum

17. An electron has a speed $v = 0.90c$. At what speed will a proton have a momentum equal to that of the electron?

18. Calculate the momentum of an electron moving with a speed of (a) $0.010c$, (b) $0.50c$, (c) $0.90c$.

19. An unstable particle at rest breaks up into two fragments of *unequal mass*. The mass of the lighter fragment is 2.50×10^{-28} kg, and that of the heavier fragment is 1.67×10^{-27} kg. If the lighter fragment has a speed of $0.893c$ after the breakup, what is the speed of the heavier fragment?

20. The nonrelativistic expression for the momentum of a particle, $p = mv$, can be used if $v \ll c$. For what speed does the use of this formula give an error in the momentum of (a) 1.00% and (b) 10.0%?

Section 26.8 Relativistic Addition of Velocities

21. An electron moves to the right with a speed of $0.90c$ relative to the laboratory frame. A proton moves to the left with a speed of $0.70c$ relative to the electron. Find the speed of the proton relative to the laboratory frame.

22. Spaceship R is moving to the right at a speed of $0.70c$ with respect to Earth. A second spaceship, L, moves to the left at the same speed with respect to Earth. What is the speed of L with respect to R?

23. A Klingon spaceship moves away from Earth at a speed of $0.800c$ (Fig. P26.23). The starship *Enterprise* pursues at a speed of $0.900c$ relative to Earth. Observers on Earth see the *Enterprise* overtaking the Klingon ship at a relative speed of $0.100c$. With what speed is the *Enterprise* overtaking the Klingon ship as seen by the crew of the *Enterprise*?

Figure P26.23

24. A spaceship travels at $0.750c$ relative to Earth. If the spaceship fires a small rocket in the forward direction, how fast (relative to the ship) must it be fired for it to travel at $0.950c$ relative to Earth?

25. **Physics⊗Now™** A rocket moves with a velocity of $0.92c$ to the right with respect to a stationary observer A. An observer B moving relative to observer A finds that the rocket is moving with a velocity of $0.95c$ to the left. What is the velocity of observer B relative to observer A? [*Hint:* Consider observer B's velocity in the frame of reference of the rocket.]

26. A pulsar is a stellar object that emits light in short bursts. Suppose a pulsar with a speed of $0.950c$ approaches Earth and a rocket with a speed of $0.995c$ heads toward the pulsar. (Both speeds are measured in Earth's frame of reference.) If the pulsar emits 10.0 pulses per second in its own frame of reference, at what rate are the pulses emitted in the rocket's frame of reference?

27. Spaceship I, which contains students taking a physics exam, approaches Earth with a speed of $0.60c$, while spaceship II, which contains instructors proctoring the exam, moves away from Earth at $0.28c$, as in Figure P26.27. If the instructors in spaceship II stop the exam after 50 min have passed *on their clock*, how long does the exam last as measured by (a) the students? (b) an observer on Earth?

Figure P26.27

Section 26.9 Relativistic Energy and the Equivalence of Mass and Energy

28. A proton moves with a speed of $0.950c$. Calculate (a) its rest energy, (b) its total energy, and (c) its kinetic energy.

29. What is the speed of a particle whose kinetic energy is equal to its own rest energy?

30. If it takes 3 750 MeV of work to accelerate a proton from rest to a speed of v, determine v.

31. What speed must a particle attain before its kinetic energy is double the value predicted by the nonrelativistic expression $KE = \frac{1}{2}mv^2$?

32. Determine the energy required to accelerate an electron from (a) $0.500c$ to $0.900c$ and (b) $0.900c$ to $0.990c$.

33. A cube of steel has a volume of 1.00 cm^3 and a mass of 8.00 g when at rest on Earth. If this cube is now given a speed $v = 0.900c$, what is its density as measured by a stationary observer? Note that relativistic density is E_R/c^2V.

34. An unstable particle with a mass equal to 3.34×10^{-27} kg is initially at rest. The particle decays into two fragments that fly off with velocities of $0.987c$ and $-0.868c$, respectively. Find the masses of the fragments. [*Hint:* Conserve both mass–energy and momentum.]

Section 26.10 Pair Production and Annihilation

35. How much total kinetic energy will an electron–positron pair have if produced by a photon of energy 3.00 MeV?

36. If an electron–positron pair with a total kinetic energy of 2.50 MeV is produced, find (a) the energy of the photon that produced the pair and (b) its frequency.

37. **Physics⊗Now™** Two photons are produced when a proton and an antiproton annihilate each other. What are the minimum frequency and the corresponding wavelength of each photon?

38. An electron moving at a speed of $0.60c$ collides head-on with a positron also moving at $0.60c$. Determine the energy and momentum of each photon produced in the process.

ADDITIONAL PROBLEMS

39. What is the speed of a proton that has been accelerated from rest through a difference of potential of (a) 500 V and (b) 5.00×10^8 V?

40. An electron has a total energy equal to five times its rest energy. (a) What is its momentum? (b) Repeat for a proton.

41. What is the momentum (in units of MeV/c) of an electron with a kinetic energy of 1.00 MeV?

42. An astronomer on Earth observes a meteoroid in the southern sky approaching Earth at a speed of $0.800c$. At the time of its discovery, the meteoroid is 20.0 lightyears from Earth. Calculate (a) the time required for the meteoroid to reach Earth, as measured by Earthbound astronomer, (b) this time as measured by a tourist on the meteoroid, and (c) the distance to Earth as measured by the tourist.

43. Ted and Mary are playing a game of catch in frame S', which is moving with a speed of $0.60c$; Jim in frame S is watching (Fig. P26.43, page 872). Ted throws the ball to Mary with a speed of $0.80c$ (according to Ted) and their separation (measured in S') is 1.8×10^{12} m. (a) According to Mary, how fast is the ball moving? (b) According to Mary, how long will it take the ball to reach her? (c) According to Jim, how far apart are Ted and Mary and how fast is the ball moving?

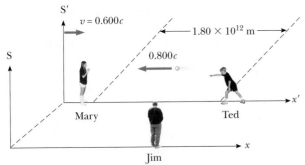

Figure P26.43

44. An alarm clock is set to sound in 10 h. At $t = 0$, the clock is placed in a spaceship moving with a speed of $0.75c$ (relative to Earth). What distance, as determined by an Earth observer, does the spaceship travel before the alarm clock sounds?

45. A radioactive nucleus moves with a speed v relative to a laboratory observer. The nucleus emits an electron in the positive x-direction with a speed of $0.70c$ relative to the decaying nucleus and a speed of $0.85c$ in the $+x$-direction relative to the laboratory observer. What is the value of v?

46. A certain quasar recedes from Earth at $v = 0.870c$. A jet of material ejected from the quasar back toward Earth moves at $0.550c$ relative to the quasar. Find the speed of the ejected material relative to Earth.

47. An astronaut wishes to visit the Andromeda galaxy, making a one-way trip that will take 30.0 years in the spaceship's frame of reference. Assume that the galaxy is 2.00 million light years away and that his speed is constant. (a) How fast must he travel relative to Earth? (b) What will be the kinetic energy of his spacecraft, which has a mass of 1.00×10^6 kg? (c) What is the cost of this energy if it is purchased at a typical consumer price for electric energy, 13.0 cents per kWh? The following approximation will prove useful:

$$\frac{1}{\sqrt{1+x}} \approx 1 - \frac{x}{2} \quad \text{for } x \ll 1$$

48. The cosmic rays of highest energy are protons that have kinetic energy on the order of 10^{13} MeV. (a) How long would it take a proton of this energy to travel across the Milky Way galaxy, having a diameter $\sim 10^5$ light years, as measured in the proton's frame? (b) From the point of view of the proton, how many kilometers across is the galaxy?

49. A spaceship of proper length 300 m takes 0.75 μs to pass an Earth observer. Determine the speed of this spaceship as measured by the Earth observer.

50. Find the kinetic energy of a 78.0-kg spacecraft launched out of the solar system with speed 106 km/s by using (a) the classical equation $KE = \frac{1}{2}mv^2$ and (b) the relativistic equation. You will need to use the approximation

$$\frac{1}{\sqrt{1-x}} \approx 1 + \frac{x}{2} \quad \text{for } x \ll 1$$

51. **Physics⊗Now™** An alien spaceship traveling $0.600c$ toward Earth launches a landing craft with an advance guard of purchasing agents. The lander travels in the same direction with a velocity of $0.800c$ relative to the spaceship. As observed on Earth, the spaceship is 0.200 light year from Earth when the lander is launched. (a) With what velocity is the lander observed to be approaching by observers on Earth? (b) What is the distance to Earth at the time of lander launch, as observed by the aliens on the mother ship? (c) How long does it take the lander to reach Earth, as observed by the aliens on the mother ship? (d) If the lander has a mass of 4.00×10^5 kg, what is its kinetic energy as observed in Earth reference frame?

52. (a) Show that a potential difference of 1.02×10^6 V would be sufficient to give an electron a speed equal to twice the speed of light if Newtonian mechanics remained valid at high speeds. (b) What speed would an electron actually acquire in falling through a potential difference of 1.02×10^6 V?

53. The muon is an unstable particle that spontaneously decays into an electron and two neutrinos. In a reference frame in which the muons are stationary, if the number of muons at $t = 0$ is N_0, the number at time t is given by $N = N_0 e^{-t/\tau}$, where τ is the mean lifetime, equal to 2.2 μs. Suppose that the muons move at a speed of $0.95c$ and that there are 5.0×10^4 muons at $t = 0$. (a) What is the observed lifetime of the muons? (b) How many muons remain after traveling a distance of 3.0 km?

54. An observer in a rocket moves toward a mirror at speed v relative to the reference frame labeled by S in Figure P26.54. The mirror is stationary with respect to S. A light pulse emitted by the rocket travels toward the mirror and is reflected back to the rocket. The front of the rocket is a distance d from the mirror (as measured by observers in S) at the moment the light pulse leaves the rocket. What is the total travel time of the pulse as measured by observers in (a) the S frame and (b) the front of the rocket?

Figure P26.54

55. A physics professor on Earth gives a final exam to her students, who are on a rocket ship traveling at speed v with respect to Earth. The moment the ship passes the professor, she signals the start of the exam. If she wishes her students to have T_0 (rocket time) to complete the exam, show that she should wait a time

$$T = T_0 \sqrt{\frac{1 - v/c}{1 + v/c}}$$

(Earth time) before sending a light signal telling them to stop. [*Hint:* Remember that it takes some time for the second light signal to travel from the professor to the students.]

56. Imagine that the entire Sun collapses to a sphere of radius R_g such that the work required to remove a small mass m from the surface would be equal to its rest energy mc^2. This radius is called the *gravitational radius* for the Sun. Find R_g. (It is believed that the ultimate fate of very massive stars is to collapse beyond their gravitational radii into black holes.)

57. A rod of length L_0 moves with a speed of v along the horizontal direction. The rod makes an angle of θ_0 with respect to the axis of a coordinate system moving with the rod. (a) Show that the length of the rod, as measured by a stationary observer, is given by

$$L = L_0 \left[1 - \left(\frac{v^2}{c^2} \right) \cos^2 \theta_0 \right]^{1/2}$$

(b) Show that the angle the rod makes with the axis, as seen by the stationary observer, is given by the expression $\tan \theta = \gamma \tan \theta_0$. These results demonstrate that the rod is both contracted and rotated. (Take the lower end of the rod to be at the origin of the moving coordinate system.)

58. The speed limit on a certain roadway is 90.0 km/h. Suppose that speeding fines are made proportional to the amount by which a vehicle's momentum exceeds the momentum the vehicle would have when traveling at the speed limit. The fine for driving at 190 km/h (that is, 100 km/h over the speed limit) is $80.0. What then will be the fine for traveling (a) at 1 090 km/h? (b) at 1 000 000 090 km/h?

59. The identical twins Speedo and Goslo join a migration from the Earth to Planet X. It is 20.0 lightyears away in a reference frame in which both planets are at rest. The twins, of the same age, depart at the same time on different spacecraft. Speedo's craft travels steadily at $0.950c$, Goslo's at $0.750c$. Calculate the age difference between the twins after Goslo's spacecraft lands on Planet X. Which twin is the older?

Color-enhanced scanning electron micrograph of the storage mite *Lepidoglyphus destructor.* These common mites grow to 0.75 mm and feed on molds, flour, and rice. They thrive at 25°C and high humidity and can trigger allergies.

© Eye of Science/Science Source/Photo Researchers, Inc.

Quantum Physics

Although many problems were resolved by the theory of relativity in the early part of the 20th century, many other problems remained unsolved. Attempts to explain the behavior of matter on the atomic level with the laws of classical physics were consistently unsuccessful. Various phenomena, such as the electromagnetic radiation emitted by a heated object (blackbody radiation), the emission of electrons by illuminated metals (the photoelectric effect), and the emission of sharp spectral lines by gas atoms in an electric discharge tube, couldn't be understood within the framework of classical physics. Between 1900 and 1930, however, a modern version of mechanics called *quantum mechanics* or *wave mechanics* was highly successful in explaining the behavior of atoms, molecules, and nuclei.

The earliest ideas of quantum theory were introduced by Planck, and most of the subsequent mathematical developments, interpretations, and improvements were made by a number of distinguished physicists, including Einstein, Bohr, Schrödinger, de Broglie, Heisenberg, Born, and Dirac. In this chapter we introduce the underlying ideas of quantum theory and the wave–particle nature of matter, and discuss some simple applications of quantum theory, including the photoelectric effect, the Compton effect, and x-rays.

27.1 BLACKBODY RADIATION AND PLANCK'S HYPOTHESIS

An object at any temperature emits electromagnetic radiation, called **thermal radiation**. Stefan's law, discussed in Section 11.5, describes the total power radiated. The spectrum of the radiation depends on the temperature and properties of the object. At low temperatures, the wavelengths of the thermal radiation are mainly in the infrared region and hence not observable by the eye. As the temperature of an object increases, the object eventually begins to glow red. At sufficiently high temperatures, it appears to be white, as in the glow of the hot tungsten filament of a lightbulb. A careful study of thermal radiation shows that it consists of a continuous distribution of wavelengths from the infrared, visible, and ultraviolet portions of the spectrum.

From a classical viewpoint, thermal radiation originates from accelerated charged particles near the surface of an object; such charges emit radiation, much as small antennas do. The thermally agitated charges can have a distribution of frequencies, which accounts for the continuous spectrum of radiation emitted by the object. By the end of the 19th century, it had become apparent that the classical theory of thermal radiation was inadequate. The basic problem was in understanding the observed distribution energy as a function of wavelength in the radiation emitted by a blackbody. By definition, a blackbody is an ideal system that absorbs *all* radiation incident on it. A good approximation of a blackbody is a small hole leading to the inside of a hollow object, as shown in Figure 27.1. The nature of the radiation emitted through the small hole leading to the cavity depends *only on the temperature* of the cavity walls, and not at all on the material composition of the object, its shape, or other factors.

Experimental data for the distribution of energy in blackbody radiation at three temperatures are shown in Active Figure 27.2 (page 876). The radiated energy varies with wavelength and temperature. As the temperature of the blackbody increases, the total amount of energy (area under the curve) it emits increases. Also, with increasing temperature, the peak of the distribution shifts to shorter wavelengths. This shift obeys the following relationship, called **Wien's displacement law**,

$$\lambda_{max} T = 0.289\,8 \times 10^{-2}\ \text{m} \cdot \text{K} \qquad [27.1]$$

where λ_{max} is the wavelength at which the curve peaks and T is the absolute temperature of the object emitting the radiation.

Figure 27.1 An opening in the cavity of a body is a good approximation of a blackbody. As light enters the cavity through the small opening, part is reflected and part is absorbed on each reflection from the interior walls. After many reflections, essentially all of the incident energy is absorbed.

TIP 27.1   Expect to Be Confused

Your life experiences take place in the macroscopic world, where quantum effects are not evident. Quantum effects can be even more bizarre than relativistic effects, but don't despair: confusion is normal and expected. As the Nobel prize-winning physicist Richard Feynman once said, "Nobody understands quantum mechanics."

Applying Physics 27.1   Star Colors

If you look carefully at stars in the night sky, you can distinguish three main colors: red, white, and blue. What causes these particular colors?

Explanation These colors result from the different surface temperatures of stars. A relatively cool star, with a surface temperature of 3 000 K, has a radiation curve like the middle curve in Active Figure 27.2 (page 876). The peak in this curve is above the visible wavelengths, 0.4 μm–0.7 μm, beyond the wavelength of red light, so significantly more radiation is emitted within the visible range at the red end than the blue end of the spectrum. Consequently, the star appears reddish in color, similar to the red glow from the burner of an electric stove.

A hotter star has a radiation curve more like the upper curve in Active Figure 27.2. In this case, the star emits significant radiation throughout the visible range, and the combination of all colors causes the star to look white. This is the case with our own Sun, with a surface temperature of 5 800 K. For very hot stars, the peak can be shifted so far below the visible range that significantly more blue radiation is emitted than red, so the star appears bluish in color.

EXAMPLE 27.1   Thermal Radiation from the Human Body

Goal Apply Wien's law.

Problem The temperature of the skin is approximately 35.0°C. At what wavelength does the radiation emitted from the skin reach its peak?

Strategy This is a matter of substitution into Wien's law, Equation 27.1.

Solution

Apply Wien's displacement law:

$$\lambda_{max} T = 0.289\,8 \times 10^{-2}\ \text{m} \cdot \text{K}$$

Solve for λ_{max}, noting that 35.0°C corresponds to an absolute temperature of 308 K:

$$\lambda_{max} = \frac{0.289\,8 \times 10^{-2}\,\text{m} \cdot \text{K}}{308\,\text{K}} = 9.41\,\mu\text{m}$$

Remark   This radiation is in the infrared region of the spectrum.

Exercise 27.1

(a) Find the wavelength corresponding to the peak of the radiation curve for the heating element of an electric oven at a temperature of 1.20×10^3 K. (Note that although this radiation peak lies in the infrared, there is enough visible radiation at this temperature to give the element a red glow.) (b) Suppose the peak in the radiation curve of a star is 780 nm. Calculate the temperature of the surface of the star.

Answers   (a) 2.42 μm;   (b) 3 700 K

ACTIVE FIGURE 27.2
Intensity of blackbody radiation versus wavelength at three different temperatures. Note that the total radiation emitted (the area under a curve) increases with increasing temperature.

Physics⊗Now™
Log into PhysicsNow at **www.cp7e.com** and go to Active Figure 27.2, where you can adjust the temperature of the blackbody and study the radiation emitted from it.

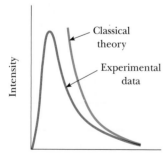

ACTIVE FIGURE 27.3
Comparison of experimental data with the classical theory of blackbody radiation. Planck's theory matches the experimental data perfectly.

Physics⊗Now™
Log into PhysicsNow at **www.cp7e.com** and go to Active Figure 27.3, where you can investigate the discrete energies emitted in the Planck model.

Attempts to use classical ideas to explain the shapes of the curves shown in Active Figure 27.2 failed. Active Figure 27.3 shows an experimental plot of the blackbody radiation spectrum (red curve), together with the theoretical picture of what this curve should look like based on classical theories (blue curve). At long wavelengths, classical theory is in good agreement with the experimental data. At short wavelengths, however, major disagreement exists between classical theory and experiment. As λ approaches zero, classical theory predicts that the amount of energy being radiated should increase. In fact, the theory erroneously predicts that the intensity should be infinite, when the experimental data shows it should approach zero. This contradiction is called the **ultraviolet catastrophe**, because theory and experiment disagree strongly in the short-wavelength, ultraviolet region of the spectrum.

In 1900 Planck developed a formula for blackbody radiation that was in complete agreement with experiments at all wavelengths, leading to a curve shown by the red line in Active Figure 27.3. Planck hypothesized that blackbody radiation was produced by submicroscopic charged oscillators, which he called *resonators*. He assumed that the walls of a glowing cavity were composed of billions of these resonators, although their exact nature was unknown. The resonators were allowed to have only certain discrete energies E_n, given by

$$E_n = nhf \qquad \text{[27.2]}$$

where n is a positive integer called a **quantum number**, f is the frequency of vibration of the resonator, and h is a constant known as **Planck's constant**, which has the value

$$h = 6.626 \times 10^{-34}\,\text{J} \cdot \text{s} \qquad \text{[27.3]}$$

Because the energy of each resonator can have only discrete values given by Equation 27.2, we say the energy is *quantized*. Each discrete energy value represents a different *quantum state*, with each value of n representing a specific quantum state. (When the resonator is in the $n = 1$ quantum state, its energy is hf; when it is in the $n = 2$ quantum state, its energy is $2\,hf$; and so on.)

The key point in Planck's theory is the assumption of quantized energy states. This is a radical departure from classical physics, the "quantum leap" that led to a totally new understanding of nature. It's shocking: it's like saying a pitched baseball can have only a fixed number of different speeds, and no speeds in between those fixed values. When Planck presented his theory, most scientists (including Planck!) didn't consider the quantum concept to be realistic; however, subsequent developments showed that a theory based on the quantum concept (rather than on classical concepts) had to be used to explain a number of other phenomena at the atomic level.

EXAMPLE 27.2   The Quantized Macroscopic Oscillator

Goal Contrast the classical and quantum oscillator.

Problem A 2.00-kg mass is attached to a spring having force constant $k = 25.0$ N/m and negligible mass. The spring is stretched 0.400 m from its equilibrium position and released. **(a)** Find the total energy and frequency of oscillation according to classical calculations. **(b)** Assume that Planck's law of energy quantization applies to any oscillator, atomic or large scale, and find the quantum number n for this system. **(c)** How much energy would be carried away in a one-quantum change?

Strategy We are given the spring constant and the oscillation amplitude, so we can find the total energy with the conservation of mechanical energy, using the point of maximum displacement. Equation 13.10 gives the frequency of a spring system, which can then be used with the quantum hypothesis, Equation 27.2, to obtain the value of the quantum number n. In part (c), a single quantum of energy is always equal to Planck's constant times the frequency.

Solution

(a) Find the energy and the classical frequency of the system.

Substitute into the classical energy when the block is at maximum amplitude:

$$E = \frac{1}{2} kA^2 = \frac{1}{2}(25.0 \text{ N/m})(0.400 \text{ m})^2 = \boxed{2.00 \text{ J}}$$

Compute the frequency of oscillation:

$$f = \frac{1}{2\pi}\sqrt{\frac{k}{m}} = \frac{1}{2\pi}\sqrt{\frac{25.0 \text{ N/m}}{2.00 \text{ kg}}} = \boxed{0.563 \text{ Hz}}$$

(b) Calculate the value of the quantum number n corresponding to the classical energy.

Solve Equation 27.2 for n:

$$E_n = nhf \quad \rightarrow \quad n = \frac{E_n}{hf}$$

$$n = \frac{2.00 \text{ J}}{(6.63 \times 10^{-34} \text{ J·s})(0.563 \text{ Hz})} = \boxed{5.36 \times 10^{33}}$$

(c) How much energy would be carried away in a one-quantum change?

Compute the difference between two adjacent energy levels and substitute:

$$\Delta E = E_{n+1} - E_n = hf$$

$$= (6.63 \times 10^{-34} \text{ J·s})(0.563 \text{ Hz}) = \boxed{3.73 \times 10^{-34} \text{ J}}$$

Remarks The energy carried away by a one-quantum change is such a small fraction of the total energy of the oscillator that we couldn't expect to measure it. Consequently, the energy of an object–spring system decreases by such small quantum transitions that the decrease in energy appears to be continuous. Quantum effects become important and measurable only on the submicroscopic level of atoms and molecules.

Exercise 27.2

A pendulum has a length of 1.50 m. Treating it as a quantum system, calculate (a) its frequency in the presence of Earth's gravitational field and (b) the energy carried away in a change of energy levels from $n = 3$ to $n = 1$.

Answers (a) 0.407 Hz   (b) 5.39×10^{-34} J

27.2   THE PHOTOELECTRIC EFFECT AND THE PARTICLE THEORY OF LIGHT

In the latter part of the 19th century, experiments showed that light incident on certain metallic surfaces caused the emission of electrons are emitted from the surfaces. This phenomenon is known as the **photoelectric effect**, and the emitted electrons are called **photoelectrons**. The first discovery of this phenomenon was made by Hertz, who was also the first to produce the electromagnetic waves predicted by Maxwell.

© Bettmann/CORBIS

MAX PLANCK, German Physicist, (1858–1947)

Planck introduced the concept of a "quantum of action" (Planck's constant *h*) in an attempt to explain the spectral distribution of blackbody radiation, which laid the foundations for quantum theory. In 1918, he was awarded the Nobel Prize for this discovery of the quantized nature of energy.

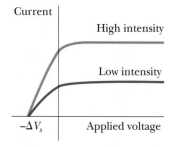

FIGURE 27.5 Photoelectric current versus applied potential difference for two light intensities. The current increases with intensity, but reaches a saturation level for large values of ΔV. At voltages equal to or less than $-\Delta V_s$, where ΔV_s is the stopping potential, the current is zero.

ACTIVE FIGURE 27.4
A circuit diagram for studying the photoelectric effect. When light strikes plate E (the emitter), photoelectrons are ejected from the plate. Electrons moving from plate E to plate C (the collector) create a current in the circuit, registered at the ammeter, A.

Physics⚛Now™
Log into PhysicsNow at **www.cp7e.com** and go to Active Figure 27.4, where you can observe the motion of electrons for various frequencies and voltages.

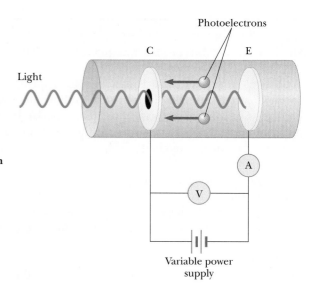

Active Figure 27.4 is a schematic diagram of a photoelectric effect apparatus. An evacuated glass tube known as a photocell contains a metal plate E (the emitter) connected to the negative terminal of a variable power supply. Another metal plate, C (the collector), is maintained at a positive potential by the power supply. When the tube is kept in the dark, the ammeter reads zero, indicating that there is no current in the circuit. However, when plate E is illuminated by light having a wavelength shorter than some particular wavelength that depends on the material used to make plate E, a current is detected by the ammeter, indicating a flow of charges across the gap between E and C. This current arises from photoelectrons emitted from the negative plate E and collected at the positive plate C.

Figure 27.5 is a plot of the photoelectric current versus the potential difference ΔV between E and C for two light intensities. At large values of ΔV, the current reaches a maximum value. In addition, the current increases as the incident light intensity increases, as you might expect. Finally, when ΔV is negative—that is, when the power supply in the circuit is reversed to make E positive and C negative—the current drops to a low value because most of the emitted photoelectrons are repelled by the now negative plate C. In this situation, only those electrons having a kinetic energy greater than the magnitude of $e\Delta V$ reach C, where e is the charge on the electron.

When ΔV is equal to or more negative than $-\Delta V_s$, the **stopping potential**, no electrons reach C and the current is zero. The stopping potential is *independent* of the radiation intensity. The maximum kinetic energy of the photoelectrons is related to the stopping potential through the relationship

$$KE_{max} = e\Delta V_s \qquad [27.4]$$

Several features of the photoelectric effect can't be explained with classical physics or with the wave theory of light:

- No electrons are emitted if the incident light frequency falls below some **cutoff frequency** f_c, which is characteristic of the material being illuminated. This is inconsistent with the wave theory, which predicts that the photoelectric effect should occur at *any* frequency, provided the light intensity is sufficiently high.
- The maximum kinetic energy of the photoelectrons is independent of light intensity. According to wave theory, light of higher intensity should carry more energy into the metal per unit time and therefore eject photoelectrons having higher kinetic energies.
- The maximum kinetic energy of the photoelectrons increases with increasing light frequency. The wave theory predicts no relationship between photoelectron energy and incident light frequency.
- Electrons are emitted from the surface almost instantaneously (less than 10^{-9} s after the surface is illuminated), even at low light intensities. Classically, we expect the photoelectrons to require some time to absorb the incident radiation before they acquire enough kinetic energy to escape from the metal.

A successful explanation of the photoelectric effect was given by Einstein in 1905, the same year he published his special theory of relativity. As part of a general paper on electromagnetic radiation, for which he received the Nobel Prize in 1921, Einstein extended Planck's concept of quantization to electromagnetic waves. He suggested that a tiny packet of light energy or **photon** would be emitted when a quantized oscillator made a jump from an energy state $E_n = nhf$ to the next lower state $E_{n-1} = (n-1)hf$. Conservation of energy would require the decrease in oscillator energy, hf, to be equal to the photon's energy E, so that

$$E = hf \qquad [27.5]$$

◀ Energy of a photon

where h is Planck's constant and f is the frequency of the light, which is equal to the frequency of Planck's oscillator.

The key point here is that the light energy lost by the emitter, hf, stays sharply localized in a tiny packet or particle called a photon. In Einstein's model, a photon is so localized that it can give *all* its energy hf to a single electron in the metal. According to Einstein, the maximum kinetic energy for these liberated photoelectrons is

$$KE_{max} = hf - \phi \qquad [27.6]$$

◀ Photoelectric effect equation

where ϕ is called the **work function** of the metal. The work function, which represents the minimum energy with which an electron is bound in the metal, is on the order of a few electron volts. Table 27.1 lists work functions for various metals.

With the photon theory of light, we can explain the previously mentioned features of the photoelectric effect that cannot be understood using concepts of classical physics:

- Photoelectrons are created by absorption of a single photon, so the energy of that photon must be greater than or equal to the work function, else no photoelectrons will be produced. This explains the cutoff frequency.
- From Equation 27.6, KE_{max} depends only on the frequency of the light and the value of the work function. Light intensity is immaterial, because absorption of a single photon is responsible for the electron's change in kinetic energy.
- Equation 27.6 is linear in the frequency, so KE_{max} increases with increasing frequency.
- Electrons are emitted almost instantaneously, regardless of intensity, because the light energy is concentrated in packets rather than spread out in waves. If the frequency is high enough, no time is needed for the electron to gradually acquire sufficient energy to escape the metal.

Experimentally, a linear relationship is observed between f and KE_{max}, as sketched in Figure 27.6. The intercept on the horizontal axis, corresponding to $KE_{max} = 0$, gives the cutoff frequency below which no photoelectrons are emitted, regardless of light intensity. The cutoff *wavelength* λ_c can be derived from Equation 27.6:

$$KE_{max} = hf_c - \phi = 0 \quad \rightarrow \quad h\frac{c}{\lambda_c} - \phi = 0$$

$$\lambda_c = \frac{hc}{\phi} \qquad [27.7]$$

where c is the speed of light. Wavelengths *greater* than λ_c incident on a material with work function ϕ don't result in the emission of photoelectrons.

TABLE 27.1

Work Functions of Selected Metals

Metal	ϕ (eV)
Na	2.46
Al	4.08
Cu	4.70
Zn	4.31
Ag	4.73
Pt	6.35
Pb	4.14
Fe	4.50

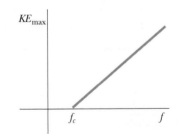

Figure 27.6 A sketch of KE_{max} versus the frequency of incident light for photoelectrons in a typical photoelectric effect experiment. Photons with frequency less than f_c don't have sufficient energy to eject an electron from the metal.

INTERACTIVE EXAMPLE 27.3   Photoelectrons from Sodium

Goal   Understand the quantization of light and its role in the photoelectric effect.

Problem   A sodium surface is illuminated with light of wavelength 0.300 μm. The work function for sodium is 2.46 eV. **(a)** Calculate the energy of each photon in electron volts, **(b)** the maximum kinetic energy of the ejected photoelectrons, and **(c)** the cutoff wavelength for sodium.

Strategy Parts (a), (b), and (c) require substitution of values into Equations 27.5, 27.6, and 27.7, respectively.

Solution
(a) Calculate the energy of each photon.

Obtain the frequency from the wavelength:

$$c = f\lambda \rightarrow f = \frac{c}{\lambda} = \frac{3.00 \times 10^8 \text{ m/s}}{0.300 \times 10^{-6} \text{ m}}$$

$$f = 1.00 \times 10^{15} \text{ Hz}$$

Use Equation 27.5 to calculate the photon's energy:

$$E = hf = (6.63 \times 10^{-34} \text{ J} \cdot \text{s})(1.00 \times 10^{15} \text{ Hz})$$
$$= 6.63 \times 10^{-19} \text{ J}$$
$$= (6.63 \times 10^{-19} \text{ J})\left(\frac{1.00 \text{ eV}}{1.60 \times 10^{-19} \text{ J}}\right) = \boxed{4.14 \text{ eV}}$$

(b) Find the maximum kinetic energy of the photoelectrons.

Substitute into Equation 27.6:

$$KE_{max} = hf - \phi = 4.14 \text{ eV} - 2.46 \text{ eV} = \boxed{1.68 \text{ eV}}$$

(c) Compute the cutoff wavelength.

Convert ϕ from electron volts to joules:

$$\phi = 2.46 \text{ eV} = (2.46 \text{ eV})(1.60 \times 10^{-19} \text{ J/eV})$$
$$= 3.94 \times 10^{-19} \text{ J}$$

Find the cutoff wavelength using Equation 27.7.

$$\lambda_c = \frac{hc}{\phi} = \frac{(6.63 \times 10^{-34} \text{ J} \cdot \text{s})(3.00 \times 10^8 \text{ m/s})}{3.94 \times 10^{-19} \text{ J}}$$
$$= 5.05 \times 10^{-7} \text{ m} = \boxed{505 \text{ nm}}$$

Remark The cutoff wavelength is in the yellow-green region of the visible spectrum.

Exercise 27.3
(a) What minimum-frequency light will eject photoelectrons from a copper surface? (b) If this frequency is tripled, find the maximum kinetic energy (in eV) of the resulting photoelectrons. (Answer in eV.)

Answers (a) 1.13×10^{15} Hz (b) 9.40 eV

Physics⊗Now™ Investigate the photoelectric effect for different materials and different wavelengths of light by logging into PhysicsNow at **www.cp7e.com** and going to Interactive Example 27.3.

Photocells

APPLICATION

Photocells

The photoelectric effect has many interesting applications using a device called the *photocell*. The photocell shown in Active Figure 27.4 produces a current in the circuit when light of sufficiently high frequency falls on the cell, but it doesn't allow a current in the dark. This device is used in streetlights: a photoelectric control unit in the base of the light activates a switch that turns off the streetlight when ambient light strikes it. Many garage-door systems and elevators use a light beam and a photocell as a safety feature in their design. When the light beam strikes the photocell, the electric current generated is sufficiently large to maintain a closed circuit. When an object or a person blocks the light beam, the current is interrupted, which signals the door to open.

27.3 X-RAYS

In 1895 at the University of Wurzburg, Wilhelm Roentgen (1845–1923) was studying electrical discharges in low-pressure gases when he noticed that a fluorescent screen glowed even when placed several meters from the gas discharge tube and even when black cardboard was placed between the tube and the screen. He concluded that the effect was caused by a mysterious type of radiation, which he called **x-rays** because of their unknown nature. Subsequent study showed that these rays traveled at or near the speed of light and that they couldn't be deflected by either electric or magnetic fields. This last fact indicated that x-rays did not consist of beams of charged particles, although the possibility that they were beams of uncharged particles remained.

In 1912 Max von Laue (1879–1960) suggested that if x-rays were electromagnetic waves with very short wavelengths, it should be possible to diffract them by using the regular atomic spacings of a crystal lattice as a diffraction grating, just as visible light is diffracted by a ruled grating. Shortly thereafter, researchers demonstrated that such a diffraction pattern could be observed, similar to that shown in Figure 27.7 for NaCl. The wavelengths of the x-rays were then determined from the diffraction data and the known values of the spacing between atoms in the crystal. X-ray diffraction has proved to be an invaluable technique for understanding the structure of matter (as discussed in more detail in the next section).

Typical x-ray wavelengths are about 0.1 nm, which is on the order of the atomic spacing in a solid. We now know that x-rays are a part of the electromagnetic spectrum, characterized by frequencies higher than those of ultraviolet radiation and having the ability to penetrate most materials with relative ease.

X-rays are produced when high-speed electrons are suddenly slowed down—for example, when a metal target is struck by electrons that have been accelerated through a potential difference of several thousand volts. Figure 27.8a shows a schematic diagram of an x-ray tube. A current in the filament causes electrons to be emitted, and these freed electrons are accelerated toward a dense metal target, such as tungsten, which is held at a higher potential than the filament.

Figure 27.9 represents a plot of x-ray intensity versus wavelength for the spectrum of radiation emitted by an x-ray tube. Note that the spectrum has are two distinct components. One component is a continuous broad spectrum that depends on the voltage applied to the tube. Superimposed on this component is a series of sharp, intense lines that depend on the nature of the target material. The accelerating voltage must exceed a certain value, called the **threshold voltage**, in order to observe these sharp lines, which represent radiation emitted by the target atoms as their electrons undergo rearrangements. We will discuss this further in Chapter 28. The continuous radiation is sometimes called **bremsstrahlung**, a German word meaning "braking radiation," because electrons emit radiation when they undergo an acceleration inside the target.

Figure 27.10 (page 882) illustrates how x-rays are produced when an electron passes near a charged target nucleus. As the electron passes close to a positively

Figure 27.7   X-ray diffraction pattern of NaCl.

(a)

(b)

Figure 27.8   (a) Diagram of an x-ray tube. (b) Photograph of an x-ray tube.

Figure 27.9   The x-ray spectrum of a metal target consists of a broad continuous spectrum plus a number of sharp lines, which are due to *characteristic x-rays*. The data shown were obtained when 35-keV electrons bombarded a molybdenum target. Note that 1 pm = 10^{-12} m = 10^{-3} nm.

Figure 27.10 An electron passing near a charged target atom experiences an acceleration, and a photon is emitted in the process.

charged nucleus contained in the target material, it is deflected from its path because of its electrical attraction to the nucleus; hence, it undergoes an acceleration. An analysis from classical physics shows that any charged particle will emit electromagnetic radiation when it is accelerated. (An example of this phenomenon is the production of electromagnetic waves by accelerated charges in a radio antenna, as described in Chapter 21.) According to quantum theory, this radiation must appear in the form of photons. Because the radiated photon shown in Figure 27.10 carries energy, the electron must lose kinetic energy because of its encounter with the target nucleus. An extreme example would consist of the electron losing all of its energy in a single collision. In this case, the initial energy of the electron ($e\Delta V$) is transformed completely into the energy of the photon (hf_{max}). In equation form,

$$e\Delta V = hf_{max} = \frac{hc}{\lambda_{min}}$$ **[27.8]**

where $e\Delta V$ is the energy of the electron after it has been accelerated through a potential difference of ΔV volts and e is the charge on the electron. This says that the shortest wavelength radiation that can be produced is

$$\lambda_{min} = \frac{hc}{e\Delta V}$$ **[27.9]**

The reason that not all the radiation produced has this particular wavelength is because many of the electrons aren't stopped in a single collision. This results in the production of the continuous spectrum of wavelengths.

Interesting insights into the process of painting and revising a masterpiece are being revealed by x-rays. Long wavelength x-rays are absorbed in varying degrees by some paints, such as those having lead, cadmium, chromium, or cobalt as a base. The x-ray interactions with the paints give contrast, because the different elements in the paints have different electron densities. Also, thicker layers will absorb more than thin layers. To examine a painting by an old master, a film is placed behind it while it is x-rayed from the front. Ghost outlines of earlier paintings and earlier forms of the final masterpiece are sometimes revealed when the film is developed.

APPLICATION

Using X-Rays to Study the Work of Master Painters

EXAMPLE 27.4   An X-Ray Tube

Goal   Calculate the minimum x-ray wavelength due to accelerated electrons.

Problem   Medical x-ray machines typically operate at a potential difference of 1.00×10^5 V. Calculate the minimum wavelength their x-ray tubes produce when electrons are accelerated through this potential difference.

Strategy   The minimum wavelength corresponds to the most energetic photons. Substitute the given potential difference into Equation 27.9.

Solution
Substitute into Equation 27.9:

$$\lambda_{min} = \frac{hc}{e\Delta V}$$

$$= \frac{(6.63 \times 10^{-34}\,\text{J}\cdot\text{s})(3.00 \times 10^8\,\text{m/s})}{(1.60 \times 10^{-19}\,\text{C})(1.00 \times 10^5\,\text{V})}$$

$$= 1.24 \times 10^{-11}\,\text{m}$$

Remarks   X-ray tubes generally operate with half the voltage with respect to Earth, $+50\,000$ V, applied to the anode, and the other half, $-50\,000$ V, applied to the cathode. This lengthens tube lifetime by reducing the probability of voltage breakthroughs.

Exercise 27.4

What potential difference would be necessary to produce gamma rays with wavelength 1.00×10^{-15} m? This wavelength is about the same size as the diameter of a proton.

Solution 1.24×10^9 V

27.4   DIFFRACTION OF X-RAYS BY CRYSTALS

In Chapter 24 we described how a diffraction grating could be used to measure the wavelength of light. In principle, the wavelength of *any* electromagnetic wave can be measured if a grating having a suitable line spacing can be found. The spacing between lines must be approximately equal to the wavelength of the radiation to be measured. X-rays are electromagnetic waves with wavelengths on the order of 0.1 nm. It would be impossible to construct a grating with such a small spacing. However, as noted in the previous section, Max von Laue suggested that the regular array of atoms in a crystal could act as a three-dimensional grating for observing the diffraction of x-rays.

One experimental arrangement for observing x-ray diffraction is shown in Figure 27.11. A narrow beam of x-rays with a continuous wavelength range is incident on a crystal such as sodium chloride. The diffracted radiation is very intense in certain directions, corresponding to constructive interference from waves reflected from layers of atoms in the crystal. The diffracted radiation is detected by a photographic film and forms an array of spots known as a *Laue pattern*. The crystal structure is determined by analyzing the positions and intensities of the various spots in the pattern.

The arrangement of atoms in a crystal of NaCl is shown in Figure 27.12. The smaller red spheres represent Na$^+$ ions, and the larger blue spheres represent Cl$^-$ ions. The spacing between successive Na$^+$ (or Cl$^-$) ions in this cubic structure, denoted by the symbol a in Figure 27.12, is approximately 0.563 nm.

A careful examination of the NaCl structure shows that the ions lie in various planes. The shaded areas in Figure 27.12 represent one example, in which the atoms lie in equally spaced planes. Now suppose an x-ray beam is incident at grazing angle θ on one of the planes, as in Figure 27.13. The beam can be reflected from both the upper and lower plane of atoms. However, the geometric construction in Figure 27.13 shows that the beam reflected from the lower surface travels farther than the beam reflected from the upper surface by a distance of $2d \sin \theta$. The two portions of the reflected beam will combine to produce constructive interference when this path difference equals some integral multiple of the wavelength λ. The condition for constructive interference is given by

$$2d \sin \theta = m\lambda \qquad (m = 1, 2, 3, \dots) \qquad \text{[27.10]} \qquad \blacktriangleleft \text{Bragg's law}$$

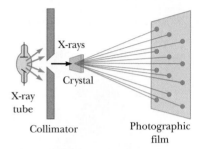

Figure 27.11 Schematic diagram of the technique used to observe the diffraction of x-rays by a single crystal. The array of spots formed on the film by the diffracted beams is called a Laue pattern. (See Fig. 27.7.)

Figure 27.12 A model of the cubic crystalline structure of sodium chloride. The blue spheres represent the Cl$^-$ ions, and the red spheres represent the Na$^+$ ions. The length of the cube edge is $a = 0.563$ nm.

Incident beam   Reflected beam

Upper plane

Lower plane

$d \sin \theta$

Figure 27.13 A two-dimensional depiction of the reflection of an x-ray beam from two parallel crystalline planes separated by a distance d. The beam reflected from the lower plane travels farther than the one reflected from the upper plane by an amount equal to $2d \sin \theta$.

Figure 27.14 An x-ray diffraction photograph of DNA taken by Rosalind Franklin. The cross pattern of spots was a clue that DNA has a helical structure.

|← — 2 nm — →|

Figure 27.15 The double-helix structure of DNA.

This condition is known as **Bragg's law**, after W. L. Bragg (1890–1971), who first derived the relationship. If the wavelength and diffraction angle are measured, Equation 27.10 can be used to calculate the spacing between atomic planes.

The method of x-ray diffraction to determine crystalline structures was thoroughly developed in England by W. H. Bragg and his son W. L. Bragg, who shared a Nobel prize in 1915 for their work. Since then, thousands of crystalline structures have been investigated. Most importantly, the technique of x-ray diffraction has been used to determine the atomic arrangement of complex organic molecules such as proteins. Proteins are large molecules containing thousands of atoms that help to regulate and speed up chemical life processes in cells. Some proteins are amazing catalysts, speeding up the slow room temperature reactions in cells by 17 orders of magnitude. In order to understand this incredible biochemical reactivity, it is important to determine the structure of these intricate molecules.

The main technique used to determine the molecular structure of proteins, DNA, and RNA is x-ray diffraction using x-rays of wavelength of about 1.0 A. This technique allows the experimenter to "see" individual atoms that are separated by about this distance in molecules. Since the biochemical x-ray diffraction sample is prepared in crystal form, the *geometry* (position of the bright spots in space) of the diffraction pattern is determined by the regular three-dimensional crystal lattice arrangement of molecules in the sample. The *intensities* of the bright diffraction spots are determined by the atoms and their electronic distributions in the fundamental building block of the crystal: the unit cell. Using complicated computational techniques, investigators can essentially deduce the molecular structure by matching the observed intensities of diffracted beams with a series of assumed atomic positions that determine the atomic structure and electron density of the molecule. Figure 27.14 shows a classic x-ray diffraction image of DNA made by Rosalind Franklin in 1952.

This and similar x-ray diffraction photos played an important role in the determination of the double-helix structure of DNA by F. H. C. Crick and J. D. Watson in 1953. A model of the famous DNA double helix is shown in Figure 27.15.

EXAMPLE 27.5   X-Ray Diffraction from Calcite

Goal   Understand Bragg's law and apply it to a crystal.

Problem   If the spacing between certain planes in a crystal of calcite ($CaCO_3$) is 0.314 nm, find the grazing angles at which first- and third-order interference will occur for x-rays of wavelength 0.070 0 nm.

Strategy   Solve Bragg's law for $\sin\theta$ and substitute, using the inverse-sine function to obtain the angle.

Solution

Find the grazing angle corresponding to $m = 1$, for first-order interference:

$$\sin\theta = \frac{m\lambda}{2d} = \frac{(0.070\ 0\ \text{nm})}{2(0.314\ \text{nm})} = 0.111$$

$$\theta = \sin^{-1}(0.111) = \boxed{6.37°}$$

Repeat the calculation for third-order interference ($m = 3$):

$$\sin\theta = \frac{m\lambda}{2d} = \frac{3(0.070\ 0\ \text{nm})}{2(0.314\ \text{nm})} = 0.334$$

$$\theta = \sin^{-1}(0.334) = \boxed{19.5°}$$

Remark   Notice there is little difference between this kind of problem and a Young's slit experiment.

Exercise 27.5
X-rays of wavelength 0.060 0 nm are scattered from a crystal with a grazing angle of 11.7°. Assume $m = 1$ for this process. Calculate the spacing between the crystal planes.

Answer   0.148 nm

27.5   THE COMPTON EFFECT

Further justification for the photon nature of light came from an experiment conducted by Arthur H. Compton in 1923. In his experiment, Compton directed an x-ray beam of wavelength λ_0 toward a block of graphite. He found that the scattered x-rays had a slightly longer wavelength λ than the incident x-rays, and hence the energies of the scattered rays were lower. The amount of energy reduction depended on the angle at which the x-rays were scattered. The change in wavelength $\Delta\lambda$ between a scattered x-ray and an incident x-ray is called the **Compton shift**.

In order to explain this effect, Compton assumed that if a photon behaves like a particle, its collision with other particles is similar to a collision between two billiard balls. Hence, the x-ray photon carries both measurable *energy* and *momentum*, and these two quantities must be conserved in a collision. If the incident photon collides with an electron initially at rest, as in Figure 27.16, the photon transfers some of its energy and momentum to the electron. As a consequence, the energy and frequency of the scattered photon are lowered and its wavelength increases. Applying relativistic energy and momentum conservation to the collision described in Figure 27.16, the shift in wavelength of the scattered photon is given by

$$\Delta\lambda = \lambda - \lambda_0 = \frac{h}{m_e c}(1 - \cos\theta) \qquad \textbf{[27.11]}$$

◀ The Compton shift formula

where m_e is the mass of the electron and θ is the angle between the directions of the scattered and incident photons. The quantity $h/m_e c$ is called the **Compton wavelength** and has a value of 0.002 43 nm. The Compton wavelength is very small relative to the wavelengths of visible light, so the shift in wavelength would be difficult to detect if visible light were used. Further, note that the Compton shift depends on the scattering angle θ and not on the wavelength. Experimental results for x-rays scattered from various targets obey Equation 27.11 and strongly support the photon concept.

Quick Quiz 27.1

An x-ray photon is scattered by an electron. The frequency of the scattered photon relative to that of the incident photon (a) increases, (b) decreases, or (c) remains the same.

Courtesy of AIP Niels Bohr Library

ARTHUR HOLLY COMPTON,
American Physicist (1892–1962)

Compton was born in Wooster, Ohio, and he attended Wooster College and Princeton University. He became director of the laboratory at the University of Chicago, where experimental work concerned with sustained chain reactions was conducted. This work was of central importance to the construction of the first atomic bomb. His discovery of the Compton effect and his work with cosmic rays led to his sharing the 1927 Nobel Prize in physics with Charles Wilson.

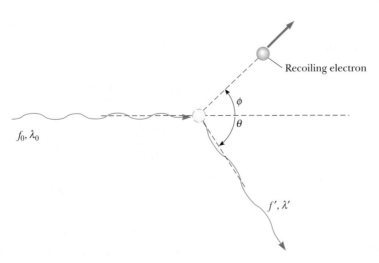

Figure 27.16 Diagram representing Compton scattering of a photon by an electron. The scattered photon has less energy (or a longer wavelength) than the incident photon.

Quick Quiz 27.2

A photon of energy E_0 strikes a free electron, with the scattered photon of energy E moving in the direction opposite that of the incident photon. In this Compton effect interaction, the resulting kinetic energy of the electron is (a) E_0 (b) E (c) $E_0 - E$ (d) $E_0 + E$ (e) None of the above

Applying Physics 27.2   Color Changes through the Compton Effect

The Compton effect involves a change in wavelength as photons are scattered through different angles. Suppose we illuminate a piece of material with a beam of light and then view the material from different angles relative to the beam of light. Will we see a color change corresponding to the change in wavelength of the scattered light?

Explanation There will be a wavelength change for visible light scattered by the material, but the change will be far too small to detect as a color change. The largest possible wavelength change, at 180° scattering, will be twice the Compton wavelength, about 0.005 nm. This represents a change of less than 0.001% of the wavelength of red light. The Compton effect is only detectable for wavelengths that are very short to begin with, so that the Compton wavelength is an appreciable fraction of the incident wavelength. As a result, the usual radiation for observing the Compton effect is in the x-ray range of the electromagnetic spectrum.

INTERACTIVE EXAMPLE 27.6   Scattering X-Rays

Goal   Understand Compton scattering and its effect on the photon's energy.

Problem   X-rays of wavelength $\lambda_0 = 0.200\,000$ nm are scattered from a block of material. The scattered x-rays are observed at an angle of 45.0° to the incident beam. (a) Calculate the wavelength of the x-rays scattered at this angle. (b) Compute the fractional change in the energy of a photon in the collision.

Solution
(a) Calculate the wavelength of the x-rays.

Substitute into Equation 27.11 to obtain the wavelength shift:

$$\Delta\lambda = \frac{h}{m_e c}(1 - \cos\theta)$$

$$= \frac{6.63 \times 10^{-34}\,\text{J}\cdot\text{s}}{(9.11 \times 10^{-31}\,\text{kg})(3.00 \times 10^8\,\text{m/s})}(1 - \cos 45.0°)$$

$$= 7.11 \times 10^{-13}\,\text{m} = 0.000\,711\,\text{nm}$$

Add this shift to the original wavelength to obtain the wavelength of the scattered photon:

$$\lambda = \Delta\lambda + \lambda_0 = \boxed{0.200\,711\,\text{nm}}$$

(b) Find the fraction of energy lost by the photon in the collision.

Rewrite the energy E in terms of wavelength, using $c = f\lambda$:

$$E = hf = h\frac{c}{\lambda}$$

Compute $\Delta E/E$ using this expression:

$$\frac{\Delta E}{E} = \frac{E_f - E_i}{E_i} = \frac{hc/\lambda_f - hc/\lambda_i}{hc/\lambda_i}$$

Cancel hc and rearrange terms:

$$\frac{\Delta E}{E} = \frac{1/\lambda_f - 1/\lambda_i}{1/\lambda_i} = \frac{\lambda_i}{\lambda_f} - 1 = \frac{\lambda_i - \lambda_f}{\lambda_f} = -\frac{\Delta\lambda}{\lambda_f}$$

Substitute values from part (a):

$$\frac{\Delta E}{E} = -\frac{0.000\,711\,\text{nm}}{0.200\,711\,\text{nm}} = \boxed{-3.54 \times 10^{-3}}$$

Remarks   It is also possible to find this answer by substituting into the energy expression at an earlier stage, but the algebraic derivation is more elegant and instructive.

Exercise 27.6
Repeat the exercise for a photon with wavelength 3.00×10^{-2} nm that scatters at an angle of $60.0°$.

Answers   (a) 3.12×10^{-2} nm   (b) $\Delta E / E = -3.88 \times 10^{-2}$

Physics⊗Now™   Study Compton scattering for different angles by logging into PhysicsNow at **www.cp7e.com** and going to Interactive Example 27.6.

27.6   THE DUAL NATURE OF LIGHT AND MATTER

Light and Electromagnetic Radiation

Phenomena such as the photoelectric effect and the Compton effect offer evidence that when light (or other forms of electromagnetic radiation) and matter interact, the light behaves as if it were composed of particles having energy hf and momentum h/λ. In other contexts, however, light acts like a wave, exhibiting interference and diffraction effects. Is light a wave or a particle?

The answer depends on the phenomenon being observed. Some experiments can be better explained with the photon concept, whereas others are best described with a wave model. The end result is that both models are needed. **Light has a dual nature, exhibiting both wave and particle characteristics**.

To understand why photons are compatible with electromagnetic waves, consider 2.5-MHz radio waves as an example. The energy of a photon having this frequency is only about 10^{-8} eV, too small to allow the photon to be detected. A sensitive radio receiver might require as many as 10^{10} of these photons to produce a detectable signal. Such a large number of photons would appear, on the average, as a continuous wave. With so many photons reaching the detector every second, we wouldn't be able to detect the individual photons striking the antenna.

Now consider what happens as we go to higher frequencies. In the visible region, it's possible to observe both the particle characteristics and the wave characteristics of light. As we mentioned earlier, a light beam shows interference phenomena (thus, it is a wave) and at the same time can produce photoelectrons (thus, it is a particle). At even higher frequencies, the momentum and energy of the photons increase. Consequently, the particle nature of light becomes more evident than its wave nature. For example, the absorption of an x-ray photon is easily detected as a single event, but wave effects are difficult to observe.

The Wave Properties of Particles

In his doctoral dissertation in 1924, Louis de Broglie postulated that, **because photons have wave and particle characteristics, perhaps all forms of matter have both properties**. This was a highly revolutionary idea with no experimental confirmation at that time. According to de Broglie, electrons, just like light, have a dual particle–wave nature.

In Chapter 26 we found that the relationship between energy and momentum for a photon, which has a rest energy of zero, is $p = E/c$. We also know from Equation 27.5 that the energy of a photon is

$$E = hf = \frac{hc}{\lambda} \qquad \textbf{[27.12]}$$

Consequently, the momentum of a photon can be expressed as

$$p = \frac{E}{c} = \frac{hc}{c\lambda} = \frac{h}{\lambda} \qquad \textbf{[27.13]}$$ ◀ Momentum of a photon

From this equation, we see that the photon wavelength can be specified by its momentum, or $\lambda = h/p$. De Broglie suggested that *all* material particles with momentum p should have a characteristic wavelength $\lambda = h/p$. Because the

LOUIS DE BROGLIE, French Physicist, (1892–1987)

De Broglie was born in Dieppe, France. At the Sorbonne in Paris, he studied history in preparation for what he hoped to be a career in the diplomatic service. The world of science is lucky that he changed his career path to become a theoretical physicist. De Broglie was awarded the Nobel Prize in 1929 for his discovery of the wave nature of electrons.

AIP Niels Bohr Library

momentum of a particle of mass m and speed v is $mv = p$, the **de Broglie wavelength** of a particle is

de Broglie's hypothesis ▶

$$\lambda = \frac{h}{p} = \frac{h}{mv}$$ [27.14]

Further, de Broglie postulated that the frequencies of matter waves (waves associated with particles having nonzero rest energy) obey the Einstein relationship for photons, $E = hf$, so that

Frequency of matter waves ▶

$$f = \frac{E}{h}$$ [27.15]

The dual nature of matter is quite apparent in Equations 27.14 and 27.15, because each contains both particle concepts (mv and E) and wave concepts (λ and f). The fact that these relationships had been established experimentally for photons made the de Broglie hypothesis that much easier to accept.

The Davisson–Germer Experiment

De Broglie's proposal in 1923 that matter exhibits both wave and particle properties was first regarded as pure speculation. If particles such as electrons had wave-like properties, then, under the correct conditions, they should exhibit diffraction effects. In 1927, three years after de Broglie published his work, C. J. Davisson (1881–1958) and L. H. Germer (1896–1971) of the United States succeeded in measuring the wavelength of electrons. Their important discovery provided the first experimental confirmation of the matter waves proposed by de Broglie.

The intent of the initial Davisson–Germer experiment was not to confirm the de Broglie hypothesis. In fact, their discovery was made by accident (as is often the case). The experiment involved the scattering of low-energy electrons (about 54 eV) from a nickel target in a vacuum. During one experiment, the nickel surface was badly oxidized because of an accidental break in the vacuum system. After the nickel target was heated in a flowing stream of hydrogen to remove the oxide coating, electrons scattered by it exhibited intensity maxima and minima at specific angles. The experimenters finally realized that the nickel had formed large crystalline regions upon heating and that the regularly spaced planes of atoms in the crystalline regions served as a diffraction grating for electron matter waves. (See Section 27.5.)

Shortly thereafter, Davisson and Germer performed more extensive diffraction measurements on electrons scattered from single-crystal targets. Their results showed conclusively the wave nature of electrons and confirmed the de Broglie relation $\lambda = h/p$. In the same year, G. P. Thomson (1892–1975) of Scotland also observed electron diffraction patterns by passing electrons through very thin gold foils. Diffraction patterns have since been observed for helium atoms, hydrogen atoms, and neutrons. Hence, the universal nature of matter waves has been established in various ways.

Quick Quiz 27.3

A nonrelativistic electron and a nonrelativistic proton are moving and have the same de Broglie wavelength. Which of the following are also the same for the two particles?
(a) speed (b) kinetic energy (c) momentum (d) frequency

Quick Quiz 27.4

We have seen two wavelengths assigned to the electron: the Compton wavelength and the de Broglie wavelength. Which is an actual *physical* wavelength associated with the electron? (a) the Compton wavelength (b) the de Broglie wavelength (c) both wavelengths (d) neither wavelength

EXAMPLE 27.7   The Electron versus the Baseball

Goal   Apply the de Broglie hypothesis to a quantum and a classical object.

Problem   (a) Compare the de Broglie wavelength for an electron ($m_e = 9.11 \times 10^{-31}$ kg) moving at a speed of 1.00×10^7 m/s with that of a baseball of mass 0.145 kg pitched at 45.0 m/s. (b) Compare these wavelengths with that of an electron traveling at $0.999c$.

Strategy   This is a matter of substitution into Equation 27.14 for the de Broglie wavelength. In part (b), the relativistic momentum must be used.

Solution

(a) Compare the de Broglie wavelengths of the electron and the baseball.

Substitute data for the electron into Equation 27.14:

$$\lambda_e = \frac{h}{m_e v} = \frac{6.63 \times 10^{-34}\,\text{J}\cdot\text{s}}{(9.11 \times 10^{-31}\,\text{kg})(1.00 \times 10^7\,\text{m/s})}$$

$$= 7.28 \times 10^{-11}\,\text{m}$$

Repeat the calculation with the baseball data:

$$\lambda_b = \frac{h}{m_b v} = \frac{6.63 \times 10^{-34}\,\text{J}\cdot\text{s}}{(0.145\,\text{kg})(45.0\,\text{m/s})} = 1.02 \times 10^{-34}\,\text{m}$$

(b) Find the wavelength for an electron traveling at $0.999c$.

Replace the momentum in Equation 27.14 with the relativistic momentum:

$$\lambda_e = \frac{h}{m_e v/\sqrt{1 - v^2/c^2}} = \frac{h\sqrt{1 - v^2/c^2}}{m_e v}$$

Substitute:

$$\lambda_e = \frac{(6.63 \times 10^{-34}\,\text{J}\cdot\text{s})\sqrt{1 - (0.999c)^2/c^2}}{(9.11 \times 10^{-31}\,\text{kg})(0.999 \cdot 3.00 \times 10^8\,\text{m/s})}$$

$$= 1.09 \times 10^{-13}\,\text{m}$$

Remarks   The electron wavelength corresponds to that of x-rays in the electromagnetic spectrum. The baseball, by contrast, has a wavelength much smaller than any aperture through which the baseball could possibly pass, so we couldn't observe any of its diffraction effects. It is generally true that the wave properties of large-scale objects can't be observed. Notice that even at extreme relativistic speeds, the electron wavelength is still far larger than the baseball's.

Exercise 27.7

Find the de Broglie wavelength of a proton ($m_p = 1.67 \times 10^{-27}$ kg) moving with a speed of 1.00×10^7 m/s.

Answer 3.97×10^{-14} m

Application: The Electron Microscope

A practical device that relies on the wave characteristics of electrons is the **electron microscope**. A *transmission* electron microscope, used for viewing flat, thin samples, is shown in Figure 27.17 (page 890). In many respects, it is similar to an optical microscope, but the electron microscope has a much greater resolving power because it can accelerate electrons to very high kinetic energies, giving them very short wavelengths. No microscope can resolve details that are significantly smaller than the wavelength of the radiation used to illuminate the object. Typically, the wavelengths of electrons are about 100 times smaller than those of the visible light used in optical microscopes. (Radiation of the same wavelength as the electrons in an electron microscope is in the x-ray region of the spectrum.)

APPLICATION

Electron Microscopes

Electron gun

Cathode

Anode

Electromagnetic lens

Electromagnetic condenser lens

Screen

Visual transmission

Vacuum

Core

Coil

Electron beam

Specimen goes here

Specimen chamber door

Projector lens

Photo chamber

(a) (b)

Figure 27.17 (a) Diagram of a transmission electron microscope for viewing a thin sectioned sample. The "lenses" that control the electron beam are magnetic deflection coils. (b) An electron microscope.

The electron beam in an electron microscope is controlled by electrostatic or magnetic deflection, which acts on the electrons to focus the beam to an image. Rather than examining the image through an eyepiece as in an optical microscope, the viewer looks at an image formed on a fluorescent screen. (The viewing screen must be fluorescent because otherwise the image produced wouldn't be visible.)

Applying Physics 27.3   X-Ray Microscopes?

Electron microscopes (Fig. 27.17) take advantage of the wave nature of particles. Electrons are accelerated to high speeds, giving them a short de Broglie wavelength. Imagine an electron microscope using electrons with a de Broglie wavelength of 0.2 nm. Why don't we design a microscope using 0.2-nm *photons* to do the same thing?

Explanation Because electrons are charged particles, they interact electrically with the sample in the microscope and scatter according to the shape and density of various portions of the sample, providing a means of viewing the sample. Photons of wavelength 0.2 nm are uncharged and in the x-ray region of the spectrum. They tend to simply pass through the thin sample without interacting.

27.7 THE WAVE FUNCTION

De Broglie's revolutionary idea that particles should have a wave nature soon moved out of the realm of skepticism to the point where it was viewed as a necessary concept in understanding the subatomic world. In 1926, the Austrian–German physicist Erwin Schrödinger proposed a wave equation that described how matter waves change in space and time. The Schrödinger wave equation represents a key element in the theory of quantum mechanics. It's as important in quantum mechanics as Newton's laws in classical mechanics. Schrödinger's equation has been successfully applied to the hydrogen atom and to many other microscopic systems.

Solving Schrödinger's equation (beyond the level of this course) determines a quantity Ψ called the **wave function**. Each particle is represented by a wave function Ψ that depends both on position and on time. Once Ψ is found, Ψ^2 gives us information on the **probability** (per unit volume) of finding the particle in any given region. To understand this, we return to Young's experiment involving coherent light passing through a double slit.

First, recall from Chapter 21 that the intensity of a light beam is proportional to the square of the electric field strength E associated with the beam: $I \propto E^2$. According to the wave model of light, there are certain points on the viewing screen where the net electric field is zero as a result of destructive interference of waves from the two slits. Because E is zero at these points, the intensity is also zero, and the screen is dark there. Likewise, at points on the screen at which constructive interference occurs, E is large, as is the intensity; hence, these locations are bright.

Now consider the same experiment when light is viewed as having a particle nature. The number of photons reaching a point on the screen per second increases as the intensity (brightness) increases. Consequently, the number of photons that strike a unit area on the screen each second is proportional to the square of the electric field, or $N \propto E^2$. From a probabilistic point of view, a photon has a high probability of striking the screen at a point at which the intensity (and E^2) is high and a low probability of striking the screen where the intensity is low.

When describing particles rather than photons, Ψ rather than E plays the role of the amplitude. Using an analogy with the description of light, we make the following interpretation of Ψ for particles: If Ψ is a wave function used to describe a single particle, the value of Ψ^2 at some location at a given time is proportional to the probability per unit volume of finding the particle at that location at that time. Adding up all the values of Ψ^2 in a given region gives the probability of finding the particle in that region.

ERWIN SCHRÖDINGER, Austrian Theoretical Physicist (1887–1961)

Schrödinger is best known as the creator of wave mechanics, a less cumbersome theory than the equivalent matrix mechanics developed by Werner Heisenberg. In 1933 Schrödinger left Germany and eventually settled at the Dublin Institute of Advanced Study, where he spent 17 happy, creative years working on problems in general relativity, cosmology, and the application of quantum physics to biology. In 1956, he returned home to Austria and his beloved Tirolean mountains, where he died in 1961.

27.8 THE UNCERTAINTY PRINCIPLE

If you were to measure the position and speed of a particle at any instant, you would always be faced with experimental uncertainties in your measurements. According to classical mechanics, no fundamental barrier to an ultimate refinement of the apparatus or experimental procedures exists. In other words, it's possible, in principle, to make such measurements with arbitrarily small uncertainty. Quantum theory predicts, however, that such a barrier does exist. In 1927, Werner Heisenberg (1901–1976) introduced this notion, which is now known as the **uncertainty principle**:

> If a measurement of the position of a particle is made with precision Δx and a simultaneous measurement of linear momentum is made with precision Δp_x, then the product of the two uncertainties can never be smaller than $h/4\pi$:
>
> $$\Delta x \, \Delta p_x \geq \frac{h}{4\pi}$$ [27.16]

In other words, **it is physically impossible to measure simultaneously the exact position and exact linear momentum of a particle.** If Δx is very small, then Δp_x is large, and vice versa.

To understand the physical origin of the uncertainty principle, consider the following thought experiment introduced by Heisenberg. Suppose you wish to measure the position and linear momentum of an electron as accurately as possible. You might be able to do this by viewing the electron with a powerful light microscope. For you to see the electron and determine its location, at least one photon of light must bounce off the electron, as shown in Figure 27.18a, and pass through the

WERNER HEISENBERG, German Theoretical Physicist (1901–1976)

Heisenberg obtained his Ph.D. in 1923 at the University of Munich, where he studied under Arnold Sommerfeld. While physicists such as de Broglie and Schrödinger tried to develop physical models of the atom, Heisenberg developed an abstract mathematical model called *matrix mechanics* to explain the wavelengths of spectral lines. Heisenberg made many other significant contributions to physics, including his famous uncertainty principle, for which he received the Nobel Prize in 1932; the prediction of two forms of molecular hydrogen; and theoretical models of the nucleus of an atom.

Figure 27.18 A thought experiment for viewing an electron with a powerful microscope. (a) The electron is viewed before colliding with the photon. (b) The electron recoils (is disturbed) as the result of the collision with the photon.

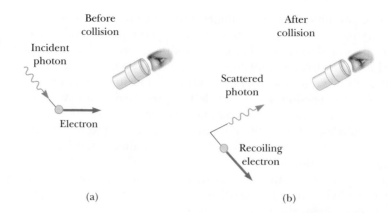

Before collision

Incident photon

Electron

(a)

After collision

Scattered photon

Recoiling electron

(b)

microscope into your eye, as shown in Figure 27.18b. When it strikes the electron, however, the photon transfers some unknown amount of its momentum to the electron. Thus, in the process of locating the electron very accurately (that is, by making Δx very small), the light that enables you to succeed in your measurement changes the electron's momentum to some undeterminable extent (making Δp_x very large).

The incoming photon has momentum h/λ. As a result of the collision, the photon transfers part or all of its momentum along the x-axis to the electron. Therefore, the *uncertainty* in the electron's momentum after the collision is as great as the momentum of the incoming photon: $\Delta p_x = h/\lambda$. Further, because the photon also has wave properties, we expect to be able to determine the electron's position to within one wavelength of the light being used to view it, so $\Delta x = \lambda$. Multiplying these two uncertainties gives

$$\Delta x \, \Delta p_x = \lambda \left(\frac{h}{\lambda} \right) = h$$

The value h represents the minimum in the product of the uncertainties. Because the uncertainty can always be greater than this minimum, we have

$$\Delta x \, \Delta p_x \geq h$$

Apart from the numerical factor $1/4\pi$ introduced by Heisenberg's more precise analysis, this inequality agrees with Equation 27.16.

Another form of the uncertainty relationship sets a limit on the accuracy with which the energy E of a system can be measured in a finite time interval Δt:

$$\Delta E \, \Delta t \geq \frac{h}{4\pi} \qquad \text{[27.17]}$$

It can be inferred from this relationship that the energy of a particle cannot be measured with complete precision in a very short interval of time. Thus, when an electron is viewed as a particle, the uncertainty principle tells us that (a) its position and velocity cannot both be known precisely at the same time and (b) its energy can be uncertain for a period given by $\Delta t = h/(4\pi \, \Delta E)$.

Applying Physics 27.4   Motion at Absolute Zero

A common, but erroneous, description of the absolute zero of temperature is "that temperature at which all molecular motion ceases." How can the uncertainty principle be used to argue against this description?

Explanation Imagine a particular molecule in a piece of material. The molecule is confined within the material, so there is a fixed uncertainty Δx in its position along one axis, corresponding to the size of that piece of material. If all molecular motion ceased at absolute zero, the given molecule's velocity, in particular, would be exactly zero, so its uncertainty in velocity would be $\Delta v = 0$, meaning its uncertainty in momentum would also be zero, since $p = mv$. The product of zero uncertainty in momentum and a nonzero uncertainty in position is zero, violating the uncertainty principle. So according to the uncertainty principle, there must be some molecular motion even at absolute zero.

EXAMPLE 27.8   Locating an Electron

Goal   Apply Heisenberg's position–momentum uncertainty principle.

Problem   The speed of an electron is measured to be 5.00×10^3 m/s to an accuracy of 0.003 00%. Find the minimum uncertainty in determining the position of this electron.

Strategy   After computing the momentum and its uncertainty, substitute into Heisenberg's uncertainty principle, Equation 27.16.

Solution

Calculate the momentum of the electron:

$$p_x = m_e v = (9.11 \times 10^{-31}\ \text{kg})(5.00 \times 10^3\ \text{m/s})$$
$$= 4.56 \times 10^{-27}\ \text{kg} \cdot \text{m/s}$$

The uncertainty in p_x is 0.003 00% of this value:

$$\Delta p_x = 0.000\ 030\ 0 p_x = (0.000\ 030\ 0)(4.56 \times 10^{-27}\ \text{kg} \cdot \text{m/s})$$
$$= 1.37 \times 10^{-31}\ \text{kg} \cdot \text{m/s}$$

Now calculate the uncertainty in position using this value of Δp_x and Equation 27.17:

$$\Delta x \Delta p_x \geq \frac{h}{4\pi} \quad \rightarrow \quad \Delta x \geq \frac{h}{4\pi \Delta p_x}$$

$$\Delta x \geq \frac{6.626 \times 10^{-34}\ \text{J} \cdot \text{s}}{4\pi(1.37 \times 10^{-31}\ \text{kg} \cdot \text{m/s})} = 0.384 \times 10^{-3}\ \text{m}$$

$$= 0.384\ \text{mm}$$

Remarks   Notice that this isn't an exact calculation: the uncertainty in position can take any value, as long as it's greater than or equal to the value given by the uncertainty principle.

Exercise 27.8

Suppose an electron is found somewhere in an atom of diameter 1.25×10^{-10} m. Estimate the uncertainty in the electron's momentum (in one dimension).

Answer $\Delta p \geq 4.22 \times 10^{-25}$ kg · m/s

EXAMPLE 27.9   Excited States of Atoms

Goal   Apply the energy–time form of the uncertainty relation.

Problem   As we'll see in the next chapter, electrons in atoms can be found in certain high states of energy called **excited states** for short periods of time. If the average time that an electron exists in one of these states is 1.00×10^{-8} s, what is the minimum uncertainty in energy of the excited state?

Strategy   Substitute values into Equation 27.17, the energy–time form of Heisenberg's uncertainty relation.

Solution

Use Equation 27.17 to obtain the minimum uncertainty in the energy:

$$\Delta E \geq \frac{h}{4\pi \Delta t} = \frac{6.63 \times 10^{-34}\ \text{J} \cdot \text{s}}{4\pi(1.00 \times 10^{-8}\ \text{s})} = 5.28 \times 10^{-27}\ \text{J}$$

$$= 3.30 \times 10^{-8}\ \text{eV}$$

Remarks   This is again an imprecise calculation, giving only a lower bound on the uncertainty.

Exercise 27.9

A muon may be considered to be an excited state of an electron, to which it decays in an average of 2.2×10^{-6} s. What's the minimum uncertainty in the muon's (rest) energy, according to the uncertainty principle?

Answer 2.40×10^{-29} J

27.9   THE SCANNING TUNNELING MICROSCOPE[1]

One of the basic phenomena of quantum mechanics—tunneling—is at the heart of a very practical device—the scanning tunneling microscope, or STM—which enables us to get highly detailed images of surfaces with a resolution comparable to the size of a single atom.

Figure 27.19 shows an image of a ring of 48 iron atoms located on a copper surface. Note the high quality of the STM image and the recognizable ring of iron atoms. What makes this image so remarkable is that its resolution—the size of the smallest detail that can be discerned—is about 0.2 nm. For an ordinary microscope, the resolution is limited by the wavelength of the waves used to make the image. An optical microscope has a resolution no better than 200 nm, about half the wavelength of visible light, and so could never show the detail displayed in Figure 27.19. Electron microscopes can have a resolution of 0.2 nm by using electron waves of that wavelength, given by the de Broglie formula $\lambda = h/p$. The electron momentum p required to give this wavelength is 10 000 eV/c, corresponding to an electron speed of 2% of the speed of light. Electrons traveling at this speed would penetrate into the interior of the sample in Figure 27.20 and so could not give us information about individual surface atoms.

The STM achieves its very fine resolution by using the basic idea shown in Figure 27.20. A conducting probe with a sharp tip is brought near the surface to be studied. Because it is attracted to the positive ions in the surface, an electron in the surface has a lower total energy than an electron in the empty space between surface and tip. The same thing is true for an electron in the probe tip, which is attracted to the positive ions in the tip. In Newtonian mechanics, this means that electrons cannot move between surface and tip because they lack the energy to escape either material. Because the electrons obey quantum mechanics, however, they can "tunnel" across the barrier of empty space. By applying a voltage between surface and tip, the electrons can be made to tunnel preferentially from surface to tip. In this way, the tip samples the distribution of electrons just above the surface.

Because of the nature of tunneling, the STM is very sensitive to the distance z from tip to surface. The reason is that in the empty space between tip and surface, the electron wave function falls off exponentially with a decay length on the order of 0.1 nm; that is, the wave function decreases by $1/e$ over that distance. For dis-

Figure 27.19   This is a photograph of a "quantum corral" consisting of a ring of 48 iron atoms located on a copper surface. The diameter of the ring is 143 nm. The photograph was obtained with a low-temperature scanning tunneling microscope (STM).

IBM Corporation Research Division

[1]This section was written by Roger A. Freedman, University of California, Santa Barbara.

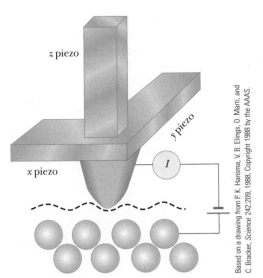

Based on a drawing from P. K. Hansma, V. B. Elings, O. Marti, and C. Bracker, *Science* 242:209, 1988, Copyright 1988 by the AAAS.

Figure 27.20   A schematic view of an STM. The tip, shown as a rounded cone, is mounted on a piezoelectric *x, y, z* scanner. A scan of the tip over the sample can reveal contours of the surface down to the atomic level. An STM image is composed of a series of scans displaced laterally from each other.

tances *z* greater than 1 nm (that is, beyond a few atomic diameters), essentially no tunneling takes place. This exponential behavior causes the current of electrons tunneling from surface to tip to depend very strongly on *z*. This sensitivity is the basis of the operation of the STM: by monitoring the tunneling current as the tip is scanned over the surface, scientists obtain a sensitive measure of the topography of the electron distribution on the surface. The result of this scan is used to make images like that in Figure 27.20. In this way the STM can measure the height of surface features to within 0.001 nm, approximately 1/100 of an atomic diameter!

The STM has, however, one serious limitation: it depends on electrical conductivity of the sample and the tip. Unfortunately, the surfaces of most materials are not electrically conductive. Even metals such as aluminum are covered with nonconductive oxides. A newer microscope—the atomic force microscope, or AFM—overcomes this limitation. It measures the force between a tip and the sample, rather than an electrical current. This force depends strongly on the tip–sample separation just as the electron tunneling current does for the STM. The AFM has comparable sensitivity for measuring topography and has become widely used for technological applications.

Perhaps the most remarkable thing about the STM is that its operation is based on a quantum mechanical phenomenon—tunneling—that was well understood in the 1920s, even though the first STM was not built until the 1980s. What other applications of quantum mechanics may yet be waiting to be discovered?

SUMMARY

Physics⊗Now™ Take a practice test by logging into PhysicsNow at **www.cp7e.com** and clicking on the Pre-Test link for this chapter.

27.1 Blackbody Radiation and Planck's Hypothesis

The characteristics of **blackbody radiation** can't be explained with classical concepts. The peak of a black-body radiation curve is given by **Wien's displacement law**;

$$\lambda_{max} T = 0.289\ 8 \times 10^{-2}\ \text{m} \cdot \text{K} \qquad \textbf{[27.1]}$$

where λ_{max} is the wavelength at which the curve peaks and T is the absolute temperature of the object emitting the radiation.

Planck first introduced the quantum concept when he assumed that the subatomic oscillators responsible for blackbody radiation could have only discrete amounts of energy given by

$$E_n = nhf \qquad \textbf{[27.2]}$$

where n is a positive integer called a **quantum number** and f is the frequency of vibration of the resonator.

27.2 The Photoelectric Effect and the Particle Theory of Light

The **photoelectric effect** is a process whereby electrons are ejected from a metal surface when light is incident on that surface. Einstein provided a successful explanation of this effect by extending Planck's quantum hypothesis to electromagnetic waves. In this model, light is viewed as a stream of particles called photons, each with energy $E = hf$, where f is the light frequency and h is **Planck's constant**. The maximum kinetic energy of the ejected photoelectrons is

$$KE_{max} = hf - \phi \qquad \textbf{[27.6]}$$

where ϕ is the **work function** of the metal.

27.3 X-Rays

27.4 Diffraction of X-Rays by Crystals

X-rays are produced when high-speed electrons are suddenly decelerated. When electrons have been accel-erated through a voltage V, the shortest-wavelength radiation that can be produced is

$$\lambda_{min} = \frac{hc}{eV} \qquad \textbf{[27.9]}$$

The regular array of atoms in a crystal can act as a diffraction grating for x-rays and for electrons. The condition for constructive interference of the diffracted rays is given by **Bragg's law**:

$$2d \sin \theta = m\lambda \qquad (m = 1, 2, 3, \ldots) \qquad \textbf{[27.10]}$$

Bragg's law bears a similarity to the equation for the diffraction pattern of a double slit.

27.5 The Compton Effect

X-rays from an incident beam are scattered at various angles by electrons in a target such as carbon. In such a scattering event, a shift in wavelength is observed for the scattered x-rays. This phenomenon is known as the **Compton shift**. Conservation of momentum and energy applied to a photon–electron collision yields the following expression for the shift in wavelength of the scattered x-rays:

$$\Delta\lambda = \lambda - \lambda_0 = \frac{h}{m_e c}(1 - \cos\theta) \qquad \textbf{[27.11]}$$

Here, m_e is the mass of the electron, c is the speed of light, and θ is the scattering angle.

27.6 The Dual Nature of Light and Matter

Light exhibits both a particle and a wave nature. De Broglie proposed that *all* matter has both a particle and a wave nature. The **de Broglie wavelength** of any particle of mass m and speed v is

$$\lambda = \frac{h}{p} = \frac{h}{mv} \qquad \textbf{[27.14]}$$

De Broglie also proposed that the frequencies of the waves associated with particles obey the Einstein relationship $E = hf$.

27.7 The Wave Function

In the theory of **quantum mechanics**, each particle is described by a quantity Ψ called the **wave function**.

The probability per unit volume of finding the particle at a particular point at some instant is proportional to Ψ^2. Quantum mechanics has been highly successful in describing the behavior of atomic and molecular systems.

Also,

$$\Delta E \, \Delta t \geq \frac{h}{4\pi} \qquad \text{[27.17]}$$

where ΔE is the uncertainty in the energy of the particle and Δt is the uncertainty in the time it takes to measure the energy.

27.8   The Uncertainty Principle

According to Heisenberg's **uncertainty principle**, it is impossible to measure simultaneously the exact position and exact momentum of a particle. If Δx is the uncertainty in the measured position and Δp_x the uncertainty in the momentum, the product $\Delta x \, \Delta p_x$ is given by

$$\Delta x \, \Delta p_x \geq \frac{h}{4\pi} \qquad \text{[27.16]}$$

CONCEPTUAL QUESTIONS

1. If you observe objects inside a very hot kiln why is it difficult to discern the shapes of the objects?

2. Why is an electron microscope more suitable than an optical microscope for "seeing" objects of atomic size?

3. Are blackbodies black?

4. Why is it impossible to simultaneously measure the position and velocity of a particle with infinite accuracy?

5. All objects radiate energy. Why, then, are we not able to see all objects in a dark room?

6. Is light a wave or a particle? Support your answer by citing specific experimental evidence.

7. A student claims that he is going to eject electrons from a piece of metal by placing a radio transmitter antenna adjacent to the metal and sending a strong AM radio signal into the antenna. The work function of a metal is typically a few electron volts. Will this work?

8. Light acts sometimes like a wave and sometimes like a particle. For the following situations, which best describes the behavior of light? Defend your answers. (a) The photoelectric effect. (b) The Compton effect. (c) Young's double-slit experiment.

9. In the photoelectric effect, explain why the stopping potential depends on the frequency of the light but not on the intensity.

10. Which has more energy, a photon of ultraviolet radiation or a photon of yellow light?

11. Why does the existence of a cutoff frequency in the photoelectric effect favor a particle theory of light rather than a wave theory?

12. What effect, if any, would you expect the temperature of a material to have on the ease with which electrons can be ejected from it via the photoelectric effect?

13. The cutoff frequency of a material is f_0. Are electrons emitted from the material when (a) light of frequency greater than f_0 is incident on the material? (b) Less than f_0?

14. The brightest star in the constellation Lyra is the bluish star Vega, whereas the brightest star in Boötes is the reddish star Arcturus. How do you account for the difference in color of the two stars?

15. If the photoelectric effect is observed in one metal, can you conclude that the effect will also be observed in another metal under the same conditions? Explain.

16. A beam of blue light and a beam of red light carry the same total amount of energy. Which beam contains the larger number of photons?

17. An x-ray photon is scattered by an electron which is initially at rest. What happens to the frequency of the scattered photon relative to that of the incident photon?

PROBLEMS

1, 2, 3 = straightforward, intermediate, challenging □ = full solution available in *Student Solutions Manual/Study Guide*
Physics⊗Now™ = coached problem with hints available at **www.cp7e.com** 🐭 = biomedical application

Section 27.1 Blackbody Radiation and Planck's Hypothesis

1. What is the surface temperature of (a) Betelgeuse, a red giant star in the constellation of Orion, which radiates with a peak wavelength of about 970 nm? (b) Rigel, a bluish-white star in Orion, radiates with a peak wavelength of 145 nm? Find the temperature of Rigel's surface.

2. (a) Lightning produces a maximum air temperature on the order of 10^4 K, while (b) a nuclear explosion produces a temperature on the order of 10^7 K. Use Wien's displacement law to find the order of magnitude of the wavelength of the thermally produced photons radiated with greatest intensity by each of these sources. Name the part of the electromagnetic spectrum where you would expect each to radiate most strongly.

3. If the surface temperature of the Sun is 5 800 K, find the wavelength that corresponds to the maximum rate of energy emission from the Sun.

4. A beam of blue light and a beam of red light each carry a total energy of 2 500 eV. If the wavelength of the red light is 690 nm and the wavelength of the blue light is 420 nm, find the number of photons in each beam.

5. Calculate the energy in electron volts of a photon having a wavelength (a) in the microwave range, 5.00 cm, (b) in the visible light range, 500 nm, and (c) in the x-ray range, 5.00 nm.

6. A certain light source is found to emit radiation whose peak value has a frequency of 1.00×10^{15} Hz. Find the temperature of the source assuming that it is a blackbody radiator.

7. An FM radio transmitter has a power output of 150 kW and operates at a frequency of 99.7 MHz. How many photons per second does the transmitter emit?

8. The threshold of dark-adapted (scotopic) vision is 4.0×10^{-11} W/m^2 at a central wavelength of 500 nm. If light with this intensity and wavelength enters the eye when the pupil is open to its maximum diameter of 8.5 mm, how many photons per second enter the eye?

9. A 1.5-kg mass vibrates at an amplitude of 3.0 cm on the end of a spring of spring constant 20.0 N/m. (a) If the energy of the spring is quantized, find its quantum num-ber. (b) If n changes by 1, find the fractional change in energy of the spring.

10. A 70.0-kg jungle hero swings at the end of a vine at a frequency of 0.50 Hz at 2.0 m/s as he moves through the lowest point on his arc. (a) Assume the energy is quantized and find the quantum number n for this system. (b) Find the energy carried away in a one-quantum change in the jungle hero's energy.

Section 27.2 The Photoelectric Effect and the Particle Theory of Light

11. Physics⊗Now™ When light of wavelength 350 nm falls on a potassium surface, electrons having a maximum kinetic energy of 1.31 eV are emitted. Find (a) the work function of potassium, (b) the cutoff wavelength, and (c) the frequency corresponding to the cutoff wavelength.

12. When a certain metal is illuminated with light of frequency 3.0×10^{15} Hz, a stopping potential of 7.0 V is required to stop the most energetic ejected electrons. What is the work function of this metal?

13. What wavelength of light would have to fall on sodium (with a work function of 2.46 eV) if it is to emit electrons with a maximum speed of 1.0×10^6 m/s?

14. Lithium, beryllium, and mercury have work functions of 2.30 eV, 3.90 eV, and 4.50 eV, respectively. If 400-nm light is incident on each of these metals, determine (a) which metals exhibit the photoelectric effect and (b) the maximum kinetic energy of the photoelectrons in each case.

15. From the scattering of sunlight, Thomson calculated that the classical radius of the electron has a value of 2.82×10^{-15} m. If sunlight having an intensity of 500 W/m^2 falls on a disk with this radius, estimate the time required to accumulate 1.00 eV of energy. Assume that light is a classical wave and that the light striking the disk is completely absorbed. How does your estimate compare with the observation that photoelectrons are promptly (within 10^{-9} s) emitted?

16. An isolated copper sphere of radius 5.00 cm, initially uncharged, is illuminated by ultraviolet light of wavelength 200 nm. What charge will the photoelectric effect induce on the sphere? The work function for copper is 4.70 eV.

17. When light of wavelength 254 nm falls on cesium, the required stopping potential is 3.00 V. If light of wavelength 436 nm is used, the stopping potential is 0.900 V. Use this information to plot a graph like that shown in Figure 27.6, and from the graph determine the cutoff frequency for cesium and its work function.

18. Ultraviolet light is incident normally on the surface of a certain substance. The binding energy of the electrons in this substance is 3.44 eV. The incident light has an intensity of 0.055 W/m^2. The electrons are photoelectrically emitted with a maximum speed of 4.2×10^5 m/s. How many electrons are emitted from a square centimeter of the surface each second? Assume that the absorption of every photon ejects an electron.

Section 27.3 X-Rays

19. The extremes of the x-ray portion of the electromagnetic spectrum range from approximately 1.0×10^{-8} m to 1.0×10^{-13} m. Find the minimum accelerating voltages required to produce wavelengths at these two extremes.

20. Calculate the minimum-wavelength x-ray that can be produced when a target is struck by an electron that has been accelerated through a potential difference of (a) 15.0 kV and (b) 100 kV.

21. What minimum accelerating voltage would be required to produce an x-ray with a wavelength of 0.030 0 nm?

Section 27.4 Diffraction of X-Rays by Crystals

22. A monochromatic x-ray beam is incident on a NaCl crystal surface with $d = 0.353$ nm. The second-order maximum in the reflected beam is found when the angle between the incident beam and the surface is 20.5°. Determine the wavelength of the x-rays.

23. Potassium iodide has an interplanar spacing of $d = 0.296$ nm. A monochromatic x-ray beam shows a first-order diffraction maximum when the grazing angle is 7.6°. Calculate the x-ray wavelength.

24. The spacing between certain planes in a crystal is known to be 0.30 nm. Find the smallest angle of incidence at which constructive interference will occur for wavelength 0.070 nm.

25. **Physics⊗Now™** X-rays of wavelength 0.140 nm are reflected from a certain crystal, and the first-order maximum occurs at an angle of 14.4°. What value does this give for the interplanar spacing of the crystal?

Section 27.5 The Compton Effect

26. X-rays are scattered from electrons in a carbon target. The measured wavelength shift is 1.50×10^{-3} nm. Calculate the scattering angle.

27. Calculate the energy and momentum of a photon of wavelength 700 nm.

28. A beam of 0.68-nm photons undergoes Compton scattering from free electrons. What are the energy and momentum of the photons that emerge at a 45° angle with respect to the incident beam?

29. A 0.001 6-nm photon scatters from a free electron. For what (photon) scattering angle will the recoiling electron and scattered photon have the same kinetic energy?

30. X-rays with an energy of 300 keV undergo Compton scattering from a target. If the scattered rays are deflected at 37.0° relative to the direction of the incident rays, find (a) the Compton shift at this angle, (b) the energy of the scattered x-ray, and (c) the kinetic energy of the recoiling electron.

31. A 0.110-nm photon collides with a stationary electron. After the collision, the electron moves forward and the photon recoils backwards. Find the momentum and kinetic energy of the electron.

32. After a 0.800-nm x-ray photon scatters from a free electron, the electron recoils with a speed equal to 1.40×10^6 m/s. (a) What was the Compton shift in the photon's wavelength? (b) Through what angle was the photon scattered?

33. A 0.45-nm x-ray photon is deflected through a 23° angle after scattering from a free electron. (a) What is the kinetic energy of the recoiling electron? (b) What is its speed?

Section 27.6 The Dual Nature of Light and Matter

34. Calculate the de Broglie wavelength of a proton moving at (a) 2.00×10^4 m/s; (b) 2.00×10^7 m/s.

35. (a) If the wavelength of an electron is 5.00×10^{-7} m, how fast is it moving? (b) If the electron has a speed of 1.00×10^7 m/s, what is its wavelength?

36. A 0.200-kg ball is released from rest at the top of a 50.0-m tall building. Find the de Broglie wavelength of the ball just before it strikes the Earth.

37. The nucleus of an atom is on the order of 10^{-14} m in diameter. For an electron to be confined to a nucleus, its de Broglie wavelength would have to be of that order of magnitude or smaller. (a) What would be the kinetic energy of an electron confined to

this region? (b) On the basis of your result in part (a), would you expect to find an electron in a nucleus? Explain.

38. After learning about de Broglie's hypothesis that particles of momentum p have wave characteristics with wavelength $\lambda = h/p$, an 80.0-kg student has grown concerned about being diffracted when passing through a 75.0-cm-wide doorway. Assume that significant diffraction occurs when the width of the diffraction aperture is less than 10.0 times the wavelength of the wave being diffracted. (a) Determine the maximum speed at which the student can pass through the doorway in order to be significantly diffracted. (b) With that speed, how long will it take the student to pass through the doorway if it is 15.0 cm thick? Compare your result with the currently accepted age of the Universe, which is 4.00×10^{17} s. (c) Should this student worry about being diffracted?

39. **Physics⊗Now™** De Broglie postulated that the relationship $\lambda = h/p$ is valid for relativistic particles. What is the de Broglie wavelength for a (relativistic) electron whose kinetic energy is 3.00 MeV?

40. A monoenergetic beam of electrons is incident on a single slit of width 0.500 nm. A diffraction pattern is formed on a screen 20.0 cm from the slit. If the distance between successive minima of the diffraction pattern is 2.10 cm, what is the energy of the incident electrons?

41. The resolving power of a microscope is proportional to the wavelength used. A resolution of 1.0×10^{-11} m (0.010 nm) would be required in order to "see" an atom. (a) If electrons were used (electron microscope), what minimum kinetic energy would be required of the electrons? (b) If photons were used, what minimum photon energy would be needed to obtain 1.0×10^{-11} m resolution?

Section 27.7 The Wave Function

Section 27.8 The Uncertainty Principle

42. A 50.0-g ball moves at 30.0 m/s. If its speed is measured to an accuracy of 0.10%, what is the minimum uncertainty in its position?

43. In the ground state of hydrogen, the uncertainty in the position of the electron is roughly 0.10 nm. If the speed of the electron is on the order of the uncertainty in its speed, how fast is the electron moving?

44. Suppose Fuzzy, a quantum mechanical duck, lives in a world in which $h = 2\pi$ J·s. Fuzzy has a mass of 2.00 kg

and is initially known to be within a pond 1.00 m wide. (a) What is the minimum uncertainty in his speed? (b) Assuming this uncertainty in speed to prevail for 5.00 s, determine the uncertainty in Fuzzy's position after this time.

45. Suppose optical radiation ($\lambda = 5.00 \times 10^{-7}$ m) is used to determine the position of an electron to within the wavelength of the light. What will be the resulting uncertainty in the electron's velocity?

46. (a) Show that the kinetic energy of a nonrelativistic particle can be written in terms of its momentum as $KE = p^2/2m$. (b) Use the results of (a) to find the minimum kinetic energy of a proton confined within a nucleus having a diameter of 1.0×10^{-15} m.

ADDITIONAL PROBLEMS

47. Figure P27.47 shows the spectrum of light emitted by a firefly. Determine the temperature of a blackbody that would emit radiation peaked at the same frequency. Based on your result, would you say firefly radiation is blackbody radiation?

Figure P27.47

48. An x-ray tube is operated at 50 000 V. (a) Find the minimum wavelength of the radiation emitted by this tube. (b) If the radiation is directed at a crystal, the first-order maximum in the reflected radiation occurs when the grazing angle is 2.5°. What is the spacing between reflecting planes in the crystal?

49. The spacing between planes of nickel atoms in a nickel crystal is 0.352 nm. At what angle does a second-order Bragg reflection occur in nickel for 11.3-keV x-rays?

50. Johnny Jumper's favorite trick is to step out of his 16th-story window and fall 50.0 m into a pool. A news reporter takes a picture of 75.0-kg Johnny just before he makes a splash, using an exposure

time of 5.00 ms. Find (a) Johnny's de Broglie wavelength at this moment, (b) the uncertainty of his kinetic energy measurement during such a period of time, and (c) the percent error caused by such an uncertainty.

51. Photons of wavelength 450 nm are incident on a metal. The most energetic electrons ejected from the metal are bent into a circular arc of radius 20.0 cm by a magnetic field with a magnitude of 2.00×10^{-5} T. What is the work function of the metal?

52. A 200-MeV photon is scattered at 40.0° by a free proton that is initially at rest. Find the energy (in MeV) of the scattered photon.

53. **Physics⊗Now™** A light source of wavelength λ illuminates a metal and ejects photoelectrons with a maximum kinetic energy of 1.00 eV. A second light source of wavelength $\lambda/2$ ejects photoelectrons with a maximum kinetic energy of 4.00 eV. What is the work function of the metal?

54. Red light of wavelength 670 nm produces photoelectrons from a certain photoemissive material. Green light of wavelength 520 nm produces photoelectrons from the same material with 1.50 times the maximum kinetic energy. What is the material's work function?

55. How fast must an electron be moving if all its kinetic energy is lost to a single x-ray photon (a) at the high end of the x-ray electromagnetic spectrum with a wavelength of 1.00×10^{-8} m; (b) at the low end of the x-ray electromagnetic spectrum with a wavelength of 1.00×10^{-13} m?

56. Show that if an electron were confined inside an atomic nucleus of diameter 2.0×10^{-15} m, it would have to be moving relativistically, while a proton confined to the same nucleus can be moving at less than one-tenth the speed of light.

57. A photon strikes a metal with a work function ϕ and produces a photoelectron with a de Broglie wavelength equal to the wavelength of the original photon. (a) Show that the energy of this photon must have been given by

$$E = \frac{\phi(m_e c^2 - \phi/2)}{m_e c^2 - \phi}$$

where m_e is the mass of the electron. [*Hint*: Begin with the conservation of energy, $E + m_e c^2 = \phi + \sqrt{(pc)^2 + (m_e c^2)^2}$.] (b) If one of these photons strikes platinum ($\phi = 6.35$ eV), determine the resulting maximum speed of the photoelectron that is emitted.

58. In a Compton scattering event, the scattered photon has an energy of 120.0 keV and the recoiling electron has a kinetic energy of 40.0 keV. Find (a) the wavelength of the incident photon, (b) the angle θ at which the photon is scattered, and (c) the recoil angle of the electron. [*Hint*: Conserve both mass–energy and relativistic momentum.]

59. A woman on a ladder drops small pellets toward a point target on the floor. (a) Show that, according to the uncertainty principle, the average distance by which she misses the target must be at least

$$\Delta x_f = \left(\frac{2\hbar}{m}\right)^{1/2}\left(\frac{2H}{g}\right)^{1/4}$$

where H is the initial height of each pellet above the floor and m is the mass of each pellet. Assume that the spread in impact points is given by $\Delta x_f = \Delta x_i + (\Delta v_x)t$. (b) If $H = 2.00$ m and $m = 0.500$ g, what is Δx_f?

60. Show that the speed of a particle having de Broglie wavelength λ and Compton wavelength $\lambda_C = h/(mc)$ is

$$v = \frac{c}{\sqrt{1 + (\lambda/\lambda_C)^2}}$$

61. (a) Find the mass of a solid iron sphere 2.00 cm in radius. (b) Assume that it is at 20°C and has emissivity 0.860. Find the power with which it is radiating electromagnetic waves. (c) If this sphere were alone in the Universe, at what rate would its temperature be changing? (d) Assume Wien's law describes the sphere. Find the wavelength λ_{max} of electromagnetic radiation it emits most strongly. Although it emits a spectrum of waves having all different wavelengths, model its whole power output as carried by photons of wavelength λ_{max}. Find (e) the energy of one photon and (f) the number of photons it emits each second. When the sphere is at thermal equilibrium with its surroundings, it emits and also absorbs photons at this rate.

ACTIVITIES

1. Use a black marker or pieces of dark electrical tape to make a very dark area on the outside of a shoebox. Poke a hole in the center of the dark area with a pencil. Now put a lid on the box, and compare the blackness of the hole with the blackness of the surrounding dark area. Based on your observation, explain why the radiation emitted from the hole is like that emitted from a black body.

2. On a clear night, go outdoors far from city lights and find the constellation Orion. Your instructor should be able to furnish you with a star chart to assist you in locating this grouping of stars. Look very carefully at the color of the two stars Betelgeuse and Rigel. (See Fig. A27.2.) Can you tell which star is hotter? Orion is visible only from November through April in the evening sky, so if Orion is not visible when you go out, compare two of the brightest stars you can see, such as Vega in the constellation Lyra and Arcturus in Boötes.

John Chumack/Photo Researchers

Figure A27.2

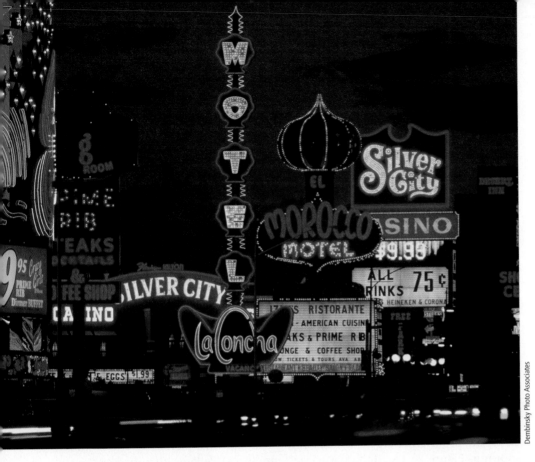

Dembinsky Photo Associates

"Neon lights," commonly used in advertising signs, consist of thin glass tubes filled with various gases, such as neon and helium. The gas atoms are excited to higher energy levels by electric discharge through the tube. When the electrons in these excited levels return to lower energy levels, the atoms emit light having a wavelength (color) that depends on the type of gas in the tube. For example, a tube filled with neon produces a red-orange color, while helium produces pink.

Atomic Physics

A large portion of this chapter concerns the hydrogen atom. Although the hydrogen atom is the simplest atomic system, it's especially important for several reasons:

- The quantum numbers used to characterize the allowed states of hydrogen can also be used to describe (approximately) the allowed states of more complex atoms. This enables us to understand the periodic table of the elements, one of the greatest triumphs of quantum mechanics.
- The hydrogen atom is an ideal system for performing precise comparisons of theory with experiment and for improving our overall understanding of atomic structure.
- Much of what is learned about the hydrogen atom with its single electron can be extended to such single-electron ions as He^+ and Li^{2+}.

In this chapter we first discuss the Bohr model of hydrogen, which helps us understand many features of that element but fails to explain finer details of atomic structure. Next we examine the hydrogen atom from the viewpoint of quantum mechanics and the quantum numbers used to characterize various atomic states. Quantum numbers aren't mere mathematical abstractions: they have physical significance, such as the role they play in the effect of a magnetic field on certain quantum states. The fact that no two electrons in an atom can have the same set of quantum numbers—the Pauli exclusion principle—is extremely important in understanding the properties of complex atoms and the arrangement of elements in the periodic table. Finally, we apply our knowledge of atomic structure to describe the mechanisms involved in the production of x-rays, the operation of a laser, and the behavior of solid-state devices such as diodes and transistors.

28.1   EARLY MODELS OF THE ATOM

The model of the atom in the days of Newton was a tiny, hard, indestructible sphere. Although this model was a good basis for the kinetic theory of gases, new models had to be devised when later experiments revealed the electronic nature of

Figure 28.1 Thomson's model of the atom, with the electrons embedded inside the positive charge like seeds in a watermelon.

SIR JOSEPH JOHN THOMSON, English Physicist (1856–1940)

Thomson, usually considered the discoverer of the electron, opened up the field of subatomic particle physics with his extensive work on the deflection of cathode rays (electrons) in an electric field. He received the 1906 Nobel prize for his discovery of the electron.

atoms. J. J. Thomson (1856–1940) suggested a model of the atom as a volume of positive charge with electrons embedded throughout the volume, much like the seeds in a watermelon (Fig. 28.1).

In 1911 Ernest Rutherford (1871–1937) and his students Hans Geiger and Ernest Marsden performed a critical experiment showing that Thomson's model couldn't be correct. In this experiment, a beam of positively charged **alpha particles** was projected against a thin metal foil, as in Figure 28.2a. The results of the experiment were astounding. Most of the alpha particles passed through the foil as if it were empty space, but a few particles deflected from their original direction of travel were scattered through large angles. Some particles were even deflected backwards, reversing their direction of travel. When Geiger informed Rutherford of these results, Rutherford wrote, "It was quite the most incredible event that has ever happened to me in my life. It was almost as incredible as if you fired a 15-inch shell at a piece of tissue paper and it came back and hit you."

Such large deflections were not expected on the basis of Thomson's model. According to that model, a positively charged alpha particle would never come close enough to a large positive charge to cause any large-angle deflections. Rutherford explained these astounding results by assuming that the positive charge in an atom was concentrated in a region that was small relative to the size of the atom. He called this concentration of positive charge the **nucleus** of the atom. Any electrons belonging to the atom were assumed to be in the relatively large volume outside the nucleus. In order to explain why electrons in this outer region of the atom were not pulled into the nucleus, Rutherford viewed them as moving in orbits about the positively charged nucleus in the same way that planets orbit the Sun, as shown in Figure 28.2b. Alpha particles themselves were later identified as the nuclei of helium atoms.

There are two basic difficulties with Rutherford's planetary model. First, an atom emits certain discrete characteristic frequencies of electromagnetic radiation and no others; the Rutherford model is unable to explain this phenomenon. Second, the electrons in Rutherford's model undergo a centripetal acceleration. According to Maxwell's theory of electromagnetism, centripetally accelerated charges revolving with frequency f should radiate electromagnetic waves of the same frequency. Unfortunately, this classical model leads to disaster when applied to the atom. As the electron radiates energy, the radius of its orbit steadily decreases and its frequency of revolution increases. This leads to an ever-increasing frequency of emitted radiation and a rapid collapse of the atom as the electron spirals into the nucleus.

28.2 ATOMIC SPECTRA

The hydrogen atom is the simplest atomic system and an especially important one to understand. Much of what we know about the hydrogen atom (which consists of one proton and one electron) can be extended directly to other single-electron

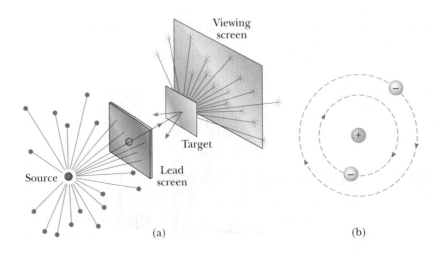

Figure 28.2 (a) Geiger and Marsden's technique for observing the scattering of alpha particles from a thin foil target. The source is a naturally occurring radioactive substance, such as radium. (b) Rutherford's planetary model of the atom.

(a)

(b)

Figure 28.3 Visible spectra. (a) Line spectra produced by emission in the visible range for the elements hydrogen, mercury, and neon. (b) The absorption spectrum for hydrogen. The dark absorption lines occur at the same wavelengths as the emission lines for hydrogen shown in (a).

K. W. Whitten, R. E. Davis, M. L. Peck, and G. G. Stanley, General Chemistry, 7th ed., Belmont, CA, Brooks/Cole, 2004.

ions such as He^+ and Li^{2+}. Further, a thorough understanding of the physics underlying the hydrogen atom can then be used to describe more complex atoms and the periodic table of the elements.

Suppose an evacuated glass tube is filled with hydrogen (or some other gas) at low pressure. If a voltage applied between metal electrodes in the tube is great enough to produce an electric current in the gas, the tube emits light having a color that depends on the gas inside. (This is how a neon sign works.) When the emitted light is analyzed with a spectrometer, discrete bright lines are observed, each having a different wavelength, or color. Such a series of spectral lines is commonly called an **emission spectrum**. The wavelengths contained in such a spectrum are characteristic of the element emitting the light (Fig. 28.3). Because no two elements emit the same line spectrum, this phenomenon represents a marvelous and reliable technique for identifying elements in a gaseous substance.

The emission spectrum of hydrogen shown in Figure 28.4 includes four prominent lines that occur at wavelengths of 656.3 nm, 486.1 nm, 434.1 nm, and 410.2 nm, respectively. In 1885 Johann Balmer (1825–1898) found that the wavelengths of these and less prominent lines can be described by the simple empirical equation

λ(nm)

364.6 410.2 434.1   486.1   656.3

Figure 28.4 The Balmer series of spectral lines for atomic hydrogen, with several lines marked with the wavelength in nanometers. The line labeled 346.6 is the shortest-wavelength line and is in the ultraviolet region of the electromagnetic spectrum. The other labeled lines are in the visible region.

$$\frac{1}{\lambda} = R_H \left(\frac{1}{2^2} - \frac{1}{n^2} \right)$$ **[28.1]**   ◀ Balmer series

where n may have integral values of 3, 4, 5, . . . , and R_H is a constant, called the **Rydberg constant**. If the wavelength is in meters, R_H has the value

$$R_H = 1.097\ 373\ 2 \times 10^7\ m^{-1}$$ **[28.2]**   ◀ Rydberg constant

The first line in the Balmer series, at 656.3 nm, corresponds to $n = 3$ in Equation 28.1, the line at 486.1 nm corresponds to $n = 4$, and so on. In addition to the Balmer series of spectral lines, a Lyman series was subsequently discovered in the far ultraviolet, with the radiated wavelengths described by a similar equation.

In addition to emitting light at specific wavelengths, an element can absorb light at specific wavelengths. The spectral lines corresponding to this process form what is known as an **absorption spectrum**. An absorption spectrum can be obtained by passing a continuous radiation spectrum (one containing all wavelengths) through a vapor of the element being analyzed. The absorption spectrum consists of a series of dark lines superimposed on the otherwise bright continuous spectrum. Each line in the absorption spectrum of a given element coincides with a line in the emission spectrum of the element. This means that if hydrogen is the

APPLICATION

Discovery of Helium

absorbing vapor, dark lines will appear at the visible wavelengths 656.3 nm, 486.1 nm, 434.1 nm, and 410.2 nm, as shown in Figures 28.3b and 28.4.

The absorption spectrum of an element has many practical applications. For example, the continuous spectrum of radiation emitted by the Sun must pass through the cooler gases of the solar atmosphere before reaching the Earth. The various absorption lines observed in the solar spectrum have been used to identify elements in the solar atmosphere, including one that was previously unknown. When the solar spectrum was first being studied, some lines were found that didn't correspond to any known element. A new element had been discovered! Because the Greek word for Sun is *helios*, the new element was named *helium*. It was later identified in underground gases on Earth. Scientists are able to examine the light from stars other than our Sun in this way, but elements other than those present on Earth have never been detected.

Applying Physics 28.1   Thermal or Spectral?

On observing a yellow candle flame, your laboratory partner claims that the light from the flame originates from excited sodium atoms in the flame. You disagree, stating that because the candle flame is hot, the radiation must be thermal in origin. Before the disagreement leads to fisticuffs, how could you determine who is correct?

Explanation A simple determination could be made by observing the light from the candle flame through a spectrometer, which is a slit and diffraction grating combination discussed in Chapter 25. If the spectrum of the light is continuous, then it's probably thermal in origin. If the spectrum shows discrete lines, it's atomic in origin. The results of the experiment show that the light is indeed thermal in origin and originates from random molecular motion in the candle flame.

Applying Physics 28.2   Auroras

At extreme northern latitudes, the aurora borealis provides a beautiful and colorful display in the night sky. A similar display occurs near the southern polar region and is called the aurora australis. What's the origin of the various colors seen in the auroras?

Explanation The aurora is due to high speed particles interacting with the Earth's magnetic field and entering the atmosphere. When these particles collide with molecules in the atmosphere, they excite the molecules in a way similar to the voltage in the spectrum tubes discussed earlier in this section. In response, the molecules emit colors of light according to the characteristic spectrum of their atomic constituents. For our atmosphere, the primary constituents are nitrogen and oxygen, which provide the red, blue, and green colors of the aurora.

28.3   THE BOHR THEORY OF HYDROGEN

At the beginning of the 20th century, scientists were perplexed by the failure of classical physics to explain the characteristics of spectra. Why did atoms of a given element emit only certain lines? Further, why did the atoms absorb only those wavelengths that they emitted? In 1913 Bohr provided an explanation of atomic spectra that includes some features of the currently accepted theory. Using the simplest atom, hydrogen, Bohr developed a model of what he thought must be the atom's structure in an attempt to explain why the atom was stable. His model of the hydrogen atom contains some classical features, as well as some revolutionary postulates that could not be justified within the framework of classical physics. The basic assumptions of the Bohr theory as it applies to the hydrogen atom are as follows:

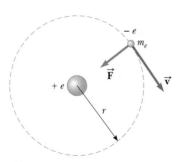

Figure 28.5 Diagram representing Bohr's model of the hydrogen atom. The orbiting electron is allowed only in specific orbits of discrete radii.

1. The electron moves in circular orbits about the proton under the influence of the Coulomb force of attraction, as in Figure 28.5. The Coulomb force produces the electron's centripetal acceleration.

2. Only certain electron orbits are stable. These are orbits in which the hydrogen atom doesn't emit energy in the form of electromagnetic radiation. Hence, the total energy of the atom remains constant, and classical mechanics can be used to describe the electron's motion.

3. Radiation is emitted by the hydrogen atom when the electron "jumps" from a more energetic initial state to a less energetic state. The "jump" can't be visualized or treated classically. In particular, the frequency f of the radiation emitted in the jump is related to the change in the atom's energy and is *independent of the frequency of the electron's orbital motion*. The frequency of the emitted radiation is given by

$$E_i - E_f = hf \qquad \text{[28.3]}$$

where E_i is the energy of the initial state, E_f is the energy of the final state, h is Planck's constant, and $E_i > E_f$.

4. The size of the allowed electron orbits is determined by a condition imposed on the electron's orbital angular momentum: the allowed orbits are those for which the electron's orbital angular momentum about the nucleus is an integral multiple of $\hbar$ (pronounced "h bar"), where $\hbar = h/2\pi$:

$$m_e v r = n\hbar \quad n = 1, 2, 3, \dots \qquad \text{[28.4]}$$

With these four assumptions, we can calculate the allowed energies and emission wavelengths of the hydrogen atom. We use the model pictured in Figure 28.5, in which the electron travels in a circular orbit of radius r with an orbital speed v.

The electrical potential energy of the atom is

$$PE = k_e \frac{q_1 q_2}{r} = k_e \frac{(-e)(e)}{r} = -k_e \frac{e^2}{r}$$

where k_e is the Coulomb constant. Assuming the nucleus is at rest, the total energy E of the atom is the sum of the kinetic and potential energy:

$$E = KE + PE = \tfrac{1}{2} m_e v^2 - k_e \frac{e^2}{r} \qquad \text{[28.5]}$$

We apply Newton's second law to the electron. We know that the electric force of attraction on the electron, $k_e e^2 / r^2$, must equal $m_e a_r$, where $a_r = v^2 / r$ is the centripetal acceleration of the electron. Thus,

$$k_e \frac{e^2}{r^2} = m_e \frac{v^2}{r} \qquad \text{[28.6]}$$

From this equation, we see that the kinetic energy of the electron is

$$\tfrac{1}{2} m_e v^2 = \frac{k_e e^2}{2r} \qquad \text{[28.7]}$$

We can combine this result with Equation 28.5 and express the energy of the atom as

$$E = -\frac{k_e e^2}{2r} \qquad \text{[28.8]}$$

◀ Energy of the hydrogen atom

where the negative value of the energy indicates that the electron is bound to the proton.

An expression for r is obtained by solving Equations 28.4 and 28.6 for v and equating the results:

$$v^2 = \frac{n^2 \hbar^2}{m_e^2 r^2} = \frac{k_e e^2}{m_e r}$$

$$r_n = \frac{n^2 \hbar^2}{m_e k_e e^2} \quad n = 1, 2, 3, \dots \qquad \text{[28.9]}$$

◀ The radii of the Bohr orbits are quantized

This equation is based on the assumption that the **electron can exist only in certain allowed orbits determined by the integer n.**

NIELS BOHR, Danish Physicist (1885–1962)

Bohr was an active participant in the early development of quantum mechanics and provided much of its philosophical framework. During the 1920s and 1930s, he headed the Institute for Advanced Studies in Copenhagen. The institute was a magnet for many of the world's best physicists and provided a forum for the exchange of ideas. When Bohr visited the United States in 1939 to attend a scientific conference, he brought news that the fission of uranium had been observed by Hahn and Strassman in Berlin. The results were the foundations of the atomic bomb developed in the United States during World War II. Bohr was awarded the 1922 Nobel Prize for his investigation of the structure of atoms and of the radiation emanating from them.

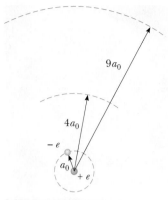

The orbit with the smallest radius, called the **Bohr radius**, a_0, corresponds to $n = 1$ and has the value

$$a_0 = \frac{\hbar^2}{m k_e e^2} = 0.052\ 9 \text{ nm} \qquad [28.10]$$

A general expression for the radius of any orbit in the hydrogen atom is obtained by substituting Equation 28.10 into Equation 28.9:

$$r_n = n^2 a_0 = n^2 (0.052\ 9 \text{ nm}) \qquad [28.11]$$

The first three Bohr orbits for hydrogen are shown in Active Figure 28.6.

Equation 28.9 may be substituted into Equation 28.8 to give the following expression for the energies of the quantum states:

$$E_n = -\frac{m_e k_e^2 e^4}{2\hbar^2}\left(\frac{1}{n^2}\right) \quad n = 1, 2, 3, \ldots \qquad [28.12]$$

If we insert numerical values into Equation 28.12, we obtain

$$E_n = -\frac{13.6}{n^2} \text{ eV} \qquad [28.13]$$

The lowest energy state, or **ground state**, corresponds to $n = 1$ and has an energy $E_1 = -m_e k_e^2 e^4 / 2\hbar^2 = -13.6$ eV. The next state, corresponding to $n = 2$, has an energy $E_2 = E_1/4 = -3.40$ eV, and so on. An energy level diagram showing the energies of these stationary states and the corresponding quantum numbers is given in Active Figure 28.7. The uppermost level shown, corresponding to $E = 0$ and $n \to \infty$, represents the state for which the electron is completely removed from the atom. In this state, the electron's KE and PE are both zero, which means that the electron is at rest infinitely far away from the proton. The minimum energy required to ionize the atom—that is, to completely remove the electron—is called the **ionization energy**. The ionization energy for hydrogen is 13.6 eV.

Equations 28.3 and 28.12 and the third Bohr postulate show that if the electron jumps from one orbit with quantum number n_i to a second orbit with quantum number, n_f, it emits a photon of frequency f given by

$$f = \frac{E_i - E_f}{h} = \frac{m_e k_e^2 e^4}{4\pi\hbar^3}\left(\frac{1}{n_f^2} - \frac{1}{n_i^2}\right) \qquad [28.14]$$

where $n_f < n_i$.

Finally, to compare this result with the empirical formulas for the various spectral series, we use Equation 28.14 and the fact that for light, $\lambda f = c$, to get

$$\frac{1}{\lambda} = \frac{f}{c} = \frac{m_e k_e^2 e^4}{4\pi c \hbar^3}\left(\frac{1}{n_f^2} - \frac{1}{n_i^2}\right) \qquad [28.15]$$

A comparison of this result with Equation 28.1 gives the following expression for the Rydberg constant:

$$R_H = \frac{m_e k_e^2 e^4}{4\pi c \hbar^3} \qquad [28.16]$$

If we insert the known values of m_e, k_e, e, c, and $\hbar$ into this expression, the resulting theoretical value for R_H is found to be in excellent agreement with the value determined experimentally for the Rydberg constant. When Bohr demonstrated this agreement, it was recognized as a major accomplishment of his theory.

In order to compare Equation 28.15 with spectroscopic data, it is convenient to express it in the form

$$\frac{1}{\lambda} = R_H\left(\frac{1}{n_f^2} - \frac{1}{n_i^2}\right) \qquad [28.17]$$

Energy level diagram:

n	E (eV)
∞	0.00
5	−0.5442
4	−0.8504
3	−1.512
2	−3.401
1	−13.606

Paschen series

Balmer series

Lyman series

ENERGY

We can use this expression to evaluate the wavelengths for the various series in the hydrogen spectrum. For example, in the Balmer series, $n_f = 2$ and $n_i = 3, 4, 5, \ldots$ (Eq. 28.1). For the Lyman series, we take $n_f = 1$ and $n_i = 2, 3, 4, \ldots$. The energy level diagram for hydrogen shown in Active Figure 28.7 indicates the origin of the spectral lines described previously. The transitions between levels are represented by vertical arrows. Note that whenever a transition occurs between a state designated by n_i to one designated by n_f (where $n_i > n_f$), a photon with a frequency $(E_i - E_f)/h$ is emitted. This can be interpreted as follows: the lines in the visible part of the hydrogen spectrum arise when the electron jumps from the third, fourth, or even higher orbit to the second orbit. Likewise, the lines of the Lyman series (in the ultraviolet) arise when the electron jumps from the second, third, or even higher orbit to the innermost ($n_f = 1$) orbit. Hence, the Bohr theory successfully predicts the wavelengths of all the observed spectral lines of hydrogen.

TIP 28.1 Energy Depends On n Only for Hydrogen

According to Equation 28.13, the energy depends only on the quantum number n. Note that this is only true for the hydrogen atom. For more complicated atoms, the energy levels depend primarily on n, but also on other quantum numbers.

INTERACTIVE EXAMPLE 28.1   The Balmer Series for Hydrogen

Goal   Calculate the wavelength, frequency, and energy of a photon emitted during an electron transition in an atom.

Problem   The Balmer series for the hydrogen atom corresponds to electronic transitions that terminate in the state with quantum number $n = 2$, as shown in Figure 28.8. **(a)** Find the longest-wavelength photon emitted in the Balmer series and determine its frequency and energy. **(b)** Find the shortest-wavelength photon emitted in the same series.

Strategy   This is a matter of substituting values into Equation 28.17. The frequency can then be obtained from $c = f\lambda$ and the energy from $E = hf$. The longest wavelength photon corresponds to the one that is emitted when the electron jumps from the $n_i = 3$ state to the $n_f = 2$ state. The shortest wavelength photon corresponds to the one that is emitted when the electron jumps from $n_i = \infty$ to the state $n_f = 2$.

Figure 28.8   (Example 28.1) Transitions responsible for the Balmer series for the hydrogen atom. All transitions terminate at the $n = 2$ level.

Solution

(a) Find the longest wavelength photon emitted in the Balmer series, and determine its frequency and energy.

Substitute into Equation 28.17, with $n_i = 3$ and $n_f = 2$:

$$\frac{1}{\lambda} = R_H \left(\frac{1}{n_f^2} - \frac{1}{n_i^2} \right) = R_H \left(\frac{1}{2^2} - \frac{1}{3^2} \right) = \frac{5R_H}{36}$$

Take the reciprocal and substitute, finding the wavelength:

$$\lambda = \frac{36}{5R_H} = \frac{36}{5(1.097 \times 10^7 \, \text{m}^{-1})} = 6.563 \times 10^{-7} \, \text{m}$$

$$= \boxed{656.3 \, \text{nm}}$$

Now use $c = f\lambda$ to obtain the frequency:

$$f = \frac{c}{\lambda} = \frac{2.998 \times 10^8 \, \text{m/s}}{6.563 \times 10^{-7} \, \text{m}} = \boxed{4.568 \times 10^{14} \, \text{Hz}}$$

Calculate the photon's energy by substituting into Equation 27.5:

$$E = hf = (6.626 \times 10^{-34} \, \text{J} \cdot \text{s})(4.568 \times 10^{14} \, \text{Hz})$$

$$= 3.027 \times 10^{-19} \, \text{J} = \boxed{1.892 \, \text{eV}}$$

(b) Find the shortest wavelength photon emitted in the Balmer series.

Substitute into Equation 28.17, with $n_i = \infty$ and $n_f = 2$.

$$\frac{1}{\lambda} = R_H \left(\frac{1}{n_f^2} - \frac{1}{n_i^2} \right) = R_H \left(\frac{1}{2^2} - \frac{1}{\infty} \right) = \frac{R_H}{4}$$

Take the reciprocal and substitute, finding the wavelength:

$$\lambda = \frac{4}{R_H} = \frac{4}{(1.097 \times 10^7 \, \text{m}^{-1})} = 3.646 \times 10^{-7} \, \text{m}$$

$$= \boxed{364.6 \, \text{nm}}$$

Remarks   The first wavelength is in the red region of the visible spectrum. We could also obtain the energy of the photon by using Equation 28.3 in the form $hf = E_3 - E_2$, where E_2 and E_3 are the energy levels of the hydrogen atom, calculated from Equation 28.13. Note that this is the lowest energy photon in the Balmer series, because it involves the smallest energy change. The second photon, the most energetic, is in the ultraviolet region.

Exercise 28.1

(a) Calculate the energy of the shortest wavelength photon emitted in the Balmer series for hydrogen. (b) Calculate the wavelength of a transition from $n = 4$ to $n = 2$.

Answers   (a) 3.40 eV (b) 486 nm

Physics⊗Now™   Investigate transitions between various states by logging into PhysicsNow at **www.cp7e.com** and going to Interactive Example 28.1.

___

Bohr's Correspondence Principle

In our study of relativity in Chapter 26, we found that Newtonian mechanics cannot be used to describe phenomena that occur at speeds approaching the speed of light. Newtonian mechanics is a special case of relativistic mechanics and applies only when v is much smaller than c. Similarly, **quantum mechanics is in agreement with classical physics when the energy differences between quantized levels are very small**. This principle, first set forth by Bohr, is called the **correspondence principle**.

For example, consider the hydrogen atom with $n > 10\,000$. For such large values of n, the energy differences between adjacent levels approach zero and the levels are nearly continuous, as Equation 28.13 shows. As a consequence, the classical model is reasonably accurate in describing the system for large values of n. According to the classical model, the frequency of the light emitted by the atom is equal to the frequency of revolution of the electron in its orbit about the nucleus. Calculations show that for $n > 10\,000$, this frequency is different from that predicted by quantum mechanics by less than 0.015%.

28.4   MODIFICATION OF THE BOHR THEORY

The Bohr theory of the hydrogen atom was a tremendous success in certain areas because it explained several features of the hydrogen spectrum that had previously defied explanation. It accounted for the Balmer series and other series; it predicted a value for the Rydberg constant that is in excellent agreement with the experimental value; it gave an expression for the radius of the atom; and it predicted the energy levels of hydrogen. Although these successes were important to scientists, it is perhaps even more significant that the Bohr theory gave us a model of what the atom looks like and how it behaves. Once a basic model is constructed, refinements and modifications can be made to enlarge on the concept and to explain finer details.

The analysis used in the Bohr theory is also successful when applied to *hydrogen-like* atoms. An atom is said to be hydrogen-like when it contains only one electron. Examples are singly ionized helium, doubly ionized lithium, triply ionized beryllium, and so forth. The results of the Bohr theory for hydrogen can be extended to hydrogen-like atoms by substituting Ze^2 for e^2 in the hydrogen equations, where Z is the atomic number of the element. For example, Equations 28.12 and 28.15 become

$$E_n = -\frac{m_e k_e^2 Z^2 e^4}{2\hbar^2}\left(\frac{1}{n^2}\right) \qquad n = 1, 2, 3, \ldots \qquad \textbf{[28.18]}$$

and

$$\frac{1}{\lambda} = \frac{m_e k_e^2 Z^2 e^4}{4\pi c \hbar^3}\left(\frac{1}{n_f^2} - \frac{1}{n_i^2}\right) \qquad\qquad \textbf{[28.19]}$$

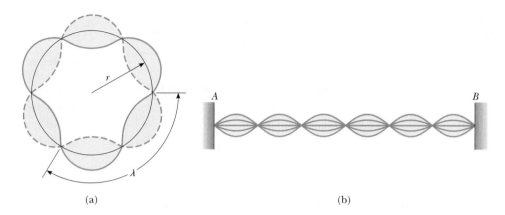

(a)   (b)

Figure 28.10 (a) Standing-wave pattern for an electron wave in a stable orbit of hydrogen. There are three full wavelengths in this orbit. (b) Standing-wave pattern for a vibrating stretched string fixed at its ends. This pattern also has three full wavelengths.

For more than a decade following Bohr's publication, no one was able to explain why the angular momentum of the electron was restricted to these discrete values. Finally, de Broglie gave a direct physical way of interpreting this condition. He assumed that an electron orbit would be stable (allowed) only if it contained an integral number of electron wavelengths. Figure 28.10a demonstrates this point when three complete wavelengths are contained in one circumference of the orbit. Similar patterns can be drawn for orbits containing one wavelength, two wavelengths, four wavelengths, five wavelengths, and so forth. These waves are analogous to standing waves on a string, discussed in Chapter 14. There, we found that strings have preferred (resonant) frequencies of vibration. Figure 28.10b shows a standing-wave pattern containing three wavelengths for a string fixed at each end. Now imagine that the vibrating string is removed from its supports at A and B and bent into a circular shape that brings those points together. The end result is a pattern such as the one shown in Figure 28.10a.

In general, the condition for a de Broglie standing wave in an electron orbit is that the circumference must contain an integral number of electron wavelengths. We can express this condition as

$$2\pi r = n\lambda \qquad n = 1, 2, 3, \ldots$$

Because the de Broglie wavelength of an electron is $\lambda = h/m_e v$, we can write the preceding equation as $2\pi r = nh/m_e v$, or

$$m_e v r = n\hbar$$

This is the same as the quantization of angular momentum condition imposed by Bohr in his original theory of hydrogen.

The electron orbit shown in Figure 28.10a contains three complete wavelengths and corresponds to the case in which the principal quantum number $n = 3$. The orbit with one complete wavelength in its circumference corresponds to the first Bohr orbit, $n = 1$; the orbit with two complete wavelengths corresponds to the second Bohr orbit, $n = 2$; and so forth.

By applying the wave theory of matter to electrons in atoms, de Broglie was able to explain the appearance of integers in the Bohr theory as a natural consequence of standing-wave patterns. This was the first convincing argument that the wave nature of matter was at the heart of the behavior of atomic systems. Although the analysis provided by de Broglie was a promising first step, gigantic strides were made subsequently with the development of Schrödinger's wave equation and its application to atomic systems.

28.6 QUANTUM MECHANICS AND THE HYDROGEN ATOM

One of the first great achievements of quantum mechanics was the solution of the wave equation for the hydrogen atom. The details of the solution are far beyond the level of this course, but we'll describe its properties and implications for atomic structure.

TABLE 28.2

Three Quantum Numbers for the Hydrogen Atom

Quantum Number	Name	Allowed Values	Number of Allowed States
n	Principal quantum number	1, 2, 3, . . .	Any number
ℓ	Orbital quantum number	0, 1, 2, . . . , $n-1$	n
m_ℓ	Orbital magnetic quantum number	$-\ell, -\ell+1, \ldots,$ 0, . . . , $\ell-1, \ell$	$2\ell+1$

According to quantum mechanics, the energies of the allowed states are in exact agreement with the values obtained by the Bohr theory (Eq. 28.12) when the allowed energies depend only on the principal quantum number n.

In addition to the principal quantum number, two other quantum numbers emerged from the solution of the wave equation: ℓ and m_ℓ. The quantum number ℓ is called the **orbital quantum number**, and m_ℓ is called the **orbital magnetic quantum number**. As pointed out in Section 28.4, these quantum numbers had already appeared in empirical modifications made to the Bohr theory. The significance of quantum mechanics is that those numbers and the restrictions placed on their values arose directly from mathematics and not from any ad hoc assumptions to make the theory consistent with experimental observation. Because we will need to make use of the various quantum numbers in the sections that follow, the allowed ranges of their values are repeated:

The value of n can range from 1 to ∞ in integer steps.
The value of ℓ can range from 0 to $n-1$ in integer steps.
The value of m_ℓ can range from $-\ell$ to ℓ in integer steps.

From these rules, it can be seen that for a given value of n, there are n possible values of ℓ, while for a given value of ℓ there are $2\ell+1$ possible values of m_ℓ. For example, if $n=1$, there is only 1 value of ℓ, $\ell=0$. Because $2\ell+1 = 2\cdot0+1 = 1$, there is only one value of m_ℓ, which is $m_\ell=0$. If $n=2$, the value of ℓ may be 0 or 1; if $\ell=0$, then $m_\ell=0$, but if $\ell=1$, then m_ℓ may be 1, 0, or -1. Table 28.2 summarizes the rules for determining the allowed values of ℓ and m_ℓ for a given value of n.

States that violate the rules given in Table 28.2 cannot exist. For instance, one state that cannot exist is the $2d$ state, which would have $n=2$ and $\ell=2$. This state is not allowed because the highest allowed value of ℓ is $n-1$, or 1 in this case. Thus, for $n=2$, $2s$ and $2p$ are allowed states, but $2d$, $2f$, . . . are not. For $n=3$, the allowed states are $3s$, $3p$, and $3d$.

In general, for a given value of n there are n^2 states with distinct pairs of values of ℓ and m_ℓ.

Quick Quiz 28.3

When the principal quantum number is $n=5$, how many different values of (a) ℓ and (b) m_ℓ are possible? (c) How many states have distinct pairs of values of ℓ and m_ℓ?

EXAMPLE 28.3 The $n=2$ Level of Hydrogen

Goal Count states and determine energy based on atomic energy level.

Problem (a) Determine the number of states with a unique set of values for ℓ and m_ℓ in the hydrogen atom for $n=2$. (b) Calculate the energies of these states.

Strategy This is a matter of counting, following the quantum rules for n, ℓ, and m_ℓ. "Unique" means that no other quantum state has the same pair of numbers for ℓ and m_ℓ. The energies are all the same because all states have the same principal quantum number, $n=2$.

Solution

(a) Determine the number of states with a unique set of values for ℓ and m_ℓ in the hydrogen atom for $n = 2$.

Determine the different possible values of ℓ for $n = 2$:

$0 \le \ell \le n - 1$, so, for $n = 2$, $0 \le \ell \le 1$ and $\ell = 0$ or 1

Find the different possible values of m_ℓ for $\ell = 0$:

$-\ell \le m_\ell \le \ell$, so $-0 \le m_\ell \le 0$ implies $m_\ell = 0$

List the distinct pairs of (ℓ, m_ℓ) for $\ell = 0$:

There is only one: $(\ell, m_\ell) = (0, 0)$.

Find the different possible values of m_ℓ for $\ell = 1$:

$-\ell \le m_\ell \le \ell$, so $-1 \le m_\ell \le 1$ implies $m_\ell = -1, 0,$ or 1

List the distinct pairs of (ℓ, m_ℓ) for $\ell = 1$:

There are three: $(\ell, m_\ell) = (1, -1), (1, 0),$ and $(1, 1)$.

Sum the results for $\ell = 0$ and $\ell = 1$:

Number of states $= 1 + 3 = \boxed{4}$

(b) Calculate the energies of these states.

The common energy of all of the states can be found with Equation 28.13:

$$E_n = -\frac{13.6 \text{ eV}}{n^2} \quad \rightarrow \quad E_2 = -\frac{13.6 \text{ eV}}{2^2} = \boxed{-3.40 \text{ eV}}$$

Remarks While these states normally have the same energy, applying a magnetic field will result in their taking slightly different energies centered around the energy corresponding to $n = 2$. As seen in the next section, there are in fact twice as many states, corresponding to a new quantum number called *spin*.

Exercise 28.3

(a) Determine the number of states with a unique pair of values for ℓ and m_ℓ in the $n = 3$ level of hydrogen. (b) Determine the energies of those states.

Answers   (a) 9   (b) $E_3 = -1.51 \text{ eV}$

28.7 THE SPIN MAGNETIC QUANTUM NUMBER

As we'll see in this section, there actually are *eight* states corresponding to $n = 2$ for hydrogen, not four as given in Example 28.3. This happens because another quantum number, m_s, the **spin magnetic quantum number**, has to be introduced to explain the splitting of each level into two.

The need for this new quantum number first came about because of an unusual feature in the spectra of certain gases, such as sodium vapor. Close examination of one of the prominent lines of sodium shows that it is, in fact, two very closely spaced lines. The wavelengths of these lines occur in the yellow region of the spectrum, at 589.0 nm and 589.6 nm. In 1925, when this doublet was first noticed, atomic theory couldn't explain it. To resolve the dilemma, Samuel Goudsmit and George Uhlenbeck, following a suggestion by the Austrian physicist Wolfgang Pauli, proposed the introduction of a fourth quantum number to describe atomic energy levels, called the *spin quantum number*.

In order to describe the spin quantum number, it's convenient (but technically incorrect) to think of the electron as spinning on its axis as it orbits the nucleus, just as the Earth spins on its axis as it orbits the Sun. Strangely, there are only two ways in which the electron can spin as it orbits the nucleus, as shown in Figure 28.11. If the direction of spin is as shown in Figure 28.11a, the electron is said to have "spin up." If the direction of spin is reversed, as in Figure 28.11b, the electron is said to have "spin down." The energy of the electron is slightly different for the two spin directions, and this energy difference accounts for the sodium doublet. The quantum numbers associated with electron spin are $m_s = \frac{1}{2}$ for the spin-up state and $m_s = -\frac{1}{2}$ for the spin-down state. As we'll see in Example 28.4, this new quantum number doubles the number of allowed states specified by the quantum numbers n, ℓ, and m_ℓ.

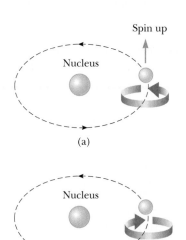

Figure 28.11 As an electron moves in its orbit about the nucleus, its spin can be either (a) up or (b) down.

Any classical description of electron spin is incorrect because quantum mechanics tells us that since the electron can't be located precisely in space, it cannot be considered to be a spinning solid object, as pictured in Figure 28.11. In spite of this conceptual difficulty, all experimental evidence supports the fact that an electron does have some intrinsic property that can be described by the spin magnetic quantum number.

The spin quantum number didn't come from the original formulation of quantum mechanics by Schrodinger (and independently, by Heisenberg). The English mathematical physicist P. A. M. Dirac developed a relativistic quantum theory in which spin appears naturally.

EXAMPLE 28.4   The Quantum Numbers for the 2p Subshell

Goal   List the distinct quantum states of a subshell by their quantum numbers, including spin.

Problem   List the unique sets of quantum numbers for electrons in the 2p subshell.

Strategy   This is again a matter following the quantum rules for n, ℓ, and m_ℓ, and now m_s as well. The 2p subshell has $n = 2$ (that's the "2" in 2p) and $\ell = 1$ (that's from the p in 2p).

Solution

Because $\ell = 1$, the magnetic quantum number can have the values -1, 0, 1, and the spin quantum number is always $+\frac{1}{2}$ or $-\frac{1}{2}$. Consequently, there are $3 \times 2 = 6$ possible sets of quantum numbers with $n = 2$ and $\ell = 1$, listed in the table at right.

n	ℓ	m_ℓ	m_s
2	1	-1	$-\frac{1}{2}$
2	1	-1	$\frac{1}{2}$
2	1	0	$-\frac{1}{2}$
2	1	0	$\frac{1}{2}$
2	1	1	$-\frac{1}{2}$
2	1	1	$\frac{1}{2}$

Remark   Remember that these quantum states are not just abstractions; they have real physical consequences, such as which electronic transitions can be made within an atom and, consequently, which wavelengths of radiation can be observed.

Exercise 28.4

(a) How many different sets of quantum numbers are there in the 3d subshell? (b) How many sets of quantum numbers are there in a 2d subshell?

Answers   (a) 10   (b) None. A 2d subshell doesn't exist because that would imply a quantum state with $n = 2$ and $\ell = 2$, impossible because $\ell \leq n - 1$.

28.8   ELECTRON CLOUDS

The solution of the wave equation, discussed in Section 27.7, yields a wave function Ψ that depends on the quantum numbers n, ℓ, and m_ℓ. We assume that we have found such a wave function Ψ and see what it may tell us about the hydrogen atom. Let $n = 1$ for the principal quantum number, which corresponds to the lowest energy state for hydrogen. For $n = 1$, the restrictions placed on the remaining quantum numbers are that $\ell = 0$ and $m_\ell = 0$.

The quantity Ψ^2 has great physical significance. If p is a point and V_p a very small volume containing that point, then $\Psi^2 V_p$ is approximately the probability of finding the electron inside the volume V_p. Figure 28.12 gives the probability per unit length of finding the electron at various distances from the nucleus in the 1s state of hydrogen. Some useful and surprising information can be extracted from this curve. First, the curve peaks at a value of $r = 0.052\ 9$ nm, the Bohr radius for the first ($n = 1$) electron orbit in hydrogen. This means that there is a maximum probability of finding the electron in a small interval centered at that distance from the nucleus. However, as the curve indicates, there is also a probability of finding the electron in a small interval centered at any other distance from the nucleus. In other words, the electron is not confined to a particular orbital distance from the nucleus, as assumed in the Bohr model. The electron may be found at various distances from the nucleus, but **the probability of finding it at a distance corresponding to the Bohr radius is a maximum**. Quantum mechanics also predicts that the wave function for the hydrogen atom in the ground state is spherically symmetric; hence the electron can be found in a spherical region surrounding the nucleus. This is in contrast to the Bohr theory, which confines the position of the electron to points in a plane. The quantum mechanical result is often interpreted by viewing the electron as a cloud surrounding the nucleus. An attempt at picturing this cloud-like behavior is shown in Figure 28.13. The densest regions of the cloud represent those locations where the electron is most likely to be found.

If a similar analysis is carried out for the $n = 2$, $\ell = 0$, state of hydrogen, a peak of the probability curve is found at $4a_0$. Likewise, for the $n = 3$, $\ell = 0$ state, the curve peaks at $9a_0$. Thus, quantum mechanics predicts a most probable electron distance to the nucleus that is in agreement with the location predicted by the Bohr theory.

28.9   THE EXCLUSION PRINCIPLE AND THE PERIODIC TABLE

Earlier, we found that the state of an electron in an atom is specified by four quantum numbers: n, ℓ, m_ℓ, and m_s. For example, an electron in the ground state of hydrogen could have quantum numbers of $n = 1$, $\ell = 0$, $m_\ell = 0$, and $m_s = \frac{1}{2}$. As it turns out, the state of an electron in any other atom may also be specified by this same set of quantum numbers. In fact, these four quantum numbers can be used to describe all the electronic states of an atom, regardless of the number of electrons in its structure.

How many electrons in an atom can have a particular set of quantum numbers? This important question was answered by Pauli in 1925 in a powerful statement known as the **Pauli exclusion principle**:

> No two electrons in an atom can ever have the same set of values for the set of quantum numbers n, ℓ, m_ℓ, and m_s.

The Pauli exclusion principle explains the electronic structure of complex atoms as a succession of filled levels with different quantum numbers increasing in energy, where the outermost electrons are primarily responsible for the chemical properties of the element. If this principle weren't valid, every electron would end up in the lowest energy state of the atom and the chemical behavior of the elements would be grossly different. Nature as we know it would not exist—and *we* would not exist to wonder about it!

As a general rule, the order that electrons fill an atom's subshell is as follows: once one subshell is filled, the next electron goes into the vacant subshell that is lowest in energy. If the atom were not in the lowest energy state available to it, it would radiate energy until it reached that state. A subshell is filled when it contains $2(2\ell + 1)$ electrons. This rule is based on the analysis of quantum numbers to be described later. Following the rule, shells and subshells can contain numbers of electrons according to the pattern given in Table 28.3.

The exclusion principle can be illustrated by an examination of the electronic arrangement in a few of the lighter atoms.

Hydrogen has only one electron, which, in its ground state, can be described by either of two sets of quantum numbers: 1, 0, 0, $\frac{1}{2}$ or 1, 0, 0, $-\frac{1}{2}$. The electronic configuration of this atom is often designated as $1s^1$. The notation $1s$ refers to a

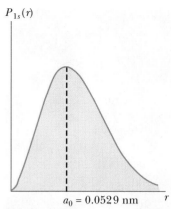

Figure 28.12 The probability per unit length of finding the electron versus distance from the nucleus for the hydrogen atom in the $1s$ (ground) state. Note that the graph has its maximum value when r equals the first Bohr radius, a_0.

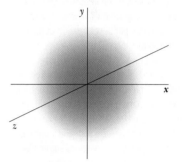

Figure 28.13 The spherical electron cloud for the hydrogen atom in its $1s$ state.

◀ The Pauli exclusion principle

TIP 28.3   The Exclusion Principle is More General

The exclusion principle stated here is a limited form of the more general exclusion principle, which states that no two *fermions* (particles with spin $1/2, 3/2, \ldots$) can be in the same quantum state.

WOLFGANG PAULI (1900–1958)

An extremely talented Austrian theoretical physicist who made important contributions in many areas of modern physics, Pauli gained public recognition at the age of 21 with a masterful review article on relativity that is still considered one of the finest and most comprehensive introductions to the subject. Other major contributions were the discovery of the exclusion principle, the explanation of the connection between particle spin and statistics, and theories of relativistic quantum electrodynamics, the neutrino hypothesis, and the hypothesis of nuclear spin.

TABLE 28.3

Number of Electrons in Filled Subshells and Shells

Shell	Subshell	Number of Electrons in Filled Subshell	Number of Electrons in Filled Shell
K ($n = 1$)	$s(\ell = 0)$	2	2
L ($n = 2$)	$s(\ell = 0)$	2	8
	$p(\ell = 1)$	6	
M ($n = 3$)	$s(\ell = 0)$	2	18
	$p(\ell = 1)$	6	
	$d(\ell = 2)$	10	
N ($n = 4$)	$s(\ell = 0)$	2	32
	$p(\ell = 1)$	6	
	$d(\ell = 2)$	10	
	$f(\ell = 3)$	14	

state for which $n = 1$ and $\ell = 0$, and the superscript indicates that one electron is present in this level.

Neutral *helium* has two electrons. In the ground state, the quantum numbers for these two electrons are 1, 0, 0, $\frac{1}{2}$ and 1, 0, 0, $-\frac{1}{2}$. No other possible combinations of quantum numbers exist for this level, and we say that the K shell is filled. The helium electronic configuration is designated as $1s^2$.

Neutral *lithium* has three electrons. In the ground state, two of these are in the $1s$ subshell and the third is in the $2s$ subshell, because the latter is lower in energy than the $2p$ subshell. Hence, the electronic configuration for lithium is $1s^2 2s^1$.

A list of electronic ground-state configurations for a number of atoms is provided in Table 28.4. In 1871 Dmitri Mendeleev (1834–1907), a Russian chemist, arranged the elements known at that time into a table according to their atomic masses and chemical similarities. The first table Mendeleev proposed contained many blank spaces, and he boldly stated that the gaps were there only because those elements had not yet been discovered. By noting the column in which these missing elements should be located, he was able to make rough predictions about their chemical properties. Within 20 years of this announcement, the elements were indeed discovered.

The elements in our current version of the periodic table are still arranged so that all those in a vertical column have similar chemical properties. For example, consider the elements in the last column: He (helium), Ne (neon), Ar (argon), Kr (krypton), Xe (xenon), and Rn (radon). The outstanding characteristic of these elements is that they don't normally take part in chemical reactions, joining with other atoms to form molecules, and are therefore classified as inert. Because of this "aloofness," they are referred to as the *noble gases*. We can partially understand their behavior by looking at the electronic configurations shown in Table 28.4, page 919. The element helium has the electronic configuration $1s^2$. In other words, one shell is filled. The electrons in this filled shell are considerably separated in energy from the next available level, the $2s$ level.

The electronic configuration for neon is $1s^2 2s^2 2p^6$. Again, the outer shell is filled and there is a large difference in energy between the $2p$ level and the $3s$ level. Argon has the configuration $1s^2 2s^2 2p^6 3s^2 3p^6$. Here, the $3p$ subshell is filled and there is a wide gap in energy between the $3p$ subshell and the $3d$ subshell. Through all the noble gases, the pattern remains the same: a noble gas is formed when either a shell or a subshell is filled, and there is a large gap in energy before the next possible level is encountered.

The elements in the first column of the periodic table are called the *alkali metals* and are highly active chemically. Referring to Table 28.4, we can understand why these elements interact so strongly with other elements. All of these alkali

TABLE 28.4

Electronic Configurations of Some Elements

Z	Symbol	Ground-State Configuration	Ionization Energy (eV)	Z	Symbol	Ground-State Configuration	Ionization Energy (eV)
1	H	$1s^1$	13.595	19	K	[Ar] $4s^1$	4.339
2	He	$1s^2$	24.581	20	Ca	$4s^2$	6.111
				21	Sc	$3d4s^2$	6.54
3	Li	[He] $2s^1$	5.390	22	Ti	$3d^24s^2$	6.83
4	Be	$2s^2$	9.320	23	V	$3d^34s^2$	6.74
5	B	$2s^22p^1$	8.296	24	Cr	$3d^54s^1$	6.76
6	C	$2s^22p^2$	11.256	25	Mn	$3d^54s^2$	7.432
7	N	$2s^22p^3$	14.545	26	Fe	$3d^64s^2$	7.87
8	O	$2s^22p^4$	13.614	27	Co	$3d^74s^2$	7.86
9	F	$2s^22p^5$	17.418	28	Ni	$3d^84s^2$	7.633
10	Ne	$2s^22p^6$	21.559	29	Cu	$3d^{10}4s^1$	7.724
				30	Zn	$3d^{10}4s^2$	9.391
11	Na	[Ne] $3s^1$	5.138	31	Ga	$3d^{10}4s^24p^1$	6.00
12	Mg	$3s^2$	7.644	32	Ge	$3d^{10}4s^24p^2$	7.88
13	Al	$3s^23p^1$	5.984	33	As	$3d^{10}4s^24p^3$	9.81
14	Si	$3s^23p^2$	8.149	34	Se	$3d^{10}4s^24p^4$	9.75
15	P	$3s^23p^3$	10.484	35	Br	$3d^{10}4s^24p^5$	11.84
16	S	$3s^23p^4$	10.357	36	Kr	$3d^{10}4s^24p^6$	13.996
17	Cl	$3s^23p^5$	13.01				
18	Ar	$3s^23p^6$	15.755				

Note: The bracket notation is used as a shorthand method to avoid repetition in indicating inner-shell electrons. Thus, [He] represents $1s^2$, [Ne] represents $1s^22s^22p^6$, [Ar] represents $1s^22s^22p^63s^23p^6$, and so on.

metals have a single outer electron in an *s* subshell. This electron is shielded from the nucleus by all the electrons in the inner shells. Consequently, it's only loosely bound to the atom and can readily be accepted by other atoms that bind it more tightly to form molecules.

The elements in the seventh column of the periodic table are called the *halogens* and are also highly active chemically. All these elements are lacking one electron in a subshell, so they readily accept electrons from other atoms to form molecules.

Quick Quiz 28.4

Krypton (atomic number 36) has how many electrons in its next to outer shell ($n = 3$)?
(a) 2   (b) 4   (c) 8   (d) 18

Applying Physics 28.3   The Periodic Table

Scanning from left to right across one row of the periodic table, the effective size of the atoms first decreases and then increases. What would cause this behavior?

Explanation Starting on the left side of the periodic table and moving toward the middle, the nuclear charge is increasing. As a result, there is an increasing Coulomb attraction between the nucleus and the electrons, and the electrons are pulled into an average position that is closer to the nucleus. From the middle of the row to the right side, the increasing number of electrons being placed in proximity to each other results in a mutual repulsion that increases the average distance from the nucleus and causes the atomic size to grow.

Figure 28.14 The x-ray spectrum of a metal target consists of a broad continuous spectrum (*bremsstrahlung*) plus a number of sharp lines that are due to *characteristic x-rays*. The data shown were obtained when 35-keV electrons bombarded a molybdenum target. Note that 1 pm = 10^{-12} m = 0.001 nm.

28.10 CHARACTERISTIC X-RAYS

X-rays are emitted when a metal target is bombarded with high-energy electrons. The x-ray spectrum typically consists of a broad continuous band and a series of intense sharp lines that are dependent on the type of metal used for the target, as shown in Figure 28.14. These discrete lines, called **characteristic x-rays**, were discovered in 1908, but their origin remained unexplained until the details of atomic structure were developed.

The first step in the production of characteristic x-rays occurs when a bombarding electron collides with an electron in an inner shell of a target atom with sufficient energy to remove the electron from the atom. The vacancy created in the shell is filled when an electron in a higher level drops down into the lower energy level containing the vacancy. The time it takes for this to happen is very short, less than 10^{-9} s. The transition is accompanied by the emission of a photon with energy equaling the difference in energy between the two levels. Typically, the energy of such transitions is greater than 1 000 eV, and the emitted x-ray photons have wavelengths in the range of 0.01 nm to 1 nm.

We assume that the incoming electron has dislodged an atomic electron from the innermost shell, the K shell. If the vacancy is filled by an electron dropping from the next higher shell, the L shell, the photon emitted in the process is referred to as the K_α line on the curve of Figure 28.14. If the vacancy is filled by an electron dropping from the M shell, the line produced is called the K_β line.

Other characteristic x-ray lines are formed when electrons drop from upper levels to vacancies other than those in the K shell. For example, L lines are produced when vacancies in the L shell are filled by electrons dropping from higher shells. An L_α line is produced as an electron drops from the M shell to the L shell, and an L_β line is produced by a transition from the N shell to the L shell.

We can estimate the energy of the emitted x-rays as follows: consider two electrons in the K shell of an atom whose atomic number is Z. Each electron partially shields the other from the charge of the nucleus, Ze, so each is subject to an effective nuclear charge $Z_{\text{eff}} = (Z - 1)e$. We can now use a modified form of Equation 28.18 to estimate the energy of either electron in the K shell (with $n = 1$). We have

$$E_K = -m_e Z_{\text{eff}}^2 \frac{k_e^2 e^4}{2\hbar^2} = -Z_{\text{eff}}^2 E_0$$

where E_0 is the ground-state energy. Substituting $Z_{\text{eff}} = Z - 1$ gives

$$E_K = -(Z - 1)^2 (13.6 \text{ eV}) \qquad \textbf{[28.20]}$$

As Example 28.5 will show, we can estimate the energy of an electron in an L or an M shell in a similar fashion. Taking the energy difference between these two levels, we can then calculate the energy and wavelength of the emitted photon.

In 1914, Henry G. J. Moseley plotted the Z values for a number of elements against $\sqrt{1/\lambda}$, where λ is the wavelength of the K_α line for each element. He found that such a plot produced a straight line, as in Figure 28.15. This is consistent with our rough calculations of the energy levels based on Equation 28.20. From his plot, Moseley was able to determine the Z values of other elements, providing a periodic chart in excellent agreement with the known chemical properties of the elements.

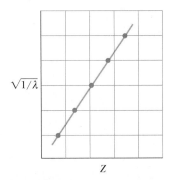

Figure 28.15 A Moseley plot of $\sqrt{1/\lambda}$ versus Z, where λ is the wavelength of the K_α x-ray line of the element of atomic number Z.

EXAMPLE 28.5   Characteristic X-Rays

Goal   Calculate the energy and wavelength of characteristic x-rays.

Problem   Estimate the energy and wavelength of the characteristic x-ray emitted from a tungsten target when an electron drops from an M shell ($n = 3$ state) to a vacancy in the K shell ($n = 1$ state).

Strategy   Develop two estimates, one for the electron in the K shell ($n = 1$) and one for the electron in the M shell ($n = 3$). For the K-shell estimate, we can use Equation 28.20. For the M shell, we need a new equation. There is one electron in the K shell (because one is missing) and 8 in the L shell, making 9 electrons shielding the nuclear charge. This means $Z_{eff} = 74 - 9$ and $E_M = -Z_{eff}^2 E_3$, where E_3 is the energy of the $n = 3$ level in hydrogen. The difference $E_M - E_K$ is the energy of the photon.

Solution

Use Equation 28.20 to estimate the energy of an electron in the K shell of tungsten, atomic number $Z = 74$:

$$E_K = -(74 - 1)^2(13.6 \text{ eV}) = -72\ 500 \text{ eV}$$

Estimate the energy of an electron in the M shell in the same way:

$$E_M = -Z_{eff}^2 E_3 = -(Z - 9)^2 \frac{E_0}{3^2} = -(74 - 9)^2 \frac{(13.6 \text{ eV})}{9}$$
$$= -6\ 380 \text{ eV}$$

Calculate the difference in energy between the M and K shells:

$$E_M - E_K = -6\ 380 \text{ eV} - (-72\ 500 \text{ eV}) = \boxed{66\ 100 \text{ eV}}$$

Find the wavelength of the emitted x-ray:

$$\Delta E = hf = h\frac{c}{\lambda} \quad \rightarrow \quad \lambda = \frac{hc}{\Delta E}$$

$$\lambda = \frac{(6.63 \times 10^{-34} \text{ J·s})(3.00 \times 10^8 \text{ m/s})}{(6.61 \times 10^4 \text{ eV})(1.60 \times 10^{-19} \text{ J/eV})}$$
$$= 1.88 \times 10^{-11} \text{ m} = \boxed{0.018\ 8 \text{ nm}}$$

Exercise 28.5

Repeat the problem for a $2p$ electron transiting from the L shell to the K shell. (For technical reasons, the L shell electron must have $\ell = 1$, so a single $1s$ electron and two $2s$ electrons shield the nucleus.)

Answer   (a) 5.54×10^4 eV   (b) $0.022\ 4$ nm

28.11 ATOMIC TRANSITIONS

We have seen that an atom will emit radiation only at certain frequencies that correspond to the energy separation between the various allowed states. Consider an atom with many allowed energy states, labeled E_1, E_2, E_3, . . . , as in Figure 28.16. When light is incident on the atom, only those photons whose energy hf matches the energy separation ΔE between two levels can be absorbed by the atom. A schematic diagram representing this **stimulated absorption process** is shown in Active Figure 28.17. At ordinary temperatures, most of the atoms in a sample are in the ground state. If a vessel containing many atoms of a gas is illuminated with a light beam containing all possible photon frequencies (that is, a continuous spectrum), only those photons of energies $E_2 - E_1$, $E_3 - E_1$, $E_4 - E_1$, and so on, can be absorbed. As a result of this absorption, some atoms are raised to various allowed higher energy levels, called **excited states**.

Once an atom is in an excited state, there is a constant probability that it will jump back to a lower level by emitting a photon, as shown in Active Figure 28.18 (page 922).

Figure 28.16 Energy level diagram of an atom with various allowed states. The lowest energy state, E_1, is the ground state. All others are excited states.

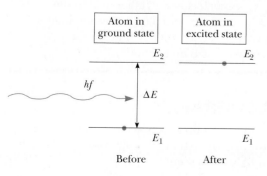

Before   After

ACTIVE FIGURE 28.17
Diagram representing the process of *stimulated absorption* of a photon by an atom. The blue dot represents an electron. The electron is transferred from the ground state to the excited state when the atom absorbs a photon of energy $hf = E_2 - E_1$.

Physics Now™

Log into PhysicsNow at **www.cp7e.com** and go to Active Figure 28.17 to observe stimulated absorption.

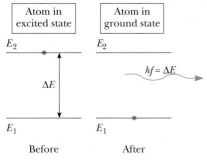

Before   After

ACTIVE FIGURE 28.18

Diagram representing the process of *spontaneous emission* of a photon by an atom that is initially in the excited state E_2. When the electron falls to the ground state, the atom emits a photon of energy $hf = E_2 - E_1$.

Physics⊗Now™

Log into PhysicsNow at **www.cp7e.com** and go to Active Figure 28.17 to observe spontaneous emission.

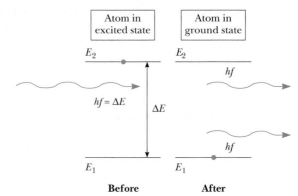

Before   After

ACTIVE FIGURE 28.19

Diagram representing the process of *stimulated emission* of a photon by an incoming photon of energy *hf*. Initially, the atom is in the excited state. The incoming photon stimulates the atom to emit a second photon of energy $hf = E_2 - E_1$.

Physics⊗Now™

Log into PhysicsNow at **www.cp7e.com** and go to Active Figure 28.17 to observe stimulated emission.

This process is known as **spontaneous emission**. Typically, an atom will remain in an excited state for only about 10^{-8} s.

A third process that is important in lasers, **stimulated emission**, was predicted by Einstein in 1917. Suppose an atom is in the excited state E_2, as in Active Figure 28.19, and a photon with energy $hf = E_2 - E_1$ is incident on it. The incoming photon increases the probability that the excited atom will return to the ground state and thereby emit a second photon having the same energy *hf*. Note that two identical photons result from stimulated emission: the incident photon and the emitted photon. *The emitted photon is exactly in phase with the incident photon.* These photons can stimulate other atoms to emit photons in a chain of similar processes. The many photons produced in this fashion are the source of the intense, coherent (in-phase) light in a laser.

Applying Physics 28.4   Streaking Meteoroids

A physics student is watching a meteor shower in the early morning hours. She notices that the streaks of light from the meteoroids entering the very high regions of the atmosphere last for as long as 2 or 3 seconds before fading. She also notices a lightning storm off in the distance. The streaks of light from the lightning fade away almost immediately after the flash, certainly in much less than 1 second. Both lightning and meteors cause the air to turn into a plasma because of the very high temperatures generated. The light is given off when the stripped electrons in the plasma recombine with the ionized atoms. Why would the light last longer for meteors than for lightning?

Explanation To answer this question, we examine the phrase "the streaks of light from the meteoroids

entering the very high regions of the atmosphere." In the very high regions of the atmosphere, the pressure is very low, so the density is also very low and the atoms of the gas are relatively far apart. Low density means that after the air is ionized by the passing meteoroid, the probability of freed electrons finding an ionized atom with which to recombine is relatively low. As a result, the recombination process occurs over a relatively long time, measured in seconds. Lightning, however, occurs in the lower regions of the atmosphere (the troposphere), where the pressure and density are relatively high. After the ionization by the lightning flash, the electrons and ionized atoms are much closer together than in the upper atmosphere. The probability of a recombination is accordingly much higher, and the time for the recombination to occur is much shorter.

28.12   LASERS AND HOLOGRAPHY

We have described how an incident photon can cause atomic transitions either upward (stimulated absorption) or downward (stimulated emission). The two processes are equally probable. When light is incident on a system of atoms, there is usually a net absorption of energy, because when the system is in thermal equilibrium, there are many more atoms in the ground state than in excited states. However, if the situation can be inverted so that there are more atoms in an excited state than in the ground state, a net emission of photons can result. Such a condition is called **population inversion**. This is the fundamental principle involved in the operation of a laser, an acronym for *light amplification by stimulated emission of radiation*. The amplification corresponds to a buildup of photons in the system as the result of a chain reaction of events. The following three conditions must be satisfied in order to achieve laser action:

1. The system must be in a state of population inversion (that is, more atoms in an excited state than in the ground state).
2. The excited state of the system must be a *metastable state*, which means its lifetime must be long compared with the otherwise usually short lifetimes of excited states. When that is the case, stimulated emission will occur before spontaneous emission.
3. The emitted photons must be confined within the system long enough to allow them to stimulate further emission from other excited atoms. This is achieved by the use of reflecting mirrors at the ends of the system. One end is totally reflecting, and the other is slightly transparent to allow the laser beam to escape.

One device that exhibits stimulated emission of radiation is the helium–neon gas laser. Figure 28.20 is an energy-level diagram for the neon atom in this system. The mixture of helium and neon is confined to a glass tube sealed at the ends by mirrors. A high voltage applied to the tube causes electrons to sweep through it, colliding with the atoms of the gas and raising them into excited states. Neon atoms are excited to state E_3^* through this process and also as a result of collisions with excited helium atoms. When a neon atom makes a transition to state E_2, it stimulates emission by neighboring excited atoms. This results in the production of coherent light at a wavelength of 632.8 nm. Figure 28.21 summarizes the steps in the production of a laser beam.

Figure 28.20 Energy-level diagram for the neon atom in a helium–neon laser. The atom emits 632.8-nm photons through stimulated emission in the transition $E_3^* \rightarrow E_2$. This is the source of coherent light in the laser.

(a)

(b)

Figure 28.21 (a) Steps in the production of a laser beam. The tube contains atoms, which represent the active medium. An external source of energy (optical, electrical, etc.) is needed to "pump" the atoms to excited energy states. The parallel end mirrors provide the feedback of the stimulating wave. (b) Photograph of the first ruby laser, showing the flash lamp surrounding the ruby rod.

Courtesy of Central Scientific Company

(a)   (b)

Figure 28.22 (a) Experimental arrangement for producing a hologram. (b) Photograph of a hologram made with a cylindrical film. Note the detail of the Volkswagen image.

Philippe Plailly/Photo Researchers, Inc.

Scientist checking the performance of an experimental laser-cutting device mounted on a robot arm. The laser is being used to cut through a metal plate.

Since the development of the first laser in 1960, laser technology has exhibited tremendous growth. Lasers that cover wavelengths in the infrared, visible, and ultraviolet regions of the spectrum are now available. Applications include the surgical "welding" of detached retinas, "lasik" surgery, precision surveying and length measurement, a potential source for inducing nuclear fusion reactions, precision cutting of metals and other materials, and telephone communication along optical fibers. These and other applications are possible because of the unique characteristics of laser light. In addition to being highly monochromatic and coherent, laser light is also highly directional and can be sharply focused to produce regions of extremely intense light energy.

Holography

One interesting application of the laser is holography: the production of three-dimensional images of objects. Figure 28.22a shows how a hologram is made. Light from the laser is split into two parts by a half-silvered mirror at B. One part of the beam reflects off the object to be photographed and strikes an ordinary photographic film. The other half of the beam is diverged by lens L_2, reflects from mirrors M_1 and M_2, and finally strikes the film. The two beams overlap to form an extremely complicated interference pattern on the film, one that can be produced only if the phase relationship of the waves is constant throughout the exposure of the film. This condition is met through the use of light from a laser, because such light is coherent. The hologram records not only the intensity of the light scattered from the object (as in a conventional photograph), but also the phase difference between the reference beam and the beam scattered from the object. Because of this phase difference, an interference pattern is formed that produces an image with full three-dimensional perspective.

A hologram is best viewed by allowing coherent light to pass through the developed film while you look back along the direction from which the beam comes. Figure 28.22b is a photograph of a hologram made using a cylindrical film.

28.13 ENERGY BANDS IN SOLIDS

In this section we trace the changes that occur in the discrete energy levels of isolated atoms when the atoms group together and form a solid. We find that in solids, the discrete levels of isolated atoms broaden into allowed energy bands

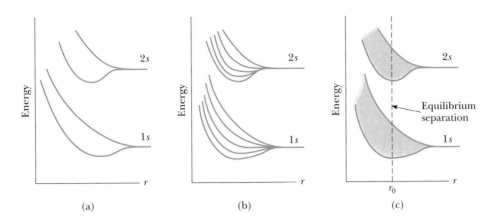

Figure 28.23   (a) Splitting of the 1s and 2s states when two atoms are brought together. (b) Splitting of the 1s and 2s states when five atoms are brought close together. (c) Formation of energy bands when a large number of sodium atoms are assembled to form a solid.

separated by forbidden gaps. The separation and electron population of the highest bands determines whether a given solid is a conductor, an insulator, or a semiconductor.

Consider two identical atoms, initially widely separated, that are brought closer and closer together. If two identical atoms are very far apart, they do not interact, and their electronic energy levels can be considered to be those of isolated atoms. Hence, the energy levels are exactly the same. As the atoms come close together, they essentially become one quantum system, and the Pauli exclusion principle demands that the electrons be in different quantum states for this single system. The exclusion principle manifests itself as a changing or splitting of electron energy levels that were identical in the widely separated atoms, as shown in Figure 28.23a. Figure 28.23b shows that with 5 atoms, each energy level in the isolated atom splits into five different, more closely spaced levels.

If we extend this argument to the large number of atoms found in solids (on the order of 10^{23} atoms/cm^3), we obtain a large number of levels so closely spaced that they may be regarded as a continuous **band** of energy levels, as in Figure 28.23c. An electron can have any energy within an allowed energy band, but cannot have an energy in the **band gap**, or the region between allowed bands. Note that the band gap energy E_g is indicated in Figure 28.23c. In practice we are only interested in the band structure of a solid at some equilibrium separation of its atoms r_0, and so we remove the distance scale on the x-axis and simply plot the allowed energy bands of a solid as a series of horizontal bands, as shown in Figure 28.24 for sodium.

Conductors and Insulators

Figure 28.24 shows that the band structure of a particular solid is quite complicated with individual atomic levels broadening by varying amounts and some levels (3s and 3p) broadening so much that they overlap. Nevertheless, it is possible to gain a qualitative understanding of whether a solid is a conductor, an insulator, or a semiconductor by considering only the structure of the upper or upper two energy bands and whether they are occupied by electrons.

Deciding whether an energy band is empty (unoccupied by electrons), partially filled, or full is carried out in basically the same way as for the energy-level population of atoms: we distribute the total number of electrons from the lowest energy levels up in a way consistent with the exclusion principle. While we omit the details of this process here, one important case is that shown in Figure 28.25a (page 926), where the highest-energy occupied band is only partially full. The other important case, where the highest occupied band is completely full, is shown in Figure 28.25b. Notice that this figure also shows that the highest filled band is called the **valence band** and the next higher empty band is called the **conduction band**. The energy band gap, which varies with the solid, is also indicated as the energy difference E_g between the top of the valence band and the bottom of the conduction band.

Figure 28.24   Energy bands of sodium. Note the energy gaps (white regions) between the allowed bands; electrons can't occupy states that lie in these forbidden gaps. Blue represents energy bands occupied by the sodium electrons when the atom is in its ground state. Gold represents energy bands that are empty. Note that the 3s and 3p levels broaden so much that they overlap.

Figure 28.25 (a) Half-filled band of a metal, an electrical conductor. (b) An electrical insulator at $T = 0$ K has a filled valence band and an empty conduction band. (c) Band structure of a semiconductor at ordinary temperatures ($T \approx 300$ K). The energy gap is much smaller than in an insulator, and many electrons occupy states in the conduction band.

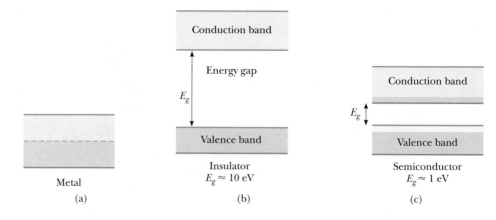

With these ideas and definitions we are now in a position to understand what determines, quantum mechanically, whether a solid will be a conductor or an insulator. When a modest voltage is applied to a good conductor, the electrons accelerate and gain energy. In quantum terms, electron energies increase *if there are higher unoccupied energy levels for electrons to jump to.* For example, electrons near the top of the partially filled band in sodium need to gain very little energy from the applied voltage to reach one of the nearby, closely spaced, empty states. Thus, it is easy for a small voltage to kick electrons into higher energy states, and charge flows easily in sodium, an excellent conductor.

Now consider the case of a material in which the highest occupied band is completely full of electrons and there is a band gap separating this filled valence band from the vacant conduction band, as in Figure 28.25b. A typical case might be diamond (carbon), in which the band gap is about 10 eV. When a voltage is applied, electrons can't easily gain energy, because there are no vacant energy states nearby to which electrons can make transitions. Because the only empty band is the conduction band, an electron must gain an amount of energy at least equal to the band gap in order for it to move through the solid. This large amount of energy can't be supplied by a modest applied voltage, so no charge flows and diamond is a good insulator. In summary then, a conductor has a highest-energy occupied band which is *partially filled*, and in an insulator, has a highest-energy occupied band which is *completely filled* with a large energy gap between the valence and conduction bands.

Semiconductors

To this point, we have completely ignored the influence of temperature on the electronic populations of energy bands. Recalling that the average thermal energy of a particle at temperature T is $3k_B T/2$, we find that an electron at room temperature has an average energy of about 0.04 eV. Because this energy is about 100 times smaller than the band gap in a typical insulator, very few electrons would have enough random thermal energy to jump the energy gap in an insulator and contribute to conduction. However things are different for a semiconductor. As we see in Figure 28.25c, a **semiconductor** is a material with a small band gap of about 1 eV whose conductivity results from appreciable thermal excitation of electrons across the gap into the conduction band at room temperature. The most commonly used semiconductors are silicon and gallium arsenide, with band gaps of 1.14 eV and 1.43 eV, respectively, at 300 K. As you might expect, the resistivity of semiconductors usually decreases with increasing temperature, because $k_B T$ becomes a larger fraction of the band gap energy.

It is interesting that the electrons in the conduction band of a semiconductor don't carry the entire current when a voltage is applied, as Figure 28.26 shows. (It might be said that conduction electrons do not constitute the "whole" story.) The missing electrons in the valence band, shown as a narrow white band in the

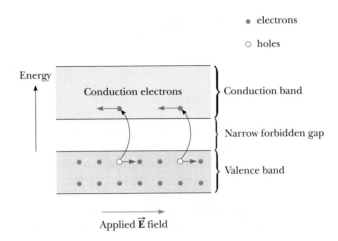

figure, provide a few empty states called **holes** for valence band electrons to fill; so some electrons in the valence band can gain energy and move towards a positive electrode and thus also carry the current. Since the valence band electrons that fill holes leave behind other holes, it is equally valid and more common to view the conduction process in the valence band as a flow of positive holes towards the negative electrode applied to a semiconductor. Thus, a pure semiconductor, such as silicon, can be viewed in a symmetric way: silicon has equal numbers of mobile electrons in the conduction band and holes in the valence band. Furthermore, when an external voltage is applied to the semiconductor, electrons move toward the positive electrode and holes move toward the negative electrode. In the next section we will look at the concepts of an electron and a hole in a simpler, more graphic way as the presence or absence of an outer-shell electron at a particular location in a crystal lattice.

When small amounts of impurities are added to a semiconductor such as silicon (about one impurity atom per 10^7 silicon atoms), both the band structure of the semiconductor and its resistivity are modified. The process of adding impurities, called **doping**, is important in making devices having well-defined regions of different resistivity. For example, when an atom containing five outer-shell electrons, such as arsenic, is added to a semiconductor such as silicon, four of the arsenic electrons form shared bonds with atoms of the semiconductor and one is left over. This extra electron is nearly free of its parent atom and has an energy level that lies in the energy gap, just below the conduction band. Such a pentavalent atom in effect donates an electron to the structure and hence is referred to as a **donor atom**. Because the spacing between the energy level of the electron of the donor atom and the bottom of the conduction band is very small (typically, about 0.05 eV), only a small amount of thermal energy is needed to cause this electron to move into the conduction band. (Recall that the average thermal energy of an electron at room temperature is $3k_BT/2 \approx 0.04$ eV). Semiconductors doped with donor atoms are called *n*-**type semiconductors**, because the charge carriers are electrons, the charge of which is *n*egative.

If a semiconductor is doped with atoms containing three outer-shell electrons, such as aluminum, the three electrons form shared bonds with neighboring semiconductor atoms, leaving an electron deficiency—a hole—where the fourth bond would be if an impurity-atom electron was available to form it. The energy level of this hole lies in the energy gap, just above the valence band. An electron from the valence band has enough energy at room temperature to fill that impurity level, leaving behind a hole in the valence band. Because a trivalent atom, in effect, accepts an electron from the valence band, such impurities are referred to as **acceptor atoms**. A semiconductor doped with acceptor impurities is known as a *p*-**type semiconductor**, because the majority of charge carriers are *p*ositively charged holes.

28.14 SEMICONDUCTOR DEVICES

The $p-n$ Junction

Now let us consider what happens when a p-semiconductor is joined to an n-semiconductor to form a $p-n$ junction. The junction consists of the three distinct regions shown in Figure 28.27a: a p-region, a depletion region, and an n-region.

The depletion region, which extends several micrometers to either side of the center of the junction, may be visualized as arising when the two halves of the junction are brought together. Mobile donor electrons from the n side nearest the junction (the blue area in Fig. 28.27a) diffuse to the p side, leaving behind immobile positive ions. At the same time, holes from the p side nearest the junction diffuse to the n side and leave behind a region (the red area in Fig. 28.27a) of fixed negative ions. The depletion region is so named because it is depleted of mobile charge carriers.

The depletion region contains an internal electric field (arising from the charges of the fixed ions) on the order of 10^4 to 10^6 V/cm. This field sweeps mobile charge out of the depletion region and keeps it truly depleted. This internal electric field creates an internal potential difference ΔV_0 that prevents further diffusion of holes and electrons across the junction and thereby ensures zero current in the junction when no external potential difference is applied.

Perhaps the most notable feature of the $p-n$ junction is its ability to pass current in only one direction. Such *diode* action is easiest to understand in terms of the potential-difference graph shown in Figure 28.27c. If an external voltage ΔV is applied to the junction such that the p side is connected to the positive terminal of a voltage source as in Figure 28.27a, the internal potential difference ΔV_0 across the junction is decreased, resulting in a current that increases exponentially with increasing forward voltage, or *forward bias*. In *reverse bias* (where the n side of the junction is connected to the positive terminal of a voltage source), the internal potential difference ΔV_0 *increases* with increasing reverse bias. This results in a very small reverse current that quickly reaches a saturation value I_0. The current–voltage relationship for an ideal diode is

$$I = I_0(e^{q\Delta V/k_B T} - 1) \qquad\qquad \text{[28.21]}$$

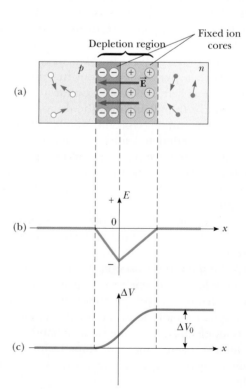

Figure 28.27   (a) Physical arrangement of a $p-n$ junction. (b) Internal electric field versus x for the $p-n$ junction. (c) Internal electric potential ΔV versus x for the $p-n$ junction. ΔV_0 represents the potential difference across the junction in the absence of an applied electric field.

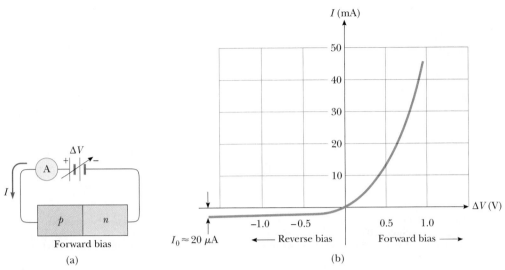

Figure 28.28 (a) Schematic of a $p-n$ junction under forward bias. (b) The characteristic curve for a real $p-n$ junction.

where q is the electron charge, k_B is Boltzmann's constant, and T is the temperature in kelvins. Figure 28.28 shows an I–ΔV plot characteristic of a real $p-n$ junction, along with a schematic of such a device under forward bias.

The most common use of the semiconductor diode is as a rectifier, a device that changes 120-V AC voltage supplied by the power company to, say the 12-V DC voltage needed by your music keyboard. We can understand how a diode rectifies a current by considering Figure 28.29a, which shows a diode connected in series with a resistor and an AC source. Because appreciable current can pass through the diode in just one direction, the alternating current in the resistor is reduced to the form shown in Figure 28.29b. The diode is said to act as a **half-wave rectifier**, because there is current in the circuit during only half of each cycle.

Figure 28.30a shows a circuit that lowers the AC voltage to 12 V with a step-down transformer and then rectifies both halves of the 12-V AC. Such a rectifier is called a **full-wave rectifier** and when combined with a step-down transformer is the most common DC power supply around the home today. A capacitor added in parallel with the load will yield an even steadier DC voltage.

The Junction Transistor

The invention of the transistor by John Bardeen (1908–1991), Walter Brattain (1902–1987), and William Shockley (1910–1989) in 1948 totally revolutionized the world of electronics. For this work, these three men shared a Nobel prize in 1956. By 1960, the transistor had replaced the vacuum tube in many electronic applications. The advent of the transistor created a multitrillion-dollar industry that produced such popular devices as pocket radios, handheld calculators, computers, television receivers, and electronic games. In this section we explain how a transistor acts as an amplifier to boost the tiny voltages and currents generated in a microphone to the ear-splitting levels required to drive a speaker.

One simple form of the transistor, called the **junction transistor**, consists of a semiconducting material in which a very narrow n region is sandwiched between two p regions. This configuration is called a **pnp transistor**. Another configuration is the **npn transistor**, which consists of a p region sandwiched between two n regions. Because the operation of the two transistors is essentially the same, we describe only the **pnp** transistor. The structure of the **pnp** transistor, together with its circuit symbol, is shown in Figure 28.31 (page 930). The outer regions are called the **emitter** and **collector**, and the narrow central region is called the **base**. The configuration contains two junctions: the emitter–base interface and the collector–base interface.

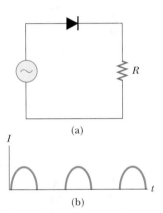

Figure 28.29 (a) A diode in series with a resistor allows current to pass in only one direction. (b) The current versus time for the circuit in (a).

Figure 28.30 (a) A full-wave rectifier circuit. (b) The current versus time in the resistor R.

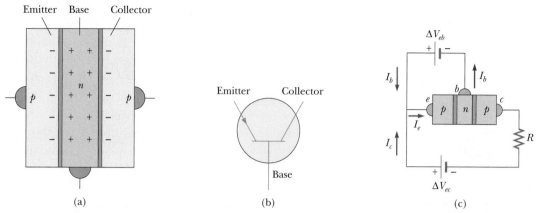

Figure 28.31   (a) The *pnp* transistor consists of an *n* region (base) sandwiched between two *p* regions (emitter and collector). (b) Circuit symbol for the *pnp* transistor. (c) A bias voltage ΔV_{eb} applied to the base as shown produces a small base current I_b that is used to control the collector current I_c in a *pnp* transistor.

Suppose a voltage is applied to the transistor so that the emitter is at a higher electric potential than the collector. (This is accomplished with the battery labeled ΔV_{ec} in Figure 28.31c.) If we think of the transistor as two diodes back to back, we see that the emitter–base junction is forward biased and the base–collector junction is reverse biased. The emitter is heavily doped relative to the base, and as a result, nearly all the current consists of holes moving across the emitter–base junction. Most of these holes do not recombine with electrons in the base because it is very narrow. Instead they are accelerated across the reverse-biased base–collector junction, producing the emitter current I_e in Figure 28.31c.

Although only a small percentage of holes recombine in the base, those that do limit the emitter current to a small value because positive charge carriers accumulating in the base prevent holes from flowing in. In order not to limit the emitter current, some of the positive charge on the base must be drawn off; this is accomplished by connecting the base to the battery labeled ΔV_{eb} in Figure 28.31c. Those positive charges that are not swept across the base–collector junction leave the base through this added pathway. **This base current I_b is very small, but a small change in it can significantly change the collector current I_c.** If the transistor is properly biased, the collector (output) current is directly proportional to the base (input) current and the transistor acts as a current amplifier. This condition may be written

$$I_c = \beta I_b$$

where β, the *current gain* factor, is typically in the range from 10 to 100. Thus, the transistor may be used to amplify a small signal. The small voltage to be amplified is placed in series with the battery V_{eb}. The input signal produces a small variation in the base current, resulting in a large change in the collector current and hence a large change in the voltage across the output resistor.

The Integrated Circuit

Invented independently by Jack Kilby (b. 1923) at Texas Instruments in late 1958 and by Robert Noyce at Fairchild Camera and Instrument in early 1959, the integrated circuit has been justly called "the most remarkable technology ever to hit mankind." Kilby's first device is shown in Figure 28.32a. Integrated circuits have indeed started a "second industrial revolution" and are found at the heart of computers, watches, cameras, automobiles, aircraft, robots, space vehicles, and all sorts of communication and switching networks.

In simplest terms, an **integrated circuit** is a collection of interconnected transistors, diodes, resistors, and capacitors fabricated on a single piece of silicon known as a chip. State-of-the-art chips easily contain several million components in

(a)   (b)

Figure 28.32   (a) Jack Kilby's first integrated circuit was tested on September 12, 1958. (b) Integrated circuits continue to shrink in size and price while simultaneously growing in capability.

a 1-cm^2 area, with the number of components per square inch having doubled every year since the integrated circuit was invented.

Integrated circuits were invented partly to solve the interconnection problem spawned by the transistor. In the era of vacuum tubes, power and size considerations of individual components set significant limits on the number of components that could be interconnected in a given circuit. With the advent of the tiny, low-power, highly reliable transistor, design limits on the number of components disappeared and were replaced by the problem of wiring together hundreds of thousands of components. The magnitude of this problem can be appreciated when we consider that second-generation computers (consisting of discrete transistors rather than integrated circuits) contained several hundred thousand components requiring more than a million hand-soldered joints to be made and tested.

In addition to solving the interconnection problem, integrated circuits possess the advantages of miniaturization and fast response, two attributes critical for high-speed computers. The fast response results from the miniaturization and close packing of components, because the response time of a circuit depends on the time it takes for electrical signals traveling at about the speed of light to pass from one component to another. This time is clearly reduced by packing components closely.

SUMMARY

Physics⊗Now™ Take a practice test by logging into PhysicsNow at **www.cp7e.com** and clicking on the Pre-Test link for this chapter.

28.3 The Bohr Theory of Hydrogen &

28.4 Modification of the Bohr Theory

The **Bohr model** of the atom is successful in describing the spectra of atomic hydrogen and hydrogenlike ions. One of the basic assumptions of the model is that the electron can exist only in certain orbits such that its an-

gular momentum mvr is an integral multiple of $\hbar$, where $\hbar$ is Planck's constant divided by 2π. Assuming circular orbits and a Coulomb force of attraction between electron and proton, the energies of the quantum states for hydrogen are

$$E_n = -\frac{m_e k_e^2 e^4}{2\hbar^2}\left(\frac{1}{n^2}\right) \qquad n = 1, 2, 3, \ldots \qquad \text{[28.12]}$$

where k_e is the Coulomb constant, e is the charge on the electron, and n is an integer called a **quantum number**.

If the electron in the hydrogen atom jumps from an orbit having quantum number n_i to an orbit having quantum number n_f, it emits a photon of frequency f, given by

$$f = \frac{m_e k_e^2 e^4}{4\pi\hbar^3}\left(\frac{1}{n_f^2} - \frac{1}{n_i^2}\right) \qquad [28.14]$$

Bohr's **correspondence principle** states that quantum mechanics is in agreement with classical physics when the quantum numbers for a system are very large.

The Bohr theory can be generalized to hydrogen-like atoms, such as singly ionized helium or doubly ionized lithium. This modification consists of replacing e^2 by Ze^2 wherever it occurs.

28.6   Quantum Mechanics and the Hydrogen Atom &

28.7   The Spin Magnetic Quantum Number

One of the many successes of quantum mechanics is that the quantum numbers n, ℓ, and m_ℓ associated with atomic structure arise directly from the mathematics of the theory. The quantum number n is called the **principal quantum number**, ℓ is the **orbital quantum number**, and m_ℓ is the **orbital magnetic quantum number**. These quantum numbers can take only certain values: $1 \le n < \infty$ in integer steps, $0 \le \ell \le n - 1$, and $-\ell \le m_\ell \le \ell$. In addition, a fourth quantum number, called the **spin magnetic quantum number** m_s, is needed to explain a fine doubling of lines in atomic spectra, with $m_s = \pm\frac{1}{2}$.

28.9   The Exclusion Principle and the Periodic Table

An understanding of the periodic table of the elements became possible when Pauli formulated the **exclusion principle**, which states that no two electrons in an atom in the same atom can have the same values for the set of quantum numbers n, ℓ, m_ℓ, and m_s. A particular set

of these quantum numbers is called a quantum state. The exclusion principle explains how different energy levels in atoms are populated. Once one subshell is filled, the next electron goes into the vacant subshell that is lowest in energy. Atoms with similar configurations in their outermost shell have similar chemical properties and are found in the same column of the periodic table.

28.10   Characteristic X-Rays

Characteristic x-rays are produced when a bombarding electron collides with an electron in an inner shell of an atom with sufficient energy to remove the electron from the atom. The vacancy is filled when an electron from a higher level drops down into the level containing the vacancy, emitting a photon in the x-ray part of the spectrum in the process.

28.11   Atomic Transitions &

28.12   Lasers and Holography

When an atom is irradiated by light of all different wavelengths, it will only absorb only wavelengths equal to the difference in energy of two of its energy levels. This phenomenon, called **stimulated absorption**, places an atom's electrons into **excited states**. Atoms in an excited state have a probability of returning to a lower level of excitation by **spontaneous emission**. The wavelengths that can be emitted are the same as the wavelengths that can be absorbed. If an atom is in an excited state and a photon with energy $hf = E_2 - E_1$ is incident on it, the probability of emission of a second photon of this energy is greatly enhanced. The emitted photon is exactly in phase with the incident photon. This process is called **stimulated emission**. The emitted and original photon can then stimulate more emission, creating an amplifying effect.

Lasers are monochromatic, coherent light sources that work on the principle of **stimulated emission** of radiation from a system of atoms.

CONCEPTUAL QUESTIONS

1. In the hydrogen atom, the quantum number n can increase without limit. Because of this, does the frequency of possible spectral lines from hydrogen also increase without limit?

2. Does the light emitted by a neon sign constitute a continuous spectrum or only a few colors? Defend your answer.

3. In an x-ray tube, if the energy with which the electrons strike the metal target is increased, the wavelengths of the characteristic x-rays do not change. Why not?

4. Must an atom first be ionized before it can emit light? Discuss.

5. Is it possible for a spectrum from an x-ray tube to show the continuous spectrum of x-rays without the presence of the characteristic x-rays?

6. Suppose that the electron in the hydrogen atom obeyed classical mechanics rather than quantum mechanics. Why should such a hypothetical atom emit a continuous spectrum rather than the observed line spectrum?

7. When a hologram is produced, the system (including light source, object, beam splitter, and so on) must be held motionless within a quarter of the light's wavelength. Why?

8. If matter has a wave nature, why is it not observable in our daily experience?

9. Discuss some consequences of the exclusion principle.

10. Can the electron in the ground state of hydrogen absorb a photon of energy less than 13.6 eV? Can it absorb a photon of energy greater than 13.6 eV? Explain.

11. Why do lithium, potassium, and sodium exhibit similar chemical properties?

12. List some ways in which quantum mechanics altered our view of the atom pictured by the Bohr theory.

13. It is easy to understand how two electrons (one with spin up, one with spin down) can fill the $1s$ shell for a helium atom. How is it possible that eight more electrons can fit into the $2s$, $2p$ level to complete the $1s2s^22p^6$ shell for a neon atom?

14. The ionization energies for Li, Na, K, Rb, and Cs are 5.390, 5.138, 4.339, 4.176, and 3.893 eV, respectively. Explain why these values are to be expected in terms of the atomic structures.

15. Why is stimulated emission so important in the operation of a laser?

16. The Bohr theory of the hydrogen atom is based upon several assumptions. Discuss these assumptions and their significance. Do any of them contradict classical physics?

17. Explain why, in the Bohr model, the total energy of the hydrogen atom is negative.

18. Consider the quantum numbers n, ℓ, m_ℓ, and m_s. (a) Which of these are integers and which are fractional? (b) Which are always positive and which can be negative? (c) If $n = 2$, what is the largest value of ℓ? (d) If $\ell = 1$, what are the possible values of m_ℓ?

19. Photon A is emitted when an electron in a hydrogen atom drops from the $n = 3$ level to the $n = 2$ level. Photon B is emitted when an electron in a hydrogen atom drops from the $n = 4$ level to the $n = 2$ level. (a) In which case is the wavelength of the emitted photon greater? (b) In which case is the energy of the emitted photon greater?

PROBLEMS

1, 2, 3 = straightforward, intermediate, challenging   ☐ = full solution available in *Student Solutions Manual/Study Guide*
Physics⊗Now™ = coached problem with hints available at **www.cp7e.com**   = biomedical application

Section 28.1 Early Models of the Atom

Section 28.2 Atomic Spectra

1. Use Equation 28.1 to calculate the wavelength of the first three lines in the Balmer series for hydrogen.

2. Show that the wavelengths for the Balmer series satisfy the equation

$$\lambda = \frac{364.5n^2}{n^2 - 4}\text{nm} \quad \text{where } n = 3, 4, 5, \dots$$

3. The "size" of the *atom* in Rutherford's model is about 1.0×10^{-10} m. (a) Determine the attractive electrostatic force between an electron and a proton separated by this distance. (b) Determine (in eV) the electrostatic potential energy of the atom.

4. The "size" of the *nucleus* in Rutherford's model of the atom is about 1.0 fm = 1.0×10^{-15} m. (a) Determine the repulsive electrostatic force between two protons separated by this distance. (b) Determine (in MeV) the electrostatic potential energy of the pair of protons.

5. Physics⊗Now™ The "size" of the atom in Rutherford's model is about 1.0×10^{-10} m. (a) Determine the speed of an electron moving about the proton using the attractive electrostatic force between an electron and a proton separated by this distance. (b) Does this speed suggest that Einsteinian relativity must be considered in studying the atom? (c) Compute the de Broglie wavelength of the electron as it moves about the proton. (d) Does this wavelength suggest that wave effects, such as diffraction and interference, must be considered in studying the atom?

6. In a Rutherford scattering experiment, an α-particle (charge $= +2e$) heads directly toward a gold nucleus (charge $= +79e$). The α-particle had a kinetic energy of 5.0 MeV when very far ($r \to \infty$) from the nucleus. Assuming the gold nucleus to be fixed in space, determine the distance of closest approach. [*Hint*: Use conservation of energy with $PE = k_e q_1 q_2 / r$.]

Section 28.3 The Bohr Theory of Hydrogen

7. A hydrogen atom is in its first excited state ($n = 2$). Using the Bohr theory of the atom, calculate (a) the radius of the orbit, (b) the linear momentum of the electron, (c) the angular momentum of the electron, (d) the kinetic energy, (e) the potential energy, and (f) the total energy.

8. For a hydrogen atom in its ground state, use the Bohr model to compute (a) the orbital speed of the electron, (b) the kinetic energy of the electron, and (c) the electrical potential energy of the atom.

9. Show that the speed of the electron in the nth Bohr orbit in hydrogen is given by

$$v_n = \frac{k_e e^2}{n\hbar}$$

10. A photon is emitted as a hydrogen atom undergoes a transition from the $n = 6$ state to the $n = 2$ state. Calculate (a) the energy, (b) the wavelength, and (c) the frequency of the emitted photon.

11. A hydrogen atom emits a photon of wavelength 656 nm. From what energy orbit to what lower energy orbit did the electron jump?

12. Following are four possible transitions for a hydrogen atom

    I. $n_i = 2$; $n_f = 5$ II. $n_i = 5$; $n_f = 3$

    III. $n_i = 7$; $n_f = 4$ IV. $n_i = 4$; $n_f = 7$

    (a) Which transition will emit the shortest-wavelength photon? (b) For which transition will the atom gain the most energy? (c) For which transition(s) does the atom lose energy?

13. What is the energy of a photon that, when absorbed by a hydrogen atom, could cause (a) an electronic transition from the $n = 3$ state to the $n = 5$ state and (b) an electronic transition from the $n = 5$ state to the $n = 7$ state?

14. A hydrogen atom initially in its ground state ($n = 1$) absorbs a photon and ends up in the state for which $n = 3$. (a) What is the energy of the absorbed photon? (b) If the atom eventually returns to the ground state, what photon energies could the atom emit?

15. Determine both the longest and the shortest wavelengths in (a) the Lyman series ($n_f = 1$) and (b) the Paschen series ($n_f = 3$) of hydrogen.

16. Show that the speed of the electron in the first (ground-state) Bohr orbit of the hydrogen atom may be expressed as

$$v = (1/137)c.$$

17. A monochromatic beam of light is absorbed by a collection of ground-state hydrogen atoms in such a way that six different wavelengths are observed when the hydrogen relaxes back to the ground state. What is the wavelength of the incident beam?

18. A particle of charge q and mass m, moving with a constant speed v, perpendicular to a constant magnetic field, B, follows a circular path. If in this case the angular momentum about the center of this circle is quantized so that $mvr = 2n\hbar$, show that the allowed radii for the particle are

$$r_n = \sqrt{\frac{2\,n\hbar}{qB}}$$

where $n = 1, 2, 3, \ldots$

19. (a) If an electron makes a transition from the $n = 4$ Bohr orbit to the $n = 2$ orbit, determine the wavelength of the photon created in the process. (b) Assuming that the atom was initially at rest, determine the recoil speed of the hydrogen atom when this photon is emitted.

20. Consider a large number of hydrogen atoms, with electrons all initially in the $n = 4$ state. (a) How many different wavelengths would be observed in the emission spectrum of these atoms? (b) What is the longest wavelength that could be observed? To which series does it belong?

21. Analyze the Earth–Sun system by following the Bohr model, where the gravitational force between Earth (mass m) and Sun (mass M) replaces the Coulomb force between the electron and proton (so that $F = GMm/r^2$ and $PE = -GMm/r$). Show that (a) the total energy of the Earth in an orbit of radius r is given by (a) $E = -GMm/2r$, (b) the radius of the nth orbit is given by $r_n = r_0 n^2$, where $r_0 = \hbar^2/GMm^2 = 2.32 \times 10^{-138}$ m, and (c) the energy of the nth orbit is given by $E_n = -E_0/n^2$, where $E_0 = G^2 M^2 m^3/2\hbar^2 = 1.71 \times 10^{182}$ J. (d) Using the Earth–Sun orbit radius of $r = 1.49 \times 10^{11}$ m, determine the value of the quan-

tum number n. (e) Should you expect to observe quantum effects in the Earth–Sun system?

22. An electron is in the nth Bohr orbit of the hydrogen atom. (a) Show that the period of the electron is $T = t_0 n^3$, and determine the numerical value of t_0. (b) On the average, an electron remains in the $n = 2$ orbit for about 10 μs before it jumps down to the $n = 1$ (ground-state) orbit. How many revolutions does the electron make before it jumps to the ground state? (c) If one revolution of the electron is defined as an "electron year" (analogous to an Earth year being one revolution of the Earth around the Sun), does the electron in the $n = 2$ orbit "live" very long? Explain. (d) How does the above calculation support the "electron cloud" concept?

23. **Physics⊗Now™** Consider a hydrogen atom. (a) Calculate the frequency f of the $n = 2 \rightarrow n = 1$ transition, and compare it with the frequency f_{orb} of the electron orbital motion in the $n = 2$ state. (b) Make the same calculation for the $n = 10\ 000 \rightarrow n = 9\ 999$ transition. Comment on the results.

24. Two hydrogen atoms collide head-on and end up with zero kinetic energy. Each then emits a 121.6-nm photon ($n = 2$ to $n = 1$ transition). At what speed were the atoms moving before the collision?

25. Two hydrogen atoms, both initially in the ground state, undergo a head-on collision. If both atoms are to be excited to the $n = 2$ level in this collision, what is the minimum speed each atom can have before the collision?

26. (a) Calculate the angular momentum of the Moon due to its orbital motion about the Earth. In your calculation, use 3.84×10^8 m as the average Earth–Moon distance and 2.36×10^6 s as the period of the Moon in its orbit. (b) If the angular momentum of the moon obeys Bohr's quantization rule ($L = n\hbar$), determine the value of the quantum number n. (c) By what fraction would the Earth–Moon radius have to be increased to increase the quantum number by 1?

Section 28.4 Modification of the Bohr Theory

Section 28.5 De Broglie Waves and the Hydrogen Atom

27. (a) Find the energy of the electron in the ground state of doubly ionized lithium, which has an atomic number $Z = 3$. (b) Find the radius of its ground-state orbit.

28. (a) Construct an energy level diagram for the He$^+$ ion, for which $Z = 2$. (b) What is the ionization energy for He$^+$?

29. The orbital radii of a hydrogen-like atom is given by the equation

$$r = \frac{n^2 \hbar^2}{Z m_e k_e e^2}.$$

What is the radius of the first Bohr orbit in (a) He$^+$, (b) Li^{2+}, and (c) Be^{3+}?

30. (a) Substitute numerical values into Equation 28.19 to find a value for the Rydberg constant for singly ionized helium, He$^+$. (b) Use the result of part (a) to find the wavelength associated with a transition from the $n = 2$ state to the $n = 1$ state of He$^+$. (c) Identify the region of the electromagnetic spectrum associated with this transition.

31. Determine the wavelength of an electron in the third excited orbit of the hydrogen atom, with $n = 4$.

32. Using the concept of standing waves, de Broglie was able to derive Bohr's stationary orbit postulate. He assumed that a confined electron could exist only in states where its de Broglie waves form standing-wave patterns, as in Figure 28.10a. Consider a particle confined in a box of length L to be equivalent to a string of length L and fixed at both ends. Apply de Broglie's concept to show that (a) the linear momentum of this particle is quantized with $p = mv = nh/2L$ and (b) the allowed states correspond to particle energies of $E_n = n^2 E_0$, where $E_0 = h^2/(8mL^2)$.

Section 28.6 Quantum Mechanics and the Hydrogen Atom

Section 28.7 The Spin Magnetic Quantum Number

33. List the possible sets of quantum numbers for electrons in the $3p$ subshell.

34. When the principal quantum number is $n = 4$, how many different values of (a) ℓ and (b) m_ℓ are possible?

35. The ρ-meson has a charge of $-e$, a spin quantum number of 1, and a mass 1 507 times that of the electron. If the electrons in atoms were replaced by ρ-mesons, list the possible sets of quantum numbers for ρ-mesons in the $3d$ subshell.

Section 28.9 The Exclusion Principle and the Periodic Table

36. (a) Write out the electronic configuration of the ground state for oxygen ($Z = 8$). (b) Write out the values for the set of quantum numbers n, ℓ, m_ℓ, and m_s for each of the electrons in oxygen.

37. Two electrons in the same atom have $n = 3$ and $\ell = 1$. (a) List the quantum numbers for the possible states of the atom. (b) How many states would be possible if the exclusion principle did not apply to the atom?

38. How many different sets of quantum numbers are possible for an electron for which (a) $n = 1$, (b) $n = 2$, (c) $n = 3$, (d) $n = 4$, and (e) $n = 5$? Check your results to show that they agree with the general rule that the number of different sets of quantum numbers is equal to $2n^2$.

39. Physics⊗Now™ Zirconium ($Z = 40$) has two electrons in an incomplete d subshell. (a) What are the values of n and ℓ for each electron? (b) What are all possible values of m_ℓ and m_s? (c) What is the electron configuration in the ground state of zirconium?

Section 28.10 Characteristic X-Rays

40. The K-shell ionization energy of copper is 8 979 eV. The L-shell ionization energy is 951 eV. Determine the wavelength of the K_α emission line of copper. What must the minimum voltage be on an x-ray tube with a copper target in order to see the K_α line?

41. The K_α x-ray is emitted when an electron undergoes a transition from the L shell ($n = 2$) to the K shell ($n = 1$). Use the method illustrated in Example 28.5 to calculate the wavelength of the K_α x-ray from a nickel target ($Z = 28$).

42. When an electron drops from the M shell ($n = 3$) to a vacancy in the K shell ($n = 1$), the measured wavelength of the emitted x-ray is found to be 0.101 nm. Identify the element.

43. The K series of the discrete spectrum of tungsten contains wavelengths of 0.018 5 nm, 0.020 9 nm, and 0.021 5 nm. The K-shell ionization energy is 69.5 keV. Determine the ionization energies of the L, M, and N shells. Sketch the transitions that produce the above wavelengths.

ADDITIONAL PROBLEMS

44. In a hydrogen atom, what is the principle quantum number of the electron orbit with a radius closest to 1.0 μm?

45. (a) How much energy is required to cause an electron in hydrogen to move from the $n = 1$ state to the $n = 2$ state? (b) If the electrons gain this energy by collision between hydrogen atoms in a high-temperature gas, find the minimum temperature of the heated hydrogen gas. The thermal energy of the heated atoms is given by $3k_B T/2$, where k_B is the Boltzmann constant.

46. A pulsed ruby laser emits light at 694.3 nm. For a 14.0-ps pulse containing 3.00 J of energy, find (a) the physical length of the pulse as it travels through space and (b) the number of photons in it. (c) If the beam has a circular cross section 0.600 cm in diameter, find the number of photons per cubic millimeter.

47. The Lyman series for a (new?) one-electron atom is observed in a distant galaxy. The wavelengths of the first four lines and the short-wavelength limit of this Lyman series are given by the energy-level diagram in Figure P28.47. Based on this information, calculate (a) the energies of the ground state and first four excited states for this one-electron atom and (b) the longest-wavelength (alpha) lines and the short-wavelength series limit in the Balmer series for this atom.

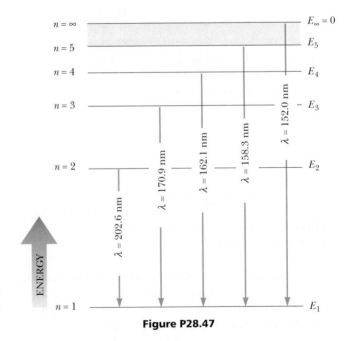

Figure P28.47

48. A dimensionless number that often appears in atomic physics is the fine-structure constant $\alpha = k_e e^2/\hbar c$, where k_e is the Coulomb constant. (a) Obtain a numerical value for $1/\alpha$. (b) In terms of α, what is the ratio of the Bohr radius a_0 to the Compton wavelength $\lambda_C = h/m_e c$? (d) In terms of α, what is the ratio of the reciprocal of the Rydberg constant $1/R_H$ to the Bohr radius?

49. Mercury's ionization energy is 10.39 eV. The three longest wavelengths of the absorption spectrum of mercury are 253.7 nm, 185.0 nm, and 158.5 nm. (a) Construct an energy-level diagram for mercury. (b) Indicate all emission lines that can occur when an electron is raised to the third level above the ground state. (c) Disregarding recoil of the mercury atom, determine the minimum speed an electron must have in order to make an inelastic collision with a mercury atom in its ground state.

50. Suppose the ionization energy of an atom is 4.100 eV. In this same atom, we observe emission lines that have wavelengths of 310.0 nm, 400.0 nm, and 1 378 nm. Use this information to construct the energy-level diagram with the least number of levels. Assume the higher energy levels are closer together.

51. A laser used in eye surgery emits a 3.00-mJ pulse in 1.00 ns, focused to a spot 30.0 μm in diameter on the retina. (a) Find (in SI units) the power per unit area at the retina. (This quantity is called the *irradiance*.) (b) What energy is delivered per pulse to an area of molecular size—say, a circular area 0.600 nm in diameter.

52. An electron has a de Broglie wavelength equal to the diameter of a hydrogen atom in its ground state. (a) What is the kinetic energy of the electron? (b) How does this energy compare with the ground-state energy of the hydrogen atom?

53. Use Bohr's model of the hydrogen atom to show that, when the atom makes a transition from the state n to the state $n - 1$, the frequency of the emitted light is given by

$$f = \frac{2\pi^2 m k_e^2 e^4}{h^3}\left(\frac{2n - 1}{(n - 1)^2 n^2}\right)$$

54. Calculate the classical frequency for the light emitted by an atom. To do so, note that the frequency of revolution is $v/2\pi r$, where r is the Bohr radius. Show that as n approaches infinity in the equation of the preceding problem, the expression given there varies as $1/n^3$ and reduces to the classical frequency. (This is an example of the correspondence principle, which requires that the classical and quantum models agree for large values of n.)

55. Physics⊗Now™ A pi meson (π^-) of charge $-e$ and mass 273 times greater than that of the electron is captured by a helium nucleus ($Z = +2$) as shown in Figure P28.55. (a) Draw an energy-level diagram (in units of eV) for this "Bohr-type" atom up to the first six energy levels. (b) When the π-meson makes a transition between two orbits, a photon is emitted that Compton scatters off a free electron initially at rest, producing a scattered photon of wavelength $\lambda' = 0.089\,929\,3$ nm at an angle of $\theta = 42.68°$, as shown on the right-hand side of Figure P28.55. Between which two orbits did the π-meson make a transition?

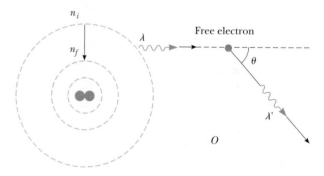

"Pi mesonic" He$^+$ atom
($Z = 2$, $m_\pi = 273 m_e$)

Figure P28.55

56. When a muon with charge $-e$ is captured by a proton, the resulting bound system forms a "muonic atom," which is the same as hydrogen, except with a muon (of mass 207 times the mass of an electron) replacing the electron. For this "muonic atom," determine (a) the Bohr radius and (b) the three lowest energy levels.

57. In this problem, you will estimate the classical lifetime of the hydrogen atom. An accelerating charge loses electromagnetic energy at a rate given by $\mathcal{P} = -2k_e q^2 a^2/(3c^3)$, where k_e is the Coulomb constant, q is the charge of the particle, a is its acceleration, and c is the speed of light in a vacuum. Assume that the electron is one Bohr radius (0.052 9 nm) from the center of the hydrogen atom. (a) Determine its acceleration. (b) Show that $\mathcal{P}$ has units of energy per unit time and determine the rate of energy loss. (c) Calculate the kinetic energy of the electron and determine how long it will take for all of this energy to be converted into electromagnetic waves, assuming that the rate calculated in part (b) remains constant throughout the electron's motion.

58. An electron in a hydrogen atom jumps from some initial Bohr orbit n_i to some final Bohr orbit n_f, as in

Figure P28.58. (a) If the photon emitted in the process is capable of ejecting a photoelectron from tungsten (work function = 4.58 eV), determine n_f. (b) If a minimum stopping potential of $V_0 = 7.51$ volts is required to prevent the photoelectron from hitting the anode, determine the value of n_i.

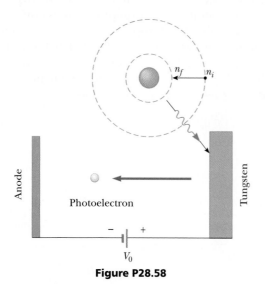

Figure P28.58

ACTIVITIES

1. With your partner not looking, use modeling clay to build one or more mounds on top of a table. Place a piece of cardboard over your mound(s), and assign your partner the task of determining the size, shape, and number of mounds without looking. He is to do this by rolling marbles at the unseen mounds and observing how they emerge. This experiment models the Rutherford scattering experiment.

2. Your instructor can probably lend you a small plastic diffraction grating to enable you to examine the spectrum of different light sources. You can use these gratings to examine a source by holding the grating very close to your eye and noting the spectrum produced by glancing out of the corner of your eye while looking at a light source. You should look at light sources such as sodium vapor lights and mercury vapor lights used in many parking lots, neon lights used in many signs, black lights, ordinary incandescent light bulbs, and so forth.

Aerial view of a nuclear power plant that generates electrical power. Energy is generated in such plants from the process of nuclear fission, in which a heavy nucleus such as ^{235}U splits into smaller particles having a large amount of kinetic energy. This surplus energy can be used to heat water into high pressure steam and drive a turbine.

Courtesy of Public Service Electric and Gas Company

29

Nuclear Physics

In 1896, the year that marks the birth of nuclear physics, Henri Becquerel (1852–1908) discovered radioactivity in uranium compounds. A great deal of activity followed this discovery as researchers attempted to understand and characterize the radiation that we now know to be emitted by radioactive nuclei. Pioneering work by Rutherford showed that the radiation was of three types, which he called *alpha, beta,* and *gamma rays.* These types are classified according to the nature of their electric charge and their ability to penetrate matter. Later experiments showed that alpha rays are helium nuclei, beta rays are electrons, and gamma rays are high-energy photons.

In 1911 Rutherford and his students Geiger and Marsden performed a number of important scattering experiments involving alpha particles. These experiments established the idea that the nucleus of an atom can be regarded as essentially a point mass and point charge and that most of the atomic mass is contained in the nucleus. Further, such studies demonstrated a wholly new type of force: the *nuclear force,* which is predominant at distances of less than about 10^{-14} m and drops quickly to zero at greater distances.

Other milestones in the development of nuclear physics include

- the first observations of nuclear reactions by Rutherford and coworkers in 1919, in which naturally occurring α particles bombarded nitrogen nuclei to produce oxygen,
- the first use of artificially accelerated protons to produce nuclear reactions, by Cockcroft and Walton in 1932,
- the discovery of the neutron by Chadwick in 1932,
- the discovery of artificial radioactivity by Joliot and Irene Curie in 1933,
- the discovery of nuclear fission by Hahn, Strassman, Meitner, and Frisch in 1938, and
- the development of the first controlled fission reactor by Fermi and his collaborators in 1942.

In this chapter we discuss the properties and structure of the atomic nucleus. We start by describing the basic properties of nuclei and follow with a discussion of the phenomenon of radioactivity. Finally, we explore nuclear reactions and the various processes by which nuclei decay.

ERNEST RUTHERFORD,
New Zealand Physicist
(1871–1937)

Rutherford was awarded the Nobel Prize in 1908 for discovering that atoms can be broken apart by alpha rays and for studying radioactivity. "On consideration, I realized that this scattering backward must be the result of a single collision, and when I made calculations I saw that it was impossible to get anything of that order of magnitude unless you took a system in which the greater part of the mass of the atom was concentrated in a minute nucleus. It was then that I had the idea of an atom with a minute massive center carrying a charge."

Definition of the unified mass unit u ▶

TIP 29.1   Mass Number is not the Atomic Mass

Don't confuse the mass number A with the atomic mass. Mass number is an integer that specifies an isotope and has no units—it's simply equal to the number of nucleons. Atomic mass is an average of the masses of the isotopes of a given element and has units of u.

29.1   SOME PROPERTIES OF NUCLEI

All nuclei are composed of two types of particles: protons and neutrons. The only exception is the ordinary hydrogen nucleus, which is a single proton. In describing some of the properties of nuclei, such as their charge, mass, and radius, we make use of the following quantities:

- the **atomic number** Z, which equals the number of protons in the nucleus,
- the **neutron number** N, which equals the number of neutrons in the nucleus,
- the **mass number** A, which equals the number of nucleons in the nucleus (*nucleon* is a generic term used to refer to either a proton or a neutron).

The symbol we use to represent nuclei is $^{A}_{Z}X$, where X represents the chemical symbol for the element. For example, $^{27}_{13}Al$ has the mass number 27 and the atomic number 13; therefore, it contains 13 protons and 14 neutrons. When no confusion is likely to arise, we often omit the subscript Z, because the chemical symbol can always be used to determine Z.

The nuclei of all atoms of a particular element must contain the same number of protons, but they may contain different numbers of neutrons. Nuclei that are related in this way are called **isotopes**. **The isotopes of an element have the same Z value, but different N and A values**. The natural abundances of isotopes can differ substantially. For example, $^{11}_{6}C$, $^{12}_{6}C$, $^{13}_{6}C$, and $^{14}_{6}C$ are four isotopes of carbon. The natural abundance of the $^{12}_{6}C$ isotope is about 98.9%, whereas that of the $^{13}_{6}C$ isotope is only about 1.1%. Some isotopes don't occur naturally, but can be produced in the laboratory through nuclear reactions. Even the simplest element, hydrogen, has isotopes: $^{1}_{1}H$, hydrogen; $^{2}_{1}H$, deuterium; and $^{3}_{1}H$, tritium.

Charge and Mass

The proton carries a single positive charge $+ e = 1.602\ 177\ 33 \times 10^{-19}$ C, the electron carries a single negative charge $- e$, and the neutron is electrically neutral. Because the neutron has no charge, it's difficult to detect. The proton is about 1 836 times as massive as the electron, and the masses of the proton and the neutron are almost equal (Table 29.1).

For atomic masses, it is convenient to define the **unified mass unit** u in such a way that the mass of one atom of the isotope ^{12}C is exactly 12 u, where $1\ u = 1.660\ 559 \times 10^{-27}$ kg. The proton and neutron each have a mass of about 1 u, and the electron has a mass that is only a small fraction of an atomic mass unit.

Because the rest energy of a particle is given by $E_R = mc^2$, it is often convenient to express the particle's mass in terms of its energy equivalent. For one atomic mass unit, we have an energy equivalent of

$$E_R = mc^2 = (1.660\ 559 \times 10^{-27}\ \text{kg})(2.997\ 92 \times 10^8\ \text{m/s})^2$$
$$= 1.492\ 431 \times 10^{-10}\ \text{J} = 931.494\ \text{MeV}$$

In calculations, nuclear physicists often express *mass* in terms of the unit MeV/c^2, where

$$1\ u = 931.494\ \text{MeV}/c^2$$

TABLE 29.1

Masses of the Proton, Neutron, and Electron in Various Units

Particle	Mass		
	kg	**u**	**MeV/c^2**
Proton	1.6726×10^{-27}	1.007 276	938.28
Neutron	1.6750×10^{-27}	1.008 665	939.57
Electron	9.109×10^{-31}	5.486×10^{-4}	0.511

The Size of Nuclei

The size and structure of nuclei were first investigated in the scattering experiments of Rutherford, discussed in Section 28.1. Using the principle of conservation of energy, Rutherford found an expression for how close an alpha particle moving directly toward the nucleus can come to the nucleus before being turned around by Coulomb repulsion.

In such a head-on collision, the kinetic energy of the incoming alpha particle must be converted completely to electrical potential energy when the particle stops at the point of closest approach and turns around (Active Fig. 29.1). If we equate the initial kinetic energy of the alpha particle to the maximum electrical potential energy of the system (alpha particle plus target nucleus), we have

$$\tfrac{1}{2}mv^2 = k_e \frac{q_1 q_2}{r} = k_e \frac{(2e)(Ze)}{d}$$

where d is the distance of closest approach. Solving for d, we get

$$d = \frac{4k_e Z e^2}{mv^2}$$

From this expression, Rutherford found that alpha particles approached to within 3.2×10^{-14} m of a nucleus when the foil was made of gold. Thus, the radius of the gold nucleus must be less than this value. For silver atoms, the distance of closest approach was 2×10^{-14} m. From these results, Rutherford concluded that the positive charge in an atom is concentrated in a small sphere, which he called the nucleus, with radius no greater than about 10^{-14} m. Because such small lengths are common in nuclear physics, a convenient unit of length is the *femtometer* (fm), sometimes called the **fermi** and defined as

$$1 \text{ fm} \equiv 10^{-15} \text{ m}$$

Since the time of Rutherford's scattering experiments, a multitude of other experiments have shown that most nuclei are approximately spherical and have an average radius given by

$$r = r_0 A^{1/3} \qquad \text{[29.1]}$$

where A is the total number of nucleons and r_0 is a constant equal to 1.2×10^{-15} m. Because the volume of a sphere is proportional to the cube of its radius, it follows from Equation 29.1 that the volume of a nucleus (assumed to be spherical) is directly proportional to A, the total number of nucleons. This relationship then suggests that **all nuclei have nearly the same density**. Nucleons combine to form a nucleus *as though* they were tightly packed spheres (Fig. 29.2).

ACTIVE FIGURE 29.1
An alpha particle on a head-on collision course with a nucleus of charge Ze. Because of the Coulomb repulsion between the like charges, the alpha particle will stop instantaneously at a distance d from the nucleus, called the distance of closest approach.

Physics⊗Now™
Log into PhysicsNow at **www.cp7e.com** and go to Active Figure 29.1, where you can adjust the atomic number of the target nucleus and the kinetic energy of the alpha particle. Then observe the approach of the alpha particle toward the nucleus.

Figure 29.2 A nucleus can be visualized as a cluster of tightly packed spheres, each of which is a nucleon.

EXAMPLE 29.1   Sizing a Neutron Star

Goal   Apply the concepts of nuclear size.

Problem   One of the end stages of stellar life is a neutron star, where matter collapses and electrons combine with protons to form neutrons. Some liken neutron stars to a single gigantic nucleus. **(a)** Approximately how many nucleons are in a neutron star with a mass of 3.00×10^{30} kg? (This is the mass number of the star.) **(b)** Calculate the radius of the star, treating it as a giant nucleus. **(c)** Calculate the density of the star, assuming the mass is distributed uniformly.

Strategy   The effective mass number of the neutron star can be found by dividing the star mass in kg by the mass of a neutron. Equation 29.1 then gives an estimate of the radius of the star, which together with the mass determines the density.

Solution
(a) Find the approximate number of nucleons in the star.

Divide the star's mass by the mass of a neutron to find A: $A = \left(\dfrac{3.00 \times 10^{30} \text{ kg}}{1.675 \times 10^{-27} \text{ kg}} \right) = 1.79 \times 10^{57}$

(b) Calculate the radius of the star, treating it as a giant atomic nucleus.

Substitute into Equation 29.1:

$$r = r_0 A^{1/3} = (1.2 \times 10^{-15} \text{ m})(1.79 \times 10^{57})^{1/3}$$
$$= 1.46 \times 10^4 \text{ m}$$

(c) Calculate the density of the star, assuming that its mass is distributed uniformly.

Substitute values into the equation for density and assume the star is a uniform sphere:

$$\rho = \frac{m}{V} = \frac{m}{\frac{4}{3}\pi r^3} = \frac{3.00 \times 10^{30} \text{ kg}}{\frac{4}{3}\pi(1.46 \times 10^4 \text{ m})^3}$$
$$= 2.30 \times 10^{17} \text{ kg/m}^3$$

Remarks This density is typical of atomic nuclei as well as of neutron stars. A ball of neutron star matter having a radius of only 1 meter would have a powerful gravity field: it could attract objects a kilometer away at an acceleration of over 50 m/s^2!

Exercise 29.1
Estimate the radius of a uranium-235 nucleus.

Answer 7.41×10^{-15} m

MARIA GOEPPERT-MAYER,
German Physicist (1906–1972)

Goeppert-Mayer was born and educated in Germany. She is best known for her development of the shell model of the nucleus, published in 1950. A similar model was simultaneously developed by Hans Jensen, a German scientist. Maria Goeppert-Mayer and Hans Jensen were awarded the Nobel Prize in physics in 1963 for their extraordinary work in understanding the structure of the nucleus.

Nuclear Stability

Given that the nucleus consists of a closely packed collection of protons and neutrons, you might be surprised that it can even exist. The very large repulsive electrostatic forces between protons should cause the nucleus to fly apart. However, nuclei are stable because of the presence of another, short-range (about 2 fm) force: the **nuclear force**, an attractive force that acts between all nuclear particles. The protons attract each other via the nuclear force, and at the same time they repel each other through the Coulomb force. The attractive nuclear force also acts between pairs of neutrons and between neutrons and protons.

The nuclear attractive force is stronger than the Coulomb repulsive force within the nucleus (at short ranges). If this were not the case, stable nuclei would not exist. Moreover, the strong nuclear force is nearly independent of charge. In other words, the nuclear forces associated with proton–proton, proton–neutron, and neutron–neutron interactions are approximately the same, apart from the additional repulsive Coulomb force for the proton–proton interaction.

There are about 260 stable nuclei; hundreds of others have been observed, but are unstable. A plot of N versus Z for a number of stable nuclei is given in Figure 29.3. Note that light nuclei are most stable if they contain equal numbers of protons and neutrons, so that $N = Z$, but heavy nuclei are more stable if $N > Z$. This difference can be partially understood by recognizing that as the number of protons increases, the strength of the Coulomb force increases, which tends to break the nucleus apart. As a result, more neutrons are needed to keep the nucleus stable, because neutrons are affected only by the attractive nuclear forces. In effect, the additional neutrons "dilute" the nuclear charge density. Eventually, when $Z = 83$, the repulsive forces between protons cannot be compensated for by the addition of neutrons. Elements that contain more than 83 protons don't have stable nuclei, but decay or disintegrate into other particles in various amounts of time. The masses and some other properties of selected isotopes are provided in Appendix B.

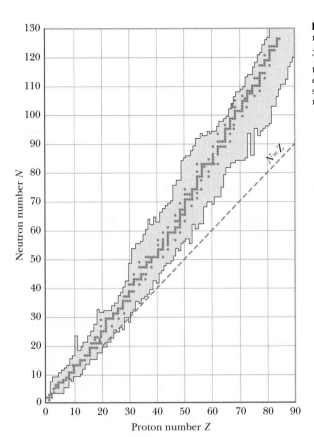

Figure 29.3 A plot of the neutron number N versus the proton number Z for the stable nuclei (solid points). The dashed straight line corresponds to the condition $N = Z$. They are centered on the so-called line of stability. The shaded area shows radioactive (unstable) nuclei.

29.2 BINDING ENERGY

The total mass of a nucleus is always less than the sum of the masses of its nucleons. Also, because mass is another manifestation of energy, **the total energy of the bound system (the nucleus) is less than the combined energy of the separated nucleons**. This difference in energy is called the **binding energy** of the nucleus and can be thought of as the energy that must be added to a nucleus to break it apart into its separated neutrons and protons.

EXAMPLE 29.2 The Binding Energy of the Deuteron

Goal Calculate the binding energy of a nucleus.

Problem The nucleus of the deuterium atom, called the deuteron, consists of a proton and a neutron. Calculate the deuteron's binding energy in MeV, given that its atomic mass—that is, *the mass of a deuterium nucleus plus an electron*—is 2.014 102 u.

Strategy Calculate the sum of the masses of the individual particles and subtract the mass of the combined particle. The masses of the neutral atoms can be used instead of the nuclei because the electron masses cancel. Use the values from Table 29.4 or Table B of the appendix. The mass of an atom given in Appendix B includes the mass of Z electrons, where Z is the atom's atomic number.

Solution

To find the binding energy, first sum the masses of the hydrogen atom and neutron and subtract the mass of the deuteron:

$$\Delta m = (m_p + m_n) - m_d$$
$$= (1.007\ 825\ \text{u} + 1.008\ 665\ \text{u}) - 2.014\ 102\ \text{u}$$
$$= 0.002\ 388\ \text{u}$$

Convert this mass difference to its equivalent in MeV:

$$E_b = (0.002\ 388\ \text{u})\frac{931.5\ \text{MeV}}{1\ \text{u}} = \boxed{2.224\ \text{MeV}}$$

Remarks This result tells us that to separate a deuteron into a proton and a neutron, it's necessary to add 2.224 MeV of energy to the deuteron to overcome the attractive nuclear force between the proton and the neutron. One way of supplying the deuteron with this energy is by bombarding it with energetic particles.

If the binding energy of a nucleus were zero, the nucleus would separate into its constituent protons and neutrons without the addition of any energy; that is, it would spontaneously break apart.

Exercise 29.2
Calculate the binding energy of $^3_2$He.

Answer   7.718 MeV

It's interesting to examine a plot of binding energy per nucleon, E_b/A, as a function of mass number for various stable nuclei (Fig. 29.4). Except for the lighter nuclei, the average binding energy per nucleon is about 8 MeV. Note that the curve peaks in the vicinity of $A = 60$, which means that nuclei with mass numbers greater or less than 60 are not as strongly bound as those near the middle of the periodic table. As we'll see later, this fact allows energy to be released in fission and fusion reactions. The curve is slowly varying for $A > 40$, which suggests that the nuclear force saturates. In other words, a particular nucleon can interact with only a limited number of other nucleons, which can be viewed as the "nearest neighbors" in the close-packed structure illustrated in Figure 29.2.

Applying Physics 29.1   Binding Nucleons and Electrons

Figure 29.4 shows a graph of the amount of energy required to remove a nucleon from the nucleus. The figure indicates that an approximately constant amount of energy is necessary to remove a nucleon above $A = 40$, whereas we saw in Chapter 28 that widely varying amounts of energy are required to remove an electron from the atom. What accounts for this difference?

Explanation In the case of Figure 29.4, the approximately constant value of the nuclear binding energy is a result of the short-range nature of the nuclear force. A given nucleon interacts only with its few nearest neighbors, rather than with all of the nucleons in the nucleus. Thus, no matter how many nucleons are present in the nucleus, pulling any one nucleon out involves separating it only from its nearest neighbors. The energy to do this, therefore, is approximately independent of how many nucleons are present. For the clearest comparison with the electron, think of averaging the energies required to strip all of the electrons out of a particular atom, from the outermost valence electron to the innermost K-shell electron. This average increases steeply with increasing atomic number. The electrical force binding the electrons to the nucleus in an atom is a long-range force. An electron in an atom interacts with all the protons in the nucleus. When the nuclear charge increases, there is a stronger attraction between the nucleus and the electrons. Therefore, as the nuclear charge increases, more energy is necessary to remove an average electron.

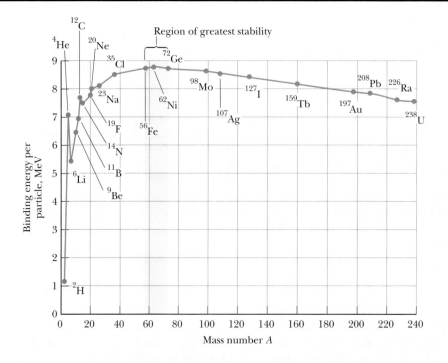

Figure 29.4   Binding energy per nucleon versus the mass number A for nuclei that are along the line of stability shown in Figure 29.3. Some representative nuclei appear as blue dots with labels. (Nuclei to the right of ^{208}Pb are unstable. The curve represents the binding energy for the most stable isotopes.)

29.3 RADIOACTIVITY

In 1896, Becquerel accidentally discovered that uranium salt crystals emit an invisible radiation that can darken a photographic plate even if the plate is covered to exclude light. After several such observations under controlled conditions, he concluded that the radiation emitted by the crystals was of a new type, one requiring no external stimulation. This spontaneous emission of radiation was soon called **radioactivity**. Subsequent experiments by other scientists showed that other substances were also radioactive.

The most significant investigations of this type were conducted by Marie and Pierre Curie. After several years of careful and laborious chemical separation processes on tons of pitchblende, a radioactive ore, the Curies reported the discovery of two previously unknown elements, both of which were radioactive. These were named polonium and radium. Subsequent experiments, including Rutherford's famous work on alpha-particle scattering, suggested that radioactivity was the result of the decay, or disintegration, of unstable nuclei.

Three types of radiation can be emitted by a radioactive substance: alpha (α) particles, in which the emitted particles are $^4_2$He nuclei; beta (β) particles, in which the emitted particles are either electrons or positrons; and gamma (γ) rays, in which the emitted "rays" are high-energy photons. A **positron** is a particle similar to the electron in all respects, except that it has a charge of $+e$. (The positron is said to be the **antiparticle** of the electron.) The symbol e^- is used to designate an electron, and e^+ designates a positron.

It's possible to distinguish these three forms of radiation by using the scheme described in Figure 29.5. The radiation from a radioactive sample is directed into a region with a magnetic field, and the beam splits into three components, two bending in opposite directions and the third not changing direction. From this simple observation it can be concluded that the radiation of the undeflected beam (the gamma ray) carries no charge, the component deflected upward contains positively charged particles (alpha particles), and the component deflected downward contains negatively charged particles (e^-). If the beam includes a positron (e^+), it is deflected upward.

The three types of radiation have quite different penetrating powers. Alpha particles barely penetrate a sheet of paper, beta particles can penetrate a few millimeters of aluminum, and gamma rays can penetrate several centimeters of lead.

The Decay Constant and Half-Life

Observation has shown that if a radioactive sample contains N radioactive nuclei at some instant, then the number of nuclei, ΔN, that decay in a small time interval Δt is proportional to N; mathematically,

$$\frac{\Delta N}{\Delta t} \propto N$$

or

$$\Delta N = -\lambda N \Delta t \qquad \text{[29.2]}$$

where λ is a constant called the **decay constant**. The negative sign signifies that N decreases with time; that is, ΔN is negative. The value of λ for any isotope determines the rate at which that isotope will decay. **The decay rate, or activity R, of a sample is defined as the number of decays per second**. From Equation 29.2, we see that the decay rate is

$$R = \left| \frac{\Delta N}{\Delta t} \right| = \lambda N \qquad \text{[29.3]}$$

◀ Decay rate

Isotopes with a large λ value decay rapidly; those with small λ decay slowly.

MARIE CURIE, Polish Scientist (1867–1934)

In 1903 Marie Curie shared the Nobel Prize in physics with her husband, Pierre, and with Becquerel for their studies of radioactive substances. In 1911 she was awarded a second Nobel Prize in chemistry for the discovery of radium and polonium. Marie Curie died of leukemia caused by years of exposure to radioactive substances. "I persist in believing that the ideas that then guided us are the only ones which can lead to the true social progress. We cannot hope to build a better world without improving the individual. Toward this end, each of us must work toward his own highest development, accepting at the same time his share of responsibility in the general life of humanity."

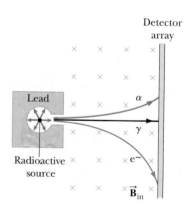

Figure 29.5 The radiation from a radioactive source, such as radium, can be separated into three components using a magnetic field to deflect the charged particles. The detector array at the right records the events. The gamma ray isn't deflected by the magnetic field.

© Richard Megna/Fundamental Photographs

The hands and numbers of this luminous watch contain minute amounts of radium salt. The radioactive decay of radium causes the phosphors to glow in the dark.

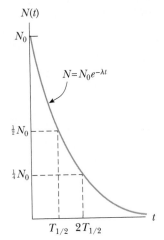

ACTIVE FIGURE 29.6
Plot of the exponential decay law for radioactive nuclei. The vertical axis represents the number of radioactive nuclei present at any time t, and the horizontal axis is time. The parameter $T_{1/2}$ is the half-life of the sample.

Physics⊗Now™

Log into PhysicsNow at **www.cp7e.com** and go to Active Figure 29.6, where you can observe the decay curves for nuclei with varying half-lives.

TIP 29.2 Two Half-Lives Don't Make a Whole-Life

A half-life is the time it takes for half of a given number of nuclei to decay. During a second half-life, half the remaining nuclei decay, so in two half-lives, three-quarters of the original material has decayed, not all of it.

A general decay curve for a radioactive sample is shown in Active Figure 29.6. It can be shown from Equation 29.2 (using calculus) that the number of nuclei present varies with time according to the equation

$$N = N_0 e^{-\lambda t} \qquad [29.4a]$$

where N is the number of radioactive nuclei present at time t, N_0 is the number present at time $t = 0$, and $e = 2.718 \ldots$ is Euler's constant. Processes that obey Equation 29.4a are sometimes said to undergo **exponential decay**.[1]

Another parameter that is useful for characterizing radioactive decay is the **half-life** $T_{1/2}$. **The half-life of a radioactive substance is the time it takes for half of a given number of radioactive nuclei to decay.** Using the concept of half-life, it can be shown that Equation 29.4a can also be written as

$$N = N_0 \left(\frac{1}{2}\right)^n \qquad [29.4b]$$

where n is the number of half-lives. The number n can take any non-negative value and need not be an integer. From the definition, it follows that n is related to time t and the half-life $T_{1/2}$ by

$$n = \frac{t}{T_{1/2}} \qquad [29.4c]$$

Setting $N = N_0/2$ and $t = T_{1/2}$ in Equation 29.4a gives

$$\frac{N_0}{2} = N_0 e^{-\lambda T_{1/2}}$$

Writing this in the form $e^{\lambda T_{1/2}} = 2$ and taking the natural logarithm of both sides, we get

$$T_{1/2} = \frac{\ln 2}{\lambda} = \frac{0.693}{\lambda} \qquad [29.5]$$

This is a convenient expression relating the half-life to the decay constant. Note that after an elapsed time of one half-life, $N_0/2$ radioactive nuclei remain (by definition); after two half-lives, half of these will have decayed and $N_0/4$ radioactive nuclei will be left; after three half-lives, $N_0/8$ will be left; and so on.

The unit of activity R is the **curie** (Ci), defined as

$$1 \text{ Ci} \equiv 3.7 \times 10^{10} \text{ decays/s} \qquad [29.6]$$

This unit was selected as the original activity unit because it is the approximate activity of 1 g of radium. The SI unit of activity is the **becquerel** (Bq):

$$1 \text{ Bq} = 1 \text{ decay/s} \qquad [29.7]$$

Therefore, 1 Ci = 3.7×10^{10} Bq. The most commonly used units of activity are the millicurie (10^{-3} Ci) and the microcurie (10^{-6} Ci).

Quick Quiz 29.1

What fraction of a radioactive sample has decayed after two half-lives have elapsed? (a) 1/4 (b) 1/2 (c) 3/4 (d) not enough information to say

Quick Quiz 29.2

Suppose the decay constant of radioactive substance A is twice the decay constant of radioactive substance B. If substance B has a half life of 4 hr, what's the half life of substance A? (a) 8 hr (b) 4 hr (c) 2 hr (d) not enough information to say

[1] Other examples of exponential decay were discussed in Chapter 18 in connection with *RC* circuits and in Chapter 20 in connection with *RL* circuits.

INTERACTIVE EXAMPLE 29.3   The Activity of Radium

Goal   Calculate the activity of a radioactive substance at different times.

Problem   The half-life of the radioactive nucleus $^{226}_{88}$Ra is 1.6×10^3 yr. If a sample initially contains 3.00×10^{16} such nuclei, determine **(a)** the initial activity in curies, **(b)** the number of radium nuclei remaining after 4.8×10^3 yr, and **(c)** the activity at this later time.

Strategy   For parts (a) and (c), find the decay constant and multiply it by the number of nuclei. Part (b) requires multiplying the initial number of nuclei by one-half for every elapsed half-life. (Essentially, this is an application of Equation 29.4b.)

Solution
(a) Determine the initial activity in curies.

Convert the half-life to seconds:

$$T_{1/2} = (1.6 \times 10^3 \text{ yr})(3.156 \times 10^7 \text{ s/yr}) = 5.0 \times 10^{10} \text{ s}$$

Substitute this value into Equation 29.5 to get the decay constant:

$$\lambda = \frac{0.693}{T_{1/2}} = \frac{0.693}{5.0 \times 10^{10} \text{ s}} = 1.4 \times 10^{-11} \text{ s}^{-1}$$

Calculate the activity of the sample at $t = 0$, using $R_0 = \lambda N_0$, where R_0 is the decay rate at $t = 0$ and N_0 is the number of radioactive nuclei present at $t = 0$:

$$R_0 = \lambda N_0 = (1.4 \times 10^{-11} \text{ s}^{-1})(3.0 \times 10^{16} \text{ nuclei})$$
$$= 4.2 \times 10^5 \text{ decays/s}$$

Convert to curies to obtain the activity at $t = 0$, using the fact that $1 \text{ Ci} = 3.7 \times 10^{10}$ decays/s:

$$R_0 = (4.2 \times 10^5 \text{ decays/s})\left(\frac{1 \text{ Ci}}{3.7 \times 10^{10} \text{ decays/s}}\right)$$
$$= 1.1 \times 10^{-5} \text{ Ci} = \boxed{11 \ \mu\text{Ci}}$$

(b) How many radium nuclei remain after 4.8×10^3 years?

Calculate the number of half-lives, n:

$$n = \frac{4.8 \times 10^3 \text{ yr}}{1.6 \times 10^3 \text{ yr/half-life}} = 3.0 \text{ half-lives}$$

Multiply the initial number of nuclei by the number of factors of one-half:

$$N = N_0 \left(\frac{1}{2}\right)^n \tag{1}$$

Substitute $N_0 = 3.0 \times 10^{16}$ and $n = 3.0$:

$$N = (3.0 \times 10^{16} \text{ nuclei})\left(\frac{1}{2}\right)^{3.0} = \boxed{3.8 \times 10^{15} \text{ nuclei}}$$

(c) Calculate the activity after 4.8×10^3 yr.

Multiply the number of remaining nuclei by the decay constant to find the activity R:

$$R = \lambda N = (1.4 \times 10^{-11} \text{ s}^{-1})(3.8 \times 10^{15} \text{ nuclei})$$
$$= 5.3 \times 10^4 \text{ decays/s}$$
$$= \boxed{1.4 \ \mu\text{Ci}}$$

Remarks   The activity is reduced by half every half-life, which is naturally the case because activity is proportional to the number of remaining nuclei. The precise number of nuclei at any time is never truly exact, because particles decay according to a probability. The larger the sample, however, the more accurate are the predictions from Equation 29.4.

Exercise 29.3
Find **(a)** the number of remaining radium nuclei after 3.2×10^3 yr and **(b)** the activity at this time.

Answer   (a) 7.5×10^{15} nuclei   (b) $2.8 \ \mu\text{Ci}$

Physics⊗Now™ Practice evaluating the parameters for the radioactive decay of various isotopes of radium by logging into PhysicsNow at **www.cp7e.com** and going to Interactive Example 29.3.

EXAMPLE 29.4   Radon Gas

Goal   Calculate the number of nuclei after an arbitrary time and the time required for a given number of nuclei to decay.

Problem   Radon $^{222}_{86}$Rn is a radioactive gas that can be trapped in the basements of homes, and its presence in high concentrations is a known health hazard. Radon has a half-life of 3.83 days. A gas sample contains 4.00×10^8 radon atoms initially. **(a)** How many atoms will remain after 14.0 days have passed if no more radon leaks in? **(b)** What is the activity of the radon sample after 14.0 days? **(c)** How long before 99% of the sample has decayed?

Strategy   The activity can be found by substitution into Equation 29.5, as before. Equation 29.4a (or Eq. 29.4b) must be used to find the number of particles remaining after 14.0 days. To obtain the time asked for in part (c), Equation 29.4a must be solved for time.

Solution

(a) How many atoms will remain after 14.0 days have passed?

Determine the decay constant from Equation 29.5:

$$\lambda = \frac{0.693}{T_{1/2}} = \frac{0.693}{3.83 \text{ days}} = 0.181 \text{ day}^{-1}$$

Now use Equation 29.4a, taking $N_0 = 4.0 \times 10^8$, and the value of λ just found to obtain the number N remaining after 14 days:

$$N = N_0 e^{-\lambda t} = (4.00 \times 10^8 \text{ atoms}) e^{-(0.181 \text{ day}^{-1})(14.0 \text{ days})}$$
$$= 3.17 \times 10^7 \text{ atoms}$$

(b) What is the activity of the radon sample after 14.0 days?

Express the decay constant in units of s^{-1}:

$$\lambda = (0.181 \text{ day}^{-1})\left(\frac{1 \text{ day}}{8.64 \times 10^4 \text{ s}}\right) = 2.09 \times 10^{-6} \text{ s}^{-1}$$

From Equation 29.3 and this value of λ, compute the activity R:

$$R = \lambda N = (2.09 \times 10^{-6} \text{ s}^{-1})(3.17 \times 10^7 \text{ atoms})$$
$$= 66.3 \text{ decays/s} = 66.3 \text{ Bq}$$

(c) How much time must pass before 99% of the sample has decayed?

Solve Equation 29.4a for t, using natural logarithms:

$$\ln(N) = \ln(N_0 e^{-\lambda t}) = \ln(N_0) + \ln(e^{-\lambda t})$$
$$\ln(N) - \ln(N_0) = \ln(e^{-\lambda t}) = -\lambda t$$
$$t = \frac{\ln(N_0) - \ln(N)}{\lambda} = \frac{\ln(N_0/N)}{\lambda}$$

Substitute values:

$$t = \frac{\ln(N_0/0.01 \, N_0)}{2.09 \times 10^{-6} \text{ s}^{-1}} = 2.20 \times 10^6 \text{ s} = 25.5 \text{ days}$$

Remarks   This kind of calculation is useful in determining how long you would have to wait for radioactivity at a given location to fall to safe levels.

Exercise 29.4

(a) Find the activity of the radon sample after 12.0 days have elapsed. **(b)** How long would it take for 85.0% of the sample to decay?

Answer   (a) 95.3 Bq   (b) 9.08×10^5 s = 10.5 days

29.4   THE DECAY PROCESSES

As stated in the previous section, radioactive nuclei decay spontaneously via alpha, beta, and gamma decay. As we'll see in this section, these processes are very different from each other.

Alpha Decay

If a nucleus emits an alpha particle ($_2^4$He), it loses two protons and two neutrons. Therefore, the neutron number N of a single nucleus decreases by 2, Z decreases by 2, and A decreases by 4. The decay can be written symbolically as

$$_Z^A X \rightarrow {}_{Z-2}^{A-4} Y + {}_2^4 He \qquad \text{[29.8]}$$

where X is called the **parent nucleus** and Y is known as the **daughter nucleus**. As examples, ^{238}U and ^{226}Ra are both alpha emitters and decay according to the schemes

$$_{92}^{238} U \rightarrow {}_{90}^{234} Th + {}_2^4 He \qquad \text{[29.9]}$$

and

$$_{88}^{226} Ra \rightarrow {}_{86}^{222} Rn + {}_2^4 He \qquad \text{[29.10]}$$

The half-life for ^{238}U decay is 4.47×10^9 years, and the half-life for ^{226}Ra decay is 1.60×10^3 years. In both cases, note that the A of the daughter nucleus is four less than that of the parent nucleus, while Z is reduced by two. The differences are accounted for in the emitted alpha particle (the ^{4}He nucleus).

The decay of ^{226}Ra is shown in Active Figure 29.7. When one element changes into another, as happens in alpha decay, the process is called **spontaneous decay** or transmutation. As a general rule, (1) the sum of the mass numbers A must be the same on both sides of the equation, and (2) the sum of the atomic numbers Z must be the same on both sides of the equation.

In order for alpha emission to occur, the mass of the parent must be greater than the combined mass of the daughter and the alpha particle. In the decay process, this excess mass is converted into energy of other forms and appears in the form of kinetic energy in the daughter nucleus and the alpha particle. Most of the kinetic energy is carried away by the alpha particle because it is much less massive than the daughter nucleus. This can be understood by first noting that a particle's kinetic energy and momentum p are related as follows:

$$KE = \frac{p^2}{2m}$$

Because momentum is conserved, the two particles emitted in the decay of a nucleus at rest must have equal, but oppositely directed, momenta. As a result, the lighter particle, with the smaller mass in the denominator, has more kinetic energy than the more massive particle.

After decay

ACTIVE FIGURE 29.7
The alpha decay of radium-226. The radium nucleus is initially at rest. After the decay, the radon nucleus has kinetic energy KE_{Rn} and momentum $\vec{p}_{Rn}$, and the alpha particle has kinetic energy KE_α and momentum $\vec{p}_\alpha$.

Log into PhysicsNow at **www.cp7e.com** and go to Active Figure 29.7, where you can observe the decay of radium-226.

Quick Quiz 29.3

If a nucleus such as ^{226}Ra that is initially at rest undergoes alpha decay, which of the following statements is true? (a) The alpha particle has more kinetic energy than the daughter nucleus. (b) The daughter nucleus has more kinetic energy than the alpha particle. (c) The daughter nucleus and the alpha particle have the same kinetic energy.

Applying Physics 29.2   Energy and Half-life

In comparing alpha decay energies from a number of radioactive nuclides, why is it found that the half-life of the decay goes down as the energy of the decay goes up?

Explanation It should seem reasonable that the higher the energy of the alpha particle, the more likely it is to escape the confines of the nucleus. The higher probability of escape translates to a faster rate of decay, which appears as a shorter half-life.

EXAMPLE 29.5   Decaying Radium

Goal   Calculate the energy released during an alpha decay.

Problem   We showed that the $^{226}_{88}$Ra nucleus undergoes alpha decay to $^{222}_{86}$Rn (Eq. 29.10). Calculate the amount of energy liberated in this decay. Take the mass of $^{226}_{88}$Ra to be 226.025 402 u, that of $^{222}_{86}$Rn to be 222.017 571 u, and that of $^{4}_{2}$He to be 4.002 602 u, as found in Appendix B.

Strategy   This is a matter of subtracting the neutral masses of the daughter particles from the original mass of the radon nucleus.

Solution

Compute the sum of the mass of the daughter particle, m_d, and the mass of the alpha particle, m_α:

$$m_d + m_\alpha = 222.017\ 571\text{ u} + 4.002\ 602\text{ u} = 226.020\ 173\text{ u}$$

Compute the loss of mass, Δm, during the decay by subtracting the previous result from M_p, the mass of the original particle:

$$\Delta m = M_p - (m_d + m_\alpha) = 226.025\ 402\text{ u} - 226.020\ 173\text{ u}$$
$$= 0.005\ 229\text{ u}$$

Convert the loss of mass Δm to its equivalent energy in MeV:

$$E = (0.005\ 229\text{ u})(931.494\text{ MeV/u}) = \boxed{4.871\text{ MeV}}$$

Remark   The potential barrier is typically higher than this value of the energy, but quantum tunneling permits the event to occur, anyway.

Exercise 29.5

Calculate the energy released when $^{8}_{4}$Be splits into two alpha particles. Beryllium-8 has an atomic mass of 8.005 305 u.

Answer   0.094 1 MeV

Beta Decay

When a radioactive nucleus undergoes beta decay, the daughter nucleus has the same number of nucleons as the parent nucleus, but the atomic number is changed by 1:

$$^{A}_{Z}X \rightarrow\ ^{A}_{Z+1}Y + e^- \qquad [29.11]$$
$$^{A}_{Z}X \rightarrow\ ^{A}_{Z-1}Y + e^+ \qquad [29.12]$$

Again, note that the nucleon number and total charge are both conserved in these decays. However, as we will see shortly, these processes are not described completely by these expressions. A typical beta decay event is

$$^{14}_{6}C \rightarrow\ ^{14}_{7}N + e^- \qquad [29.13]$$

The emission of electrons from a *nucleus* is surprising, because, in all our previous discussions, we stated that the nucleus is composed of protons and neutrons only. This apparent discrepancy can be explained by noting that the emitted electron is created in the nucleus by a process in which a neutron is transformed into a proton. This process can be represented by the equation

$$^{1}_{0}n \rightarrow\ ^{1}_{1}p + e^- \qquad [29.14]$$

Consider the energy of the system of Equation 29.13 before and after decay. As with alpha decay, energy must be conserved in beta decay. The next example illustrates how to calculate the amount of energy released in the beta decay of $^{14}_{6}C$.

EXAMPLE 29.6   The Beta Decay of Carbon-14

Goal   Calculate the energy released in a beta decay.

Problem   Find the energy liberated in the beta decay of $^{14}_{6}C$ to $^{14}_{7}N$, as represented by Equation 29.13. That equation refers to nuclei, while Appendix B gives the masses of neutral atoms. Adding six electrons to both sides of Equation 29.13 yields

$$^{14}_{6}C\text{ atom} \rightarrow\ ^{14}_{7}N\text{ atom}$$

Strategy As in preceding problems, finding the released energy involves computing the difference in mass between the resultant particle(s) and the initial particle(s) and converting to MeV.

Solution

Obtain the masses of $^{14}_{6}C$ and $^{14}_{7}N$ from Appendix B and compute the difference between them:

$$\Delta m = m_C - m_N = 14.003\ 242\ u - 14.003\ 074\ u = 0.000\ 168\ u$$

Convert the mass difference to MeV:

$$E = (0.000\ 168\ u)(931.494\ MeV/u) = \boxed{0.156\ MeV}$$

Remarks The calculated energy is generally more than the energy observed in this process. The discrepancy led to a crisis in physics, because it appeared that energy wasn't conserved. As discussed below, this crisis was resolved by the discovery that another particle was also produced in the reaction.

Exercise 29.6

Calculate the maximum energy liberated in the beta decay of radioactive potassium to calcium: $^{40}_{19}K \rightarrow {}^{40}_{20}Ca$.

Answer 1.31 MeV

From Example 29.6, we see that the energy released in the beta decay of ^{14}C is approximately 0.16 MeV. As with alpha decay, we expect the electron to carry away virtually all of this energy as kinetic energy because, apparently, it is the lightest particle produced in the decay. As Figure 29.8 shows, however, only a small number of electrons have this maximum kinetic energy, represented as KE_{max} on the graph; most of the electrons emitted have kinetic energies lower than that predicted value. If the daughter nucleus and the electron aren't carrying away this liberated energy, then where has the energy gone? As an additional complication, further analysis of beta decay shows that the principles of conservation of both angular momentum and linear momentum appear to have been violated!

In 1930 Pauli proposed that a third particle must be present to carry away the "missing" energy and to conserve momentum. Later, Enrico Fermi developed a complete theory of beta decay and named this particle the **neutrino** ("little neutral one") because it had to be electrically neutral and have little or no mass. Although it eluded detection for many years, the neutrino (ν) was finally detected experimentally in 1956. The neutrino has the following properties:

◀ Properties of the neutrino

- Zero electric charge
- A mass much smaller than that of the electron, but probably not zero. (Recent experiments suggest that the neutrino definitely has mass, but the value is uncertain—perhaps less than 1 eV/c^2.)
- A spin of $\frac{1}{2}$
- Very weak interaction with matter, making it difficult to detect

With the introduction of the neutrino, we can now represent the beta decay process of Equation 29.13 in its correct form:

$$^{14}_{6}C \rightarrow {}^{14}_{7}N + e^- + \bar{\nu} \qquad [29.15]$$

The bar in the symbol $\bar{\nu}$ indicates an **antineutrino**. To explain what an antineutrino is, we first consider the following decay:

$$^{12}_{7}N \rightarrow {}^{12}_{6}C + e^+ + \nu \qquad [29.16]$$

TIP 29.3 Mass Number of the Electron

Another notation that is sometimes used for an electron is $_{-1}^{0}e$. This notation does not imply that the electron has zero rest energy. The mass of the electron is much smaller than that of the lightest nucleon, so we can approximate it as zero when we study nuclear decays and reactions.

(a)

(b)

Figure 29.8 (a) Distribution of beta particle energies in a typical beta decay. All energies are observed up to a maximum value. (b) In contrast, the energies of alpha particles from an alpha decay are discrete.

Here, we see that when ^{12}N decays into ^{12}C, a particle is produced which is identical to the electron except that it has a positive charge of $+e$. This particle is called a **positron**. Because it is like the electron in all respects except charge, the positron is said to be the **antiparticle** of the electron. We will discuss antiparticles further in Chapter 30; for now, it suffices to say that, **in beta decay, an electron and an antineutrino are emitted or a positron and a neutrino are emitted**.

Unlike beta decay, which results in a daughter particle with a variety of possible kinetic energies, alpha decays come in discrete amounts, as seen in Figure 29.8b. This is because the two daughter particles have momenta with equal magnitude and opposite direction and are each composed of a fixed number of nucleons.

Gamma Decay

Very often a nucleus that undergoes radioactive decay is left in an excited energy state. The nucleus can then undergo a second decay to a lower energy state—perhaps even to the ground state—by emitting one or more high-energy photons. The process is similar to the emission of light by an atom. An atom emits radiation to release some extra energy when an electron "jumps" from a state of high energy to a state of lower energy. Likewise, the nucleus uses essentially the same method to release any extra energy it may have following a decay or some other nuclear event. In nuclear de-excitation, the "jumps" that release energy are made by protons or neutrons in the nucleus as they move from a higher energy level to a lower level. The photons emitted in the process are called **gamma rays**, which have very high energy relative to the energy of visible light.

A nucleus may reach an excited state as the result of a violent collision with another particle. However, it's more common for a nucleus to be in an excited state as a result of alpha or beta decay. The following sequence of events typifies the gamma decay processes:

$$^{12}_{5}B \rightarrow\ ^{12}_{6}C^* + e^- + \bar{\nu} \qquad [29.17]$$

$$^{12}_{6}C^* \rightarrow\ ^{12}_{6}C + \gamma \qquad [29.18]$$

Equation 29.17 represents a beta decay in which ^{12}B decays to ^{12}C*, where the asterisk indicates that the carbon nucleus is left in an excited state following the decay. The excited carbon nucleus then decays to the ground state by emitting a gamma ray, as indicated by Equation 29.18. Note that gamma emission doesn't result in any change in either Z or A.

Practical Uses of Radioactivity

Carbon Dating   The beta decay of ^{14}C given by Equation 29.15 is commonly used to date organic samples. Cosmic rays (high-energy particles from outer space) in the upper atmosphere cause nuclear reactions that create ^{14}C from ^{14}N. In fact, the ratio of ^{14}C to ^{12}C (by numbers of nuclei) in the carbon dioxide molecules of our atmosphere has a constant value of about 1.3×10^{-12}, as determined by measuring carbon ratios in tree rings. All living organisms have the same ratio of ^{14}C to ^{12}C because they continuously exchange carbon dioxide with their surroundings. When an organism dies, however, it no longer absorbs ^{14}C from the atmosphere, so the ratio of ^{14}C to ^{12}C decreases as the result of the beta decay of ^{14}C. It's therefore possible to determine the age of a material by measuring its activity per unit mass as a result of the decay of ^{14}C. Through carbon dating, samples of wood, charcoal, bone, and shell have been identified as having lived from 1 000 to 25 000 years ago. This knowledge has helped researchers reconstruct the history of living organism—including human—during that time span.

A particularly interesting example is the dating of the Dead Sea Scrolls. This group of manuscripts was first discovered by a young Bedouin boy in a cave at Qumran near the Dead Sea in 1947. Translation showed the manuscripts to be religious documents, including most of the books of the Old Testament. Because of their historical and religious significance, scholars wanted to know their age. Carbon dating applied to fragments of the scrolls and to the material in which

ENRICO FERMI, Italian Physicist (1901–1954)

Fermi was awarded the Nobel Prize in 1938 for producing the transuranic elements by neutron irradiation and for his discovery of nuclear reactions bought about by slow neutrons. He made many other outstanding contributions to physics, including his theory of beta decay, the free-electron theory of metals, and the development of the world's first fission reactor in 1942. Fermi was truly a gifted theoretical and experimental physicist. He was also well known for his ability to present physics in a clear and exciting manner. "Whatever Nature has in store for mankind, unpleasant as it may be, men must accept, for ignorance is never better than knowledge."

APPLICATION

Carbon Dating of the Dead Sea Scrolls

they were wrapped established that they were about 1950 years old. The scrolls are now stored at the Israel museum in Jerusalem.

Smoke Detectors   Smoke detectors are frequently used in homes and industry for fire protection. Most of the common ones are the ionization-type that use radioactive materials. (See Fig. 29.9.) A smoke detector consists of an ionization chamber, a sensitive current detector, and an alarm. A weak radioactive source ionizes the air in the chamber of the detector, which creates charged particles. A voltage is maintained between the plates inside the chamber, setting up a small but detectable current in the external circuit. As long as the current is maintained, the alarm is deactivated. However, if smoke drifts into the chamber, the ions become attached to the smoke particles. These heavier particles do not drift as readily as do the lighter ions, which causes a decrease in the detector current. The external circuit senses this decrease in current and sets off the alarm.

Radon Detection   Radioactivity can also affect our daily lives in harmful ways. Soon after the discovery of radium by the Curies, it was found that the air in contact with radium compounds becomes radioactive. It was then shown that this radioactivity came from the radium itself, and the product was therefore called "radium emanation." Rutherford and Soddy succeeded in condensing this "emanation," confirming that it was a real substance: the inert, gaseous element now called **radon** (Rn). Later, it was discovered that the air in uranium mines is radioactive because of the presence of radon gas. The mines must therefore be well ventilated to help protect the miners. Finally, the fear of radon pollution has moved from uranium mines into our own homes. (See Example 29.4.) Because certain types of rock, soil, brick, and concrete contain small quantities of radium, some of the resulting radon gas finds its way into our homes and other buildings. The most serious problems arise from leakage of radon from the ground into the structure. One practical remedy is to exhaust the air through a pipe just above the underlying soil or gravel directly to the outdoors by means of a small fan or blower.

APPLICATION
Smoke Detectors

Figure 29.9   An ionization-type smoke detector. Smoke entering the chamber reduces the detected current, causing the alarm to sound.

APPLICATION
Radon Pollution

Applying Physics 29.3   Radioactive Dating of the Iceman

In 1991, a German tourist discovered the well-preserved remains of a man trapped in a glacier in the Italian Alps. (See Fig. 29.10.) Radioactive dating of a sample of bone from this hunter–gatherer, dubbed the "Iceman," revealed an age of 5 300 years. Why did scientists date the sample using the isotope ^{14}C, rather than ^{11}C, a beta emitter with a half-life of 20.4 min?

Explanation ^{14}C has a long half-life of 5 730 years, so the fraction of ^{14}C nuclei remaining after one half-life is high enough to accurately measure changes in the sample's activity. The ^{11}C isotope, which has a very short half-life, is not useful, because its activity decreases to a vanishingly small value over the age of the sample, making it impossible to detect.

   If a sample to be dated is not very old—say, about 50 years—then you should select the isotope of some other element with half-life comparable to the age of the sample. For example, if the sample contained hydrogen, you could measure the activity of ^{3}H (tritium), a beta emitter of half-life 12.3 years. As a general rule, the expected age of the sample should be long enough to measure a change in activity, but not so long that its activity can't be detected.

Hanny Paul/Gamma Liaison

Figure 29.10   (Applying Physics 29.3) The body of an ancient man (dubbed the Iceman) was exposed by a melting glacier in the Alps.

To use radioactive dating techniques, we need to recast some of the equations already introduced. We start by multiplying both sides of Equation 29.4 by λ:

$$\lambda N = \lambda N_0 e^{-\lambda t}$$

From Equation 29.3, we have $\lambda N = R$ and $\lambda N_0 = R_0$. Substitute these expressions into the above equation and divide through by R_0:

$$\frac{R}{R_0} = e^{-\lambda t}$$

R is the present activity and R_0 was the activity when the object in question was part of a living organism. We can solve for time by taking the natural logarithm of both sides of the foregoing equation:

$$\ln\left(\frac{R}{R_0}\right) = \ln(e^{-\lambda t}) = -\lambda t$$

$$t = -\frac{\ln\left(\dfrac{R}{R_0}\right)}{\lambda} \qquad \textbf{[29.19]}$$

EXAMPLE 29.7   Should We Report This to Homicide?

Goal   Apply the technique of carbon-14 dating.

Problem   A 50.0-g sample of carbon is taken from the pelvis bone of a skeleton and is found to have a carbon-14 decay rate of 200.0 decays/min. It is known that carbon from a living organism has a decay rate of 15.0 decays/min · g and that ^{14}C has a half-life of 5 730 yr $= 3.01 \times 10^9$ min. Find the age of the skeleton.

Strategy   Calculate the original activity and the decay constant, and then substitute those numbers and the current activity into Equation 29.19.

Solution

Calculate the original activity R_0 from the decay rate and the mass of the sample:

$$R_0 = \left(15.0\ \frac{\text{decays}}{\text{min} \cdot \text{g}}\right)(50.0\ \text{g}) = 7.50 \times 10^2\ \frac{\text{decays}}{\text{min}}$$

Find the decay constant from Equation 29.5:

$$\lambda = \frac{0.693}{T_{1/2}} = \frac{0.693}{3.01 \times 10^9\ \text{min}} = 2.30 \times 10^{-10}\ \text{min}^{-1}$$

R is given, so now we substitute all values into Equation 29.19 to find the age of the skeleton:

$$t = -\frac{\ln\left(\dfrac{R}{R_0}\right)}{\lambda} = -\frac{\ln\left(\dfrac{200.0\ \text{decays/min}}{7.50 \times 10^2\ \text{decays/min}}\right)}{2.30 \times 10^{-10}\ \text{min}^{-1}}$$

$$= \frac{1.32}{2.30 \times 10^{-10}\ \text{min}^{-1}}$$

$$= 5.74 \times 10^9\ \text{min} = \boxed{1.09 \times 10^4\ \text{yr}}$$

Remark   For much longer periods, other radioactive substances with longer half-lives must be used to develop estimates.

Exercise 29.7

A sample of carbon of mass 7.60 g taken from an animal jawbone has an activity of 4.00 decays/min. How old is the jawbone?

Answer 2.77×10^4 yr

APPLICATION

Carbon-14 Dating of the Shroud of Turin

Carbon-14 and the Shroud of Turin

Since the Middle Ages, many people have marveled at a 14-foot-long, yellowing piece of linen found in Turin, Italy, purported to be the burial shroud of Jesus Christ (Fig. 29.11). The cloth bears a remarkable, full-size likeness of a crucified body, with wounds on the head that could have been caused by a crown of thorns and another wound in the side that could have been the cause of death. Skepticism over the authenticity of the shroud has existed since its first public showing in 1354; in fact, a French bishop declared it to be a fraud at the time. Because of its controversial nature, religious bodies have taken a neutral stance on its authenticity.

In 1978 the bishop of Turin allowed the cloth to be subjected to scientific analysis, but notably missing from these tests was carbon-14 dating. The reason for this omission was that, at the time, carbon-dating techniques required a piece of cloth about the size of a handkerchief. In 1988 the process had been refined to the point that pieces as small as one square inch were sufficient, and at that time permission was granted to allow the dating to proceed. Three labs were selected for the testing, and each was given four pieces of material. One of these was a piece of the shroud, and the other three pieces were control pieces similar in appearance to the shroud.

The testing procedure consisted of burning the cloth to produce carbon dioxide, which was then converted chemically to graphite. The graphite sample was subjected to carbon-14 analysis, and in the end all three labs agreed amazingly well on the age of the shroud. The average of their results gave a date for the cloth of A.D. 1 320 ± 60 years, with an assurance that the cloth could not be older than A.D. 1 200. Carbon-14 dating has thus unraveled the most important mystery concerning the shroud, but others remain. For example, investigators have not yet been able to explain how the image was imprinted.

Figure 29.11   The Shroud of Turin as it appears in a photographic negative image.

29.5   NATURAL RADIOACTIVITY

Radioactive nuclei are generally classified into two groups: (1) unstable nuclei found in nature, which give rise to what is called **natural radioactivity**, and (2) nuclei produced in the laboratory through nuclear reactions, which exhibit **artificial radioactivity**.

Three series of naturally occurring radioactive nuclei exist (Table 29.2). Each starts with a specific long-lived radioactive isotope with half-life exceeding that of any of its descendants. The fourth series in Table 29.2 begins with ^{237}Np, a transuranic element (an element having an atomic number greater than that of uranium) not found in nature. This element has a half-life of "only" 2.14×10^6 yr.

The two uranium series are somewhat more complex than the ^{232}Th series (Fig. 29.12). Also, there are several other naturally occurring radioactive isotopes, such as ^{14}C and ^{40}K, that are not part of either decay series.

Natural radioactivity constantly supplies our environment with radioactive elements that would otherwise have disappeared long ago. For example, because the Solar System is about 5×10^9 years old, the supply of ^{226}Ra (with a half-life of only 1 600 yr) would have been depleted by radioactive decay long ago were it not for the decay series that starts with ^{238}U, with a half-life of 4.47×10^9 yr.

Figure 29.12   Decay series beginning with ^{232}Th.

29.6   NUCLEAR REACTIONS

It is possible to change the structure of nuclei by bombarding them with energetic particles. Such changes are called **nuclear reactions**. Rutherford was the first to observe nuclear reactions, using naturally occurring radioactive sources for the

TABLE 29.2

The Four Radioactive Series

Series	Starting Isotope	Half-life (years)	Stable End Product
Uranium	$^{238}_{92}$U	4.47×10^9	$^{206}_{82}$Pb
Actinium	$^{235}_{92}$U	7.04×10^8	$^{207}_{82}$Pb
Thorium	$^{232}_{90}$Th	1.41×10^{10}	$^{208}_{82}$Pb
Neptunium	$^{237}_{93}$Np	2.14×10^6	$^{209}_{83}$Bi

bombarding particles. He found that protons were released when alpha particles were allowed to collide with nitrogen atoms. The process can be represented symbolically as

$$\ _{2}^{4}\text{He} + \ _{7}^{14}\text{N} \ \rightarrow \ X + \ _{1}^{1}\text{H} \qquad\qquad \textbf{[29.20]}$$

This equation says that an alpha particle ($_{2}^{4}\text{He}$) strikes a nitrogen nucleus and produces an unknown product nucleus (X) and a proton ($_{1}^{1}\text{H}$). Balancing atomic numbers and mass numbers, as we did for radioactive decay, enables us to conclude that the unknown is characterized as $_{8}^{17}\text{X}$. Because the element with atomic number 8 is oxygen, we see that the reaction is

$$\ _{2}^{4}\text{He} + \ _{7}^{14}\text{N} \ \rightarrow \ _{8}^{17}\text{O} + \ _{1}^{1}\text{H} \qquad\qquad \textbf{[29.21]}$$

This nuclear reaction starts with two stable isotopes—helium and nitrogen—and produces two different stable isotopes—hydrogen and oxygen.

Since the time of Rutherford, thousands of nuclear reactions have been observed, particularly following the development of charged-particle accelerators in the 1930s. With today's advanced technology in particle accelerators and particle detectors, it is possible to achieve particle energies of at least 1 000 GeV = 1 TeV. These high-energy particles are used to create new particles whose properties are helping to solve the mysteries of the nucleus (and indeed, of the Universe itself).

Quick Quiz 29.4

Which of the following are possible reactions?

(a) $_{0}^{1}\text{n} + \ _{92}^{235}\text{U} \ \rightarrow \ _{54}^{140}\text{Xe} + \ _{38}^{94}\text{Sr} + 2(_{0}^{1}\text{n})$

(b) $_{0}^{1}\text{n} + \ _{92}^{235}\text{U} \ \rightarrow \ _{50}^{132}\text{Sn} + \ _{42}^{101}\text{Mo} + 3(_{0}^{1}\text{n})$

(c) $_{0}^{1}\text{n} + \ _{94}^{239}\text{Pu} \ \rightarrow \ _{53}^{127}\text{I} + \ _{41}^{93}\text{Nb} + 3(_{0}^{1}\text{n})$

EXAMPLE 29.8   The Discovery of the Neutron

Goal   Balance a nuclear reaction to determine an unknown decay product.

Problem   A nuclear reaction of significant note occurred in 1932 when Chadwick, in England, bombarded a beryllium target with alpha particles. Analysis of the experiment indicated that the following reaction occurred:

$$\ _{2}^{4}\text{He} + \ _{4}^{9}\text{Be} \ \rightarrow \ _{6}^{12}\text{C} + \ _{Z}^{A}\text{X}$$

What is $_{Z}^{A}\text{X}$ in this reaction?

Strategy   Balancing mass numbers and atomic numbers yields the answer.

Solution

Write an equation relating the atomic masses on either side: $4 + 9 = 12 + A \ \rightarrow \ A = 1$

Write an equation relating the atomic numbers: $2 + 4 = 6 + Z \ \rightarrow \ Z = 0$

Identify the particle: $_{Z}^{A}\text{X} = \ _{0}^{1}\text{n}$ (a neutron)

Remarks   This experiment was the first to provide positive proof of the existence of neutrons.

Exercise 29.8

Identify the unknown particle in this reaction:

$$\ _{2}^{4}\text{He} + \ _{7}^{14}\text{N} \ \rightarrow \ _{8}^{17}\text{O} + \ _{Z}^{A}\text{X}$$

Answer $_{Z}^{A}\text{X} = \ _{1}^{1}\text{H}$ (a neutral hydrogen atom)

EXAMPLE 29.9   Synthetic Elements

Goal   Construct equations for a series of radioactive decays.

Problem   **(a)** A beam of neutrons is directed at a target of $^{238}_{92}$U. The reaction products are a gamma ray and another isotope. What is the isotope? **(b)** The isotope $^{239}_{92}$U is radioactive and undergoes beta decay. Write the equation symbolizing this decay and identify the resulting isotope.

Strategy   Balance the mass numbers and atomic numbers on both sides of the equations.

Solution

(a) Identify the isotope produced by the reaction of a neutron with a target of $^{238}_{92}$U, with production of a gamma ray.

Write an equation for the reaction in terms of the unknown isotope:

$$^{1}_{0}n + {}^{238}_{92}U \rightarrow {}^{A}_{Z}X + \gamma$$

Write and solve equations for the atomic mass and atomic number:

$$A = 1 + 238 = 239; \quad Z = 0 + 92 = 92$$

Identify the isotope:

$$^{A}_{Z}X = {}^{239}_{92}U$$

(b) Write the equation for the beta decay of $^{239}_{92}$U, identifying the resulting isotope.

Write an equation for the decay of $^{239}_{92}$U by beta emission in terms of the unknown isotope:

$$^{239}_{92}U \rightarrow {}^{A}_{Z}Y + e^{-} + \bar{\nu}$$

Write and solve equations for the atomic mass and charge conservation (the electron counts as -1 on the right):

$$A = 239; \quad 92 = Z - 1 \rightarrow Z = 93$$

Identify the isotope:

$$^{A}_{Z}Y = {}^{239}_{93}Np \text{ (neptunium)}$$

Remarks   The interesting feature of these reactions is the fact that uranium is the element with the greatest number of protons (92) which exists in nature in any appreciable amount. The reactions in parts (a) and (b) do occur occasionally in nature; hence, minute traces of neptunium and plutonium are present. In 1940, however, researchers bombarded uranium with neutrons to produce plutonium and neptunium. These two elements were the first elements made in the laboratory. Since then, the list of synthetic elements has been extended to include those up to atomic number 112. Recently, elements 113 and 115 have been observed, but as of this writing, their existence has not yet been confirmed.

Exercise 29.9

The isotope $^{238}_{93}$U is also radioactive and decays by beta emission. What is the end product?

Answer $^{239}_{94}$Pu

Q Values

We have just examined some nuclear reactions for which mass numbers and atomic numbers must be balanced in the equations. We will now consider the energy involved in these reactions, because energy is another important quantity that must be conserved.

We illustrate this procedure by analyzing the following nuclear reaction:

$$^{2}_{1}H + {}^{14}_{7}N \rightarrow {}^{12}_{6}C + {}^{4}_{2}He \qquad\qquad \text{[29.22]}$$

The total mass on the left side of the equation is the sum of the mass of ^2_1H (2.014 102 u) and the mass of $^{14}_7N$ (14.003 074 u), which equals 16.017 176 u. Similarly, the mass on the right side of the equation is the sum of the mass of $^{12}_6C$ (12.000 000 u) plus the mass of ^4_2He (4.002 602 u), for a total of 16.002 602 u. Thus, the total mass before the reaction is greater than the total mass after the reaction. The mass difference in the reaction is equal to 16.017 176 u − 16.002 602 u = 0.014 574 u. This "lost" mass is converted to the kinetic energy of the nuclei present after the reaction. In energy units, 0.014 574 u is equivalent to 13.576 MeV of kinetic energy carried away by the carbon and helium nuclei.

The energy required to balance the equation is called the Q value of the reaction. In Equation 29.22, the Q value is 13.576 MeV. Nuclear reactions in which there is a release of energy—that is, positive Q values—are said to be **exothermic reactions**.

The energy balance sheet isn't complete, however: We must also consider the kinetic energy of the incident particle before the collision. As an example, assume that the deuteron in Equation 29.22 has a kinetic energy of 5 MeV. Adding this to our Q value, we find that the carbon and helium nuclei have a total kinetic energy of 18.576 MeV following the reaction.

Now consider the reaction

$$^4_2He + ^{14}_7N \rightarrow ^{17}_8O + ^1_1H \qquad [29.23]$$

Before the reaction, the total mass is the sum of the masses of the alpha particle and the nitrogen nucleus: 4.002 602 u + 14.003 074 u = 18.005 676 u. After the reaction, the total mass is the sum of the masses of the oxygen nucleus and the proton: 16.999 133 u + 1.007 825 u = 18.006 958 u. In this case, the total mass after the reaction is *greater* than the total mass before the reaction. The mass deficit is 0.001 282 u, equivalent to an energy deficit of 1.194 MeV. This deficit is expressed by the negative Q value of the reaction, −1.194 MeV. Reactions with negative Q values are called **endothermic reactions**. Such reactions won't take place unless the incoming particle has at least enough kinetic energy to overcome the energy deficit.

At first it might appear that the reaction in Equation 29.23 can take place if the incoming alpha particle has a kinetic energy of 1.194 MeV. In practice, however, the alpha particle must have more energy than this. If it has an energy of only 1.194 MeV, energy is conserved but careful analysis shows that momentum isn't. This can be understood by recognizing that the incoming alpha particle has some momentum before the reaction. However, if its kinetic energy is only 1.194 MeV, the products (oxygen and a proton) would be created with zero kinetic energy and thus zero momentum. It can be shown that in order to conserve both energy and momentum, the incoming particle must have a minimum kinetic energy given by

$$KE_{min} = \left(1 + \frac{m}{M}\right)|Q| \qquad [29.24]$$

where m is the mass of the incident particle, M is the mass of the target, and the absolute value of the Q value is used. For the reaction given by Equation 29.23, we find that

$$KE_{min} = \left(1 + \frac{4.002\ 602}{14.003\ 074}\right)|-1.194\ \text{MeV}| = 1.535\ \text{MeV}$$

This minimum value of the kinetic energy of the incoming particle is called the **threshold energy**. The nuclear reaction shown in Equation 29.23 won't occur if the incoming alpha particle has a kinetic energy of less than 1.535 MeV, but can occur if its kinetic energy is equal to or greater than 1.535 MeV.

Quick Quiz 29.5

If the Q value of an endothermic reaction is −2.17 MeV, then the minimum kinetic energy needed in the reactant nuclei if the reaction is to occur must be (a) equal to 2.17 MeV, (b) greater than 2.17 MeV, (c) less than 2.17 MeV, or (d) exactly half of 2.17 MeV.

29.7 MEDICAL APPLICATIONS OF RADIATION

Radiation Damage in Matter

Radiation absorbed by matter can cause severe damage. The degree and kind of damage depend on several factors, including the type and energy of the radiation and the properties of the absorbing material. Radiation damage in biological organisms is due primarily to ionization effects in cells. The normal function of a cell may be disrupted when highly reactive ions or radicals are formed as the result of ionizing radiation. For example, hydrogen and hydroxyl radicals produced from water molecules can induce chemical reactions that may break bonds in proteins and other vital molecules. Large acute doses of radiation are especially dangerous because damage to a great number of molecules in a cell may cause the cell to die. Also, cells that do survive the radiation may become defective, which can lead to cancer.

In biological systems, it is common to separate radiation damage into two categories: somatic damage and genetic damage. **Somatic damage** is radiation damage to any cells except the reproductive cells. Such damage can lead to cancer at high radiation levels or seriously alter the characteristics of specific organisms. **Genetic damage** affects only reproductive cells. Damage to the genes in reproductive cells can lead to defective offspring. Clearly, we must be concerned about the effect of diagnostic treatments, such as x-rays and other forms of exposure to radiation.

Several units are used to quantify radiation exposure and dose. The **roentgen** (R) is defined as **that amount of ionizing radiation which will produce 2.08 × 10⁹ ion pairs in 1 cm³ of air under standard conditions.** Equivalently, the roentgen is **that amount of radiation which deposits 8.76 × 10⁻³ J of energy into 1 kg of air.**

For most applications, the roentgen has been replaced by the **rad** (an acronym for *r*adiation *a*bsorbed *d*ose), defined as follows: **One rad is that amount of radiation which deposits 10⁻² J of energy into 1 kg of absorbing material.**

Although the rad is a perfectly good physical unit, it's not the best unit for measuring the degree of biological damage produced by radiation, because the degree of damage depends not only on the dose, but also on the *type* of radiation. For example, a given dose of alpha particles causes about 10 times more biological damage than an equal dose of x-rays. The **RBE** (*r*elative *b*iological *e*ffectiveness) factor is defined as **the number of rads of x-radiation or gamma radiation that produces the same biological damage as 1 rad of the radiation being used.** The RBE factors for different types of radiation are given in Table 29.3. Note that the values are only approximate because they vary with particle energy and the form of damage.

Finally, the **rem** (*r*oentgen *e*quivalent in *m*an) is defined as the product of the dose in rads and the RBE factor:

$$\text{Dose in rem} = \text{dose in rads} \times \text{RBE}$$

According to this definition, 1 rem of any two kinds of radiation will produce the same amount of biological damage. From Table 29.3, we see that a dose of 1 rad of fast neutrons represents an effective dose of 10 rem and that 1 rad of x-radiation is equivalent to a dose of 1 rem.

TABLE 29.3

RBE Factors for Several Types of Radiation

Radiation	RBE Factor
X-rays and gamma rays	1.0
Beta particles	1.0–1.7
Alpha particles	10–20
Slow neutrons	4–5
Fast neutrons and protons	10
Heavy ions	20

Low-level radiation from natural sources, such as cosmic rays and radioactive rocks and soil, delivers a dose of about 0.13 rem/year per person. The upper limit of radiation dose recommended by the U.S. government (apart from background radiation and exposure related to medical procedures) is 0.5 rem/year. Many occupations involve higher levels of radiation exposure, and for individuals in these occupations, an upper limit of 5 rem/year has been set for whole-body exposure. Higher upper limits are permissible for certain parts of the body, such as the hands and forearms. An acute whole-body dose of 400 to 500 rem results in a mortality rate of about 50%. The most dangerous form of exposure is ingestion or inhalation of radioactive isotopes, especially those elements the body retains and concentrates, such as ^{90}Sr. In some cases, a dose of 1000 rem can result from ingesting 1 mCi of radioactive material.

Sterilizing objects by exposing them to radiation has been going on for at least 25 years, but in recent years the methods used have become safer to use and more economical. Most bacteria, worms, and insects are easily destroyed by exposure to gamma radiation from radioactive cobalt. There is no intake of radioactive nuclei by an organism in such sterilizing processes, as there is in the use of radioactive tracers. The process is highly effective in destroying Trichinella worms in pork, salmonella bacteria in chickens, insect eggs in wheat, and surface bacteria on fruits and vegetables that can lead to rapid spoilage. Recently, the procedure has been expanded to include the sterilization of medical equipment while in its protective covering. Surgical gloves, sponges, sutures, and so forth are irradiated while packaged. Also, bone, cartilage, and skin used for grafting is often irradiated to reduce the chance of infection.

Tracing

Radioactive particles can be used to trace chemicals participating in various reactions. One of the most valuable uses of radioactive tracers is in medicine. For example, ^{131}I is an artificially produced isotope of iodine. (The natural, nonradioactive isotope is ^{127}I.) Iodine, a necessary nutrient for our bodies, is obtained largely through the intake of seafood and iodized salt. The thyroid gland plays a major role in the distribution of iodine throughout the body. In order to evaluate the performance of the thyroid, the patient drinks a small amount of radioactive sodium iodide. Two hours later, the amount of iodine in the thyroid gland is determined by measuring the radiation intensity in the neck area.

A medical application of the use of radioactive tracers occurring in emergency situations is that of locating a hemorrhage inside the body. Often the location of the site cannot easily be determined, but radioactive chromium can identify the location with a high degree of precision. Chromium is taken up by red blood cells and carried uniformly throughout the body. However, the blood will be dumped at a hemorrhage site, and the radioactivity of that region will increase markedly.

The tracer technique is also useful in agricultural research. Suppose the best method of fertilizing a plant is to be determined. A certain material in the fertilizer, such as nitrogen, can be tagged with one of its radioactive isotopes. The fertilizer is then sprayed onto one group of plants, sprinkled on the ground for a second group, and raked into the soil for a third. A Geiger counter is then used to track the nitrogen through the three types of plants.

Tracing techniques are as wide ranging as human ingenuity can devise. Present applications range from checking the absorption of fluorine by teeth to checking contamination of food-processing equipment by cleansers to monitoring deterioration inside an automobile engine. In the last case, a radioactive material is used in the manufacture of the pistons, and the oil is checked for radioactivity to determine the amount of wear on the pistons.

Computed Axial Tomography (CAT) Scans

The normal x-ray of a human body has two primary disadvantages when used as a source of clinical diagnosis. First, it is difficult to distinguish between various types of tissue in the body because they all have similar x-ray absorption properties.

Second, a conventional x-ray absorption picture is indicative of the average amount of absorption along a particular direction in the body, leading to somewhat obscured pictures. To overcome these problems, an instrument called a CAT scanner was developed in England in 1973; the device is capable of producing pictures of much greater clarity and detail than previously possible.

The operation of a CAT scanner can be understood by considering the following hypothetical experiment: suppose a box consists of four compartments, labeled A, B, C, and D, as in Figure 29.13a. Each compartment has a different amount of absorbing material from any other compartment. What set of experimental procedures will enable us to determine the relative amounts of material in each compartment? The following steps outline one method that will provide this information: first, a beam of x-rays is passed through compartments A and C, as in Figure 29.13b. The intensity of the exiting radiation is reduced by absorption by some number that we assign as 8. (The number 8 could mean, for example, that the intensity of the exiting beam is reduced by eight-tenths of 1% from its initial value.) Because we don't know which of the compartments, A or C, was responsible for this reduction in intensity, half the loss is assigned to each compartment, as in Figure 29.13c. Next, a beam of x-rays is passed through compartments B and D, as in Figure 29.13b. The reduction in intensity for this beam is 10, and again we assign half the loss to each compartment. We now redirect the x-ray source so that it sends one beam through compartments A and B and another through compartments C and D, as in Figure 29.13d. Once more, we measure the absorption. Suppose the absorption through compartments A and B in this experiment is measured to be 7 units. On the basis of our first experiment, we would have guessed it would be 9 units: 4 by compartment A and 5 by compartment B. Thus, we have reduced the guessed absorption for each compartment by 1 unit, so that the sum is 7 rather than 9, giving the numbers shown in Figure 29.13e. Likewise, when the beam is passed through compartments C and D, as in Figure 29.13d, we may find the total absorption to be 11 as compared to our first experiment of 9. In this case, we add 1 unit of absorption to each compartment to give a sum of 11, as in Figure 29.13e. This somewhat crude procedure could be improved by measuring the absorption along other paths. However, these simple measurements are sufficient to enable us to conclude that compartment D contains the most absorbing material and A the least. A visual representation of these results can be obtained by assigning, to each compartment, a shade of gray corresponding to the particular number

APPLICATION

CAT Scans

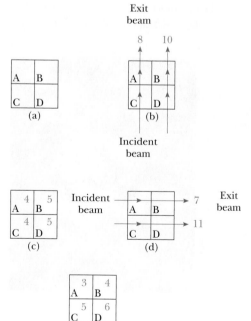

Figure 29.13 An experimental procedure for determining the relative amounts of x-ray absorption by four different compartments in a box.

associated with the absorption. In this example, compartment D would be very dark and compartment A would be very light.

The steps outlined previously are representative of how a CAT scanner produces images of the human body. A thin slice of the body is subdivided into perhaps 10 000 compartments, rather than 4 as in our simple example. The function of the CAT scanner is to determine the relative absorption in each of these 10 000 compartments and to display a picture of its calculations in various shades of gray. Note that "CAT" stands for **computed axial tomography**. The term *axial* is used because the slice of the body to be analyzed corresponds to a plane perpendicular to the head-to-toe axis. *Tomos* is the Greek word for slice and *graph* is the Greek word for picture. In a typical diagnosis, the patient is placed in the position shown in Figure 29.14 and a narrow beam of x-rays is sent through the plane of interest. The emerging x-rays are detected and measured by photomultiplier tubes behind the patient. The x-ray tube is then rotated a few degrees, and the intensity is recorded again. An extensive amount of information is obtained by rotating the beam through 180° at intervals of about 1° per measurement, resulting in a set of numbers assigned to each of the 10 000 "compartments" in the slice. These numbers are then converted by the computer to a photograph in various shades of gray for the segment of the body that is under observation.

A brain scan of a patient can now be made in about 2 s, and a full-body scan requires about 6 s. The final result is a picture containing much greater quantitative information and clarity than a conventional x-ray photograph. Because CAT scanners use x-rays, which are an ionizing form of radiation, the technique presents a modest health risk to the patient being diagnosed.

Magnetic Resonance Imaging (MRI)

At the heart of magnetic resonance imaging (MRI) is the fact that when a nucleus having a magnetic moment is placed in an external magnetic field, its moment precesses about the magnetic field with a frequency that is proportional to the field. For example, a proton, with a spin of 1/2, can occupy one of two energy states when placed in an external magnetic field. The lower energy state corresponds to the case in which the spin is aligned with the field, whereas the higher energy state corresponds to the case in which the spin is opposite the field. Transitions between these two states can be observed with a technique known as **nuclear magnetic resonance**. A DC magnetic field is applied to align the magnetic moments, and a second, weak oscillating magnetic field is applied perpendicular to the DC field. When the frequency of the oscillating field is adjusted to match the precessional frequency of the magnetic moments, the nuclei will "flip" between

(a)   (b)

Figure 29.14   (a) CAT scanner detector assembly. (b) Photograph of a patient undergoing a CAT scan in a hospital.

Figure 29.15 Computer-enhanced MRI images of (a) a normal human brain and (b) a human brain with a glioma tumor.

(a)   (b)

the two spin states. These transitions result in a net absorption of energy by the spin system, which can be detected electronically.

In MRI, image reconstruction is obtained using spatially varying magnetic fields and a procedure for encoding each point in the sample being imaged. Some MRI images taken on a human head are shown in Figure 29.15. In practice, a computer-controlled pulse-sequencing technique is used to produce signals that are captured by a suitable processing device. The signals are then subjected to appropriate mathematical manipulations to provide data for the final image. The main advantage of MRI over other imaging techniques in medical diagnostics is that it causes minimal damage to cellular structures. Photons associated with the rf signals used in MRI have energies of only about 10^{-7} eV. Because molecular bond strengths are much larger (on the order of 1 eV), the rf photons cause little cellular damage. In comparison, x-rays or γ-rays have energies ranging from 10^4 to 10^6 eV and can cause considerable cellular damage.

29.8 RADIATION DETECTORS

Although most medical applications of radiation require instruments to make quantitative measurements of radioactive intensity, we have not yet explained how such instruments operate. Various devices have been developed to detect the energetic particles emitted when a radioactive nucleus decays. The **Geiger counter** (Fig. 29.16) is perhaps the most common device used to detect radioactivity. It can be considered the prototype of all counters that use the ionization of a medium as the basic detection process. A Geiger counter consists of a thin wire electrode aligned along the central axis of a cylindrical metallic tube filled with a gas at low pressure. The wire is maintained at a high positive voltage of about 1 000 V relative to the tube. When an energetic charged particle or gamma-ray photon enters the tube through a thin window at one end, some of the gas atoms are ionized. The electrons removed from these atoms are attracted toward the wire electrode,

Figure 29.16 (a) Diagram of a Geiger counter. The voltage between the wire electrode and the metallic tube is usually about 1 000 V. (b) A Geiger counter.

and in the process they ionize other atoms in their path. This sequential ionization results in an *avalanche* of electrons that produces a current pulse. After the pulse has been amplified, it can either be used to trigger an electronic counter or delivered to a loudspeaker that clicks each time a particle is detected. Although a Geiger counter reliably detects the presence and quantity of radiation, it cannot be used to measure the energy of the detected radiation.

A **semiconductor diode detector** is essentially a reverse biased $p-n$ junction. As an energetic particle passes through the junction, it produces electron–hole pairs that are separated by the internal electric field. This movement of electrons and holes creates a brief pulse of current that is measured with an electronic counter. In a typical device, the duration of the pulse is 10^{-8} s.

A **scintillation counter** usually uses a solid or liquid material having atoms that are easily excited by radiation. The excited atoms then emit photons of visible light when they return to their ground state. Common materials used as scintillators are transparent crystals of sodium iodide and certain plastics. If the scintillator material is attached to one end of a device called a **photomultiplier** (PM) tube, as shown in Figure 29.17, the photons emitted by the scintillator can be converted to an electrical signal. The PM tube consists of numerous electrodes, called *dynodes*, whose electric potentials increase in succession along the length of the tube. Between the top of the tube and the scintillator material is a plate called a *photocathode*. When photons leaving the scintillator hit this plate, electrons are emitted because of the photoelectric effect. As one of these emitted electrons strikes the first dynode, the electron has sufficient kinetic energy to eject several other electrons from the surface of the dynode. When these electrons are accelerated to the second dynode, many more electrons are ejected, and a multiplication process occurs. The end result is 1 million or more electrons striking the last dynode. Hence, one particle striking the scintillator produces a sizable electrical pulse at the PM output, and this pulse is sent to an electronic counter.

Both the scintillator and the semiconductor diode detector are much more sensitive than a Geiger counter, mainly because of the higher mass density of the detecting medium. Both can also be used to measure particle energy from the height of the pulses produced.

Track detectors are various devices used to view the tracks or paths of charged particles directly. High-energy particles produced in particle accelerators may have energies ranging from 10^9 to 10^{12} eV. The energy of such particles can't be measured with the small detectors already mentioned. Instead, their energy and

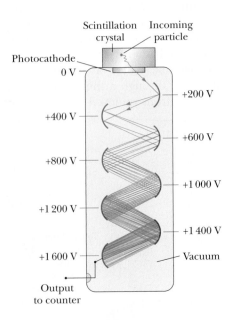

Figure 29.17 Diagram of a scintillation counter connected to a photomultiplier tube.

(a)   (b)

Figure 29.18 (a) Artificially colored photograph showing tracks of particles that have passed through a bubble chamber. (b) This research scientist is studying a photograph of particle tracks made in a bubble chamber at Fermilab.

momentum are found from the curvature of their paths in a magnetic field of known magnitude and direction.

A **photographic emulsion** is the simplest example of a track detector. A charged particle ionizes the atoms in an emulsion layer. The path of the particle corresponds to a family of points at which chemical changes have occurred in the emulsion. When the emulsion is developed, the particle's track becomes visible.

A **cloud chamber** contains a gas that has been supercooled to just below its usual condensation point. An energetic charged particle passing through ionizes the gas along its path. The ions serve as centers for condensation of the supercooled gas. The track can be seen with the naked eye and can be photographed. A magnetic field can be applied to determine the charges of the radioactive particles, as well as their momentum and energy.

A device called a **bubble chamber**, invented in 1952 by D. Glaser, uses a liquid (usually liquid hydrogen) maintained near its boiling point. Ions produced by incoming charged particles leave bubblelike tracks, which can be photographed (Fig. 29.18). Because the density of the liquid in a bubble chamber is much higher than the density of the gas in a cloud chamber, the bubble chamber has a much higher sensitivity.

A **wire chamber** consists of thousands of closely spaced parallel wires that collect the electrons created by a passing ionizing particle. A second grid, with wires perpendicular to the first, allows the x,y position of the particle in the plane of the two sets of wires to be determined. Finally several such x,y grids arranged parallel to each other along the z-axis can be used to determine the particle's track in three dimensions. Wire chambers form a part of most detectors used at high-energy accelerator labs and provide electronic readouts to a computer for rapid reconstruction and display of tracks.

SUMMARY

Physics⊗Now™ Take a practice test by logging into PhysicsNow at **www.cp7e.com** and clicking on the Pre-Test link for this chapter.

29.1   Some Properties of Nuclei &
29.2   Binding Energy

Nuclei are represented symbolically as $^A_Z X$, where X represents the chemical symbol for the element. The quantity A

is the **mass number**, which equals the total number of nucleons (neutrons plus protons) in the nucleus. The quantity Z is the **atomic number**, which equals the number of protons in the nucleus. Nuclei that contain the same number of protons but different numbers of neutrons are called **isotopes**. In other words, isotopes have the same Z value but different A values.

Most nuclei are approximately spherical, with an average radius given by

$$r = r_0 A^{1/3} \qquad [29.1]$$

where A is the mass number and r_0 is a constant equal to 1.2×10^{-15} m.

The total mass of a nucleus is always less than the sum of the masses of its individual nucleons. This mass difference Δm, multiplied by c^2, gives the **binding energy** of the nucleus.

29.3 Radioactivity

The spontaneous emission of radiation by certain nuclei is called **radioactivity**. There are three processes by which a radioactive substance can decay: alpha (α) decay, in which the emitted particles are $^4_2$He nuclei; beta (β) decay, in which the emitted particles are electrons or positrons; and gamma (γ) decay, in which the emitted particles are high-energy photons.

The **decay rate**, or **activity**, R, of a sample is given by

$$R = \left| \frac{\Delta N}{\Delta t} \right| = \lambda N \qquad \text{[29.3]}$$

where N is the number of radioactive nuclei at some instant and λ is a constant for a given substance called the **decay constant**.

Nuclei in a radioactive substance decay in such a way that the number of nuclei present varies with time according to the expression

$$N = N_0 e^{-\lambda t} \qquad \text{[29.4a]}$$

where N is the number of radioactive nuclei present at time t, N_0 is the number at time $t = 0$, and $e = 2.718 \ldots$ is the base of the natural logarithms.

The **half-life** $T_{1/2}$ of a radioactive substance is the time required for half of a given number of radioactive nuclei to decay. The half-life is related to the decay constant by

$$T_{1/2} = \frac{0.693}{\lambda} \qquad \text{[29.5]}$$

29.4 The Decay Processes

If a nucleus decays by alpha emission, it loses two protons and two neutrons. A typical alpha decay is

$$^{238}_{92}\text{U} \rightarrow ^{234}_{90}\text{Th} + ^4_2\text{He} \qquad \text{[29.9]}$$

Note that in this decay, as in all radioactive decay processes, the sum of the Z values on the left equals the sum of the Z values on the right; the same is true for the A values.

A typical beta decay is

$$^{14}_6\text{C} \rightarrow ^{14}_7\text{N} + e^- + \overline{\nu} \qquad \text{[29.15]}$$

When a nucleus undergoes beta decay, an **antineutrino** is emitted along with an electron, or a **neutrino** along with a positron. A neutrino has zero electric charge and a small mass (which may be zero) and interacts weakly with matter.

Nuclei are often in an excited state following radioactive decay, and they release their extra energy by emitting a high-energy photon called a **gamma ray** (γ). A typical gamma ray emission is

$$^{12}_6\text{C}^* \rightarrow ^{12}_6\text{C} + \gamma \qquad \text{[29.18]}$$

where the asterisk indicates that the carbon nucleus was in an excited state before gamma emission.

29.6 Nuclear Reactions

Nuclear reactions can occur when a bombarding particle strikes another nucleus. A typical nuclear reaction is

$$^4_2\text{He} + ^{14}_7\text{N} \rightarrow ^{17}_8\text{O} + ^1_1\text{H} \qquad \text{[29.21]}$$

In this reaction, an alpha particle strikes a nitrogen nucleus, producing an oxygen nucleus and a proton. As in radioactive decay, atomic numbers and mass numbers balance on the two sides of the arrow.

Nuclear reactions in which energy is released are said to be **exothermic reactions** and are characterized by positive Q values. Reactions with negative Q values, called **endothermic reactions**, cannot occur unless the incoming particle has at least enough kinetic energy to overcome the energy deficit. In order to conserve both energy and momentum, the incoming particle must have a minimum kinetic energy, called the **threshold energy**, given by

$$KE_{\text{min}} = \left(1 + \frac{m}{M} \right) |Q| \qquad \text{[29.24]}$$

where m is the mass of the incident particle and M is the mass of the target atom.

CONCEPTUAL QUESTIONS

1. Isotopes of a given element have different physical properties, such as mass, but the same chemical properties. Why is this?

2. If a heavy nucleus that is initially at rest undergoes alpha decay, which has more kinetic energy after the decay, the alpha particle or the daughter nucleus?

3. A student claims that a heavy form of hydrogen decays by alpha emission. How do you respond?

4. Explain the main differences between alpha, beta, and gamma rays.

5. In beta decay, the energy of the electron or positron emitted from the nucleus lies somewhere in a relatively large range of possibilities. In alpha decay, however, the alpha particle energy can only have discrete values. Why is there this difference?

6. If film is kept in a box, alpha particles from a radioactive source outside the box cannot expose the film, but beta particles can. Explain.

7. In positron decay, a proton in the nucleus becomes a neutron, and the positive charge is carried away by the positron. But a neutron has a larger rest energy than a proton. How is this possible?

8. An alpha particle has twice the charge of a beta particle. Why does the former deflect less than the latter when passing between electrically charged plates, assuming they both have the same speed?

9. Can carbon-14 dating be used to measure the age of a stone?

10. Pick any beta-decay process and show that the neutrino must have zero charge.

11. Why do heavier elements require more neutrons in order to maintain stability?

12. Suppose it could be shown that the intensity of cosmic rays was much greater 10 000 years ago. How would this affect the ages we assign to ancient samples of once-living matter?

13. Compare and contrast a photon and a neutrino.

14. Why is carbon dating unable to provide accurate estimates of very old materials?

15. Two samples of the same radioactive nuclide are prepared. Sample A has twice the intial activity of sample B. How does the half-life of A compare with the half-life of B? After each has passed through five half-lives, what is the ratio of their activities?

16. (a) Describe what happens to the number of protons and neutrons in a nucleus when the nucleus undergoes alpha decay. (b) Repeat for beta decay.

PROBLEMS

1, 2, 3 = straightforward, intermediate, challenging   □ = full solution available in *Student Solutions Manual/Study Guide*

Physics⊗Now™ = coached problem with hints available at **www.cp7e.com**   = biomedical application

Table 29.4 will be useful for many of these problems. A more complete list of atomic masses is given in Appendix B.

Section 29.1 Some Properties of Nuclei

1. Compare the nuclear radii of the following nuclides: $^2_1$H, $^{60}_{27}$Co, $^{197}_{79}$Au, $^{239}_{94}$Pu.

2. What is the order of magnitude of the number of protons in your body? Of the number of neutrons? Of the number of electrons?

3. Using the result of Example 29.1, find the radius of a sphere of nuclear matter that would have a mass equal to that of the Earth. The Earth has a mass of 5.98×10^{24} kg and average radius of 6.37×10^6 m.

4. Consider the hydrogen atom to be a sphere of radius equal to the Bohr radius, 0.53×10^{-10} m, and calculate the approximate value of the ratio of the nuclear density to the atomic density.

5. An alpha particle ($Z = 2$, mass 6.64×10^{-27} kg) approaches to within 1.00×10^{-14} m of a carbon nucleus ($Z = 6$). What are (a) the maximum Coulomb

TABLE 29.4

Some Atomic Masses

Element	Atomic Mass (u)	Element	Atomic Mass (u)
($^0_{-1}$e)	0.000 549	$^{23}_{11}$Na	22.989 770
($^0_1$n)	1.008 665	$^{23}_{12}$Mg	22.994 127
$^1_1$H	1.007 825	$^{27}_{13}$Al	26.981 538
$^2_1$H	2.014 102	$^{30}_{15}$P	29.978 310
$^4_2$He	4.002 602	$^{40}_{20}$Ca	39.962 591
$^7_3$Li	7.016 003	$^{42}_{20}$Ca	41.958 622
$^9_4$Be	9.012 174	$^{43}_{20}$Ca	42.958 770
$^{10}_5$B	10.012 936	$^{56}_{26}$Fe	55.934 940
$^{12}_6$C	12.000 000	$^{64}_{30}$Zn	63.929 144
$^{13}_6$C	13.003 355	$^{64}_{29}$Cu	63.929 599
$^{14}_7$N	14.003 074	$^{93}_{41}$Nb	92.906 377
$^{15}_7$N	15.000 108	$^{197}_{79}$Au	196.966 543
$^{15}_8$O	15.003 065	$^{202}_{80}$Hg	201.970 617
$^{17}_8$O	16.999 131	$^{216}_{84}$Po	216.001 790
$^{18}_8$O	17.999 160	$^{220}_{86}$Rn	220.011 401
$^{18}_9$F	18.000 937	$^{234}_{90}$Th	234.043 583
$^{20}_{10}$Ne	19.992 435	$^{238}_{92}$U	238.050 784

force on the alpha particle, (b) the acceleration of the alpha particle at this time, and (c) the potential energy of the alpha particle at the same time?

6. Singly ionized carbon atoms are accelerated through 1 000 V and passed into a mass spectrometer to determine the isotopes present. (See Chapter 19.) The magnetic field strength in the spectrometer is 0.200 T. (a) Determine the orbital radii for the ^{12}C and the ^{13}C isotopes as they pass through the field. (b) Show that the ratio of the radii may be written in the form

$$\frac{r_1}{r_2} = \sqrt{\frac{m_1}{m_2}}$$

and verify that your radii in part (a) satisfy this formula.

7. Physics⊗Now™ (a) Find the speed an alpha particle requires to come within 3.2×10^{-14} m of a gold nucleus. (b) Find the energy of the alpha particle in MeV.

8. Find the radius of a nucleus of (a) $^4_2$He and (b) $^{238}_{92}$U.

Section 29.2 Binding Energy

9. Calculate the average binding energy per nucleon of $^{93}_{41}$Nb and $^{197}_{79}$Au.

10. Calculate the binding energy per nucleon for (a) ^{2}H, (b) ^{4}He, (c) ^{56}Fe, and (d) ^{238}U.

11. A pair of nuclei for which $Z_1 = N_2$ and $Z_2 = N_1$ are called *mirror isobars*. (The atomic and neutron numbers are interchangeable.) Binding-energy measurements on such pairs can be used to obtain evidence of the charge independence of nuclear forces. Charge independence means that the proton–proton, proton–neutron, and neutron–neutron forces are approximately equal. Calculate the difference in binding energy for the two mirror nuclei $^{15}_8$O and $^{15}_7$N.

12. The peak of the stability curve occurs at ^{56}Fe. This is why iron is prominent in the spectrum of the Sun and stars. Show that ^{56}Fe has a higher binding energy per nucleon than its neighbors ^{55}Mn and ^{59}Co. Compare your results with Figure 29.4.

13. Two nuclei having the same mass number are known as *isobars*. Calculate the difference in binding energy per nucleon for the isobars $^{23}_{11}$Na and $^{23}_{12}$Mg. How do you account for this difference?

14. Calculate the binding energy of the last neutron in the $^{43}_{20}$Ca nucleus. [*Hint*: You should compare the mass of $^{43}_{20}$Ca with the mass of $^{42}_{20}$Ca plus the mass of a neutron. The mass of $^{42}_{20}$Ca = 41.958 622 u.]

Section 29.3 Radioactivity

15. The half-life of an isotope of phosphorus is 14 days. If a sample contains 3.0×10^{16} such nuclei, determine its activity. Express your answer in curies.

16. A drug tagged with $^{99}_{43}$Tc (half-life = 6.05 h) is prepared for a patient. If the original activity of the sample was 1.1×10^4 Bq, what is its activity after it has sat on the shelf for 2.0 h?

17. The half-life of ^{131}I is 8.04 days. (a) Calculate the decay constant for this isotope. (b) Find the number of ^{131}I nuclei necessary to produce a sample with an activity of 0.50 μCi.

18. After 2.00 days, the activity of a sample of an unknown type of radioactive material has decreased to 84.2% of the initial activity. (a) What is the half-life of this material? (b) Can you identify it by using the table of isotopes in Appendix B?

19. Suppose that you start with 1.00×10^{-3} g of a pure radioactive substance and 2.0 h later determine that only 0.25×10^{-3} g of the substance remains. What is the half-life of this substance?

20. Radon gas has a half-life of 3.83 days. If 3.00 g of radon gas is present at time $t = 0$, what mass of radon will remain after 1.50 days have passed?

21. Many smoke detectors use small quantities of the isotope ^{241}Am in their operation. The half-life of ^{241}Am is 432 yr. How long will it take for the activity of this material to decrease to 1.00×10^{-3} of the original activity?

22. After a plant or animal dies, its ^{14}C content decreases with a half-life of 5 730 yr. If an archaeologist finds an ancient firepit containing partially consumed firewood, and the ^{14}C content of the wood is only 12.5% that of an equal carbon sample from a present-day tree, what is the age of the ancient site?

23. Physics⊗Now™ A freshly prepared sample of a certain radioactive isotope has an activity of 10.0 mCi. After 4.00 h, the activity is 8.00 mCi. (a) Find the decay constant and half-life of the isotope. (b) How many atoms of the isotope were contained in the freshly prepared sample? (c) What is the sample's activity 30 h after it is prepared?

24. A building has become accidentally contaminated with radioactivity. The longest-lived material in the building is strontium-90. (The atomic mass of $^{90}_{38}\text{Sr}$ is 89.907 7.) If the building initially contained 5.0 kg of this substance, and the safe level is less than 10.0 counts/min, how long will the building be unsafe?

Section 29.4 The Decay Processes

25. Complete the following radioactive decay formulas:

$$^{212}_{83}\text{Bi} \rightarrow ? + ^4_2\text{He}$$
$$^{95}_{36}\text{Kr} \rightarrow ? + e^-$$
$$? \rightarrow ^4_2\text{He} + ^{140}_{58}\text{Ce}$$

26. Complete the following radioactive decay formulas:

$$^{12}_5\text{B} \rightarrow ? + e^-$$
$$^{234}_{90}\text{Th} \rightarrow ^{230}_{88}\text{Ra} + ?$$
$$? \rightarrow ^{14}_7\text{N} + e^-$$

27. Complete the following radioactive decay formulas:

$$? \rightarrow e^+ + ^{40}_{19}\text{K}$$
$$^{94}_{44}\text{Ru} \rightarrow ^4_2\text{He} + ?$$
$$^{144}_{60}\text{Nd} \rightarrow ? + ^{140}_{58}\text{Ce}$$

28. Figure P29.28 shows the steps by which $^{235}_{92}\text{U}$ decays to $^{207}_{82}\text{Pb}$. Enter the correct isotope symbol in each square.

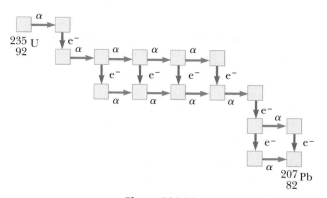

Figure P29.28

29. The mass of ^{56}Fe is 55.934 9 u and the mass of ^{56}Co is 55.939 9 u. Which isotope decays into the other and by what process?

30. Find the energy released in the alpha decay of $^{238}_{92}\text{U}$. The following mass value will be useful: $^{234}_{90}\text{Th}$ has a mass of 234.043 583 u.

31. Determine which of the following suggested decays can occur spontaneously:

(a) $^{40}_{20}\text{Ca} \rightarrow e^+ + ^{40}_{19}\text{K}$; (b) $^{144}_{60}\text{Nd} \rightarrow ^4_2\text{He} + ^{140}_{58}\text{Ce}$

32. $^{66}_{28}\text{Ni}$ (mass = 65.929 1 u) undergoes beta decay to $^{66}_{29}\text{Cu}$ (mass = 65.9289 u). (a) Write the complete decay formula for this process. (b) Find the maximum kinetic energy of the emerging electrons.

33. An ^3H nucleus beta decays into ^3He by creating an electron and an antineutrino according to the reaction

$$^3_1\text{H} \rightarrow ^3_2\text{He} + e^- + \bar{\nu}$$

Use Appendix B to determine the total energy released in this reaction.

34. A piece of charcoal used for cooking is found at the remains of an ancient campsite. A 1.00-kg sample of carbon from the wood has an activity of 2.00×10^3 decays per minute. Find the age of the charcoal. [*Hint:* Living material has an activity of 15.0 decays/minute per gram of carbon present.]

35. A wooden artifact is found in an ancient tomb. Its carbon-14 ($^{14}_6\text{C}$) activity is measured to be 60.0% of that in a fresh sample of wood from the same region. Assuming the same amount of ^{14}C was initially present in the wood from which the artifact was made, determine the age of the artifact.

36. A living specimen in equilibrium with the atmosphere contains one atom of ^{14}C (half-life = 5 730 yr) for every 7.70×10^{11} stable carbon atoms. An archeological sample of wood (cellulose, $C_{12}H_{22}O_{11}$) contains 21.0 mg of carbon. When the sample is placed inside a shielded beta counter with 88.0% counting efficiency, 837 counts are accumulated in one week. Assuming that the cosmic-ray flux and the Earth's atmosphere have not changed appreciably since the sample was formed, find the age of the sample.

Section 29.6 Nuclear Reactions

37. Physics⊗Now™ The first known reaction in which the product nucleus was radioactive (achieved in 1934) was one in which $^{27}_{13}\text{Al}$ was bombarded with alpha particles. Produced in the reaction were a neutron and a product nucleus. (a) What was the product nucleus? (b) Find the Q value of the reaction.

38. Complete the following nuclear reactions:

$$? + ^{14}_7\text{N} \rightarrow ^1_1\text{H} + ^{17}_8\text{O}$$
$$^7_3\text{Li} + ^1_1\text{H} \rightarrow ^4_2\text{He} + ?$$

39. Identify the unknown particles X and X′ in the following nuclear reactions:

$$X + {}_{2}^{4}\text{He} \rightarrow {}_{12}^{24}\text{Mg} + {}_{0}^{1}\text{n}$$
$$ {}_{92}^{235}\text{U} + {}_{0}^{1}\text{n} \rightarrow {}_{38}^{90}\text{Sr} + X + 2{}_{0}^{1}\text{n}$$
$$ 2{}_{1}^{1}\text{H} \rightarrow {}_{1}^{2}\text{H} + X + X′$$

40. The first nuclear reaction utilizing particle accelerators was performed by Cockcroft and Walton. Accelerated protons were used to bombard lithium nuclei, producing the following reaction:

$$ {}_{1}^{1}\text{H} + {}_{3}^{7}\text{Li} \rightarrow {}_{2}^{4}\text{He} + {}_{2}^{4}\text{He}$$

Since the masses of the particles involved in the reaction were well known, these results were used to obtain an early proof of the Einstein mass–energy relation. Calculate the Q value of the reaction.

41. (a) Suppose ${}_{5}^{10}\text{B}$ is struck by an alpha particle, releasing a proton and a product nucleus in the reaction. What is the product nucleus? (b) An alpha particle and a product nucleus are produced when ${}_{6}^{13}\text{C}$ is struck by a proton. What is the product nucleus?

42. (a) Determine the product of the reaction ${}_{3}^{7}\text{Li} + {}_{2}^{4}\text{He} \rightarrow ? + \text{n}$. (b) What is the Q value of the reaction?

43. Natural gold has only one isotope: ${}_{79}^{197}\text{Au}$. If gold is bombarded with slow neutrons, e^- particles are emitted. (a) Write the appropriate reaction equation. (b) Calculate the maximum energy of the emitted beta particles. The mass of ${}_{80}^{198}\text{Hg}$ is 197.966 75 u.

44. Find the threshold energy that an incident neutron must have to produce the reaction

$$ {}_{0}^{1}\text{n} + {}_{2}^{4}\text{He} \rightarrow {}_{1}^{2}\text{H} + {}_{1}^{3}\text{H}$$

45. When ${}^{18}\text{O}$ is struck by a proton, ${}^{18}\text{F}$ and another particle are produced. (a) What is the other particle? (b) The reaction has a Q value of -2.453 MeV, and the atomic mass of ${}^{18}\text{O}$ is 17.999 160 u. What is the atomic mass of ${}^{18}\text{F}$?

Section 29.7 Medical Applications of Radiation

46. In terms of biological damage, how many rad of heavy ions is equivalent to 100 rad of x-rays?

47. A person whose mass is 75.0 kg is exposed to a whole-body dose of 25.0 rads. How many joules of energy are deposited in the person's body?

48. A 200-rad dose of radiation is administered to a patient in an effort to combat a cancerous growth. Assuming all of the energy deposited is absorbed by the growth, (a) calculate the amount of energy delivered per unit mass. (b) Assuming the growth has a mass of 0.25 kg and a specific heat equal to that of water, calculate its temperature rise.

49. A "clever" technician decides to heat some water for his coffee with an x-ray machine. If the machine produces 10 rad/s, how long will it take to raise the temperature of a cup of water by 50°C. Ignore heat losses during this time.

50. An x-ray technician works 5 days per week, 50 weeks per year. Assume that the technician takes an average of eight x-rays per day and receives a dose of 5.0 rem/yr as a result. (a) Estimate the dose in rem per x-ray taken. (b) How does this result compare with the amount of low-level background radiation the technician is exposed to?

51. Physics⊗Now™ A patient swallows a radiopharmaceutical tagged with phosphorus-32 (${}_{15}^{32}\text{P}$), a β^- emitter with a half-life of 14.3 days. The average kinetic energy of the emitted electrons is 700 keV. If the initial activity of the sample is 1.31 MBq, determine (a) the number of electrons emitted in a 10-day period, (b) the total energy deposited in the body during the 10 days, and (c) the absorbed dose if the electrons are completely absorbed in 100 g of tissue.

52. A particular radioactive source produces 100 mrad of 2-MeV gamma rays per hour at a distance of 1.0 m. (a) How long could a person stand at this distance before accumulating an intolerable dose of 1 rem? (b) Assuming the gamma radiation is emitted uniformly in all directions, at what distance would a person receive a dose of 10 mrad/h from this source?

ADDITIONAL PROBLEMS

53. A 200.0-mCi sample of a radioactive isotope is purchased by a medical supply house. If the sample has a half-life of 14.0 days, how long will it keep before its activity is reduced to 20.0 mCi?

54. A sample of organic material is found to contain 18 g of carbon. The investigators believe the material to be 20 000 years old, based on samples of pottery found at the site. If so, what is the expected activity of the organic material? Take data from Example 29.7.

55. Deuterons that have been accelerated are used to bombard other deuterium nuclei, resulting in the reaction

$$ {}_{1}^{2}\text{H} + {}_{1}^{2}\text{H} \rightarrow {}_{2}^{3}\text{He} + {}_{0}^{1}\text{n}$$

Does this reaction require a threshold energy? If so, what is its value?

56. Free neutrons have a characteristic half-life of 12 min. What fraction of a group of free neutrons at a thermal energy of 0.040 eV will decay before traveling a distance of 10.0 km?

57. A by-product of some fission reactors is the isotope $^{239}_{94}$Pu, an alpha emitter having a half-life of 24 120 yr. The reaction involved is

$$^{239}_{94}\text{Pu} \rightarrow {}^{235}_{92}\text{U} + \alpha$$

Consider a sample of 1.00 kg of pure $^{239}_{94}$Pu at $t = 0$. Calculate (a) the number of $^{239}_{94}$Pu nuclei present at $t = 0$ and (b) the initial activity in the sample. (c) How long does the sample have to be stored if a "safe" activity level is 0.100 Bq?

58. (a) Find the radius of the $^{12}_{6}$C nucleus. (b) Find the force of repulsion between a proton at the surface of a $^{12}_{6}$C nucleus and the remaining five protons. (c) How much work (in MeV) has to be done to overcome this electrostatic repulsion in order to put the last proton into the nucleus? (d) Repeat (a), (b), and (c) for $^{238}_{92}$U.

59. In a piece of rock from the Moon, the ^{87}Rb content is assayed to be 1.82×10^{10} atoms per gram of material and the ^{87}Sr content is found to be 1.07×10^{9} atoms per gram. (The relevant decay is ^{87}Rb $\rightarrow$ ^{87}Sr + e$^-$. The half-life of the decay is 4.8×10^{10} yr.) (a) Determine the age of the rock. (b) Could the material in the rock actually be much older? What assumption is implicit in using the radioactive-dating method?

60. Many radioisotopes have important industrial, medical, and research applications. One such radioisotope is ^{60}Co, which has a half-life of 5.2 yr and decays by the emission of a beta particle (energy 0.31 MeV) and two gamma photons (energies 1.17 MeV and 1.33 MeV). A scientist wishes to prepare a ^{60}Co sealed source that will have an activity of at least 10 Ci after 30 months of use. What is the minimum initial mass of ^{60}Co required?

61. A medical laboratory stock solution is prepared with an initial activity due to ^{24}Na of 2.5 mCi/ml, and 10.0 ml of the stock solution is diluted at $t_0 = 0$ to a working solution whose total volume is 250 ml. After 48 h, a 5.0-ml sample of the working solution is monitored with a counter. What is the measured activity? (Note that 1 ml = 1 milliliter.)

62. A theory of nuclear astrophysics is that all the heavy elements such as uranium are formed in supernova explosions of massive stars, which immediately release the elements into space. If we assume that at the time of an explosion there were equal amounts of ^{235}U and ^{238}U, how long ago were the elements that formed our Earth released, given that the present ^{235}U/^{238}U ratio is 0.007? (The half-lives of ^{235}U and ^{238}U are 0.70×10^{9} yr and 4.47×10^{9} yr, respectively.)

63. A fission reactor is hit by a nuclear weapon, causing 5.0×10^{6} Ci of ^{90}Sr ($T_{1/2} = 28.7$ yr) to evaporate into the air. The ^{90}Sr falls out over an area of 10^{4} km^2. How long will it take the activity of the ^{90}Sr to reach the agriculturally "safe" level of 2.0 μCi/m^2?

64. After the sudden release of radioactivity from the Chernobyl nuclear reactor accident in 1986, the radioactivity of milk in Poland rose to 2 000 Bq/L due to iodine-131, with a half-life of 8.04 days. Radioactive iodine is particularly hazardous, because the thyroid gland concentrates iodine. The Chernobyl accident caused a measurable increase in thyroid cancers among children in Belarus. (a) For comparison, find the activity of milk due to potassium. Assume that 1 liter of milk contains 2.00 g of potassium, of which 0.011 7% is the isotope ^{40}K, which has a half-life of 1.28×10^{9} yr. (b) After what length of time would the activity due to iodine fall below that due to potassium?

65. During the manufacture of a steel engine component, radioactive iron (^{59}Fe) is included in the total mass of 0.20 kg. The component is placed in a test engine when the activity due to the isotope is 20.0 μCi. After a 1 000-h test period, oil is removed from the engine and is found to contain enough ^{59}Fe to produce 800 disintegrations/min per liter of oil. The total volume of oil in the engine is 6.5 L. Calculate the total mass worn from the engine component per hour of operation. (The half-life of ^{59}Fe is 45.1 days.)

66. After determining that the Sun has existed for hundreds of millions of years, but before the discovery of nuclear physics, scientists could not explain why the Sun has continued to burn for such a long time. For example, if it were a coal fire, the Sun would have burned up in about 3 000 yr. Assume that the Sun, whose mass is 1.99×10^{30} kg, originally consisted entirely of hydrogen and that its total power output is 3.76×10^{26} W. (a) If the energy-generating mechanism of the Sun is the transforming of hydrogen into helium via the net reaction

$$4{}^{1}_{1}\text{H} + 2e^- \rightarrow {}^{4}_{2}\text{He} + 2\nu + \gamma$$

calculate the energy (in joules) given off by this reaction. (b) Determine how many hydrogen atoms constitute the Sun. Take the mass of one hydrogen atom to be 1.67×10^{-27} kg. (c) Assuming that the total power output remains constant, after what time will all the hydrogen be converted into helium, making the Sun die? The actual projected lifetime of the Sun is about 10 billion years, because only the hydrogen in a relatively small core is available as a fuel. (Only in the Sun's core are temperatures and densities high enough for the fusion reaction to be self-sustaining).

ACTIVITIES

1. This experiment will take a little longer to do than most that we have suggested, but the time spent is worthwhile to help you understand the concept of half-life. Obtain a box of sugar cubes and with a pencil make a mark on one side of each of about 200 cubes. Each of these cubes will represent the nucleus of a radioactive substance. Thus, at $t = 0$, you have 200 undecayed nuclei. Now, put the 200 marked cubes in a box and roll them out on a table, just as you would roll dice. Next, count and remove any cubes that have landed marked-side up. These cubes represent nuclei that emitted radiation during the roll. They are no longer radioactive and thus do not participate in the rest of the action. Record the number of undecayed cubes remaining as the number of undecayed nuclei at $t = 1$ roll.

    Continue rolling, counting, and removing until you have completed 12 to 15 rolls. By then, you should have only a few cubes remaining. Plot a graph of undecayed cubes versus the roll number and from this determine the "half-roll" of the cubes.

2. Use a nail to punch a hole in the bottom of a large tin can. Hold the can beneath a faucet and adjust the water flow from the faucet to a fine constant stream. Although water flows from the hole at the bottom, you will note that the level of the water in the can rises. As it does so, however, the flow of water leaving the can increases due to increased water pressure caused by the greater depth of water. Unless the flow of water is too great, an equilibrium point will be reached at which the amount of water flowing out of the can each second exactly equals the amount flowing in each second. When this happens, the level of water in the can is constant. As noted in the text, carbon-14 is continually being produced in the atmosphere and is also continually disappearing as it decays into nitrogen. What is the analogy between water entering the can, remaining in the can, and flowing out of the can and the behavior of carbon-14 in the atmosphere?

This photo shows scientist Melissa Douglas and part of the Z machine, an inertial-electrostatic confinement fusion apparatus at Sandia National Laboratories. In the device, giant capacitors discharge through a grid of tungsten wires finer than human hairs, vaporizing them. The tungsten ions implode inward at a million miles an hour. Braking strongly in the grip of a "Z-pinch," they emit powerful x-rays that compress a deuterium pellet, causing collisions between the deuterium atoms that lead to fusion events.

Sandia National Laboratories. Photo by Randy Montoya

Nuclear Energy and Elementary Particles

In this concluding chapter we discuss the two means by which energy can be derived from nuclear reactions: fission, in which a nucleus of large mass number splits into two smaller nuclei, and fusion, in which two light nuclei fuse to form a heavier nucleus. In either case, there is a release of large amounts of energy, which can be used destructively through bombs or constructively through the production of electric power. We end our study of physics by examining the known subatomic particles and the fundamental interactions that govern their behavior. We also discuss the current theory of elementary particles, which states that all matter in nature is constructed from only two families of particles: quarks and leptons. Finally, we describe how such models help us understand the evolution of the Universe.

30.1 NUCLEAR FISSION

Nuclear fission occurs when a heavy nucleus, such as ^{235}U, splits, or fissions, into two smaller nuclei. In such a reaction, **the total mass of the products is less than the original mass of the heavy nucleus**.

Nuclear fission was first observed in 1939 by Otto Hahn and Fritz Strassman, following some basic studies by Fermi. After bombarding uranium ($Z = 92$) with neutrons, Hahn and Strassman discovered two medium-mass elements, barium and lanthanum, among the reaction products. Shortly thereafter, Lise Meitner and Otto Frisch explained what had happened: the uranium nucleus had split into two nearly equal fragments after absorbing a neutron. This was of considerable interest to physicists attempting to understand the nucleus, but it was to have even more far-reaching consequences. Measurements showed that about 200 MeV of energy is released in each fission event, and this fact was to affect the course of human history.

The fission of ^{235}U by slow (low-energy) neutrons can be represented by the sequence of events

$$\frac{1}{0}n + \frac{235}{92}U \rightarrow \frac{236}{92}U^* \rightarrow X + Y + neutrons \qquad [30.1]$$

where ^{236}U* is an intermediate state that lasts only for about 10^{-12} s before splitting into nuclei X and Y, called **fission fragments**. There are many combinations of X and Y that satisfy the requirements of conservation of energy and charge. In the fission of uranium, about 90 different daughter nuclei can be formed. The process also results in the production of several (typically two or three) neutrons per fission event. On the average, 2.47 neutrons are released per event.

A typical reaction of this type is

$$\frac{1}{0}n + \frac{235}{92}U \rightarrow \frac{141}{56}Ba + \frac{92}{36}Kr + 3\frac{1}{0}n \qquad [30.2]$$

The fission fragments, barium and krypton, and the released neutrons have a great deal of kinetic energy following the fission event.

The breakup of the uranium nucleus can be compared to what happens to a drop of water when excess energy is added to it. All of the atoms in the drop have energy, but not enough to break up the drop. However, if enough energy is added to set the drop vibrating, it will undergo elongation and compression until the amplitude of vibration becomes large enough to cause the drop to break apart. In the uranium nucleus, a similar process occurs (Fig. 30.1). The sequence of events is as follows:

Sequence of events in a nuclear fission process ▶

1. The ^{235}U nucleus captures a thermal (slow-moving) neutron.
2. The capture results in the formation of ^{236}U*, and the excess energy of this nucleus causes it to undergo violent oscillations.
3. The ^{236}U* nucleus becomes highly elongated, and the force of repulsion between protons in the two halves of the dumbbell-shaped nucleus tends to increase the distortion.
4. The nucleus splits into two fragments, emitting several neutrons in the process.

The energy released in a typical fission process Q can be estimated. From Figure 29.4, we see that the binding energy per nucleon is about 7.2 MeV for heavy nuclei (those having a mass number of approximately 240) and about 8.2 MeV for nuclei of intermediate mass. This means that the nucleons in the fission fragments are more tightly bound, and therefore have less mass, than the nucleons in the original heavy nucleus. The decrease in mass per nucleon appears as released energy when fission occurs. The amount of energy released is (8.2 − 7.2) MeV per nucleon. Assuming a total of 240 nucleons, we find that the energy released per fission event is

$$Q = (240 \ nucleons)(8.2 \ MeV/nucleon - 7.2 \ MeV/nucleon) = 240 \ MeV$$

This is a large amount of energy relative to the amount released in chemical processes. For example, the energy released in the combustion of one molecule of the octane used in gasoline engines is about one hundred-millionth the energy released in a single fission event!

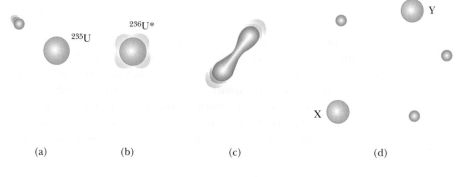

Figure 30.1 A nuclear fission event as described by the liquid-drop model of the nucleus. (a) A slow neutron approaches a ^{235}U nucleus. (b) The neutron is absorbed by the ^{235}U nucleus, changing it to ^{236}U*, which is a ^{236}U nucleus in an excited state. (c) The nucleus deforms and oscillates like a liquid drop. (d) The nucleus undergoes fission, resulting in two lighter nuclei X and Y, along with several neutrons.

(a)   (b)   (c)   (d)

Applying Physics 30.1   Unstable Products

If a heavy nucleus were to fission into just two product nuclei, they would be very unstable. Why is this?

Explanation   According to Figure 29.3, the ratio of the number of neutrons to the number of protons increases with Z. As a result, when a heavy nucleus splits in a fission reaction to two lighter nuclei, the lighter nuclei tend to have too many neutrons. This leads to instability, as the nucleus returns to the curve in Figure 29.3 by decay processes that reduce the number of neutrons.

EXAMPLE 30.1   The Fission of Uranium

Goal   Balance a nuclear equation to determine details of the fission fragments.

Problem   When ^{235}U is struck by a neutron, there are various possible fission fragments. Determine the number of neutrons produced when the fission fragments are ^{140}Xe and ^{94}Sr (isotopes of xenon and strontium).

Strategy   This is something like balancing chemical equations: the atomic numbers and mass numbers should balance on either side of the equation.

Solution

Write the equation describing the process, with an unknown number x of neutrons:

$$^{1}_{0}n + {}^{235}_{92}U \rightarrow {}^{140}_{54}Xe + {}^{94}_{38}Sr + x({}^{1}_{0}n)$$

The atomic numbers balance already, as they should. Write an equation relating the mass numbers:

$$1 + 235 = 140 + 94 + x \rightarrow \boxed{x = 2}$$

Remark   In this case, the number of protons balanced automatically. If that were not the case, there might be other possible daughter particles, such as protons or helium nuclei (also called alpha particles).

Exercise 30.1

Find the number of neutrons released if the two major fragments are ^{132}Sn and ^{101}Mo.

Answer   Three neutrons

Quick Quiz 30.1

In the first atomic bomb, the energy released was equivalent to about 30 kilotons of TNT, where a ton of TNT releases an energy of about 4.0×10^9 J. Estimate the amount of mass converted into energy in this event. (a) $1 \, \mu g$ (b) $1 \, mg$ (c) $1 \, g$ (d) $1 \, kg$ (e) 20 kilotons

---

EXAMPLE 30.2   A Fission-Powered World

Goal   Relate raw material to energy output.

Problem   **(a)** Calculate the total energy released if 1.00 kg of ^{235}U undergoes fission, taking the disintegration energy per event to be $Q = 208$ MeV (a more accurate value than the estimate given previously). **(b)** How many kilograms of ^{235}U would be needed to satisfy the world's annual energy consumption (about 4×10^{20} J)?

Strategy   In part (a), use the concept of a mole and Avogadro's number to obtain the total number of nuclei. Multiplying by the energy per reaction then gives the total energy released. Part (b) requires some light algebra.

Solution

(a) Calculate the total energy released from 1.00 kg of ^{235}U.

Find the total number of nuclei in 1.00 kg of uranium:

$$N = \left(\frac{6.02 \times 10^{23} \text{ nuclei/mol}}{235 \text{ g/mol}} \right) (1.00 \times 10^3 \text{ g})$$

$$= 2.56 \times 10^{24} \text{ nuclei}$$

Multiply N by the energy yield per nucleus, obtaining the total disintegration energy:

$$E = NQ = (2.56 \times 10^{24} \text{ nuclei})\left(208 \frac{\text{MeV}}{\text{nucleus}}\right)$$

$$= 5.32 \times 10^{26} \text{ MeV}$$

(b) How many kilograms would provide for the world's annual energy needs?

Set the energy per kilogram, E_{kg}, times the number of kilograms, N_{kg}, equal to the total annual energy consumption. Solve for N_{kg}:

$$E_{kg} N_{kg} = E_{tot}$$

$$N_{kg} = \frac{E_{tot}}{E_{kg}} = \frac{4 \times 10^{20} \text{ J}}{(5.32 \times 10^{32} \text{ eV/kg})(1.60 \times 10^{-19} \text{ J/eV})}$$

$$= 5 \times 10^{6} \text{ kg}$$

Remark   The calculation implicitly assumes perfect conversion to usable power, which is never the case in real systems. There are enough known uranium deposits to provide the world's current energy requirements for a few hundred years.

Exercise 30.2
How long can one kilogram of U-235 keep a 100-watt lightbulb burning if all its released energy is converted to electrical energy?

Answer $\sim 30\ 000$ yr

30.2   NUCLEAR REACTORS

We have seen that neutrons are emitted when ^{235}U undergoes fission. These neutrons can in turn trigger other nuclei to undergo fission, with the possibility of a chain reaction (Active Fig. 30.2). Calculations show that if the chain reaction isn't controlled, it will proceed too rapidly and possibly result in the sudden release of an enormous amount of energy (an explosion), even from only 1 g of ^{235}U. If the energy in 1 kg of ^{235}U were released, it would equal that released by the detonation

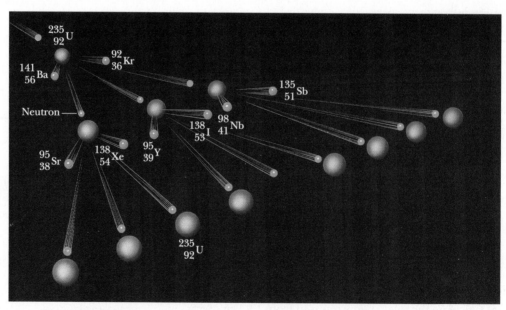

ACTIVE FIGURE 30.2
A nuclear chain reaction initiated by the capture of a neutron.

Physics⊗Now™
Log into PhysicsNow at **www.cp7e.com** and go to Active Figure 30.2 to observe a chain reaction.

Painting of the world's first nuclear reactor. Because of wartime secrecy, there are no photographs of the completed reactor, which was composed of layers of graphite interspersed with uranium. A self-sustained chain reaction was first achieved on December 2, 1942. Word of the success was telephoned immediately to Washington with this message: "The Italian navigator has landed in the New World and found the natives very friendly." The historic event took place in an improvised laboratory in the racquet court under the west stands of the University of Chicago's Stagg Field. The Italian navigator was Fermi.

of about 20 000 tons of TNT! An uncontrolled fission reaction, of course, is the principle behind the first nuclear bomb.

A nuclear reactor is a system designed to maintain what is called a **self-sustained chain reaction**. This important process was first achieved in 1942 by a group led by Fermi at the University of Chicago, with natural uranium as the fuel. Most reactors in operation today also use uranium as fuel. Natural uranium contains only about 0.7% of the ^{235}U isotope, with the remaining 99.3% being the ^{238}U isotope. This is important to the operation of a reactor because ^{238}U almost never undergoes fission. Instead, it tends to absorb neutrons, producing neptunium and plutonium. For this reason, reactor fuels must be artificially enriched so that they contain several percent of the ^{235}U isotope.

Earlier we mentioned that an average of about 2.5 neutrons are emitted in each fission event of ^{235}U. In order to achieve a self-sustained chain reaction, one of these neutrons must be captured by another ^{235}U nucleus and cause it to undergo fission. A useful parameter for describing the level of reactor operation is the **reproduction constant K, defined as the average number of neutrons from each fission event that will cause another event**. As we have seen, K can have a maximum value of 2.5 in the fission of uranium. In practice, however, K is less than this because of several factors, which we soon discuss.

A self-sustained chain reaction is achieved when $K = 1$. Under this condition, the reactor is said to be **critical**. When K is less than one, the reactor is subcritical and the reaction dies out. When K is greater than one the reactor is said to be supercritical, and a runaway reaction occurs. In a nuclear reactor used to furnish power to a utility company, it is necessary to maintain a K value close to one.

The basic design of a nuclear reactor is shown in Figure 30.3. The fuel elements consist of enriched uranium. The functions of the remaining parts of the reactor and some aspects of its design are described next.

Figure 30.3 Cross section of a reactor core showing the control rods, fuel elements containing enriched fuel, and moderating material, all surrounded by a radiation shield.

Neutron Leakage

In any reactor, a fraction of the neutrons produced in fission will leak out of the core before inducing other fission events. If the fraction leaking out is too large, the reactor will not operate. The percentage lost is large if the reactor is very small because leakage is a function of the ratio of surface area to volume. Therefore, a critical requirement of reactor design is choosing the correct surface-area-to-volume ratio so that a sustained reaction can be achieved.

Regulating Neutron Energies

The neutrons released in fission events are highly energetic, with kinetic energies of about 2 MeV. It is found that slow neutrons are far more likely than fast neutrons to produce fission events in ^{235}U. Further, ^{238}U doesn't absorb slow neutrons. In order for the chain reaction to continue, therefore, the neutrons must be slowed down. This is accomplished by surrounding the fuel with a substance called a **moderator**.

In order to understand how neutrons are slowed down, consider a collision between a light object and a massive one. In such an event, the light object rebounds from the collision with most of its original kinetic energy. However, if the collision is between objects having masses that are nearly the same, the incoming projectile transfers a large percentage of its kinetic energy to the target. In the first nuclear reactor ever constructed, Fermi placed bricks of graphite (carbon) between the fuel elements. Carbon nuclei are about 12 times more massive than neutrons, but after about 100 collisions with carbon nuclei, a neutron is slowed sufficiently to increase its likelihood of fission with ^{235}U. In this design the carbon is the moderator; most modern reactors use heavy water (D_2O) as the moderator.

Neutron Capture

In the process of being slowed down, neutrons may be captured by nuclei that do not undergo fission. The most common event of this type is neutron capture by ^{238}U. The probability of neutron capture by ^{238}U is very high when the neutrons have high kinetic energies and very low when they have low kinetic energies. The slowing down of the neutrons by the moderator serves the dual purpose of making them available for reaction with ^{235}U and decreasing their chances of being captured by ^{238}U.

Control of Power Level

It is possible for a reactor to reach the critical stage ($K = 1$) after all neutron losses described previously are minimized. However, a method of control is needed to adjust K to a value near one. If K were to rise above this value, the heat produced in the runaway reaction would melt the reactor. To control the power level, control rods are inserted into the reactor core. (See Fig. 30.3.) These rods are made of materials such as cadmium that are highly efficient in absorbing neutrons. By adjusting the number and position of the control rods in the reactor core, the K value can be varied and any power level within the design range of the reactor can be achieved.

A diagram of a pressurized-water reactor is shown in Figure 30.4. This type of reactor is commonly used in electric power plants in the United States. Fission

Figure 30.4 Main components of a pressurized-water nuclear reactor.

events in the reactor core supply heat to the water contained in the primary (closed) system, which is maintained at high pressure to keep it from boiling. This water also serves as the moderator. The hot water is pumped through a heat exchanger, and the heat is transferred to the water contained in the secondary system. There the hot water is converted to steam, which drives a turbine–generator to create electric power. Note that the water in the secondary system is isolated from the water in the primary system in order to prevent contamination of the secondary water and steam by radioactive nuclei from the reactor core.

Reactor Safety[1]

The safety aspects of nuclear power reactors are often sensationalized by the media and misunderstood by the public. The 1979 near disaster of Three Mile Island in Pennsylvania and the accident at the Chernobyl reactor in the Ukraine rightfully focused attention on reactor safety. Yet the safety record in the United States is enviable. The records show no fatalities attributed to commercial nuclear power generation in the history of the United States nuclear industry.

Commercial reactors achieve safety through careful design and rigid operating procedures. Radiation exposure and the potential health risks associated with such exposure are controlled by three layers of containment. The fuel and radioactive fission products are contained inside the reactor vessel. Should this vessel rupture, the reactor building acts as a second containment structure to prevent radioactive material from contaminating the environment. Finally, the reactor facilities must be in a remote location to protect the general public from exposure should radiation escape the reactor building.

According to the *Oak Ridge National Laboratory Review*, "The health risk of living within 8 km (5 miles) of a nuclear reactor for 50 years is no greater than the risk of smoking 1.4 cigarettes, drinking 0.5 liters of wine, traveling 240 km by car, flying 9 600 km by jet, or having one chest x-ray in a hospital. Each of these activities is estimated to increase a person's chances of dying in any given year by one in a million."

Another potential danger in nuclear reactor operations is the possibility that the water flow could be interrupted. Even if the nuclear fission chain reaction were stopped immediately, residual heat could build up in the reactor to the point of melting the fuel elements. The molten reactor core would melt its way to the bottom of the reactor vessel and conceivably into the ground below—the so-called China syndrome. Although it might appear that this deep underground burial site would be an ideal safe haven for a radioactive blob, there would be danger of a steam explosion should the molten mass encounter water. This nonnuclear explosion could spread radioactive material to the areas surrounding the power plant. To prevent such an unlikely chain of events, nuclear reactors are designed with emergency core cooling systems, requiring no power, that automatically flood the reactor with water in the event of a loss of coolant. The emergency cooling water moderates heat build-up in the core, which in turn prevents the melting of the reactor vessel.

A continuing concern in nuclear fission reactors is the safe disposal of radioactive material when the reactor core is replaced. This waste material contains long-lived, highly radioactive isotopes and must be stored for long periods of time in such a way that there is no chance of environmental contamination. At present, sealing radioactive wastes in waterproof containers and burying them in deep salt mines seems to be the most promising solution.

Transportation of reactor fuel and reactor wastes poses additional safety risks. However, neither the waste nor the fuel of nuclear power reactors can be used to construct a nuclear bomb.

Accidents during transportation of nuclear fuel could expose the public to harmful levels of radiation. The Department of Energy requires stringent crash

[1]The authors are grateful to Professor Gene Skluzacek of the University of Nebraska at Omaha for rewriting this section.

tests on all containers used to transport nuclear materials. Container manufacturers must demonstrate that their containers will not rupture, even in high-speed collisions.

The safety issues associated with nuclear power reactors are complex and often emotional. All sources of energy have associated risks. Coal, for example, exposes workers to health hazards (including radioactive radon) and produces atmospheric pollution (including greenhouse gases). In each case, the risks must be weighed against the benefits and the availability of the energy source.

30.3 NUCLEAR FUSION

Figure 29.4 shows that the binding energy of light nuclei (those having a mass number lower than 20) is much smaller than the binding energy of heavier nuclei. This suggests a process that is the reverse of fission. **When two light nuclei combine to form a heavier nucleus, the process is called nuclear fusion**. Because the mass of the final nucleus is less than the masses of the original nuclei, there is a loss of mass, accompanied by a release of energy. Although fusion power plants have not yet been developed, a worldwide effort is under way to harness the energy from fusion reactions in the laboratory.

This photograph of the Sun, taken on December 19, 1973, during the third and final manned Skylab mission, shows one of the most spectacular solar flares ever recorded, spanning more than 588 000 km (365 000 mi) across the solar surface. Several active regions can be seen on the eastern side of the disk. The photograph was taken in the light of ionized helium by the extreme ultraviolet spectroheliograph instrument of the U.S. Naval Research Laboratory.

Fusion in the Sun

All stars generate their energy through fusion processes. About 90% of stars, including the Sun, fuse hydrogen, whereas some older stars fuse helium or other heavier elements. Stars are born in regions of space containing vast clouds of dust and gas. Recent mathematical models of these clouds indicate that star formation is triggered by shock waves passing through a cloud. These shock waves are similar to sonic booms and are produced by events such as the explosion of a nearby star, called a *supernova*. The shock wave compresses certain regions of the cloud, causing them to collapse under their own gravity. As the gas falls inward toward the center, the atoms gain speed, which causes the temperature of the gas to rise. Two conditions must be met before fusion reactions in the star can sustain its energy needs: (1) The temperature must be high enough (about 10^7 K for hydrogen) to allow the kinetic energy of the positively charged hydrogen nuclei to overcome their mutual Coulomb repulsion as they collide, and (2) the density of nuclei must be high enough to ensure a high rate of collision.

It's interesting that temperatures inside stars like the Sun are not sufficient to allow colliding protons to overcome Coulomb repulsion. In a certain percentage of collisions, the nuclei pass through the barrier anyway, an example of *quantum tunneling*. So a quantum effect is key in making sunshine.

When fusion reactions occur at the core of a star, the energy that is liberated eventually becomes sufficient to prevent further collapse of the star under its own gravity. The star then continues to live out the remainder of its life under a balance between the inward force of gravity pulling it toward collapse and the outward force due to thermal effects and radiation pressure.

The **proton–proton cycle** is a series of three nuclear reactions that are believed to be the stages in the liberation of energy in the Sun and other stars rich in hydrogen. An overall view of the proton–proton cycle is that four protons combine to form an alpha particle and two positrons, with the release of 25 MeV of energy in the process.

The specific steps in the proton–proton cycle are

$$^1_1\text{H} + ^1_1\text{H} \rightarrow ^2_1\text{D} + e^+ + \nu$$

and

$$^1_1\text{H} + ^2_1\text{D} \rightarrow ^3_2\text{He} + \gamma \qquad \textbf{[30.3]}$$

where D stands for deuterium, the isotope of hydrogen having one proton and one neutron in the nucleus. (It can also be written as ^2_1H.) The second reaction is followed by either hydrogen–helium fusion or helium–helium fusion:

$$_1^1H + _2^3He \rightarrow _2^4He + e^+ + \nu$$

or

$$_2^3He + _2^3He \rightarrow _2^4He + 2(_1^1H)$$

The energy liberated is carried primarily by gamma rays, positrons, and neutrinos, as can be seen from the reactions. The gamma rays are soon absorbed by the dense gas, thus raising its temperature. The positrons combine with electrons to produce gamma rays, which in turn are also absorbed by the gas within a few centimeters. The neutrinos, however, almost never interact with matter; hence, they escape from the star, carrying about 2% of the energy generated with them. These energy-liberating fusion reactions are called **thermonuclear fusion reactions**. The hydrogen (fusion) bomb, first exploded in 1952, is an example of an uncontrolled thermonuclear fusion reaction.

Fusion Reactors

APPLICATION
Fusion Reactors

The enormous amount of energy released in fusion reactions suggests the possibility of harnessing this energy for useful purposes on Earth. A great deal of effort is under way to develop a sustained and controllable thermonuclear reactor—a fusion power reactor. Controlled fusion is often called the ultimate energy source because of the availability of its fuel source: water. For example, if deuterium, the isotope of hydrogen consisting of a proton and a neutron, were used as the fuel, 0.06 g of it could be extracted from 1 gal of water at a cost of about four cents. Such rates would make the fuel costs of even an inefficient reactor almost insignificant. An additional advantage of fusion reactors is that comparatively few radioactive by-products are formed. As noted in Equation 30.3, the end product of the fusion of hydrogen nuclei is safe, nonradioactive helium. Unfortunately, a thermonuclear reactor that can deliver a net power output over a reasonable time interval is not yet a reality, and many problems must be solved before a successful device is constructed.

We have seen that the Sun's energy is based, in part, on a set of reactions in which ordinary hydrogen is converted to helium. Unfortunately, the proton–proton interaction is not suitable for use in a fusion reactor because the event requires very high pressures and densities. The process works in the Sun only because of the extremely high density of protons in the Sun's interior. In fact, even at the densities and temperatures that exist at the center of the Sun, the average proton takes 14 billion years to react!

The fusion reactions that appear most promising in the construction of a fusion power reactor involve deuterium (D) and tritium (T), which are isotopes of hydrogen. These reactions are

$$_1^2D + _1^2D \rightarrow _2^3He + _0^1n \qquad Q = 3.27 \text{ MeV}$$
$$_1^2D + _1^2D \rightarrow _1^3T + _1^1H \qquad Q = 4.03 \text{ MeV} \qquad \textbf{[30.4]}$$

and

$$_1^2D + _1^3T \rightarrow _2^4He + _0^1n \qquad Q = 17.59 \text{ MeV}$$

where the Q values refer to the amount of energy released per reaction. As noted earlier, deuterium is available in almost unlimited quantities from our lakes and oceans and is very inexpensive to extract. Tritium, however, is radioactive ($T_{1/2} = 12.3$ yr) and undergoes beta decay to ^3He. For this reason, tritium doesn't occur naturally to any great extent and must be artificially produced.

One of the major problems in obtaining energy from nuclear fusion is the fact that the Coulomb repulsion force between two charged nuclei must be overcome before they can fuse. The fundamental challenge is to give the two nuclei enough kinetic energy to overcome this repulsive force. This can be accomplished by heating the fuel to extremely high temperatures (about 10^8 K, far greater than the interior temperature of the Sun). As you might expect, such high temperatures are not easy to obtain in a laboratory or a power plant. At these high temperatures, the atoms are ionized and the system consists of a collection of electrons and nuclei, commonly referred to as a *plasma*.

In addition to the high temperature requirements, there are two other critical factors that determine whether or not a thermonuclear reactor will function: the **plasma ion density** n and the **plasma confinement time** τ—the time the interacting ions are maintained at a temperature equal to or greater than that required for the reaction to proceed. The density and confinement time must both be large enough to ensure that more fusion energy will be released than is required to heat the plasma.

Lawson's criterion states that a net power output in a fusion reactor is possible under the following conditions:

Lawson's criterion ▶

$$n\tau \geq 10^{14} \text{ s/cm}^3 \text{ Deuterium–tritium interaction} \qquad \text{[30.5]}$$
$$n\tau \geq 10^{16} \text{ s/cm}^3 \text{ Deuterium–deuterium interaction}$$

The problem of plasma confinement time has yet to be solved. How can a plasma be confined at a temperature of 10^8 K for times on the order of 1 s? The basic plasma-confinement technique under investigation is discussed following Example 30.3.

EXAMPLE 30.3   Astrofuel on the Moon

Goal   Calculate the energy released in a fusion reaction.

Problem   Find the energy released in the reaction of helium-3 with deuterium:

$$^3_2\text{He} + {}^2_1\text{D} \rightarrow {}^4_2\text{He} + {}^1_1\text{H}$$

Strategy   The energy released is the difference between the mass energy of the reactants and the products.

Solution

Add the masses on the left-hand side, and subtract the masses on the right, obtaining Δm in atomic mass units:

$$\begin{aligned}\Delta m &= m_{\text{He-3}} + m_D - m_{\text{He-4}} - m_H \\ &= 3.016\ 029 \text{ u} + 2.014\ 102 \text{ u} - 4.002\ 602 \text{ u} - 1.007\ 825 \text{ u} \\ &= 0.019\ 704 \text{ u}\end{aligned}$$

Convert the mass difference to an equivalent amount of energy in MeV:

$$E = (0.019\ 704 \text{ u})\left(\frac{931.5 \text{ MeV}}{1 \text{ u}}\right) = \boxed{18.35 \text{ MeV}}$$

Remarks   This is a large amount of energy per reaction. Helium-3 is rare on Earth, but plentiful on the Moon, where it has become trapped in the fine dust of the lunar regolith. Helium-3 has the advantage of producing more protons than neutrons (some neutrons are still produced by side reactions, such as D–D), but has the disadvantage of a higher ignition temperature. If fusion power plants using helium-3 became a reality, studies indicate that it would be economically advantageous to mine helium-3 robotically and return it to Earth. The energy return per dollar would be far greater than for mining coal or drilling for oil!

Exercise 30.3

Find the energy yield in the fusion of two helium-3 nuclei:

$$^3_2\text{He} + {}^3_2\text{He} \rightarrow {}^4_2\text{He} + 2({}^1_1\text{H})$$

Answer   12.9 MeV

Magnetic Field Confinement

Most fusion experiments use magnetic field confinement to contain a plasma. One device, called a **tokamak**, has a doughnut-shaped geometry (a toroid), as shown in Figure 30.5a. This device, first developed in the former Soviet Union, uses a

(a)

(b)

(c)

Figure 30.5 (a) Diagram of a tokamak used in the magnetic confinement scheme. The plasma is trapped within the spiraling magnetic field lines as shown. (b) Interior view of the Tokamak Fusion Test Reactor (TFTR) vacuum vessel located at the Princeton Plasma Physics Laboratory, Princeton University, Princeton, New Jersey. (c) The National Spherical Torus Experiment (NSTX) that began operation in March 1999.

combination of two magnetic fields to confine the plasma inside the doughnut. A strong magnetic field is produced by the current in the windings, and a weaker magnetic field is produced by the current in the toroid. The resulting magnetic field lines are helical, as shown in the figure. In this configuration, the field lines spiral around the plasma and prevent it from touching the walls of the vacuum chamber.

In order for the plasma to reach ignition temperature, some form of auxiliary heating is necessary. A successful and efficient auxiliary heating technique that has been used recently is the injection of a beam of energetic neutral particles into the plasma.

When it was in operation, the Tokamak Fusion Test Reactor (TFTR) at Princeton reported central ion temperatures of 510 million degrees Celsius, more than 30 times hotter than the center of the Sun. TFTR $n\tau$ values for the D–T reaction were well above 10^{13} s/cm^3 and close to the value required by Lawson's criterion. In 1991, reaction rates of 6×10^{17} D–T fusions per second were reached in the JET tokamak at Abington, England.

One of the new generations of fusion experiments is the National Spherical Torus Experiment (NSTX) shown in Figure 30.5c. Rather than generating the donut-shaped plasma of a tokamak, the NSTX produces a spherical plasma that has a hole through its center. The major advantage of the spherical configuration is its ability to confine the plasma at a higher pressure in a given magnetic field. This approach could lead to the development of smaller and more economical fusion reactors.

There are a number of other methods of creating fusion events. In inertial laser confinement fusion, the fuel is put into the form of a small pellet and then collapsed by ultrahigh-power lasers. Fusion can also take place in a device the size of a TV set, and in fact was invented by Philo Farnsworth, one of the fathers of electronic television. In this method, called inertial electrostatic confinement, positively charged particles are rapidly attracted towards a negatively charged grid. Some of the positive particles then collide and fuse.

An international collaborative effort involving four major fusion programs is currently under way to build a fusion reactor called the International Thermonuclear Experimental Reactor (ITER). This facility will address the remaining technological and scientific issues concerning the feasibility of fusion power. The design is completed, and site and construction negotiations are under way. If the planned device works as expected, the Lawson number for ITER will be about six times greater than the current record holder, the JT-60U tokamak in Japan.

30.4   ELEMENTARY PARTICLES

The word "atom" is from the Greek word *atomos*, meaning "indivisible." At one time, atoms were thought to be the indivisible constituents of matter; that is, they were regarded as elementary particles. Discoveries in the early part of the 20th century revealed that the atom is not elementary, but has protons, neutrons, and electrons as its constituents. Until 1932, physicists viewed these three constituent particles as elementary because, with the exception of the free neutron, they are highly stable. The theory soon fell apart, however, and beginning in 1937, many new particles were discovered in experiments involving high-energy collisions between known particles. These new particles are characteristically unstable and have very short half-lives, ranging between 10^{-23} s and 10^{-6} s. So far more than 300 of them have been cataloged.

Until the 1960s, physicists were bewildered by the large number and variety of subatomic particles being discovered. They wondered whether the particles were like animals in a zoo or whether a pattern could emerge that would provide a better understanding of the elaborate structure in the subnuclear world. In the last 30 years, physicists have made tremendous advances in our knowledge of the structure of matter by recognizing that all particles (with the exception of electrons, photons, and a few others) are made of smaller particles called *quarks*. Protons and neutrons, for example, are not truly elementary but are systems of tightly bound quarks. The quark model has reduced the bewildering array of particles to a manageable number and has predicted new quark combinations that were subsequently found in many experiments.

30.5   THE FUNDAMENTAL FORCES OF NATURE

The key to understanding the properties of elementary particles is to be able to describe the forces between them. All particles in nature are subject to four fundamental forces: strong, electromagnetic, weak, and gravitational.

The **strong force** is responsible for the tight binding of quarks to form neutrons and protons and for the nuclear force, a sort of residual strong force, binding neutrons and protons into nuclei. This force represents the "glue" that holds the nucleons together and is the strongest of all the fundamental forces. It is a very short-range force and is negligible for separations greater than about 10^{-15} m (the approximate size of the nucleus). The **electromagnetic force**, which is about 10^{-2} times the strength of the strong force, is responsible for the binding of atoms and molecules. It is a long-range force that decreases in strength as the inverse square of the separation between interacting particles. The **weak force** is a short-range nuclear force that tends to produce instability in certain nuclei. It is responsible for beta decay, and its strength is only about 10^{-6} times that of the strong force. (As we discuss later, scientists now believe that the weak and electromagnetic forces are two manifestations of a single force called the *electroweak* force). Finally, the **gravitational force** is a long-range force with a strength only about 10^{-43} times that of the strong force. Although this familiar interaction is the force that holds the planets, stars, and galaxies together, its effect on elementary

TABLE 30.1

Particle Interactions

Interaction (Force)	Relative Strength[a]	Range of Force	Mediating Field Particle
Strong	1	Short (≈ 1 fm)	Gluon
Electromagnetic	10^{-2}	Long ($\propto 1/r^2$)	Photon
Weak	10^{-6}	Short ($\approx 10^{-3}$ fm)	$W^{\pm}$ and Z bosons
Gravitational	10^{-43}	Long ($\propto 1/r^2$)	Graviton

[a] For two quarks separated by 3×10^{-17} m.

particles is negligible. The gravitational force is by far the weakest of all the fundamental forces.

Modern physics often describes the forces between particles in terms of the actions of field particles or quanta. In the case of the familiar electromagnetic interaction, the field particles are photons. In the language of modern physics, the electromagnetic force is *mediated* (carried) by photons, which are the quanta of the electromagnetic field. Likewise, the strong force is mediated by field particles called *gluons*, the weak force is mediated by particles called the W and Z *bosons*, and the gravitational force is thought to be mediated by quanta of the gravitational field called *gravitons*. All of these field quanta have been detected except for the graviton, which may never be found directly because of the weakness of the gravitational field. These interactions, their ranges, and their relative strengths are summarized in Table 30.1.

30.6   POSITRONS AND OTHER ANTIPARTICLES

In the 1920s, the theoretical physicist Paul Adrien Maurice Dirac (1902–1984) developed a version of quantum mechanics that incorporated special relativity. Dirac's theory successfully explained the origin of the electron's spin and its magnetic moment. But it had one major problem: its relativistic wave equation required solutions corresponding to negative energy states even for free electrons, and if negative energy states existed, we would expect a normal free electron in a state of positive energy to make a rapid transition to one of these lower states, emitting a photon in the process. Normal electrons would not exist and we would be left with a universe of photons and electrons locked up in negative energy states.

Dirac circumvented this difficulty by postulating that all negative energy states are normally filled. The electrons that occupy the negative energy states are said to be in the "Dirac sea" and are not directly observable when all negative energy states are filled. However, if one of these negative energy states is vacant, leaving a hole in the sea of filled states, the hole can react to external forces and therefore can be observed as the electron's positive antiparticle. The general and profound implication of Dirac's theory is that **for every particle, there is an antiparticle with the same mass as the particle, but the opposite charge**. For example, the electron's antiparticle, the *positron*, has a mass of 0.511 MeV/c^2 and a positive charge of 1.6×10^{-19} C. As noted in Chapter 29, we usually designate an antiparticle with a bar over the symbol for the particle. For example, $\bar{p}$ denotes the antiproton and $\bar{\nu}$ the antineutrino. In this book, the notation e^+ is preferred for the positron.

The positron was discovered by Carl Anderson in 1932, and in 1936 he was awarded the Nobel prize for his achievement. Anderson discovered the positron while examining tracks created by electron-like particles of positive charge in a cloud chamber. (These early experiments used cosmic rays—mostly energetic protons passing through interstellar space—to initiate high-energy reactions on the order of several GeV.) In order to discriminate between positive and negative charges, the cloud chamber was placed in a magnetic field, causing moving charges to follow curved paths. Anderson noted that some of the

Courtesy AIP Emilio Segre Visual Archives

PAUL ADRIEN MAURICE DIRAC
(1902–1984)

Dirac was instrumental in the understanding of antimatter and in the unification of quantum mechanics and relativity. He made numerous contributions to the development of quantum physics and cosmology, and won the Nobel Prize for physics in 1933.

electronlike tracks deflected in a direction corresponding to a positively charged particle.

Since Anderson's initial discovery, the positron has been observed in a number of experiments. Perhaps the most common process for producing positrons is pair production, introduced in Chapter 26. In this process, a gamma ray with sufficiently high energy collides with a nucleus, creating an electron–positron pair. Because the total rest energy of the pair is $2m_e c^2 = 1.02$ MeV, the gamma ray must have at least this much energy to create such a pair.

Practically every known elementary particle has a distinct antiparticle. Among the exceptions are the photon and the neutral pion (π^0), which are their own antiparticles. Following the construction of high-energy accelerators in the 1950s, many of these antiparticles were discovered. They included the antiproton $\bar{p}$, discovered by Emilio Segrè and Owen Chamberlain in 1955, and the antineutron $\bar{n}$, discovered shortly thereafter.

The process of electron–positron annihilation is used in the medical diagnostic technique of positron emission tomography (PET). The patient is injected with a glucose solution containing a radioactive substance that decays by positron emission. Examples of such substances are oxygen-15, nitrogen-13, carbon-11, and fluorine-18. The radioactive material is carried to the brain. When a decay occurs, the emitted positron annihilates with an electron in the brain tissue, resulting in two gamma ray photons. With the assistance of a computer, an image can be created of the sites in the brain at which the glucose accumulates.

The images from a PET scan can point to a wide variety of disorders in the brain, including Alzheimer's disease. In addition, because glucose metabolizes more rapidly in active areas of the brain, the PET scan can indicate which areas of the brain are involved in various processes such as language, music, and vision.

APPLICATION

Positron Emission Tomography

HIDEKI YUKAWA, Japanese Physicist (1907–1981)

Yukawa was awarded the Nobel Prize in 1949 for predicting the existence of mesons. This photograph of Yukawa at work was taken in 1950 in his office at Columbia University.

30.7   MESONS AND THE BEGINNING OF PARTICLE PHYSICS

Physicists in the mid-1930s had a fairly simple view of the structure of matter. The building blocks were the proton, the electron, and the neutron. Three other particles were known or postulated at the time: the photon, the neutrino, and the positron. These six particles were considered the fundamental constituents of matter. Although the accepted picture of the world was marvelously simple, no one was able to provide an answer to the following important question: Because the many protons in proximity in any nucleus should strongly repel each other due to their like charges, what is the nature of the force that holds the nucleus together? Scientists recognized that this mysterious nuclear force must be much stronger than anything encountered up to that time.

The first theory to explain the nature of the nuclear force was proposed in 1935 by the Japanese physicist Hideki Yukawa (1907–1981), an effort that later earned him the Nobel prize. In order to understand Yukawa's theory, it is useful to first note that **two atoms can form a covalent chemical bond by the exchange of electrons.** Similarly, in the modern view of electromagnetic interactions, **charged particles interact by exchanging a photon.** Yukawa used this same idea to explain the nuclear force by proposing a new particle that is exchanged by nucleons in the nucleus to produce the strong force. Further, he demonstrated that the range of the force is inversely proportional to the mass of this particle, and predicted that the mass would be about 200 times the mass of the electron. Because the new particle would have a mass between that of the electron and the proton, it was called a **meson** (from the Greek *meso*, meaning "middle").

In an effort to substantiate Yukawa's predictions, physicists began looking for the meson by studying cosmic rays that enter the Earth's atmosphere. In 1937, Carl Anderson and his collaborators discovered a particle with mass 106 MeV/c^2, about 207 times the mass of the electron. However, subsequent experiments showed that the particle interacted very weakly with matter and hence could not be the carrier of the nuclear force. This puzzling situation inspired several theoreticians to propose that there are two mesons with slightly different masses, an idea that was confirmed in 1947 with the discovery of the pi meson (π), or simply *pion*, by Cecil Frank Powell (1903–1969) and Guiseppe P. S. Occhialini (1907–1993). The lighter meson discovered earlier by Anderson, now called a *muon* (μ), has only weak and electromagnetic interactions and plays no role in the strong interaction.

The pion comes in three varieties, corresponding to three charge states: π^+, π^-, and π^0. The π^+ and π^- particles have masses of 139.6 MeV/c^2, and the π^0 has a mass of 135.0 MeV/c^2. Pions and muons are highly unstable particles. For example, the π^-, which has a lifetime of about 2.6×10^{-8} s, decays into a muon and an antineutrino. The muon, with a lifetime of 2.2 μs, then decays into an electron, a neutrino, and an antineutrino. The sequence of decays is

$$\pi^- \rightarrow \mu^- + \overline{\nu} \qquad \text{[30.6]}$$

$$\mu^- \rightarrow e^- + \nu + \overline{\nu}$$

The interaction between two particles can be understood in general with a simple illustration called a *Feynman diagram*, developed by Richard P. Feynman (1918–1988). Figure 30.6 is a Feynman diagram for the electromagnetic interaction between two electrons. In this simple case, a photon is the field particle that mediates the electromagnetic force between the electrons. The photon transfers energy and momentum from one electron to the other in the interaction. Such a photon, called a *virtual photon*, can never be detected directly because it is absorbed by the second electron very shortly after being emitted by the first electron. The existence of a virtual photon might be expected to violate the law of conservation of energy, but it does not because of the time–energy uncertainty principle. Recall that the uncertainty principle says that the energy is uncertain or not conserved by an amount ΔE for a time Δt such that $\Delta E \, \Delta t \approx \hbar$.

Now consider the pion exchange between a proton and a neutron via the nuclear force (Fig. 30.7). The energy needed to create a pion of mass m_π is given by $\Delta E = m_\pi c^2$. Again, the existence of the pion is allowed in spite of conservation of energy if this energy is surrendered in a short enough time Δt, the time it takes the pion to transfer from one nucleon to the other. From the uncertainty principle, $\Delta E \, \Delta t \approx \hbar$, we get

$$\Delta t \approx \frac{\hbar}{\Delta E} = \frac{\hbar}{m_\pi c^2} \qquad \text{[30.7]}$$

Because the pion can't travel faster than the speed of light, the maximum distance d it can travel in a time Δt is $c \, \Delta t$. Using Equation 30.7 and $d = c \, \Delta t$, we find this maximum distance to be

$$d \approx \frac{\hbar}{m_\pi c} \qquad \text{[30.8]}$$

The measured range of the nuclear force is about 1.5×10^{-15} m. Using this value for d in Equation 30.8, the rest energy of the pion is calculated to be

$$m_\pi c^2 = \frac{\hbar c}{d} = \frac{(1.05 \times 10^{-34} \, \text{J} \cdot \text{s})(3.00 \times 10^8 \, \text{m/s})}{1.5 \times 10^{-15} \, \text{m}}$$

$$= 2.1 \times 10^{-11} \, \text{J} \cong 130 \, \text{MeV}$$

This corresponds to a mass of 130 MeV/c^2 (about 250 times the mass of the electron), which is in good agreement with the observed mass of the pion.

The concept we have just described is quite revolutionary. In effect, it says that a proton can change into a proton plus a pion, as long as it returns to its original state in a very short time. High-energy physicists often say that a nucleon undergoes "fluctuations" as it emits and absorbs pions. As we have seen, these fluctuations are a consequence of a combination of quantum mechanics (through the uncertainty principle) and special relativity (through Einstein's energy–mass relation $E = mc^2$).

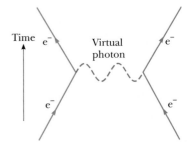

Figure 30.6 Feynman diagram representing a photon mediating the electromagnetic force between two electrons.

© Shelly Gazin/CORBIS

RICHARD FEYNMAN, American Physicist (1918–1988)

Feynman, together with Julian S. Schwinger and Shinichiro Tomonaga, won the 1965 Nobel Prize for physics for fundamental work in the principles of quantum electrodynamics. His many important contributions to physics include work on the first atomic bomb in the Manhattan project, the invention of simple diagrams to represent particle interactions graphically, the theory of the weak interaction of subatomic particles, a reformulation of quantum mechanics, and the theory of superfluid helium. Later he served on the commission investigating the *Challenger* tragedy, demonstrating the problem with the O-rings by dipping a scale-model O-ring in his glass of ice water and then shattering it with a hammer. He also contributed to physics education through the magnificent three-volume text *The Feynman Lectures on Physics*.

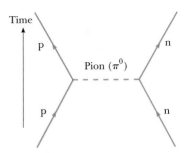

Figure 30.7 Feynman diagram representing a proton interacting with a neutron via the strong force. In this case, the pion mediates the nuclear force.

This section has dealt with the early theory of Yukawa of particles that mediate the nuclear force, pions, and the mediators of the electromagnetic force, photons. Although his model led to the modern view, it has been superseded by the more basic quark–gluon theory, as explained in Sections 30.12 and 30.13.

30.8 CLASSIFICATION OF PARTICLES

Hadrons

All particles other than photons can be classified into two broad categories, hadrons and leptons, according to their interactions. Particles that interact through the strong force are called *hadrons*. There are two classes of hadrons, known as *mesons* and *baryons*, distinguished by their masses and spins. All mesons are known to decay finally into electrons, positrons, neutrinos, and photons. The pion is the lightest of known mesons, with a mass of about 140 MeV/c^2 and a spin of 0. Another is the K meson, with a mass of about 500 MeV/c^2 and spin 0 also.

Baryons have masses equal to or greater than the proton mass (the name *baryon* means "heavy" in Greek), and their spin is always a non-integer value (1/2 or 3/2). Protons and neutrons are baryons, as are many other particles. With the exception of the proton, all baryons decay in such a way that the end products include a proton. For example, the baryon called the Ξ hyperon first decays to a Λ^0 in about 10^{-10} s. The Λ^0 then decays to a proton and a π^- in about 3×10^{-10} s.

Today it is believed that hadrons are composed of quarks. (Later, we will have more to say about the quark model.) Some of the important properties of hadrons are listed in Table 30.2.

TABLE 30.2

Some Particles and Their Properties

Category	Particle Name	Symbol	Anti-particle	Mass (MeV/c^2)	B	L_e	L_μ	L_τ	S	Lifetime(s)	Principal Decay Modes[a]
Leptons	Electron	e^-	e^+	0.511	0	+1	0	0	0	Stable	
	Electron–neutrino	ν_e	$\bar{\nu}_e$	$< 7\text{eV}/c^2$	0	+1	0	0	0	Stable	
	Muon	μ^-	μ^+	105.7	0	0	+1	0	0	2.20×10^{-6}	$e^- \bar{\nu}_e \nu_\mu$
	Muon–neutrino	ν_μ	$\bar{\nu}_\mu$	< 0.3	0	0	+1	0	0	Stable	
	Tau	τ^-	τ^+	1 784	0	0	0	+1	0	$< 4 \times 10^{-13}$	$\mu^- \bar{\nu}_\mu \nu_\tau, e^- \bar{\nu}_e \nu_\tau$
	Tau–neutrino	ν_τ	$\bar{\nu}_\tau$	< 30	0	0	0	+1	0	Stable	
Hadrons											
Mesons	Pion	π^+	π^-	139.6	0	0	0	0	0	2.60×10^{-8}	$\mu^+ \nu_\mu$
		π^0	Self	135.0	0	0	0	0	0	0.83×10^{-16}	2γ
	Kaon	K^+	K^-	493.7	0	0	0	0	+1	1.24×10^{-8}	$\mu^+ \nu_\mu, \pi^+ \pi^0$
		K^0_S	$\bar{K}^0_S$	497.7	0	0	0	0	+1	0.89×10^{-10}	$\pi^+ \pi^-, 2\pi^0$
		K^0_L	$\bar{K}^0_L$	497.7	0	0	0	0	+1	5.2×10^{-8}	$\pi^\pm e^\mp \bar{\nu}_e, 3\pi^0$ $\pi^\pm \mu^\mp \bar{\nu}_\mu$
	Eta	η	Self	548.8	0	0	0	0	0	$< 10^{-18}$	$2\gamma, 3\pi$
		η'	Self	958	0	0	0	0	0	2.2×10^{-21}	$\eta \pi^+ \pi^-$
Baryons	Proton	p	$\bar{p}$	938.3	+1	0	0	0	0	Stable	
	Neutron	n	$\bar{n}$	939.6	+1	0	0	0	0	920	$pe^- \bar{\nu}_e$
	Lambda	Λ^0	$\bar{\Lambda}^0$	1 115.6	+1	0	0	0	-1	2.6×10^{-10}	$p\pi^-, n\pi^0$
	Sigma	Σ^+	$\bar{\Sigma}^-$	1 189.4	+1	0	0	0	-1	0.80×10^{-10}	$p\pi^0, n\pi^+$
		Σ^0	$\bar{\Sigma}^0$	1 192.5	+1	0	0	0	-1	6×10^{-20}	$\Lambda^0 \gamma$
		Σ^-	$\bar{\Sigma}^+$	1 197.3	+1	0	0	0	-1	1.5×10^{-10}	$n\pi^-$
	Xi	Ξ^0	$\bar{\Xi}^0$	1 315	+1	0	0	0	-2	2.9×10^{-10}	$\Lambda^0 \pi^0$
		Ξ^-	Ξ^+	1 321	+1	0	0	0	-2	1.64×10^{-10}	$\Lambda^0 \pi^-$
	Omega	Ω^-	Ω^+	1 672	+1	0	0	0	-3	0.82×10^{-10}	$\Xi^0 \pi^0, \Lambda^0 K^-$

[a] Notations in this column, such as $p\pi^-$, $n\pi^0$ mean two possible decay modes. In this case, the two possible decays are $\Lambda^0 \to p + \pi^-$ and $\Lambda^0 \to n + \pi^0$.

Leptons

Leptons (from the Greek *leptos*, meaning "small" or "light") are a group of particles that participate in the weak interaction. All leptons have a spin of 1/2. Included in this group are electrons, muons, and neutrinos, which are less massive than the lightest hadron. Although hadrons have size and structure, leptons appear to be truly elementary, with no structure down to the limit of resolution of experiment (about 10^{-19} m).

Unlike hadrons, the number of known leptons is small. Currently, scientists believe there are only six leptons (each having an antiparticle): the electron, the muon, the tau, and a neutrino associated with each:

$$\begin{pmatrix} e^- \\ \nu_e \end{pmatrix} \quad \begin{pmatrix} \mu^- \\ \nu_\mu \end{pmatrix} \quad \begin{pmatrix} \tau^- \\ \nu_\tau \end{pmatrix}$$

The tau lepton, discovered in 1975, has a mass about twice that of the proton.

Although neutrinos have masses of about zero, there is strong indirect evidence that the electron neutrino has a nonzero mass of about 3 eV/c^2, or 1/180 000 of the electron mass. A firm knowledge of the neutrino's mass could have great significance in cosmological models and in our understanding of the future of the Universe.

30.9   CONSERVATION LAWS

A number of conservation laws are important in the study of elementary particles. Although the two described here have no theoretical foundation, they are supported by abundant empirical evidence.

Baryon Number

The law of conservation of baryon number tells us that whenever a baryon is created in a reaction or decay, an antibaryon is also created. This information can be quantified by assigning a baryon number: $B = +1$ for all baryons, $B = -1$ for all antibaryons, and $B = 0$ for all other particles. Thus, the **law of conservation of baryon number** states that whenever a nuclear reaction or decay occurs, the sum of the baryon numbers before the process equals the sum of the baryon numbers after the process.

◀ Conservation of baryon number

Note that if the baryon number is absolutely conserved, the proton must be absolutely stable: if it were not for the law of conservation of baryon number, the proton could decay into a positron and a neutral pion. However, such a decay has never been observed. At present, we can only say that the proton has a half-life of at least 10^{31} years. (The estimated age of the Universe is about 10^{10} years.) In one recent version of a so-called grand unified theory (GUT), physicists have predicted that the proton is actually unstable. According to this theory, the baryon number (sometimes called the *baryonic charge*) is not absolutely conserved, whereas electric charge is always conserved.

EXAMPLE 30.4   Checking Baryon Numbers

Goal   Use conservation of baryon number to determine whether a given reaction can occur.

Problem   Determine whether the following reaction can occur based on the law of conservation of baryon number.

$$p + n \;\longrightarrow\; p + p + n + \bar{p}$$

Strategy   Count the baryons on both sides of the reaction, recalling that that $B = +1$ for baryons and $B = -1$ for antibaryons.

Solution

Count the baryons on the left: The neutron and proton are both baryons; hence, $1 + 1 = 2$.

Count the baryons on the right:

There are three baryons and one antibaryon, so
$$1 + 1 + 1 + (-1) = 2.$$

Remark Baryon number is conserved in this reaction, so it can occur, provided the incoming proton has sufficient energy.

Exercise 30.4

Can the following reaction occur, based on the law of conservation of baryon number?

$$p + n \;\rightarrow\; p + p + \bar{p}$$

Answer No. (Show this by computing the baryon number on both sides and finding that they're not equal.)

Lepton Number

Conservation of lepton number ▶

There are three conservation laws involving lepton numbers, one for each variety of lepton. The **law of conservation of electron-lepton number** states that the sum of the electron-lepton numbers before a reaction or decay must equal the sum of the electron-lepton numbers after the reaction or decay. The electron and the electron neutrino are assigned a positive electron-lepton number $L_e = +1$, the antileptons e^+ and $\bar{\nu}_e$ are assigned the electron-lepton number $L_e = -1$, and all other particles have $L_e = 0$. For example, consider neutron decay:

Neutron decay ▶

$$n \;\rightarrow\; p^+ + e^- + \bar{\nu}_e$$

Before the decay, the electron-lepton number is $L_e = 0$; after the decay, it is $0 + 1 + (-1) = 0$, so the electron-lepton number is conserved. It's important to recognize that the baryon number must also be conserved. This can easily be seen by noting that before the decay $B = +1$, whereas after the decay $B = +1 + 0 + 0 = +1$.

Similarly, when a decay involves muons, the muon-lepton number L_μ is conserved. The μ^- and the ν_μ are assigned $L_\mu = +1$, the antimuons μ^+ and $\bar{\nu}_\mu$ are assigned $L_\mu = -1$, and all other particles have $L_\mu = 0$. Finally, the tau-lepton number L_τ is conserved, and similar assignments can be made for the τ lepton and its neutrino.

EXAMPLE 30.5   Checking Lepton Numbers

Goal Use conservation of lepton number to determine whether a given process is possible.

Problem Determine which of the following decay schemes can occur on the basis of conservation of lepton number.

$$\mu^- \;\rightarrow\; e^- + \bar{\nu}_e + \nu_\mu \tag{1}$$

$$\pi^+ \;\rightarrow\; \mu^+ + \nu_\mu + \nu_e \tag{2}$$

Strategy Count the leptons on either side and see if the numbers are equal.

Solution
Because decay 1 involves both a muon and an electron, L_μ and L_e must both be conserved. Before the decay, $L_\mu = +1$ and $L_e = 0$. After the decay, $L_\mu = 0 + 0 + 1 = +1$ and $L_e = +1 - 1 + 0 = 0$. Both lepton numbers are conserved, and on this basis, the decay mode is possible.

Before decay 2 occurs, $L_\mu = 0$ and $L_e = 0$. After the decay, $L_\mu = -1 + 1 + 0 = 0$, but $L_e = +1$. This decay isn't possible because the electron-lepton number is not conserved.

Exercise 30.5

Determine whether the decay $\mu^- \rightarrow e^- + \bar{\nu}_e$ can occur.

Answer No. (Show this by computing muon-lepton numbers on both sides and showing they're not equal.)

Quick Quiz 30.2

Which of the following reactions cannot occur?
(a) $p + p \rightarrow p + p + \bar{p}$ (b) $n \rightarrow p + e^- + \bar{\nu}_e$
(c) $\mu^- \rightarrow e^- + \bar{\nu}_e + \nu_\mu$ (d) $\pi^- \rightarrow \mu^- + \bar{\nu}_\mu$

Quick Quiz 30.3

Which of the following reactions cannot occur?
(a) $p + \bar{p} \rightarrow 2\gamma$ (b) $\gamma + p \rightarrow n + \pi^0$
(c) $\pi^0 + n \rightarrow K^+ + \Sigma^-$ (d) $\pi^+ + p \rightarrow K^+ + \Sigma^+$

Quick Quiz 30.4

Suppose a claim is made that the decay of a neutron is given by $n \rightarrow p^+ + e^-$. Which of the following conservation laws are necessarily violated by this proposed decay scheme? (a) energy (b) linear momentum (c) electric charge (d) lepton number (e) baryon number

30.10   STRANGE PARTICLES AND STRANGENESS

Many particles discovered in the 1950s were produced by the nuclear interaction of pions with protons and neutrons in the atmosphere. A group of these particles, namely the K, Λ, and Σ particles, was found to exhibit unusual properties in their production and decay and hence were called *strange particles*.

One unusual property of strange particles is that they are always produced in pairs. For example, when a pion collides with a proton, two neutral strange particles are produced with high probability (Fig. 30.8) following the reaction

$$\pi^- + p^+ \quad \rightarrow \quad K^0 + \Lambda^0$$

On the other hand, the reaction $\pi^- + p^+ \rightarrow K^0 + n$ has never occurred, even though it violates no known conservation laws and the energy of the pion is sufficient to initiate the reaction.

The second peculiar feature of strange particles is that although they are produced by the strong interaction at a high rate, they do not decay into particles that interact via the strong force at a very high rate. Instead, they decay very slowly, which is characteristic of the weak interaction. Their half-lives are in the range from 10^{-10} s to 10^{-8} s; most other particles that interact via the strong force have lifetimes on the order of 10^{-23} s.

To explain these unusual properties of strange particles, a law called *conservation of strangeness* was introduced, together with a new quantum number S called **strangeness**. The strangeness numbers for some particles are given in Table 30.2. The production of strange particles in pairs is explained by assigning $S = +1$ to one of the particles and $S = -1$ to the other. All nonstrange particles are assigned strangeness $S = 0$. The **law of conservation of strangeness** states that whenever a nuclear reaction or decay occurs, the sum of the strangeness numbers before the process must equal the sum of the strangeness numbers after the process.

The slow decay of strange particles can be explained by assuming that the strong and electromagnetic interactions obey the law of conservation of strangeness, whereas the weak interaction does not. Because the decay reaction involves the loss of one strange particle, it violates strangeness conservation and hence proceeds slowly via the weak interaction.

Courtesy Lawrence Berkeley Laboratory, University of California

Figure 30.8 This drawing represents tracks of many events obtained by analyzing a bubble-chamber photograph. The strange particles Λ^0 and K^0 are formed (at the bottom) as the π^- interacts with a proton according to the interaction $\pi^- + p \rightarrow \Lambda^0 + K^0$. (Note that the neutral particles leave no tracks, as is indicated by the dashed lines.) The Λ^0 and K^0 then decay according to the interactions $\Lambda^0 \rightarrow \pi + p$ and $K^0 \rightarrow \pi + \mu^- + \nu_\mu$.

◀ Conservation of strangeness number

Applying Physics 30.2   Breaking Conservation Laws

A student claims to have observed a decay of an electron into two neutrinos traveling in opposite directions. What conservation laws would be violated by this decay?

Explanation   Several conservation laws are violated. Conservation of electric charge is violated because the negative charge of the electron has disappeared. Conservation of electron lepton number is also violated, because there is one lepton before the decay and two afterward. If both neutrinos were electron-neutrinos, electron lepton number conservation

would be violated in the final state. However, if one of the product neutrinos were other than an electron-neutrino, then another lepton conservation law would be violated, because there were no other leptons in the initial state.

Other conservation laws are obeyed by this decay. Energy can be conserved—the rest energy of the electron appears as the kinetic energy (and possibly some small rest energy) of the neutrinos. The opposite directions of the velocities of the two neutrinos allow for the conservation of momentum. Conservation of baryon number and conservation of other lepton numbers are also upheld in this decay.

EXAMPLE 30.6   Is Strangeness Conserved?

Goal   Apply conservation of strangeness to determine whether a process can occur.

Problem   Determine whether the following reactions can occur on the basis of conservation of strangeness:

$$\pi^0 + n \;\rightarrow\; K^+ + \Sigma^- \tag{1}$$

$$\pi^- + p \;\rightarrow\; \pi^- + \Sigma^+ \tag{2}$$

Strategy   Count strangeness on each side of a given process. If strangeness is conserved, the reaction is possible.

Solution

In the first process, the neutral pion and neutron both have strangeness of zero, so $S_{\text{initial}} = 0 + 0 = 0$. Because the strangeness of the K^+ is $S = +1$, and the strangeness of the Σ^- is $S = -1$, the total strangeness of the final state is $S_{\text{final}} = +1 - 1 = 0$. Strangeness is conserved and the reaction is allowed.

In the second process, the initial state has strangeness $S_{\text{initial}} = 0 + 0 = 0$, but the final state has strangeness $S_{\text{final}} = 0 + (-1) = -1$. Strangeness is not conserved and the reaction isn't allowed.

Exercise 30.6

Does the reaction $p^+ + \pi^- \rightarrow K^0 + \Lambda^0$ obey the law of conservation of strangeness? Show why or why not.

Answer   Yes. (Show this by computing the strangeness on both sides.)

30.11   THE EIGHTFOLD WAY

Quantities such as spin, baryon number, lepton number, and strangeness are labels we associate with particles. Many classification schemes that group particles into families based on such labels have been proposed. First, consider the first eight baryons listed in Table 30.2, all having a spin of 1/2. The family consists of the proton, the neutron, and six other particles. If we plot their strangeness versus their charge using a sloping coordinate system, as in Figure 30.9a, a fascinating pattern emerges: six of the baryons form a hexagon, and the remaining two are at the hexagon's center. (Particles with spin quantum number 1/2 or 3/2 are called fermions.)

Now consider the family of mesons listed in Table 30.2 with spins of zero. (Particles with spin quantum number 0 or 1 are called bosons.) If we count both particles and antiparticles, there are nine such mesons. Figure 30.9b is a plot of strangeness versus charge for *this* family. Again, a fascinating hexagonal pattern emerges. In this case, the particles on the perimeter of the hexagon lie opposite their antiparticles, and the remaining three (which form their own antiparticles)

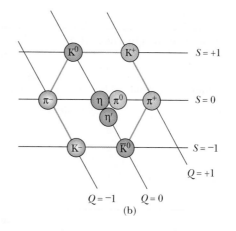

Figure 30.9 (a) The hexagonal eightfold-way pattern for the eight spin $-\frac{1}{2}$ baryons. This strangeness versus charge plot uses a horizontal axis for the strangeness values S, but a sloping axis for the charge number Q. (b) The eightfold-way pattern for the nine spin-zero mesons.

are at its center. These and related symmetric patterns, called the **eightfold way**, were proposed independently in 1961 by Murray Gell-Mann and Yuval Ne'eman.

The groups of baryons and mesons can be displayed in many other symmetric patterns within the framework of the eightfold way. For example, the family of spin-3/2 baryons contains ten particles arranged in a pattern like the tenpins in a bowling alley. After the pattern was proposed, one of the particles was missing—it had yet to be discovered. Gell-Mann predicted that the missing particle, which he called the *omega minus* (Ω^-), should have a spin of 3/2, a charge of -1, a strangeness of -3, and a mass of about 1 680 MeV/c^2. Shortly thereafter, in 1964, scientists at the Brookhaven National Laboratory found the missing particle through careful analyses of bubble chamber photographs and confirmed all its predicted properties.

The patterns of the eightfold way in the field of particle physics have much in common with the periodic table. Whenever a vacancy (a missing particle or element) occurs in the organized patterns, experimentalists have a guide for their investigations.

30.12   QUARKS

As we have noted, leptons appear to be truly elementary particles because they have no measurable size or internal structure, are limited in number, and do not seem to break down into smaller units. Hadrons, on the other hand, are complex particles with size and structure. Further, we know that hadrons decay into other hadrons and are many in number. Table 30.2 lists only those hadrons that are stable against hadronic decay; hundreds of others have been discovered. These facts strongly suggest that hadrons cannot be truly elementary but have some substructure.

The Original Quark Model

In 1963 Gell-Mann and George Zweig independently proposed that hadrons have an elementary substructure. According to their model, all hadrons are composite systems of two or three fundamental constituents called **quarks**, which rhymes with "forks" (though some rhyme it with "sharks"). Gell-Mann borrowed the word *quark* from the passage "Three quarks for Muster Mark" in James Joyce's book *Finnegans Wake*. In the original model there were three types of quarks designated by the symbols u, d, and s. These were given the arbitrary names *up*, *down*, and *sideways* (or, now more commonly, *strange*).

An unusual property of quarks is that they have fractional electronic charges, as shown—along with other properties—in Table 30.3 (page 994). Associated with each quark is an antiquark of opposite charge, baryon number, and strangeness. The compositions of all hadrons known when Gell-Mann and Zweig presented their models could be completely specified by three simple rules:

Photo courtesy of Michael R. Dressler

MURRAY GELL-MANN,
American Physicist (1929–)

Gell-Mann was awarded the Nobel Prize in 1969 for his theoretical studies dealing with subatomic particles.

TABLE 30.3

Properties of Quarks and Antiquarks

Quarks

Name	Symbol	Spin	Charge	Baryon Number	Strangeness	Charm	Bottomness	Topness
Up	u	$\frac{1}{2}$	$+\frac{2}{3}e$	$\frac{1}{3}$	0	0	0	0
Down	d	$\frac{1}{2}$	$-\frac{1}{3}e$	$\frac{1}{3}$	0	0	0	0
Strange	s	$\frac{1}{2}$	$-\frac{1}{3}e$	$\frac{1}{3}$	-1	0	0	0
Charmed	c	$\frac{1}{2}$	$+\frac{2}{3}e$	$\frac{1}{3}$	0	$+1$	0	0
Bottom	b	$\frac{1}{2}$	$-\frac{1}{3}e$	$\frac{1}{3}$	0	0	$+1$	0
Top	t	$\frac{1}{2}$	$+\frac{2}{3}e$	$\frac{1}{3}$	0	0	0	$+1$

Antiquarks

Name	Symbol	Spin	Charge	Baryon Number	Strangeness	Charm	Bottomness	Topness
Anti-up	$\bar{u}$	$\frac{1}{2}$	$-\frac{2}{3}e$	$-\frac{1}{3}$	0	0	0	0
Anti-down	$\bar{d}$	$\frac{1}{2}$	$+\frac{1}{3}e$	$-\frac{1}{3}$	0	0	0	0
Anti-strange	$\bar{s}$	$\frac{1}{2}$	$+\frac{1}{3}e$	$-\frac{1}{3}$	$+1$	0	0	0
Anti-charmed	$\bar{c}$	$\frac{1}{2}$	$-\frac{2}{3}e$	$-\frac{1}{3}$	0	-1	0	0
Anti-bottom	$\bar{b}$	$\frac{1}{2}$	$+\frac{1}{3}e$	$-\frac{1}{3}$	0	0	-1	0
Anti-top	$\bar{t}$	$\frac{1}{2}$	$-\frac{2}{3}e$	$-\frac{1}{3}$	0	0	0	-1

TABLE 30.4

Quark Composition of Several Hadrons

Particle	Quark Composition
Mesons	
π^+	$\bar{d}u$
π^-	$\bar{u}d$
K^+	$\bar{s}u$
K^-	$\bar{u}s$
K^0	$\bar{s}d$
Baryons	
p	uud
n	udd
Λ^0	uds
Σ^+	uus
Σ^0	uds
Σ^-	dds
Ξ^0	uss
Ξ^-	dss
Ω^-	sss

1. Mesons consist of one quark and one antiquark, giving them a baryon number of 0, as required.
2. Baryons consist of three quarks.
3. Antibaryons consist of three antiquarks.

Table 30.4 lists the quark compositions of several mesons and baryons. Note that just two of the quarks, u and d, are contained in all hadrons encountered in ordinary matter (protons and neutrons). The third quark, s, is needed only to construct strange particles with a strangeness of either $+1$ or -1. Active Figure 30.10 is a pictorial representation of the quark compositions of several particles.

Applying Physics 30.3   Conservation of Meson Number?

We have seen a law of conservation of lepton number and a law of conservation of baryon number. Why isn't there a law of conservation of meson number?

Explanation We can argue this from the point of view of creating particle–antiparticle pairs from available energy. If energy is converted to the rest energy of a lepton–antilepton pair, then there is no net change in lepton number, because the lepton has a lepton number of $+1$ and the antilepton -1. Energy could also be transformed into the rest energy of a

baryon–antibaryon pair. The baryon has baryon number $+1$, the antibaryon -1, and there is no net change in baryon number.

But now suppose energy is transformed into the rest energy of a quark–antiquark pair. By definition in quark theory, a quark–antiquark pair is a meson. There was no meson before, and now there's a meson, so already there is violation of conservation of meson number. With more energy, we can create more mesons, with no restriction from a conservation law other than that of energy.

Charm and Other Recent Developments

Although the original quark model was highly successful in classifying particles into families, there were some discrepancies between predictions of the model and certain experimental decay rates. Consequently, a fourth quark was proposed by several physicists in 1967. The fourth quark, designated by c, was given a property called **charm**. A charmed quark would have the charge $+2e/3$, but its charm would distinguish it from the other three quarks. The new quark would have a charm $C = +1$, its antiquark would have a charm $C = -1$, and all other quarks would have $C = 0$, as indicated in Table 30.3. Charm, like strangeness, would be conserved in strong and electromagnetic interactions but not in weak interactions.

In 1974 a new heavy meson called the J/ψ particle (or simply, ψ) was discovered independently by a group led by Burton Richter at the Stanford Linear Accelerator (SLAC) and another group led by Samuel Ting at the Brookhaven National Laboratory. Richter and Ting were awarded the Nobel Prize in 1976 for this work. The J/ψ particle didn't fit into the three-quark model, but had the properties of a combination of a charmed quark and its antiquark ($c\bar{c}$). It was much heavier than the other known mesons ($\sim 3\,100$ MeV/c^2) and its lifetime was much longer than those of other particles that decay via the strong force. In 1975, researchers at Stanford University reported strong evidence for the existence of the tau (τ) lepton, with a mass of 1 784 MeV/c^2. Such discoveries led to more elaborate quark models and the proposal of two new quarks, named *top* (t) and *bottom* (b). To distinguish these quarks from the old ones, quantum numbers called *topness* and *bottomness* were assigned to these new particles and are included in Table 30.3. In 1977 researchers at the Fermi National Laboratory, under the direction of Leon Lederman, reported the discovery of a very massive new meson Υ with composition $b\bar{b}$. In March of 1995, researchers at Fermilab announced the discovery of the top quark (supposedly the last of the quarks to be found) having mass 173 GeV/c^2.

You are probably wondering whether such discoveries will ever end. How many "building blocks" of matter really exist? The numbers of different quarks and leptons have implications for the primordial abundance of certain elements, so at present it appears there may be no further fundamental particles. Some properties of quarks and leptons are given in Table 30.5.

Despite extensive experimental efforts, no isolated quark has ever been observed. Physicists now believe that quarks are permanently confined inside ordinary particles because of an exceptionally strong force that prevents them from escaping. This force, called the color force (which will be discussed in Section 30.13), increases with separation distance (similar to the force of a spring). The great strength of the force between quarks has been described by one author as follows:[2]

> Quarks are slaves of their own color charge, . . . bound like prisoners of a chain gang. . . . Any locksmith can break the chain between two prisoners, but no locksmith is expert enough to break the gluon chains between quarks. Quarks remain slaves forever.

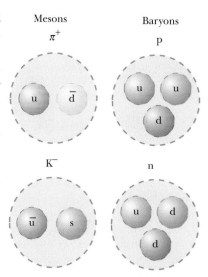

ACTIVE FIGURE 30.10
Quark compositions of two mesons and two baryons. Note that the mesons on the left contain two quarks, and the baryons on the right contain three quarks.

Physics⊗Now™

Log into PhysicsNow at **www.cp7e.com** and go to Active Figure 30.10 to observe the quark compositions for the mesons and baryons.

TABLE 30.5

The Fundamental Particles and Some of Their Properties

Particle	Rest Energy	Charge
Quarks		
u	360 MeV	$+\frac{2}{3}e$
d	360 MeV	$-\frac{1}{3}e$
c	1500 MeV	$+\frac{2}{3}e$
s	540 MeV	$-\frac{1}{3}e$
t	173 GeV	$+\frac{2}{3}e$
b	5 GeV	$-\frac{1}{3}e$
Leptons		
e^-	511 keV	$-e$
μ^-	107 MeV	$-e$
τ^-	1784 MeV	$-e$
ν_e	<30 eV	0
ν_μ	<0.5 MeV	0
ν_τ	<250 MeV	0

[2]Harald Fritzsch, *Quarks: The Stuff of Matter* (London: Allen Lane, 1983).

Computers at Fermilab create a pictorial representation such as this of the paths of particles after a collision.

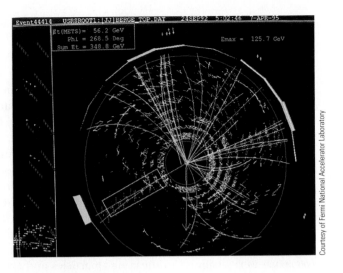

Courtesy of Fermi National Accelerator Laboratory

30.13 COLORED QUARKS

Shortly after the theory of quarks was proposed, scientists recognized that certain particles had quark compositions that were in violation of the Pauli exclusion principle. Because all quarks have spins of $1/2$, they are expected to follow the exclusion principle. One example of a particle that violates the exclusion principle is the Ω^- (sss) baryon, which contains three s quarks having parallel spins, giving it a total spin of $3/2$. Other examples of baryons that have identical quarks with parallel spins are the Δ^{++} (uuu) and the Δ^- (ddd). To resolve this problem, Moo-Young Han and Yoichiro Nambu suggested in 1965 that quarks possess a new property called **color** or **color charge**. This "charge" property is similar in many respects to electric charge, except that it occurs in three varieties, labeled *red*, *green*, and *blue*! (The antiquarks are labeled *anti-red*, *anti-green*, and *anti-blue*.) To satisfy the exclusion principle, all three quarks in a baryon must have different colors. Just as a combination of actual colors of light can produce the neutral color white, a combination of three quarks with different colors is also "white," or colorless. A meson consists of a quark of one color and an antiquark of the corresponding anticolor. The result is that baryons and mesons are always colorless (or white).

Although the concept of color in the quark model was originally conceived to satisfy the exclusion principle, it also provided a better theory for explaining certain experimental results. For example, the modified theory correctly predicts the lifetime of the π^0 meson. The theory of how quarks interact with each other by means of color charge is called **quantum chromodynamics**, or QCD, to parallel quantum electrodynamics (the theory of interactions among electric charges). In QCD, the quark is said to carry a **color charge**, in analogy to electric charge. The strong force between quarks is often called the **color force**. The force is carried by massless particles called **gluons** (which are analogous to photons for the electromagnetic force). According to QCD, there are eight gluons, all with color charge. When a quark emits or absorbs a gluon, its color changes. For example, a blue quark that emits a gluon may become a red quark, and a red quark that absorbs this gluon becomes a blue quark. The color force between quarks is analogous to the electric force between charges: Like colors repel and opposite colors attract. Therefore, two red quarks repel each other, but a red quark will be attracted to an anti-red quark. The attraction between quarks of opposite color to form a meson ($q\overline{q}$) is indicated in Figure 30.11a.

Different-colored quarks also attract each other, but with less intensity than opposite colors of quark and antiquark. For example, a cluster of red, blue, and green quarks all attract each other to form baryons, as indicated in Figure 30.11b. Every baryon contains three quarks of three different colors.

Although the color force between two color-neutral hadrons (such as a proton and a neutron) is negligible at large separations, the strong color force between their constituent quarks does not exactly cancel at small separations of about 1 fm. **This residual strong force is in fact the nuclear force that binds protons and**

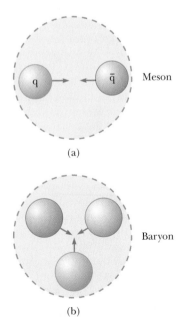

Figure 30.11 (a) A green quark is attracted to an anti-green quark to form a meson with quark structure ($q\overline{q}$). (b) Three different-colored quarks attract each other to form a baryon.

(a) Yukawa's pion model

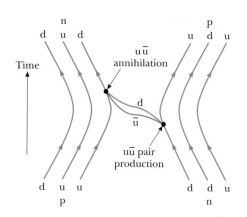

(b) Quark model

Figure 30.12 (a) A nuclear interaction between a proton and a neutron explained in terms of Yukawa's pion exchange model. Because the pion carries charge, the proton and neutron switch identities. (b) The same interaction explained in terms of quarks and gluons. Note that the exchanged $\bar{u}d$ quark pair makes up a π^- meson.

neutrons to form nuclei. It is similar to the residual electromagnetic force that binds neutral atoms into molecules. According to QCD, a more basic explanation of nuclear force can be given in terms of quarks and gluons, as shown in Figure 30.12, which shows contrasting Feynman diagrams of the same process. Each quark within the neutron and proton is continually emitting and absorbing virtual gluons and creating and annihilating virtual $(q\bar{q})$ pairs. When the neutron and proton approach within 1 fm of each other, these virtual gluons and quarks can be exchanged between the two nucleons, and such exchanges produce the nuclear force. Figure 30.12b depicts one likely possibility or contribution to the process shown in Figure 30.12a: a down quark emits a virtual gluon (represented by a wavy line in Fig. 30.12b), which creates a $u\bar{u}$ pair. Both the recoiling d quark and the $\bar{u}$ are transmitted to the proton where the $\bar{u}$ annihilates a proton u quark (with the creation of a gluon) and the d is captured.

30.14 ELECTROWEAK THEORY AND THE STANDARD MODEL

Recall that the weak interaction is an extremely short range force having an interaction distance of approximately 10^{-18} m (Table 30.1). Such a short-range interaction implies that the quantized particles which carry the weak field (the spin one W^+, W^-, and Z^0 bosons) are extremely massive, as is indeed the case. These amazing bosons can be thought of as structureless, pointlike particles as massive as krypton atoms! The weak interaction is responsible for the decay of the c, s, b, and t quarks into lighter, more stable u and d quarks, as well as the decay of the massive μ and τ leptons into (lighter) electrons. **The weak interaction is very important because it governs the stability of the basic particles of matter.**

A mysterious feature of the weak interaction is its lack of symmetry, especially when compared to the high degree of symmetry shown by the strong, electromagnetic, and gravitational interactions. For example, the weak interaction, unlike the strong interaction, is not symmetric under mirror reflection or charge exchange. (*Mirror reflection* means that all the quantities in a given particle reaction are exchanged as in a mirror reflection—left for right, an inward motion toward the mirror for an outward motion, etc. *Charge exchange* means that all the electric charges in a particle reaction are converted to their opposites—all positives to negatives and vice versa.) When we say that the weak interaction is not symmetric, we mean that the reaction with all quantities changed occurs less frequently than the direct reaction. For example, the decay of the K^0, which is governed by the weak interaction, is not symmetric under charge exchange because the reaction $K^0 \rightarrow \pi^- + e^+ + \nu_e$ occurs much more frequently than the reaction $K^0 \rightarrow \pi^+ + e^- + \bar{\nu}_e$.

In 1979, Sheldon Glashow, Abdus Salam, and Steven Weinberg won a Nobel prize for developing a theory called the **electroweak theory** that unified the electromagnetic and weak interactions. This theory postulates that the weak and electromagnetic interactions have the same strength at very high particle energies,

Figure 30.13 The Standard Model of particle physics.

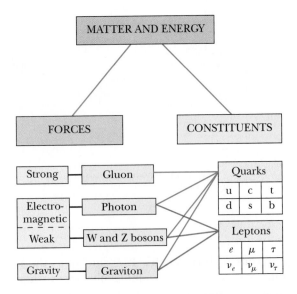

and are different manifestations of a single unifying electroweak interaction. The photon and the three massive bosons ($W^\pm$ and Z^0) play a key role in the electroweak theory. The theory makes many concrete predictions, but perhaps the most spectacular is the prediction of the masses of the W and Z particles at about $82\ \text{GeV}/c^2$ and $93\ \text{GeV}/c^2$, respectively. A 1984 Nobel Prize was awarded to Carlo Rubbia and Simon van der Meer for their work leading to the discovery of these particles at just those energies at the CERN Laboratory in Geneva, Switzerland.

The combination of the electroweak theory and QCD for the strong interaction form what is referred to in high energy physics as the **Standard Model**. Although the details of the Standard Model are complex, its essential ingredients can be summarized with the help of Figure 30.13. The strong force, mediated by gluons, holds quarks together to form composite particles such as protons, neutrons, and mesons. Leptons participate only in the electromagnetic and weak interactions. The electromagnetic force is mediated by photons, and the weak force is mediated by W and Z bosons. Note that all fundamental forces are mediated by bosons (particles with spin 1) whose properties are given, to a large extent, by symmetries involved in the theories.

However, the Standard Model does not answer all questions. A major question is why the photon has no mass while the W and Z bosons do. Because of this mass difference, the electromagnetic and weak forces are quite distinct at low energies, but become similar in nature at very high energies, where the rest energies of the W and Z bosons are insignificant fractions of their total energies. This behavior during the transition from high to low energies, called **symmetry breaking**, doesn't answer the question of the origin of particle masses. To resolve that problem, a hypothetical particle called the **Higgs boson** has been proposed which provides a mechanism for breaking the electroweak symmetry and bestowing different particle masses on different particles. The Standard Model, including the Higgs mechanism, provides a logically consistent explanation of the massive nature of the W and Z bosons. Unfortunately, the Higgs boson has not yet been found, but physicists know that its mass should be less than $1\ \text{TeV}/c^2$ ($10^{12}\ \text{eV}$).

In order to determine whether the Higgs boson exists, two quarks of at least 1 TeV of energy must collide, but calculations show that this requires injecting 40 TeV of energy within the volume of a proton. Scientists are convinced that because of the limited energy available in conventional accelerators using fixed targets, it is necessary to build colliding-beam accelerators called **colliders**. The concept of a collider is straightforward. In such a device, particles with equal masses and kinetic energies, traveling in opposite directions in an accelerator ring, collide head-on to produce the required reaction and the formation of new particles. Because the total momentum of the interacting particles is zero, all of their kinetic energy is available for the reaction. The Large Electron–Positron (LEP) collider at CERN, near Geneva, Switzerland, and the Stanford Linear Collider in California collide both electrons and positrons. The Super Proton Synchrotron at CERN accelerates

A view from inside the Large Electron–Positron (LEP) collider tunnel, which is 27 km in circumference.

protons and antiprotons to energies of 270 GeV, and the world's highest-energy proton acclerator, the Tevatron, at the Fermi National Laboratory in Illinois, produces protons at almost 1 000 GeV (or 1 TeV). CERN has started construction of the Large Hadron Collider (LHC), a proton–proton collider that will provide a center-of-mass energy of 14 TeV and allow an exploration of Higgs-boson physics. The accelerator is being constructed in the same 27-km circumference tunnel as CERN's LEP collider, and construction is expected to be completed in 2005.

Following the success of the electroweak theory, scientists attempted to combine it with QCD in a **grand unification theory** (GUT). In this model, the electroweak force was merged with the strong color force to form a grand unified force. One version of the theory considers leptons and quarks as members of the same family that are able to change into each other by exchanging an appropriate particle. Many GUT theories predict that protons are unstable and will decay with a lifetime of about 10^{31} years, a period far greater than the age of the Universe. As yet, proton decays have not been observed.

Applying Physics 30.4   Head-on Collisions

Consider a car making a head-on collision with an identical car moving in the opposite direction at the same speed. Compare that collision to one in which one of the cars collides with a second car that is at rest. In which collision is there a larger transformation of kinetic energy to other forms? How does this idea relate to producing exotic particles in collisions?

Explanation   In the head-on collision with both cars moving, conservation of momentum causes most, if not all, of the kinetic energy to be transformed to other forms. In the collision between a moving car and

a stationary car, the cars are still moving after the collision in the direction of the moving car, but with reduced speed. Thus, only part of the kinetic energy is transformed to other forms. This suggests the advantage of using colliding beams to produce exotic particles, as opposed to firing a beam into a stationary target. When particles moving in opposite directions collide, all of the kinetic energy is available for transformation into other forms—in this case, the creation of new particles. When a beam is fired into a stationary target, only part of the energy is available for transformation, so particles of higher mass cannot be created.

30.15   THE COSMIC CONNECTION

In this section we describe one of the most fascinating theories in all of science—the Big Bang theory of the creation of the Universe—and the experimental evidence that supports it. This theory of cosmology states that the Universe had a beginning and that this beginning was so cataclysmic that it is impossible to look back beyond it. According to the theory, the Universe erupted from an infinitely dense singularity about 15 to 20 billion years ago. The first few minutes after the Big Bang saw such extremes of energy that it is believed that all four interactions of physics were unified and all matter was contained in an undifferentiated "quark soup."

The evolution of the four fundamental forces from the Big Bang to the present is shown in Figure 30.14. During the first 10^{-43} s (the ultrahot epoch, with

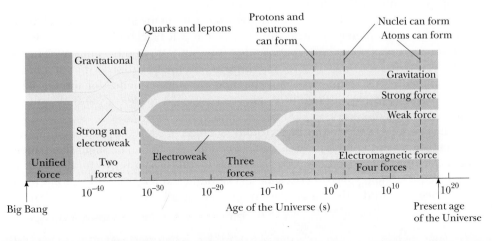

Figure 30.14   A brief history of the Universe from the Big Bang to the present. The four forces became distinguishable during the first microsecond. Following this, all the quarks combined to form particles that interact via the strong force. The leptons remained separate, however, and exist as individually observable particles to this day.

GEORGE GAMOW
(1904–1968)

Gamow and two of his students, Ralph Alpher and Robert Herman, were the first to take the first half hour of the Universe seriously. In a mostly overlooked paper published in 1948, they made truly remarkable cosmological predictions. They correctly calculated the abundances of hydrogen and helium after the first half hour (75% H and 25% He) and predicted that radiation from the Big Bang should still be present and have an apparent temperature of about 5 K.

$T \approx 10^{32}$ K), it is presumed that the strong, electroweak, and gravitational forces were joined to form a completely unified force. In the first 10^{-35} s following the Big Bang (the hot epoch, with $T \approx 10^{29}$ K), gravity broke free of this unification and the strong and electroweak forces remained as one, described by a grand unification theory. This was a period when particle energies were so great ($> 10^{16}$ GeV) that very massive particles as well as quarks, leptons, and their antiparticles, existed. Then, after 10^{-35} s, the Universe rapidly expanded and cooled (the warm epoch, with $T \approx 10^{29}$ to 10^{15} K), the strong and electroweak forces parted company, and the grand unification scheme was broken. As the Universe continued to cool, the electroweak force split into the weak force and the electromagnetic force about 10^{-10} s after the Big Bang.

After a few minutes, protons condensed out of the hot soup. For half an hour the Universe underwent thermonuclear detonation, exploding like a hydrogen bomb and producing most of the helium nuclei now present. The Universe continued to expand, and its temperature dropped. Until about 700 000 years after the Big Bang, the Universe was dominated by radiation. Energetic radiation prevented matter from forming single hydrogen atoms because collisions would instantly ionize any atoms that might form. Photons underwent continuous Compton scattering from the vast number of free electrons, resulting in a Universe that was opaque to radiation. By the time the Universe was about 700 000 years old, it had expanded and cooled to about 3 000 K, and protons could bind to electrons to form neutral hydrogen atoms. Because the energies of the atoms were quantized, far more wavelengths of radiation were not absorbed by atoms than were, and the Universe suddenly became transparent to photons. Radiation no longer dominated the Universe, and clumps of neutral matter grew steadily—first atoms, followed by molecules, gas clouds, stars, and finally galaxies.

Observation of Radiation from the Primordial Fireball

In 1965 Arno A. Penzias (b. 1933) and Robert W. Wilson (b. 1936) of Bell Laboratories made an amazing discovery while testing a sensitive microwave receiver. A pesky signal producing a faint background hiss was interfering with their satellite communications experiments. In spite of their valiant efforts, the signal remained. Ultimately it became clear that they were observing microwave background radiation (at a wavelength of 7.35 cm) representing the leftover "glow" from the Big Bang.

The microwave horn that served as their receiving antenna is shown in Figure 30.15. The intensity of the detected signal remained unchanged as the antenna was pointed in different directions. The fact that the radiation had equal strengths in all directions suggested that the entire Universe was the source of this radiation. Evicting a flock of pigeons from the 20-foot horn and cooling the microwave detector both failed to remove the signal. Through a casual conversation, Penzias and Wilson discovered that a group at Princeton had predicted the residual radiation from the Big Bang and were planning an experiment to confirm the theory. The excitement in the scientific community was high when Penzias and Wilson announced that they had already observed an excess microwave background compatible with a 3-K blackbody source.

Figure 30.15 Robert W. Wilson (*left*) and Arno A. Penzias (*right*), with Bell Telephone Laboratories' horn-reflector antenna.

Because Penzias and Wilson made their measurements at a single wavelength, they did not completely confirm the radiation as 3-K blackbody radiation. Subsequent experiments by other groups added intensity data at different wavelengths, as shown in Figure 30.16. The results confirm that the radiation is that of a blackbody at 2.9 K. This figure is perhaps the most clear-cut evidence for the Big Bang theory. The 1978 Nobel Prize in physics was awarded to Penzias and Wilson for their important discovery.

The discovery of the cosmic background radiation produced a problem, however: the radiation was too uniform. Scientists believed there had to be slight fluctuations in this background in order for such objects as galaxies to form. In 1989, NASA launched a satellite called the Cosmic Background Explorer (COBE, pronounced KOH-bee) to study this radiation in greater detail. In 1992, George Smoot

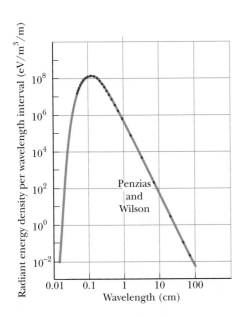

Figure 30.16 Theoretical blackbody (brown curve) and measured radiation spectra (blue points) of the Big Bang. Most of the data were collected from the Cosmic Background Explorer (COBE) satellite. The datum of Wilson and Penzias is indicated.

(b. 1945) at the Lawrence Berkeley Laboratory found that the background was not perfectly uniform, but instead contained irregularities corresponding to temperature variations of 0.000 3 K. It is these small variations that provided nucleation sites for the formation of the galaxies and other objects we now see in the sky.

30.16 PROBLEMS AND PERSPECTIVES

While particle physicists have been exploring the realm of the very small, cosmologists have been exploring cosmic history back to the first microsecond of the Big Bang. Observation of the events that occur when two particles collide in an accelerator is essential in reconstructing the early moments in cosmic history. Perhaps the key to understanding the early Universe is first to understand the world of elementary particles. Cosmologists and particle physicists find that they have many common goals and are joining efforts to study the physical world at its most fundamental level.

Our understanding of physics at short and long distances is far from complete. Particle physics is faced with many questions: why is there so little antimatter in the Universe? Do neutrinos have a small mass, and if so, how much do they contribute to the "dark matter" holding the universe together gravitationally? How can we understand the latest astronomical measurements, which show that the expansion of the universe is accelerating and that there may be a kind of "antigravity force" acting between widely separated galaxies? Is it possible to unify the strong and electroweak theories in a logical and consistent manner? Why do quarks and leptons form three similar but distinct families? Are muons the same as electrons (apart from their different masses), or do they have subtle differences that have not been detected? Why are some particles charged and others neutral? Why do quarks carry a fractional charge? What determines the masses of the fundamental particles? The questions go on and on. Because of the rapid advances and new discoveries in the related fields of particle physics and cosmology, by the time you read this book some of these questions may have been resolved and others may have emerged.

An important question that remains is whether leptons and quarks have a substructure. If they do, one could envision an infinite number of deeper structure levels. However, if leptons and quarks are indeed the ultimate constituents of matter, as physicists today tend to believe, we should be able to construct a final theory of the structure of matter, as Einstein dreamed of doing. In the view of many physicists, the end of the road is in sight, but how long it will take to reach that goal is anyone's guess.

SUMMARY

30.1 Nuclear Fission &
30.2 Nuclear Reactors

In **nuclear fission**, the total mass of the products is always less than the original mass of the reactants. Nuclear fission occurs when a heavy nucleus splits, or fissions, into two smaller nuclei. The lost mass is transformed into energy, electromagnetic radiation, and the kinetic energy of daughter particles.

A **nuclear reactor** is a system designed to maintain a self-sustaining chain reaction. Nuclear reactors using controlled fission events are currently being used to generate electric power. A useful parameter for describing the level of reactor operation is the reproduction constant K, which is the average number of neutrons from each fission event that will cause another event. A self-sustaining reaction is achieved when $K = 1$.

30.3 Nuclear Fusion

In nuclear fusion, two light nuclei combine to form a heavier nucleus. This type of nuclear reaction occurs in the Sun, assisted by a quantum tunneling process that helps particles get through the Coulomb barrier.

Controlled fusion events offer the hope of plentiful supplies of energy in the future. The nuclear fusion reactor is considered by many scientists to be the ultimate energy source because its fuel is water. **Lawson's criterion** states that a fusion reactor will provide a net output power if the product of the plasma ion density n and the plasma confinement time τ satisfies the following relationships:

$$n\tau \geq 10^{14}\,\text{s/cm}^3 \quad \text{Deuterium–tritium interaction} \qquad [30.5]$$
$$n\tau \geq 10^{16}\,\text{s/cm}^3 \quad \text{Deuterium–deuterium interaction}$$

30.5 The Fundamental Forces of Nature

There are four fundamental forces of nature: the **strong** (hadronic), **electromagnetic**, **weak**, and **gravitational** forces. The strong force is the force between nucleons that keeps the nucleus together. The weak force is responsible for beta decay. The electromagnetic and weak forces are now considered to be manifestations of a single force called the **electroweak** force.

Every fundamental interaction is said to be mediated by the exchange of field particles. The electromagnetic interaction is mediated by the photon, the weak interaction by the $W^\pm$ and Z^0 bosons, the gravitational interaction by gravitons, and the strong interaction by gluons.

30.6 Positrons and Other Antiparticles

An antiparticle and a particle have the same mass, but opposite charge, and may also have other properties with opposite values, such as lepton number and baryon number. It is possible to produce particle–antiparticle pairs in nuclear reactions if the available energy is greater than $2mc^2$, where m is the mass of the particle (or antiparticle).

30.8 Classification of Particles

Particles other than photons are classified as hadrons or leptons. **Hadrons** interact primarily through the strong force. They have size and structure and hence are not elementary particles. There are two types of hadrons: *baryons* and *mesons*. Mesons have a baryon number of zero and have either zero or integer spin. Baryons, which generally are the most massive particles, have nonzero baryon numbers and spins of 1/2 or 3/2. The neutron and proton are examples of baryons.

Leptons have no known structure, down to the limits of current resolution (about 10^{-19} m). Leptons interact only through the weak and electromagnetic forces. There are six leptons: the electron, e^-; the muon, μ^-; the tau, τ^-; and their associated neutrinos, ν_e, ν_μ, and ν_τ.

30.9 Conservation Laws &
30.10 Strange Particles and Strangeness

In all reactions and decays, quantities such as energy, linear momentum, angular momentum, electric charge, baryon number, and lepton number are strictly conserved. Certain particles have properties called **strangeness** and **charm**. These unusual properties are conserved only in those reactions and decays that occur via the strong force.

30.12 Quarks &
30.13 Colored Quarks

Recent theories postulate that all hadrons are composed of smaller units known as **quarks** which have fractional electric charges and baryon numbers of 1/3 and come in six "flavors": up, down, strange, charmed, top, and bottom. Each baryon contains three quarks, and each meson contains one quark and one antiquark.

According to the theory of **quantum chromodynamics**, quarks have a property called **color**, and the strong force between quarks is referred to as the **color force**. The color force increases as the distance between particles increases, so quarks are confined and are never observed in isolation. When two bound quarks are widely separated, a new quark–antiquark pair forms between them, and the single particle breaks into two new particles, each composed of a quark–antiquark pair.

30.15 The Cosmic Connection

Observation of background microwave radiation by Penzias and Wilson strongly confirmed that the Universe started with a Big Bang about 15 billion years ago and has been expanding ever since. The background radiation is equivalent to that of a blackbody at a temperature of about 3 K.

The cosmic microwave background has very small irregularities, corresponding to temperature variations of 0.000 3 K. Without these irregularities acting as nucleation sites, particles would never have clumped together to form galaxies and stars.

CONCEPTUAL QUESTIONS

1. If high-energy electrons with de Broglie wavelengths smaller than the size of the nucleus are scattered from nuclei, the behavior of the electrons is consistent with scattering from very massive structures much smaller in size than the nucleus, namely, quarks. How is this similar to a classic experiment that detected small structures in an atom?

2. What factors make a fusion reaction difficult to achieve?

3. Doubly charged baryons are known to exist. Why are there no doubly charged mesons?

4. Why would a fusion reactor produce less radioactive waste than a fission reactor?

5. Atoms didn't exist until hundreds of thousands of years after the Big Bang. Why?

6. Particles known as resonances have very short half-lives, on the order of 10^{-23} s. Would you guess they are hadrons or leptons?

7. Describe the quark model of hadrons, including the properties of quarks.

8. In the theory of quantum chromodynamics, quarks come in three colors. How would you justify the statement "All baryons and mesons are colorless?"

9. Describe the properties of baryons and mesons and the important differences between them.

10. Identify the particle decays in Table 30.2 that occur by the electromagnetic interaction. Justify your answer.

11. Kaons all decay into final states that contain no protons or neutrons. What is the baryon number of kaons?

12. When an electron and a positron meet at low speeds in free space, why are *two* 0.511-MeV gamma rays produced, rather than *one* gamma ray with an energy of 1.02 MeV?

13. Two protons in a nucleus interact via the strong interaction. Are they also subject to a weak interaction?

14. Why is a neutron stable inside the nucleus? (In free space, the neutron decays in 900 s.)

15. An antibaryon interacts with a meson. Can a baryon be produced in such an interaction? Explain.

16. Why is water a better shield against neutrons than lead or steel is?

17. How many quarks are there in (a) a baryon, (b) an antibaryon, (c) a meson, and (d) an antimeson? How do you account for the fact that baryons have half-integral spins and mesons have spins of 0 or 1? [*Hint:* quarks have spin $\frac{1}{2}$.]

18. A typical chemical reaction is one in which a water molecule is formed by combining hydrogen and oxygen. In such a reaction, about 2.5 eV of energy is released. Compare this reaction to a nuclear event such as $_0^1 n + {}_{92}^{235} U \rightarrow {}_{53}^{136} I + {}_{39}^{98} Y + 2 {}_0^1 n$. Would you expect the energy released in this nuclear event to be much greater, much less, or about the same as that released in the chemical reaction? Explain.

19. The neutral ρ meson decays by the strong interaction into two pions according to $\rho^0 \rightarrow \pi^+ + \pi^-$, with a half-life of about 10^{-23} s. The neutral K meson also decays into two pions according to $K^0 \rightarrow \pi^+ + \pi^-$, but with a much longer half-life of about 10^{-10} s. How do you explain these observations?

PROBLEMS

1, 2, 3 = straightforward, intermediate, challenging   □ = full solution available in *Student Solutions Manual/Study Guide*

Physics⊗Now™ = coached problem with hints available at **www.cp7e.com**   🔬 = biomedical application

Section 30.1 Nuclear Fission
Section 30.2 Nuclear Reactors

1. If the average energy released in a fission event is 208 MeV, find the total number of fission events required to operate a 100-W lightbulb for 1.0 h.

2. Find the energy released in the fission reaction

$$n + {}_{92}^{235} U \quad \rightarrow \quad {}_{40}^{98} Zr + {}_{52}^{135} Te + 3n$$

The atomic masses of the fission products are 97.912 0 u for $_{40}^{98}$Zr and 134.908 7 u for $_{52}^{135}$ Te.

3. Find the energy released in the following fission reaction:

$$_0^1 n + {}_{92}^{235} U \quad \rightarrow \quad {}_{38}^{88} Sr + {}_{54}^{136} Xe + 12 {}_0^1 n$$

4. Strontium-90 is a particularly dangerous fission product of ^{235}U because it is radioactive and it substitutes for calcium in bones. What other direct fission products would accompany it in the neutron-induced fission of ^{235}U? [*Note*: This reaction may release two, three, or four free neutrons.]

5. Assume that ordinary soil contains natural uranium in amounts of 1 part per million by mass. (a) How much uranium is in the top 1.00 meter of soil on a 1-acre (43 560-ft²) plot of ground, assuming the specific gravity of soil is 4.00? (b) How much of the isotope ^{235}U, appropriate for nuclear reactor fuel, is in this soil? [*Hint*: See Appendix B for the percent abundance of $_{92}^{235}$U.]

6. A typical nuclear fission power plant produces about 1.00 GW of electrical power. Assume that the plant has an overall efficiency of 40.0% and

that each fission produces 200 MeV of thermal energy. Calculate the mass of ^{235}U consumed each day.

7. Physics⊗Now™ Suppose that the water exerts an average frictional drag of 1.0×10^5 N on a nuclear-powered ship. How far can the ship travel per kilogram of fuel if the fuel consists of enriched uranium containing 1.7% of the fissionable isotope ^{235}U and the ship's engine has an efficiency of 20%? (Assume 208 MeV is released per fission event.)

8. It has been estimated that the Earth contains 1.0×10^9 tons of natural uranium that can be mined economically. If all the world's energy needs $(7.0 \times 10^{12}$ J/s$)$ were supplied by ^{235}U fission, how long would this supply last? [*Hint*: See Appendix B for the percent abundance of $_{92}^{235}$U.]

9. An all-electric home uses approximately 2 000 kWh of electric energy per month. How much uranium-235 would be required to provide this house with its energy needs for 1 year? (Assume 100% conversion efficiency and 208 MeV released per fission.)

Section 30.3 Nuclear Fusion

10. Find the energy released in the fusion reaction

$$_1^1 H + {}_1^2 H \quad \rightarrow \quad {}_2^3 He + \gamma$$

11. When a star has exhausted its hydrogen fuel, it may fuse other nuclear fuels. At temperatures above 1.0×10^8 K, helium fusion can occur. Write the equations for the following processes: (a) Two alpha

particles fuse to produce a nucleus A and a gamma ray. What is nucleus A? (b) Nucleus A absorbs an alpha particle to produce a nucleus B and a gamma ray. What is nucleus B? (c) Find the total energy released in the reactions given in (a) and (b). [*Note*: The mass of ${}^{8}_{4}\text{Be} = 8.005\ 305$ u.]

12. Another series of nuclear reactions that can produce energy in the interior of stars is the cycle described below. This cycle is most efficient when the central temperature in a star is above 1.6×10^7 K. Because the temperature at the center of the Sun is only 1.5×10^7 K, the following cycle produces less than 10% of the Sun's energy. (a) A high-energy proton is absorbed by ${}^{12}\text{C}$. Another nucleus, A, is produced in the reaction, along with a gamma ray. Identify nucleus A. (b) Nucleus A decays through positron emission to form nucleus B. Identify nucleus B. (c) Nucleus B absorbs a proton to produce nucleus C and a gamma ray. Identify nucleus C. (d) Nucleus C absorbs a proton to produce nucleus D and a gamma ray. Identify nucleus D. (e) Nucleus D decays through positron emission to produce nucleus E. Identify nucleus E. (f) Nucleus E absorbs a proton to produce nucleus F plus an alpha particle. What is nucleus F? [*Note*: If nucleus F is not ${}^{12}\text{C}$—that is, the nucleus you started with—you have made an error and should review the sequence of events.]

13. If an all-electric home uses approximately 2 000 kWh of electric energy per month, how many fusion events described by the reaction ${}^{2}_{1}\text{H} + {}^{3}_{1}\text{H} \rightarrow {}^{4}_{2}\text{He} + {}^{1}_{0}\text{n}$ would be required to keep this home running for one year?

14. To understand why plasma containment is necessary, consider the rate at which an unconfined plasma would be lost. (a) Estimate the rms speed of deuterons in a plasma at 4.00×10^8 K. (b) Estimate the order of magnitude of the time such a plasma would remain in a 10-cm cube if no steps were taken to contain it.

15. The oceans have a volume of 317 million cubic miles and contain 1.32×10^{21} kg of water. Of all the hydrogen nuclei in this water, 0.030 0% of the mass is deuterium. (a) If all of these deuterium nuclei were fused to helium via the first reaction in Equation 30.4, determine the total amount of energy that could be released. (b) The present world electric power consumption is about 7.00×10^{12} W. If consumption were 100 times greater, how many years would the energy supply calculated in part (a) last?

Section 30.6 Positrons and Other Antiparticles

16. Two photons are produced when a proton and an antiproton annihilate each other. What is the minimum frequency and corresponding wavelength of each photon?

17. **Physics⊗Now™** A photon produces a proton–antiproton pair according to the reaction $\gamma \rightarrow \text{p} + \bar{\text{p}}$. What is the minimum possible frequency of the photon? What is its wavelength?

18. A photon with an energy of 2.09 GeV creates a proton–antiproton pair in which the proton has a kinetic energy of 95.0 MeV. What is the kinetic energy of the antiproton?

Section 30.7 Mesons and the Beginning of Particle Physics

19. When a high-energy proton or pion traveling near the speed of light collides with a nucleus, it travels an average distance of 3.0×10^{-15} m before interacting with another particle. From this information, estimate the time for the strong interaction to occur.

20. Calculate the order of magnitude of the range of the force that might be produced by the virtual exchange of a proton.

21. One of the mediators of the weak interaction is the Z^0 boson, which has a mass of 96 GeV/c^2. Use this information to find an approximate value for the range of the weak interaction.

22. If a π^0 at rest decays into two γ's, what is the energy of each of the γ's?

Section 30.9 Conservation Laws
Section 30.10 Strange Particles and Strangeness

23. Each of the following reactions is forbidden. Determine a conservation law that is violated for each reaction.
    (a) $\text{p} + \bar{\text{p}} \rightarrow \mu^+ + e^-$
    (b) $\pi^- + \text{p} \rightarrow \text{p} + \pi^+$
    (c) $\text{p} + \text{p} \rightarrow \text{p} + \pi^+$
    (d) $\text{p} + \text{p} \rightarrow \text{p} + \text{p} + \text{n}$
    (e) $\gamma + \text{p} \rightarrow \text{n} + \pi^0$

24. For the following two reactions, the first may occur but the second cannot. Explain.

$$K^0 \rightarrow \pi^+ + \pi^- \quad \text{(can occur)}$$
$$\Lambda^0 \rightarrow \pi^+ + \pi^- \quad \text{(cannot occur)}$$

25. Physics⊗Now™ Identify the unknown particle on the left side of the reaction

$$? + p \rightarrow n + \mu^+$$

26. Determine the type of neutrino or antineutrino involved in each of the following processes:
(a) $\pi^+ \rightarrow \pi^0 + e^+ + ?$
(b) $? + p \rightarrow \mu^- + p + \pi^+$
(c) $\Lambda^0 \rightarrow p + \mu^- + ?$
(d) $\tau^+ \rightarrow \mu^+ + ? + ?$

27. The following reactions or decays involve one or more neutrinos. Supply the missing neutrinos.
(a) $\pi^- \rightarrow \mu^- + ?$
(b) $K^+ \rightarrow \mu^+ + ?$
(c) $? + p \rightarrow n + e^+$
(d) $? + n \rightarrow p + e^-$
(e) $? + n \rightarrow p + \mu^-$
(f) $\mu^- \rightarrow e^- + ? + ?$

28. Determine which of the reactions below can occur. For those that cannot occur, determine the conservation law (or laws) that each violates:
(a) $p \rightarrow \pi^+ + \pi^0$
(b) $p + p \rightarrow p + p + \pi^0$
(c) $p + p \rightarrow p + \pi^+$
(d) $\pi^+ \rightarrow \mu^+ + \nu_\mu$
(e) $n \rightarrow p + e^- + \bar{\nu}_e$
(f) $\pi^+ \rightarrow \pi^+ + n$

29. Which of the following processes are allowed by the strong interaction, the electromagnetic interaction, the weak interaction, or no interaction at all?
(a) $\pi^- + p \rightarrow 2\eta^0$
(b) $K^- + n \rightarrow \Lambda^0 + \pi^-$
(c) $K^- \rightarrow \pi^- + \pi^0$
(d) $\Omega^- \rightarrow \Xi^- + \pi^0$
(e) $\eta^0 \rightarrow 2\gamma$

30. A K^0 particle at rest decays into a π^+ and a π^-. What will be the speed of each of the pions? The mass of the K^0 is 497.7 MeV/c^2 and the mass of each pion is 139.6 MeV/c^2.

31. Determine whether or not strangeness is conserved in the following decays and reactions:
(a) $\Lambda^0 \rightarrow p + \pi^-$
(b) $\pi^- + p \rightarrow \Lambda^0 + K^0$
(c) $\bar{p} + p \rightarrow \bar{\Lambda}^0 + \Lambda^0$

(d) $\pi^- + p \rightarrow \pi^- + \Sigma^+$
(e) $\Xi^- \rightarrow \Lambda^0 + \pi^-$
(f) $\Xi^0 \rightarrow p + \pi^-$

32. Fill in the missing particle. Assume that (a) occurs via the strong interaction while (b) and (c) involve the weak interaction.
(a) $K^+ + p \rightarrow \underline{\quad} + p$
(b) $\Omega^- \rightarrow \underline{\quad} + \pi^-$
(c) $K^+ \rightarrow \underline{\quad} + \mu^+ + \nu_\mu$

33. Identify the conserved quantities in the following processes:
(a) $\Xi^- \rightarrow \Lambda^0 + \mu^- + \nu_\mu$
(b) $K^0 \rightarrow 2\pi^0$
(c) $K^- + p \rightarrow \Sigma^0 + n$
(d) $\Sigma^0 \rightarrow \Lambda^0 + \gamma$
(e) $e^+ + e^- \rightarrow \mu^+ + \mu^-$
(f) $\bar{p} + n \rightarrow \bar{\Lambda}^0 + \Sigma^-$

Section 30.12 Quarks
Section 30.13 Colored Quarks

34. The quark composition of the proton is uud, while that of the neutron in udd. Show that the charge, baryon number, and strangeness of these particles equal the sums of these numbers for their quark constituents.

35. Find the number of electrons, and of each species of quark, in 1 L of water.

36. The quark compositions of the K^0 and Λ^0 particles are $d\bar{s}$ and uds, respectively. Show that the charge, baryon number, and strangeness of these particles equal the sums of these numbers for the quark constituents.

37. Identify the particles corresponding to the following quark states: (a) suu; (b) $\bar{u}$d; (c) $\bar{s}$d; (d) ssd.

38. What is the electrical charge of the baryons with the quark compositions (a) $\bar{u}\bar{u}$d and (b) $\bar{u}$d$\bar{d}$? What are these baryons called?

39. Analyze the first three of the following reactions at the quark level, and show that each conserves the net number of each type of quark; then, in the last reaction, identify the mystery particle:
(a) $\pi^- + p \rightarrow K^0 + \Lambda^0$
(b) $\pi^+ + p \rightarrow K^+ + \Sigma^+$
(c) $K^- + p \rightarrow K^+ + K^0 + \Omega^-$
(d) $p + p \rightarrow K^0 + p + \pi^+ + ?$

40. **Physics⊗Now™** Assume binding energies can be neglected. Find the masses of the u and d quarks from the masses of the proton and neutron.

ADDITIONAL PROBLEMS

41. A Σ^0 particle traveling through matter strikes a proton and a Σ^+, and a gamma ray, as well as a third particle, emerges. Use the quark model of each to determine the identity of the third particle.

42. It was stated in the text that the reaction $\pi^- + p^+ \rightarrow K^0 + \Lambda^0$ occurs with high probability, whereas the reaction $\pi^- + p^+ \rightarrow K^0 + n$ never occurs. Analyze these reactions at the quark level and show that the first conserves the net number of each type of quark while the second does not.

43. Two protons approach each other with equal and opposite velocities. Find the minimum kinetic energy of each of the protons if they are to produce a π^+ meson at rest in the reaction

$$p + p \rightarrow p + n + \pi^+$$

44. Name at least one conservation law that prevents each of the following reactions from occurring:
(a) $\pi^- + p \rightarrow \Sigma^+ + \pi^0$
(b) $\mu^- \rightarrow \pi^- + \nu_e$
(c) $p \rightarrow \pi^+ + \pi^+ + \pi^-$

45. Find the energy released in the fusion reaction

$$H + {}_2^3He \rightarrow {}_2^4He + e^+ + \nu$$

46. Occasionally, high-energy muons collide with electrons and produce two neutrinos according to the reaction $\mu^+ + e^- \rightarrow 2\nu$. What kind of neutrinos are these?

47. Each of the following decays is forbidden. For each process, determine a conservation law that is violated:
(a) $\mu^- \rightarrow e^- + \gamma$
(b) $n \rightarrow p + e^- + \nu_e$
(c) $\Lambda^0 \rightarrow p + \pi^0$
(d) $p \rightarrow e^+ + \pi^0$
(e) $\Xi^0 \rightarrow n + \pi^0$

48. Two protons approach each other with 70.4 MeV of kinetic energy and engage in a reaction in which a proton and a positive pion emerge at rest. What third particle, obviously uncharged and therefore difficult to detect, must have been created?

49. The atomic bomb dropped on Hiroshima on August 6, 1945, released 5×10^{13} J of energy (equivalent to that from 12 000 tons of TNT). Estimate (a) the number of ${}_{92}^{235}U$ nuclei fissioned and (b) the mass of this ${}_{92}^{235}U$.

50. A Σ^0 particle at rest decays according to $\Sigma^0 \rightarrow \Lambda^0 + \gamma$. Find the gamma-ray energy. [*Hint*: remember to conserve momentum.]

51. If baryon number is not conserved, then one possible mechanism by which a proton can decay is

$$p \rightarrow e^+ + \gamma$$

Show that this reaction violates the conservation of baryon number. (b) Assuming that the reaction occurs and that the proton is initially at rest, determine the energy and momentum of the photon after the reaction. [*Hint*: recall that energy and momentum must be conserved in the reaction.] (c) Determine the speed of the positron after the reaction.

52. Classical general relativity views the space–time manifold as a deterministic structure completely well defined down to arbitrarily small distances. On the other hand, quantum general relativity forbids distances smaller than the Planck length $L = (\hbar G/c^3)^{1/2}$. (a) Calculate the value of L. The answer suggests that, after the Big Bang (when all the known Universe was reduced to a singularity), nothing could be observed until that singularity grew larger than the Planck length, L. Since the size of the singularity grew at the speed of light, we can infer that during the time it took for light to travel the Planck length, no observations were possible. (b) Determine this time (known as the Planck time T), and compare it to the ultra-hot epoch discussed in the text. (c) Does your answer to part (b) suggest that we may never know what happened between the time $t = 0$ and the time $t = T$?

53. (a) Show that about 1.0×10^{10} J would be released by the fusion of the deuterons in 1.0 gal of water. Note that 1 out of every 6 500 hydrogen atoms is a deuteron. (b) The average energy consumption rate of a person living in the United States is about 1.0×10^4 J/s (an average power of 10 kW). At this rate, how long would the energy needs of one person be supplied by the fusion of the deuterons in 1.0 gal of water? Assume that the energy released per deuteron is 1.64 MeV.

54. Calculate the mass of ${}^{235}U$ required to provide the total energy requirements of a nuclear submarine

during a 100-day patrol, assuming a constant power demand of 100 000 kW, a conversion efficiency of 30%, and an average energy released per fission of 208 MeV.

55. A 2.0-MeV neutron is emitted in a fission reactor. If it loses one-half of its kinetic energy in each collision with a moderator atom, how many collisions must it undergo in order to achieve thermal energy (0.039 eV)?

"QUARKS. NEUTRINOS. MESONS. ALL THOSE DAMN PARTICLES YOU CAN'T SEE. THAT'S WHAT DROVE ME TO DRINK. BUT NOW I CAN SEE THEM."

Mathematical Review

A.1 MATHEMATICAL NOTATION

Many mathematical symbols are used throughout this book. You are no doubt familiar with some, such as the symbol $=$ to denote the equality of two quantities.

The symbol $\propto$ denotes a proportionality. For example, $y \propto x^2$ means that y is proportional to the square of x.

The symbol $<$ means *is less than,* and $>$ means *is greater than.* For example, $x > y$ means x is greater than y.

The symbol $\ll$ means *is much less than,* and $\gg$ means *is much greater than.*

The symbol $\approx$ indicates that two quantities are *approximately equal* to each other.

The symbol $\equiv$ means *is defined as.* This is a stronger statement than a simple $=$.

It is convenient to use the notation Δx (read as "delta x") to indicate the *change in the quantity x.* (Note that Δx does not mean "the product of Δ and x.") For example, suppose that a person out for a morning stroll starts measuring her distance away from home when she is 10 m from her doorway. She then moves along a straight-line path and stops strolling 50 m from the door. Her change in position during the walk is $\Delta x = 50$ m $- 10$ m $= 40$ m or, in symbolic form,

$$\Delta x = x_f - x_i$$

In this equation x_f is the *final position* and x_i is the *initial position.*

We often have occasion to add several quantities. A useful abbreviation for representing such a sum is the Greek letter Σ (capital sigma). Suppose we wish to add a set of five numbers represented by x_1, x_2, x_3, x_4, and x_5. In the abbreviated notation, we would write the sum as

$$x_1 + x_2 + x_3 + x_4 + x_5 = \sum_{i=1}^{5} x_i$$

where the subscript i on x represents any one of the numbers in the set. For example, if there are five masses in a system, m_1, m_2, m_3, m_4, and m_5, the total mass of the system $M = m_1 + m_2 + m_3 + m_4 + m_5$ could be expressed as

$$M = \sum_{i=1}^{5} m_i$$

Finally, the magnitude of a quantity x, written $|x|$, is simply the absolute value of that quantity. The sign of $|x|$ is always positive, regardless of the sign of x. For example, if $x = -5$, $|x| = 5$; if $x = 8$, $|x| = 8$.

A.2 SCIENTIFIC NOTATION

Many quantities that scientists deal with often have very large or very small values. For example, the speed of light is about 300 000 000 m/s and the ink required to make the dot over an i in this textbook has a mass of about 0.000 000 001 kg. Obviously, it is cumbersome to read, write, and keep track of numbers such as these. We avoid this problem by using a method dealing with powers of the number 10:

$$10^0 = 1$$
$$10^1 = 10$$
$$10^2 = 10 \times 10 = 100$$
$$10^3 = 10 \times 10 \times 10 = 1\,000$$
$$10^4 = 10 \times 10 \times 10 \times 10 = 10\,000$$
$$10^5 = 10 \times 10 \times 10 \times 10 \times 10 = 100\,000$$

and so on. The number of zeros corresponds to the power to which 10 is raised, called the **exponent** of 10. For example, the speed of light, 300 000 000 m/s, can be expressed as 3×10^8 m/s.

For numbers less than one, we note the following:

$$10^{-1} = \frac{1}{10} = 0.1$$

$$10^{-2} = \frac{1}{10 \times 10} = 0.01$$

$$10^{-3} = \frac{1}{10 \times 10 \times 10} = 0.001$$

$$10^{-4} = \frac{1}{10 \times 10 \times 10 \times 10} = 0.000\ 1$$

$$10^{-5} = \frac{1}{10 \times 10 \times 10 \times 10 \times 10} = 0.000\ 01$$

In these cases, the number of places the decimal point is to the left of the digit 1 equals the value of the (negative) exponent. Numbers that are expressed as some power of 10 multiplied by another number between 1 and 10 are said to be in **scientific notation.** For example, the scientific notation for 5 943 000 000 is 5.943×10^9 and that for 0.000 083 2 is 8.32×10^{-5}.

When numbers expressed in scientific notation are being multiplied, the following general rule is very useful:

$$10^n \times 10^m = 10^{n+m} \qquad \text{[A.1]}$$

where n and m can be *any* numbers (not necessarily integers). For example, $10^2 \times 10^5 = 10^7$. The rule also applies if one of the exponents is negative. For example, $10^3 \times 10^{-8} = 10^{-5}$.

When dividing numbers expressed in scientific notation, note that

$$\frac{10^n}{10^m} = 10^n \times 10^{-m} = 10^{n-m} \qquad \text{[A.2]}$$

EXERCISES

With help from the above rules, verify the answers to the following:

1. $86\ 400 = 8.64 \times 10^4$
2. $9\ 816\ 762.5 = 9.816\ 762\ 5 \times 10^6$
3. $0.000\ 000\ 039\ 8 = 3.98 \times 10^{-8}$
4. $(4.0 \times 10^8)(9.0 \times 10^9) = 3.6 \times 10^{18}$
5. $(3.0 \times 10^7)(6.0 \times 10^{-12}) = 1.8 \times 10^{-4}$
6. $\dfrac{75 \times 10^{-11}}{5.0 \times 10^{-3}} = 1.5 \times 10^{-7}$
7. $\dfrac{(3 \times 10^6)(8 \times 10^{-2})}{(2 \times 10^{17})(6 \times 10^5)} = 2 \times 10^{-18}$

A.3   ALGEBRA

A. Some Basic Rules

When algebraic operations are performed, the laws of arithmetic apply. Symbols such as x, y, and z are frequently used to represent quantities that are not specified, what are called the **unknowns.**

First, consider the equation

$$8x = 32$$

If we wish to solve for x, we can divide (or multiply) each side of the equation by the same factor without destroying the equality. In this case, if we divide both sides by 8, we have

$$\frac{8x}{8} = \frac{32}{8}$$

$$x = 4$$

Next consider the equation

$$x + 2 = 8$$

In this type of expression, we can add or subtract the same quantity from each side. If we subtract 2 from each side, we get

$$x + 2 - 2 = 8 - 2$$

$$x = 6$$

In general, if $x + a = b$, then $x = b - a$.

Now consider the equation

$$\frac{x}{5} = 9$$

If we multiply each side by 5, we are left with x on the left by itself and 45 on the right:

$$\left(\frac{x}{5}\right)(5) = 9 \times 5$$

$$x = 45$$

In all cases, **whatever operation is performed on the left side of the equality must also be performed on the right side.**

The following rules for multiplying, dividing, adding, and subtracting fractions should be recalled, where a, b, and c are three numbers:

	Rule	Example
Multiplying	$\left(\frac{a}{b}\right)\left(\frac{c}{d}\right) = \frac{ac}{bd}$	$\left(\frac{2}{3}\right)\left(\frac{4}{5}\right) = \frac{8}{15}$
Dividing	$\frac{(a/b)}{(c/d)} = \frac{ad}{bc}$	$\frac{2/3}{4/5} = \frac{(2)(5)}{(4)(3)} = \frac{10}{12}$
Adding	$\frac{a}{b} \pm \frac{c}{d} = \frac{ad \pm bc}{bd}$	$\frac{2}{3} - \frac{4}{5} = \frac{(2)(5) - (4)(3)}{(3)(5)} = -\frac{2}{15}$

EXERCISES

In the following exercises, solve for x:

ANSWERS

1. $a = \dfrac{1}{1 + x}$ $x = \dfrac{1 - a}{a}$

2. $3x - 5 = 13$ $x = 6$

3. $ax - 5 = bx + 2$ $x = \dfrac{7}{a - b}$

4. $\dfrac{5}{2x + 6} = \dfrac{3}{4x + 8}$ $x = -\dfrac{11}{7}$

B. Powers

When powers of a given quantity x are multiplied, the following rule applies:

$$x^n x^m = x^{n+m} \tag{A.3}$$

For example, $x^2 x^4 = x^{2+4} = x^6$.

When dividing the powers of a given quantity, note that

$$\frac{x^n}{x^m} = x^{n-m} \qquad \text{[A.4]}$$

For example, $x^8/x^2 = x^{8-2} = x^6$.

A power that is a fraction, such as $\frac{1}{3}$, corresponds to a root as follows:

$$x^{1/n} = \sqrt[n]{x} \qquad \text{[A.5]}$$

For example, $4^{1/3} = \sqrt[3]{4} = 1.587\,4$. (A scientific calculator is useful for such calculations.)

Finally, any quantity x^n that is raised to the mth power is

$$(x^n)^m = x^{nm} \qquad \text{[A.6]}$$

TABLE A.1

Rules of Exponents

$x^0 = 1$

$x^1 = x$

$x^n x^m = x^{n+m}$

$x^n/x^m = x^{n-m}$

$x^{1/n} = \sqrt[n]{x}$

$(x^n)^m = x^{nm}$

Table A.1 summarizes the rules of exponents.

EXERCISES

Verify the following:

1. $3^2 \times 3^3 = 243$
2. $x^5 x^{-8} = x^{-3}$
3. $x^{10}/x^{-5} = x^{15}$
4. $5^{1/3} = 1.709\,975$ (Use your calculator.)
5. $60^{1/4} = 2.783\,158$ (Use your calculator.)
6. $(x^4)^3 = x^{12}$

C. Factoring

Some useful formulas for factoring an equation are

$$ax + ay + az = a(x + y + z) \qquad \text{common factor}$$
$$a^2 + 2ab + b^2 = (a + b)^2 \qquad \text{perfect square}$$
$$a^2 - b^2 = (a + b)(a - b) \qquad \text{differences of squares}$$

D. Quadratic Equations

The general form of a quadratic equation is

$$ax^2 + bx + c = 0 \qquad \text{[A.7]}$$

where x is the unknown quantity and a, b, and c are numerical factors referred to as **coefficients** of the equation. This equation has two roots, given by

$$x = \frac{-b \pm \sqrt{b^2 - 4ac}}{2a} \qquad \text{[A.8]}$$

If $b^2 \geq 4ac$, the roots will be real.

EXAMPLE

The equation $x^2 + 5x + 4 = 0$ has the following roots corresponding to the two signs of the square root term:

$$x = \frac{-5 \pm \sqrt{5^2 - (4)(1)(4)}}{2(1)} = \frac{-5 \pm \sqrt{9}}{2} = \frac{-5 \pm 3}{2}$$

that is,

$$x_+ = \frac{-5 + 3}{2} = -1 \qquad x_- = \frac{-5 - 3}{2} = -4$$

where x_+ refers to the root corresponding to the positive sign and x_- refers to the root corresponding to the negative sign.

EXERCISES
Solve the following quadratic equations:

ANSWERS

1. $x^2 + 2x - 3 = 0$	$x_+ = 1$	$x_- = -3$
2. $2x^2 - 5x + 2 = 0$	$x_+ = 2$	$x_- = 1/2$
3. $2x^2 - 4x - 9 = 0$	$x_+ = 1 + \sqrt{22}/2$	$x_- = 1 - \sqrt{22}/2$

E. Linear Equations

A linear equation has the general form

$$y = ax + b \qquad \text{[A.9]}$$

where a and b are constants. This equation is referred to as being linear because the graph of y versus x is a straight line, as shown in Figure A.1. The constant b, called the **intercept**, represents the value of y at which the straight line intersects the y axis. The constant a is equal to the **slope** of the straight line. If any two points on the straight line are specified by the coordinates (x_1, y_1) and (x_2, y_2), as in Figure A.1, then the slope of the straight line can be expressed

$$\text{Slope} = \frac{y_2 - y_1}{x_2 - x_1} = \frac{\Delta y}{\Delta x} \qquad \text{[A.10]}$$

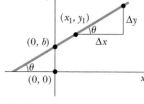

Figure A.1

Note that a and b can have either positive or negative values. If $a > 0$, the straight line has a positive slope, as in Figure A.1. If $a < 0$, the straight line has a *negative* slope. In Figure A.1, both a and b are positive. Three other possible situations are shown in Figure A.2: $a > 0$, $b < 0$; $a < 0$, $b > 0$; and $a < 0$, $b < 0$.

EXERCISES

1. Draw graphs of the following straight lines:
   (a) $y = 5x + 3$ **(b)** $y = -2x + 4$ **(c)** $y = -3x - 6$
2. Find the slopes of the straight lines described in Exercise 1.
   Answers: **(a)** 5   **(b)** -2 **(c)** -3
3. Find the slopes of the straight lines that pass through the following sets of points: **(a)** $(0, -4)$ and $(4, 2)$, **(b)** $(0, 0)$ and $(2, -5)$, and **(c)** $(-5, 2)$ and $(4, -2)$
   Answers: **(a)** $\frac{3}{2}$ **(b)** $-\frac{5}{2}$ **(c)** $-\frac{4}{9}$

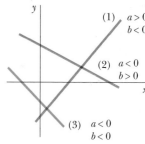

Figure A.2

F. Solving Simultaneous Linear Equations

Consider an equation such as $3x + 5y = 15$, which has two unknowns, x and y. Such an equation does not have a unique solution. That is, $(x = 0, y = 3)$, $(x = 5, y = 0)$ and $(x = 2, y = \frac{9}{5})$ are all solutions to this equation.

If a problem has two unknowns, a unique solution is possible only if we have *two* independent equations. In general, if a problem has n unknowns, its solution requires n independent equations. In order to solve two simultaneous equations involving two unknowns, x and y, we solve one of the equations for x in terms of y and substitute this expression into the other equation.

EXAMPLE

Solve the following two simultaneous equations:

$$\textbf{(1)} \quad 5x + y = -8 \qquad \textbf{(2)} \quad 2x - 2y = 4$$

Solution   From (2), we find that $x = y + 2$. Substitution of this into (1) gives

$$5(y + 2) + y = -8$$
$$6y = -18$$
$$y = -3$$
$$x = y + 2 = -1$$

Alternate Solution   Multiply each term in (1) by the factor 2 and add the result to (2):

$$10x + 2y = -16$$
$$\underline{2x - 2y = 4}$$
$$12x = -12$$
$$x = -1$$
$$y = x - 2 = -3$$

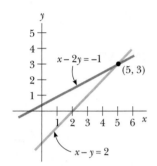

Figure A.3

Two linear equations with two unknowns can also be solved by a graphical method. If the straight lines corresponding to the two equations are plotted in a conventional coordinate system, the intersection of the two lines represents the solution. For example, consider the two equations

$$x - y = 2$$
$$x - 2y = -1$$

These are plotted in Figure A.3. The intersection of the two lines has the coordinates $x = 5$, $y = 3$. This represents the solution to the equations. You should check this solution by the analytical technique discussed above.

EXERCISES

Solve the following pairs of simultaneous equations involving two unknowns:

ANSWERS

1. $x + y = 8$ $x = 5, y = 3$
    $x - y = 2$
2. $98 - T = 10a$ $T = 65, a = 3.27$
    $T - 49 = 5a$
3. $6x + 2y = 6$ $x = 2, y = -3$
    $8x - 4y = 28$

G. Logarithms

Suppose that a quantity x is expressed as a power of some quantity a:

$$x = a^y \qquad \text{[A.11]}$$

The number a is called the **base** number. The **logarithm** of x with respect to the base a is equal to the exponent to which the base must be raised in order to satisfy the expression $x = a^y$:

$$y = \log_a x \qquad \text{[A.12]}$$

Conversely, the **antilogarithm** of y is the number x:

$$x = \text{antilog}_a y \qquad \text{[A.13]}$$

In practice, the two bases most often used are base 10, called the *common* logarithm base, and base $e = 2.718 \ldots$, called the *natural* logarithm base. When common logarithms are used,

$$y = \log_{10} x \qquad (\text{or } x = 10^y) \qquad \text{[A.14]}$$

When natural logarithms are used,

$$y = \ln_e x \qquad (\text{or } x = e^y) \qquad \qquad \textbf{[A.15]}$$

For example, $\log_{10} 52 = 1.716$, so that $\text{antilog}_{10} 1.716 = 10^{1.716} = 52$. Likewise, $\ln_e 52 = 3.951$, so $\text{antiln}_e 3.951 = e^{3.951} = 52$.

In general, note that you can convert between base 10 and base e with the equality

$$\ln_e x = (2.302\ 585)\log_{10} x \qquad \qquad \textbf{[A.16]}$$

Finally, some useful properties of logarithms are

$$\log(ab) = \log a + \log b \qquad \ln e = 1$$
$$\log(a/b) = \log a - \log b \qquad \ln e^a = a$$
$$\log(a^n) = n \log a \qquad \ln\left(\frac{1}{a}\right) = -\ln a$$

A.4 GEOMETRY

Table A.2 gives the areas and volumes for several geometric shapes used throughout this text:

TABLE A.2

Useful Information for Geometry

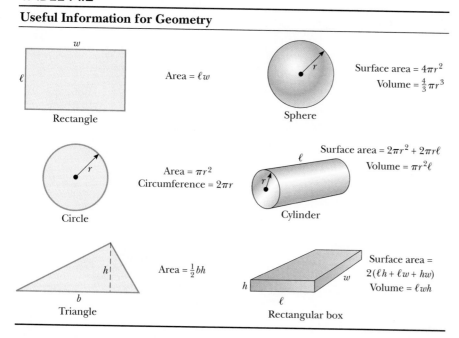

Rectangle — Area $= \ell w$

Sphere — Surface area $= 4\pi r^2$; Volume $= \frac{4}{3}\pi r^3$

Circle — Area $= \pi r^2$; Circumference $= 2\pi r$

Cylinder — Surface area $= 2\pi r^2 + 2\pi r\ell$; Volume $= \pi r^2 \ell$

Triangle — Area $= \frac{1}{2}bh$

Rectangular box — Surface area $= 2(\ell h + \ell w + hw)$; Volume $= \ell w h$

A.5 TRIGONOMETRY

Some of the most basic facts concerning trigonometry are presented in Chapter 1, and we encourage you to study the material presented there if you are having trouble with this branch of mathematics. In addition to the discussion of Chapter 1, certain useful trig identities that can be of value to you follow.

$$\sin^2\theta + \cos^2\theta = 1$$
$$\sin\theta = \cos(90° - \theta)$$
$$\cos\theta = \sin(90° - \theta)$$
$$\sin 2\theta = 2\sin\theta\cos\theta$$
$$\cos 2\theta = \cos^2\theta - \sin^2\theta$$
$$\sin(\theta \pm \phi) = \sin\theta\cos\phi \pm \cos\theta\sin\phi$$
$$\cos(\theta \pm \phi) = \cos\theta\cos\phi \mp \sin\theta\sin\phi$$

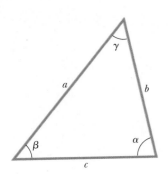

Figure A.4

The following relationships apply to *any* triangle, as shown in Figure A.4:

$$\alpha + \beta + \gamma = 180°$$

$$a^2 = b^2 + c^2 - 2bc \cos \alpha$$

Law of cosines $\quad b^2 = a^2 + c^2 - 2ac \cos \beta$

$$c^2 = a^2 + b^2 - 2ab \cos \gamma$$

Law of sines $\quad \dfrac{a}{\sin \alpha} = \dfrac{b}{\sin \beta} = \dfrac{c}{\sin \gamma}$

An Abbreviated Table of Isotopes

Atomic Number Z	Element	Symbol	Chemical Atomic Mass (u)	Mass Number (* Indicates Radioactive) A	Atomic Mass (u)	Percent Abundance	Half-Life (If Radioactive) $T_{1/2}$
0	(Neutron)	n		1*	1.008 665		10.4 min
1	Hydrogen	H	1.007 94	1	1.007 825	99.988 5	
	Deuterium	D		2	2.014 102	0.011 5	
	Tritium	T		3*	3.016 049		12.33 yr
2	Helium	He	4.002 602	3	3.016 029	0.000 137	
				4	4.002 603	99.999 863	
3	Lithium	Li	6.941	6	6.015 122	7.5	
				7	7.016 004	92.5	
4	Beryllium	Be	9.012 182	7*	7.016 929		53.3 days
				9	9.012 182	100	
5	Boron	B	10.811	10	10.012 937	19.9	
				11	11.009 306	80.1	
6	Carbon	C	12.010 7	10*	10.016 853		19.3 s
				11*	11.011 434		20.4 min
				12	12.000 000	98.93	
				13	13.003 355	1.07	
				14*	14.003 242		5 730 yr
7	Nitrogen	N	14.006 7	13*	13.005 739		9.96 min
				14	14.003 074	99.632	
				15	15.000 109	0.368	
8	Oxygen	O	15.999 4	15*	15.003 065		122 s
				16	15.994 915	99.757	
				18	17.999 160	0.205	
9	Fluorine	F	18.998 403 2	19	18.998 403	100	
10	Neon	Ne	20.179 7	20	19.992 440	90.48	
				22	21.991 385	9.25	
11	Sodium	Na	22.989 77	22*	21.994 437		2.61 yr
				23	22.989 770	100	
				24*	23.990 963		14.96 h
12	Magnesium	Mg	24.305 0	24	23.985 042	78.99	
				25	24.985 837	10.00	
				26	25.982 593	11.01	
13	Aluminum	Al	26.981 538	27	26.981 539	100	
14	Silicon	Si	28.085 5	28	27.976 926	92.229 7	
15	Phosphorus	P	30.973 761	31	30.973 762	100	
				32*	31.973 907		14.26 days
16	Sulfur	S	32.066	32	31.972 071	94.93	
				35*	34.969 032		87.5 days
17	Chlorine	Cl	35.452 7	35	34.968 853	75.78	
				37	36.965 903	24.22	
18	Argon	Ar	39.948	40	39.962 383	99.600 3	
19	Potassium	K	39.098 3	39	38.963 707	93.258 1	
				40*	39.963 999	0.011 7	1.28×10^9 yr

(Continued)

Atomic Number Z	Element	Symbol	Chemical Atomic Mass (u)	Mass Number (* Indicates Radioactive) A	Atomic Mass (u)	Percent Abundance	Half-Life (If Radioactive) $T_{1/2}$
20	Calcium	Ca	40.078	40	39.962 591	96.941	
21	Scandium	Sc	44.955 910	45	44.955 910	100	
22	Titanium	Ti	47.867	48	47.947 947	73.72	
23	Vanadium	V	50.941 5	51	50.943 964	99.750	
24	Chromium	Cr	51.996 1	52	51.940 512	83.789	
25	Manganese	Mn	54.938 049	55	54.938 050	100	
26	Iron	Fe	55.845	56	55.934 942	91.754	
27	Cobalt	Co	58.933 200	59	58.933 200	100	
				60*	59.933 822		5.27 yr
28	Nickel	Ni	58.693 4	58	57.935 348	68.076 9	
				60	59.930 790	26.223 1	
29	Copper	Cu	63.546	63	62.929 601	69.17	
				65	64.927 794	30.83	
30	Zinc	Zn	65.39	64	63.929 147	48.63	
				66	65.926 037	27.90	
				68	67.924 848	18.75	
31	Gallium	Ga	69.723	69	68.925 581	60.108	
				71	70.924 705	39.892	
32	Germanium	Ge	72.61	70	69.924 250	20.84	
				72	71.922 076	27.54	
				74	73.921 178	36.28	
33	Arsenic	As	74.921 60	75	74.921 596	100	
34	Selenium	Se	78.96	78	77.917 310	23.77	
				80	79.916 522	49.61	
35	Bromine	Br	79.904	79	78.918 338	50.69	
				81	80.916 291	49.31	
36	Krypton	Kr	83.80	82	81.913 485	11.58	
				83	82.914 136	11.49	
				84	83.911 507	57.00	
				86	85.910 610	17.30	
37	Rubidium	Rb	85.467 8	85	84.911 789	72.17	
				87*	86.909 184	27.83	4.75×10^{10} yr
38	Strontium	Sr	87.62	86	85.909 262	9.86	
				88	87.905 614	82.58	
				90*	89.907 738		29.1 yr
39	Yttrium	Y	88.905 85	89	88.905 848	100	
40	Zirconium	Zr	91.224	90	89.904 704	51.45	
				91	90.905 645	11.22	
				92	91.905 040	17.15	
				94	93.906 316	17.38	
41	Niobium	Nb	92.906 38	93	92.906 378	100	
42	Molybdenum	Mo	95.94	92	91.906 810	14.84	
				95	94.905 842	15.92	
				96	95.904 679	16.68	
				98	97.905 408	24.13	
43	Technetium	Tc		98*	97.907 216		4.2×10^6 yr
				99*	98.906 255		2.1×10^5 yr

Atomic Number Z	Element	Symbol	Chemical Atomic Mass (u)	Mass Number (* Indicates Radioactive) A	Atomic Mass (u)	Percent Abundance	Half-Life (If Radioactive) $T_{1/2}$
44	Ruthenium	Ru	101.07	99	98.905 939	12.76	
				100	99.904 220	12.60	
				101	100.905 582	17.06	
				102	101.904 350	31.55	
				104	103.905 430	18.62	
45	Rhodium	Rh	102.905 50	103	102.905 504	100	
46	Palladium	Pd	106.42	104	103.904 035	11.14	
				105	104.905 084	22.33	
				106	105.903 483	27.33	
				108	107.903 894	26.46	
				110	109.905 152	11.72	
47	Silver	Ag	107.868 2	107	106.905 093	51.839	
				109	108.904 756	48.161	
48	Cadmium	Cd	112.411	110	109.903 006	12.49	
				111	110.904 182	12.80	
				112	111.902 757	24.13	
				113*	112.904 401	12.22	9.3×10^{15} yr
				114	113.903 358	28.73	
49	Indium	In	114.818	115*	114.903 878	95.71	4.4×10^{14} yr
50	Tin	Sn	118.710	116	115.901 744	14.54	
				118	117.901 606	24.22	
				120	119.902 197	32.58	
51	Antimony	Sb	121.760	121	120.903 818	57.21	
				123	122.904 216	42.79	
52	Tellurium	Te	127.60	126	125.903 306	18.84	
				128*	127.904 461	31.74	$>8 \times 10^{24}$ yr
				130*	129.906 223	34.08	$\leq 1.25 \times 10^{21}$ yr
53	Iodine	I	126.904 47	127	126.904 468	100	
				129*	128.904 988		1.6×10^{7} yr
54	Xenon	Xe	131.29	129	128.904 780	26.44	
				131	130.905 082	21.18	
				132	131.904 145	26.89	
				134	133.905 394	10.44	
				136*	135.907 220	8.87	$\geq 2.36 \times 10^{21}$ yr
55	Cesium	Cs	132.905 45	133	132.905 447	100	
56	Barium	Ba	137.327	137	136.905 821	11.232	
				138	137.905 241	71.698	
57	Lanthanum	La	138.905 5	139	138.906 349	99.910	
58	Cerium	Ce	140.116	140	139.905 434	88.450	
				142*	141.909 240	11.114	$>5 \times 10^{16}$ yr
59	Praseodymium	Pr	140.907 65	141	140.907 648	100	
60	Neodymium	Nd	144.24	142	141.907 719	27.2	
				144*	143.910 083	23.8	2.3×10^{15} yr
				146	145.913 112	17.2	
61	Promethium	Pm		145*	144.912 744		17.7 yr
62	Samarium	Sm	150.36	147*	146.914 893	14.99	1.06×10^{11} yr
				149*	148.917 180	13.82	$>2 \times 10^{15}$ yr
				152	151.919 728	26.75	
				154	153.922 205	22.75	

(Continued)

Atomic Number Z	Element	Symbol	Chemical Atomic Mass (u)	Mass Number (* Indicates Radioactive) A	Atomic Mass (u)	Percent Abundance	Half-Life (If Radioactive) $T_{1/2}$
63	Europium	Eu	151.964	151	150.919 846	47.81	
				153	152.921 226	52.19	
64	Gadolinium	Gd	157.25	156	155.922 120	20.47	
				158	157.924 100	24.84	
				160	159.927 051	21.86	
65	Terbium	Tb	158.925 34	159	158.925 343	100	
66	Dysprosium	Dy	162.50	162	161.926 796	25.51	
				163	162.928 728	24.90	
				164	163.929 171	28.18	
67	Holmium	Ho	164.930 32	165	164.930 320	100	
68	Erbium	Er	167.6	166	165.930 290	33.61	
				167	166.932 045	22.93	
				168	167.932 368	26.78	
69	Thulium	Tm	168.934 21	169	168.934 211	100	
70	Ytterbium	Yb	173.04	172	171.936 378	21.83	
				173	172.938 207	16.13	
				174	173.938 858	31.83	
71	Lutecium	Lu	174.967	175	174.940 768	97.41	
72	Hafnium	Hf	178.49	177	176.943 220	18.60	
				178	177.943 698	27.28	
				179	178.945 815	13.62	
				180	179.946 549	35.08	
73	Tantalum	Ta	180.947 9	181	180.947 996	99.988	
74	Tungsten (Wolfram)	W	183.84	182	181.948 206	26.50	
				183	182.950 224	14.31	
				184*	183.950 933	30.64	$>3 \times 10^{17}$ yr
				186	185.954 362	28.43	
75	Rhenium	Re	186.207	185	184.952 956	37.40	
				187*	186.955 751	62.60	4.4×10^{10} yr
76	Osmium	Os	190.23	188	187.955 836	13.24	
				189	188.958 145	16.15	
				190	189.958 445	26.26	
				192	191.961 479	40.78	
77	Iridium	Ir	192.217	191	190.960 591	37.3	
				193	192.962 924	62.7	
78	Platinum	Pt	195.078	194	193.962 664	32.967	
				195	194.964 774	33.832	
				196	195.964 935	25.242	
79	Gold	Au	196.966 55	197	196.966 552	100	
80	Mercury	Hg	200.59	199	198.968 262	16.87	
				200	199.968 309	23.10	
				201	200.970 285	13.18	
				202	201.970 626	29.86	
81	Thallium	Tl	204.383 3	203	202.972 329	29.524	
				205	204.974 412	70.476	
		(Th C″)		208*	207.982 005		3.053 min
		(Ra C″)		210*	209.990 066		1.30 min

Atomic Number Z	Element	Symbol	Chemical Atomic Mass (u)	Mass Number (* Indicates Radioactive) A	Atomic Mass (u)	Percent Abundance	Half-Life (If Radioactive) $T_{1/2}$
82	Lead	Pb	207.2	204*	203.973 029	1.4	$\geq 1.4 \times 10^{17}$ yr
				206	205.974 449	24.1	
				207	206.975 881	22.1	
				208	207.976 636	52.4	
		(Ra D)		210*	209.984 173		22.3 yr
		(Ac B)		211*	210.988 732		36.1 min
		(Th B)		212*	211.991 888		10.64 h
		(Ra B)		214*	213.999 798		26.8 min
83	Bismuth	Bi	208.980 38	209	208.980 383	100	
		(Th C)		211*	210.987 258		2.14 min
84	Polonium	Po					
		(Ra F)		210*	209.982 857		138.38 days
		(Ra C′)		214*	213.995 186		164 μs
85	Astatine	At		218*	218.008 682		1.6 s
86	Radon	Rn		222*	222.017 570		3.823 days
87	Francium	Fr					
		(Ac K)		223*	223.019 731		22 min
88	Radium	Ra		226*	226.025 403		1 600 yr
		(Ms Th$_1$)		228*	228.031 064		5.75 yr
89	Actinium	Ac		227*	227.027 747		21.77 yr
90	Thorium	Th	232.038 1				
		(Rd Th)		228*	228.028 731		1.913 yr
		(Th)		232*	232.038 050	100	1.40×10^{10} yr
91	Protactinium	Pa	231.035 88	231*	231.035 879		32.760 yr
92	Uranium	U	238.028 9	232*	232.037 146		69 yr
				233*	233.039 628		1.59×10^5 yr
		(Ac U)		235*	235.043 923	0.720 0	7.04×10^8 yr
				236*	236.045 562		2.34×10^7 yr
		(UI)		238*	238.050 783	99.274 5	4.47×10^9 yr
93	Neptunium	Np		237*	237.048 167		2.14×10^6 yr
94	Plutonium	Pu		239*	239.052 156		2.412×10^4 yr
				242*	242.058 737		3.73×10^6 yr
				244*	244.064 198		8.1×10^7 yr

[a]Chemical atomic masses are from T. B. Coplen, "Atomic Weights of the Elements 1999," a technical report to the International Union of Pure and Applied Chemistry, and published in *Pure and Applied Chemistry*, 73(4), 667–683, 2001. Atomic masses of the isotopes are from G. Audi and A. H. Wapstra, "The 1995 Update to the Atomic Mass Evaluation," *Nuclear Physics*, A595, vol. 4, 409–480, December 25, 1995. Percent abundance values are from K. J. R. Rosman and P. D. P. Taylor, "Isotopic Compositions of the Elements 1999", a technical report to the International Union of Pure and Applied Chemistry, and published in *Pure and Applied Chemistry*, 70(1), 217–236, 1998.

APPENDIX C
Some Useful Tables

TABLE C.1

Mathematical Symbols Used in the Text and Their Meaning

Symbol	Meaning
$=$	is equal to
$\neq$	is not equal to
$\equiv$	is defined as
$\propto$	is proportional to
$>$	is greater than
$<$	is less than
$\gg$	is much greater than
$\ll$	is much less than
$\approx$	is approximately equal to
$\sim$	is on the order of magnitude of
Δx	change in x or uncertainty in x
Σx_i	sum of all quantities x_i
$\lvert x \rvert$	absolute value of x (always a positive quantity)

TABLE C.2

Standard Symbols for Units

Symbol	Unit	Symbol	Unit
A	ampere	kcal	kilocalorie
Å	angstrom	kg	kilogram
atm	atmosphere	km	kilometer
Bq	bequerel	kmol	kilomole
Btu	British thermal unit	L	liter
C	coulomb	lb	pound
°C	degree Celsius	ly	light year
cal	calorie	m	meter
cm	centimeter	min	minute
Ci	curie	mol	mole
d	day	N	newton
deg	degree (angle)	nm	nanometer
eV	electronvolt	Pa	pascal
°F	degree Fahrenheit	rad	radian
F	farad	rev	revolution
ft	foot	s	second
G	Gauss	T	tesla
g	gram	u	atomic mass unit
H	henry	V	volt
h	hour	W	watt
hp	horsepower	Wb	weber
Hz	hertz	yr	year
in.	inch	μm	micrometer
J	joule	Ω	ohm
K	kelvin		

TABLE C.3

The Greek Alphabet

Alpha	A	α	Nu	N	ν
Beta	B	β	Xi	Ξ	ξ
Gamma	Γ	γ	Omicron	O	o
Delta	Δ	δ	Pi	Π	π
Epsilon	E	ϵ	Rho	P	ρ
Zeta	Z	ζ	Sigma	Σ	σ
Eta	H	η	Tau	T	τ
Theta	Θ	θ	Upsilon	Y	υ
Iota	I	ι	Phi	Φ	ϕ
Kappa	K	κ	Chi	X	χ
Lambda	Λ	λ	Psi	Ψ	ψ
Mu	M	μ	Omega	Ω	ω

TABLE C.4

Physical Data Often Used[a]

Average Earth-Moon distance	3.84×10^8 m
Average Earth-Sun distance	1.496×10^{11} m
Equatorial radius of Earth	6.38×10^6 m
Density of air (20°C and 1 atm)	1.20 kg/m^3
Density of water (20°C and 1 atm)	1.00×10^3 kg/m^3
Free-fall acceleration	9.80 m/s^2
Mass of Earth	5.98×10^{24} kg
Mass of Moon	7.36×10^{22} kg
Mass of Sun	1.99×10^{30} kg
Standard atmospheric pressure	1.013×10^5 Pa

[a] These are the values of the constants as used in the text.

TABLE C.5

Some Fundamental Constants[a]

Quantity	Symbol	Value[b]
Atomic mass unit	u	$1.660\ 540\ 2(10) \times 10^{-27}$ kg $931.494\ 32(28)$ MeV/c^2
Avogadro's number	N_A	$6.022\ 136\ 7(36) \times 10^{23}$ (mol)$^{-1}$
Bohr radius	$a_0 = \dfrac{\hbar^2}{m_e e^2 k_e}$	$0.529\ 177\ 249(24) \times 10^{-10}$ m
Boltzmann's constant	$k_B = R/N_A$	$1.380\ 658(12) \times 10^{-23}$ J/K
Compton wavelength	$\lambda_C = \dfrac{h}{m_e c}$	$2.426\ 310\ 58(22) \times 10^{-12}$ m
Coulomb constant	$k_e = \dfrac{1}{4\pi\epsilon_0}$	$8.987\ 551\ 787 \times 10^9$ N·m^2/C^2 (exact)
Electron mass	m_e	$9.109\ 389\ 7(54) \times 10^{-31}$ kg $5.485\ 799\ 03(13) \times 10^{-4}$ u $0.510\ 999\ 06(15)$ MeV/c^2
Electron volt	eV	$1.602\ 177\ 33(49) \times 10^{-19}$ J
Elementary charge	e	$1.602\ 177\ 33(49) \times 10^{-19}$ C
Gas constant	R	$8.314\ 510(70)$ J/K·mol
Gravitational constant	G	$6.672\ 59(85) \times 10^{-11}$ N·m^2/kg^2
Hydrogen ionization energy	$-E_1 = \dfrac{m_e e^4 k_e^2}{2\hbar^2} = \dfrac{e^2 k_e}{2a_0}$	$13.605\ 698(40)$ eV
Neutron mass	m_n	$1.674\ 928\ 6(10) \times 10^{-27}$ kg $1.008\ 664\ 904(14)$ u $939.565\ 63(28)$ MeV/c^2
Permeability of free space	μ_0	$4\pi \times 10^{-7}$ T·m/A (exact)
Permittivity of free space	$\epsilon_0 = 1/\mu_0 c^2$	$8.854\ 187\ 817 \times 10^{-12}$ C^2/N·m^2 (exact)
Planck's constant	h $\hbar = h/2\pi$	$6.626\ 075(40) \times 10^{-34}$ J·s $1.054\ 572\ 66(63) \times 10^{-34}$ J·s
Proton mass	m_p	$1.672\ 623(10) \times 10^{-27}$ kg $1.007\ 276\ 470(12)$ u $938.272\ 3(28)$ MeV/c^2
Rydberg constant	R_H	$1.097\ 373\ 153\ 4(13) \times 10^7$ m^{-1}
Speed of light in vacuum	c	$2.997\ 924\ 58 \times 10^8$ m/s (exact)

[a] These constants are the values recommended in 1986 by CODATA, based on a least-squares adjustment of data from different measurements. For a more complete list, see Cohen, E. Richard, and Barry N. Taylor, *Rev. Mod. Phys.* **59**:1121, 1987.

[b] The numbers in parentheses for the values below represent the uncertainties in the last two digits.

APPENDIX D
SI Units

TABLE D.1

SI Base Units

Base Quantity	SI Base Unit	
	Name	**Symbol**
Length	meter	m
Mass	kilogram	kg
Time	second	s
Electric current	ampere	A
Temperature	kelvin	K
Amount of substance	mole	mol
Luminous intensity	candela	cd

TABLE D.2

Derived SI Units

Quantity	Name	Symbol	Expression in Terms of Base Units	Expression in Terms of Other SI Units
Plane angle	radian	rad	m/m	
Frequency	hertz	Hz	s^{-1}	
Force	newton	N	$kg \cdot m/s^2$	J/m
Pressure	pascal	Pa	$kg/m \cdot s^2$	N/m^2
Energy: work	joule	J	$kg \cdot m^2/s^2$	$N \cdot m$
Power	watt	W	$kg \cdot m^2/s^3$	J/s
Electric charge	coulomb	C	$A \cdot s$	
Electric potential (emf)	volt	V	$kg \cdot m^2/A \cdot s^3$	$W/A, J/C$
Capacitance	farad	F	$A^2 \cdot s^4/kg \cdot m^2$	C/V
Electric resistance	ohm	Ω	$kg \cdot m^2/A^2 \cdot s^3$	V/A
Magnetic flux	weber	Wb	$kg \cdot m^2/A \cdot s^2$	$V \cdot s, T \cdot m^2$
Magnetic field intensity	tesla	T	$kg/A \cdot s^2$	Wb/m^2
Inductance	henry	H	$kg \cdot m^2/A^2 \cdot s^2$	Wb/A

MCAT Skill Builder Study Guide

VECTORS

EXAMPLE 1

A hiker walks due north at 4.00 km/h from his campsite. A grizzly bear, starting 3.00 km due east of the campsite, walks at 7.00 km/h in a direction 30° west of north. After one hour, how far apart are they?

Solution

Write the components of the position vector of the hiker after one hour, choosing the campsite as the origin:

$H_x = 0 \qquad H_y = 4.00 \text{ m}$

The bear travels at an angle of $30° + 90° = 120°$ with respect to due east, where due east corresponds to the positive x-direction. Write the components of the position vector of the bear after one hour, again relative the campsite:

$B_x = 3.00 \text{ km} + (7.00 \text{ km}) \cos 120° = -0.500 \text{ km}$

$B_y = (7.00 \text{ km}) \sin 120° = 6.06 \text{ km}$

Subtracting the bear's position vector from the hiker's position vector will result in a vector $\vec{R}$ pointing from the bear to the hiker:

$R_x = H_x - B_x = 0 - (-0.500 \text{ km}) = 0.500 \text{ km}$

$R_y = H_y - B_y = 4.00 \text{ km} - 6.06 \text{ km} = -2.06 \text{ km}$

Calculate the magnitude of the resultant vector, obtaining the distance between the hiker and the bear:

$R = (R_x^2 + R_y^2)^{1/2} = [(0.500 \text{ km})^2 + (-2.06 \text{ km})^2]^{1/2}$

$= 2.12 \text{ km}$

EXAMPLE 2

Two vectors $\vec{A}$ and $\vec{B}$ are not parallel to each other (nor "antiparallel"—pointing in opposite directions). If $\vec{R} = \vec{A} + \vec{B}$, which of the following must be true of the magnitudes A, B, and R? (a) $R > A + B$ (b) $R = A + B$ (c) $R < A + B$

Conceptual Solution

Because the vectors $\vec{A}$ and $\vec{B}$ are not parallel, they can be arranged to form two sides of a triangle. By the geometric law of vector addition, the resultant vector $\vec{R}$ forms the third side of the same triangle. The shortest distance between two points is a straight line, so the distance along R is shorter than the sum of the lengths of A and B. Hence the correct answer is **(c)**.

MULTIPLE CHOICE PROBLEMS

1. A car travels due east for a distance of 3 mi and then due north for an additional 4 mi before stopping. What is the shortest straight-line distance between the starting and ending points of this trip? (a) 3 mi (b) 4 mi (c) 5 mi (d) 7 mi

2. In the previous example, what is the angle α of the shortest path relative to due north? (a) $\alpha = \text{arc cos } 3/5$ (b) $\alpha = \text{arc sin } 5/3$ (c) $\alpha = \text{arc sin } 4/3$ (d) $\alpha = \text{arc tan } 3/4$

3. A vector $\vec{A}$ makes an angle of 60° with the x-axis of a Cartesian coordinate system. Which of the following statements is true of the indicated magnitudes? (a) $A_x > A_y$ (b) $A_y > A_x$ (c) $A_y > A$ (d) $A_x > A$

4. Force is a vector quantity measured in units of newtons, N. What must be the angle between two concurrently acting forces of 5 N and 3 N, respectively, if the resultant vector has a magnitude of 8 N? (a) 0° (b) 45° (c) 90° (d) 180°

5. Two forces act concurrently on an object. Both vectors have the same magnitude of 10 N and act at right angles to each other. What is the closest estimate of their resultant? (a) 0 N (b) 14 N (c) 20 N (d) 100 N

Answers

1. **(c).** The two legs of the trip are perpendicular, therefore the shortest distance is given by the hypotenuse of the corresponding right triangle. The value is given by the Pythagorean theorem: $R = (3^2 + 4^2)^{1/2} = 5$ mi

2. **(d).** The angle α that gives the direction of the hypotenuse relative to the y-axis is given by:

$$\tan \alpha = \text{opposite side/adjacent side} = 3/4 \quad \rightarrow \quad \alpha = \text{arc tan } 3/4$$

3. **(b).** Choices (c) and (d) can be eliminated immediately because the projection of a vector can never be greater than the magnitude of the vector itself. As the angle between the vector and the axis increases towards 90°, the magnitude of the projection decreases toward zero. The angle between the vector and the x-axis is 60°, which means the angle between the vector and the y-axis must be 30°. Because the angle between the vector and the y-axis is smaller than the angle between the vector and the x-axis, the projection onto the y-axis, A_y, must be greater than the projection onto the x-axis, A_x.

4. **(a).** The only way the vector sum of a 3 N and a 5 N force can equal 8 N is if both forces act in the same direction. The angle between them must therefore be zero degrees.

5. **(b).** The two vectors form the legs of a right triangle. The resultant vector is the hypotenuse. Choices (a) and (c) can be eliminated immediately because they require that the two forces be anti-parallel or parallel respectively. Choice (d) can also be eliminated since it's too large to be the hypotenuse. The magnitude of the resultant can be calculated with the Pythagorean theorem:

$$R = (10^2 + 10^2)^{1/2} = (200)^{1/2} = 10(2)^{1/2} = 10(1.41) \approx 14 \text{ N}$$

MOTION

EXAMPLE 1

A motorist traveling 24.0 m/s slams on his brakes and comes to rest in 6.00 s. What is the car's acceleration assuming it's constant, and how far does the car travel during this time?

Solution

Apply the kinematics equation for velocity:

$$v = at + v_0 \tag{1}$$

Solve Equation (1) for the acceleration a and substitute $v = 0$, $t = 6.00$ s, and $v_0 = 24.0$ m/s:

$$a = \frac{v - v_0}{t} = \frac{0 - 24.0 \text{ m/s}}{6.00 \text{ s}} = \boxed{-4.00 \text{ m/s}^2}$$

The displacement can then be found with the time-independent equation:

$$v^2 - v_0^2 = 2a\Delta x \tag{2}$$

Solve Equation (2) for Δx and substitute values:

$$\Delta x = \frac{v^2 - v_0^2}{2a} = \frac{0 - (24.0 \text{ m/s})^2}{2(-4.00 \text{ m/s}^2)} = \boxed{72.0 \text{ m}}$$

EXAMPLE 2

A student on top of a building releases a rock from rest. When the rock is halfway to the ground, he releases a golf ball from rest. As both objects fall, does the distance between them increase, decrease, or stay the same?

Conceptual Solution

At the instant the golf ball is released the rock is traveling faster, hence will always travel faster before it reaches the ground, because the velocities of both objects are changing at the same rate. Because the rock is always falling faster, the distance between it and the golf ball will increase.

EXAMPLE 3

A baseball player throws a ball at an angle θ above the horizontal and at speed v_0. Neglecting air drag, what is the ball's speed when it returns to the same height at which it was thrown? (a) The ball's speed will be the same. (b) The ball's speed will be larger. (c) The ball's speed will be smaller.

Conceptual Solution

The ball's speed will be the same, so the correct answer is **(a)**. As there is no acceleration in the horizontal direction, the velocity in the x-direction is constant, and only the y-direction need be considered. By symmetry, the time taken to reach maximum height is the same time taken to fall back to the original height. If the ball's velocity in the y-direction changes from v_{0y} to 0 on the way to maximum height, then it will change from 0 to $-v_{0y}$ on returning to its original height. Hence the ball's velocity components will be the same as before, except for a sign change in the y-component, which doesn't affect the speed.

Quantitative Solution

Let the point of release correspond to $x = 0$, $y = 0$. Set the equation of displacement in the y-direction equal to zero:

$$\Delta y = \tfrac{1}{2}at^2 + v_{0y}t = 0 \tag{1}$$

Solve Equation (1), obtaining the times when $\Delta y = 0$ (i.e., at the start and end of the trajectory.):

$$t = 0, \ \frac{2v_{0y}}{a}$$

Substitute these two solutions into the equation for the velocity in the y-direction:

$$v_y = at + v_{0y}$$

Case 1: $t = 0$

$$v_y = a\cdot 0 + v_{0y} = v_{0y}$$

Case 2: $t = \dfrac{-2v_{0y}}{a}$

$$v_y = a\cdot\left(\frac{-2v_{0y}}{a}\right) + v_{0y} = -v_{0y}$$

The speed at any time is given by the following expression:

$$v = (v_{0x}{}^2 + v_{0y}{}^2)^{1/2}$$

Because the y-component is squared, the negative sign in Case 2 makes no difference and the speeds at the two times in question are the same, again giving answer **(a)**.

EXAMPLE 4

A merry-go-round has a radius of 2.00 m and turns once every 5.00 s. What magnitude net force is required to hold a 35.0 kg boy in place at the rim?

Solution

First, calculate the speed of an object at the rim, which is the distance around the rim divided by the time for one rotation:

$$v = \frac{d}{t} = \frac{2\pi r}{t} = \frac{2\pi(2.00 \text{ m})}{5.00 \text{ s}} = 2.51 \text{ m/s}$$

A body traveling in a circle at uniform speed requires a centripetal acceleration of $a_c = v^2/r$. The force producing this acceleration can be found by substituting this expression into Newton's second law:

$$F = ma_c = m\frac{v^2}{r} = (35.0 \text{ kg})\frac{(2.51 \text{ m/s})^2}{2.00 \text{ m}}$$

$$= 1.10 \times 10^2 \text{ N}$$

MULTIPLE CHOICE PROBLEMS

1. A bird flies 4.0 m due north in 2.0 s and then flies 2.0 m due west in 1.0 s. What is the bird's average speed? (a) 2.0 m/s (b) 4.0 m/s (c) 8.0 m/s (d) $2\sqrt{5}/3$ m/s

2. Applying the brakes to a car traveling at 45 km/h provides an acceleration of 5.0 m/s^2 in the opposite direction. How long will it take the car to stop? (a) 0.40 s (b) 2.5 s (c) 5.0 s (d) 9.0 s

3. A ball rolls down a long inclined plane with a uniform acceleration of magnitude 1.0 m/s^2. If its speed at some instant of time is 10 m/s, what will be its speed 5.0 seconds later? (a) 5 m/s (b) 10 m/s (c) 15 m/s (d) 16 m/s

4. A ball rolls down an inclined plane with a uniform acceleration of 1.00 m/s^2 down the incline. If its initial velocity is 1.00 m/s down the incline, how far will it travel along the incline in 10.0 s? (a) 10.0 m (b) 12.0 m (c) 60.0 m (d) 1.00×10^2 m

5. An object with an initial velocity of 25.0 m/s accelerates uniformly for 10.0 s to a final velocity of 75.0 m/s. What is the magnitude of its acceleration? (a) 3.00 m/s^2 (b) 5.00 m/s^2 (c) 25.0 m/s^2 (d) 50.0 m/s^2

6. A rock is dropped from a height of 19.6 m above the ground. How long does it take the rock to hit the ground? (a) 2.0 s (b) 4.0 s (c) 4.9 s (d) 9.8 s

7. A spacecraft hovering above the surface of a distant planet releases a probe to explore the planet's surface. The probe falls freely a distance of 40.0 m during the first 4.00 s after its release. What is the magnitude of the acceleration due to gravity on this planet? (a) 4.00 m/s^2 (b) 5.00 m/s^2 (c) 10.0 m/s^2 (d) 16.0 m/s^2

8. A ball is dropped from the roof of a very tall building. What is its speed after falling for 5.00 s? (a) 1.96 m/s (b) 9.80 m/s (c) 49.0 m/s (d) 98.0 m/s

9. A quarterback throws a football with a velocity of 7.00 m/s at an angle of 15.0° with the horizontal. How far away should the receiver be? (a) 1.25 m (b) 2.50 m (c) 5.00 m (d) 6.25 m

10. A bomber is flying parallel to the ground with a speed of 500 km/h at an altitude of 1 960 m when it drops a bomb. How long does it take for the bomb to hit the ground? (a) 0.500 s (b) 20.0 s (c) 50.0 s (d) 2.00×10^2 s

11. A ball is thrown horizontally with a speed of 6.0 m/s. What is its speed after 3.0 seconds of flight? (a) 30 m/s (b) 15.8 m/s (c) 18 m/s (d) 4.9 m/s

12. A 70-kg woman standing at the equator rotates with the Earth around its axis at a tangential speed of about 5×10^2 m/s. If the radius of the Earth is approximately 6×10^6 m, which is the best estimate of the centripetal acceleration experienced by the woman? (a) 4×10^{-2} m/s^2 (b) 4 m/s^2 (c) 10^{-4} m/s^2 (d) 24 m/s^2

13. A ball rolls with uniform speed around a flat horizontal circular track. If the speed of the ball is doubled, its centripetal acceleration is (a) quadrupled (b) doubled (c) halved (d) unchanged

ANSWERS

1. **(a)**. The average speed is the total distance traveled divided by the elapsed time.
$$v = d/\Delta t = 6.0 \text{ m}/3.0 \text{ s} = 2.0 \text{ m/s}$$

2. **(b)**. $t = (v_f - v_i)/a$. The final velocity is 0 km/h and the acceleration is -5.0 m/s^2. (The negative sign appears because the acceleration is anti-parallel to the velocity.) Remember to convert from kilometers to meters and from hours to seconds.
$$t = \frac{\Delta v}{a} = \frac{0 - (45 \text{ km/h})(10^3 \text{ m}/1 \text{ km})(1 \text{ h}/3\,600 \text{ s})}{-5.0 \text{ m/s}^2} = 2.5 \text{ s}$$

3. **(c)**. $v_f = v_i + at = 10 \text{ m/s} + (1.0 \text{ m/s}^2)(5.0 \text{ s}) = 15 \text{ m/s}$

4. **(c)**. $d = v_i t + \frac{1}{2}at^2 = (1.00 \text{ m/s})(10.0 \text{ s}) + \frac{1}{2}(1.00 \text{ m/s}^2)(10.0 \text{ s})^2 = 60.0 \text{ m}$

5. **(b)**. $a = \Delta v/\Delta t = (75.0 \text{ m/s} - 25.0 \text{ m/s})/10.0 \text{ s} = 5.00 \text{ m/s}^2$

6. **(a)**. This is free fall with zero initial velocity and initial height of $y_0 = 19.6$ m. The displacement equation becomes $\Delta y = y - y_0 = -\frac{1}{2}gt^2$. Solving for time t with $y = 0$ gives:
$$t = (2y_0/g)^{1/2} = (2 \times 19.6 \text{ m}/9.8 \text{ m/s}^2)^{1/2} = (4.0 \text{ s}^2)^{1/2} = 2.0 \text{ s}$$

7. **(b)**. This is a free-fall problem. The object starts from rest, $v_0 = 0$. Rearrange $\Delta y = -\frac{1}{2}gt^2$ and solve for the acceleration of gravity, g:
$$g = -2(y - y_0)/t^2 = -2(-40.0 \text{ m} - 0)/16.0 \text{ s}^2 = 5.00 \text{ m/s}^2$$

8. **(c)**. The free-fall velocity starting with an initial velocity of zero is: $v = -gt = -(9.8 \text{ m/s}^2)(5.00 \text{ s}) = -49 \text{ m/s}$

9. **(b).** The initial velocity in the y-direction is $v_{0y} = v_0 \sin \theta = (7.00 \text{ m/s}) \sin 15.0° = 1.81 \text{ m/s}$. Maximum height is reached when the velocity in the y-direction is zero: $v = 0 = -gt + v_{0y} \rightarrow 0 = -(9.80 \text{ m/s}^2)t + 1.81 \text{ m/s} \rightarrow t = 0.185 \text{ s}$. Doubling this gives the total time of flight. The range is then $\Delta x = (v_0 \cos \theta)t = (7.00 \text{ m/s})(\cos 15.0°)(0.370 \text{ s}) = 2.50 \text{ m}$

10. **(b).** The horizontal and the vertical components of velocity are independent. The vertical component is due only to the acceleration of gravity:

$$y = \tfrac{1}{2} gt^2 \quad \rightarrow \quad t = (2y/g)^{1/2}$$
$$t = [2(1\,960 \text{ m})/(9.80 \text{ m/s}^2)]^{1/2} = (4.00 \times 10^2 \text{ s}^2)^{1/2} = 20.0 \text{ s}$$

11. **(a).** The horizontal component of velocity is $v_x = 6.00 \text{ m/s}$, and the vertical component is $v_y = gt = (9.80 \text{ m/s}^2)(3.00 \text{ s}) = 29.4 \text{ m/s}$, so

$$v = (v_x^2 + v_y^2)^{1/2} = [(6.00 \text{ m/s})^2 + (29.4 \text{ m/s})^2]^{1/2} = 3.00 \text{ m/s}.$$

12. **(a).** The woman is moving with constant speed equal to the rotation of the Earth on its axis in a circle equal to the radius of the Earth. The centripetal acceleration is:

$$a_c = v^2/R = (500 \text{ m/s})^2/6 \times 10^6 \text{ m} = (25 \times 10^4 \text{ m}^2/\text{s}^2)/6 \times 10^6 \text{ m} \approx 4 \times 10^{-2} \text{ m/s}^2$$

13. **(a).** $a_c = v^2/R$. The acceleration and the square of the velocity are directly proportional. Because the velocity is squared, doubling the velocity quadruples the acceleration.

FORCE

EXAMPLE 1

A net force F accelerates an object of mass m at 6 m/s². The same net force is applied to a body with mass $3m$. What is the acceleration of the more massive body?

Conceptual Solution
According to the Newton's second law, the acceleration imparted to an object by a force is inversely proportional to the object's mass. The acceleration of an object three times as massive is therefore $\tfrac{1}{3}(6 \text{ m/s}^2) = \boxed{2 \text{ m/s}^2}$.

Quantitative Solution
Write Newton's second law for each of the objects:

$$ma_m = F \tag{1}$$

$$Ma_M = F \tag{2}$$

Divide Equation (2) by Equation (1):

$$\frac{Ma_M}{ma_m} = \frac{F}{F} = 1 \tag{3}$$

Substitute $M = 3m$ into Equation (3) and solve for a_M:

$$\frac{(3m)a_M}{ma_m} = 1 \quad \rightarrow \quad a_M = \tfrac{1}{3}a_m = \tfrac{1}{3} \cdot (6 \text{ m/s}^2)$$

$$a_M = \boxed{2 \text{ m/s}^2}$$

EXAMPLE 2

A block of mass m is attached by a horizontal string of negligible mass to the left side of a second block of mass M, with $m < M$. A second such string is attached to the right side of mass M, and the system is pulled in the positive x-direction by a constant force F. If the surface is frictionless, what can be said about the tension T in the string connecting the two blocks? (a) $T = F$ (b) $T < F$ (c) $T > F$

Conceptual Solution
The answer is **(b).** The acceleration of the block m is the same as that of the system of both, which has total mass $m + M$. The tension T must accelerate only the mass m, while the force F must provide the same acceleration to a system of greater mass, $m + M$. It therefore follows that $F > T$.

Quantitative Solution

In this problem the y-component of Newton's second law gives only the normal forces acting on the blocks in terms of the gravity force. As there is no friction, these forces don't affect the acceleration. Write the x-component of Newton's second law for the system:

$$(m + M)a = F \tag{1}$$

Write the x-component of Newton's second law for the less massive block:

$$ma = T \tag{2}$$

Divide Equation (1) by Equation (2), canceling the acceleration, which is the same for both:

$$\frac{(m + M)a}{ma} = \frac{F}{T} \quad \rightarrow \quad \frac{(m + M)}{m} = \frac{F}{T} \tag{3}$$

Inspecting Equation (3), it's clear that $F > T$, because $m + M > m$, and again the answer is **(b)**.

EXAMPLE 3

A block with mass m is started at the top of an incline with coefficient of kinetic friction μ_k and allowed to slide to the bottom. A second block with mass $M = 2m$ is also allowed to slide down the incline. If μ_k is the same for both blocks, how does the acceleration a_M of the second block compare to the acceleration a_m of the first block? (a) $a_M = 2a_m$ (b) $a_M = a_m$ (c) $a_M = \frac{1}{2}a_m$

Conceptual Solution

The accelerations are the same, because the force of gravity, the force of friction, and the ma side of Newton's second law when applied in this physical context are all proportional to the mass. Hence a different mass should not affect the acceleration of the body, and the answer is **(b)**.

EXAMPLE 4

Planet A has twice the mass and twice the radius of Planet B. On which planet would an astronaut have a greater weight? (a) Planet A (b) Planet B (c) His weight is unaffected.

Conceptual Solution

Weight as measured on a given planet is the magnitude of the gravitational force at the surface of that planet. In Newton's law of gravitation, the gravitational force is directly proportional to the mass of each of two bodies but inversely proportional to the square of the distance between them. Hence in considering only the masses, planet A exerts twice the gravitational force of planet B. Planet A also has twice the radius of planet B, however, and the inverse square of 2 is $\frac{1}{4}$. Overall, therefore, Planet A has a weaker gravitational acceleration at its surface than Planet B by a factor of one-half, and the weight of an astronaut will be greater on planet B [answer **(b)**].

Quantitative Solution

Write Newton's law of gravitation for the weight of an astronaut of mass m on planet A:

$$w_A = \frac{mM_A G}{r_A{}^2} \tag{1}$$

Write the same law for the weight of the astronaut on Planet B:

$$w_B = \frac{mM_B G}{r_B{}^2} \tag{2}$$

Divide Equation (2) by Equation (1):

$$\frac{w_B}{w_A} = \frac{\dfrac{mM_B G}{r_B{}^2}}{\dfrac{mM_A G}{r_A{}^2}} = \frac{M_B r_A{}^2}{M_A r_B{}^2}$$

Now substitute $r_A = 2r_B$ and $M_A = 2M_B$:

$$\frac{w_B}{w_A} = \frac{M_B(2r_B)^2}{(2M_B)r_B^2} = 2$$

Solve for w_B:

$$w_B = \boxed{2w_A} \ [\text{answer } \textbf{(b)}]$$

MULTIPLE CHOICE PROBLEMS

1. Which of the following will result from the application of a non-zero net force on an object? (a) The velocity of the object will remain constant. (b) The velocity of the object will remain constant, but the direction in which the object moves will change. (c) The velocity of the object will change. (d) None of the above.

2. Body A has a mass that is twice as great as that of body B. If a force acting on body A is half the value of a force acting on body B, which statement is true? (a) The acceleration of A will be twice that of B. (b) The acceleration of A will be half that of B. (c) The acceleration of A will be equal to that of B. (d) The acceleration of A will be one fourth that of B.

3. Which of the following is a statement of Newton's second law of motion? (a) For every action there is an equal and opposite reaction. (b) Force and the acceleration it produces are directly proportional, with mass as the constant of proportionality. (c)A body at rest tends to remain at rest unless acted upon by a force. (d) None of the above.

4. If a body has an acceleration of zero, which statement is *most* true? (a) The body must be at rest. (b) The body may be at rest. (c) The body must slow down. (d) The body may speed up.

5. What is the weight of a 2.00-kg body on or near the surface of the Earth? (a) 4.90 N (b) 16.0 lb (c) 19.6 N (d) $64.0 \text{ kg} \cdot \text{m/s}^2$

6. Two objects of equal mass are separated by a distance of 2 m. If the mass of one object is doubled, the force of gravity between the two objects will be (a) Half as great (b) Twice as great (c) One fourth as great (d) Four times as great

7. The distance between a spaceship and the center of the Earth increases from one Earth radius to three Earth radii. What happens to the force of gravity acting on the spaceship? (a) It becomes 1/9 as great. (b) It becomes 9 times as great. (c) It becomes 1/3 as great. (d)It becomes 3 times as great.

8. A 100-kg astronaut lands on a planet with a radius three times that of Earth and a mass nine times that of Earth. The acceleration due to gravity, g, experienced by the astronaut will be (a) Nine times the value of g on Earth. (b) Three times the value of g on Earth. (c) The same value of g as on Earth. (d) One third the value of g on Earth.

9. The measured weight of a person standing on a scale and riding in an elevator will be the greatest when (a) The elevator rises at a constant velocity. (b) The elevator accelerates upward. (c) The elevator falls at a constant velocity. (d) None of the above, because weight is constant.

10. A block with mass 25.0 kg rests on a rough level surface. A horizontal force of 1.50×10^2 N is applied to it, and an acceleration of 4.00 m/s^2 is subsequently observed. What is the magnitude of the force of friction acting on the block? (a) 25.0 N (b) 1.00×10^2 N (c) 75.0 N (d) 50.0 N

ANSWERS

1. **(c).** Force produces acceleration, which results in a change in velocity.

2. **(d).** Obtain the ratio of the accelerations from Newton's Second Law, $F = ma$:

$$\frac{m_A a_A}{m_B a_B} = \frac{F_A}{F_B} \quad \rightarrow \quad \frac{(2m_B)a_A}{m_B a_B} = \frac{(\frac{1}{2}F_B)}{F_B} \quad \rightarrow \quad \frac{a_A}{a_B} = \frac{1}{4}$$

3. **(b).** Choice (a) is a statement of the third law, and choice (c) a statement of the first law.

4. **(b).** Choices (c) and (d) can be eliminated immediately because changing velocity (either its magnitude or direction) requires the application of a force. Choice (a) can be eliminated because the first law applies to both bodies at rest and those moving with uniform velocity.

5. **(c).** $w = F = mg = (2.00 \text{ kg})(9.80 \text{ m/s}^2) = 19.6 \text{ N}$

6. **(b).** In the law of universal gravitation, the gravitational force is directly proportional to the product of the masses. Doubling one of the masses doubles their product and therefore doubles the force.

7. **(a).** Choices (b) and (d) can be eliminated because the force of gravitational attraction must decrease with increasing distance. Because $F_{grav} = Gm_1m_2/r^2$, when r is tripled, the force decreases by a factor of 9.

8. **(c).** See Example 4. Gravitational force is proportional to mass and inversely proportional to the radius squared, so nine times the mass and three times the radius yields a factor of $9/3^2 = 1$. There is no discernible difference in weight, because the two effects exactly cancel each other.

9. **(b).** The weight measured on the scale is essentially the magnitude of the normal force. Apply Newton's second law to the passenger and solve for the normal force:

$$ma = n - mg \quad \rightarrow \quad n = ma + mg$$

Inspecting the equation for the normal force, it's clear that a positive acceleration results in a larger normal force, hence a larger reading on the scale.

10. **(d).** Use Newton's second law:

$$ma = F_{app} - F_{fric}$$
$$(25.0 \text{ kg})(4.00 \text{ m/s}^2) = 1.50 \times 10^2 \text{ N} - F_{fric}$$
$$F_{fric} = 50.0 \text{ N}$$

EQUILIBRIUM

EXAMPLE 1

A uniform steel beam with length L and mass $M = 60.0$ kg rests on two pivots. The first pivot is located at the left end of the beam and exerts a normal force n_1 at that point. A second pivot is two-thirds of the distance from the left end to the right end. Calculate the normal force n_2 exerted on the beam by the second pivot. (a) 441 N (b) 264 N (c) 372 N (d) 188 N

Strategy

In equilibrium, the sum of the torques is zero. Let τ_{n_1} be the torque exerted by the normal force n_1 associated with the first pivot, τ_{n_2} the torque exerted by the normal force n_2 associated with the second pivot, and τ_M the torque exerted by the force of gravity on the beam. Choose the pivot at the left end as the point around which to calculate torques. Use $\tau = rF \sin \theta$ for each torque, where θ is the angle between the position vector of $\vec{r}$ and the force $\vec{F}$. The sign of a given torque is positive if the force tends to rotate the beam counterclockwise, and negative if the force tends to rotate the beam clockwise.

Solution

In equilibrium, the sum of the torques is zero:

$$\sum \tau_i = \tau_{n_1} + \tau_{n_2} + \tau_M = 0 \tag{1}$$

The torque due to the normal force n_1 is zero because its lever arm is zero. The weight of the uniform beam may be considered to act at the middle of the beam, its center of gravity. Equation (1) becomes:

$$n_2(2L/3) \sin 90° - Mg(L/2) \sin 90° = 0 \tag{2}$$

Notice that for n_2 the torque is positive, whereas the gravitational torque is negative (see the Strategy). Cancel the factors of L in Equation (2) and solve for n_2:

$$n_2 = \tfrac{3}{4}Mg = \tfrac{3}{4}(60.0 \text{ kg})(9.80 \text{ m/s}^2) = 441 \text{ N}$$

$$= \boxed{441 \text{ N}} \text{ [answer (a)]}$$

EXAMPLE 2

Ice skater A has mass M and is at rest on a frozen pond, holding a backpack of mass m. She tosses the backpack to ice skater B, who also has mass M and is initially at rest. If the surface of the pond is considered frictionless, which skater has the larger subsequent speed? (Changes of momentum in the vertical direction can be ignored, here.) (a) skater A (b) skater B (c) Both skaters have the same subsequent speed.

Conceptual Solution

By conservation of momentum, the final momentum of skater A is equal and opposite the momentum of the backpack. Skater B acquires this same momentum when catching the backpack, but also increases her mass by m to $M + m$. Because momentum is proportional to both mass and velocity, skater B, with greater mass, has a smaller subsequent speed. Thus, the answer is **(a).**

Quantitative Solution

Apply conservation of momentum to Skater A. The initial momentum equals the final momentum:

$$p_i = p_f$$

The initial total momentum is zero. Skater A's final momentum is MV_A, whereas the backpack has momentum mv:

$$0 = MV_A + mv$$

Solve for the velocity V_A of skater A:

$$V_A = -\frac{mv}{M} \tag{1}$$

Apply conservation of momentum to skater B. The initial momentum of the backpack, mv, equals the final momentum of the system of skater and backpack:

$$mv = (m + M)V_B$$

Solve for the velocity V_B of the backpack-skater B system:

$$V_B = \frac{mv}{m + M} \tag{2}$$

Divide the magnitude of Equation (1) by Equation (2):

$$\frac{|V_A|}{|V_B|} = \frac{\left|-\dfrac{mv}{M}\right|}{\left|\dfrac{mv}{m + M}\right|} = \frac{m + M}{M} > 1 \tag{3}$$

By Equation (3), $|V_A|/|V_B| > 1$, so $|V_A| > |V_B|$ [answer **(a)**].

MULTIPLE CHOICE PROBLEMS

1. A nonuniform bar 8.00 m long is placed on a pivot 2.00 m from the right end of the bar, which is also the lighter end. The center of gravity of the bar is located 2.00 m from the heavier left end. If a weight of $W = 5.00 \times 10^2$ N on the right end balances the bar, what must be the weight w of the bar? (a) 1.25×10^2 N (b) 2.50×10^2 N (c) 5.00×10^2 N (d) 1.00×10^3 N

2. A rod of negligible mass is 10 m in length. If a 30-kg object is suspended from the left end of the rod and a 20-kg object from the right end, where must the pivot point be placed to ensure equilibrium? (a) 4 m from the 30-kg object (b) 4 m from the 20-kg object (c) 8 m from the 30-kg object (d) 5 m from the 20-kg object

3. A car with a mass of 8.00×10^2 kg is stalled on a road. A truck with a mass of 1 200 kg comes around the curve at 20.0 m/s and hits the car. The two vehicles remain locked together after the collision. What is their combined velocity after the impact? (a) 3.0 m/s (b) 6.0 m/s (c) 12 m/s (d) 24 m/s

4. A car of mass 1.00×10^3 kg traveling at 5.00 m/s overtakes and collides with a truck of mass 3.00×10^3 kg traveling in the same direction at 1.00 m/s. During the collision, the two vehicles remain in contact and continue to move as one unit. What is the speed of the coupled vehicles right after impact? (a) 2.00 m/s (b) 4.00 m/s (c) 5.00 m/s (d) 6.00 m/s

5. A pistol with a mass of 0.20 kg fires a 0.50-gram bullet with an initial velocity of 1.0×10^2 m/s. What is the magnitude of the recoil velocity of the pistol? (a) 0.25 m/s (b) 0.50 m/s (c) 1.0 m/s (d) 10 m/s

6. A 0.20-kg ball is bounced against a wall. It hits the wall with a speed of 20.0 m/s and rebounds in an elastic collision. What is the magnitude of the change in momentum of the ball? (a) 0 kg·m/s (b) 4.0 kg·m/s (c) 8.0 kg·m/s (d) 10 kg·m/s

7. A tennis ball is hit with a tennis racket and the change in the momentum of the ball is 4.0 kg·m/s. If the collision time of the ball and racket is 0.01 s, what is the magnitude of the force exerted by the ball on the racket? (a) 2.5×10^{-3} N (b) 4×10^{-2} N (c) 4.0 N (d) 400 N

8. A 30-kg cart traveling due north at 5 m/s collides with a 50-kg cart that had been traveling due south. Both carts immediately come to rest after the collision. What must have been the speed of the southbound cart? (a) 3 m/s (b) 5 m/s (c) 6 m/s (d) 10 m/s

Answers

1. (b). At equilibrium,

$$\sum \vec{\tau}_i = \vec{\tau}_p + \vec{\tau}_B + \vec{\tau}_W = 0 \tag{1}$$

The torque $\vec{\tau}_p$ exerted by the pivot is zero because the torques are computed around that point, meaning a zero lever arm for the force exerted by the pivot. The center of gravity is 4.00 m from the pivot point, and the gravitational force exerts a counterclockwise torque $\vec{\tau}_B$ on the bar as if all the mass were concentrated at that point. The torque due to the weight of the bar is therefore equal to $+rF = +(4.00 \text{ m})w_B$. The weight W on the right side gives a clockwise torque $\vec{\tau}_W = -(2.00 \text{ m})W$. Hence from Equation (1):

$$0 + w_B(4.00 \text{ m}) - (5.00 \times 10^2 \text{ N})(2.00 \text{ m}) = 0 \quad \rightarrow \quad w_B = 2.50 \times 10^2 \text{ N}$$

2. (a). The pivot point must be closer to the heavier weight, otherwise the torques due to the weights could not be equal in magnitude and opposite in direction. Given this, choices (b) and (c) can be eliminated immediately. Choice (d) can also be eliminated because it puts the pivot at the center of the rod, which would only give equilibrium if the two weights were identical. The choice must be (a), which can be confirmed using the second condition of equilibrium. If x is the distance of the pivot from the 30-kg mass, then $10 - x$ is the distance from the pivot to the 20-kg mass. Apply the second condition of equilibrium, $\Sigma \tau_i = 0$:

$$(30 \text{ kg})g(x) - (20 \text{ kg})g(10 - x) = 0 \quad \rightarrow \quad x = 4 \text{ m}$$

3. (c). Conservation of momentum ensures that the total momentum after the collision must equal the total momentum before the collision. If A is the car of mass 8.00×10^2 kg and B is the 1 200-kg truck, then

$$m_A v_{A \text{ initial}} + m_B v_{B \text{ initial}} = m_A v_{A \text{ final}} + m_B v_{B \text{ final}}$$

Because the car is stalled, its initial velocity is zero and $m_A v_{A \text{ initial}}$ is zero. After collision the vehicles are joined and $v_{A \text{ final}} = v_{B \text{ final}} = v_{\text{final}}$. The equation becomes:

$$m_B v_{B \text{ initial}} = (m_A + m_B)v_{\text{final}}$$

Solve for v_{final}:

$v_{\text{final}} = m_B v_{B \text{ initial}}/(m_A + m_B) = (1\ 200 \text{ kg})(20.0 \text{ m/s})/(1\ 200 \text{ kg} + 8.00 \times 10^2 \text{ kg})$
      $= 12 \text{ m/s}.$

4. (a). Conservation of momentum gives $p_{\text{before}} = p_{\text{after}}$. The total momentum before the collision is the sum of momenta for the two vehicles. After the collision, p is the momentum associated with their coupled masses:

$$(1.00 \times 10^3 \text{ kg})(5.00 \text{ m/s}) + (3.00 \times 10^3 \text{ kg})(1.00 \text{ m/s}) = (4.00 \times 10^3 \text{ kg})v_{\text{final}}$$
$$v_{\text{final}} = (5.00 \times 10^3 + 3.00 \times 10^3)(\text{kg·m /s})/(4.00 \times 10^3 \text{ kg}) = 2.00 \text{ m/s}$$

5. (a). The momentum is conserved because there are no external forces acting on the pistol-bullet system during the firing. The initial momentum, however, is zero. The final momentum is the sum of the momentum of the bullet and the recoil momentum of the pistol:

$$p_{\text{before}} = p_{\text{after}} \quad \rightarrow \quad 0 = m_{\text{pistol}}v_{\text{pistol}} + m_{\text{bullet}}v_{\text{bullet}}$$
$$v_{\text{pistol}} = -(0.50 \text{ g})(1.0 \times 10^2 \text{ m/s})/0.20 \times 10^3 \text{ g} = -0.25 \text{ m/s}$$

6. (c). The change in momentum is given by $\Delta p = mv_f - mv_i$, where v_i is the initial velocity and v_f is the rebound velocity. Because the collision is elastic, the ball rebounds in the opposite direction but with the same speed with which it hit the wall. Therefore, $v_f = -v_i$, and the change in momentum is:

$$\Delta p = m(-v_i) - mv_i = -2mv_i = -2(0.20 \text{ kg})(20.0 \text{ m/s}) = -8.0 \text{ kg·m/s}$$

The magnitude of a number is its absolute value: $|-8.0 \text{ kg·m/s}| = 8.0 \text{ kg·m/s}$

7. (d). Force is related to momentum by the impulse-momentum equation, $F\Delta t = \Delta p$. Therefore:

$$F = \Delta p/\Delta t = (4 \text{ kg·m/s})/0.01 \text{ s} = 400 \text{ kg·m/s}^2 = 400 \text{ N}$$

8. (a). Momentum is conserved. Since both carts come to rest, the total momentum after the collision must be zero, which means the total momentum prior to the collision was also zero. If the two carts are labeled A and B, then:

$$m_A v_A = m_B v_B \quad \rightarrow \quad v_B = (30 \text{ kg})(5 \text{ m/s})/50 \text{ kg} = 3 \text{ m/s}$$

WORK

EXAMPLE 1

A car traveling at speed v can stop in distance d due to kinetic friction. At twice the speed, what stopping distance is required? (a) d (b) $2d$ (c) $4d$ (d) $8d$

Conceptual Solution

The answer is **(c)**. Answer (a) can be eliminated immediately, because at double the velocity the car's energy must be greater than before, hence the stopping distance must be greater. Because kinetic energy is proportional to v^2, doubling the velocity quadruples the energy. Work done by friction is linear in distance d, so four times the distance is required. This can more readily be seen through a proportion.

Quantitative Solution

Apply the work-energy theorem:

$$W = \Delta KE = KE_f - KE_i \tag{1}$$

Assume the car is traveling in the positive x-direction. Substitute expressions for work done by a kinetic friction force with magnitude F_k performed over a displacement Δx, and for kinetic energy, noting that $KE_f = 0$:

$$-F_k \Delta x = 0 - \tfrac{1}{2} m v^2 \tag{2}$$

Multiply both sides by -1 and drop the zero term:

$$F_k \Delta x = \tfrac{1}{2} m v^2 \tag{3}$$

Denote the two possible different displacements by d_1 and d_2, the two velocities by v_1 and v_2. Substitute into Equation (3), creating one equation for the first case and a similar equation for the second case:

$$F_k d_1 = \tfrac{1}{2} m v_1{}^2 \tag{4}$$
$$F_k d_2 = \tfrac{1}{2} m v_2{}^2 \tag{5}$$

Now divide equation (5) by Equation (4), canceling terms F_k, m, and $\tfrac{1}{2}$ that don't change:

$$\frac{F_k d_2}{F_k d_1} = \frac{\tfrac{1}{2} m v_2{}^2}{\tfrac{1}{2} m v_1{}^2} \quad \rightarrow \quad \frac{d_2}{d_1} = \frac{v_2{}^2}{v_1{}^2} = \frac{(2v)^2}{v^2} = 4$$

It follows, then, that answer **(c)** is correct:

$$d_2 = \boxed{4d_1} \, .$$

EXAMPLE 2

A truck of mass M and a car of mass m are traveling at the same speed. They both slam on their brakes, locking them and skidding along the road, which has friction coefficient μ_k. The car stops after going a distance d. Assuming μ_k is the same for both vehicles, which of the following is true of the distance D that the truck travels before stopping? (a) $D = d$ (b) $D > d$ (c) $D < d$

Conceptual Solution

Kinetic energy is directly proportional to mass, and because the truck is more massive than the car, its kinetic energy is greater, hence it might be thought a greater braking distance would be required. The friction force, however, is also directly proportional to mass, so the friction force is proportionately greater for the truck than the car. All other things being equal, the truck and car stop in the same distance, $D = d$, which is answer **(a).**

Quantitative Solution

Write the expression for the force of friction, where n is the normal force:

$$F_k = \mu_k n = \mu_k m g \tag{1}$$

CONCEPTUAL QUESTIONS

1. An ellipsoid. The dimension in the direction of motion would be measured to be less than D.

3. This scenario is not possible with light. Light waves are described by the principles of special relativity. As you detect the light wave ahead of you and moving away from you (which would be a pretty good trick—think about it!), its velocity relative to you is c. Thus, you will not be able to catch up to the light wave.

5. No. The principle of relativity implies that nothing can travel faster than the speed of light in a *vacuum*, which is 3.00×10^8 m/s.

7. The light from the quasar moves at 3.00×10^8 m/s. The speed of light is independent of the motion of the source or the observer.

9. For a wonderful fictional exploration of this question, get a "Mr. Tompkins" book by George Gamow. All of the relativity effects would be obvious in our lives. Time dilation and length contraction would both occur. Driving home in a hurry, you would push on the gas pedal not to increase your speed very much, but to make the blocks shorter. Big Doppler shifts in wave frequencies would make red lights look green as you approached and make car horns and radios useless. High-speed transportation would be both very expensive, requiring huge fuel purchases, as well as dangerous, since a speeding car could knock down a building. When you got home, hungry for lunch, you would find that you had missed dinner; there would be a five-day delay in transit when you watch a live TV program originating in Australia. Finally, we would not be able to see the Milky Way, since the fireball of the Big Bang would surround us at the distance of Rigel or Deneb.

11. A photon transports energy. The relativistic equivalence of mass and energy means that is enough to give it momentum.

13. Your assignment: measure the length of a rod as it slides past you. Mark the position of its front end on the floor and have an assistant mark the position of the back end. Then measure the distance between the two marks. This distance will represent the length of the rod only if the two marks were made simultaneously in your frame of reference.

PROBLEMS

1. (a) $t_{OB} = 1.67 \times 10^3$ s, $t_{OA} = 2.04 \times 10^3$ s
   (b) $t_{BO} = 2.50 \times 10^3$ s, $t_{AO} = 2.04 \times 10^3$ s
   (c) $\Delta t = 90$ s
3. 5.0 s
5. (a) 20 m (b) 19 m (c) $0.31c$
7. (a) 1.3×10^{-7} s (b) 38 m (c) 7.6 m
9. (a) 2.2 μs (b) 0.65 km
11. $0.950c$
13. Yes, with 19 m to spare
15. (a) 39.2 μs (b) Accurate to one digit
17. 3.3×10^5 m/s
19. $0.285c$
21. $0.54c$ to the right
23. $0.357c$
25. $0.998c$ toward the right
27. (a) 54 min (b) 52 min
29. $c(\sqrt{3}/2)$
31. $0.786c$

33. 18.4 g/cm^3
35. 1.98 MeV
37. 2.27×10^{23} Hz, 1.32 fm for each photon
39. (a) 3.10×10^5 m/s (b) $0.758c$
41. 1.42 MeV/c
43. (a) $0.80c$ (b) 7.5×10^3 s (c) 1.4×10^{12} m, $0.38c$
45. $0.37c$ in the $+ x$-direction
47. (a) $v/c = 1 - 1.12 \times 10^{-10}$ (b) 6.00×10^{27} J
   (c) $\$2.17 \times 10^{20}$
49. $0.80c$
51. (a) $0.946c$ (b) 0.160 ly (c) 0.114 yr (d) 7.50×10^{22} J
53. (a) 7.0 μs (b) 1.1×10^4 muons
59. 5.45 yr; Goslo is older.

Chapter 27

QUICK QUIZZES

1. (b)
2. (c)
3. (c)
4. (b)

CONCEPTUAL QUESTIONS

1. The shape of an object is normally determined by observing the light reflecting from its surface. In a kiln, the object will be very hot and will be glowing red. The emitted radiation is far stronger than the reflected radiation, and the thermal radiation emitted is only slightly dependent on the material from which the object is made. Unlike reflected light, the emitted light comes from all surfaces with equal intensity, so contrast is lost and the shape of the object is harder to discern.

3. The "blackness" of a blackbody refers to its ideal property of absorbing all radiation incident on it. If an observed room temperature object in everyday life absorbs all radiation, we describe it as (visibly) black. The black appearance, however, is due to the fact that our eyes are sensitive only to visible light. If we could detect infrared light with our eyes, we would see the object emitting radiation. If the temperature of the blackbody is raised, Wien's law tells us that the emitted radiation will move into the visible range of the spectrum. Thus, the blackbody could appear as red, white, or blue, depending on its temperature.

5. All objects do radiate energy, but at room temperature this energy is primarily in the infrared region of the electromagnetic spectrum, which our eyes cannot detect. (Pit vipers have sensory organs that *are* sensitive to infrared radiation; thus, they can seek out their warm-blooded prey in what we would consider absolute darkness.

7. Most metals have cutoff frequencies corresponding to photons in or near the visible range of the electromagnetic spectrum. AM radio wave photons have far too little energy to eject electrons from the metal.

9. We can picture higher frequency light as a stream of photons of higher energy. In a collision, one photon can give all of its energy to a single electron. The kinetic energy of such an electron is measured by the stopping potential. The reverse voltage (stopping voltage) required to stop the current is proportional to the frequency of the incoming light. More intense light consists of more photons striking a unit area each second, but atoms are so small that one emitted electron never gets a "kick" from more than one photon. Increasing the intensity of the

3. What happens to the speed of a body if its kinetic energy is doubled? (a) It increases by a factor of $2^{1/2}$. (b) It is doubled. (c) It is halved. (d) It increases by a factor of 4.

4. A ball with a mass of 1.0 kg sits at the top of a 30° incline plane that is 20.0 meters long. If the potential energy of the ball relative the bottom of the incline is 98 J at the top of the incline, what is its potential energy once it rolls halfway down the incline? (a) 0 J (b) 49 J (c) 98 J (d) 196 J

5. How much work is done by a horizontal force of 20.0 N when applied to a mass of 0.500-kg over a distance of 10.0 m? (a) 5.00 J (b) 10.0 J (c) 49.0 J (d) 2.00×10^2 J

6. A body located 10.0 meters above the surface of the Earth has a gravitational potential energy of 4.90×10^2 J relative to the Earth's surface. What is the new gravitational potential energy if the body drops to a height of 7.00 m above the Earth? (a) 70.0 J (b) 147 J (c) 281 J (d) 343 J

7. Cart A has a mass of 1.0 kg and a constant velocity of 3.0 m/s. Cart B has a mass of 1.5 kg and a constant velocity of 2.0 m/s. Which of the following statements is true? (a) Cart A has the greater kinetic energy. (b) Cart B has the greater kinetic energy. (c) Cart A has the greater acceleration. (d) Cart B has the greater acceleration.

8. The work done in raising a body must (a) increase the kinetic energy of the body. (b) decrease the total mechanical energy of the body. (c) decrease the internal energy of the body. (d) increase the gravitational potential energy.

9. What is the average power output of a 50.0-kg woman who climbs a 2.00-m step ladder at constant speed in 10.0 seconds? (a) 10.0 W (b) 49.0 W (c) 98.0 W (d) 2.50×10^2 W.

Answers

1. **(d).** The kinetic energy is directly proportional to the square of the velocity. If the velocity is tripled, then its square becomes $(3v)^2 = 9v^2$, and *KE* must increase by a factor of nine.

2. **(d).** The instantaneous power is the product of the force acting on the plane and the speed, $\mathscr{P} = Fv$. Solving for the required force gives:

$$F = \mathscr{P}/v = 500 \times 10^3 \, \text{W}/(100 \, \text{m/s}) = 5 \times 10^3 \, \text{N}$$

3. **(a).** Let KE_i be the initial kinetic energy, and KE_f the final kinetic energy. Make a proportion:

$$\frac{KE_f}{KE_i} = \frac{\frac{1}{2} mv_f^2}{\frac{1}{2} mv_i^2} \quad \rightarrow \quad \frac{(2KE_i)}{KE_i} = \frac{\frac{1}{2} mv_f^2}{\frac{1}{2} mv_i^2} \quad \rightarrow \quad 2 = \frac{v_f^2}{v_i^2}$$

Solve for v_f, getting $v_f = \sqrt{2} v_i$.

4. **(b).** Because the height is decreasing, the gravitational potential energy is decreasing and must have a value less than its initial reading of 98 J. Choice (d) can be eliminated because its value is higher than the initial value of the potential energy, as can choice (c), because its value is the same as the initial value of the potential energy. Choice (a) can be eliminated because by the problem statement, the gravitational potential energy is zero at the bottom of the ramp. That leaves (b) as the only option.

Quantitatively: $PE = mgh = (1.0 \, \text{kg})(9.8 \, \text{m/s}^2)((10.0 \, \text{m}) \cdot \sin(30°)) = 49 \, \text{J}$

5. **(d).** $W = F\Delta s = (20.0 \, \text{N})(10.0 \, \text{m}) = 2.00 \times 10^2 \, \text{J}$

6. **(d).** The gravitational potential energy and the height of the body above the ground are directly proportional: $PE = mgh$. Reducing the height by a factor of 7/10 must reduce the gravitational potential energy by the same factor: $(4.90 \times 10^2 \, \text{J})(0.700) = 343 \, \text{J}$.

7. **(a).** Because the velocity is constant, the acceleration must be zero and we can eliminate choices (c) and (d). Solving $KE = \frac{1}{2} mv^2$ for the two carts yields:

$$KE_A = \tfrac{1}{2}(1.0 \, \text{kg})(3.0 \, \text{m/s})^2 = 4.5 \, \text{J} \qquad KE_B = \tfrac{1}{2}(1.5 \, \text{kg})(4.0 \, \text{m/s})^2 = 3.0 \, \text{J}$$

8. **(d).** By definition, gravitational potential energy changes with the height of a body above the ground.

9. **(c).** The average power is the work done on an object divided by the length of the time interval, $\mathscr{P} = W/\Delta t$. The work done here is against gravity. In the period under consideration, the woman must increase her mechanical energy by an amount equal to her change in gravitational potential energy, so $\mathscr{P} = mgh/\Delta t = (50.0 \, \text{kg})(9.80 \, \text{m/s}^2)(2.00 \, \text{m})/10.0 \, \text{s} = 98.0 \, \text{W}$.

MATTER

EXAMPLE 1

Tank A is filled with water, while in an identical tank B the bottom half is filled with water and the top half with oil. For which tank is the pressure at the bottom of the tank the greatest? (a) Tank A (b) Tank B (c) The pressure in both tanks is the same.

Conceptual Solution

The oil in tank B floats on top of the water because it's less dense than water. The total weight of the fluid in tank B must therefore be less than an equal volume of water, alone. Consequently, the fluid weight per unit area is greater at the bottom of tank A, and the answer is **(a).**

EXAMPLE 2

A steel cable with cross-sectional area A and length L stretches by length ΔL_1 when suspending a given weight. A second steel cable with half the cross-sectional area and three times the length is used to support the same weight, and stretches by an amount ΔL_2. What is the ratio $\Delta L_2/\Delta L_1$? (a) 2 (b) 4 (c) 6 (d) 8

Conceptual Solution

Tensile stress, a force per unit area, is proportional to tensile strain, which is the fractional change in length of an object, or the change in length divided by the original length. The amount the cable stretches is thus inversely proportional to the cross-sectional area and directly proportional to the length. Hence the second cable, being three times as long, should stretch three times as much under a given load. With only half the cross-sectional area, it should stretch further by a factor of $(\frac{1}{2})^{-1}$, or 2. Taken together, the second cable should stretch six times as much as the first shorter and thicker cable, so $\Delta L_2/\Delta L_1 = 6$, which is answer **(c).**

Quantitative Solution

Write the equation relating tensile stress and strain for the first cable:

$$\frac{F_1}{A_1} = Y\frac{\Delta L_1}{L_1} \tag{1}$$

Rewrite Equation (1) for the second cable, noting that Y, Young's Modulus, is constant for a given material

$$\frac{F_2}{A_2} = Y\frac{\Delta L_2}{L_2} \tag{2}$$

Divide Equation (2) by Equation (1), canceling Y:

$$\frac{\frac{F_2}{A_2}}{\frac{F_1}{A_1}} = \frac{\frac{\Delta L_2}{L_2}}{\frac{\Delta L_1}{L_1}} \rightarrow \frac{F_2 A_1}{F_1 A_2} = \frac{\Delta L_2}{\Delta L_1}\frac{L_1}{L_2} \tag{3}$$

Cancel F_2 and F_1 in Equation (3), as they are equal in this case, rearrange the equation, and substitute $L_2 = 3L_1$ and $A_2 = \frac{1}{2}A_1$:

$$\frac{\Delta L_2}{\Delta L_1} = \frac{A_1}{A_2}\frac{L_2}{L_1} = \frac{A_1}{\frac{1}{2}A_1}\frac{3L_1}{L_1} = \frac{3}{\frac{1}{2}} = \boxed{6}$$

The answer is **(c),** as expected.

EXAMPLE 3

An object floats with three-quarters of its volume submerged in water. What is its density? ($\rho_{water} = 1.00 \times 10^3$ kg/m³) (a) 1.00×10^3 kg/m³ (b) 8.70×10^2 kg/m³ (c) 7.50×10^2 kg/m³ (d) 1.25×10^3 kg/m³

Conceptual Solution

Choice (d) can be eliminated immediately, because the object would sink if it were more dense than water. If choice (a) were true, then the object would be floating but would be completely submerged. These considerations leave only choices (b) and (c). The buoyant force depends on the amount of fluid displaced, and this force must exactly balance the gravitational force, or weight. Because the object is only three-quarters submerged, its weight must be only three-quarters that of a similar volume of water. It follows that the density of the object must be three-quarters that of water, which is answer **(c).**

Quantitative Solution

Neglecting the buoyancy of air, two forces act on the object: the buoyant force F_B, which is equal to the weight of displaced fluid, and the gravitational force:

$$\Sigma \vec{F} = \vec{F}_B + \vec{F}_{grav} = 0$$

Substitute the expressions for the buoyant and gravitational forces:

$$F_B - mg = 0 \tag{1}$$

By definition, the mass of the object is given by $m = \rho_{obj} V$, where ρ_{obj} is the object's density and V is its volume. The buoyant force is the weight of the displaced water: $F_B = \rho_{water} V_{sub} g$, where V_{sub} is the volume of water displaced. Substitute these expressions into Equation (1) and solve for ρ_{obj}:

$$\rho_{water} V_{sub} g - \rho_{obj} V g = 0 \tag{2}$$

Solve Equation (2) for ρ_{obj}, obtaining answer **(c):**

$$\rho_{obj} = \frac{V_{sub}}{V} \rho_{water} = (0.750)(1.00 \times 10^3 \text{ kg/m}^3)$$

$$= 7.50 \times 10^2 \text{ kg/m}^3$$

EXAMPLE 4

At point 1, water flows smoothly at speed v_1 through a horizontal pipe with radius r_1. The pipe then narrows to half that radius at point 2. What can be said of the speed v_1 of the water at point 1 compared to its speed v_2 at point 2? (a) $v_1 = v_2$ (b) $v_1 > v_2$ (c) $v_1 < v_2$

Conceptual Solution

The volume flow rate is proportional to the velocity and cross-sectional area of the pipe. A larger radius at point 1 means a larger cross-sectional area. Because water is essentially incompressible, the flow rate must be the same in both sections of pipe, so the larger cross-section at point 1 results in a smaller fluid velocity, and the answer is **(c).**

Quantitative Solution

Apply the equation of continuity for an incompressible fluid:

$$A_1 v_1 = A_2 v_2 \tag{1}$$

Substitute an expression for the area on each side of Equation (1):

$$\pi r_1^2 v_1 = \pi r_2^2 v_2 \tag{2}$$

Solve Equation (2) for v_1, and substitute $r_2 = \frac{1}{2} r_1$, obtaining answer **(c):**

$$v_1 = \frac{r_2^2}{r_1^2} v_2 = \frac{r_2^2}{(2r_2)^2} v_2 = \frac{1}{4} v_2$$

MULTIPLE CHOICE PROBLEMS

1. A diver is swimming 10.0 m below the surface of the water in a reservoir. There is no current, the air has a pressure of 1.00 atmosphere, and the density of the water is 1.00×10^3 kilograms per cubic meter. What is the pressure as measured by the diver? (a) 1.10 atm (b) 1.99×10^5 Pa (c) 11.0 atm (d) 1.01×10^5 Pa

2. The aorta of a 70.0-kg man has a cross-sectional area of 3.00 cm^2 and carries blood with a speed of 30.0 cm/s. What is the average volume flow rate? (a) 10.0 cm/s (b) 33.0 cm^3/s (c) 10.0 cm^2/s (d) 90.0 cm^3/s

3. At 20.0°C the density of water is 1.00 g/cm³. What is the density of a body that has a weight of 0.980 N in air but registers an apparent weight of only 0.245 N on a spring scale when fully immersed in water? (a) 0.245 g/cm³ (b) 0.735 g/cm³ (c) 1.33 g/cm³ (d) 4.00 g/cm³

4. Two insoluble bodies, A and B, appear to lose the same amount of weight when submerged in alcohol. Which statement is most applicable? (a) Both bodies have the same mass in air. (b) Both bodies have the same volume. (c) Both bodies have the same density. (d) Both bodies have the same weight in air.

5. The bottom of each foot of an 80-kg man has an area of about 400 cm². What is the effect of his wearing snowshoes with an area of about 0.400 m²? (a) The pressure exerted on the snow becomes 10 times as great. (b) The pressure exerted on the snow becomes 1/10 as great. (c) The pressure exerted on the snow remains the same. (d) The force exerted on the snow is 1/10 as great.

6. In a hydraulic lift, the surface of the input piston is 10 cm² and that of the output piston is 3 000 cm². What is the work done if a 100 N force applied to the input piston raises the output piston by 2.0 m? (a) 20 kJ (b) 30 kJ (c) 40 kJ (d) 60 kJ

7. The Young's modulus for steel is 2.0×10^{11} N/m². What is the stress on a steel rod that is 100 centimeters long and 20 millimeters in diameter when it is stretched by a force of 6.3×10^3 N? (a) 2.01×10^7 N/m² (b) 12.6×10^{12} N/m² (c) 3.15×10^8 N/m² (d) 4.0×10^{11} N/m²

Answers

1. (b). The fluid is at rest (no currents) so this is a hydrostatic pressure calculation. In SI units 1 atm = 1.01×10^5 Pa. Choice (d) can be eliminated because the pressure below the water must be greater than the pressure at the surface.

$$P_{\text{diver}} = P_{\text{atm}} + \rho g h = (1.01 \times 10^5 \text{ Pa}) + (1.00 \times 10^3 \text{ kg/m}^3)(9.80 \text{ m/s}^2)(10.0 \text{ m})$$
$$= 1.99 \times 10^5 \text{ Pa}$$

2. (d). The answer follows from the definition:

$$\text{volume flow rate} = vA = (30 \text{ cm/s})(3 \text{ cm}^2) = 90 \text{ cm}^3/\text{s}$$

3. (c). This problem is an application of Archimedes' Principle. The apparent loss of weight of a submerged body equals the weight of the fluid displaced. The weight w of the displaced water is $w = 0.980$ N $- 0.245$ N $= 0.735$ N. Using the definition of density, $\rho_{\text{water}} g = w/V$, so the volume occupied by this weight of water is $V = 0.735$ N$/\rho_{\text{water}} g$. The volume of the body must equal the volume of the water displaced. The density of the body is: $\rho = m/V = (0.980 \text{ N}/0.735 \text{ N})\rho_{\text{water}} = 1.33$ g/cm³

4. (b). According to Archimedes' Principle, the apparent weight lost is equal to the weight of the displaced fluid; therefore, both must have the same volume because the volume of the fluid equals the volume of the body.

5. (b). The force exerted by the man is his weight and it is assumed to be constant. This eliminates choice (d). For a constant force, pressure and area are inversely proportional. The area of the snowshoes is ten times the area of the foot so that the pressure associated with the snowshoes is the inverse of 10 or 1/10 the pressure exerted by the foot.

6. (d). By Pascal's principle, the pressure at the input and output pistons is the same, so

$$\frac{F_{\text{in}}}{A_{\text{in}}} = \frac{F_{\text{out}}}{A_{\text{out}}} \quad \rightarrow \quad F_{\text{out}} = \frac{A_{\text{out}}}{A_{\text{in}}} F_{\text{in}}$$

$$= \left(\frac{3\,000 \text{ cm}^2}{10 \text{ cm}^2}\right) 100 \text{ N} = 3 \times 10^4 \text{ N}$$

Work is a force times a displacement:

$$W = F\Delta s = (3 \times 10^4 \text{ N})(2 \text{ m}) = 6 \times 10^4 \text{ J} = 60 \text{ kJ}$$

7. (a). Stress is force per unit area, so neither Young's modulus nor the length of the rod are needed to solve the problem.

$$\text{Stress} = F/A = (6.30 \times 10^3 \text{ N})/\pi(1 \times 10^{-2} \text{ m})^2 = 2.01 \times 10^7 \text{ N/m}^2$$

WAVES

EXAMPLE 1

A transverse wave travels at speed v_1 and has twice the frequency and one-quarter the wavelength of a second transverse wave. (These could be waves on two different strings, for example.) How does the speed v_1 of the first wave compare to the speed v_2 of the second wave? (a) $v_1 = v_2$ (b) $v_1 = 2v_2$ (c) $v_1 = \frac{1}{2}v_2$.

Solution

The speed of a wave is proportional to both the frequency and wavelength:

$$v = f\lambda \tag{1}$$

Make a ratio of the v_2 to v_1 using Equation 1, and substitute $f_1 = 2f_2$ and $\lambda_1 = \lambda_2/4$:

$$\frac{v_1}{v_2} = \frac{f_1\lambda_1}{f_2\lambda_2} = \frac{(2f_2)(\lambda_2/4)}{f_2\lambda_2} = \frac{2}{4} = \frac{1}{2}$$

Solve for v_1, obtaining answer (c):

$$v_1 = \boxed{\tfrac{1}{2}v_2}$$

Example 2

A block of mass m oscillates at the end of a horizontal spring on a frictionless surface. At maximum extension, an identical block drops onto the top of the first block and sticks to it. How does the new period T_{new} compare to the original period, T_0? (a) $T_{new} = \sqrt{2}T_0$ (b) $T_{new} = T_0$ (c) $T_{new} = 2T_0$ (d) $T_{new} = \frac{1}{2}T_0$

Conceptual Solution

An increased mass would increase the inertia of the system without augmenting the mechanical energy, so a mass of $2m$ should move more slowly than a mass of m. This in turn would lengthen the period, eliminating choices (b) and (d). The period is proportional to the square root of the mass, so doubling the mass increases the period by a factor of the square root of two, which is answer (a).

Quantitative Solution

Apply the equation for the period of a mass-spring system:

$$T = 2\pi\sqrt{\frac{m}{k}} \tag{1}$$

Using Equation (1), make a ratio of the new period T_{new} to the old period, T_0, canceling common terms and substituting $m_{new} = 2m_0$:

$$\frac{T_{new}}{T_0} = \frac{2\pi\sqrt{\dfrac{m_{new}}{k}}}{2\pi\sqrt{\dfrac{m_0}{k}}} = \frac{\sqrt{m_{new}}}{\sqrt{m_0}} = \sqrt{\frac{2m_0}{m_0}} = \sqrt{2} \tag{2}$$

Solving Equation (2) for the new period T_{new} yields answer (a):

$$T_{new} = \boxed{\sqrt{2}\,T_0} \ [\text{answer (a)}]$$

EXAMPLE 3

A simple pendulum swings back and forth 36 times in 68 s. What is its length, if the local acceleration of gravity is 9.80 m/s^2? (a) 0.067 7 m (b) 0.356 m (c) 1.25 m (d) 0.887 m

Solution

First, find the period T of the motion:

$$T = \frac{68 \text{ s}}{36} = 1.89 \text{ s}$$

Write the equation for the period of a pendulum and solve it for the length L, obtaining answer (d):

$$T = 2\pi\sqrt{\frac{L}{g}} \quad \rightarrow \quad L = \frac{gT^2}{4\pi^2} = \frac{(9.80 \text{ m/s}^2)(1.89 \text{ s})^2}{4\pi^2}$$

$$= \boxed{0.887 \text{ m}} \ [\text{answer (d)}]$$

MULTIPLE CHOICE PROBLEMS

1. A simple pendulum has a period of 4.63 s at a place on the Earth where the acceleration of gravity is 9.82 m/s^2. At a different location, the period increases to 4.64 seconds. What is the value of g at this second point? (a) 9.78 m/s^2 (b) 9.82 m/s^2 (c) 9.86 m/s^2 (d) Cannot be determined without knowing the length of the pendulum.

2. What is the wavelength of a transverse wave having a speed of 15 m/s and a frequency of 5.0 Hz? (a) 3.0 m (b) 10 m (c) 20 m (d) 45 m

3. What is the optimum difference in phase for maximum destructive interference between two waves of the same frequency? (a) 360° (b) 270° (c) 180° (d) 90°

4. Standing waves can be formed if coincident waves have (a) the same direction of propagation. (b) the same frequency. (c) different amplitudes. (d) different wavelengths.

5. A simple pendulum with a length L has a period of 2 s. In order for the pendulum to have a period of 4 s, we must (a) halve the length. (b) quarter the length. (c) double the length. (d) quadruple the length.

6. If a simple pendulum 12 m long has a frequency of 0.25 Hz, what will be the period of a second pendulum at the same location if its length is 3.0 m? (a) 2.0 s (b) 3.0 s (c) 4.0 s (d) 6.0 s

7. A pendulum clock runs too slowly (i.e., is losing time). Which of the following adjustments could rectify the problem? (a) The weight of the bob should be decreased so it can move faster. (b) The length of the wire holding the bob should be shortened. (c) The amplitude of the swing should be reduced so the path covered is shorter. (d) None of the above.

8. A 20.0-kg object placed on a frictionless floor is attached to a wall by a spring. A 5.00-N force horizontally displaces the object 1.00 m from its equilibrium position. What is the period of oscillation of the object? (a) 2.00 s (b) 6.08 s (c) 12.6 s (d) 16.4 s

Answers

1. **(a).** The answer can be determined without doing a numerical solution. Rearrange $T = 2\pi(L/g)^{1/2}$ to $Tg^{1/2} = $ constant. The period is inversely related to the square root of the acceleration due to gravity. Because T has increased, $g^{1/2}$ and hence g must decrease. Choice (a) is the only value of g that is less than the original 9.82 m/s^2. Quantitatively:

$$g_2 = (T_1^2 g_1/T_2^2) = (4.63 \text{ s})^2(9.82 \text{ m/s}^2)/(4.64 \text{ s})^2 = 9.78 \text{ m/s}^2$$

2. **(a).** Wavelength is velocity divided by frequency. The formula does not depend upon the type of wave involved.

$$\lambda = v/f = (15 \text{ m/s})/5.0 \text{ s}^{-1} = 3.0 \text{ m}$$

3. **(c).** Two waves are completely out of phase when their antinodes coincide so that each crest on one wave coincides with a trough on the other. This occurs when the waves differ in phase by 180°.

4. **(b).** In standing waves, the nodes are stationary. This can be accomplished when two waves with the same frequency travel in opposite directions.

5. **(d).** In a pendulum, the period and the square root of the length are directly proportional:

$$T = 2\pi(L/g)^{1/2} \text{ so that } T/L^{1/2} = \text{constant}$$

To double the period, you must double the square root of the length. To double the square root of the length, you must quadruple the length:

$$(4L)^{1/2} = 4^{1/2} L^{1/2} = 2L^{1/2}$$

6. **(a).** The frequency is the reciprocal of the period: $f = 1/T$, so the first pendulum has period $T_1 = 4.0$ s. For pendulums, the period and the square root of the length are directly proportional, so the ratio of the two periods is:

$$T_2/T_1 = (L_2/L_1)^{1/2} = (12/3)^{1/2} = (4.0)^{1/2} = 2.0$$

It follows that $T_2 = 2.0$ s.

7. **(b).** The period of a pendulum is directly related to the square root of the length of the cord holding the bob. It is independent of the mass and amplitude.

8. **(c).** The force constant is $k = F/\Delta x = 5.0 \text{ N}/1.0 \text{ m} = 5.0 \text{ N} \cdot \text{m}^{-1}$. The period is

$$T = 2\pi(m/k)^{1/2} = 2\pi(20.0 \text{ kg}/5.00 \text{ N} \cdot \text{m}^{-1})^{1/2} = 12.6 \text{ s}$$

SOUND

EXAMPLE 1

When a given sound intensity doubles, by how much does the sound intensity level (or decibel level) increase?

Solution

Write the expression for a difference in decibel level, $\Delta\beta$:

$$\Delta\beta = \beta_2 - \beta_1 = 10\log\left(\frac{I_2}{I_0}\right) - 10\log\left(\frac{I_1}{I_0}\right)$$

Factor and apply the logarithm rule $\log a - \log b = \log(a/b) = \log(a\cdot b^{-1})$:

$$\Delta\beta = 10\left(\log\left(\frac{I_2}{I_0}\right) - \log\left(\frac{I_1}{I_0}\right)\right) = 10\log\left(\frac{I_2}{I_0}\cdot\frac{I_0}{I_1}\right)$$

$$= 10\log\left(\frac{I_2}{I_1}\right)$$

Substitute $I_2 = 2I_1$ and evaluate the expression:

$$\Delta\beta = 10\log\left(\frac{2I_1}{I_1}\right) = 10\log(2) = \boxed{3.01\text{ dB}}$$

EXAMPLE 2

A sound wave passes from air into water. What can be said of its wavelength in water as compared to air? (a) The wavelengths are the same. (b) The wavelength in air is greater than in water. (c) The wavelength in air is less than in water.

Conceptual Solution

The frequency of the sound doesn't change in going from one type of media to the next because it's caused by periodic variations of pressure in time. The wavelength must change, however, because the speed of sound changes and during a single period the sound wave will travel a different distance, which corresponds to a single wavelength. The speed of sound in water is greater than in air, so in a single period the sound wave will travel a greater distance. Hence the wavelength of a given sound wave is greater in water than in air, which is **(c).** This result can be made quantitative by using $v = f\lambda$, where v is the wave speed, f the frequency, and λ the wavelength.

EXAMPLE 3

A sound wave emitted from a sonar device reflects off a submarine that is traveling away from the sonar source. How does the frequency of the reflected wave, f_R, compare to the frequency of the source, f_S? (a) $f_R = f_S$ (b) $f_R < f_S$ (c) $f_R > f_S$

Solution

The reflected wave will have a lower frequency [answer **(b)**]. To see this, consider pressure maximums impinging sequentially on the submarine. The first pressure maximum hits the submarine and reflects, but the second reaches the submarine when it has moved further away. The distance between consecutive maximums in the reflected waves is thereby increased; hence the wavelength also increases. Because $v = f\lambda$ and the speed isn't affected, the frequency must decrease.

MULTIPLE CHOICE PROBLEMS

1. The foghorn of a ship echoes off an iceberg in the distance. If the echo is heard 5.00 seconds after the horn is sounded, and the air temperature is $-15.0°C$, how far away is the iceberg? (a) 224 m (b) 805 m (c) 827 m (d) 930 m

2. What is the sound level of a wave with an intensity of 10^{-3} W/m^2? (a) 30 dB (b) 60 dB (c) 90 dB (d) 120 dB

3. At $0°C$, approximately how long does it take sound to travel 5.00 km through air? (a) 15 s (b) 30 s (c) 45 s (d) 60 s

4. If the speed of a transverse wave of a violin string is 12.0 m/s and the frequency played is 4.00 Hz, what is the wavelength of the sound in air? (Use 343 m/s for the speed of sound.) (a) 48.0 m (b) 12.0 m (c) 3.00 m (d) 85.8 m

5. If two identical sound waves interact in phase, the resulting wave will have a (a) shorter period. (b) larger amplitude. (c) higher frequency. (d) greater velocity.

6. What is the speed of a longitudinal sound wave in a steel rod if Young's modulus for steel is 2.0×10^{11} N/m^2 and the density of steel is 8.0×10^3 kg/m^3? (a) 4.0×10^{-8} m/s (b) 5.0×10^3 m/s (c) 25×10^6 m/s (d) 2.5×10^9 m/s

7. If two frequencies emitted from two sources are 48 and 54 Hz, how many beats per second are heard? (a) 3 (b) 6 (c) 9 (d) 12

8. The frequency registered by a detector is higher than the frequency emitted by the source. Which of the statements below **must** be true? (a) The source must be moving away from the detector. (b) The source must be moving toward the detector. (c) The distance between the source and the detector must be decreasing. (d) The detector must be moving away from the source.

Answers

1. (b). The normal speed of sound at 0°C in air is 331 m/s. The speed of sound in air at various temperatures is given by

$$v = (331 \text{ m/s}) \sqrt{\frac{T}{273\ K}} = (331 \text{ m/s}) \sqrt{\frac{273\ K - 15.0\ K}{273\ K}}$$

$$= 322 \text{ m/s}$$

Calculate the distance traveled in half of five seconds:

$$d = vt = (322 \text{ m/s})(2.50 \text{ s}) = 805 \text{ m}.$$

2. (c). Substitute: $\beta = 10 \log I/I_0 = 10 \log (1 \times 10^{-3} \text{ W/m}^2/10^{-12} \text{ W/m}^2) = 90$ dB.

3. (a). At 0°C the speed of sound in air is 331 m/s, so $t = (5.00 \times 10^3 \text{ m})/(331 \text{ m/s}) \sim 15$ s

4. (d). The frequency is the same for the string as for the sound wave the string produces in the surrounding air, but the wavelength differs. The speed of sound is equal to the product of the frequency and wavelength, so $\lambda = v/f = (343 \text{ m/s})/4.00 \text{ s}^{-1} = 85.8$ m.

5. (b). Two waves are in phase if their crests and troughs coincide. The amplitude of the resulting wave is the algebraic sum of the amplitudes of the two waves being superposed at that point, so the amplitude is doubled.

6. (b). The solution requires substituting into an expression for the velocity of sound through a rod of solid material having Young's modulus Y:

$$v = (Y/\rho)^{1/2} = (2.0 \times 10^{11} \text{ kg} \cdot \text{m/s}^2)/(8.0 \times 10^3 \text{ kg/m}^3)^{1/2}$$
$$= 25 \times 10^6 \text{ m}^2/\text{s}^2)^{1/2} = 5.0 \times 10^3 \text{ m/s}$$

7. (b). $f_{\text{beat}} = f_1 - f_2 = 54 - 48 = 6$ beats

8. (c). Movement of a sound source relative an observer is the Doppler effect. Because the frequency is shifted to a higher value, the source and the detector must be getting closer together—effectively shortening the wavelength from the observer's point of view. This observation eliminates choices (a) and (d). Choice (b) can be eliminated because it isn't necessarily true: the same effect kind of effect occurs when the source is held steady and the detector moves towards it.

LIGHT

EXAMPLE 1

Which substance will have smaller critical angle for total internal reflection, glass with an index of refraction of 1.50, or diamond with an index of refraction of 2.42? (a) glass (b) diamond (c) The critical angles are the same for both.

Conceptual Solution

A larger index of refraction indicates a slower speed of light inside the material, which in turn means a larger refraction angle—a larger bending towards the normal on entering the material from air, and a larger bending away from the normal on passing from the material into air. (Recall that a normal line is perpendicular to the surface of the material.) For total internal reflection to occur, a refraction angle of 90° must be possible, which occurs whenever the refracting medium has a lower index of refraction than the incident medium. Diamond bends light more than glass, so the incident ray can be closer to the normal and still be bent enough so the angle of refraction results in total internal reflection. Closer to the normal means a smaller critical angle, so the answer is **(b).**

Quantitative Solution

Write Snell's law for the diamond-air interface:

$$n_D \sin \theta_D = n_A \sin \theta_A$$

Compute the critical angle for diamond, using $n_A = 1.00$ for air and $\theta_A = 90°$, together with $n_D = 2.42$:

$$2.42 \sin \theta_D = 1.00 \quad \rightarrow \quad \theta_D = \sin^{-1}\left(\frac{1.00}{2.42}\right) = 24.4°$$

Repeat the calculation for glass:

$$1.50 \sin \theta_G = 1.00 \quad \rightarrow \quad \theta_G = \sin^{-1}\left(\frac{1.00}{1.50}\right) = 41.8°$$

The calculation explicitly shows that diamond has a smaller critical angle than glass (answer **b**).

EXAMPLE 2

A patient's near point is 85.0 cm. What focal length prescription lens will allow the patient to see objects clearly that are at a distance of 25.0 cm from the eye? Neglect the eye-lens distance.

Solution

Use the thin lens equation:

$$\frac{1}{f} = \frac{1}{p} + \frac{1}{q}$$

An object at distance $p = 25.0$ cm must form an image at the patient's near point of 85.0 cm. The image must be virtual, so $q = -85.0$ cm:

$$\frac{1}{f} = \frac{1}{25.0 \text{ cm}} + \frac{1}{-85.0 \text{ cm}} = 2.82 \times 10^{-2} \text{ cm}^{-1}$$

$$f = 35.4 \text{ cm}$$

EXAMPLE 3

Two photons traveling in vacuum have different wavelengths. Which of the following statements is true? (a) The photon with a smaller wavelength has greater energy. (b) The photon with greater wavelength has greater energy. (c) The energy of all photons is the same. (d) The photon with the greater wavelength travels at a lesser speed.

Solution

Answer (d) can be eliminated immediately, because the speed of light in vacuum is the same for all wavelengths of light. The energy of a photon, or particle of light, is given by $E = hf$, and consequently is proportional to the frequency f, which in turn is inversely proportional to the wavelength because $f = c/\lambda$, where c is the speed of light. A smaller wavelength photon has a greater frequency, and therefore a greater energy, so answer **(a)** is true.

MULTIPLE CHOICE PROBLEMS

1. Glass has an index of refraction of 1.50. What is the frequency of light that has a wavelength of 5.00×10^2 nm in glass? (a) 1.00 Hz (b) 2.25 Hz (c) 4.00×10^{14} Hz (d) 9.00×10^{16} Hz

2. Water has an index of refraction of 1.33. If a plane mirror is submerged in water, what can be said of the angle of reflection θ if light strikes the mirror with an angle of incidence of 30°? (a) $\theta < 30°$ (b) $\theta = 30°$ (c) $30° < \theta$ (d) No light is reflected because 30° is the critical angle for water.

3. The index of refraction for water is 1.33 and that for glass is 1.50. A light ray strikes the water-glass boundary with an incident angle of 30.0° on the water side. Which of the following is the refraction angle in the glass? (a) 26.3° (b) 34.7° (c) 30.0° (d) 60.0°

4. Light is incident on a prism at an angle of 90° relative to its surface. The index of refraction of the prism material is 1.50. Which of the following statements is most accurate about the angle of refraction θ? (a) $0° < \theta < 45°$ (b) $45° < \theta < 90°$ (c) $\theta = 0°$ (d) $90° < \theta$

5. White light incident on an air-glass interface is split into a spectrum within the glass. Which color light has the greatest angle of refraction? (a) red light (b) violet light (c) yellow light (d) The angle is the same for all wavelengths.

6. A real object is placed 10.0 cm from a converging lens that has a focal length of 6.00 cm. Which statement is most accurate? (a) The image is real, upright, and enlarged. (b) The image is real, inverted, and enlarged. (c) The image is real, upright, and reduced. (d) The image is real, inverted, and reduced.

7. What is the focal length of a lens that forms a virtual image 30.0 cm from the lens when a real object is placed 15.0 cm from the lens? (a) 10.0 cm (b) 15.0 cm (c) 30.0 cm (d) 45.0 cm

8. What is the magnification of a lens that forms an image 20.0 cm to its right when a real object is placed 10.0 cm to its left? (a) 0.500 (b) 1.00 (c) 1.50 (d) -2.00

9. The human eye can respond to light with a total energy of as little as 10^{-18} J. If red light has a wavelength of 600 nm, what is the minimum number of red light photons that the eye can perceive? (a) 1 (b) 2 (c) 3 (d) 5

10. Which phenomenon occurs for transverse waves but not for longitudinal waves? (a) reflection (b) refraction (c) diffraction (d) polarization

Answers

1. **(c).** The velocity of light in glass can be found from the definition of refractive index, $n = c/v$, wavelength and frequency are related to velocity by the general wave relation, $v = f\lambda$. Therefore:

$$f = v/\lambda = c/n\lambda = (3.00 \times 10^8 \text{ m/s})/(1.50)(5.00 \times 10^{-7} \text{ m}) = 4.00 \times 10^{14} \text{ Hz}$$

2. **(b).** The law of reflection is independent of the medium involved. The angle of reflection is always equal to the angle of incidence.

3. **(a).** Snell's law is given by $n_1 \sin \theta_1 = n_2 \sin \theta_2$, hence: $1.33 \sin 30.0° = 1.50 \sin \theta_2$. From this expression, we see that $\sin \theta_2 < \sin 30.0°$, so θ_2 must be less than 30.0°. Therefore, (a) is the only reasonable choice.

4. **(c).** Snell's law is $n_1 \sin \theta_1 = n_2 \sin \theta_2$. Because the incident rays are normal to the prism surface, $\theta_1 = 0°$, and $\sin 0° = 0$. Consequently, $0 = n_2 \sin \theta_2 \rightarrow \theta_2 = 0°$

5. **(b).** The greater the frequency of light, the greater its energy and the faster its speed through any material medium. From Snell's Law the velocity and $\sin \theta$ with respect to the normal are inversely proportional. Of the choices, violet light has the highest frequency and therefore, the highest velocity, and the greatest angle of refraction.

6. **(b).** By the thin lens equation, $1/f = 1/p + 1/q$, hence with $p = 10.0$ cm and $f = 6.00$ cm, substitution results in $q = 15.0$ cm. The image distance is positive, hence the image is real. The magnification is given by $M = -q/p = -15.0 \text{ cm}/10.0 = -1.50$. Because M is negative, the image is inverted, whereas $|M| > 1$ means the image is enlarged.

7. **(c).** The distance is given by the thin lens equation, $1/f = 1/p + 1/q$. Because the image formed is virtual, the sign of q is negative. Therefore:

$$\frac{1}{f} = \frac{1}{p} + \frac{1}{q} = \frac{1}{15.0 \text{ cm}} + \frac{1}{-30.0 \text{ cm}} = \frac{1}{30.0 \text{ cm}} \rightarrow f = 30.0 \text{ cm}$$

8. **(d).** Magnification is given by $M = -q/p = -20.0 \text{ cm}/10.0 \text{ cm} = -2.00$. The sign of q is positive because the image is real. The negative value of M means the image is inverted.

9. (c). $E = hc/\lambda = (6.63 \times 10^{-34}\,\text{J s})(3.00 \times 10^8\,\text{m/s})/(6.00 \times 10^{-7}\,\text{m}) = 3.31 \times 10^{-19}\,\text{J}$.
This is the energy of each red photon. The number of such photons needed to produce a total of $10^{-18}\,\text{J}$ of energy is:

$$(10^{-18}\,\text{J})/(3.31 \times 10^{-19}\,\text{J/photon}) \sim 3\text{ photons}$$

10. (d). Polarization can only occur with transverse waves because the motion must be perpendicular to the direction of propagation.

ELECTROSTATICS

EXAMPLE 1

Two protons, each of charge $q_p = e$, exert an electric force of magnitude F_{p-p} on each other when they are a distance r_p apart. A pair of alpha particles, each of charge $q_\alpha = 2e$, exert an electric force $F_{\alpha-\alpha} = \frac{1}{4}F_{p-p}$ on each other. What is the distance between the alpha particles, r_α, in terms of the distance between the protons, r_p? (a) $r_\alpha = 4r_p$ (b) $r_\alpha = 2r_p$ (c) $r_\alpha = r_p$ (d) More information is needed.

Solution

Use Coulomb's law to find an expression for the force between the protons:

$$F_{p-p} = \frac{k_e q_p q_p}{r_p^2} = \frac{k_e e^2}{r_p^2} \qquad (1)$$

Use Coulomb's law to find and expression for the force between the alpha particles:

$$F_{\alpha-\alpha} = \frac{k_e q_\alpha q_\alpha}{r_\alpha^2} = \frac{k_e (2e)^2}{r_\alpha^2} = \frac{4k_e e^2}{r_\alpha^2} \qquad (2)$$

Divide Equation (2) by Equation (1) and cancel common terms:

$$\frac{F_{p-p}}{F_{\alpha-\alpha}} = \frac{\dfrac{k_e e^2}{r_p^2}}{\dfrac{4k_e e^2}{r_\alpha^2}} = \frac{r_\alpha^2}{4r_p^2} \qquad (3)$$

Now substitute $F_{\alpha-\alpha} = \frac{1}{4}F_{p-p}$ and solve for r_α^2:

$$\frac{r_\alpha^2}{4r_p^2} = \frac{F_{p-p}}{F_{\alpha-\alpha}} = \frac{F_{p-p}}{\frac{1}{4}F_{p-p}} = 4 \quad \rightarrow \quad r_\alpha^2 = 16r_p^2$$

Take square roots, obtaining r_α in terms of r_p:

$$r_\alpha = \boxed{4r_p}$$

The distance between the alpha particles is four times the distance between the protons, which is answer **(a)**.

EXAMPLE 2

Sphere A has twice the radius of a second, very distant sphere B. Let the electric potential at infinity be taken as zero. If the electric potential at the surface of sphere A is the same as at the surface of sphere B, what can be said of the charge Q_A on sphere A as compared to charge Q_B on B? (a) $Q_A = 2Q_B$ (b) $Q_A = Q_B$ (c) $Q_A = Q_B/2$

Conceptual Solution

By Gauss's law, a spherical distribution of charge creates an electric field outside the sphere as if all the charge were concentrated as a point charge at the center of the sphere. The electric potential due to a point charge is proportional to the charge Q and inversely proportional to the distance from that charge. Twice the radius reduces the electric potential at the surface of A by a factor of one-half. The charge of sphere A must be twice that of B so that the electric potentials will be the same for both spheres. Hence the answer is **(a)**.

Quantitative Solution

Write the equation for the electric potential of a point charge q:

$$V = \frac{kQ}{r} \qquad (1)$$

Make a ratio of Equation (1) for charge A and charge B, respectively:

$$\frac{V_A}{V_B} = \frac{\dfrac{kQ_A}{r_A}}{\dfrac{kQ_B}{r_B}} = \frac{Q_A r_B}{Q_B r_A} \qquad \textbf{(2)}$$

Substitute $V_A = V_B$ and $r_A = 2r_B$ into Equation (2) and solve for Q_A, again obtaining [anwer **(a)**]:

$$1 = \frac{Q_A r_B}{Q_B (2r_B)} \quad \rightarrow \quad Q_A = \boxed{2Q_B} \;\; [\text{answer \textbf{(a)}}]$$

EXAMPLE 3

How much work is required to bring a proton with charge 1.6×10^{-19} C and an alpha particle with charge 3.2×10^{-19} C from rest at a great distance (effectively infinity) to rest positions a distance of 1.00×10^{-15} m away from each other?

Solution

Use the work-energy theorem:

$$W = \Delta KE + \Delta PE = KE_f - KE_i + PE_f - PE_i$$

The velocities are zero both initially and finally, so the kinetic energies are zero. Substitute values into the potential energy and find the necessary work to assemble the configuration:

$$W = 0 - 0 + \frac{k_e q_p q_\alpha}{r} - 0$$

$$= \frac{(9.00 \times 10^9 \text{ kg} \cdot \text{m}^3/\text{C}^2 \cdot \text{s}^2)}{1.00 \times 10^{-15} \text{ m}} \times (1.60 \times 10^{-19} \text{ C})(3.20 \times 10^{-19} \text{ C})$$

$$= \boxed{4.61 \times 10^{-13} \text{ J}}$$

EXAMPLE 4

A fixed constant electric field E accelerates a proton from rest through a displacement Δs. A fully-ionized lithium atom with three times the charge of the proton accelerates through the same constant electric field and displacement. Which of the following is true of the kinetic energies of the particles? (a) The kinetic energies of the two particles are the same. (b) The kinetic energy of the proton is larger. (c) The kinetic energy of the lithium ion is larger.

Solution

The work done by the electric field on a particle of charge q is given by $W = F\Delta s = qE\Delta s$. The electric field E and displacement Δs are the same for both particles so the field does three times as much work on the lithium ion. The Work-Energy Theorem for this physical context is $W = \Delta KE$, so the lithium ion's kinetic energy is three times that of the proton, and the answer is **(c)**.

MULTIPLE CHOICE PROBLEMS

1. What is the potential difference between point A and point B if 10.0 J of work is required to move a charge of 4.00 C from one point to the other? (a) 0.400 V (b) 2.50 V (c) 14.0 V (d) 40.0 V

2. How much work would have to be done by a nonconservative force in moving an electron through a positive potential difference of 2.0×10^6 V? Assume the electron is at rest both initially and at its final position. (a) 3.2×10^{-13} J (b) -8.0×10^{-26} J (c) 1.25×10^{25} J (d) -3.2×10^{-13} J

3. Two electrically neutral materials are rubbed together. One acquires a net positive charge. The other must have (a) lost electrons. (b) gained electrons. (c) lost protons. (d) gained protons.

4. What is the magnitude of the charge on a body that has an excess of 20 electrons? (a) 3.2×10^{-18} C (b) 1.6×10^{-18} C (c) 3.2×10^{-19} C (d) 2.4×10^{-19} C

5. Two point charges, A and B, with charges of 2.00×10^{-4} C and -4.00×10^{-4} C, respectively, are separated by a distance of 6.00 m. What is the magnitude of the electrostatic force exerted on charge A? (a) 2.20×10^{-9} N (b) 1.30 N (c) 20.0 N (d) 36.0 N

6. Two point charges, A and B, are separated by 10.0 m. If the distance between them is reduced to 5.00 m, the force exerted on each (a) decreases to one-half its original value. (b) increases to twice its original value. (c) decreases to one-quarter of its original value. (d) increases to four times its original value.

7. Sphere A with a net charge of $+3.0 \times 10^{-3}$ C is touched to a second sphere B, which has a net charge of -9.0×10^{-3} C. The two spheres, which are exactly the same size and composition, are then separated. The net charge on sphere A is now: (a) $+3.0 \times 10^{-3}$ C (b) -3.0×10^{-3} C (c) -6.0×10^{-3} C (d) -9.0×10^{-3} C

8. If the charge on a particle in an electric field is reduced to half its original value, the force exerted on the particle by the field is (a) doubled. (b) halved. (c) quadrupled. (d) unchanged.

9. In the figure below, points A, B, and C are at various distances from a given point charge.

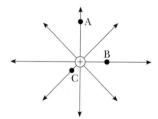

Which statement is most accurate? The electric field strength is (a) greatest at point A. (b) greatest at point B. (c) greatest at point C. (d) the same at all three points.

10. The electrostatic force between two point charges is F. If the charge of one point charge is doubled and that of the other charge is quadrupled, the force becomes which of the following? (a) $F/2$ (b) $2F$ (c) $4F$ (d) $8F$

Answers

1. (b). Potential difference between two points in an electric field is the work per unit charge required to move a charge between the two points:

$$\Delta V = W/q = 10.0 \, \text{J}/4.00 \, \text{C} = 2.50 \, \text{V}$$

2. (d). If the only effect on the particle is a change of position, negative work must be done, because otherwise negative charges gain kinetic energy on moving through a positive potential difference.

$$W = \Delta KE + q\Delta V = 0 + (-1.6 \times 10^{-19} \, \text{C})(2.0 \times 10^6 \, \text{V})$$
$$= -3.2 \times 10^{-13} \, \text{J}$$

3. (b). Protons are fixed in the nucleus and cannot be transferred by friction. Electrons can be transferred by friction. Therefore, net charges are due to the transfer of electrons between two bodies. Conservation of charge means that if there is a net positive charge, one body must have lost electrons and the other body must have gained the electrons.

4. (a). The elementary charge $e = 1.6 \times 10^{-19}$ C, so the total charge of 20 electrons has a magnitude of $20(1.6 \times 10^{-19} \, \text{C}) = 3.2 \times 10^{-18}$ C.

5. (c). Apply Coulomb's law: $F = k(q_A q_B)/r^2$

$$F = (9.00 \times 10^9 \, \text{N m}^2/\text{C}^2)(2.00 \times 10^{-4} \, \text{C})(-4.00 \times 10^{-4} \, \text{C})/(6.00 \, \text{m})^2 = -20.0 \, \text{N},$$

so the magnitude is 20.0 N.

3. (a), (e); (a), (e)

4. (a) False (b) False (c) True (d) False

5. (a)

CONCEPTUAL QUESTIONS

1. An ellipsoid. The dimension in the direction of motion would be measured to be less than D.

3. This scenario is not possible with light. Light waves are described by the principles of special relativity. As you detect the light wave ahead of you and moving away from you (which would be a pretty good trick—think about it!), its velocity relative to you is c. Thus, you will not be able to catch up to the light wave.

5. No. The principle of relativity implies that nothing can travel faster than the speed of light in a *vacuum*, which is 3.00×10^8 m/s.

7. The light from the quasar moves at 3.00×10^8 m/s. The speed of light is independent of the motion of the source or the observer.

9. For a wonderful fictional exploration of this question, get a "Mr. Tompkins" book by George Gamow. All of the relativity effects would be obvious in our lives. Time dilation and length contraction would both occur. Driving home in a hurry, you would push on the gas pedal not to increase your speed very much, but to make the blocks shorter. Big Doppler shifts in wave frequencies would make red lights look green as you approached and make car horns and radios useless. High-speed transportation would be both very expensive, requiring huge fuel purchases, as well as dangerous, since a speeding car could knock down a building. When you got home, hungry for lunch, you would find that you had missed dinner; there would be a five-day delay in transit when you watch a live TV program originating in Australia. Finally, we would not be able to see the Milky Way, since the fireball of the Big Bang would surround us at the distance of Rigel or Deneb.

11. A photon transports energy. The relativistic equivalence of mass and energy means that is enough to give it momentum.

13. Your assignment: measure the length of a rod as it slides past you. Mark the position of its front end on the floor and have an assistant mark the position of the back end. Then measure the distance between the two marks. This distance will represent the length of the rod only if the two marks were made simultaneously in your frame of reference.

PROBLEMS

1. (a) $t_{OB} = 1.67 \times 10^3$ s, $t_{OA} = 2.04 \times 10^3$ s
   (b) $t_{BO} = 2.50 \times 10^3$ s, $t_{AO} = 2.04 \times 10^3$ s
   (c) $\Delta t = 90$ s

3. 5.0 s

5. (a) 20 m (b) 19 m (c) 0.31c

7. (a) 1.3×10^{-7} s (b) 38 m (c) 7.6 m

9. (a) 2.2 μs (b) 0.65 km

11. 0.950c

13. Yes, with 19 m to spare

15. (a) 39.2 μs (b) Accurate to one digit

17. 3.3×10^5 m/s

19. 0.285c

21. 0.54c to the right

23. 0.357c

25. 0.998c toward the right

27. (a) 54 min (b) 52 min

29. $c(\sqrt{3}/2)$

31. 0.786c

33. 18.4 g/cm^3

35. 1.98 MeV

37. 2.27×10^{23} Hz, 1.32 fm for each photon

39. (a) 3.10×10^5 m/s (b) 0.758c

41. 1.42 MeV/c

43. (a) 0.80c (b) 7.5×10^3 s (c) 1.4×10^{12} m, 0.38c

45. 0.37c in the $+ x$-direction

47. (a) $v/c = 1 - 1.12 \times 10^{-10}$ (b) 6.00×10^{27} J
   (c) \$$2.17 \times 10^{20}$

49. 0.80c

51. (a) 0.946c (b) 0.160 ly (c) 0.114 yr (d) 7.50×10^{22} J

53. (a) 7.0 μs (b) 1.1×10^4 muons

59. 5.45 yr; Goslo is older.

Chapter 27

QUICK QUIZZES

1. (b)

2. (c)

3. (c)

4. (b)

CONCEPTUAL QUESTIONS

1. The shape of an object is normally determined by observing the light reflecting from its surface. In a kiln, the object will be very hot and will be glowing red. The emitted radiation is far stronger than the reflected radiation, and the thermal radiation emitted is only slightly dependent on the material from which the object is made. Unlike reflected light, the emitted light comes from all surfaces with equal intensity, so contrast is lost and the shape of the object is harder to discern.

3. The "blackness" of a blackbody refers to its ideal property of absorbing all radiation incident on it. If an observed room temperature object in everyday life absorbs all radiation, we describe it as (visibly) black. The black appearance, however, is due to the fact that our eyes are sensitive only to visible light. If we could detect infrared light with our eyes, we would see the object emitting radiation. If the temperature of the blackbody is raised, Wien's law tells us that the emitted radiation will move into the visible range of the spectrum. Thus, the blackbody could appear as red, white, or blue, depending on its temperature.

5. All objects do radiate energy, but at room temperature this energy is primarily in the infrared region of the electromagnetic spectrum, which our eyes cannot detect. (Pit vipers have sensory organs that *are* sensitive to infrared radiation; thus, they can seek out their warm-blooded prey in what we would consider absolute darkness.

7. Most metals have cutoff frequencies corresponding to photons in or near the visible range of the electromagnetic spectrum. AM radio wave photons have far too little energy to eject electrons from the metal.

9. We can picture higher frequency light as a stream of photons of higher energy. In a collision, one photon can give all of its energy to a single electron. The kinetic energy of such an electron is measured by the stopping potential.

than R_3, again because of a denominator larger than one, and so R_P is less than R_{combo}. The purely parallel combination therefore yields the least resistance, and the answer is **(c)**.

EXAMPLE 2

Two resistors are to dissipate as much energy as possible when a fixed voltage difference is placed across them. Should they be installed in parallel or in series?

Conceptual Solution

The power dissipated by a resistor is proportional to the voltage difference squared and inversely proportional to the resistance. The smallest possible resistance will therefore result in the largest power output. Consequently, the two resistors should be placed in parallel, which results in the lowest combined resistance.

EXAMPLE 3

A certain off-the-shelf resistor of resistivity ρ has resistance R. Suppose it is commercially desirable to design a new resistor that has one-third the length and one-fourth the cross-sectional area of the existing resistor, but have the same overall resistance. What should the resistivity ρ_n of the new resistor be, in terms of the resistivity ρ of the original device?

Solution

Write an expression for the resistance R_n of the new resistor, in terms of its resistivity, cross-sectional area A, and its length L:

$$R_n = \frac{\rho_n L_n}{A_n} \tag{1}$$

Divide Equation (1) by the same expression for the original resistor:

$$\frac{R_n}{R} = \frac{\frac{\rho_n L_n}{A_n}}{\frac{\rho L}{A}} = \frac{\rho_n L_n A}{\rho L A_n} \tag{2}$$

Substitute $A_n = A/4$, $R_n = R$, and $L_n = L/3$ into Equation (2):

$$1 = \frac{\rho_n (L/3) A}{\rho L (A/4)} = \frac{4\rho_n}{3\rho} \tag{3}$$

Solve Equation (3) for the new resistivity, ρ_n:

$$\rho_n = \boxed{\tfrac{3}{4}\rho}$$

MULTIPLE CHOICE PROBLEMS

1. Three resistors of resistance 1.0 Ω, 2.0 Ω, and 3.0 Ω, respectively, are in series. If a potential difference of 12 V is put across the combination, what is the resulting current in the circuit? (a) 0.50 A (b) 2.0 A (c) 6.0 A (d) 12 A

2. If the length of a conducting wire with resistance R is doubled, the resistance of the longer wire will be (a) $R/4$ (b) $R/2$ (c) $2R$ (d) $4R$

3. If a 2.0 Ω resistor and a 6.0 Ω resistor are connected in parallel, what is their combined resistance? (a) 1.5 Ω (b) 4.0 Ω (c) 8.0 Ω (d) 12 Ω

4. If all the components of an electric circuit are connected in series, which of the following physical quantities must be the same at all points in the circuit? (a) voltage (b) current (c) resistance (d) power

5. The current in a conductor is 3.0 A when it is connected across a 6.0-V battery. How much power is delivered? (a) 0.50 W (b) 2.0 W (c) 9.0 W (d) 18 W

6. A 12-Ω resistor is connected across a 6.0-V dc source. How much energy is delivered to the resistor in half an hour? (a) 1.5×10^{-3} kWh (b) 2.0×10^{-3} kWh (c) 3.0×10^{-3} kWh (d) 12×10^{-3} kWh

7. A battery with an emf of 6.20 V carries a current of 20.0 A. If the internal resistance of the battery is 0.01 Ω, what is the terminal voltage? (a) 1.24 V (b) 6.00 V (c) 6.40 V (d) 31.0 V

8. Devices A and B are connected in parallel to a battery. If the resistance R_A of device A is four times as great as the resistance R_B of device B, then what is true of I_A and I_B, the currents in devices A and B? (a) $I_A = 2I_B$ (b) $I_A = I_B/2$ (c) $I_A = 4I_B$ (d) $I_A = I_B/4$

9. A wire has resistance R. What is the resistance of a wire of the same substance that has the same length but twice the cross-sectional area? (a) $2R$ (b) $R/2$ (c) $4R$ (d) $R/4$

10. What must be the reading in the ammeter A for the circuit section shown?

(a) 0 A (b) 6.0 A (c) 8.0 A (d) 12 A

11. What is the current in a wire if 5.00 C of charge passes through the wire in 0.500 s? (a) 1.00 A (b) 2.50 A (c) 5.00 A (d) 10.0 A

Answers

1. **(b).** The three resistors are connected in series so $R_{tot} = \Sigma R = 6.0\ \Omega$. From Ohm's law, $V_{tot}/R_{tot} = I_{tot} = 12\ V/6.0\ \Omega = 2.0\ A$.

2. **(c).** Resistance is directly proportional to length and inversely proportional to the cross-sectional area of the conductor. If the area remains constant, then doubling the length will double the resistance.

3. **(a).** The resistors are in parallel:

$$R_{eq} = R_1 R_2 / (R_1 + R_2) = 12\ \Omega^2/8.0\ \Omega = 1.5\ \Omega$$

In fact, the equivalent resistance of parallel resistors is always less than the smallest resistance in the combination. This means that answers (b), (c), and (d) could have been eliminated immediately.

4. **(b).** A series circuit has only one path for current so it must be the same at all points in the circuit.

5. **(d).** Power $= I\Delta V = 18\ W$

6. **(a).** The energy used by a load is the product of the power it uses per unit time and the length of time it is operated:

$$E = W = \mathcal{P}\ \Delta t = (\Delta V)^2 \Delta t/R = (6.0\ V)^2\ (0.50\ h)/12\ \Omega$$
$$= 1.5\ Wh = 1.5 \times 10^{-3}\ kWh$$

Note the conversion from watt-hours to kilowatt-hours.

7. **(b).** The product Ir is the potential drop occurring within the battery: $Ir = (20.0\ A)(0.01\ \Omega) = 0.20\ V$. Because the battery is producing current and not being recharged, the terminal voltage will be less than the emf by the internal potential drop. This eliminates choices (c) and (d) immediately: $V = E - Ir = 6.20\ V - 0.20\ V = 6.00\ V$

8. **(d).** More current will follow the circuit branch with less resistance. Use Kirchoff's loop law around the loop consisting of the 2 parallel resistances and the fact that $R_A = 4R_B$:

$$\Sigma \Delta V_i = 0 \quad \rightarrow \quad I_A R_A - I_B R_B = 0 \quad \rightarrow$$

$$\frac{I_A}{I_B} = \frac{R_B}{R_A} = \frac{R_B}{4R_B} = \frac{1}{4}$$

Hence one-fourth as much current goes through resistor A.

9. **(b).** Resistance is inversely proportional to the area; doubling the area reduces the resistance by half.

10. **(d).** Kirchhoff's current rule says that the current entering a junction must equal the current leaving the junction. The current entering junction 1 is 6.0 A which subsequently enters junction 2 from above, so the total current entering junction 2 is 12 A and the current then leaving 2 and entering the ammeter is 12 A.

11. **(d).** By definition, the current is the amount of charge that passes a point in the circuit in a given time: $I = Q/\Delta t = 5.00 \text{ C}/0.500 \text{ s} = 10.0 \text{ A}$

ATOMS

EXAMPLE 1

Tritium is an isotope of hydrogen with a half-life of $t_{1/2} = 12.33$ yr. How long would it take for 1.60×10^2 g of tritium to decay to 20.0 g? (a) 6.17 y (b) 26.7 y (c) 37.0 y (d) 74.0 y

Solution

Calculate the number n of half-lives required. The equation can be solved by inspection (logarithms are ordinarily required):

$$\left(\frac{1}{2}\right)^n = \frac{20.0 \text{ g}}{1.60 \times 10^2 \text{ g}} = \frac{1}{8.00} \quad \rightarrow \quad n = 3$$

Multiply the number of half-lives by the length of the half-life to find the necessary time interval in question, verifying that the answer is **(c):**

$$\Delta t = nt_{1/2} = 3 \ (12.33 \text{ yr}) = \boxed{37.0 \text{ yr}}$$

EXAMPLE 2

What is the mass number of a carbon atom having 6 protons and 8 neutrons? (a) 6 (b) 8 (c) 14

Solution

Don't confuse mass number with the atomic number, Z, which is the number of protons in the nucleus. The mass number A is the number of nucleons in the nucleus. To calculate the mass number, just add the number of protons and neutrons together: $6 + 8 = 14$, which is answer **(c).**

EXAMPLE 3

Which of the following particles is given by $^A_Z X$ in the following reaction? (a) ^7_3Li (b) ^7_4Be (c) $^{14}_7\text{N}$ (d) $^{10}_6\text{C}$

$$^1_1\text{H} + ^A_Z\text{X} \rightarrow ^4_2\text{He} + ^4_2\text{He}$$

Solution

Equate the sum of the mass numbers on both sides of the reaction: $1 + A = 4 + 4 = 8 \rightarrow A = 7$

By charge conservation, the number of protons must also be the same on both sides: $1 + Z = 2 + 2 = 4 \rightarrow Z = 3$

Based on these two results, the correct answer is **(a)**, ^7_3Li .

MULTIPLE CHOICE PROBLEMS

1. In the nuclear reaction below, what particle does X represent?

$$^{22}_{11}\text{Na} \rightarrow ^{22}_{10}\text{Ne} + \text{X} + \nu_e$$

(a) an α-particle. (b) a β-particle. (c) a positron. (d) a γ-photon.

2. What is the atomic number of the daughter nuclide in the following reaction?

$$^{30}_{15}\text{P} \rightarrow ^{A}_{Z}\text{Si} + e^+ + \nu_e$$

(a) 14 (b) 16 (c) 30 (d) 31

3. If $^{13}_{7}\text{N}$ has a half-life of about 10.0 min, how long will it take for 20 g of the isotope to decay to 2.5 g? (a) 5 min (b) 10 min (c) 20 min (d) 30 min

4. A certain radionuclide decays by emitting an α-particle. What is the difference between the atomic numbers of the parent and the daughter nuclides? (a) 1 (b) 2 (c) 4 (d) 6

5. What is the difference in mass number between the parent and daughter nuclides after a β-decay process? (a) -1 (b) 0 (c) 1 (d) 2

6. A nitrogen atom has 7 protons and 6 neutrons. What is its mass number? (a) 1 (b) 6 (c) 7 (d) 13

7. Which one of the following is an isotope of $^{182}_{63}\text{X}$? (a) $^{182}_{63}\text{X}$ (b) $^{182}_{62}\text{X}$ (c) $^{180}_{64}\text{X}$ (d) $^{180}_{62}\text{X}$

8. In the following nuclear equation, what is X?

$$^{14}_{7}\text{N} + ^{4}_{2}\text{He} \rightarrow ^{17}_{8}\text{O} + \text{X}$$

(a) a proton (b) a positron (c) a β-particle (d) an α-particle

9. What is the number of neutrons in $^{140}_{54}\text{Xe}$? (a) 54 (b) 86 (c) 140 (d) 194

10. Radon-222 has a half-life of about 4 days. If a sample of ^{222}Rn gas in a container is initially doubled, the half-life will be (a) halved. (b) doubled. (c) quartered. (d) unchanged.

11. A radionuclide decays completely. In the reaction flask, the only gas that is found is helium, which was not present when the flask was sealed. The decay process was probably (a) β-decay. (b) α-decay. (c) γ-decay. (d) positron emission.

12. What is the half-life of a radionuclide if 1/16 of its initial mass is present after 2 h? (a) 15 min (b) 30 min (c) 45 min (d) 60 min

13. The half-life of $^{22}_{11}\text{Na}$ is 2.6 y. If X grams of this sodium isotope are initially present, how much is left after 13 y? (a) $X/32$ (b) $X/13$ (c) $X/8$ (d) $X/5$

Answers

1. **(c).** Notice that the mass number is unchanged, whereas the atomic number has been reduced by one. This implies that a proton has changed into a neutron. The emitting particle must have a charge equal to a proton but an atomic mass number of zero. The positron is the only choice that has both of these attributes.

2. **(a).** The atomic number is Z. Conservation of charge means: $15 = Z + 1$. Therefore, $Z = 14$.

3. **(d).** First find the number of half lives, then multiply by the value of the half-life to get the elapsed time:

$$\frac{2.5\text{ g}}{20\text{ g}} = \frac{1}{8} = \left(\frac{1}{2}\right)^3 \rightarrow n = 3$$

$$\Delta t = nt_{1/2} = 3(10\text{ min}) = 30\text{ min}$$

4. **(b).** An α-particle is a $^{4}_{2}\text{He}$ helium nucleus. In α-decay, two protons are effectively removed.

5. **(b).** β-decay emits a high-energy electron, $^{0}_{-1}\text{e}$. In the process, a neutron decays into a proton plus the emitted electron and an antineutrino. The number of nucleons remains unchanged, with the proton replacing the neutron in the sum of nucleons.

6. **(d).** The mass number is the sum of neutrons and pro-tons $= 6 + 7 = 13$ nucleons.

7. **(c).** Isotopes of an element have the same atomic number but different numbers of neutrons, so their mass numbers are different. The atomic number of X is 63.

8. **(a).** The mass numbers must be the same on both sides of the reaction. If A is the mass number of X, then $14 + 4 = 17 + A$, so $A = 1$. As for atomic number, $7 + 2 = 8 + Z$. Therefore, $Z = 1$. This describes a proton, $^{1}_{1}\text{H}$.

9. **(b).** The number of neutrons is the mass number minus the atomic number: $A - Z = 140 - 54 = 86$.

10. **(d).** The half-life is a constant that depends on the identity of the nuclide, not on the amount of nuclide present.

11. **(b).** The α-particle is a helium nucleus. Each α-particle then acquires two electrons to form a neutral helium atom.

12. **(b).** In every half-life, the mass decreases to half its previous value: $1/16 = 1/2^4$. It takes four half-lives to decay down to $1/16$ the original mass. Each must be 30 min long because the entire process takes two hours.

13. **(a).** In 13 y, there will be 5 half-lives of 2.6 y each ($5 \times 2.6 = 13$). The isotope decreases to $1/2^5 = 1/32$ of its original amount.

THOMSON

™

PETERSON'S

Answers to Quick Quizzes, Odd-Numbered Conceptual Questions and Problems

Chapter 15

QUICK QUIZZES
1. (b)
2. (b)
3. (c)
4. (a)
5. (c) and (d)
6. (a)
7. (c)
8. (b)
9. (d)
10. (b) and (d)

CONCEPTUAL QUESTIONS
1. Electrons have been removed from the object.
3. The configuration shown is inherently unstable. The negative charges repel each other. If there is any slight rotation of one of the rods, the repulsion can result in further rotation away from this configuration. There are three conceivable final configurations shown below. Configuration (a) is stable: If the positive upper ends are pushed towards each other, their mutual repulsion will move the system back to the original configuration. Configuration (b) is an equilibrium configuration, but it is unstable: If the lower ends are moved towards each other, their mutual attraction will be larger than that of the upper ends, and the configuration will shift to (c), another possible stable configuration.

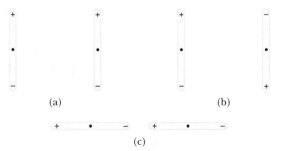

(a)   (b)

(c)

Figure Q15.3

5. Move an object A with a net positive charge so it is near, but not touching, a neutral metallic object B that is insulated from the ground. The presence of A will polarize B, causing an excess negative charge to exist on the side nearest A and an excess positive charge of equal magnitude to exist on the side farthest from A. While A is still near B, touch B with your hand. Additional electrons will then flow from ground, through your body and onto B. With A continuing to be near but not in contact with B, remove your hand from B, thus trapping the excess electrons on B. When A is now removed, B is left with excess electrons, or a net negative charge. By means of mutual repulsion, this negative charge will now spread uniformly over the entire surface of B.

7. An object's mass decreases very slightly (immeasurably) when it is given a positive charge, because it loses electrons. When the object is given a negative charge, its mass increases slightly because it gains electrons.

9. Electric field lines start on positive charges and end on negative charges. Thus, if the fair-weather field is directed into the ground, the ground must have a negative charge.

11. The two charged plates create a region with a uniform electric field between them, directed from the positive toward the negative plate. Once the ball is disturbed so as to touch one plate (say, the negative one), some negative charge will be transferred to the ball and it will be acted upon by an electric force that will accelerate it to the positive plate. Once the ball touches the positive plate, it will release its negative charge, acquire a positive charge, and accelerate back to the negative plate. The ball will continue to move back and forth between the plates until it has transferred all their net charge, thereby making both plates neutral.

13. The electric shielding effect of conductors depends on the fact that there are two kinds of charge: positive and negative. As a result, charges can move within the conductor so that the combination of positive and negative charges establishes an electric field that exactly cancels the external field within the conductor and any cavities inside the conductor. There is only one type of gravitation charge, however, because there is no negative mass. As a result, gravitational shielding is not possible.

15. The electric field patterns of each of these three configurations do not have sufficient symmetry to make the calculations practical. Gauss's law is useful only for calculating the electric fields of highly symmetric charge distributions, such as uniformly charged spheres, cylinders, and sheets.

17. No, the wall is not positively charged. The balloon induces a charge of opposite sign in the wall, causing the balloon and the wall to be attracted to each other. The balloon eventually falls because its charge slowly diminishes after leaking to ground. Some of the balloon's charge could also be lost due to positive ions in the surrounding atmosphere, which would tend to neutralize the negative charges on the balloon.

19. When the comb is nearby, charges separate on the paper, and the paper is attracted to the comb. After contact, charges from the comb are transferred to the paper, so that it has the same type of charge as the comb. The paper is thus repelled.

21. The attraction between the ball and the object could be an attraction of unlike charges, or it could be an attraction between a charged object and a neutral object as a result of polarization of the molecules of the neutral object. Two additional experiments could help us determine whether the object is charged. First, a known neutral ball could be brought near the object, and if there is an attraction, the object is negatively charged. Another possibility is to bring a known negatively charged ball near the

object. In that case, if there is a repulsion, then the object is negatively charged. If there is an attraction, then the object is neutral.

PROBLEMS

1. 1.1×10^{-8} N (attractive)
3. 91 N (repulsion)
5. (a) 36.8 N (b) 5.54×10^{27} m/s^2
7. 5.12×10^5 N
9. (a) 2.2×10^{-5} N (attraction)
   (b) 9.0×10^{-7} N (repulsion)
11. 1.38×10^{-5} N at 77.5° below the negative x-axis
13. 0.872 N at 30.0° below the positive x-axis
15. 7.2 nC
17. 1.5×10^{-3} C
19. 7.20×10^5 N/C (downward)
21. 1.2×10^4 N/C
23. (a) 6.12×10^{10} m/s^2 (b) 19.6 μs (c) 11.8 m
    (d) 1.20×10^{-15} J
25. zero
27. 1.8 m to the left of the -2.5-μC charge
33. (a) 0 (b) 5 μC inside, -5 μC outside (c) 0 inside,
    -5 μC outside (d) 0 inside, -5 μC outside
35. 1.3×10^{-3} C
37. (a) 4.8×10^{-15} N (b) 2.9×10^{12} m/s^2
39. (a) 858 N·m^2/C (b) 0 (c) 657 N·m^2/C
41. 4.1×10^6 N/C
43. (a) 0 (b) $k_e q/r^2$ outward
47. 57.5 N
49. 24 N/C in the positive x-direction
51. (a) $E = 2k_e qb (a^2 + b^2)^{-3/2}$ in the positive x-direction
    (b) $E = k_e Qb(a^2 + b^2)^{-3/2}$ in the positive x-direction
53. (a) 0 (b) 7.99×10^7 N/C (outward)
    (c) 0 (d) 7.34×10^6 N/C (outward)
55. 3.55×10^5 N·m^2/C
57. 4.4×10^5 N/C
59. (a) 10.9 nC (b) 5.44×10^{-3} N
61. $\sim 10^{-7}$ C
63. (a) 1.00×10^3 N/C (b) 3.37×10^{-8} s (c) accelerate at 1.76×10^{14} m/s^2 in the direction opposite that of the electric field

Chapter 16

QUICK QUIZZES

1. (b)
2. (b), (d)
3. (d)
4. (c)
5. (a)
6. (c)
7. (a) C decreases. (b) Q stays the same. (c) E stays the same. (d) ΔV increases. (e) The energy stored increases.
8. (a) C increases. (b) Q increases. (c) E stays the same. (d) ΔV remains the same. (e) The energy stored increases.
9. (a)

CONCEPTUAL QUESTIONS

1. (a) The proton moves in a straight line with constant acceleration in the direction of the electric field. (b) As its velocity increases, its kinetic energy increases and the electric potential energy associated with the proton decreases.

3. The work done in pulling the capacitor plates farther apart is transferred into additional electric energy stored in the capacitor. The charge is constant and the capacitance decreases, but the potential difference between the plates increases, which results in an increase in the stored electric energy.

5. If the power line makes electrical contact with the metal of the car, it will raise the potential of the car to 20 kV. It will also raise the potential of your body to 20 kV, because you are in contact with the car. In itself, this is not a problem. If you step out of the car, however, your body at 20 kV will make contact with the ground, which is at zero volts. As a result, a current will pass through your body and you will likely be injured. Thus, it is best to stay in the car until help arrives.

7. If two points on a conducting object were at different potentials, then free charges in the object would move and we would not have static conditions, in contradiction to the initial assumption. (Free positive charges would migrate from locations of higher to locations of lower potential. Free electrons would rapidly move from locations of lower to locations of higher potential.) All of the charges would continue to move until the potential became equal everywhere in the conductor.

9. The capacitor often remains charged long after the voltage source is disconnected. This residual charge can be lethal. The capacitor can be safely handled after discharging the plates by short-circuiting the device with a conductor, such as a screwdriver with an insulating handle.

11. Field lines represent the direction of the electric force on a positive test charge. If electric field lines were to cross, then, at the point of crossing, there would be an ambiguity regarding the direction of the force on the test charge, because there would be two possible forces there. Thus, electric field lines cannot cross. It is possible for equipotential surfaces to cross. (However, equipotential surfaces at different potentials cannot intersect.) For example, suppose two identical positive charges are at diagonally opposite corners of a square and two negative charges of equal magnitude are at the other two corners. Then the planes perpendicular to the sides of the square at their midpoints are equipotential surfaces. These two planes cross each other at the line perpendicular to the square at its center.

13. You should use a dielectric-filled capacitor whose dielectric constant is very large. Further, you should make the dielectric as thin as possible, keeping in mind that dielectric breakdown must also be considered.

15. (a) ii (b) i

17. It would make no difference at all. An electron volt is the kinetic energy gained by an electron in being accelerated through a potential difference of 1 V. A proton accelerated through 1 V would have the same kinetic energy, because it carries the same charge as the electron (except for the sign). The proton would be moving in the opposite direction and more slowly after accelerating through 1 V, due to its opposite charge and its larger mass, but it would still gain 1 electron volt, or 1 proton volt, of kinetic energy.

PROBLEMS

1. (a) 6.40×10^{-19} J (b) -6.40×10^{-19} J (c) -4.00 V
3. 1.4×10^{-20} J

5. 1.7×10^6 N/C
7. (a) 1.13×10^5 N/C (b) 1.80×10^{-14} N
   (c) 4.38×10^{-17} J
9. (a) 0.500 m (b) 0.250 m
11. (a) 1.44×10^{-7} V (b) -7.19×10^{-8} V
13. (a) 2.67×10^6 V (b) 2.13×10^6 V
15. (a) 103 V (b) -3.85×10^{-7} J; positive work must be done to separate the charges.
17. -11.0 kV
19. 2.74×10^{-14} m
21. 0.719 m, 1.44 m, 2.88 m. No. The equipotentials are not uniformly spaced. Instead, the radius of an equipotenial is inversely proportional to the potential.
23. (a) 1.1×10^{-8} F (b) 27 C
25. (a) 11.1 kV/m toward the negative plate (b) 3.74 pF
    (c) 74.7 pC and -74.7 pC
27. (a) 90.4 V (b) 9.04×10^4 V/m
29. (a) 13.3 μC on each (b) 20.0 μC, 40.0 μC
31. (a) 2.00 μF (b) $Q_3 = 24.0$ μC, $Q_4 = 16.0$ μC,
    $Q_2 = 8.00$ μC, $(\Delta V)_2 = (\Delta V)_4 = 4.00$ V, $(\Delta V)_3 = 8.00$ V
33. (a) 5.96 μF (b) $Q_{20} = 89.5$ μC, $Q_6 = 63.2$ μC,
    $Q_3 = Q_{15} = 26.3$ μC
35. $Q_1 = 16.0$ μC, $Q_5 = 80.0$ μC, $Q_8 = 64.0$ μC,
    $Q_4 = 32.0$ μC
37. (a) $Q_{25} = 1.25$ mC, $Q_{40} = 2.00$ mC (b) $Q'_{25} = 288$ μC,
    $Q'_{40} = 462$ μC, $\Delta V = 11.5$ V
39. $Q'_1 = 3.33$ μC, $Q'_2 = 6.67$ μC
41. 83.6 μC
43. 2.55×10^{-11} J
45. 3.2×10^{10} J
47. $\kappa = 4.0$
49. (a) 8.13 nF (b) 2.40 kV
51. (a) volume 9.09×10^{-16} m^3, area 4.54×10^{-10} m^2
    (b) 2.01×10^{-13} F (c) 2.01×10^{-14} C,
    1.26×10^5 electronic charges
55. 4.29 μF
57. 6.25 μF
59. 4.47 kV
61. 0.75 mC on C_1, 0.25 mC on C_2
65. 50 N

Chapter 17

QUICK QUIZZES

1. (d)
2. (b)
3. (c), (d)
4. (b)
5. (b)
6. (a)
7. (b)
8. (a)

CONCEPTUAL QUESTIONS

1. Charge. Because an ampere is a unit of current (1 A = 1 C/s) and an hour is a unit of time (1 h = 3 600 s), then 1 A·h = 3 600 C.
3. The gravitational force pulling the electron to the bottom of a piece of metal is much smaller than the electrical repulsion pushing the electrons apart. Thus, free electrons stay distributed throughout the metal. The concept of charges residing on the surface of a metal is true for a

metal with an excess charge. The number of free electrons in an electrically neutral piece of metal is the same as the number of positive ions—the metal has zero net charge.

5. A voltage is not something that "surges through" a completed circuit. A voltage is a potential difference that is applied across a device or a circuit. It would be more correct to say "1 ampere of electricity surged through the victim's body." Although this amount of current would have disastrous results on the human body, a value of 1 (ampere) doesn't sound as exciting for a newspaper article as 10 000 (volts). Another possibility is to write "10 000 volts of electricity were applied across the victim's body," which still doesn't sound quite as exciting.
7. We would conclude that the conductor is nonohmic.
9. The shape, dimensions, and the resistivity affect the resistance of a conductor. Because temperature and impurities affect the conductor's resistivity, these factors also affect resistance.
11. The radius of wire B is the square root of three times the radius of wire A. Therefore the cross-sectional area of B three times larger than that of A.
13. The drift velocity might increase steadily as time goes on, because collisions between electrons and atoms in the wire would be essentially nonexistent and the conduction electrons would move with constant acceleration. The current would rise steadily without bound also, because I is proportional to the drift velocity.
15. Once the switch is closed, the line voltage is applied across the bulb. As the voltage is applied across the cold filament when it is first turned on, the resistance of the filament is low, the current is high, and a relatively large amount of power is delivered to the bulb. As the filament warms, its resistance rises and the current decreases. As a result, the power delivered to the bulb decreases. The large current spike at the beginning of the bulb's operation is the reason that lightbulbs often fail just after they are turned on.

PROBLEMS

1. 3.00×10^{20} electrons move past in the direction opposite to the current.
3. 2.00 C
5. 1.05 mA
7. 27 yr
9. (a) n is unaffected (b) v_d is doubled
11. 32 V is 200 times larger than 0.16 V
13. 0.17 mm
15. (a) 30 Ω (b) 4.7×10^{-4} $\Omega \cdot$m
17. silver ($\rho = 1.59 \times 10^{-8}$ $\Omega \cdot$m)
19. 256 Ω
21. 1.98 A
23. 26 mA
25. (a) 5.89×10^{-2} Ω (b) 5.45×10^{-2} Ω
27. (a) 3.0 A (b) 2.9 A
29. (a) 1.2 Ω (b) 8.0×10^{-4} (a 0.080% increase)
31. 5.00 A, 24.0 Ω
33. 18 bulbs
35. 11.2 min
37. 34.4 Ω
39. 1.6 cm
41. 295 metric tons/h
43. 26 cents
45. 23 cents
47. $1.2

49. 1.1 km
51. $1.47 \times 10^{-6} \ \Omega \cdot m$; differs by 2.0% from value in Table 17.1
53. (a) $3.06 (b) No. The circuit must be able to handle at least 26 A.
55. (a) 667 A (b) 50.0 km
57. $3.77 \times 10^{28}/m^3$
59. (a) 144 Ω (b) 26 m (c) To fit the required length into a small space. (d) 25 m
61. 37 MΩ
63. 0.48 kg/s
65. (a) $2.6 \times 10^{-5} \ \Omega$ (b) 76 kg
67. (a) 470 W (b) 1.60 mm or more (c) 2.93 mm or more

Chapter 18

QUICK QUIZZES

1. (a), (d)
2. (b)
3. (a)
4. *Parallel*: (a) unchanged (b) unchanged (c) increase (d) decrease
5. *Series*: (a) decrease (b) decrease (c) decrease (d) increase
6. (c)

CONCEPTUAL QUESTIONS

1. No. When a battery serves as a source and supplies current to a circuit, the conventional current flows through the battery from the negative terminal to the positive one. However, when a source having a larger emf than the battery is used to charge the battery, the conventional current is forced to flow through the battery from the positive terminal to the negative one.

3. The total amount of energy delivered by the battery will be less than *W*. Recall that a battery can be considered an ideal, resistanceless battery in series with the internal resistance. When the battery is being charged, the energy delivered to it includes the energy necessary to charge the ideal battery, plus the energy that goes into raising the temperature of the battery due to I^2r heating in the internal resistance. This latter energy is not available during discharge of the battery, when part of the reduced available energy again transforms into internal energy in the internal resistance, further reducing the available energy below *W*.

5. The starter in the automobile draws a relatively large current from the battery. This large current causes a significant voltage drop across the internal resistance of the battery. As a result, the terminal voltage of the battery is reduced, and the headlights dim accordingly.

7. An electrical appliance has a given resistance. Thus, when it is attached to a power source with a known potential difference, a definite current will be drawn, and the device can therefore be labeled with both the voltage and the current. Batteries, however, can be applied to a number of devices. Each device will have a different resistance, so the current will vary with the device. As a result, only the voltage of the battery can be specified.

9. Connecting batteries in parallel does not increase the emf. A high-current device connected to two batteries in parallel can draw currents from both batteries. Thus, connecting the batteries in parallel increases the possible current output and, therefore, the possible power output.

11. The lightbulb will glow for a very short while as the capacitor is being charged. Once the capacitor is almost totally charged, the current in the circuit will be nearly zero and the bulb will not glow.

13. The bird is resting on a wire of fixed potential. In order to be electrocuted, a large potential difference is required between the bird's feet. The potential difference between the bird's feet is too small to harm the bird.

15. The junction rule is a statement of conservation of charge. It says that the amount of charge that enters a junction in some time interval must equal the charge that leaves the junction in that time interval. The loop rule is a statement of conservation of energy. It says that the increases and decreases in potential around a closed loop in a circuit must add to zero.

17. A few of the factors involved are as follows: the conductivity of the string (is it wet or dry?); how well you are insulated from ground (are you wearing thick rubber- or leather-soled shoes?); the magnitude of the potential difference between you and the kite; and the type and condition of the soil under your feet.

19. She will not be electrocuted if she holds onto only one high-voltage wire, because she is not completing a circuit. There is no potential difference across her body as long as she clings to only one wire. However, she should release the wire immediately once it breaks, because she will become part of a closed circuit when she reaches the ground or comes into contact with another object.

21. (a) The intensity of each lamp increases because lamp C is short circuited and there is current (which increases) only in lamps A and B. (b) The intensity of lamp C goes to zero because the current in this branch goes to zero. (c) The current in the circuit increases because the total resistance decreases from $3R$ (with the switch open) to $2R$ (after the switch is closed). (d) The voltage drop across lamps A and B increases, while the voltage drop across lamp C becomes zero. (e) The power dissipated increases from $\mathcal{E}^2/3R$ (with the switch open) to $\mathcal{E}^2/2R$ (after the switch is closed).

23. The statement is false. The current in each bulb is the same, because they are connected in series. The bulb that glows brightest has the larger resistance and hence dissipates more power

PROBLEMS

1. 4.92 Ω
3. 73.8 W. Your circuit diagram will consist of two 0.800-Ω resistors in series with the 192-Ω resistance of the bulb.
5. (a) 17.1 Ω (b) 1.99 A for 4.00 Ω and 9.00 Ω, 1.17 A for 7.00 Ω, 0.818 A for 10.0 Ω
7. $2.5R$
9. (a) 0.227 A (b) 5.68 V
11. 55 Ω
13. 0.43 A
15. (a) Connect two 50-Ω resistors in parallel, and then connect this combination in series with a 20-Ω resistor. (b) Connect two 50-Ω resistors in parallel, connect two 20-Ω resistors in parallel, and then connect these two combinations in series with each other.
17. 0.846 A downwards in the 8.00-Ω resistor; 0.462 A downwards in the middle branch; 1.31 A upwards in the right-hand branch

19. (a) 3.00 mA (b) -19.0 V (c) 4.50 V
21. 10.7 V
23. (a) 0.385 mA, 3.08 mA, 2.69 mA
  (b) 69.2 V, with c at the higher potential
25. $I_1 = 3.5$ A, $I_2 = 2.5$ A, $I_3 = 1.0$ A
27. $I_{30} = 0.353$ A, $I_5 = 0.118$ A, $I_{20} = 0.471$ A
29. $\Delta V_2 = 3.05$ V, $\Delta V_3 = 4.57$ V, $\Delta V_4 = 7.38$ V, $\Delta V_5 = 1.62$ V
31. (a) 12 s (b) 1.2×10^{-4} C
33. 1.3×10^{-4} C
35. 0.982 s
37. (a) heater, 10.8 A; toaster, 8.33 A; grill, 12.5 A
  (b) $I_{total} = 31.6$ A, so a 30-A breaker is insufficient.
39. (a) 6.25 A (b) 750 W
41. (a) 1.2×10^{-9} C, 7.3×10^9 K$^+$ ions. Not large, only
  $1e/290$ A^2
  (b) 1.7×10^{-9} C, 1.0×10^{10} Na$^+$ ions (c) 0.83 μA
  (d) 7.5×10^{-12} J
43. 11 nW
45. 7.5 Ω
47. (a) 15 Ω
  (b) $I_1 = 1.0$ A, $I_2 = I_3 = 0.50$ A, $I_4 = 0.30$ A, and
    $I_5 = 0.20$ A
  (c) $(\Delta V)_{ac} = 6.0$ V, $(\Delta V)_{ce} = 1.2$ V, $(\Delta V)_{ed} = (\Delta V)_{fd} =$
    1.8 V, $(\Delta V)_{cd} = 3.0$ V, $(\Delta V)_{db} = 6.0$ V
  (d) $\mathscr{P}_{ac} = 6.0$ W, $\mathscr{P}_{ce} = 0.60$ W, $\mathscr{P}_{ed} = 0.54$ W, $\mathscr{P}_{fd} = 0.36$ W,
    $\mathscr{P}_{cd} = 1.5$ W, $\mathscr{P}_{db} = 6.0$ W
49. (a) 12.4 V (b) 9.65 V
51. $I_1 = 0$, $I_2 = I_3 = 0.50$ A,
53. 112 V, 0.200 Ω
55. (a) $R_x = R_2 - \frac{1}{4}R_1$
  (b) $R_x = 2.8$ Ω (inadequate grounding)
59. $\mathscr{P} = \dfrac{(144 \text{ V}^2)R}{(R + 10.0 \ \Omega)^2}$

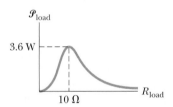

61. (a) 5.68 V (b) 0.227 A
63. 0.395 A; 1.50 V

Chapter 19

QUICK QUIZZES
1. (b)
2. (c)
3. (c)
4. (a)
5. (b)

CONCEPTUAL QUESTIONS
1. The set should be oriented such that the beam is moving either toward the east or toward the west.
3. The proton moves in a circular path upwards on the page. After completing half a circle, it exits the field and moves in a straight-line path back in the direction from whence it came. An electron will behave similarly, but the direction of traversal of the circle is downward, and the radius of the circular path is smaller.

5. The magnetic force on a moving charged particle is always perpendicular to the particle's direction of motion. There is no magnetic force on the charge when it moves parallel to the direction of the magnetic field. However, the force on a charged particle moving in an electric field is never zero and is always parallel to the direction of the field. Therefore, by projecting the charged particle in different directions, it is possible to determine the nature of the field.
7. The magnetic field produces a magnetic force on the electrons moving toward the screen that produce the image. This magnetic force deflects the electrons to regions on the screen other than the ones to which they are supposed to go. The result is a distorted image.
9. Such levitation could never occur. At the North Pole, where Earth's magnetic field is directed downward, toward the equivalent of a buried south pole, a coffin would be repelled if its south magnetic pole were directed downward. However, equilibrium would be only transitory, as any slight disturbance would upset the balance between the magnetic force and the gravitational force.
11. If you were moving along with the electrons, you would measure a zero current for the electrons, so they would not produce a magnetic field according to your observations. However, the fixed positive charges in the metal would now be moving backwards relative to you, creating a current equivalent to the forward motion of the electrons when you were stationary. Thus, you would measure the same magnetic field as when you were stationary, but it would be due to the positive charges presumed to be moving from your point of view.
13. A compass does not detect currents in wires near light switches, for two reasons. The first is that, because the cable to the light switch contains two wires, one carrying current to the switch and the other carrying it away from the switch, the net magnetic field would be very small and would fall off rapidly with increasing distance. The second reason is that the current is alternating at 60 Hz. As a result, the magnetic field is oscillating at 60 Hz also. This frequency would be too fast for the compass to follow, so the effect on the compass reading would average to zero.
15. The levitating wire is stable with respect to vertical motion: If it is displaced upward, the repulsive force weakens, and the wire drops back down. By contrast, if it drops lower, the repulsive force increases, and it moves back up. The wire is not stable, however, with respect to lateral movement: If it moves away from the vertical position directly over the lower wire, the repulsive force will have a sideways component that will push the wire away.
 In the case of the attracting wires, the hanging wire is not stable with respect to vertical movement. If it rises, the attractive force increases, and the wire moves even closer to the upper wire. If the hanging wire falls, the attractive force weakens, and the wire falls farther. If the wire moves to the right, it moves farther from the upper wire and the attractive force decreases. Although there is a restoring force component pulling it back to the left, the vertical force component is not strong enough to hold the wire up, and it falls.
17. Each coil of the Slinky® will become a magnet, because a coil acts as a current loop. The sense of rotation of the current is the same in all coils, so each coil becomes a magnet with the same orientation of poles. Thus, all of the coils attract, and the Slinky® will compress.

19. There is no net force on the wires, but there is a torque. To understand this distinction, imagine a fixed vertical wire and a free horizontal wire (see the figure below). The vertical wire carries an upward current and creates a magnetic field that circles the vertical wire, itself. To the right, the magnetic field of the vertical wire points into the page, while on the left side it points out of the page, as indicated. Each segment of the horizontal wire (of length ℓ) carries current that interacts with the magnetic field according to the equation $F = BI\ell \sin \theta$. Apply the right-hand rule on the right side: point the fingers of your right hand in the direction of the horizontal current and curl them into the page in the direction of the magnetic field. Your thumb points downward, the direction of the force on the right side of the wire. Repeating the process on the left side gives a force upward on the left side of the wire. The two forces are equal in magnitude and opposite in direction, so the net force is zero, but they create a net torque around the point where the wires cross.

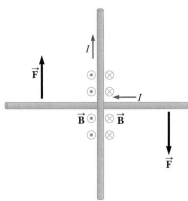

21. (a) The field is into the page. (b) The beam would deflect upwards.

PROBLEMS

1. (a) horizontal and due east (b) horizontal and 30° N of E (c) horizontal and due east (d) zero force
3. (a) into the page (b) toward the right (c) toward the bottom of the page
5. $F_g = 8.93 \times 10^{-30}$ N (downward), $F_e = 1.60 \times 10^{-17}$ N (upward), $F_m = 4.80 \times 10^{-17}$ N (downward)
7. 2.83×10^7 m/s west
9. 0.021 T in the $-y$-direction
11. 8.0×10^{-3} T in the $+z$-direction
13. (a) into the page (b) toward the right (c) toward the bottom of the page
15. 7.50 N
17. 0.131 T (downward)
19. 0.20 T directed out of the page
21. ab: 0, bc: 0.040 0 N in $-x$-direction, cd: 0.040 0 N in the $-z$-direction da: 0.056 6 N parallel to the xz-plane and at 45° to both the $+x$- and the $+z$-directions
23. 9.05×10^{-4} N·m, tending to make the left-hand side of the loop move toward you and the right-hand side move away.
25. (a) 3.97° (b) 3.39×10^{-3} N·m
27. 6.56×10^{-2} T
31. 1.77 cm
33. $r = 3R/4$
35. 20.0 μT
37. 2.4 mm

39. 20.0 μT toward bottom of page
41. 0.167 μT out of the page
43. (a) 4.00 m (b) 7.50 nT (c) 1.26 m (d) zero
45. 4.5 mm
47. 31.8 mA
49. 2.26×10^{-4} N away from the center, zero torque
51. 1.7 N·m
53. (a) 0.500 μT out of the page (b) 3.89 μT parallel to xy-plane and at 59.0° clockwise from $+x$-direction
55. 2.13 cm
57. (a) 1.33 m/s (b) the sign of the emf is independent of the charge
59. 1.41×10^{-6} N
61. 13.0 μT toward the bottom of the page
63. 53 μT toward the bottom of the page, 20 μT toward the bottom of the page, and 0
65. (a) -8.00×10^{-21} kg·m/s (b) 8.90°
67. 1.29 kW
69. (a) 12.0 cm to the left of wire 1 (b) 2.40 A, downward

Chapter 20

QUICK QUIZZES

1. b, c, a
2. (a)
3. (b)
4. (c)
5. (b)

CONCEPTUAL QUESTIONS

1. According to Faraday's law, an emf is induced in a wire loop if the magnetic flux through the loop changes with time. In this situation, an emf can be induced either by rotating the loop around an arbitrary axis or by changing the shape of the loop.
3. As the spacecraft moves through space, it is apparently moving from a region of one magnetic field strength to a region of a different magnetic field strength. The changing magnetic field through the coil induces an emf and a corresponding current in the coil.
5. If the bar were moving to the left, the magnetic force on the negative charges in the bar would be upward, causing an accumulation of negative charge on the top and positive charges at the bottom. Hence, the electric field in the bar would be upward, as well.
7. If, for any reason, the magnetic field should change rapidly, a large emf could be induced in the bracelet. If the bracelet were not a continuous band, this emf would cause high-voltage arcs to occur at any gap in the band. If the bracelet were a continuous band, the induced emf would produce a large induced current and result in resistance heating of the bracelet.
11. As the aluminum plate moves into the field, eddy currents are induced in the metal by the changing magnetic field at the plate. The magnetic field of the electromagnet interacts with this current, producing a retarding force on the plate that slows it down. In a similar fashion, as the plate leaves the magnetic field, a current is induced, and once again there is an upward force to slow the plate.
13. The energy stored in an inductor carrying a current I is equal to $PE_L = (1/2)LI^2$. Therefore, doubling the current will quadruple the energy stored in the inductor.

15. If an external battery is acting to increase the current in the inductor, an emf is induced in a direction to oppose the increase of current. Likewise, if we attempt to reduce the current in the inductor, the emf that is set up tends to support the current. Thus, the induced emf always acts to oppose the change occurring in the circuit, or it acts in the "back" direction to the change.

17. (a) clockwise (b) The net force exerted on the bar must be zero because it moves at constant speed. The component of the gravitational force down the incline is balanced by a component of the magnetic force up the incline.

19. from left to right

PROBLEMS
1. $5.9 \times 10^{-2}\,\text{T}\cdot\text{m}^2$
3. $7.71 \times 10^{-1}\,\text{T}\cdot\text{m}^2$
5. (a) $\Phi_{B,net} = 0$ (b) 0
7. (a) $3.1 \times 10^{-3}\,\text{T}\cdot\text{m}^2$ (b) $\Phi_{B,net} = 0$
9. 0.18 T
11. 94 mV
13. 2.7 T/s
15. (a) $4.0 \times 10^{-6}\,\text{T}\cdot\text{m}^2$ (b) 18 μV
17. 10.2 μV
19. 0.763 V
21. (a) toward the east (b) $4.58 \times 10^{-4}\,\text{V}$
23. (a) from left to right (b) from right to left
25. (a) $F = N^2 B^2 w^2 v/R$ to the left (b) 0
    (c) $F = N^2 B^2 w^2 v/R$ to the left
27. into the page
29. (a) from right to left (b) from right to left (c) from left to right (d) from left to right
31. $1.9 \times 10^{-11}\,\text{V}$
33. (a) 18.1 μV (b) 0
35. (a) 60 V (b) 57 V (c) 0.13 s
37. 20 mV
39. (a) 2.0 mH (b) 38 A/s
43. 12 mH
45. 1.92 Ω
47. 0.140 J
49. (a) 18 J (b) 7.2 J
51. negative ($V_a < V_b$)
53. (a) 20.0 ms (b) 37.9 V (c) 1.52 mV (d) 51.8 mA
55. 1.20 μC
57. (a) 0.500 A (b) 2.00 W (c) 2.00 W
59. 115 kV
61. (a) 0.157 mV (end B is positive) (b) 5.89 mV (end A is positive)
63. (a) 9.00 A (b) 10.8 N (c) b is at the higher potential (d) No
65. $v_t = \dfrac{mgR}{B^2 \ell^2}$

Chapter 21

ANSWERS TO QUICK QUIZZES
1. (c)
2. (b)
3. (b)
4. (b), (c)
5. (b), (d)

CONCEPTUAL QUESTIONS
1. For best reception, the length of the antenna should be parallel to the orientation of the oscillating electric field. Because of atmospheric variations and reflections of the wave before it arrives at your location, the orientation of this field may be in different directions for different stations.

3. The primary coil of the transformer is an inductor. When an AC voltage is applied, the back emf due to the inductance will limit the current in the coil. If DC voltage is applied, there is no back emf, and the current can rise to a higher value. It is possible that this increased current will deliver so much energy to the resistance in the coil that its temperature rises to the point at which insulation on the wire can burn.

5. An antenna that is a conducting line responds to the electric field of the electromagnetic wave—the oscillating electric field causes an electric force on electrons in the wire along its length. The movement of electrons along the wire is detected as a current by the radio and is amplified. Thus, a line antenna must have the same orientation as the broadcast antenna. A loop antenna responds to the magnetic field in the radio wave. The varying magnetic field induces a varying current in the loop (by Faraday's law), and this signal is amplified. The loop should be in the vertical plane containing the line of sight to the broadcast antenna, so the magnetic field lines go through the area of the loop.

7. The flashing of the light according to Morse code is a drastic amplitude modulation—the amplitude is changing from a maximum to zero. In this sense, it is similar to the on-and-off binary code used in computers and compact disks. The carrier frequency is that of the light, on the order of 10^{14} HZ. The frequency of the signal depends on the skill of the signal operator, but it is on the order of a single hertz, as the light is flashed on and off. The broadcasting antenna for this modulated signal is the filament of the lightbulb in the signal source. The receiving antenna is the eye.

9. The sail should be as reflective as possible, so that the maximum momentum is transferred to the sail from the reflection of sunlight.

11. Suppose the extraterrestrial looks around your kitchen. Lightbulbs and the toaster glow brightly in the infrared. Somewhat fainter are the back of the refrigerator and the back of the television set, while the television screen is dark. The pipes under the sink show the same weak glow as the walls, until you turn on the faucets. Then the pipe on the right gets darker and that on the left develops a gleam that quickly runs up along its length. The food on the plates shines, as does human skin, the same color for all races. Clothing is dark as a rule, but your seat and the chair seat glow alike after you stand up. Your face appears lit from within, like a jack-o'-lantern; your nostrils and the openings of your ear canals are bright; brighter still are the pupils of your eyes.

13. Radio waves move at the speed of light. They can travel around the curved surface of the Earth, bouncing between the ground and the ionosphere, which has an altitude that is small compared with the radius of the Earth. The distance across the lower 48 states is approximately 5 000 km, requiring a travel time that is equal to $(5 \times 10^6\,\text{m})/(3 \times 10^8\,\text{m/s}) \sim 10^{-2}$ s. Likewise, radio waves take only 0.07 s to travel halfway around the Earth. In other words, a speech can be heard on the other side of the world (in the form of radio waves) before it is heard at the back of the room (in the form of sound waves).

15. No. The wire will emit electromagnetic waves only if the current varies in time. The radiation is the result of accelerating charges, which can occur only when the current is not constant.

17. The resonance frequency is determined by the inductance and the capacitance in the circuit. If both L and C are doubled, the resonance frequency is reduced by a factor of two.

19. It is far more economical to transmit power at a high voltage than at a low voltage because the I^2R loss on the transmission line is significantly lower at high voltage. Transmitting power at high voltage permits the use of step-down transformers to make "low" voltages and high currents available to the end user.

21. No. A voltage is induced in the secondary coil only if the flux through the core changes with time.

PROBLEMS

1. (a) 141 V (b) 20.0 A (c) 28.3 A (d) 2.00 kW
3. 70.7 V, 2.95 A
5. 6.76 W
9. 4.0×10^2 Hz
11. 17 μF
15. 3.14 A
17. 0.450 T·m^2
19. (a) 0.361 A (b) 18.1 V (c) 23.9 V (d) $-53.0°$
21. (a) 1.4 kΩ (b) 0.10 A (c) 51° (d) voltage leads current
23. (a) 89.6 V (b) 108 V
25. 1.88 V
27. (a) 103 V (b) 150 V (c) 127 V (d) 23.6 V
29. (a) 208 Ω (b) 40.0 Ω (c) 0.541 H
31. (a) 1.8×10^2 Ω (b) 0.71 H
33. 2.29 μH
35. C_{min} = 4.9 nF, C_{max} = 51 nF
37. 0.242 J
39. 0.18% is lost
41. (a) 1.1×10^3 kW (b) 3.1×10^2 A (c) 8.3×10^3 A
43. 1 000 km; there will always be better use for tax money.
45. $f_{red} = 4.55 \times 10^{14}$ Hz, $f_{IR} = 3.19 \times 10^{14}$ Hz, $E_{max,f}/E_{max,i} = 0.57$
47. 2.94×10^8 m/s
49. $E_{max} = 1.01 \times 10^3$ V/m, $B_{max} = 3.35 \times 10^{-6}$ T
51. (a) 188 m to 556 m (b) 2.78 m to 3.4 m
53. 5.2×10^{13} Hz, 5.8 μm
55. $4.299\,999\,84 \times 10^{14}$ Hz; -1.6×10^7 Hz (the frequency decreases)
57. 99.6 mH
59. 1.7 cents
61. (a) resistor and inductor (b) R = 10 Ω, L = 30 mH
63. (a) 6.7×10^{-16} T (b) 5.3×10^{-17} W/m^2 (c) 1.7×10^{-14} W
65. (a) 0.536 N (b) 8.93×10^{-5} m/s^2 (c) 33.9 days
67. 4.47×10^{-9} J

Chapter 22

QUICK QUIZZES

1. (a)
2. Beams 2 and 4 are reflected; beams 3 and 5 are refracted.
3. (b)
4. (c)

CONCEPTUAL QUESTIONS

1. Sound radiated upward at an acute angle with the horizontal is bent back toward Earth by refraction. This means that the sound can reach the listener by this path as well as by a direct path. Thus, the sound is louder.

3. The color will not change, for two reasons. First, despite the popular statement that color depends on wavelength, it actually depends on the *frequency* of the light, which does not change under water. Second, when the light enters the eye, it travels through the fluid within. Thus, even if color did depend on wavelength, the important wavelength is that of the light in the ocular fluid, which does not depend on the medium through which the light traveled to reach the eye.

5. (a) Away from the normal (b) increases (c) remains the same

7. No, the information in the catalog is incorrect. The index of refraction is given by $n = c/v$, where c is the speed of light in a vacuum and v is the speed of light in the material. Because light travels faster in a vacuum than in any other material, it is impossible for the index of refraction of any material to have a value less than 1.

9. There is no dependence of the angle of reflection on wavelength, because the light does not enter deeply into the material during reflection—it reflects from the surface.

11. On the one hand, a ball covered with mirrors sparkles by reflecting light from its surface. On the other hand, a faceted diamond lets in light at the top, reflects it by total internal reflection in the bottom half, and sends the light out through the top again. Because of its high index of refraction, the critical angle for diamond in air for total internal reflection, namely $\theta_c = \sin^{-1}(n_{air}/n_{diamond})$, is small. Thus, light rays enter through a large area and exit through a very small area with a much higher intensity. When a diamond is immersed in carbon disulfide, the critical angle is increased to $\theta_c = \sin^{-1}(n_{carbon\ disulfide}/n_{diamond})$. As a result, the light is emitted from the diamond over a larger area and appears less intense.

13. The index of refraction of water is 1.333, quite different from that of air, which has an index of refraction of about 1. The boundary between the air and water is therefore easy to detect, because of the differing diffraction effects above and below the boundary. (Try looking at a glass half full of water.) The index of refraction of liquid helium, however, happens to be much closer to that of air. Consequently, the defractive differences above and below the helium-air boundary are harder to see.

15. The diamond acts like a prism, dispersing the light into its spectral components. Different colors are observed as a consequence of the manner in which the index of refraction varies with the wavelength.

17. Light travels through a vacuum at a speed of 3×10^8 m/s. Thus, an image we see from a distant star or galaxy must have been generated some time ago. For example, the star Altair is 16 lightyears away; if we look at an image of Altair today, we know only what Altair looked like 16 years ago. This may not initially seem significant; however, astronomers who look at other galaxies can get an idea of what galaxies looked like when they were much younger. Thus, it does make sense to speak of "looking backward in time."

PROBLEMS

1. 3.00×10^8 m/s
3. 114 rad/s for a maximum intensity of returning light
5. (b) 3.000×10^8 m/s

7. 19.5° above the horizontal
9. (a) 1.52 (b) 417 nm (c) 4.74×10^{14} Hz
   (d) 1.98×10^8 m/s
11. (a) 584 nm (b) 1.12
13. 111°
15. (a) 1.559×10^8 m/s (b) 329.1 nm (c) 4.738×10^{14} Hz
17. five times from the right-hand mirror and six times from the left
19. 0.388 cm
21. $\theta = 30.4°$, $\theta' = 22.3°$
23. 6.39 ns
25. $\theta = \tan^{-1}(n_g)$
27. 3.39 m
29. $\theta_{\text{red}} = 48.22°$, $\theta_{\text{blue}} = 47.79°$
31. (a) $\theta_{1i} = 30°$, $\theta_{1r} = 19°$, $\theta_{2i} = 41°$, $\theta_{2r} = 77°$
    (b) First surface: $\theta_{\text{reflection}} = 30°$;
    second surface: $\theta_{\text{reflection}} = 41°$
33. (a) 31.3° (b) 44.2° (c) 49.8°
35. (a) 33.4° (b) 53.4°
37. (a) 40.8° (b) 60.6°
39. 1.000 08
41. (a) 10.7° (b) air (c) Sound falling on the wall from most directions is 100% reflected.
43. 27.5°
45. 22.0°
47. (a) 53.1° (b) $\geq 38.7°$
49. (a) 38.5° (b) ≥ 1.44
53. 24.7°
55. 1.93
59. $\theta = \sin^{-1}\left(\sqrt{n^2 - 1}\ \sin\phi - \cos\phi\right)$
61. (a) 1.20 (b) 3.40 ns

Chapter 23

QUICK QUIZZES
1. At C.
2. (c)
3. (a) False (b) False (c) True
4. (b)
5. An infinite number
6. (a) False (b) True (c) False

CONCEPTUAL QUESTIONS
1. You will not be able to focus your eyes on both the picture and your image at the same time. To focus on the picture, you must adjust your eyes so that an object several centimeters away (the picture) is in focus. Thus, you are focusing on the mirror surface. But, your image in the mirror is as far behind the mirror as you are in front of it. Thus, you must focus your eyes beyond the mirror, twice as far away as the picture to bring the image into focus.
3. A single flat mirror forms a virtual image of an object due to two factors. First, the light rays from the object are necessarily diverging from the object, and second, the lack of curvature of the flat mirror cannot convert diverging rays to converging rays. If another optical element is first used to cause light rays to converge, then the flat mirror can be placed in the region in which the converging rays are present, and it will change the direction of the rays so that the real image is formed at a different location. For example, if a real image is formed by a convex lens, and the flat mirror is placed between the lens and the image position, the image formed by the mirror will be real.

5. The ultrasonic range finder sends out a sound wave and measures the time for the echo to return. Using this information, the camera calculates the distance to the subject and sets the camera lens. When the camera is facing a mirror, the ultrasonic signal reflects from the mirror surface and the camera adjusts its focus so that the mirror surface is at the correct focusing distance from the camera. But your image in the mirror is twice this distance from the camera, so it is blurry.
7. Light rays diverge from the position of a virtual image just as they do from an actual object. Thus, a virtual image can be as easily photographed as any object can. Of course, the camera would have to be placed near the axis of the lens or mirror in order to intercept the light rays.
9. We consider the two trees to be two separate objects. The far tree is an object that is farther from the lens than the near tree. Thus, the image of the far tree will be closer to the lens than the image of the near tree. The screen must be moved closer to the lens to put the far tree in focus.
11. If a converging lens is placed in a liquid having an index of refraction larger than that of the lens material, the direction of refractions at the lens surfaces will be reversed, and the lens will diverge light. A mirror depends only on reflection which is independent of the surrounding material, so a converging mirror will be converging in any liquid.
13. This is a possible scenario. When light crosses a boundary between air and ice, it will refract in the same manner as it does when crossing a boundary of the same shape between air and glass. Thus, a converging lens may be made from ice as well as glass. However, ice is such a strong absorber of infrared radiation that it is unlikely you will be able to start a fire with a small ice lens.
15. The focal length for a mirror is determined by the law of reflection from the mirror surface. The law of reflection is independent of the material of which the mirror is made and of the surrounding medium. Thus, the focal length depends only on the radius of curvature and not on the material. The focal length of a lens depends on the indices of refraction of the lens material and surrounding medium. Thus, the focal length of a lens depends on the lens material.
17. (a) all signs are positive (b) f and p are positive, q is negative
19. (c) the image becomes fuzzy and disappears

PROBLEMS
1. on the order of 10^{-9} s younger
3. 10.0 ft, 30.0 ft, 40.0 ft
5. 0.268 m behind the mirror; virtual, upright, and diminished; $M = 0.026\ 8$
7. (a) 13.3 cm in front of mirror, real, inverted, $M = -0.333$
   (b) 20.0 cm in front of mirror, real, inverted, $M = -1.00$
   (c) No image is formed. Parallel rays leave the mirror.
9. Behind the worshipper, 3.33 m from the deepest point in the niche.
11. 5.00 cm
13. 1.0 m
15. 8.05 cm
17. -20.0 cm
19. (a) concave with focal length $f = 0.83$ m
    (b) Object must be 1.0 m in front of the mirror.
21. 38.2 cm below the upper surface of the ice

23. 3.8 mm
25. $n = 2.00$
27. 20.0 cm
29. (a) 40.0 cm beyond the lens, real, inverted, $M = -1.00$
    (b) No image is formed. Parallel rays leave the lens.
    (c) 20.0 cm in front of the lens, virtual, upright, $M = +2.00$
31. (a) 13.3 cm in front of the lens, virtual, upright, $M = +1/3$
    (b) 10.0 cm in front of the lens, virtual, upright, $M = +1/2$
    (c) 6.67 cm in front of the lens, virtual, upright, $M = +2/3$
33. (a) either 9.63 cm or 3.27 cm (b) 2.10 cm
35. (a) 39.0 mm (b) 39.5 mm
37. at distance $2|f|$ in front of lens
39. 40.0 cm
41. 30.0 cm to the left of the second lens, $M = -3.00$
43. 7.47 cm in front of the second lens; 1.07 cm; virtual, upright
45. from 0.224 m to 18.2 m
47. real image, 5.71 cm in front of the mirror
49. 38.6°
51. 160 cm to the left of the lens, inverted, $M = -0.800$
53. $q = 10.7$ cm
55. 32.0 cm to the right of the second surface (real image)
57. (a) 20.0 cm to the right of the second lens; $M = -6.00$
    (b) inverted
    (c) 6.67 cm to the right of the second lens; $M = -2.00$;
        inverted
59. (a) 1.99
    (b) 10.0 cm to the left of the lens
    (c) inverted
61. (a) 5.45 m to the left of the lens
    (b) 8.24 m to the left of the lens
    (c) 17.1 m to the left of the lens
    (d) by surrounding the lens with a medium having a
        refractive index greater than that of the lens material.
63. (a) 263 cm (b) 79.0 cm

Chapter 24

QUICK QUIZZES

1. (c)
2. (b)
3. (b)
4. The compact disc.

CONCEPTUAL QUESTIONS

1. You will *not* see an interference pattern from the automobile headlights, for two reasons. The first is that the headlights are not coherent sources and are therefore incapable of producing sustained interference. Also, the headlights are so far apart in comparison to the wavelengths emitted that, even if they were made into coherent sources, the interference maxima and minima would be too closely spaced to be observable.

3. The result of the double slit is to redistribute the energy arriving at the screen. Although there is no energy at the location of a dark fringe, there is four times as much energy at the location of a bright fringe as there would be with only a single narrow slit. The total amount of energy arriving at the screen is twice as much as with a single slit, as it must be according to the law of conservation of energy.

5. One of the materials has a higher index of refraction than water, and the other has a lower index. The material with the higher index will appear black as it approaches zero thickness. There will be a 180° phase change for the light

reflected from the upper surface, but no such phase change for the light reflected from the lower surface, because the index of refraction for water on the other side is lower than that of the film. Thus, the two reflections will be out of phase and will interfere destructively. The material with index of refraction lower than water will have a phase change for the light reflected from both the upper and the lower surface, so that the reflections from the zero-thickness film will be back in phase and the film will appear bright.

7. For incidence normal to the film, the extra path length followed by the reflected ray is twice the thickness of the film. For destructive interference, this must be a distance of half a wavelength of the light in the material of the film. For a film in air, no 180° phase change will occur in these reflections, so the thickness of the film must be one-quarter wavelength, which is the same as the condition for constructive interference of reflected light. This means that the transmitted light is a minimum when the reflected light is a maximum, and vice versa.

9. Since the light reflecting at the lower surface of the film undergoes a 180° phase change, while light reflecting from the upper surface of the film does not undergo such a change, the central spot (where the film has near zero thickness) will be dark. If the observed rings are not circular, the curved surface of the lens does not have a true spherical shape.

11. For regional communication at the Earth's surface, radio waves are typically broadcast from currents oscillating in tall vertical towers. These waves have vertical planes of polarization. Light originates from the vibrations of atoms or electronic transitions within atoms, which represent oscillations in all possible directions. Thus, light generally is not polarized.

13. Yes. In order to do this, first measure the radar reflectivity of the metal of your airplane. Then choose a light, durable material that has approximately half the radar reflectivity of the metal in your plane. Measure its index of refraction, and place onto the metal a coating equal in thickness to one-quarter of 3 cm, divided by that index. Sell the plane quick, and then you can sell the supposed enemy new radars operating at 1.5 cm, which the coated metal will reflect with extra-high efficiency.

15. If you wish to perform an interference experiment, you need monochromatic coherent light. To obtain it, you must first pass light from an ordinary source through a prism or diffraction grating to disperse different colors into different directions. Using a single narrow slit, select a single color and make that light diffract to cover both slits for a Young's experiment. The procedure is much simpler with a laser because its output is already monochromatic and coherent.

17. Strictly speaking, the ribs do act as a diffraction grating, but the separation distance of the ribs is so much larger than the wavelength of the x-rays that there are no observable effects.

19. As the edge of the Moon cuts across the light from the star, edge diffraction effects occur. Thus, as the edge of the Moon moves relative to the star, the observed light from the star proceeds through a series of maxima and minima.

21. Larger. From Brewster's law, $n = \tan \theta_p$, we see that the angle increases as n increases.

PROBLEMS

1. 1.58 cm
3. (a) 2.6 mm (b) 2.62 mm
5. (a) 36.2° (b) 5.08 cm (c) 5.08×10^{14} Hz
7. (a) 55.7 m (b) 124 m
9. 75.0 m
11. 11.3 m
13. 148 m
15. 91.9 nm
17. 550 nm
19. 0.500 cm
21. (a) 238 nm (b) λ will increase (c) 328 nm
23. 4.35 μm
25. 4.75 μm
27. No, the wavelengths intensified are 276 nm, 138 nm, 92.0 nm, . . .
29. 4.22 mm
31. (a) 1.1 m (b) 1.7 mm
33. 1.20 mm, 1.20 mm
35. (a) 479 nm, 647 nm, 698 nm (b) 20.5°, 28.3°, 30.7°
37. 5.91° in first order; 13.2° in second order; and 26.5° in third order
39. 44.5 cm
41. 9.13 cm
43. (a) 25.6° (b) 19.0°
45. (a) 1.11 (b) 42.0°
47. (a) 56.7° (b) 48.8°
49. 31.2°
53. 6.89 units
55. (a) 413.7 nm, 409.7 nm (b) 8.6°
57. 0.156 mm
59. 2.50 mm
61. Any positive integral multiple of 210 nm
63. (a) 16.6 m (b) 8.28 m
65. 127 m
67. 0.350 mm
69. 115 nm

Chapter 25

QUICK QUIZZES

1. (c)
2. (a)

CONCEPTUAL QUESTIONS

1. The observer is *not* using the lens as a simple magnifier. For a lens to be used as a simple magnifier, the object distance must be less than the focal length of the lens. Also, a simple magnifier produces a virtual image at the normal near point of the eye, or at an image distance of about $q = -25$ cm. With a large object distance and a relatively short image distance, the magnitude of the magnification by the lens would be considerably less than one. Most likely, the lens in this example is part of a lens combination being used as a telescope.

3. The image formed on the retina by the lens and cornea is already inverted.

5. There will be an effect on the interference pattern—it will be distorted. The high temperature of the flame will change the index of refraction of air for the arm of the interferometer in which the match is held. As the index of refraction varies randomly, the wavelength of the light

in that region will also vary randomly. As a result, the effective difference in length between the two arms will fluctuate, resulting in a wildly varying interference pattern.

7. Large lenses are difficult to manufacture and machine with accuracy. Also, their large weight leads to sagging, which produces a distorted image. In reflecting telescopes, light does not pass through glass; hence, problems associated with chromatic aberrations are eliminated. Large-diameter reflecting telescopes are also technically easier to construct. Some designs use a rotating pool of mercury as the reflecting surface.

9. In order for someone to see an object through a microscope, the wavelength of the light in the microscope must be smaller than the size of the object. An atom is much smaller than the wavelength of light in the visible spectrum, so an atom can never be seen with the use of visible light.

11. farsighted; converging

PROBLEMS

1. 30.0 cm beyond the lens, $M = -1/5$
3. 177 m
5. $f/1.4$
7. $f/8.0$
9. 40.0 cm
11. 23.2 cm
13. (a) -2.00 diopters (b) 17.6 cm
15. $+17.0$ diopters
17. (a) 5.8 cm (b) $m = 4.3$
19. (a) 4.07 cm (b) $m = +7.14$
21. (a) $|M| = 1.22$ (b) $\theta/\theta_0 = 6.08$
23. 2.1 cm
25. $m = -115$
27. $f_o = 90$ cm, $f_e = 2.0$ cm
29. (b) $-fh/p$ (c) -1.07 mm
31. (a) $m = 1.50$ (b) $m = 1.90$
33. 492 km
35. 0.40 μrad
37. 9.1×10^7 km
39. 9.8 km
41. No. A resolving power of 2.0×10^5 is needed, and that available is only 1.8×10^5.
43. 50.4 μm
45. 40
47. 98 fringe shifts
49. (a) $+2.67$ diopters (b) 0.16 diopter too low
51. (a) $+44.6$ diopters (b) 3.03 diopters
53. (a) 1.0×10^3 lines (b) 3.3×10^2 lines
55. $m = 10.7$
57. (a) $m = 4.0$ (b) $m = 3.0$

Chapter 26

QUICK QUIZZES

1. (a)
2. No. From your perspective you're at rest with respect to the cabin, so you will measure yourself as having your normal length, and will require a normal-sized cabin.
3. (a), (e); (a), (e)
4. (a) False (b) False (c) True (d) False
5. (a)

CONCEPTUAL QUESTIONS

1. An ellipsoid. The dimension in the direction of motion would be measured to be less than D.

3. This scenario is not possible with light. Light waves are described by the principles of special relativity. As you detect the light wave ahead of you and moving away from you (which would be a pretty good trick—think about it!), its velocity relative to you is c. Thus, you will not be able to catch up to the light wave.

5. No. The principle of relativity implies that nothing can travel faster than the speed of light in a *vacuum*, which is 3.00×10^8 m/s.

7. The light from the quasar moves at 3.00×10^8 m/s. The speed of light is independent of the motion of the source or the observer.

9. For a wonderful fictional exploration of this question, get a "Mr. Tompkins" book by George Gamow. All of the relativity effects would be obvious in our lives. Time dilation and length contraction would both occur. Driving home in a hurry, you would push on the gas pedal not to increase your speed very much, but to make the blocks shorter. Big Doppler shifts in wave frequencies would make red lights look green as you approached and make car horns and radios useless. High-speed transportation would be both very expensive, requiring huge fuel purchases, as well as dangerous, since a speeding car could knock down a building. When you got home, hungry for lunch, you would find that you had missed dinner; there would be a five-day delay in transit when you watch a live TV program originating in Australia. Finally, we would not be able to see the Milky Way, since the fireball of the Big Bang would surround us at the distance of Rigel or Deneb.

11. A photon transports energy. The relativistic equivalence of mass and energy means that is enough to give it momentum.

13. Your assignment: measure the length of a rod as it slides past you. Mark the position of its front end on the floor and have an assistant mark the position of the back end. Then measure the distance between the two marks. This distance will represent the length of the rod only if the two marks were made simultaneously in your frame of reference.

PROBLEMS

1. (a) $t_{OB} = 1.67 \times 10^3$ s, $t_{OA} = 2.04 \times 10^3$ s
   (b) $t_{BO} = 2.50 \times 10^3$ s, $t_{AO} = 2.04 \times 10^3$ s
   (c) $\Delta t = 90$ s
3. 5.0 s
5. (a) 20 m (b) 19 m (c) 0.31c
7. (a) 1.3×10^{-7} s (b) 38 m (c) 7.6 m
9. (a) 2.2 μs (b) 0.65 km
11. 0.950c
13. Yes, with 19 m to spare
15. (a) 39.2 μs (b) Accurate to one digit
17. 3.3×10^5 m/s
19. 0.285c
21. 0.54c to the right
23. 0.357c
25. 0.998c toward the right
27. (a) 54 min (b) 52 min
29. $c(\sqrt{3}/2)$
31. 0.786c

33. 18.4 g/cm^3
35. 1.98 MeV
37. 2.27×10^{23} Hz, 1.32 fm for each photon
39. (a) 3.10×10^5 m/s (b) 0.758c
41. 1.42 MeV/c
43. (a) 0.80c (b) 7.5×10^3 s (c) 1.4×10^{12} m, 0.38c
45. 0.37c in the $+ x$-direction
47. (a) $v/c = 1 - 1.12 \times 10^{-10}$ (b) 6.00×10^{27} J
   (c) \$$2.17 \times 10^{20}$
49. 0.80c
51. (a) 0.946c (b) 0.160 ly (c) 0.114 yr (d) 7.50×10^{22} J
53. (a) 7.0 μs (b) 1.1×10^4 muons
59. 5.45 yr; Goslo is older.

Chapter 27

QUICK QUIZZES
1. (b)
2. (c)
3. (c)
4. (b)

CONCEPTUAL QUESTIONS
1. The shape of an object is normally determined by observing the light reflecting from its surface. In a kiln, the object will be very hot and will be glowing red. The emitted radiation is far stronger than the reflected radiation, and the thermal radiation emitted is only slightly dependent on the material from which the object is made. Unlike reflected light, the emitted light comes from all surfaces with equal intensity, so contrast is lost and the shape of the object is harder to discern.

3. The "blackness" of a blackbody refers to its ideal property of absorbing all radiation incident on it. If an observed room temperature object in everyday life absorbs all radiation, we describe it as (visibly) black. The black appearance, however, is due to the fact that our eyes are sensitive only to visible light. If we could detect infrared light with our eyes, we would see the object emitting radiation. If the temperature of the blackbody is raised, Wien's law tells us that the emitted radiation will move into the visible range of the spectrum. Thus, the blackbody could appear as red, white, or blue, depending on its temperature.

5. All objects do radiate energy, but at room temperature this energy is primarily in the infrared region of the electromagnetic spectrum, which our eyes cannot detect. (Pit vipers have sensory organs that *are* sensitive to infrared radiation; thus, they can seek out their warm-blooded prey in what we would consider absolute darkness.

7. Most metals have cutoff frequencies corresponding to photons in or near the visible range of the electromagnetic spectrum. AM radio wave photons have far too little energy to eject electrons from the metal.

9. We can picture higher frequency light as a stream of photons of higher energy. In a collision, one photon can give all of its energy to a single electron. The kinetic energy of such an electron is measured by the stopping potential. The reverse voltage (stopping voltage) required to stop the current is proportional to the frequency of the incoming light. More intense light consists of more photons striking a unit area each second, but atoms are so small that one emitted electron never gets a "kick" from more than one photon. Increasing the intensity of the light will generally increase the size of the current, but will not change the energy of the individual electrons that are ejected. Thus, the stopping potential remains constant.

11. Wave theory predicts that the photoelectric effect should occur at any frequency, provided that the light intensity is high enough. However, as seen in photoelectric experiments, the light must have sufficiently high frequency for the effect to occur.

13. (a) Electrons are emitted only if the photon frequency is greater than the cutoff frequency.

15. No. Suppose that the incident light frequency at which you first observed the photoelectric effect is above the cutoff frequency of the first metal, but less than the cutoff frequency of the second metal. In that case, the photoelectric effect would not be observed at all in the second metal.

17. The frequency of the scattered photon must decrease, because some of its energy is transferred to the electron.

PROBLEMS

1. (a) $\approx 3\,000$ K (b) $\approx 20\,000$ K
3. 500 nm
5. (a) 2.49×10^{-5} eV (b) 2.49 eV (c) 249 eV
7. 2.27×10^{30} photons/s
9. (a) 2.3×10^{31} (b) $\Delta E/E = 4.3 \times 10^{-32}$
11. (a) 2.24 eV (b) 555 nm (c) 5.41×10^{14} Hz
13. 234 nm
15. 148 days, incompatible with observation
17. 4.8×10^{14} Hz, 2.0 eV
19. 1.2×10^2 V and 1.2×10^7 V, respectively
21. 41.4 kV
23. 0.078 nm
25. 0.281 nm
27. 1.78 eV, 9.47×10^{-28} kg $\cdot$ m/s
29. 70°
31. 1.18×10^{-23} kg $\cdot$ m/s, 478 eV
33. (a) 1.2 eV (b) 6.5×10^5 m/s
35. (a) 1.46 km/s (b) 7.28×10^{-11} m
37. (a) $\approx 10^2$ MeV (b) No. With kinetic energy much larger than the magnitude of the negative potential energy, the electron would immediately escape.
39. 3.58×10^{-13} m
41. (a) 15 keV (b) 1.2×10^2 keV
43. $\sim 10^6$ m/s
45. 116 m/s
47. $\approx 5\,200$ K; clearly, a firefly is not at that temperature, so this cannot be blackbody radiation.
49. 18.2°
51. 1.36 eV
53. 2.00 eV
55. (a) $0.022\,0c$ (b) $0.999\,2c$
57. (b) 3.72 km/s
59. (b) 5.19×10^{-16} m
61. (a) 0.263 kg (b) 1.81 W
    (c) $-0.015\,3°C/s = -0.919°C/min$ (d) 9.89 μm
    (e) 2.01×10^{-20} J (f) 8.98×10^{19} photon/s

Chapter 28

QUICK QUIZZES

1. (b)
2. (a)

3. (a) 5 (b) 9 (c) 25
4. (d)

CONCEPTUAL QUESTIONS

1. If the energy of the hydrogen atom were proportional to n (or any power of n), then the energy would become infinite as n grew to infinity. But the energy of the atom is inversely proportional to n^2. Thus, as n grows to infinity, the energy of the atom approaches a value that is above the ground state by a finite amount, namely, the ionization energy 13.6 eV. As the electron falls from one bound state to another, its energy loss is always less than the ionization energy. The energy and frequency of any emitted photon are finite.

3. The characteristic x-rays originate from transitions within the atoms of the target, such as an electron from the L shell making a transition to a vacancy in the K shell. The vacancy is caused when an accelerated electron in the x-ray tube supplies energy to the K shell electron to eject it from the atom. If the energy of the bombarding electrons were to be increased, the K shell electron will be ejected from the atom with more remaining kinetic energy. But the energy difference between the K and L shell has not changed, so the emitted x-ray has exactly the same wavelength.

5. A continuous spectrum without characteristic x-rays is possible. At a low accelerating potential difference for the electron, the electron may not have enough energy to eject an electron from a target atom. As a result, there will be no characteristic x-rays. The change in speed of the electron as it enters the target will result in the continuous spectrum.

7. The hologram is an interference pattern between light scattered from the object and the reference beam. If anything moves by a distance comparable to the wavelength of the light (or more), the pattern will wash out. The effect is just like making the slits vibrate in Young's experiment, to make the interference fringes vibrate wildly so that a photograph of the screen displays only the average intensity everywhere.

9. If the Pauli exclusion principle were not valid, the elements and their chemical behavior would be grossly different, because every electron would end up in the lowest energy level of the atom. All matter would therefore be nearly alike in its chemistry and composition, since the shell structures of each element would be identical. Most materials would have a much higher density, and the spectra of atoms and molecules would be very simple, resulting in the existence of less color in the world.

11. The three elements have similar electronic configurations, with filled inner shells plus a single electron in an s orbital. Since atoms typically interact through their unfilled outer shells, and since the outer shells of these atoms are similar, the chemical interactions of the three atoms are also similar.

13. Each of the eight electrons must have at least one quantum number different from each of the others. They can differ (in m_s) by being spin-up or spin-down. They can differ (in ℓ) in angular momentum and in the general shape of the wave function. Those electrons with $\ell = 1$ can differ (in m_ℓ) in orientation of angular momentum.

15. Stimulated emission is the reason laser light is coherent and tends to travel in a well-defined parallel beam. When a photon passing by an excited atom stimulates that atom to emit a photon, the emitted photon is in phase with the

original photon and travels in the same direction. As this process is repeated many times, an intense, parallel beam of coherent light is produced. Without stimulated emission, the excited atoms would return to the ground state by emitting photons at random times and in random directions. The resulting light would not have the useful properties of laser light.

17. The atom is a bound system. The atomic electron does not have enough kinetic energy to escape from its electrical attraction to the nucleus. The electrical potential energy of the atom is negative and is greater than the kinetic energy, so the total energy of the atom is negative.

19. (a) The wavelength of photon A is greater. (b) The energy of photon B is greater.

PROBLEMS

1. 656 nm, 486 nm, and 434 nm
3. (a) 2.3×10^{-8} N (b) -14 eV
5. (a) 1.6×10^6 m/s (b) No, $v/c = 5.3 \times 10^{-3} << 1$
   (c) 0.46 nm (d) Yes. The wavelength is roughly the same size as the atom.
7. (a) 0.212 nm (b) 9.95×10^{-25} kg·m/s
   (c) 2.11×10^{-34} J·s
   (d) 3.40 eV (e) -6.80 eV (f) -3.40 eV
11. $E = -1.51$ eV ($n = 3$) to $E = -3.40$ eV ($n = 2$)
13. (a) 0.967 eV (b) 0.266 eV
15. (a) 122 nm, 91.1 nm (b) 1.87×10^3 nm, 820 nm
17. 97.2 nm
19. (a) 488 nm (b) 0.814 m/s
21. (d) $n = 2.53 \times 10^{74}$ (e) No. At such large quantum numbers, the allowed energies are essentially continuous.
23. (a) 2.47×10^{14} Hz, $f_{orb} = 8.23 \times 10^{14}$ Hz
    (b) 6.59×10^3 Hz, $f_{orb} = 6.59 \times 10^3$ Hz. For large n, classical theory and quantum theory approach each other in their results.
25. 4.42×10^4 m/s
27. (a) -122 eV (b) 1.76×10^{-11} m
29. (a) 0.026 5 nm (b) 0.017 6 nm (c) 0.013 2 nm
31. 1.33 nm
33. $n = 3, \ell = 1, m_\ell = +1, m_s = \pm 1/2; n = 3, \ell = 1, m_\ell = 0, m_s = \pm 1/2; n = 3, \ell = 1, m_\ell = -1, m_s = \pm 1/2$
35. Fifteen possible states, as summarized in the following table:

n	3	3	3	3	3	3	3	3	3	3	3	3	3	3	3
ℓ	2	2	2	2	2	2	2	2	2	2	2	2	2	2	2
m_ℓ	+2	+2	+2	+1	+1	+1	0	0	0	-1	-1	-1	-2	-2	-2
m_s	+1	0	-1	+1	0	-1	+1	0	-1	+1	0	-1	+1	0	-1

37. (a) 30 possible states (b) 36
39. (a) $n = 4$ and $\ell = 2$ (b) $m_\ell = (0, \pm 1, \pm 2)$, $m_s = \pm 1/2$
    (c) $1s^2 2s^2 2p^6 3s^2 3p^6 3d^{10} 4s^2 4p^6 4d^2 5s^2 = $ [Kr] $4d^2 5s^2$
41. 0.160 nm
43. L shell: 11.7 keV; M shell: 10.0 keV; N shell: 2.30 keV
45. (a) 10.2 eV (b) 7.88×10^4 K
47. (a) -8.18 eV, -2.04 eV, -0.904 eV, -0.510 eV, -0.325 eV
    (b) 1.09×10^3 nm and 609 nm
49. The four lowest energies are -10.39 eV, -5.502 eV, -3.687 eV, and -2.567 eV (b) The wavelengths of the emission lines are 158.5 nm, 185.0 nm, 253.7 nm, 422.5 nm, 683.2 nm, and 1 107 nm
    (c) 1.31×10^6 m/s

51. (a) 4.24×10^{15} W/m² (b) 1.20×10^{-12} J
55. (a) $E_n = (-1.49 \times 10^4 \text{ eV})/n^2$ (b) $n = 4 \rightarrow n = 1$
57. (a) 9.03×10^{22} m/s² (b) -4.63×10^{-8} W
    (c) $\sim 10^{-11}$ s

Chapter 29

QUICK QUIZZES

1. (c)
2. (c)
3. (a)
4. (a) and (b)
5. (b)

CONCEPTUAL QUESTIONS

1. Isotopes of a given element correspond to nuclei with different numbers of neutrons. This will result in a variety of different physical properties for the nuclei, including the obvious one of mass. The chemical behavior, however, is governed by the element's electrons. All isotopes of a given element have the same number of electrons and, therefore, the same chemical behavior.

3. An alpha particle contains two protons and two neutrons. Because a hydrogen nucleus contains only one proton, it cannot emit an alpha particle.

5. In alpha decay, there are only two final particles: the alpha particle and the daughter nucleus. There are also two conservation principles: of energy and of momentum. As a result, the alpha particle must be ejected with a discrete energy to satisfy both conservation principles. However, beta decay is a three-particle decay: the beta particle, the neutrino (or antineutron), and the daughter nucleus. As a result, the energy and momentum can be shared in a variety of ways among the three particles while still satisfying the two conservation principles. This allows a continuous range of energies for the beta particle.

7. The larger rest energy of the neutron means that a free proton in space will not spontaneously decay into a neutron and a positron. When the proton is in the nucleus, however, the important question is that of the total rest energy of the nucleus. If it is energetically favorable for the nucleus to have one less proton and one more neutron, then the decay process will occur to achieve this lower energy.

9. Carbon dating cannot generally be used to estimate the age of a stone, because the stone was not alive to take up carbon from the environment. Only the ages of artifacts that were once alive can be estimated with carbon dating.

11. The protons, although held together by the nuclear force, are repelled by the electrostatic force. If enough protons were placed together in a nucleus, the electrostatic force would overcome the nuclear force, which is based on the number of particles, and cause the nucleus to fission.

    The addition of neutrons prevents such fission. The neutron does not increase the electrical force, being electrically neutral, but does contribute to the nuclear force.

13. The photon and the neutrino are similar in that both particles have zero charge and very little mass. (The photon has zero mass, but recent evidence suggests that certain kinds of neutrinos have a very small mass.) Both must travel at the speed of light and are capable of transferring both energy and momentum. They differ in that the pho-

ton has spin (intrinsic angular momentum) $\hbar$ and is involved in electromagnetic interactions, while the neutrino has spin $\hbar/2$, and is closely related to beta decays.

15. Since the two samples are of the same radioactive nuclide, they have the same half-life; the 2:1 difference in activity is due to a 2:1 difference in the mass of each sample. After 5 half lives, each will have decreased in mass by a power of $2^5 = 32$. However, since this simply means that the mass of each is 32 times smaller, the ratio of the masses will still be $(2/32) : (1/32)$, or 2:1. Therefore, the ratio of their activities will *always* be 2:1.

PROBLEMS

1. $A = 2$, $r = 1.5$ fm; $A = 60$, $r = 4.7$ fm; $A = 197$, $r = 7.0$ fm; $A = 239$, $r = 7.4$ fm
3. 1.8×10^2 m
5. (a) 27.6 N (b) 4.16×10^{27} m/s^2 (c) 1.73 MeV
7. (a) 1.9×10^7 m/s (b) 7.1 MeV
9. 8.66 MeV/nucleon for $^{93}_{41}$Nb, 7.92 MeV/nucleon for $^{197}_{79}$Au
11. 3.54 MeV
13. 0.210 MeV/nucleon greater for $^{23}_{11}$Na, attributable to less proton repulsion
15. 0.46 Ci
17. (a) 9.98×10^{-7} s^{-1} (b) 1.9×10^{10} nuclei
19. 1.0 h
21. 4.31×10^3 yr
23. (a) 5.58×10^{-2} h^{-1}, 12.4 h (b) 2.39×10^{13} nuclei (c) 1.9 mCi
25. $^{208}_{81}$Tl, $^{95}_{37}$Rb, $^{144}_{60}$Nd
27. $^{40}_{20}$Ca, $^{94}_{42}$Mo, $^{4}_{2}$He
29. e^+ decay, $^{56}_{27}$Co $\rightarrow$ $^{56}_{26}$Fe $+ e^+ + \nu$
31. (a) cannot occur spontaneously
    (b) can occur spontaneously
33. 18.6 keV
35. 4.22×10^3 yr
37. (a) $^{30}_{15}$P (b) -2.64 MeV
39. (a) $^{21}_{10}$Ne (b) $^{144}_{54}$Xe (c) X $= e^+$, X$' = \nu$
41. (a) $^{13}_{6}$C (b) $^{10}_{5}$B
43. (a) $^{197}_{79}$Au $+ n \rightarrow$ $^{198}_{80}$Hg $+ e^- + \bar{\nu}$ (b) 7.88 MeV
45. (a) $^1_0$n (b) Fluoride mass $= 18.000\ 953$ u
47. 18.8 J
49. 24 d
51. (a) 8.97×10^{11} electrons (b) 0.100 J (c) 100 rad
53. 46.5 d
55. $Q = 3.27$ MeV > 0, no threshold energy required
57. (a) 2.52×10^{24} (b) 2.29×10^{12} Bq (c) 1.07×10^6 yr
59. (a) 4.0×10^9 yr (b) It could be no older. The rock could be younger if some ^{87}Sr were initially present.
61. 54 μCi
63. 2.3×10^2 yr
65. 4.4×10^{-8} kg/h

Chapter 30

QUICK QUIZZES

1. (c)
2. (a)
3. (b)
4. (d)

CONCEPTUAL QUESTIONS

1. The experiment described is a nice analogy to the Rutherford scattering experiment. In the Rutherford experiment, alpha particles were scattered from atoms and the scattering was consistent with a small structure in the atom containing the positive charge.

3. The largest charge quark is $2e/3$, so a combination of only two particles, a quark and an antiquark forming a meson, could not have an electric charge of $+2e$. Only particles containing three quarks, each with a charge of $2e/3$, can combine to produce a total charge of $2e$.

5. Until about 700 000 years after the Big Bang, the temperature of the Universe was high enough for any atoms that formed to be ionized by ambient radiation. Once the average radiation energy dropped below the hydrogen ionization energy of 13.6 eV, hydrogen atoms could form and remain as neutral atoms for relatively long period of time.

7. In the quark model, all hadrons are composed of smaller units called quarks. Quarks have a fractional electric charge and a baryon number of $\frac{1}{3}$. There are six flavors of quarks: up (u), down (d), strange (s), charmed (c), top (t), and bottom (b). All baryons contain three quarks, and all mesons contain one quark and one antiquark. Section 30.12 has a more detailed discussion of the quark model.

9. Baryons and mesons are hadrons, interacting primarily through the strong force. They are not elementary particles, being composed of either three quarks (baryons) or a quark and an antiquark (mesons). Baryons have a nonzero baryon number with a spin of either $\frac{1}{2}$ or $\frac{3}{2}$. Mesons have a baryon number of zero and a spin of either 0 or 1.

11. All stable particles other than protons and neutrons have baryon number zero. Since the baryon number must be conserved, and the final states of the kaon decay contain no protons or neutrons, the baryon number of all kaons must be *zero*.

13. Yes, but the strong interaction predominates.

15. Unless the particles have enough kinetic energy to produce a baryon–antibaryon pair, the answer is *no*. Antibaryons have a baryon number of -1, baryons have a baryon number of $+1$, and mesons have a baryon number of 0. If such an interaction were to occur and produce a baryon, the baryon number would not be conserved.

17. Baryons and antibaryons contain three quarks, while mesons and antimesons contain two quarks. Quarks have a spin of 1/2; thus, three quarks in a baryon can only combine to form a net spin that is half-integral. Likewise, two quarks in a meson can only combine to form a net spin of 0 or 1.

19. For the first decay, the half-life is characteristic of the strong interaction, so the ρ^0 must have $S = 0$, and strangeness is conserved. The second decay must occur via the weak interaction.

PROBLEMS

1. 1.1×10^{16} fissions
3. 126 MeV
5. (a) 16.2 kg (b) 117 g
7. 2.9×10^3 km ($\approx 1\ 800$ miles)
9. 1.01 g
11. (a) $^8_4$Be (b) $^{12}_6$C (c) 7.27 MeV
13. 3.07×10^{22} events/yr
15. (a) 3.44×10^{30} J (b) 1.56×10^8 yr
17. (a) 4.53×10^{23} Hz (b) 0.622 fm

19. $\sim 10^{-23}$ s

21. $\sim 10^{-18}$ m

23. (a) conservation of electron-lepton number and conservation of muon-lepton number (b) conservation of charge (c) conservation of baryon number (d) conservation of baryon number (e) conservation of charge

25. $\overline{\nu}_\mu$

27. (a) $\overline{\nu}_\mu$ (b) ν_μ (c) $\overline{\nu}_e$ (d) ν_e (e) ν_μ (f) ν_μ and $\overline{\nu}_e$

29. (a) not allowed; violates conservation of baryon number (b) strong interaction (c) weak interaction (d) weak interaction (e) electromagnetic interaction

31. (a) not conserved (b) conserved (c) conserved (d) not conserved (e) not conserved (f) not conserved

33. (a) charge, baryon number, L_e, L_τ (b) charge, baryon number, L_e, L_μ, L_τ (c) charge, L_e, L_μ, L_τ, strangeness number (d) charge, baryon number, L_e, L_μ, L_τ, strangeness number (e) charge, baryon number, L_e, L_μ, L_τ, strangeness number (f) charge, baryon number, L_e, L_μ, L_τ, strangeness number

35. 3.34×10^{26} electrons, 9.36×10^{26} up quarks, 8.70×10^{26} down quarks

37. (a) Σ^+ (b) π^- (c) K^0 (d) Ξ^-

39.

Reaction	At Quark Level	Net Quarks (before and after)
$\pi^- + p \;\rightarrow\; K^0 + \Lambda^0$	$\overline{u}d + uud \;\rightarrow\; d\overline{s} + uds$	1 up, 2 down, 0 strange
$\pi^+ + p \;\rightarrow\; K^+ + \Sigma^+$	$u\overline{d} + uud \;\rightarrow\; u\overline{s} + uus$	3 up, 0 down, 0 strange
$K^- + p \;\rightarrow$ $K^+ + K^0 + \Omega^-$	$\overline{u}s + uud \;\rightarrow$ $u\overline{s} + d\overline{s} + sss$	1 up, 1 down, 1 strange

(d) The mystery particle is a Λ^0 or a Σ^0.

41. a neutron, udd

43. 70.45 MeV

45. 18.8 MeV

47. (a) electron-lepton and muon-lepton numbers not conserved (b) electron-lepton number not conserved (c) charge not conserved (d) baryon and electron-lepton numbers not conserved (e) strangeness violated by 2 units

49. (a) 2×10^{24} nuclei (b) ≈ 0.6 kg

51. (a) 1 baryon before and zero baryons after decay. Baryon number is not conserved. (b) 469 MeV, 469 MeV$/c$ (c) 0.999 999 $4c$

53. (b) 12 days

55. 26 collisions

Credits

Photographs

This page constitutes an extension of the copyright page. We have made every effort to trace the ownership of all copyrighted material and to secure permission from copyright holders. In the event of any question arising as to the use of any material, we will be pleased to make the necessary corrections in future printings. Thanks are due to the following authors, publishers, and agents for permission to use the material indicated.

Chapter 15. **497:** © Keith Kent/Science Photo Library/Photo Researchers, Inc. **498:** © American Philosophical Society/AIP **501:** top, © 1968 Fundamental Photographs; bottom, Photo courtesy of AIP Niels Bohr Library, E. Scott Barr Collection **510:** Photo courtesy of Harold M. Waage, Princeton University **511:** top, Photo courtesy of Harold M. Waage, Princeton University; bottom, Photo courtesy of Harold M. Waage, Princeton University **513:** top, Photo courtesy of Harold M. Waage, Princeton University; middle right, Photo courtesy of Harold M. Waage, Princeton University

Chapter 16. **531:** Courtesy of Resonance Research Corporation **543:** bottom middle, Riei O'Harra/Black Star/PNI; bottom right, Greig Cranna/Stock, Boston/PNI **546:** Photo courtesy of Harold M. Waage, Princeton University **557:** © Loren Winters/Visuals Unlimited **558:** a. Paul Silverman/Fundamental Photographs; b. George Semple

Chapter 17. **568:** Telegraph Colour Library/FPG/Getty Images **572:** Michael Dalton, Fundamental Photographs **574:** © Bettmann/CORBIS **575:** Courtesy of Henry Leap and Jim Lehman **578:** Courtesy of Central Scientific Company **580:** Courtesy of IBM Research Laboratory **585:** Courtesy of Medtronic, Inc.

Chapter 18. **592:** © Lester Lefkowitz/Corbis **593:** George Semple **602:** AIP ESVA, W.F. Meggers Collection **610:** George Semple **613:** Juergen Berger, Max-Planck Institute/Science Photo Library/Photo Researchers, Inc. **616:** middle left, Superstock; middle right, Courtesy of Henry Leap and Jim Lehman

Chapter 19. **624:** Johnny Johnson/Stone/Getty **625:** Courtesy of Central Scientific Company **626:** a–c. Henry Leap and Jim Lehman **631:** Courtesy of Henry Leap and Jim Lehman **637:** Courtesy of American Honda Motor Co., Inc. **640:** middle left, North Wind Picture Archives; bottom, © Richard Megna, Fundamental Photographs **641:** Leonard de Selva/CORBIS **645:** © Richard Megna, Fundamental Photographs **646:** Courtesy of Henry Leap and Jim Lehman **650:** © Loren Winters/Visuals Unlimited **651:** top right, Courtesy of Central Scientific Company; middle right, Courtesy of Central Scientific Company

Chapter 20. **660:** PhotoDisc/Getty Images **661:** By kind permission of the President and Council of the Royal Society **666:** Charles D. Winters **667:** Courtesy of PedsLink Pediatric Healthcare Resources, Newport Beach, CA **674:** Luis Castaneda/The Image Bank/Getty Images **678:** North Wind Picture Archives **681:** Courtesy of Saab **688:** bottom left, Courtesy of CENCO

Chapter 21. **693:** © Bettmann/Corbis **701:** Bettmann/Corbis **704:** Ryan Williams/International Stock Photography **707:** George Semple **708:** top, North Wind Photo Archives; bottom, Bettmann/Corbis **713:** John Neal/Photo Researchers, Inc. **715:** Ron Chapple/Getty Images **717:** a. NASA/CXC/SAO; b. Palomar Observatory; c. WM Keck Observatory; d, VLA/NRAO **720:** George Semple

Chapter 22. **726:** Peter Bowater/Photo Researchers, Inc. **727:** Courtesy of Rijksmuseum voor de Geschiedenis der Natuurwetenschappen. Courtesy AIP Niels Bohr Library **728:** bottom left and right, Photographs Courtesy of Henry Leap and Jim Lehman **729:** top left and right, Photos by Charles D. Winters **731:** top right, Courtesy of Henry Leap and Jim Lehman **735:** bottom left, Courtesy of Sony Disc Manufacturing **737:** top right, David Parker/SPL/Photo Researchers, Inc.; bottom right, Courtesy of Bausch & Lomb **738:** Courtesy of PASCO Scientific **739:** Mark D. Phillips/Photo Researchers, Inc. **740:** Courtesy of Sabina Zigman/Benjamin Cardozo High School **742:** Courtesy of Henry Leap and Jim Lehman **744:** bottom left, Dennis O'Clair/Tony Stone Images/Getty Images; bottom right, Hank Morgan/Photo Researchers, Inc.

Chapter 23. **754:** Don Hammond/CORBIS **757:** Courtesy of Henry Leap and Jim Lehman **759:** Courtesy of Jim Lehman, James Madison University **761:** top right, middle right, bottom right, Photos courtesy of David Rogers **762:** © Junebug Clark 1988/Photo Researchers, Inc. **768:** John M. Dunay IV, Fundamental Photographs, NYC **770:** top left, Photos Courtesy of Henry Leap and Jim Lehman **778:** a–c. Photos by Norman Goldberg **780:** middle right, Richard Megna/Fundamental Photographs, NYC **784:** Photo © Michael Levin/Opti-Gone Associates

Chapter 24. **786:** RO-MA/Index Stock Imagery **787:** M. Cagnet, M. Francon, J. C. Thierr **788:** Richard Megna/Fundamental Photographs, NYC **789:** John S. Shelton **792:** Dr. Jeremy Burgess/Science Photo Library/Photo Researchers, Inc. **793:** top right, Peter Aprahamian/Science Photo Library/Photo Researchers, Inc.; bottom middle, Courtesy of Bausch & Lomb Optical Company; bottom right, From

Physical Science Study Committee, *College Physics,* Lexington Mass., D. C. Heath and Co., 1968 **794:** Richard Megna, Fundamental Photographs **796:** Courtesy of Sony Disc Manufacturing **798:** top left, Courtesy of P. M. Rinard, from *Am. J. Phys.,* 44:70, 1976; top right, From M. Cagnet, M. Francon, and J. C. Thierr, *Atlas of Optical Phenomena,* Berlin, Springer-Verlag, 1962, plate 18 **803:** © Kristen Brochmann/Fundamental Photographs **806:** a–c. Photos courtesy of Henry Leap **809:** bottom left, Sepp Seitz 1981; bottom right, Peter Aprahamian/Science Photo Library/Photo Researchers, Inc.

Chapter 25. **819:** © Denis Scott/CORBIS **821:** top right, From Lennart Nilson, in collaboration with Jan Lindberg, *Behold Man: A Photographic Journey of Discovery Inside the Body,* Boston, Little, Brown & Co., 1974 **827:** bottom right, © Tony Freeman/Photo Edit **829:** NASA **830:** top right, © Tony Freeman/Photo Edit; bottom, Courtesy of Palomar Observatory/California Institute of Technology **832:** a–c. Photos from M. Cagnet, M. Francon, and J. C. Thierr **833:** From M. Cagnet, M. Francon, and J. C. Thierr, *Atlas of Optical Phenomena,* Berlin, Springer-Verlag, 1962, plate 34

Chapter 26. **843:** Courtesy of the Archives, California Institute of Technology **849:** AIP Niels Bohr Library **861:** Lawrence Berkeley Laboratory/Science Photo Library/Photo Researchers, Inc.

Chapter 27. **874:** © Eye of Science/Science Source/Photo Researchers, Inc. **878:** © Bettmann/CORBIS **881:** Courtesy of GE Medical Systems **884:** Science Source/Photo Researchers, Inc. **885:** Courtesy of AIP Niels Bohr Library **887:** AIP Niels Bohr Library **890:** top right, © David Parker/Photo Researchers, Inc. **891:** top right, AIP Emilio Segré Visual Archives; bottom right, Courtesy of the University of Hamburg **894:** IBM Corporation Research Division **902:** John Chumack/Photo Researchers, Inc.

Chapter 28. **903:** Dembinsky Photo Associates **904:** Stock Montage, Inc. **907:** top right, Princeton University/Courtesy of AIP Emilio Segré Visual Archives **918:** CERN/Courtesy of AIP Emilio Segré Visual Archives **923:** Courtesy of HRL Laboratories LLC, Malibu, CA **924:** top right, Courtesy of Central Scientific Company; middle left, Philippe Plailly/Photo Researchers, Inc. **931:** top left, Courtesy of Texas Instruments, Inc.; top right, Courtesy of Intel Corporation

Chapter 29. **939:** Courtesy of Public Service Electric and Gas Company **940:** North Wind Picture Archives **942:** Courtesy of Louise Barker/AIP Niels Bohr Library **945:** FPG International **946:** © Richard Megna/Fundamental Photographs **952:** National Accelerator Laboratory **953:** Henry Paul/Gamma Liaison **955:** Santi Visali/The IMAGE Bank **962:** Jay Freis/The Image Bank **963:** top left, SBHA/Getty Images; top right, Scott Camazine/Science Source/Photo Researchers, Inc.; bottom right, © Hank Morgan/Photo Researchers **965:** top left, Photo Researchers, Inc./Science Photo Library; top right, Dan McCoy/Rainbow

Chapter 30. **973:** Sandia National Laboratories/Photo by Randy Montoya **977:** Courtesy of Chicago Historical Society **980:** NASA **983:** top left, Courtesy of Princeton Plasma Physics Laboratory; top right, Courtesy of Princeton University **985:** Courtesy AIP Emilio Segré Visual Archives **986:** UPI/Corbis-Bettmann **987:** © Shelly Gazin/CORBIS **993:** Photo courtesy of Michael R. Dressler **996:** Courtesy of Fermi National Accelerator Laboratory **998:** Courtesy of CERN **1000:** top left, Courtesy of AIP Emilio Segré Visual Archives; bottom left, AT&T Bell Laboratories

Tables and Illustrations

This page constitutes an extension of the copyright page. We have made every effort to trace the ownership of all copyrighted material and to secure permission from copyright holders. In the event of any question arising as to the use of any material, we will be pleased to make the necessary corrections in future printings. Thanks are due to the following authors, publishers, and agents for permission to use the material indicated.

Chapter 25. **821:** Art by Robert Demarest, from Cecie Starr and Beverly McMillan, *Human Biology* 5th ed. Copyright © 2003, Brooks/Cole, a division of Thomson Learning, Inc.

Chapter 27. **894–895:** This section was written by Roger A. Freedman, University of California, Santa Barbara **895:** Based on a drawing from P. K. Hansma, V. B. Elings, O. Marti, and C. Bracker, *Science* 242:209, 1988, Copyright 1988 by the AAAS

Chapter 28. **905:** K. W. Whitten, R. E. Davis, M. L. Peck, and G. G. Stanley, *General Chemistry,* 7th ed., Belmont, CA, Brooks/Cole, 2004

Chapter 30. **1008:** © 2005 Sidney Harris

Index

PHYSICAL CONSTANTS

Quantity	Symbol	Value	SI unit
Speed of light in vacuum	c	3.00×10^8	m/s
Permittivity of free space	ϵ_0	8.85×10^{-12}	$C^2/N \cdot m^2$
Coulomb constant, $1/4\pi\epsilon_0$	k_e	8.99×10^9	$N \cdot m^2/C^2$
Permeability of free space	μ_0	1.26×10^{-6} ($4\pi \times 10^{-7}$ exactly)	$T \cdot m/A$
Elementary charge	e	1.60×10^{-19}	C
Planck's constant	h	6.63×10^{-34}	$J \cdot s$
	$\hbar = h/2\pi$	1.05×10^{-34}	$J \cdot s$
Electron mass	m_e	9.11×10^{-31}	kg
		5.49×10^{-4}	u
Proton mass	m_p	$1.672\,65 \times 10^{-27}$	kg
		$1.007\,276$	u
Neutron mass	m_n	$1.674\,95 \times 10^{-27}$	kg
		$1.008\,665$	u
Avogadro's number	N_A	6.02×10^{23}	mol^{-1}
Universal gas constant	R	8.31	$J/mol \cdot K$
Boltzmann's constant	k_B	1.38×10^{-23}	J/K
Stefan-Boltzmann constant	σ	5.67×10^{-8}	$W/m^2 \cdot K^4$
Molar volume of ideal gas at STP	V	22.4	L/mol
		2.24×10^{-2}	m^3/mol
Rydberg constant	R_H	1.10×10^7	m^{-1}
Bohr radius	a_0	5.29×10^{-11}	m
Electron Compton wavelength	$h/m_e c$	2.43×10^{-12}	m
Gravitational constant	G	6.67×10^{-11}	$N \cdot m^2/kg^2$
Standard free-fall acceleration	g	9.80	m/s^2
Radius of Earth (at equator)	R_E	6.38×10^6	m
Mass of Earth	M_E	5.98×10^{24}	kg
Radius of Moon	R_M	1.74×10^6	m
Mass of Moon	M_M	7.36×10^{22}	kg

The values presented in this table are those used in computations in the text. Generally, the physical constants are known to much better precision.